VOLUME EIGHTY

ADVANCES IN INORGANIC CHEMISTRY

Biomedical Applications of Inorganic Photochemistry

ADVISORY BOARD

Academic Press is an imprint of Elsevier
50 Hampshire Street, 5th Floor, Cambridge, MA 02139, United States
525 B Street, Suite 1650, San Diego, CA 92101, United States
The Boulevard, Langford Lane, Kidlington, Oxford, OX5 1GB, United Kingdom
125 London Wall, London, EC2Y 5AS, United Kingdom

First edition 2022

ISBN: 978-0-323-99171-1
ISSN: 0898-8838

For information on all Academic Press publications
visit our website at https://www.elsevier.com/books-and-journals

Publisher: Zoe Kruze
Acquisitions Editor: Sam Mahfoudh
Developmental Editor: Federico Paulo Mendoza
Production Project Manager: Vijayaraj Purushothaman
Cover Designer: Vicky Pearson

Typeset by STRAIVE, India

VOLUME EIGHTY

ADVANCES IN INORGANIC CHEMISTRY

Biomedical Applications of Inorganic Photochemistry

Edited by

PETER C. FORD

Department of Chemistry and Biochemistry
University of California Santa Barbara, CA, United States

RUDI van ELDIK

University of Erlangen–Nuremberg, Germany
Nicolaus Copernicus University in Torun, Poland

ELSEVIER

ACADEMIC PRESS

An imprint of Elsevier

Contents

Contributors *xi*
Preface *xv*

1. Leveraging the photophysical properties of rhenium(I) tricarbonyl complexes for biomedical applications **1**

Justin J. Wilson

1. Introduction 2
2. Properties and applications of $Re(CO)_3$ complexes 4
3. $Re(CO)_3$ complexes as photoluminescent imaging agents 6
4. Simple diimine $Re(CO)_3$ complexes exhibit anticancer activity 8
5. Pyridylimine $Re(CO)_3$ complexes 10
6. Isonitrile-containing $Re(CO)_3$ complexes 14
7. Photoactivation of $Re(CO)_3$ anticancer agents 22
8. Conclusions and future outlook 27

Acknowledgments 28
References 28

2. Strategic design of photofunctional transition metal complexes for cancer diagnosis and therapy **35**

Lawrence Cho-Cheung Lee and Kenneth Kam-Wing Lo

1. Introduction 37
2. Transition metal complexes for targeted cancer diagnosis and therapy 38
3. Activatable transition metal complexes for cancer diagnosis and therapy 55
4. Conclusions and future perspective 88

Acknowledgments 90
References 90

3. Advances in the design of photoactivated platinum anticancer complexes **95**

Huayun Shi and Peter J. Sadler

1. Introduction 96
2. Discovery of photoactive Pt(IV) anticancer complexes 99
3. Mechanism of action 103
4. Axial derivatives of Pt(IV) complexes 108
5. Pt(IV) azide–acetylene derivatives via "click" reactions 112

6. Nanocarriers 114
Acknowledgments 124
References 124

4. Adventures in the photo-uncaging of small molecule bioregulators 129

Peter C. Ford

1. Introduction 131
2. PhotoNORMs and PhotoCORMs 133
3. Enhancing photoreaction SMB release rates with antennas 143
4. Moving toward the phototherapeutic window 148
5. Biological targeting 159
6. Summary 163
Acknowledgments 164
References 164

5. Light-triggered unconventional therapies with engineered inorganic nanoparticles 171

Aurore Fraix and Salvatore Sortino

1. Introduction 172
2. Unconventional phototherapy: Basic concepts 175
3. Nonmetallic inorganic NPs 179
4. Noble metal-based NPs 189
5. Concluding remarks and future perspectives 199
Acknowledgments 200
References 200

6. Photoactive manganese carbonyl complexes with *fac*-{Mn(CO)$_3$} moiety: Design, application, and potential as prodrugs in CO therapy 205

Indranil Chakraborty and Pradip K. Mascharak

1. Introduction 206
2. Chemical and photochemical properties of the *fac*-{Mn(CO)$_3$} moiety 207
3. Moving photosensitivity to visible region of light; Smart design of photoCORMs 208
4. Water-soluble photoCORMs for chemotherapeutic applications 216
5. Trackable photoCORMs that allow visualization of the CO donor within cellular matrix 217
6. Turn-ON photoCORMs that confirm CO delivery at the target site 220
7. PhotoCORMs with various coligands that confer specific drug actions 222

8. CO-releasing materials and platform design 224
9. Conclusion and future perspectives 227
Acknowledgment 229
References 229

7. Mechanistic insight into photoactivation of small inorganic molecules from the biomedical applications perspectives 233

Agnieszka Kyzioł, Łukasz Orzeł, Ilona Gurgul, Olga Mazuryk, Przemysław Łabuz, and Grażyna Stochel

1. Introduction 234
2. Reactive small inorganic molecules (RSIM) 234
3. RSIM photogeneration in biological systems - Pathways and mechanisms 236
4. Photoresponsive materials - Examples and development trends 246
5. Conclusions and future directions 275
Acknowledgments 277
References 277

8. Ruthenium complexes for photoactivated dual activity: Drug delivery and singlet oxygen generation 285

Sean J. Steinke, Jeremy J. Kodanko, and Claudia Turro

1. Introduction 286
2. Photodynamic therapy 289
3. Photochemotherapy 298
4. Dual activity 306
5. Concluding remarks 316
Acknowledgment 317
References 317

9. Photochemical biosignaling with ruthenium complexes 321

Oscar Filevich and Roberto Etchenique

1. Introduction 322
2. Advantages of ruthenium caged compounds 324
3. Inorganic caged compounds and clinical applications 332
4. Value in diversity 333
5. The fifth ligand itch 337
6. Three-dimensional localization 341
7. *Cis*, *trans* and other structures 347
8. A brief gaze into the future 349
Acknowledgments 350
References 350

10. Ruthenium phthalocyanines in nitric oxide modulation and singlet oxygen release: Selectivity and cytotoxic effect on cancer cell lines 355

Renata Galvão de Lima, Rafaella Rebecchi Rios,

Antonio Eduardo da Hora Machado, and Roberto Santana da Silva

1. Introduction 356
2. Spectroscopic characteristics 359
3. Molecular orbital calculations 362
4. Light irradiation process 365
5. In vitro studies 373
6. Conclusions and future perspectives 377

Acknowledgment 377

References 377

11. Photoactive organometallic compounds as antimicrobial agents 381

Ashwene Rajagopal, Jack Biddulph, Leila Tabrizi,

Deirdre Fitzgerald-Hughes, and Mary T. Pryce

1. Introduction 382
2. Group 4 and 5 organometallic complexes 386
3. Group 6 organometallic complexes 387
4. Group 7 organometallic complexes 391
5. Group 8 organometallic complexes 398
6. Group 9 organometallic compounds 399
7. Group 10 and 11 organometallic complexes 401
8. Summary and future perspectives 404

Acknowledgment 405

References 405

12. Mitochondria-targeting transition metal complexes 411

Ryan J. Morris and Massimiliano Massi

1. Introduction 413
2. Iridium 419
3. Platinum 464
4. Rhenium 470
5. Gold 476
6. Copper 479
7. Ruthenium 482
8. Osmium 486
9. Vanadium 489

10. Palladium 492
11. Rhodium 493
12. Zinc 495
13. Silver 496
14. Concluding remarks and future perspectives 497
Acknowledgments 499
References 499

Index *511*

Contributors

Jack Biddulph
School of Chemical Sciences, Dublin City University, Dublin, Ireland

Indranil Chakraborty
Department of Chemistry and Biochemistry, University of California, Santa Cruz, CA, United States

Roberto Santana da Silva
Faculdade de Ciências Farmacêuticas de Ribeirão Preto da Universidade de São Paulo, Ribeirão Preto – SP, Brazil

Renata Galvão de Lima
Instituto de Ciências Exatas e Naturais – ICENP, Universidade Federal de Uberlândia, Ituiutaba, MG, Brazil

Roberto Etchenique
Departamento de Química Inorgánica, Analítica y Química Física, INQUIMAE, Facultad de Ciencias Exactas y Naturales, Universidad de Buenos Aires, Intendente Güiraldes 2160, Ciudad Universitaria, Pabellón 2, Buenos Aires, Argentina

Oscar Filevich
Laboratorio de Neuroingeniería, Instituto de Tecnologías Emergentes y Ciencias Aplicadas (ITECA), UNSAM-CONICET, Escuela de Ciencia y Tecnología (ECyT), Universidad Nacional de San Martín, San Martín, Buenos Aires, Argentina

Deirdre Fitzgerald-Hughes
Department of Clinical Microbiology, Royal College of Surgeons in Ireland, Dublin, Ireland

Peter C. Ford
Department of Chemistry and Biochemistry, University of California, Santa Barbara, CA, United States

Aurore Fraix
PhotoChemLab, Department of Drug and Health Sciences, University of Catania, Catania, Italy

Ilona Gurgul
Faculty of Chemistry, Jagiellonian University, Kraków, Poland

Jeremy J. Kodanko
Department of Chemistry, Wayne State University, Detroit, MI, United States

Agnieszka Kyzioł
Faculty of Chemistry, Jagiellonian University, Kraków, Poland

Przemysław Łabuz
Faculty of Chemistry, Jagiellonian University, Kraków, Poland

Lawrence Cho-Cheung Lee
Department of Chemistry, City University of Hong Kong; Laboratory for Synthetic Chemistry and Chemical Biology Limited, Hong Kong, PR China

Kenneth Kam-Wing Lo
Department of Chemistry; State Key Laboratory of Terahertz and Millimeter Waves, City University of Hong Kong, Hong Kong, PR China

Antonio Eduardo da Hora Machado
Instituto de Física, Programa de Pós-Graduação em Ciências Exatas e Tecnológicas, Universidade Federal de Catalão, Catalão, GO; Instituto de Química, Laboratório de Fotoquímica e Ciência de Materiais, Universidade Federal de Uberlândia, Uberlândia, MG, Brazil

Pradip K. Mascharak
Department of Chemistry and Biochemistry, University of California, Santa Cruz, CA, United States

Massimiliano Massi
School of Molecular and Life Sciences, Curtin University, Perth, Australia

Olga Mazuryk
Faculty of Chemistry, Jagiellonian University, Kraków, Poland

Ryan J. Morris
School of Molecular and Life Sciences, Curtin University, Perth, Australia

Łukasz Orzeł
Faculty of Chemistry, Jagiellonian University, Kraków, Poland

Mary T. Pryce
School of Chemical Sciences, Dublin City University, Dublin, Ireland

Ashwene Rajagopal
School of Chemical Sciences, Dublin City University; Department of Clinical Microbiology, Royal College of Surgeons in Ireland, Dublin, Ireland

Rafaella Rebecchi Rios
Faculdade de Ciências Farmacêuticas de Ribeirão Preto da Universidade de São Paulo, Ribeirão Preto – SP, Brazil

Peter J. Sadler
Department of Chemistry, University of Warwick, Coventry, United Kingdom

Huayun Shi
Department of Chemistry, University of Warwick, Coventry, United Kingdom

Salvatore Sortino
PhotoChemLab, Department of Drug and Health Sciences, University of Catania, Catania, Italy

Sean J. Steinke
Department of Chemistry and Biochemistry, The Ohio State University, Columbus, OH, United States

Grażyna Stochel
Faculty of Chemistry, Jagiellonian University, Kraków, Poland

Leila Tabrizi
School of Chemical Sciences, Dublin City University, Dublin, Ireland

Claudia Turro
Department of Chemistry and Biochemistry, The Ohio State University, Columbus, OH, United States

Justin J. Wilson
Department of Chemistry and Chemical Biology, Cornell University, Ithaca, NY, United States

Preface

The idea of a thematic volume of *Advances in Inorganic Chemistry* devoted to "Biomedical Applications of Inorganic Photochemistry" was initiated in discussions between the *AIC* series editor Rudi van Eldik and the volume editor Peter C. Ford in early 2021. With much gratitude to the contributors, we are pleased to be able to present these 12 chapters covering various aspects of this exciting topic (pun intended). As one can see from the list of authors, the subject has received increasing attention from a broad international community of chemists with contributions from Argentina, Australia, Brazil, England, Hong Kong, Ireland, Italy, Poland, and the United States to this volume alone.

Metal complexes have long been of interest from a medical perspective,[a] and most inorganic chemists are quite aware that "cisplatin" was a known compound (*cis*-$Pt(NH_3)_2Cl_2$) for more than a century prior to its approval in 1978 for cancer chemotherapy. Photochemistry adds new dimensions. For example, electronic excitation can release (uncage) a small molecule bioregulator or chemotherapeutic agent or transform a benign precursor complex into a bioactive form. Precise spatial-temporal control of such uncaging is highly desirable, especially for therapeutic applications, and the photoactivation allows control of the timing, location, and intensity of release. The precision from such treatment has value as an investigative tool in the laboratory and, potentially, in the therapy of disease states. Alternatively, the metal complex or its photoproduct may be a photosensitizer for the transformation of dioxygen into singlet oxygen ($^1O_2$) or into other reactive oxygen species (ROS). Such photodynamic therapy (PDT) has promise for treatment of malignant tumors and for killing antibiotic-resistant bacteria or fungi, although a notable limitation in PDT treatment of cancer is the hypoxic nature of solid tumors. Furthermore, luminescent complexes and/or photoproducts may provide imaging of targeted sites. The combination of therapeutic release or PDT and imaging is sometimes given the adjective "theranostic," although this term is perhaps more commonly used in nuclear medicine.

[a] See, for example, "Medicinal Chemistry", *Advances in Inorganic Chemistry*, Vol. 75, Sadler, P. J.; van Eldik, R., Eds, pp. 2–496.

Although photo-uncaging can define the *location*, *timing*, and *dosage* of delivery, it is desirable to design molecular systems or conjugates that are specific to certain tissue types (e.g., tumors) or certain cellular structures (e.g., membranes or mitochondria). Inorganic and organometallic complexes having a wide range of metal centers and oxidation states, plus multiple coordination sites with ligands having different electronic, steric, and excited-state properties, provide an enormous range of systems that can be tailored and tuned to the mission at hand. With *in vivo* applications in mind, the complexes also need to be compatible with the mechanisms for delivery to targeted tissues, possibly via injection or by infusion via the circulatory system or by recruiting carriers such as vesicles or macrophages. As we will see in this volume, the medical opportunities for such coordination compounds are tremendous. However, to better codify the success in this field, there is a need to universally recognize standardized procedures and tests to which new compounds and delivery methods are subjected, preferably overseen by an independent laboratory to evaluate claims.

Although some metal compounds have indeed entered the clinical world, much of the research, especially with regard to photochemical applications, continues to be exploratory. When considering medical applications, one needs to consider the intended task and whether the target is topical or in deep tissue. The latter requirement introduces limitations on the necessary excitation wavelength owing to the poor tissue transmission of light, except at longer visible and near-infrared wavelengths. As seen in several chapters, this topic has been addressed by applying multiphoton excitation methods.[b] Among other questions to be considered are whether the initial complex or its photoproduct are selective for accumulation in certain tissues or cell types and whether they are physiologically compatible or toxic. The chapters in this volume address many of the issues encountered in such studies and hopefully provide valuable resources for both established and new researchers in this important intellectual endeavor.

Chapter 1 by Wilson introduces mechanistic studies from his Cornell University Laboratory concerned with the photochemistry of rhenium(I) tricarbonyl complexes and potential applications of various fac-$Re(CO)_3$ derivatives with regard to anticancer activity. The rich photoluminescence properties of these anticancer agents provide theranostic opportunities.

[b] Ford, P. C. *Acc. Chem. Res.* **2008,** *41*, 190–200.

A similar theme is evident in Chapter 2 by Lee and Lo, which describes development of photofunctional transition metal complexes for cancer diagnosis and therapy at CU Hong Kong. The luminactive chromophores are largely Re(I), ruthenium(II), or iridium(III) complexes suitably decorated with ligands that target specific cellular types. *In vitro* and *in vivo* studies are described, and limitations with regard to the excitation wavelengths required, selectivity, and biocompatibility are discussed.

Chapter 3 by Shi and Sadler directs attention to studies at the University of Warwick with regard to photoactivated platinum anticancer complexes and the mechanisms by which they are effective in the clinical use of photoactivated chemotherapy. Much of the focus is on six-coordinate diazido Pt(IV) complexes that undergo fragmentation to Pt(II) products plus azido radicals, both of which species may have physiological consequences.

Chapter 4 by Ford reports long-term studies at UC Santa Barbara and elsewhere on photochemical uncaging of the small molecule bioregulators nitric oxide (NO) and carbon monoxide (CO). Particular emphasis is paid to strategies for employing tissue transmitting deep-red or NIR excitation for this purpose and for utilizing biological mechanisms such as macrophages to deliver payloads to specific targets.

Chapter 5 by Fraix and Sortino of the University of Catania reports investigations of inorganic (including carbon, silica, and noble metal) nanoparticles for the photo release of therapeutic agents for cancer and bacterial diseases. The therapeutic mechanisms for their approach include PDT, NO-PDT, and photothermal therapy, with the last of these being facilitated by the plasmonic absorption of NIR light of gold nanoparticles with various shapes.

Chapter 6 by Chakraborty and Mascharak focuses on studies at UC Santa Cruz to develop photoactivated CO-releasing moieties based on the *fac*-$Mn(CO)_3$ unit. They use density functional theory (DFT) and time-dependent DFT to obtain insights into electronic transitions and help design molecular systems that can be photoactivated at longer visible wavelengths.

Chapter 7 by Kyzioł, Orzeł, Gurgul, Mazuryk, Łabuz, and Stochel of Jagiellonian University reviews mechanisms for the photogeneration of reactive small inorganic molecules (RSIMs) such as reactive oxygen, nitrogen, sulfur, chlorine, and carbon species. *In vivo* strategies leading to spatially and temporally controlled RSIM generation are discussed with representative examples.

Chapter 8 by Steinke, Kodanko, and Turro reports on collaborative studies from Wayne State University and Ohio State University primarily utilizing Ru(II) complexes of polydentate aromatic nitrogen heterocyclic ligands as well as other ligands. By tuning the excited-state energies with the ligands, they have developed dual-function complexes that not only photo-uncage a chemotherapeutic ligand but also retain the ability to sensitize PDT for treatment of cancerous tumors.

Chapter 9 by Filevich and Etchenique describes further applications of Ru(II) polypyridine complexes and, in this case, as Ru-based phototriggers to uncage neurotransmitters for temporal-spatial neural stimulation to test the hypotheses regarding neural dynamics and brain circuitry. These Ru(II) complexes can be activated at much longer excitation wavelengths than organic phototriggers and by two-photon excitation with NIR light to enhance the spatial precision of stimuli.

Chapter 10 by Lima, Rios, Machado, and Silva reports NO photo-uncaging coupled to photogeneration of reactive oxygen and nitrogen species by Ru(II) phthalocyanine (Ru-Pc) complexes. Phthalocyanines generally present maximum absorption maxima in the therapeutic window range with extremely high molar extinction coefficients and thereby can serve as effective photosensitizers with long-wavelength single-photon excitation.

Chapter 11 by Rajagopal, Biddulph, Tabrizi, Fitzgerald-Hughes, and Pryce addresses a critical health issue, namely, development of photo-activated drugs for the treatment of antibiotic-resistant bacterial infections, which are estimated to put millions of people in jeopardy in the foreseeable future. These authors describe organometallic compounds that can release CO or generate ROS, such as $^1O_2$, for antimicrobial therapies.

Chapter 12 by Morris and Massi is a comprehensive review of transition metal-based photosensitizers that are designed to target the mitochondria and the potential applications of these as diagnostic and/or therapeutic agents. The specificity of certain compounds for cancer cells and the toxicity and phototoxicity consequences with various normal and cancer cell lines are evaluated.

In summary, we once again thank all of the authors for their contributions to this volume, and we look forward to continuing contributions from them and many other researchers in extending the tools of photochemistry and photophysics to biomedical diagnostics and therapies.

Peter C. Ford
Co-editor of this volume
Distinguished Professor of Chemistry
University of California Santa Barbara, United States

Rudi van Eldik
Editor of *Advances in Inorganic Chemistry*
Emeritus Professor of Inorganic Chemistry
University of Erlangen–Nuremberg, Germany
Research Professor of Inorganic Chemistry
Nicolaus Copernicus University, Torun, Poland

CHAPTER ONE

Leveraging the photophysical properties of rhenium(I) tricarbonyl complexes for biomedical applications

Justin J. Wilson*
Department of Chemistry and Chemical Biology, Cornell University, Ithaca, NY, United States
*Corresponding author: e-mail address: jjw275@cornell.edu

Contents

1. Introduction 2
2. Properties and applications of $Re(CO)_3$ complexes 4
3. $Re(CO)_3$ complexes as photoluminescent imaging agents 6
4. Simple diimine $Re(CO)_3$ complexes exhibit anticancer activity 8
5. Pyridylimine $Re(CO)_3$ complexes 10
6. Isonitrile-containing $Re(CO)_3$ complexes 14
7. Photoactivation of $Re(CO)_3$ anticancer agents 22
8. Conclusions and future outlook 27
Acknowledgments 28
References 28

Abstract

Rhenium(I) tricarbonyl ($Re(CO)_3$) complexes have been extensively studied for their photochemical and catalytic properties. In recent years, it has been recognized that complexes of this class can also be applied favorably for various diagnostic and therapeutic biomedical applications. In particular, their rich photophysical properties make them amenable for use as optical imaging agents, as well as photoactivated or photodynamic therapeutic agents. Despite the recent rapid expansion of research in this area over the last 5 years, this chapter presents only a narrow, detailed focus on contributions to this field by the Wilson Laboratory in the Department of Chemistry and Chemical Biology at Cornell University. This chapter begins with a brief description of the photophysical properties of $Re(CO)_3$ complexes that make them valuable for biomedical applications. Subsequently, a discussion on the earliest and most prominent examples of how these complexes were used for imaging is presented. The rest of the chapter focuses specifically on anticancer $Re(CO)_3$ complexes developed by the Wilson Lab. These examples highlight mechanistic studies that show how these compounds are distinct from

Advances in Inorganic Chemistry, Volume 80
ISSN 0898-8838
https://doi.org/10.1016/bs.adioch.2022.04.005

conventional platinum anticancer agents and how the ligand choice can affect biological activities. Furthermore, the use of the rich photophysical properties of these anticancer agents for both imaging and photoactivation is discussed.

Abbreviations

ABC ATP-binding cassette
ATF4 activating transcription factor 4
bisim bidentate diimine ligand
bpy 2,2′-bipyridine
CHOP C/EBP homologous protein
DAPTA 3,7-diacetyl-1,3,7-triaza-5-phosphabicyclo[3.3.1]nonane
DFT density functional theory
DMSO dimethylsulfoxide
eIF2α eukaryotic translation initiation factor 2 subunit 1
ER endoplasmic reticulum
FPR formyl peptide receptor
GSH glutathione
H&E hematoxylin and eosin
HPLC high-performance liquid chromatography
Hsp90 heat shock protein 90
ic internal conversion
IC_{50} 50% growth inhibitory concentration
isc intersystem crossing
LF ligand field
MDR multidrug resistance
MLCT metal-to-ligand charge transfer
MTT 3-(4,5-dimethylthiazol-2-yl)-2,5-diphenyltetrazolium bromide
neo neocuproine
PACT photoactivated chemotherapy
PDT photodynamic therapy
Pgp P-glycoprotein
PTA 1,3,5-triaza-7-phosphaadamantane
ROS reactive oxygen species
RT-PCR reverse transcriptase-polymerase chain reaction
SAR structure-activity relationship
THP tris(hydroxymethyl)phosphine
UV ultraviolet
XRF X-ray fluorescence

1. Introduction

The unique properties of transition metal coordination and organometallic complexes, relative to organic molecules, have facilitated their use in a wide range of applications. This aspect is apparent in the field

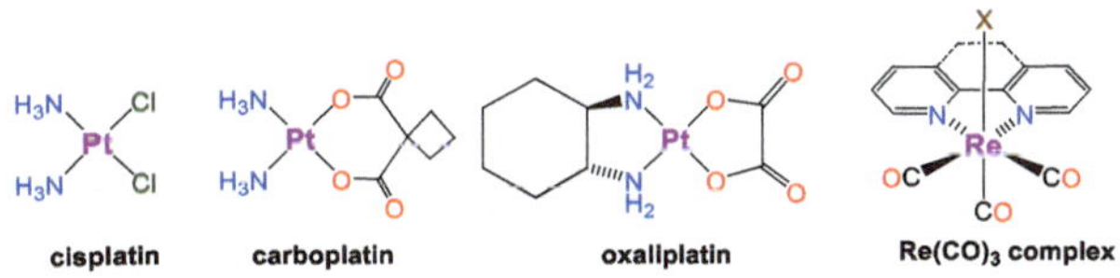

Fig. 1 Structures of the clinically used Pt anticancer drugs cisplatin, carboplatin, and oxaliplatin and the general structure of a $Re(CO)_3$ complex.

of medicine, where metal-based compounds have experienced clinical success as both diagnostic and therapeutic agents.[1–5] With respect to therapeutic applications, the platinum (Pt) compounds cisplatin, carboplatin, and oxaliplatin are among the most widely used and well known metal-containing drugs (Fig. 1).[6] The success of these chemotherapeutic agents has motivated extensive research to develop new Pt-based drug candidates.[7–9] Despite these significant efforts, the advancement of new Pt compounds to clinical trials and beyond has been met with challenges, including dose-limiting toxic side effects and acquired resistance.[10,11]

Driven to overcome the limitations of conventional Pt-based drugs, inorganic chemists have investigated complexes of other metal ions for their biomedical applications. Most notably, researchers in this field have recognized the potential use of transition metal photochemistry for this application.[12–15] In this regard, polypyridyl complexes of the precious metals ruthenium (Ru), osmium (Os), iridium (Ir), and rhenium (Re) have been the subject of most of these studies.[16–22] These complexes exhibit rich photophysical properties that have enabled their use as photoluminescent imaging agents as well as photoactivated and photodynamic therapeutic agents. Among these metal complexes, Re(I) tricarbonyl ($Re(CO)_3$) compounds (Fig. 1) have undergone an expansive growth in their investigation for biomedical applications within the last several years.[23–26] Researchers have found that this class of compounds has a number of favorable properties for both imaging and therapeutic applications. With respect to therapeutic applications, early studies have shown that $Re(CO)_3$ complexes typically exhibit ligand substitution rates that are comparable to the clinically used Pt anticancer agents.[27–29] As such, these complexes have the potential to bind to and interact with intracellular targets on a biologically relevant timescale.[28,30–34] Furthermore, when coupled with appropriate supporting ligands, the $Re(CO)_3$ core possesses valuable photoluminescent properties that can be leveraged for fluorescence microscopy imaging.[35–37] In addition to optical imaging, the intense CO stretching modes of these complexes have also been used in a wide range of vibrational microscopy techniques for visualizing their cellular uptake and intracellular distribution.[38,39]

Lastly, the ability to synthesize analogous tricarbonyl complexes of technetium (Tc), the lighter congener of Re, has enabled the use of its radioisotope ^{99m}Tc for nuclear imaging applications.[40–43] Collectively, both the therapeutic potential and diverse imaging modalities accessible to $Re(CO)_3$ complexes have led to a surge in their biomedical applications.

In this Chapter, we will discuss recent developments from our laboratory in the use of $Re(CO)_3$ complexes in biology. A broad overview of the properties of these complexes followed by some of their applications in cellular imaging, anticancer therapy, and phototoxicity studies will be presented. Given the significant expansion of this field, the scope of this chapter will be focused on work carried out specifically by our lab with several key examples from other research teams when relevant.

2. Properties and applications of $Re(CO)_3$ complexes

The $Re(CO)_3$ core is a robust organometallic fragment that has been incorporated into a number of different molecules. $Re(CO)_3$ complexes bearing aromatic diimine ligands, first reported in 1941,[44] possess a number of interesting photochemical properties that make them useful for different applications. For example, the low-energy empty π^* orbitals of these ligands render metal-to-ligand charge transfer (MLCT) excited states to be accessible in the near ultraviolet (UV) and visible region of the spectrum.[45,46] Coupled with efficient intersystem crossing afforded by the large spin-orbit coupling of the heavy Re center, these compounds are often photoluminescent through the triplet MLCT (^{3}MLCT) excited state. The rich photochemistry of these complexes has been the subject of intense investigation since the 1960s and has been critical to developing a number of fundamental principles in the field of inorganic photochemistry.[47–51] In addition, the catalytic properties of these complexes have been heavily investigated. In particular, $Re(CO)_3$ complexes are efficient photocatalysts and electrocatalysts for the reduction of CO_2 and H^+.[52,53] A large body of work has probed the mechanisms of these transformations as catalyzed by $Re(CO)_3$ complexes, and the effects of ligand modifications on their catalytic efficiencies.[54–58]

The potential biological applications of this photochemistry are dictated by the fate of the rapidly generated ^{3}MLCT excited state (Fig. 2). If this state is energetically far from low-lying ligand field (LF) states, radiative decay becomes the dominant pathway for relaxation, leading to compounds that are suitable for optical imaging applications. Depending on the supporting ligands, however, the ^{3}LF state can be sufficiently lowered in energy, such

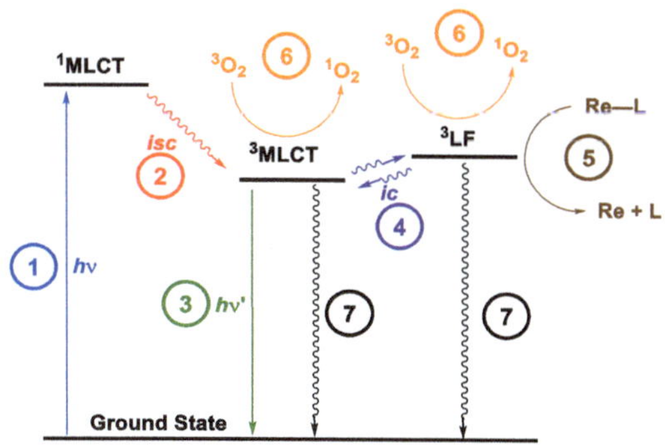

Fig. 2 Simplified Jablonski diagram demonstrating the potential fates of the ^{3}MLCT excited state of diimine $Re(CO)_3$ complexes that depend on the nature of the supporting ligands: (1) Diimine $Re(CO)_3$ complexes absorb light to populate the ^{1}MLCT state. (2) Intersystem crossing (isc) occurs rapidly to the ^{3}MLCT state. Once in the ^{3}MLCT state, (3) these complexes can relax to the ground state by emitting a photon, a process that would make them valuable as optical imaging agents. Alternatively, (4) internal conversion (ic) to thermally accessible ^{3}LF states may be possible. From the ^{3}LF state, (5) photosubstitution can occur readily, favoring the use of $Re(CO)_3$ complexes as PACT agents. Similarly, in either the ^{3}MLCT and ^{3}LF states, (6) energy transfer to ground state $^3O_2$ can take place, leading to the generation of cytotoxic $^1O_2$. $Re(CO)_3$ complexes that undergo this pathway can be leveraged as PDT agents. Lastly, (7) non-radiative relaxation to the ground state may be the dominant process. $Re(CO)_3$ complexes of this type may still be biologically active in the absence of light and useful for anticancer applications.

that it is thermally accessible from the ^{3}MLCT excited state. In these cases, excitation of the $Re(CO)_3$ complexes can lead to efficient photosubstitution reactions via population of the dissociative ^{3}LF excited states.[59] Complexes of this type have the potential to act as photoactivated chemotherapeutic (PACT) agents, or compounds that trigger light-induced cancer cell death without generating singlet oxygen ($^1O_2$). The use of $Re(CO)_3$ complexes as photodynamic therapeutic (PDT) agents, compounds that generate cytotoxic $^1O_2$, is also possible, as both the ^{3}MLCT and ^{3}LF excited states typically have sufficient energy to convert ground state triplet oxygen ($^3O_2$) to $^1O_2$.[22] The efficiency of this process is dictated by the energy difference between the Re and O_2 excited states, the respective lifetimes of the Re excited states, and the concentration of O_2. Compounds with effective $^1O_2$-sensitization capabilities can be leveraged as PDT agents, which elicit cancer cell death via the generation of this potent reactive oxygen species (ROS). Lastly, if non-radiative decay via other pathways occurs efficiently, the ^{3}MLCT returns to the ground state. Compounds for which this pathway is the dominant process may still be biologically active in the absence of light. Importantly, the relative energies and accessibilities of non-radiative decay pathways can be largely controlled by the nature of supporting ligands on these complexes. As such, $Re(CO)_3$ complexes provide a robust and diverse framework to capitalize on different photobiological applications by varying the types of ligands used.

3. $Re(CO)_3$ complexes as photoluminescent imaging agents

The earliest studies involving $Re(CO)_3$ in biological environments sought to leverage their rich photophysical properties for imaging applications. In a landmark investigation in 2004, a $Re(CO)_3$ complex bearing a tridentate bis(quinolyl) amine ligand was conjugated to a peptide, fMLF(SAACQ)G, that binds to the formyl peptide receptor (FPR). This receptor is overexpressed on neutrophils and has been used in nuclear medicine as a means of tracking white blood cells. This complex fMLF $[(SAACQ\text{-}Re(CO)_3)^+]G$ (Fig. 3) was successfully used for fluorescence imaging and flow cytometry of human leukocytes.[60] Furthermore, its corresponding isostructural ^{99m}Tc complex was also prepared as a nuclear imaging partner. Shortly after this initial report, a second study on the optical imaging applications of $Re(CO)_3$ complexes of the formula $[Re(bisim)L(CO)_3]^+$, where bisim is a bidentate diimine and L is a monodentate pyridine ligand, were reported in 2007.[61] In this work, both the bisim ligand and the monodentate pyridine analogues were systematically modified to investigate their effects on the photophysical properties and intracellular distribution of the resulting complexes. These compounds were imaged in a parasitic flagellate *Spironucleus vortens*. The researchers found that ligand modifications not only altered the emission energies and quantum yields of the complexes, but also affected their toxicity and susceptibility to photobleaching in the parasite. The complex bearing a sulfonated phenanthroline and hydroxymethyl pyridine ligand was found to possess optimal cell uptake

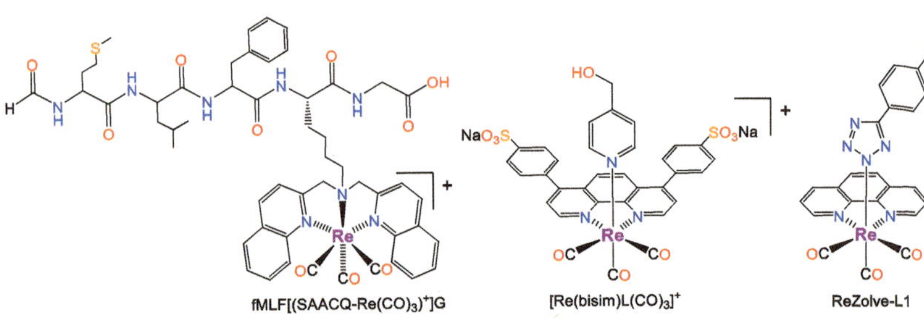

Fig. 3 Structures of several of the earliest used and most prominent $Re(CO)_3$ biological imaging agents.

with low toxicity, suggesting its value as an optical imaging agent (Fig. 3). Together, these two initial studies highlighted the promise and viability of $Re(CO)_3$ complexes bearing tridentate and a mixture of bidentate and monodentate ligands. Many subsequent efforts have been undertaken to explore these two classes of $Re(CO)_3$ complexes for optical imaging applications. One of the most significant outcomes of these efforts has been the development and commercialization of the $Re(CO)_3$ tetrazolato complex with the trade name ReZolve-L1™ (Fig. 3).[62] This neutral $Re(CO)_3$ complex selectively stains lipid droplets, enabling their imaging by two-photon fluorescence microscopy and Fourier transform infrared microspectroscopy in human adipocytes and adipose cells from *Drosophila*.[63–65] Importantly, this compound was applied as a tool to study the distinct lipid profiles of human prostate cancer cell lines (Fig. 4).[66] Furthermore, an analogue of

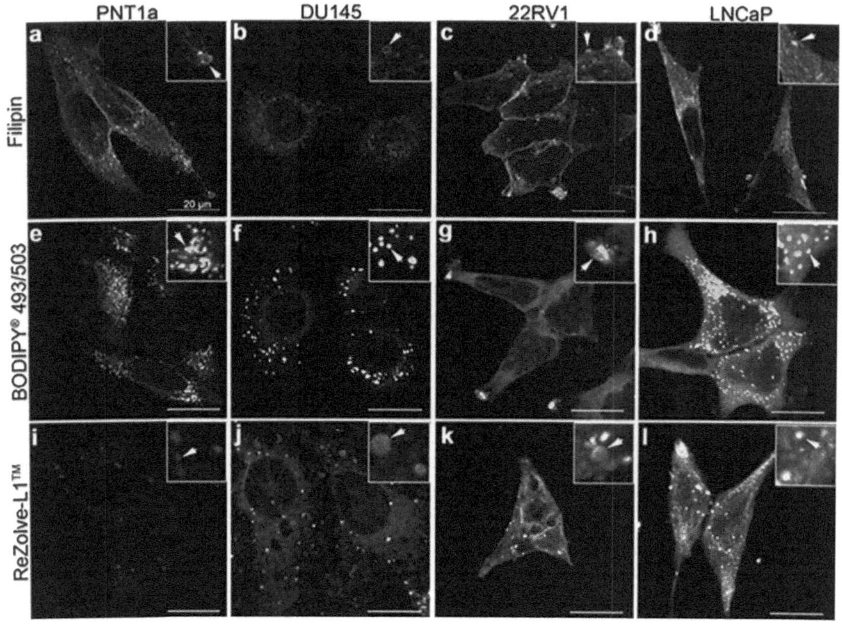

Fig. 4 Fluorescence microscopy images of the non-malignant prostate cell line PNT1a, and the prostate cancer cell lines DU145, 22RV1, and LNCaP treated with either Filipin III (a stain for free cholesterol), BODIPY® 493/503 (a stain for cholesteryl esters and triacylglycerides), or ReZolve-L1 ™. Cells in panels (A–H) were fixed with 4% paraformaldehyde, and those in panels (I–K) were imaged live. The scale bars are 20 μm. These images show how ReZolve-L1 ™ is able to identify distinct lipid droplet profiles within the prostate cancer cells lines. *Reproduced with permission from Sorvina, A.; Bader, C.A.; Caporale, C.; Carter, E.A.; Johnson, I.R.D.; Parkinson-Lawrence, E.J.; Simpson, P.V.; Wright, P.J.; Stagni, S.; Lay, P.A.; Massi, M.; Brooks, D.A.; Plush, S.E.* Oncotarget ***2018,*** 9, *35541–35552. Copyright 2018 Impact Journals, LLC.*

ReZolve-L1™ bearing a pyridyl tetrazolato ligand was found to selectively localize to the endoplasmic reticulum (ER),[67] thus demonstrating how the imaging applications of these $Re(CO)_3$ complexes can be modified with appropriate ligand choices. Given that our laboratory has focused mostly on the therapeutic applications of $Re(CO)_3$, this topic will not be discussed in greater detail and readers are referred to other comprehensive review articles on their optical imaging applications.[14,35,36,39,68–70]

4. Simple diimine $Re(CO)_3$ complexes exhibit anticancer activity

Based on the favorable photoluminescent properties and ligand substitution kinetics of $Re(CO)_3$ complexes, our lab was interested in the possibility of using these complexes as theranostic anticancer agents.[71] This interest was further bolstered by prior studies that demonstrated some $Re(CO)_3$ complexes to possess cytotoxic and antitumor activities in the absence of light irradiation.[72,73] In our initial efforts towards discovering new $Re(CO)_3$ anticancer agents, we prepared a small library of these complexes bearing different diimine ligands based on both 2,2′-bipyridine (bpy) and 1,10-phenathroline (phen) to carry out a simple in vitro screening of their anticancer activities.[74] To improve aqueous solubility and biocompatibility, we used these compounds in their cationic aquated forms, where the axial ligand is a neutral aqua ligand. Our screening of this small compound library revealed them to inhibit cancer cell growth within the micromolar concentration range. Notably, one compound identified, Neo-Re or $[Re(CO)_3(neo)(OH_2)]^+$ where neo = neocuproine (Fig. 5), was more potent than the reference metallodrug cisplatin in HeLa cells.

The potent anticancer properties of this complex in HeLa cells prompted us to carry out additional biological studies to understand its mechanism

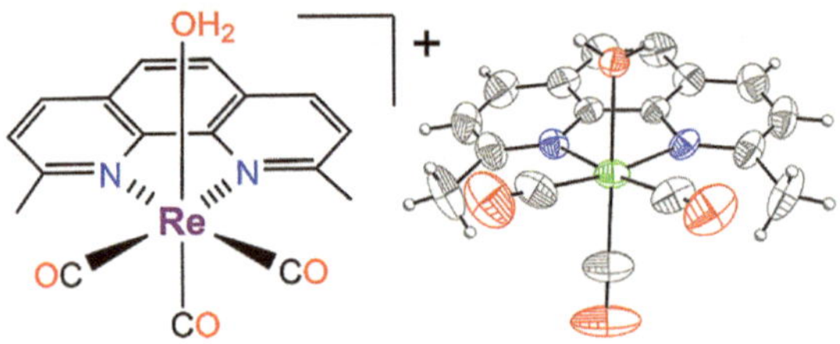

$[Re(CO)_3(neo)(OH_2)]^+$
Neo-Re

Fig. 5 Structure of Neo-Re, a cytotoxic $Re(CO)_3$ anticancer agent.

of action. The cross-resistance of this compound with cisplatin was assessed by testing its in vitro anticancer activity in both wild-type A2780 and cisplatin-resistant A2780CP70 ovarian cancer cells, as well as wild-type KB-3-1 and cisplatin-resistant KBCP20 cervical cancer cells. Although Neo-Re was less potent in the wild-type cell lines compared to cisplatin, it was more active against the cisplatin-resistant variants and exhibited no differential activities between the two cell line pairs. As such, these data clearly showed that Neo-Re could circumvent the platinum-resistance mechanisms within these cell lines. This latter result highlights the fact that these compounds act via a distinct mechanism of action.

To further study its mechanism of action, we carried out a series of biological studies to probe for canonical modes of cell death. Intriguingly, a combination of flow cytometric studies, western blots, and cytotoxicity assays in the presence of different cell death pathway inhibitors failed to paint a clear picture of a canonical form of cell death that Neo-Re could be inducing in cancer cells. In terms of other phenotypic responses of cancer cells treated with Neo-Re, the formation of enlarged endosomes and lysosomes within the cytoplasm occurred (Fig. 6). The apparent perturbation of these organelles is consistent with our cellular uptake studies that showed it to enter cells via endocytosis. Although a specific molecular target has not yet been conclusively identified, a COMPARE analysis[75] of data obtained from an NCI-60 screen[76] showed that the spectrum of activity of this compound correlated highly with other compounds that are known inhibitors of the chaperone heat shock protein 90 (Hsp90).[77] A large subset of the known Hsp90 inhibitors interact with this protein via covalent interactions with key cysteine residues.[78] Similarly, the labile aqua ligand present in Neo-Re provides an opportunity for this compound to form covalent interactions with potential protein targets. Our work and those of others has shown that these $Re(CO)_3$ have a strong affinity for sulfur donors like cysteine and glutathione (GSH),[74,79–81] suggesting that their biological activities may arise in part from these types of interactions. This latter concept was demonstrated with a similar library of compounds that was shown to inhibit a key protease responsible for the virulence of the SARS-CoV-2 virus by forming covalent adducts with a cysteine residue.[81]

Based on the promising in vitro anticancer activity of Neo-Re, the in vivo antitumor properties of this compound were assessed in mice bearing an ovarian cancer patient-derived xenograft.[82] These mice were injected twice weekly via tail vein injection with either 10, 20, or 40 mg/kg of Neo-Re. These doses were able to slow tumor growth over a period of

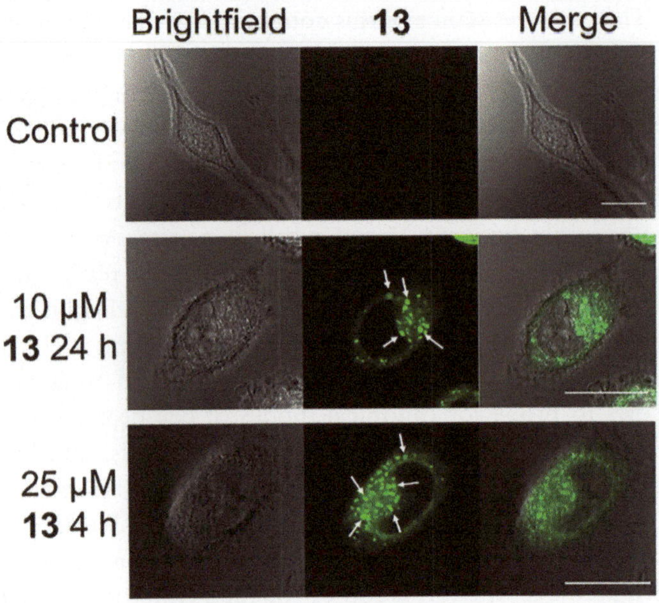

Fig. 6 Confocal fluorescence microscope images of HeLa cells in the absence or presence of Neo-Re (**13** in this figure). The photoluminescence of this compound was imaged directly, enabling the detection of vacuolization of the cytosol as indicated by the white arrows. Scale bars are 20 μm. *Reproduced with permission from Knopf, K. M.; Murphy, B. L.; MacMillan, S. N.; Baskin, J. M.; Barr, M. P.; Boros, E.; Wilson, J. J.* J. Am. Chem. Soc. ***2017,*** 139, *14302–14314. Copyright 2017 American Chemical Society.*

30 days. Notably, no differences in tumor growth were observed between the three administered doses, suggesting that higher doses beyond 10 mg/kg afford no additional beneficial biological effects. The body weight of all three mice cohorts did not decrease over the duration of this experiment, indicating that Neo-Re is well tolerated. This latter aspect was verified via histological studies, which showed no morphological abnormalities in sensitive organs like the kidneys and liver. Overall, these in vivo studies underscore the potential for compounds like Neo-Re as anticancer agents.

5. Pyridylimine $Re(CO)_3$ complexes

Based on the promising anticancer activities of the simple bpy and phen complexes discussed above, we sought alternative bidentate ligands that could also confer $Re(CO)_3$ complexes with cytotoxic activity. In this regard, pyridylimine ligands, which can be synthesized via a simple condensation reaction between an amine and a picolinaldehyde, provide a versatile platform by which the properties of these compounds could be systematically modified. Indeed, pioneering work by Ziegler and coworkers in this area has shown that this condensation reaction can be a straightforward means of attaching biologically relevant groups, like peptides, amino acids and fluorophores, to the $Re(CO)_3$ core.[83–86] Furthermore, pyridylimine ligands have been previously studied for their use in a wide range of different metal-based anticancer agents.[87] In particular, the facile assembly of these ligands from various picolinaldehydes and amines was used to generate a combinatorial library of different Ru arene complexes within a 96-well plate, which was then screened directly for anticancer activity.[88,89] Importantly, this work was able to identify novel Ru-based anticancer agents that induce ER stress.[90–92]

Motivated by these prior studies, we sought a combinatorial synthesis of $Re(CO)_3$ pyridylimine complexes for anticancer activity screening.[93] Like the prior studies involving Ru arene complexes, we varied both the picolinaldehyde and amine components along the columns and rows of a 96-well plate. In each well, $Re(CO)_5Cl$ was added to act as the template for the condensation reaction and to form complexes of the general formula $[Re(CO)_3(\text{pyridylimine})Cl]$ directly in each well. Upon mixing of these three components in each well in dimethyl sulfoxide (DMSO), the reaction required microwave irradiation to proceed (Fig. 7). Each well was then assayed by high-performance liquid chromatography (HPLC) to assess the formation and purities of the pyridylimine $Re(CO)_3$ complexes. The majority of wells displayed >80% conversion to the pyridylimine-containing product, demonstrating the success of this approach.

Following the generation of this compound library, 10 μM concentrations of these samples were transferred to a 96-well plate containing either HeLa, A2780, or A2780CP70 cancer cells to assess their effects on cell viability via the 3-(4,5-dimethylthiazol-2-yl)-2,5-diphenyltetrazolium bromide (MTT) assay. Of the 96 compounds prepared, only three were significantly cytotoxic (Fig. 7). These three compounds share the common dimethylamino functional group on the pyridylimine ligand, suggesting that this feature is critical for anticancer activity within this class of compounds. Following this initial screen, they were independently synthesized and isolated at >95% purity. The assessment of the in vitro anticancer activity of these purified compounds revealed them to also possess potent cytotoxic activity in cancer cell lines, verifying that the intact Re complex, rather than impurities or ligand fragments, is responsible for the biological properties observed from our combinatorial screening efforts.

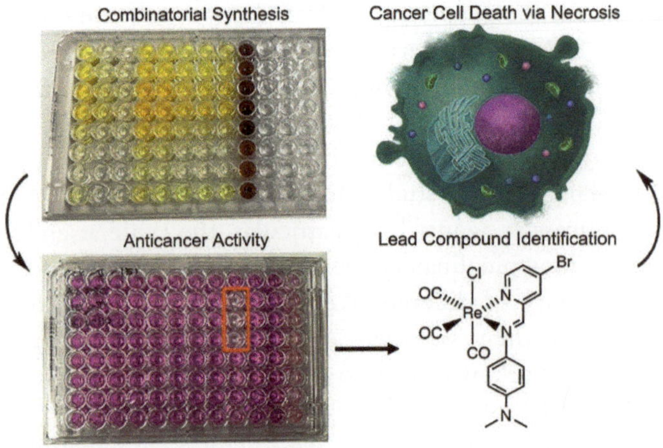

Fig. 7 Combinatorial synthetic strategy for identifying anticancer-active $Re(CO)_3$ pyridylimine complexes. (Top left) A mixture of different amines and picolinaldehydes with $Re(CO)_5Cl$ were subjected to microwave irradiation to generate a library of $[Re(CO)_3(pyridylimine)Cl]$ complexes. (Bottom left) These complexes were then transferred directly to a 96-well plate containing cancer cells and screened for cytotoxic activity with the MTT assay. (Bottom right) From these screening efforts, three potent $[Re(pyridylimine)Cl(CO)_3]$ complexes were identified. The most effective compound, Re-9D (shown), was synthesized and purified independently. (Top right) The biological mechanism of action of Re-9D was studied, revealing it to induce cell death via necrosis. *Reproduced with permission from Konkankit, C.C.; Vaughn, B.A.; MacMillan, S.N.; Boros, E.; Wilson, J. J.* Inorg. Chem. ***2019,*** 58, *3895–3909. Copyright 2019 American Chemical Society.*

Mechanistically, these pyridylimine $Re(CO)_3$ complexes are distinct from both cisplatin and Neo-Re. They are equally potent in both wild-type and cisplatin-resistant ovarian cancer cells, reflecting a lack of cross resistance with cisplatin. Both brightfield and fluorescence microscopy was used to further probe the mechanism of action of the most potent compound in this class Re-9D (Fig. 7), revealing distinct differences from Neo-Re as well. Within several hours of compound treatment, the outer membrane of cancer cells becomes permeabilized and cytoplasmic expulsion occurs. This rapid outer membrane permeabilization was also detected with the fluorescent dye propidium iodide, which can only enter cells with compromised membranes (Fig. 8). The membrane permeabilization and cytoplasmic expulsion induced by Re-9D suggest that it causes canonical necrotic cell death in cancer cells, in contrast to Neo-Re.

Building upon these studies, we targeted alternative pyridylimine complexes that may possess potent cytotoxic activities. We rationalized that

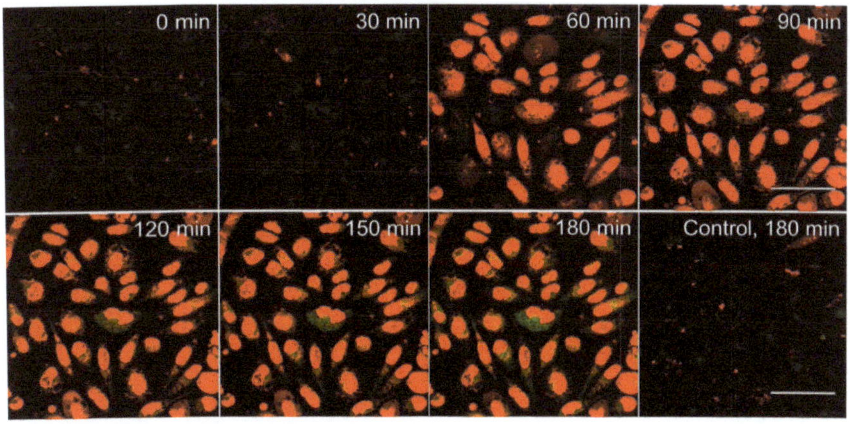

Fig. 8 A2780CP70 cells treated with 60 μM Re-9D and propidium iodide imaged by confocal fluorescence microscopy over 180 min. The influx and nuclear staining of propidium iodide is caused by the necrotic rupture of the cell membranes upon exposure to Re-9D. Scale bars are 100 μm. *Reproduced from Konkankit, C.C.; Vaughn, B.A.; MacMillan, S.N.; Boros, E.; Wilson, J.* J. Inorg. Chem. ***2019,*** 58, *3895–3909. Copyright 2019, American Chemical Society.*

increasing the compound lipophilicity would enhance the cellular uptake and cytotoxic activities of this class of compounds. This phenomenon has previously been exploited to develop highly active platinum-based drug candidates[94–98] and has also been explored in the context of different $Re(CO)_3$ complexes primarily for imaging applications.[86,99,100] Using this modular condensation chemistry, picolinaldehyde and amines with different alkyl chain lengths, ranging from 2 to 12 carbons, were attached to the $Re(CO)_3$ core to make a series of complexes with varying lipophilicities (Fig. 9).[101] The water-octanol partition coefficients (log P values) of these complexes ranged from 1.59 to 2.95, demonstrating that these values could be tuned with the length of the alkyl chain. Unexpectedly, the 50% growth inhibitory concentration (IC_{50}) values of all six of these complexes in HeLa cells were identical after a 48-h incubation period (Fig. 10A). However, the use of different incubation times with these compounds enabled us to detect significant variations on their effects on cell viability. In particular, compounds with longer alkyl chains reduced cell viability more effectively over shorter incubation time periods, but these differential cytotoxic effects were not apparent over longer time frames (Fig. 10B). This result highlights the significance of testing the effect of time in biological studies and how physiochemical property differences, like lipophilicity, may only be apparent at early time points. Several other studies have also

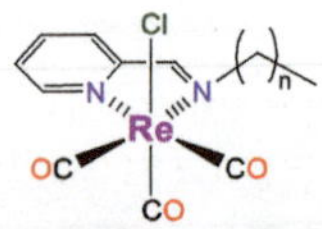

n = 1 (**Re-C2**), n = 2 (**Re-C3**)
n = 3 (**Re-C4**), n = 4 (**Re-C5**)
n = 5 (**Re-C5**), n = 11 (**Re-C12**)

Fig. 9 Structures of $Re(CO)_3$ pyridylimine complexes bearing different alkyl chain substituents.

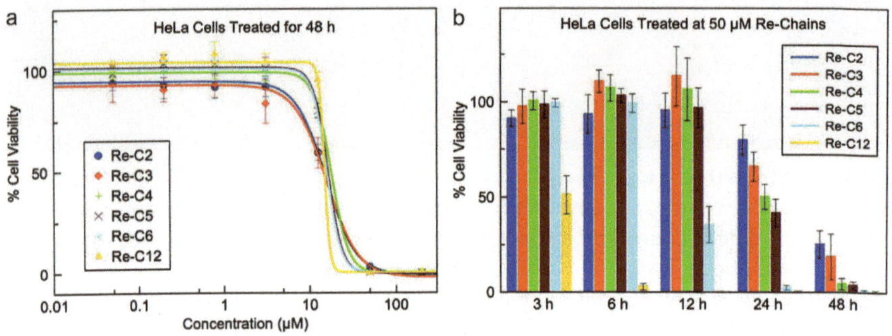

Fig. 10 (A) Dose-response curves of HeLa cells treated with the $Re(CO)_3$ pyridylimine alkyl chain complexes, named in Fig. 9, for a 48-h incubation period. No significant differences in cytotoxic activity over this time point were observed. (B) Single-dose 50 μM-treatment of HeLa cells with the $Re(CO)_3$ pyridylimine alkyl chain complexes for variable incubation times. Notably, compounds with longer alkyl chains induce their cytotoxic effects more rapidly. *Reproduced with permission from Konkankit, C.C.; Vaughn, B.A.; Huang, Z.; Boros, E.; Wilson, J. J.* Dalton Trans. ***2020,*** 49, *16062–16066. Copyright 2020 Royal Society of Chemistry.*

begun to investigate the kinetics of cellular uptake of metal-based drug candidates as an important parameter in determining their viability and use in different therapeutic and diagnostic applications.[102,103]

6. Isonitrile-containing $Re(CO)_3$ complexes

Isonitriles are π-acceptor ligands that share many common features with CO. Although they can generate metal complexes with similar chemical properties as CO, a potential advantage of isonitriles is that they can be modified by altering the substituents on the nitrogen atom. This amenability to structural diversity has led to the use of these ligands for different biomedical applications. Most notably, the clinically used perfusion imaging agent, Cardiolite®, is a homoleptic ^{99m}Tc isonitrile complex.[104] Furthermore, $Re(CO)_3$ complexes bearing these ligands have rich photophysical properties that could be leveraged for imaging applications.[105,106] Based on these favorable properties, we began to explore the use of isonitriles as ligands for new cytotoxic $Re(CO)_3$ complexes.

In our initial screening studies, we identified a tricarbonyl rhenium isonitrile polpyridyl (TRIP) complex of the formula $[Re(CO)_3(neo)(isc)]^+$, where isc = *p*-tolylisonitrile, that exhibits potent in vitro anticancer activity (Fig. 11).[107] In A2780 ovarian cancer cells, the cytotoxicity of this compound exceeds that of cisplatin, suggesting that it could be a valuable therapeutic agent. Furthermore, like the other $Re(CO)_3$ compounds described above, TRIP is not cross-resistant with cisplatin, a property that is reflected by its equitoxic effects in both wild-type A2780 and cisplatin-resistant A2780CP70 ovarian cancer cell lines. A key difference between TRIP and the other $Re(CO)_3$ complexes discussed above is that it contains no labile ligands, a property that was verified with X-ray fluorescence (XRF) microscopy. For these XRF studies, an analogue of TRIP containing an iodine (I) on the axial isonitrile ligand (I-TRIP) was used.[108] HeLa cells treated with I-TRIP were imaged by XRF microscopy, and these imaging studies showed that Re and I co-localized with one another, indicating that the isonitrile ligand does not dissociate from the Re coordination sphere intracellularly (Fig. 12). These results are in contrast to the $Re(CO)_3$ aqua and chlorido complexes described above, where the axial ligands are known to undergo facile ligand substitution reactions with biological nucleophiles,

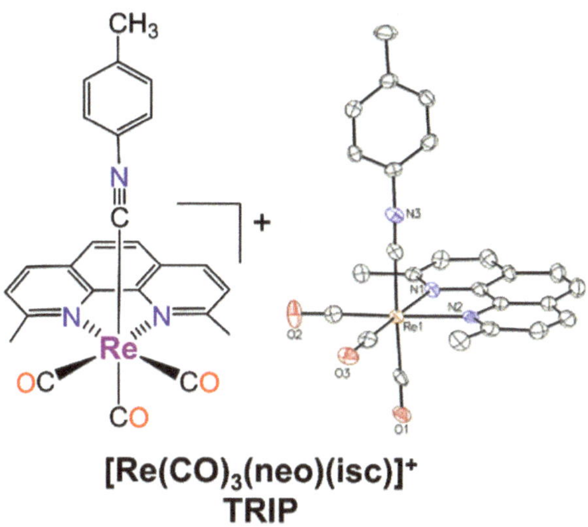

Fig. 11 Structure of the isonitrile-$Re(CO)_3$ complex TRIP.

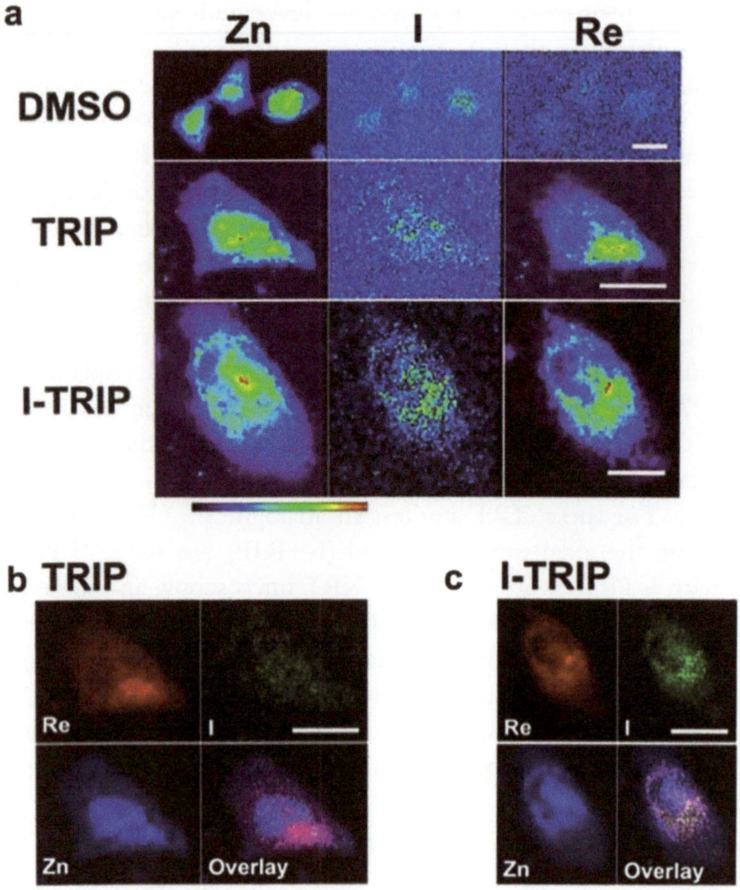

Fig. 12 (A) XRF maps displaying the elemental distribution of Zn, I, and Re within HeLa cells treated with either DMSO (0.06% v/v), 2 mM TRIP, or 3 mM I-TRIP. Correlation analysis on the co-localization of Re, I, and Zn distribution maps for cells treated with (B) TRIP and (C) I-TRIP. The scale bar is 20 μm. *Reproduced with permission from Konkankit, C.C.; Lovett, J.; Harris, H.H.; Wilson, J.* J. Chem. Commun. ***2020,*** 56, *6515–6518. Copyright 2020 Royal Society of Chemistry.*

but are similar to those obtained for iodine-labeled analogues of the tetrazolate-bearing $Re(CO)_3$ imaging agent ReZolve™, for which the axial tetrazolate ligand also remained bound in cells.[109] Collectively, these studies show how $Re(CO)_3$ complexes can be altered to afford either covalent inhibitors or intact structural inhibitors.[110]

Given the different chemical reactivity of TRIP, its biological mechanism of action was of particular interest. Fluorescence microscopy studies revealed this compound to induce mitochondrial fission 30 min post-treatment, which was then related to the ability of this compound to trigger apoptotic and autophagic cell death. Moreover, because mitochondrial fission and autophagy are frequently related to ER stress,[111] we investigated the role of this phenomenon in the mechanism of TRIP. Western blots of cell lysates from cancer cells treated with TRIP revealed this compound to induce phosphorylation of the transcription factor eIF2α. This transcription factor is a regulator of the unfolded protein response (UPR), a mechanism of cellular response to ER stress.[112] We further showed that the downstream consequences of eIF2α phosphorylation, namely the inhibition of translation and the upregulation of pro-apoptotic stress response factors ATF4 and CHOP,[113] also occurred. Collectively, these data support the fact that TRIP is capable of causing ER stress in cancer cells, which in turn leads to translation inhibition and cell death. To investigate the mechanisms by which this compound induces ER stress, we used the dye thioflavin T, which elicits a fluorescent response in the presence of protein aggregates, to show that TRIP increases the misfolded protein burden in cancer cells within a short 30 min time period. Based on these data, a mechanistic hypothesis for TRIP was formulated and is summarized in Fig. 13.

As noted in Section 1, a key limitation of the Pt-based drugs arises from the ability of cancer cells to develop resistance to them.[114] Although we showed that TRIP is not cross-resistant with cisplatin, we considered the possibility that ovarian cancer cells may be able to develop alternative resistance mechanisms to circumvent the biological properties of this $Re(CO)_3$ complex. In addition to highlighting potential deficiencies of TRIP, the identification of resistance pathways is valuable in gaining a deeper mechanistic insight on this compound, as drug resistance mechanisms have previously been correlated to specific biological targets for novel therapeutic agents.[115,116] The treatment of A2780 ovarian cancer cells with TRIP at doses below its IC_{50} value for over a year afforded a resistant cell line A2780TR, which required a ninefold higher concentration of TRIP to elicit an equivalent cytotoxic effect compared to the wild-type A2780 cell line.[117] Although the Pt-based drugs cisplatin, carboplatin, and oxaliplatin were not cross-resistant in this cell line, the organic drug taxol exhibited a remarkable 180-fold decrease in potency in the TRIP-resistant cell line.

The significant cross-resistance between TRIP and taxol indicated that taxol-resistance mechanisms were also likely a key factor in the observed TRIP resistance mechanisms. In this context, one of the most predominant

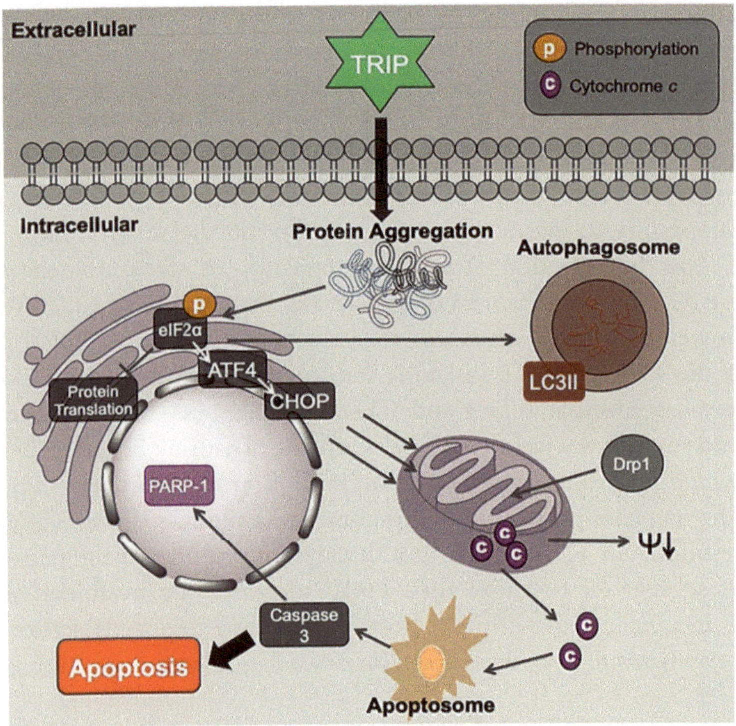

Fig. 13 Summary of the proposed mechanism of anticancer activity of TRIP. *Reproduced with permission from King, A.P.; Marker, S.C.; Swanda, R.V.; Woods, J.J.; Qian, S.-B.; Wilson, J.* J. Chem. A Eur. J. ***2019,*** *25, 9206–9210. Copyright 2019 Wiley-VCH.*

mechanisms of resistance to taxol is the upregulation of the ATP-binding cassette (ABC) transporter P-glycoprotein (Pgp).[118] The upregulation of Pgp, which effluxes a wide range of different organic drugs from cells, is a common component of the multidrug resistance (MDR) pathway that is prevalent among drug-treated cancer cells and tumors.[119] Consistent with this hypothesis, we observed that TRIP-resistance could be reversed by the Pgp inhibitor verapamil and that cellular uptake of TRIP, as measured by both fluorescence microscopy and inductively coupled plasma optical emission spectroscopy, was significantly diminished in the TRIP-resistant cell line. Lastly, Pgp-encoding mRNA levels in the A2780 and A2780TR cell lines were determined with reverse transcriptase-polymerase chain reaction (RT-PCR). These studies revealed that the resistant cell line contains 800 times more Pgp-encoding mRNA than the wild-type cell line. Thus, these data demonstrate that Pgp is an important efflux transporter for TRIP that mediates cancer cell resistance to this compound. Although Pgp has long been implicated in cancer resistance to organic drugs, these studies support the role of this transporter in mediating inorganic and organometallic complex drug resistance mechanisms. For example, in addition to our work with TRIP, earlier studies have shown that the ^{99m}Tc imaging agent Cardiolite®,[120] the Pt-based drug oxaliplatin,[121] the anti-arthritic gold drug auronafin,[122] the anti-metastatic Ru drug candidate KP1019,[123] and cyclometalated Ru and Os ER stress-inducers[124] are also substrates of Pgp. Thus, the role of Pgp in xenobiotic efflux processes may extend beyond conventional organic drugs and be more significant for the detoxification of metal-containing chemical compounds.

Based on the promising anticancer activity of TRIP and its novel mechanism of protein aggregation and ER stress induction, we next sought to find improved analogues and develop structure-activity relationships (SARs) for this compound class. A small library of 11 $Re(CO)_3$ complexes containing different diimine and isonitrile ligands was synthesized (Fig. 14). These ligands were chosen to modify both the lipophilic and electronic properties of the complexes and to assess their importance in mediating their cytotoxic activities in HeLa cells. Although no strong dependence on the lipophilicity was observed, several correlations with the electron-donating strengths of the ligands were observed. For example, both the CO and isonitrile CN vibrational stretching frequencies correlated with

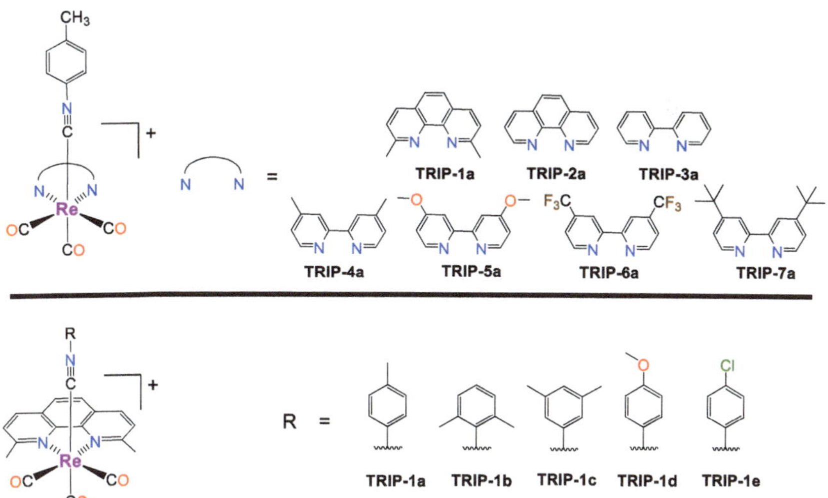

Fig. 14 Structures and naming scheme of the 11 different TRIP analogues investigated for the SARs in Ref. 125. TRIP-1a is the same compound named TRIP, shown in Fig. 11.

the cytotoxic activities of these complexes. Specifically, lower energy stretching frequencies, a property that will arise in more electron-rich metal centers, were indicative of more potent compounds. In addition to probing the cytotoxic properties of this compound library, we also investigated their mechanisms of action. Importantly, we found that this compound library triggered the same phenotypic responses in cancer cells as TRIP (Fig. 15). Specifically, these compounds upregulated CHOP, caused

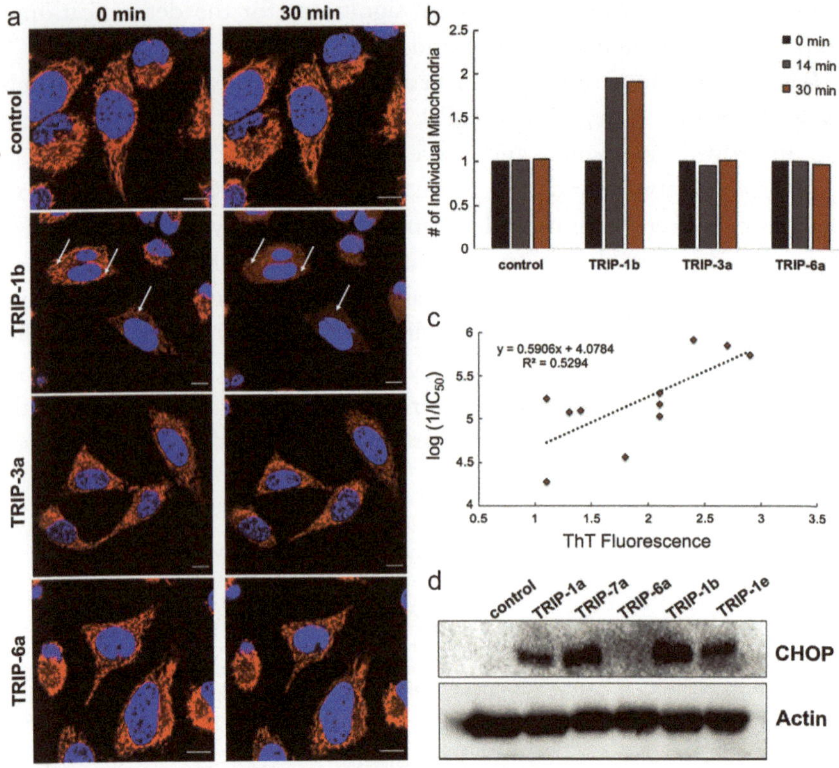

Fig. 15 Biological mechanistic studies and SARs of the TRIP analogs, which are named as shown in Fig. 14. (A) HeLa cells treated with 10 μM of the indicated TRIP analog stained with MitoTracker red and Hoechst. Arrows indicate signs of mitochondrial fission after a 30-min treatment period. Scale bars are 10 μm. (B) Quantification of the number of mitochondria resulting from fission after treatment with 10 μM of the indicated TRIP analog. (C) Plot of cytotoxic activity, log(1/IC_{50}), of the TRIP analogs vs intracellular thioflavin T fluorescence in HeLa cells, showing a correlation between these two properties. (D) CHOP expression levels shown by western blot in HeLa cells treated with the indicated TRIP analog for 24 h. *Reproduced with permission from Marker, S.C.; King, A.P.; Granja, S.; Vaughn, B.; Woods, J.J.; Boros, E.; Wilson, J. J.* Inorg. Chem. ***2020,*** *59, 10285–10303. Copyright 2020 American Chemical Society.*

mitochondrial fission, and induced intracellular protein aggregation. Furthermore, the extent of protein aggregation caused by these compounds was correlated directly with their cytotoxic activities, confirming that this phenotypic response is a critical component of their mechanism of action. The observation that all of these compounds within the library exhibited the same biological mechanism of action is important, especially in the context of developing rational SAR. An interesting problem in the field of medicinal inorganic chemistry is that often times simple ligand modifications can dramatically alter the mechanism of action.[90,126–128] These significant mechanistic changes can make it difficult to improve biological activity and pharmacokinetic properties as the resulting biological effects may be different.

Following the promising in vitro studies with TRIP and its analogues, the in vivo antitumor properties and biodistribution of this compound were assessed. For the biodistribution studies, a ^{99m}Tc analogue of TRIP, called TTIP, was prepared to determine its suitability as a diagnostic partner for this therapeutic agent. Both compounds overall showed very similar biodistribution patterns with pronounced accumulation in the kidneys and heart. The kidney uptake can be explained by the renal clearance of these compounds, whereas the heart accumulation is consistent with the lipophilic cationic nature of these compounds that favors accumulation in the mitochondria-rich cardiac cells and the high expression levels of Pgp within this tissue. Following these biodistribution studies, the in vivo antitumor activity of TRIP (TRIP-1a) was assessed in mice bearing ovarian cancer A2780 xenografts. The mice were treated with TRIP twice weekly via tail vein injection. As shown in Fig. 16, the administered dose of 20 mg/kg was able to slow tumor growth in a statistically significant manner, resulting in prolonged survival of the mice. In addition, only minimal changes in body weight were observed, and histological analysis with hematoxylin and eosin (H&E) staining revealed no detectable morphological changes in organ composition. Thus, TRIP appears to be well-tolerated in vivo. In addition to our mouse study of Neo-Re,[82] described above in Section 4, there have only been a limited number of investigations of this type using $Re(CO)_3$ complexes, but they collectively reveal the promising in vivo antitumor potential of this class of compounds.[129–134] Altogether, our in-depth studies of TRIP and its analogues as promising anticancer agents highlight the increasingly recognized importance of metal complexes as novel ER stress-inducers and protein translation inhibitors.[135,136]

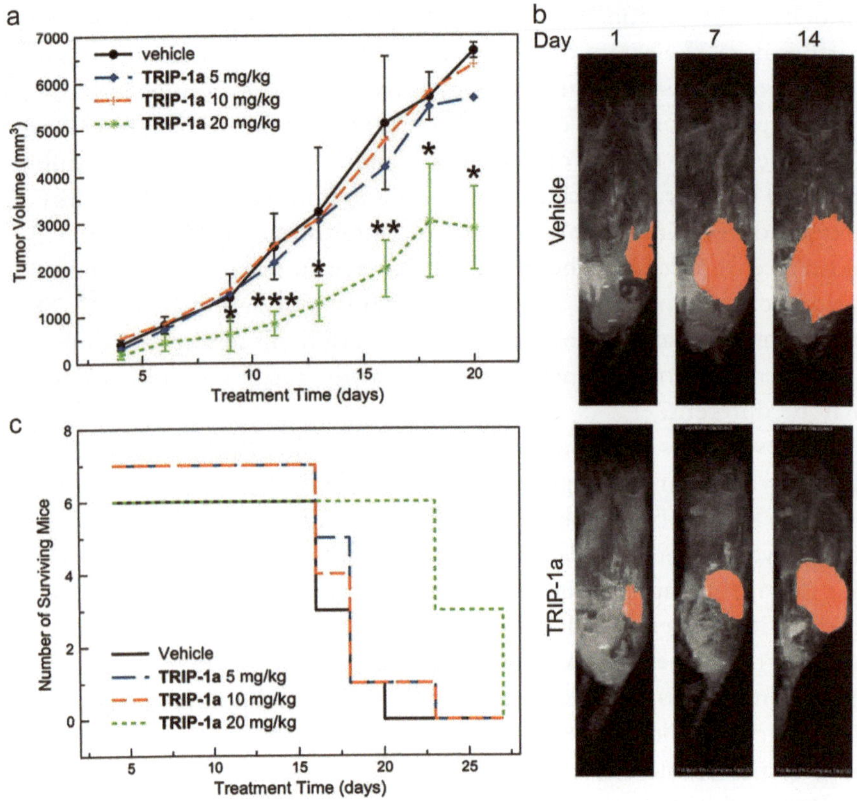

Fig. 16 In vivo antitumor efficacy of TRIP (TRIP-1a) in mice bearing A2780 ovarian cancer xenografts. (A) Tumor volume of mice treated with the indicated doses of TRIP ($* = P < 0.05$, $** = P < 0.01$, $*** = P < 0.005$). (B) Magnetic resonance images of mice treated with either the vehicle control or TRIP. The tumor xenograft is marked in red. (C) Kaplan-Meier survival plot of mice treated with TRIP over the duration of the study. *Reproduced with permission from Marker, S.C.; King, A.P.; Granja, S.; Vaughn, B.; Woods, J.J.; Boros, E.; Wilson, J.* J. Inorg. Chem. ***2020,*** 59, *10285–10303. Copyright 2020 American Chemical Society.*

7. Photoactivation of $Re(CO)_3$ anticancer agents

In addition to providing valuable applications in imaging, the rich photophysical properties of the late 2nd and 3rd row transition metal complexes have also been harnessed for developing photoactivated anticancer agents. The majority of complexes of this type are based on Ru and Ir[16,137] with substantially fewer examples for Re.[21] Despite the paucity of photoactivated Re-based anticancer agents reported to date, several early studies highlighted the value of these complexes for this application. It has been demonstrated that $Re(CO)_3$ complexes bearing suitable diimine or quinolylamine ligands generate $^1O_2$ upon photoirradiation, which can trigger photoactivated cell death in cancer cells.[138–141] Notably, by changing the nature of the ligands, photoactivation with biocompatible red light could be achieved.[142]

Although the generation of $^1O_2$ is a highly effective means of inducing cell death via PDT, this type of photoactivation requires high concentrations of oxygen, which is often depleted in the hypoxic tumor microenvironment. By contrast, the concept of PACT uses ligand photosubstitution processes to generate cytotoxic compounds. PACT has been developed extensively in the context of Ru polypyridyl complexes for which the control of ligand substitution reactions by light has been thoroughly studied.[143–145] Compared to Ru, the photolabile ^{3}LF states of $Re(CO)_3$ complexes are higher in energy, making photosubstitution much rarer for this compound class. In 2000, the group of Ishitani discovered that $Re(CO)_3$ complexes of the general formula $[Re(CO)_3(NN)(PR_3)]^+$, where NN is a diimine ligand, are photolabile, ejecting the CO trans to the phosphine upon 365-nm light irradiation.[146] Although the ^{3}LF excited state is still higher in energy than the initially populated ^{3}MLCT state, it is thermally accessible at room temperature.[147] The photochemistry of this class of compounds was leveraged to design light-activated CO-releasing molecules as tools to study the biological properties of this small gasotransmitter.[148–150]

In studying this photochemistry, we recognized that the aqua-$Re(CO)_2$ byproduct $[Re(CO)_2(OH_2)(NN)(PR_3)]^+$ generated upon light irradiation may also have biological activity, given its structural similarity to the cytotoxic Neo-Re compound discussed above in Section 3. To investigate this hypothesis, we synthesized a series of compounds with different diimine and phosphine ligands (Fig. 17).[151] The three phosphine ligands tris(hydroxymethyl)phosphine (THP), 1,3,5-triaza-7-phosphaadamantane (PTA), and 3,7-diacetyl-1,3,7-triaza-5-phosphabicyclo[3.3.1]nonane (DAPTA) were chosen for this study because they are known to confer their coordination compounds with good water solubility.[152–154] These compounds were administered to HeLa cells at a single 200 μM dose, and their cytotoxicities were evaluated in either the absence or presence of 365-nm light irradiation for 30 min with the MTT assay. With the exception of those bearing the PTA phosphine, these compounds are non-toxic in the dark, but can elicit effective cell-killing properties upon exposure to light (Fig. 18). The lack of phototoxicity of the PTA complexes could be

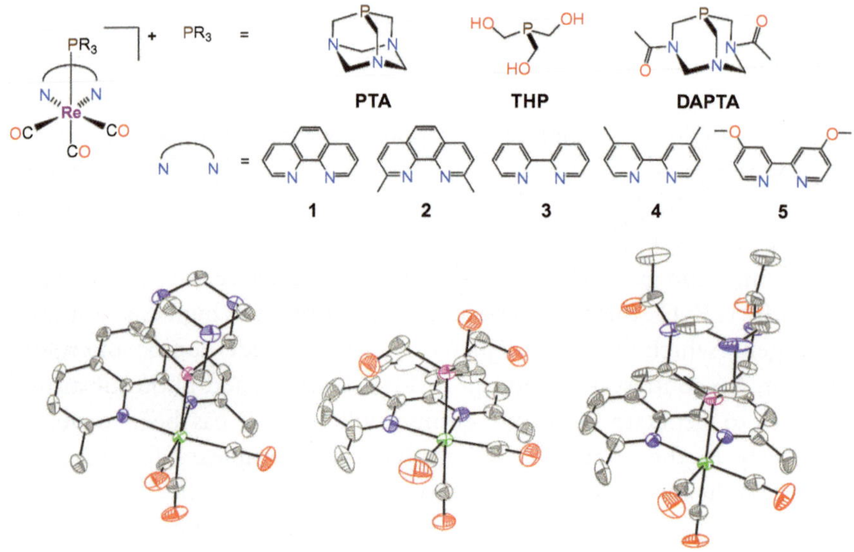

PTA-2 THP-2 DAPTA-2

Fig. 17 Structures of the photoactive $Re(CO)_3$ anticancer complexes investigated in Ref. 151.

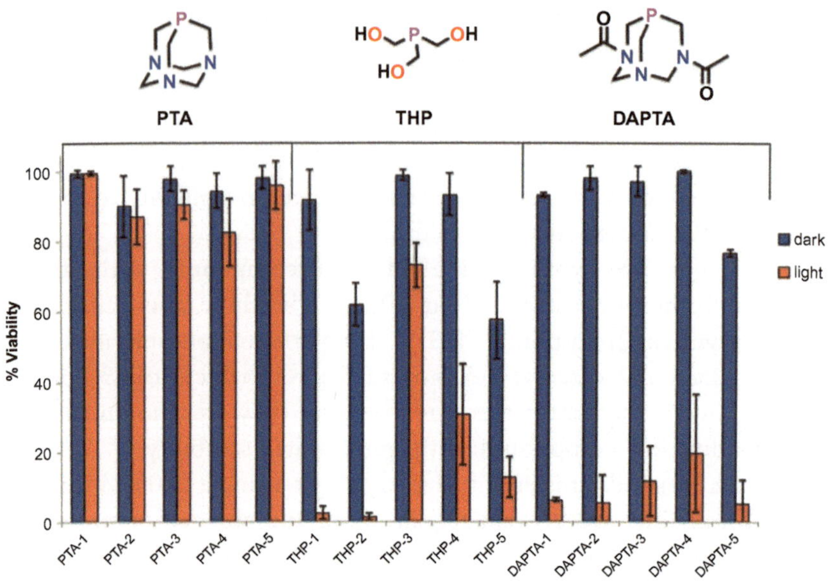

Fig. 18 Cell viability of HeLa cells treated with 200 μM of the indicated compound, named as shown in Fig. 17. After treatment, cells were either kept in the dark or irradiated with 365-nm light for 30 min at a photon flux of 2.4×10^{-10} einsteins/s. *Reproduced with permission from Marker, S.C.; MacMillan, S.N.; Zipfel, W.R.; Li, Z.; Ford, P.C.; Wilson, J. J.* Inorg. Chem. ***2018,*** 57, *1311–1331. Copyright 2018 American Chemical Society.*

a consequence of photoinduced electron transfer quenching of their excited states by the nitrogen lone pairs on this ligand. Complexes with the DAPTA phosphine show the most promising phototoxic effects. Thus, proper tuning of the ligand coordination sphere to modulate excited state energies can lead to effective photoactivated Re-based anticancer agents, as verified by recent computational studies on this subject.[155]

The means by which these compounds induce their light-activated anticancer activities was further assessed with the most potent phototoxic compound, $[Re(CO)_3(phen)(DAPTA)]^+$ (DAPTA-1). Its photoproduct, the aqua complex $[Re(CO)_2(H_2O)(phen)(DAPTA)]^+$ (DAPTA-1A) was synthesized and isolated independently, and its cytotoxic activity in HeLa cells was measured. The IC_{50} value of this compound was found to be approximately 40 μM, roughly an order of magnitude less potent than DAPTA-1 upon photoirradiation. These results indicated that the phototoxic effects of DAPTA-1 cannot be attributed solely to the Re photoproducts. It was further determined that these compounds can sensitize the formation of $^1O_2$ with quantum yields up to 70%, thus suggesting another mode by which these compounds can kill cells in the presence of light. Lastly, the cytotoxic and anticancer properties of CO have been well documented,[156] and may also be a contributing factor for these complexes. These compounds are therefore most likely acting in a multi-pronged manner with respect to their abilities to induce phototoxic effects.

We have also explored other approaches to leverage the photochemical properties of $Re(CO)_3$ complexes to design photoactivated drug candidates. Towards these efforts and motivated by other heterometallic multinuclear $Re(CO)_3$ complexes[157] and Pt(IV) complexes with photoactive ligands,[158–160] we have developed multimetallic complexes comprising one Pt(IV) and two $Re(CO)_3$ centers.[161] The Pt(IV) core can act as a prodrug for the conventional Pt(II)-based anticancer agents upon its reduction,[9] whereas the $Re(CO)_3$ centers were chosen for both their ER stress-inducing and photochemical properties. We hypothesized that these conjugates would elicit a dual mechanism of cytotoxicity, with the Pt fragment forming covalent DNA crosslinks and the $Re(CO)_3$ compounds causing ER stress. Conjugates containing Pt(IV) analogues of cisplatin and oxaliplatin, CisPtRe and OxaliPtRe, were synthesized and compared to the mononuclear $Re(CO)_3$ analogue, TRIP-suc (Fig. 19). Upon treatment of CisPtRe or OxaliPtRe with 20 equiv. of the biological reducing agent GSH over 24 h, partial reduction of the Pt(IV) center with concomitant release of the axial ligand TRIP-suc is observed. Consistent with previous comparative studies on the reduction of Pt(IV) analogs of oxaliplatin and

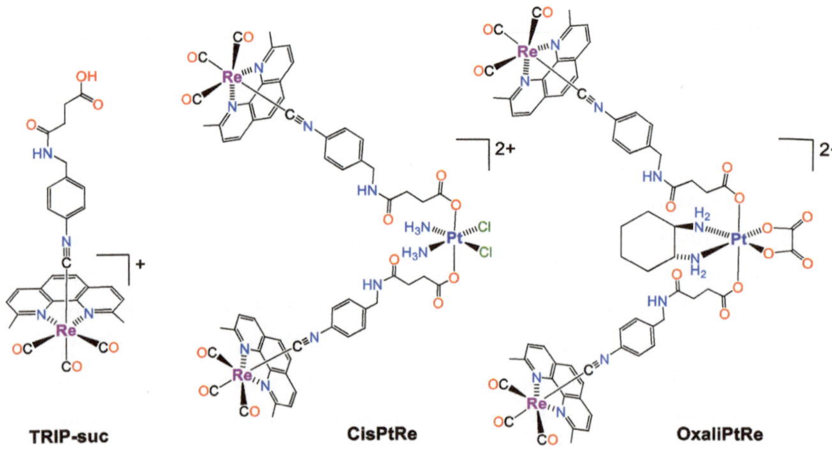

Fig. 19 Structures of the Pt(IV)-Re(I) conjugates investigated in Ref. 161.

cisplatin,[162] the extent of reduction of OxaliPtRe (<5%) was significantly smaller than that of CisPtRe (35%) over the same time period. In addition to being activated by chemical reduction, these compounds can undergo a photochemical reaction. Irradiation with 365-nm light triggers the photoreduction of the Pt(IV) center, releasing both the Pt(II) analogues and TRIP-suc. Importantly, the photoreduction occurs on a much faster timescale than chemical reduction by GSH, suggesting that these compounds could be useful as photoactivated drug candidates. The mechanism of photoreduction of these compounds was investigated by density functional theory (DFT) (Fig. 20). In this case, light irradiation of the TRIP-suc axial ligands induces a transition from the d(π) orbitals to the ligand π* orbitals, yielding a $Re(CO)_3$-based MLCT excited state. The Pt(IV) d(σ*) orbitals energetically reside between the Re d(π) and ligand π* orbitals. Therefore, once populated the ligand π* orbitals of the Re center can transfer the electron to the Pt(IV) d(σ*) orbitals, leading to photoreduction and ligand dissociation from the Pt(IV) center. Based on this photochemical activation process, we assessed the cytotoxicity of these compounds both in the dark and upon light irradiation to determine their viabilities as light-activated drug candidates. Only mild increases in cytotoxicity (2–3 fold) were observed upon light irradiation. The low phototoxicity indices suggested that other factors, such as cell uptake and intracellular localization, may also play an important role in mediating their cytotoxic activities. These compounds show that $Re(CO)_3$ centers can play a role that is distinct from $^1O_2$-generation and photoinduced ligand substitution

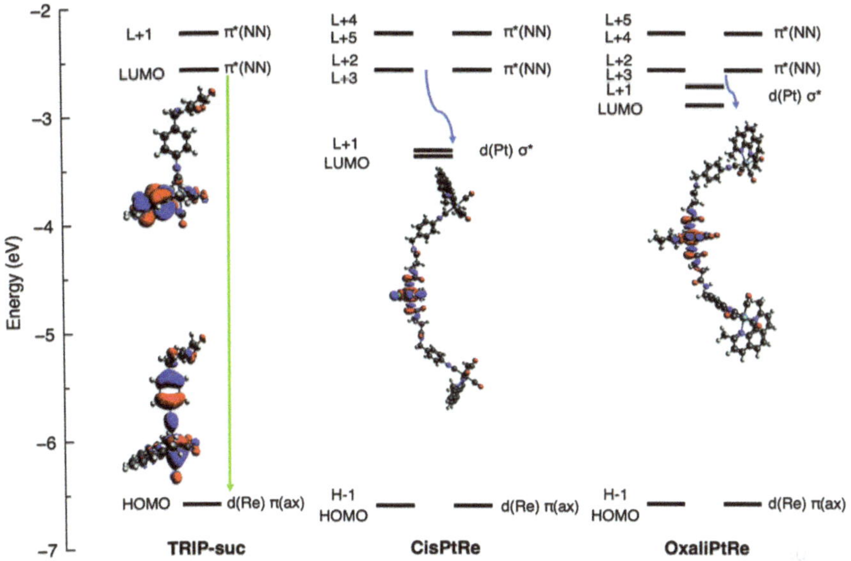

Fig. 20 Frontier Kohn-Sham molecular orbitals, as calculated by DFT, for TRIP-suc, CisPtRe, and OxaliPtRe. The green arrow represents an emissive transition from the ^{3}MLCT state. The blue arrows represent photoinduced electron transfer non-radiative decay pathways through the intermediary unoccupied Pt σ* d orbitals. *Reproduced with permission from Huang, Z.; King, A.P.; Lovett, J.; Lai, B.; Woods, J.J.; Harris, H.H.; Wilson, J.* J. Chem. Commun. ***2021,*** 57, *11189–11192. Copyright 2021 Royal Society of Chemistry.*

pathways discussed prior. Specifically, CisPtRe and OxaliPtRe demonstrate that the $Re(CO)_3$ core can act as a photoreductant, providing a novel means of photoactivation.

8. Conclusions and future outlook

The biological applications of $Re(CO)_3$ complexes are rapidly expanding as researchers recognize how the novel photophysical properties of these systems can be leveraged in unique manners. Although many researchers have explored these complexes, the examples described above highlight specific case studies and contributions from our own lab. Through these efforts, we have demonstrated several important concepts and points of interest that may motivate continued investigations in this area. Specifically, the rich spectroscopic properties of this compound class have proven to be valuable for understanding their mechanisms of action. For example, the fluorescence microscopy studies of Neo-Re in cells revealed its ability to perturb lysosomal and endosomal function. In addition, the modular coordination chemistry of $Re(CO)_3$ complexes has enabled a diverse range of easily accessible compounds that can be obtained via straightforward ligand substitution or metal ion-templated condensation reactions. This principle was demonstrated in our combinatorial synthesis of pyridylimine $Re(CO)_3$ complexes. Lastly, the photochemistry of this compound class can be leveraged for a range of photobiological properties. Compounds that sensitize $^1O_2$, undergo photosubstitution, and act as photoreductants can be accessed with appropriate ligand modifications, and these features can be used for triggering photoinduced cell death. In conjunction with our work and the great contributions by numerous other research groups over the last 5 years, $Re(CO)_3$ complexes have proven to be valuable tools for therapy and diagnosis.

Acknowledgments

My former and current students Dr. A. Paden King, Dr. Sierra C. Marker, Dr. Chilaluck C. Konkankit, Mr. Kevin M. Knopf, and Mr. Tracky Huang are thanked for their invaluable contributions to this project and research area. Early research in our lab in this area was supported by the Department of Defense Ovarian Cancer Research Program under award no. W81XWH-17-1-0097.

References

1. Guo, Z.; Sadler, P. J. *Angew. Chem. Int. Ed.* **1999**, *38*, 1512–1531.
2. Chellan, P.; Sadler, P. J. *Phil. Trans. R. Soc. A* **2015**, *373*, 20140182.
3. Anthony, E. J.; Bolitho, E. M.; Bridgewater, H. E.; Carter, O. W. L.; Donnelly, J. M.; Imberti, C.; Lant, E. C.; Lermyte, F.; Needham, R. J.; Palau, M.; Sadler, P. J.; Shi, H.; Wang, F.-X.; Zhang, W.-Y.; Zhang, Z. *Chem. Sci.* **2020**, *11*, 12888–12917.
4. Imberti, C.; Sadler, P. J. *Adv. Inorg. Chem.* **2020**, *75*, 3–56.
5. Farrell, N. P. *Adv. Inorg. Chem.* **2020**, *75*, 57–86.
6. Johnstone, T. C.; Suntharalingam, K.; Lippard, S. J. *Chem. Rev.* **2016**, *116*, 3436–3486.
7. Wang, Z.; Deng, Z.; Zhu, G. *Dalton Trans.* **2019**, *48*, 2536–2544.
8. Khoury, A.; Deo, K. M.; Aldrich-Wright, J. R. *J. Inorg. Biochem.* **2020**, *207*, 111070.
9. Xu, Z.; Wang, Z.; Deng, Z.; Zhu, G. *Coord. Chem. Rev.* **2021**, *442*, 213991.
10. Peng, K.; Liang, B.-B.; Liu, W.; Mao, Z.-W. *Coord. Chem. Rev.* **2021**, *449*, 214210.
11. Ivanova, S. *Pharmacia* **2022**, *69*, e78813.
12. Monro, S.; Colón, K. L.; Yin, H.; Roque, J., III; Konda, P.; Gujar, S.; Thummel, R. P.; Lilge, L.; Cameron, C. G.; McFarland, S. A. *Chem. Rev.* **2019**, *119*, 797–828.
13. Imberti, C.; Zhang, P.; Huang, H.; Sadler, P. J. *Angew. Chem. Int. Ed.* **2020**, *59*, 61–73.
14. Xu, G.-X.; Mak, E. C.-L.; Lo, K. K.-W. *Inorg. Chem. Front.* **2021**, *8*, 4553–4579.
15. Karges, J. *Angew. Chem. Int. Ed.* **2022**, *61*, e202112236.
16. Mari, C.; Gasser, G. *Chimia* **2015**, *69*, 176–181.
17. Mari, C.; Pierroz, V.; Ferrari, S.; Gasser, G. *Chem. Sci.* **2015**, *6*, 2660–2686.
18. Huang, H.; Banerjee, S.; Sadler, P. J. *ChemBioChem* **2018**, *19*, 1574–1589.
19. Huang, T.; Yu, Q.; Liu, S.; Huang, W.; Zhao, Q. *Dalton Trans.* **2018**, *47*, 7628–7633.

20. Zhang, P.; Huang, H. *Dalton Trans.* **2018**, *47*, 14841–14854.
21. Liew, H. S.; Mai, C.-W.; Zulkefeli, M.; Madheswaran, T.; Kiew, L. V.; Delsuc, N.; Low, M. L. *Molecules* **2020**, *25*, 4176.
22. Wolcan, E. *Inorg. Chim. Acta* **2020**, *509*, 119650.
23. Leonidova, A.; Gasser, G. *ACS Chem. Biol.* **2014**, *9*, 2180–2193.
24. Bauer, E. B.; Haase, A. A.; Reich, R. M.; Crans, D. C.; Kühn, F. E. *Coord. Chem. Rev.* **2019**, *393*, 79–117.
25. Collery, P.; Desmaele, D.; Veena, V. *Curr. Pharm. Des.* **2019**, *25*, 3306–3322.
26. Schindler, K.; Zobi, F. *Molecules* **2022**, *27*, 539.
27. Salignac, B.; Grundler, P. V.; Cayemittes, S.; Frey, U.; Scopelliti, R.; Merbach, A. E.; Hedinger, R.; Hegetschweiler, K.; Alberto, R.; Prinz, U.; Raabe, G.; Kölle, U.; Hall, S. *Inorg. Chem.* **2003**, *42*, 3516–3526.
28. Zobi, F.; Spingler, B.; Fox, T.; Alberto, R. *Inorg. Chem.* **2003**, *42*, 2818–2820.
29. Schutte-Smith, M.; Marker, S. C.; Wilson, J. J.; Visser, H. G. *Inorg. Chem.* **2020**, *59*, 15888–15897.
30. Oriskovich, T. A.; White, P. S.; Thorp, H. H. *Inorg. Chem.* **1995**, *34*, 1629–1631.
31. Zobi, F.; Spingler, B.; Alberto, R. *ChemBioChem* **2005**, *6*, 1397–1405.
32. Binkley, S. L.; Leeper, T. C.; Rowlett, R. S.; Herrick, R. S.; Ziegler, C. J. *Metallomics* **2011**, *3*, 909–916.
33. Zobi, F.; Spingler, B. *Inorg. Chem.* **2012**, *51*, 1210–1212.
34. Santoro, G.; Blacque, O.; Zobi, F. *Metallomics* **2012**, *4*, 253–259.
35. Lo, K. K.-W.; Zhang, K. Y.; Li, S. P.-Y. *Eur. J. Inorg. Chem.* **2011**, 3551–3568.
36. Lee, L. C.-C.; Leung, K.-K.; Lo, K. K.-W. *Dalton Trans.* **2017**, *46*, 16357–16380.
37. Schindler, K.; Zobi, F. *Chimia* **2021**, *75*, 837–844.
38. Clède, S.; Policar, C. *Chem. A Eur. J.* **2015**, *21*, 942–958.
39. Hostachy, S.; Policar, C.; Delsuc, N. *Coord. Chem. Rev.* **2017**, *351*, 172–188.
40. Alberto, R.; Ortner, K.; Wheatley, N.; Schibli, R.; Schubiger, A. P. *J. Am. Chem. Soc.* **2001**, *123*, 3135–3136.
41. Pitchumony, T. S.; Banevicius, L.; Janzen, N.; Zubieta, J.; Valliant, J. F. *Inorg. Chem.* **2013**, *52*, 13521–13528.
42. Yazdani, A.; Janzen, N.; Banevicius, L.; Czorny, S.; Valliant, J. F. *Inorg. Chem.* **2015**, *54*, 1728–1736.
43. Dilworth, J. R.; Parrott, S. J. *Chem. Soc. Rev.* **1998**, *27*, 43–55.
44. Hieber, W.; Fuchs, H. *Z. Anorg. Allg. Chem.* **1941**, *248*, 269–275.
45. Vogler, A.; Kunkely, H. *Coord. Chem. Rev.* **2000**, *200–202*, 991–1008.
46. Kumar, A.; Sun, S.-S.; Lees, A. J. *Top. Organomet. Chem.* **2010**, *29*, 1–35.
47. Wrighton, M.; Morse, D. L. *J. Am. Chem. Soc.* **1974**, *96*, 998–1003.
48. Wrighton, M. S.; Morse, D. L.; Pdungsap, L. *J. Am. Chem. Soc.* **1975**, *97*, 2073–2079.
49. Giordano, P. J.; Wrighton, M. S. *J. Am. Chem. Soc.* **1979**, *101*, 2888–2897.
50. Fredericks, S. M.; Luong, J. C.; Wrighton, M. S. *J. Am. Chem. Soc.* **1979**, *101*, 7415–7417.
51. Summers, D. P.; Luong, J. C.; Wrighton, M. S. *J. Am. Chem. Soc.* **1981**, *103*, 5238–5241.
52. Grice, K. A.; Kubiak, C. P. *Adv. Inorg. Chem.* **2014**, *66*, 163–188.
53. Zarkadoulas, A.; Koutsouri, E.; Kefalidi, C.; Mitsopoulou, C. A. *Coord. Chem. Rev.* **2015**, *304–305*, 55–72.
54. Kurz, P.; Probst, B.; Spingler, B.; Alberto, R. *Eur. J. Inorg. Chem.* **2006**, 2966–2974.
55. Chabolla, S. A.; Dellamary, E. A.; Machan, C. W.; Tezcan, F. A.; Kubiak, C. P. *Inorg. Chim. Acta* **2014**, *422*, 109–113.
56. Vollmer, M. V.; Machan, C. W.; Clark, M. L.; Antholine, W. E.; Agarwal, J.; Schaefer, H. F., III; Kubiak, C. P.; Walensky, J. R. *Organometallics* **2015**, *34*, 3–12.
57. Clark, M. L.; Cheung, P. L.; Lessio, M.; Carter, E. A.; Kubiak, C. P. *ACS Catal.* **2018**, *8*, 2021–2029.

58. Shirley, H.; Sexton, T. M.; Liyanage, N. P.; Palmer, C. Z.; McNamara, L. E.; Hammer, N. I.; Tschumper, G. S.; Delcamp, J. H. *Eur. J. Inorg. Chem.* **2020**, 1844–1851.
59. Sato, S.; Ishitani, O. *Coord. Chem. Rev.* **2015**, *282–283*, 50–59.
60. Stephenson, K. A.; Banerjee, S. R.; Besanger, T.; Sogbein, O. O.; Levadala, M. K.; McFarlane, N.; Lemon, J. A.; Boreham, D. R.; Maresca, K. P.; Brennan, J. D.; Babich, J. W.; Zubieta, J.; Valliant, J. F. *J. Am. Chem. Soc.* **2004**, *126*, 8598–8599.
61. Amoroso, A. J.; Coogan, M. P.; Dunne, J. E.; Fernández-Moreira, V.; Hess, J. B.; Hayes, A. J.; Lloyd, D.; Millet, C.; Pope, S. J. A.; Williams, C. *Chem. Commun.* **2007**, 3066–3068.
62. Gillam, T. A.; Sweetman, M. J.; Bader, C. A.; Morrison, J. L.; Hayball, J. D.; Brooks, D. A.; Plush, S. E. *Coord. Chem. Rev.* **2018**, *375*, 234–255.
63. Bader, C. A.; Brooks, R. D.; Ng, Y. S.; Sorvina, A.; Werrett, M. V.; Wright, P. J.; Anwer, A. G.; Brooks, D. A.; Stagni, S.; Muzzioli, S.; Silberstein, M.; Skelton, B. W.; Goldys, E. M.; Plush, S. E.; Shandala, T.; Massi, M. *RSC Adv.* **2014**, *4*, 16345–16351.
64. Bader, C. A.; Carter, E. A.; Safitri, A.; Simpson, P. V.; Wright, P.; Stagni, S.; Massi, M.; Lay, P. A.; Brooks, D. A.; Plush, S. E. *Mol. BioSyst.* **2016**, *12*, 2064–2068.
65. Bader, C. A.; Shandala, T.; Carter, E. A.; Ivask, A.; Guinan, T.; Hickey, S. M.; Werrett, M. V.; Wright, P. J.; Simpson, P. V.; Stagni, S.; Voelcker, N. H.; Lay, P. A.; Massi, M.; Plush, S. E.; Brooks, D. A. *PLoS One* **2016**, *11*, e0161557.
66. Sorvina, A.; Bader, C. A.; Caporale, C.; Carter, E. A.; Johnson, I. R. D.; Parkinson-Lawrence, E. J.; Simpson, P. V.; Wright, P. J.; Stagni, S.; Lay, P. A.; Massi, M.; Brooks, D. A.; Plush, S. E. *Oncotarget* **2018**, *9*, 35541–35552.
67. Bader, C. A.; Sorvina, A.; Simpson, P. V.; Wright, P. J.; Stagni, S.; Plush, S. E.; Massi, M.; Brooks, D. A. *FEBS Lett.* **2016**, *590*, 3051–3060.
68. Balasingham, R. G.; Coogan, M. P.; Thorp-Greenwood, F. L. *Dalton Trans.* **2011**, *40*, 11663–11674.
69. Lo, K. K.-W. *Acc. Chem. Res.* **2015**, *48*, 2985–2995.
70. Yip, A. M.-H.; Lo, K. K.-W. *Coord. Chem. Rev.* **2018**, *361*, 138–163.
71. Tan, C.-P.; Zhong, Y.-M.; Ji, L.-N.; Mao, Z.-W. *Chem. Sci.* **2021**, *12*, 2357–2367.
72. Orsa, D. K.; Haynes, G. K.; Pramanik, S. K.; Iwunze, M. O.; Greco, G. E.; Ho, D. M.; Krause, J. A.; Hill, D. A.; Williams, R. J.; Mandal, S. K. *Inorg. Chem. Commun.* **2008**, *11*, 1054–1056.
73. Orsa, D. K.; Nettles, C. R.; Pramanik, S. K.; Iwunze, M. O.; Greco, G. E.; Krause, J. A.; Mandal, S. K. Cytotoxic effects of rhenium (I) carbonyl complexes on prostate cancer cell lines. In *Handbook of Prostate Cancer Cell Research*; Nova Science Publishers, Inc.: Hauppauge, NY, 2009; pp. 323–362.
74. Knopf, K. M.; Murphy, B. L.; MacMillan, S. N.; Baskin, J. M.; Barr, M. P.; Boros, E.; Wilson, J. J. *J. Am. Chem. Soc.* **2017**, *139*, 14302–14314.
75. Zaharevitz, D. W.; Holbeck, S. L.; Bowerman, C.; Svetlik, P. A. *J. Mol. Graphics Modell.* **2002**, *20*, 297–303.
76. Shoemaker, R. H. *Nat. Rev. Cancer* **2006**, *6*, 813–823.
77. Zuehlke, A. D.; Moses, M. A.; Neckers, L. *Phil. Trans. R. Soc. B* **2018**, *373*, 20160527.
78. Li, L.; Wang, L.; You, Q.-D.; Xu, X.-L. *J. Med. Chem.* **2020**, *63*, 1798–1822.
79. Capper, M. S.; Garcia, A. E.; Macia, N.; Lai, B.; Lin, J.-B.; Nomura, M.; Alihosseinzadeh, A.; Ponnurangam, S.; Heyne, B.; Shemanko, C. S.; Jalilehvand, F. *J. Biol. Inorg. Chem.* **2020**, *25*, 759–776.
80. Capper, M. S.; Garcia, A. E.; Lai, B.; Wang, B. O.; Gelfand, B. S.; Shemanko, C. S.; Jalilehvand, F. *Dalton Trans.* **2021**, *50*, 5968–5977.
81. Karges, J.; Kalaj, M.; Gembicky, M.; Cohen, S. M. *Angew. Chem. Int. Ed.* **2021**, *60*, 10716–10723.

82. Konkankit, C. C.; King, A. P.; Knopf, K. M.; Southard, T. L.; Wilson, J. J. *ACS Med. Chem. Lett.* **2019**, *10*, 822–827.
83. Herrick, R. S.; Wrona, I.; McMicken, N.; Jones, G.; Ziegler, C. J.; Shaw, J. *J. Organomet. Chem.* **2004**, *689*, 4848–4855.
84. Costa, R.; Chanawanno, K.; Engle, J. T.; Baroody, B.; Herrick, R. S.; Ziegler, C. J. *J. Organomet. Chem.* **2013**, *734*, 25–31.
85. Chanawanno, K.; Kondeti, V.; Caporoso, J.; Paruchuri, S.; Leeper, T. C.; Herrick, R. S.; Ziegler, C. J. *Dalton Trans.* **2016**, *45*, 4729–4735.
86. Schrage, B. R.; Frisinger, B. R.; Schmidtke Sobeck, S. J.; Ziegler, C. *J. Dalton Trans.* **2021**, *50*, 1069–1075.
87. Parveen, S. *Appl. Organomet. Chem.* **2020**, *34*, e5687.
88. Chow, M. J.; Licona, C.; Wong, D. Y. Q.; Pastorin, G.; Gaiddon, C.; Ang, W. H. *J. Med. Chem.* **2014**, *57*, 6043–6059.
89. Chow, M. J.; Alfiean, M.; Pastorin, G.; Gaiddon, C.; Ang, W. H. *Chem. Sci.* **2017**, *8*, 3641–3649.
90. Chow, M. J.; Licona, C.; Pastorin, G.; Mellitzer, G.; Ang, W. H.; Gaiddon, C. *Chem. Sci.* **2016**, *7*, 4117–4124.
91. Chow, M. J.; Babak, M. V.; Wong, D. Y. Q.; Pastorin, G.; Gaiddon, C.; Ang, W. H. *Mol. Pharmaceutics* **2016**, *13*, 2543–2554.
92. Chow, M. J.; Babak, M. V.; Tan, K. W.; Cheong, M. C.; Pastorin, G.; Gaiddon, C.; Ang, W. H. *Mol. Pharmaceutics* **2018**, *15*, 3020–3031.
93. Konkankit, C. C.; Vaughn, B. A.; MacMillan, S. N.; Boros, E.; Wilson, J. *J. Inorg. Chem.* **2019**, *58*, 3895–3909.
94. Ammar, A. A.; Raveendran, R.; Gibson, D.; Nassar, T.; Benita, S. *J. Med. Chem.* **2016**, *59*, 9035–9046.
95. Zanellato, I.; Bonarrigo, I.; Colangelo, D.; Gabano, E.; Ravera, M.; Alessio, M.; Osella, D. *J. Inorg. Biochem.* **2014**, *140*, 219–227.
96. Johnstone, T. C.; Lippard, S. *J. Inorg. Chem.* **2013**, *52*, 9915–9920.
97. Zheng, Y.-R.; Suntharalingam, K.; Johnstone, T. C.; Yoo, H.; Lin, W.; Brooks, J. G.; Lippard, S. J. *J. Am. Chem. Soc.* **2014**, *136*, 8790–8798.
98. Awuah, S. G.; Zheng, Y.-R.; Bruno, P. M.; Hemann, M. T.; Lippard, S. J. *J. Am. Chem. Soc.* **2015**, *137*, 14854–14857.
99. Clède, S.; Lambert, F.; Saint-Fort, R.; Plamont, M.-A.; Bertrand, H.; Vessières, A.; Policar, C. *Chem. A Eur. J.* **2014**, *20*, 8714–8722.
100. Hallett, A. J.; Placet, E.; Prieux, R.; McCafferty, D.; Platts, J. A.; Lloyd, D.; Isaacs, M.; Hayes, A. J.; Coles, S. J.; Pitak, M. B.; Marchant, S.; Marriott, S. N.; Allemann, R. K.; Dervisi, A.; Fallis, I. A. *Dalton Trans.* **2018**, *47*, 14241–14253.
101. Konkankit, C. C.; Vaughn, B. A.; Huang, Z.; Boros, E.; Wilson, J. J. *Dalton Trans.* **2020**, *49*, 16062–16066.
102. Ballesta, A.; Billy, F.; Coverdale, J. P. C.; Song, J.-I.; Sanchez-Cano, C.; Romero-Canelón, I.; Sadler, P. J. *Metallomics* **2019**, *11*, 1648–1656.
103. Clède, S.; Sandt, C.; Dumas, P.; Policar, C. *Appl. Spectrosc.* **2020**, *74*, 63–71.
104. Kronauge, J. F.; Mindiola, D. J. *Organometallics* **2016**, *35*, 3432–3435.
105. Sacksteder, L.; Lee, M.; Demas, J. N.; DeGraff, B. A. *J. Am. Chem. Soc.* **1993**, *115*, 8230–8238.
106. Favale, J. M., Jr.; Danilov, E. O.; Yarnell, J. E.; Castellano, F. N. *Inorg. Chem.* **2019**, *58*, 8750–8762.
107. King, A. P.; Marker, S. C.; Swanda, R. V.; Woods, J. J.; Qian, S.-B.; Wilson, J. J. *Chem. A Eur. J.* **2019**, *25*, 9206–9210.
108. Konkankit, C. C.; Lovett, J.; Harris, H. H.; Wilson, J. J. *Chem. Commun.* **2020**, *56*, 6515–6518.

109. Wedding, J. L.; Harris, H. H.; Bader, C. A.; Plush, S. E.; Mak, R.; Massi, M.; Brooks, D. A.; Lai, B.; Vogt, S.; Werrett, M. V.; Simpson, P. V.; Skelton, B. W.; Stagni, S. *Metallomics* **2017**, *9*, 382–390.
110. Morrison, C. N.; Prosser, K. E.; Stokes, R. W.; Cordes, A.; Metzler-Nolte, N.; Cohen, S. M. *Chem. Sci.* **2020**, *11*, 1216–1225.
111. Senft, D.; Ronai, Z. A. *Trends Biochem. Sci.* **2015**, *40*, 141–148.
112. Wek, R. C.; Cavener, D. R. *Antioxid. Redox Signal.* **2007**, *9*, 2357–2371.
113. Almanza, A.; Carlesso, A.; Chintha, C.; Creedican, S.; Doultsinos, D.; Leuzzi, B.; Luís, A.; McCarthy, N.; Montibeller, L.; More, S.; Papaioannou, A.; Püschel, F.; Sassano, M. L.; Skoko, J.; Agostinis, P.; de Belleroche, J.; Eriksson, L. A.; Fulda, S.; Gorman, A. M.; Healy, S.; Kozlov, A.; Muñoz-Pinedo, C.; Rehm, M.; Chevet, E.; Samali, A. *FEBS J.* **2019**, *286*, 241–278.
114. Ohmichi, M.; Hayakawa, J.; Tasaka, K.; Kurachi, H.; Murata, Y. *Trends Pharmacol. Sci.* **2005**, *26*, 113–116.
115. Wacker, S. A.; Houghtaling, B. R.; Elemento, O.; Kapoor, T. M. *Nat. Chem. Biol.* **2012**, *8*, 235–237.
116. Kapoor, T. M.; Miller, R. M. *Trends Pharmacol. Sci.* **2017**, *38*, 1100–1109.
117. Marker, S. C.; King, A. P.; Swanda, R. V.; Vaughn, B.; Boros, E.; Qian, S.-B.; Wilson, J. J. *Angew. Chem. Int. Ed.* **2020**, *59*, 13391–13400.
118. Mechetner, E.; Kyshtoobayeva, A.; Zonis, S.; Kim, H.; Stroup, R.; Garcia, R.; Parker, R. J.; Fruehauf, J. P. *Clin. Cancer Res.* **1998**, *4*, 389–398.
119. Ling, V. *Cancer Chemother. Pharmacol.* **1997**, *40*, S3–S8.
120. Piwnica-Worms, D.; Chiu, M. L.; Budding, M.; Kronauge, J. F.; Kramer, R. A.; Croop, J. M. *Cancer Res.* **1993**, *53*, 977–984.
121. Hsu, H.-H.; Chen, M.-C.; Baskaran, R.; Lin, Y.-M.; Day, C. H.; Lin, Y. J.; Tu, C.-C.; Padma, V. V.; Kuo, W.-W.; Huang, C.-Y. *J. Cell. Physiol.* **2018**, *233*, 5458–5467.
122. Landini, I.; Lapucci, A.; Pratesi, A.; Massai, L.; Napoli, C.; Perrone, G.; Pinzani, P.; Messori, L.; Mini, E.; Nobili, S. *Oncotarget* **2017**, *8*, 96062–96078.
123. Heffeter, P.; Pongratz, M.; Steiner, E.; Chiba, P.; Jakupec, M. A.; Elbling, L.; Marian, B.; Körner, W.; Sevelda, F.; Micksche, M.; Keppler, B. K.; Berger, W. *J. Pharmacol. Exp. Ther.* **2005**, *312*, 281–289.
124. Licona, C.; Delhorme, J.-B.; Riegel, G.; Vidimar, V.; Cerón-Camacho, R.; Boff, B.; Venkatasamy, A.; Tomasetto, C.; da Silva Figueiredo Celestino Gomes, P.; Rognan, D.; Freund, J.-N.; Le Lagadec, R.; Pfeffer, M.; Gross, I.; Mellitzer, G.; Gaiddon, C. *Inorg. Chem. Front.* **2020**, *7*, 678–688.
125. Marker, S. C.; King, A. P.; Granja, S.; Vaughn, B.; Woods, J. J.; Boros, E.; Wilson, J. *J. Inorg. Chem.* **2020**, *59*, 10285–10303.
126. Suntharalingam, K.; Johnstone, T. C.; Bruno, P. M.; Lin, W.; Hemann, M. T.; Lippard, S. J. *J. Am. Chem. Soc.* **2013**, *135*, 14060–14063.
127. Ye, R.-R.; Tan, C.-P.; Chen, M.-H.; Hao, L.; Ji, L.-N.; Mao, Z.-W. *Chem. A Eur. J.* **2016**, *22*, 7800–7809.
128. Li, C.; Ip, K.-W.; Man, W.-L.; Song, D.; He, M.-L.; Yiu, S.-M.; Lau, T.-C.; Zhu, G. *Chem. Sci.* **2017**, *8*, 6865–6870.
129. Collery, P.; Mohsen, A.; Kermagoret, A.; Corre, S.; Bastian, G.; Tomas, A.; Wei, M.; Santoni, F.; Guerra, N.; Desmaële, D.; d'Angelo, J. *Invest. New Drugs* **2015**, *33*, 848–860.
130. Wang, F.-X.; Liang, J.-H.; Zhang, H.; Wang, Z.-H.; Wan, Q.; Tan, C.-P.; Ji, L.-N.; Mao, Z.-W. *ACS Appl. Mater. Interfaces* **2019**, *11*, 13123–13133.
131. He, L.; Pan, Z.-Y.; Qin, W.-W.; Li, Y.; Tan, C.-P.; Mao, Z.-W. *Dalton Trans.* **2019**, *48*, 4398–4404.
132. Capper, M. S.; Packman, H.; Rehkämper, M. *ChemBioChem* **2020**, *21*, 2111–2115.

133. Domenichini, A.; Casari, I.; Simpson, P. V.; Desai, N. M.; Chen, L.; Dustin, C.; Edmands, J. S.; van der Vliet, A.; Mohammadi, M.; Massi, M.; Falasca, M. *J. Exp. Clin. Cancer Res.* **2020**, *39*, 276.
134. Collery, P.; Lagadec, P.; Krossa, I.; Cohen, C.; Antomarchi, J.; Varlet, D.; Lucio, M.; Guigonis, J.-M.; Scimeca, J.-C.; Schmid-Antomarchi, H.; Schmid-Alliana, A. *Trace Elem. Med. Biol.* **2022**, *71*, 126931.
135. King, A. P.; Wilson, J. J. *Chem. Soc. Rev.* **2020**, *49*, 8113–8136.
136. Ortega, E.; Vigueras, G.; Ballester, F. J.; Ruiz, J. *Coord. Chem. Rev.* **2021**, *446*, 214129.
137. Guan, R.; Xie, L.; Ji, L.; Chao, H. *Eur. J. Inorg. Chem.* **2020**, 3978–3986.
138. Leonidova, A.; Pierroz, V.; Rubbiani, R.; Heier, J.; Ferrari, S.; Gasser, G. *Dalton Trans.* **2014**, *43*, 4287–4294.
139. Leonidova, A.; Pierroz, V.; Adams, L. A.; Barlow, N.; Ferrari, S.; Graham, B.; Gasser, G. *ACS Med. Chem. Lett.* **2014**, *5*, 809–814.
140. Kastl, A.; Dieckmann, S.; Wähler, K.; Völker, T.; Kastl, L.; Merkel, A. L.; Vultur, A.; Shannan, B.; Harms, K.; Ocker, M.; Parak, W. J.; Herlyn, M.; Meggers, E. *ChemMedChem* **2013**, *8*, 924–927.
141. Mion, G.; Gianferrara, T.; Bergamo, A.; Gasser, G.; Pierroz, V.; Rubbiani, R.; Vilar, R.; Leczkowska, A.; Alessio, E. *ChemMedChem* **2015**, *10*, 1901–1914.
142. Wähler, K.; Ludewig, A.; Szabo, P.; Harms, K.; Meggers, E. *Eur. J. Inorg. Chem.* **2014**, 807–811.
143. Knoll, J. D.; Turro, C. *Coord. Chem. Rev.* **2015**, *282–283*, 110–126.
144. Bonnet, S. *Dalton Trans.* **2018**, *47*, 10330–10343.
145. Chen, Y.; Bai, L.; Zhang, P.; Zhao, H.; Zhou, Q. *Molecules* **2021**, *26*, 5679.
146. Koike, K.; Tanabe, J.; Toyama, S.; Tsubaki, H.; Sakamoto, K.; Westwell, J. R.; Johnson, F. P. A.; Hori, H.; Saitoh, H.; Ishitani, O. *Inorg. Chem.* **2000**, *39*, 2777–2783.
147. Koike, K.; Okoshi, N.; Hori, H.; Takeuchi, K.; Ishitani, O.; Tsubaki, H.; Clark, I. P.; George, M. W.; Johnson, F. P. A.; Turner, J. J. *J. Am. Chem. Soc.* **2002**, *124*, 11448–11455.
148. Pierri, A. E.; Pallaoro, A.; Wu, G.; Ford, P. C. *J. Am. Chem. Soc.* **2012**, *134*, 18197–18200.
149. Chakraborty, I.; Jimenez, J.; Sameera, W. M. C.; Kato, M.; Mascharak, P. K. *Inorg. Chem.* **2017**, *56*, 2863–2873.
150. Chakraborty, I.; Carrington, S. J.; Roseman, G.; Mascharak, P. K. *Inorg. Chem.* **2017**, *56*, 1534–1545.
151. Marker, S. C.; MacMillan, S. N.; Zipfel, W. R.; Li, Z.; Ford, P. C.; Wilson, J. *J. Inorg. Chem.* **2018**, *57*, 1311–1331.
152. Katti, K. V.; Gali, H.; Smith, C. J.; Berning, D. E. *Acc. Chem. Res.* **1999**, *32*, 9–17.
153. Darensbourg, D. J.; Ortiz, C. G.; Kamplain, J. W. *Organometallics* **2004**, *23*, 1747–1754.
154. Phillips, A. D.; Gonsalvi, L.; Romerosa, A.; Vizza, F.; Peruzzini, M. *Coord. Chem. Rev.* **2004**, *248*, 955–993.
155. Álvarez, D.; Menéndez, M. I.; López, R. *Inorg. Chem.* **2022**, *61*, 439–455.
156. Kourti, M.; Jiang, W. G.; Cai, J. *Oxid. Med. Cell. Longevity* **2017**, *2017*, 9326454.
157. Huang, Z.; Wilson, J. J. *Eur. J. Inorg. Chem.* **2021**, 1312–1324.
158. Wang, Z.; Wang, N.; Cheng, S.-C.; Xu, K.; Deng, Z.; Chen, S.; Xu, Z.; Xie, K.; Tse, M.-K.; Shi, P.; Hirao, H.; Ko, C.-C.; Zhu, G. *Chem* **2019**, *5*, 3151–3165.
159. Yao, H.; Chen, S.; Deng, Z.; Tse, M.-K.; Matsuda, Y.; Zhu, G. *Inorg. Chem.* **2020**, *59*, 11823–11833.
160. Karges, J.; Yempala, T.; Tharaud, M.; Gibson, D.; Gasser, G. *Angew. Chem. Int. Ed.* **2020**, *59*, 7069–7075.
161. Huang, Z.; King, A. P.; Lovett, J.; Lai, B.; Woods, J. J.; Harris, H. H.; Wilson, J. J. *Chem. Commun.* **2021**, *57*, 11189–11192.
162. Wexselblatt, E.; Gibson, D. *J. Inorg. Biochem.* **2012**, *117*, 220–229.

CHAPTER TWO

Strategic design of photofunctional transition metal complexes for cancer diagnosis and therapy

Lawrence Cho-Cheung Lee[a,b] **and Kenneth Kam-Wing Lo**[a,c,*]
[a]Department of Chemistry, City University of Hong Kong, Hong Kong, PR China
[b]Laboratory for Synthetic Chemistry and Chemical Biology Limited, Hong Kong, PR China
[c]State Key Laboratory of Terahertz and Millimeter Waves, City University of Hong Kong, Hong Kong, PR China
*Corresponding author: e-mail address: bhkenlo@cityu.edu.hk

Contents

1. Introduction 37
2. Transition metal complexes for targeted cancer diagnosis and therapy 38
   2.1 Glucose transporters 39
   2.2 Biotin receptors 42
   2.3 Integrins 44
   2.4 G-Protein-coupled receptors 45
   2.5 Epidermal growth factor receptor 48
   2.6 Carbonic anhydrase 48
   2.7 Cyclooxygenase 50
   2.8 Sialylated glycoproteins and glycolipids 51
3. Activatable transition metal complexes for cancer diagnosis and therapy 55
   3.1 Hypoxia 56
   3.2 pH 62
   3.3 Polarity 65
   3.4 Viscosity 66
   3.5 Redox status 67
   3.6 Enzymes 72
   3.7 Mismatched DNA molecules 76
   3.8 Bioorthogonal chemical reporters 76
4. Conclusions and future perspective 88
Acknowledgments 90
References 90

Advances in Inorganic Chemistry, Volume 80
ISSN 0898-8838
https://doi.org/10.1016/bs.adioch.2022.06.002

Abstract

Cancer is one of the most important public health problems and poses a serious threat to human life. The development of innovative diagnostic and therapeutic agents for rapid detection and effective treatment of cancer has been an important area of biomedical research. Over the past decade, there has been enormous interest in the utilization of photofunctional transition metal complexes for bioimaging and photodynamic therapy due to their rich photophysical and photochemical properties, in particular the environment-sensitive and long-lived emission and high reactive oxygen species (ROS)-generation efficiency. In this chapter, we summarize the recent advances in the development of photofunctional transition metal complexes for cancer diagnosis and therapy, with an emphasis on the molecular designs that exploit the interesting photophysical and photochemical behavior of the complexes. The design strategies are classified into two categories: (1) targeting specific membrane proteins overexpressed on the surface of cancer cells to increase the cellular uptake of the complexes; and (2) activating the emission and ROS-photosensitization properties of the complexes in response to specific cancer-associated stimuli. The current challenges and future perspective on the translation of transition metal complexes from the laboratory to the clinic as diagnostics and therapeutics in cancer patients are also discussed.

Abbreviations

Ac_4ManNAz	1,3,4,6-tetra-*O*-acetyl-*N*-azidoacetyl-D-mannosamine
BCN-OH	(1*R*,8*S*,9*s*)-bicyclo[6.1.0]non-4-yn-9-ylmethanol
***bis*-BCN**	1,13-*bis*-((1*R*,8*S*,9*s*)-bicyclo[6.1.0]non-4-yn-9-ylmethyloxycarbonylamino)-4,7,10-trioxatridecane
BRs	biotin receptors
CA	carbonic anhydrase
CHO	Chinse hamster ovary
COX	cyclooxygenase
CT	charge transfer
CuAAC	copper(I)-catalyzed azide–alkyne cycloaddition
DIBO	dibenzocyclooctyne
DNP	2,4-dinitrophenyl
DNPS	2,4-dinitrophenylsulfonamide
EGFR	epidermal growth factor receptor
EPR	enhanced permeability and retention
ER	endoplasmic reticulum
FRET	Förster resonance energy transfer
GLUTs	glucose transporters
GRPR	gastrin-releasing peptide receptor
GSH	glutathione
HE	high-energy
HLF	human lymphatic fibroblast
HOMO	highest occupied molecular orbital
ICP-MS	inductively coupled plasma-mass spectrometry
IEDDA	inverse electron-demand Diels–Alder
IL	intraligand
ILCT	intraligand charge transfer

LE	low-energy
LSCM	laser-scanning confocal microscopy
LUMO	lowest unoccupied molecular orbital
MCTSs	multicellular tumor spheroids
MMR	mismatch repair
NADH	reduced nicotinamide adenine dinucleotide
NADPH	reduced nicotinamide adenine dinucleotide phosphate
Neu5Ac	*N*-acetylneuraminic acid
NIR	near-infrared
NLCT	amine-to-ligand charge transfer
PBA	phenylboronic acid
PDI	perylene diimide
PDT	photodynamic therapy
PEG	poly(ethylene glycol)
PET	photoinduced electron transfer
PI	photocytotoxicity index
PLIM	phosphorescence lifetime imaging microscopy
PVP	poly(*N*-vinylpyrrolidone)
ROS	reactive oxygen species
SPAAC	strain-promoted azide–alkyne cycloaddition
SPANC	strain-promoted alkyne–nitrone cycloaddition
SPSAC	strain-promoted sydnone–alkyne cycloaddition
SST	somatostatin
SSTRs	somatostatin receptors
TCO-OH	*trans*-cyclooct-4-en-1-ol
TNBC	triple-negative breast cancer
TPA	two-photon absorption

1. Introduction

Cancer is a global health issue and is among the leading causes of death worldwide.[1] Thus, the diagnosis and treatment of cancer remain one of the major focuses in biomedical research. In the past decade, there has been enormous interest in the development of photofunctional transition metal complexes for cancer diagnosis and therapy.[2–6] The high structural diversity and synthetic versatility of transition metal complexes allow facile manipulation of their photophysical and photochemical properties for diagnostic and therapeutic applications, respectively, through the use of ligands with various structural, electronic, and coordinating properties.[7] The lipophilicity, aqueous solubility, and membrane permeability of complexes can be readily modulated through rational molecular engineering. Phosphorescent transition metal complexes are attractive candidates for optical imaging as they usually display intense, long-lived, and environment-sensitive emission with large Stokes' shifts, which minimize possible self-quenching effects at high local probe concentrations (e.g., in tumors). The long emission lifetimes (in the sub-microsecond to microsecond time regime) enable the elimination of interference caused by the shorter-lived autofluorescence (on the nanosecond timescale) using time-resolved techniques such as phosphorescence lifetime imaging microscopy (PLIM).[8] Also, the high photostability of complexes allows their continuous exposure to irradiation and thus enables real-time imaging in living systems. Additionally, many transition metal complexes are efficient photosensitizers for reactive oxygen species (ROS) such as superoxide anion radical ($O_2^{\bullet-}$) and singlet oxygen ($^1O_2$) via a Type I or Type II photoreaction, respectively, which renders them attractive candidates for photodynamic therapy (PDT).[9] Furthermore, the absence of certain transition metals in living systems allows the cellular accumulation levels of the complexes to be quantitatively determined by inductively coupled plasma-mass spectrometry (ICP-MS). This eliminates possible artifacts in laser-scanning confocal microscopy (LSCM) since the emission intensity may not always be proportional to the probe concentration inside the cells due to environmental effects.[10] These remarkable features have led to the successful development of photofunctional transition metal complexes for diagnostic and therapeutic applications.

In this chapter, we summarize the recent development of photofunctional transition metal complexes for cancer diagnosis and therapy, with an emphasis on the molecular designs that exploit the interesting photophysical and photochemical properties of the complexes. The design strategies are classified into two categories: (1) targeting specific membrane proteins overexpressed on the surface of cancer cells to increase the cellular uptake of the complexes; and (2) activating the emission and ROS-photosensitization properties of the complexes in response to specific cancer-associated stimuli. The current challenges and future perspective on the translation of transition metal complexes from the laboratory to the clinic for diagnostic and therapeutic applications are also discussed.

2. Transition metal complexes for targeted cancer diagnosis and therapy

Plasma membrane proteins play important roles in solute transport, signal transduction, and cell–cell and cell–extracellular matrix interactions. The expression levels and post-translational modifications of these proteins in cancer cells are altered to promote cancer cell proliferation, migration, and adhesion.[11] Thus, the integration of transition metal complexes with a cancer-targeting motif, which binds to specific cell-surface protein with high affinity, is a useful strategy to realize selective cancer imaging and therapy.

2.1 Glucose transporters

Glucose is the most important carbohydrate in cellular metabolism and serves as an energy source for the growth of cells. The transport of glucose across the plasma membrane is facilitated by the membrane-bound glucose transporters (GLUTs).[12] The increased glucose uptake and metabolism in cancer cells is associated with the overexpression of GLUTs.[13]

A rhenium(I) complex appended with a D-glucose moiety (**1**) has been designed as a phosphorescent glucose uptake indicator.[14] Temperature dependence and chemical inhibition experiments indicate the involvement of GLUTs in the cellular uptake of the glucose complex, which is supported by the reduced uptake of the complex by human cervical adenocarcinoma HeLa cells in the presence of D-glucose or 2-deoxy-D-glucose but not L-glucose. An iridium(III) complex bearing a D-glucose pendant (**2**) displays similar GLUT-mediated cellular uptake.[15] ICP-MS measurements reveal that the uptake of the glucose complex by HeLa cells is *ca.* 4 times higher than its D-galactose counterpart. Remarkably, both the rhenium(I) and iridium(III) glucose complexes exhibit more efficient uptake by transformed cell lines (HeLa and human breast adenocarcinoma MCF-7) than non-transformed cell lines (human embryonic kidney HEK293T and mouse embryonic fibroblast NIH/3T3) due to the overexpression of GLUTs in transformed cells.[14,15] Ruthenium(II) complexes have been modified with a D-glucose unit (**3–5**) for cancer-targeted, two-photon-induced PDT.[16] The complexes are efficiently taken up by cancer cells (HeLa, human lung carcinoma A549 and its cisplatin-resistant derivative A549R, and human hepatocellular carcinoma HepG2) over normal human liver LO2 cells and accumulated in the mitochondria of the cells, which allow them to show higher photocytotoxicity toward cancer cells ($IC_{50,light}$ = 2.1–14.2 μM) than normal cells ($IC_{50,light}$ = 20.4–25.2 μM) through the generation of $^1O_2$ upon irradiation to cause mitochondrial dysfunction. The strong two-photon absorption (TPA) ($\delta_{810nm} = 128$ GM), high $^1O_2$-photosensitization capability ($\Phi_\Delta = 0.85$), and preferential enrichment in cancer cells and mitochondria enable complex **3** to achieve complete tumor ablation in mice bearing HeLa-derived tumor xenografts upon two-photon excitation ($\lambda_{ex} = 810$ nm).

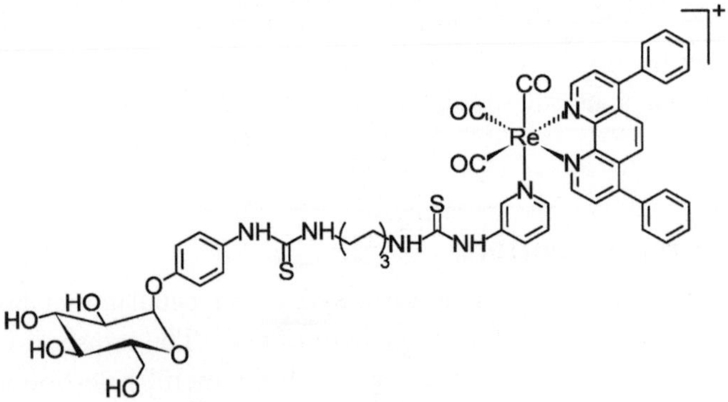

1

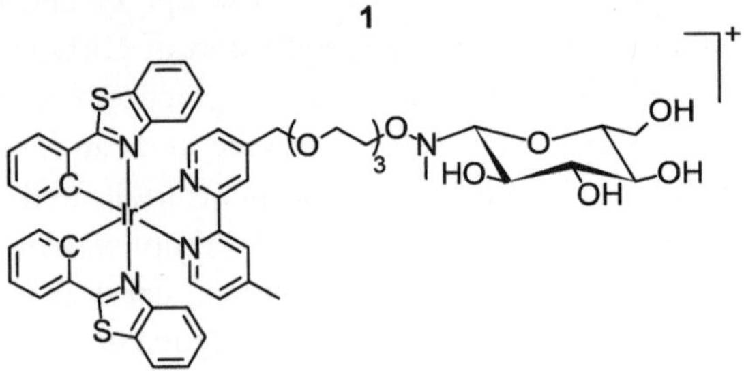

2

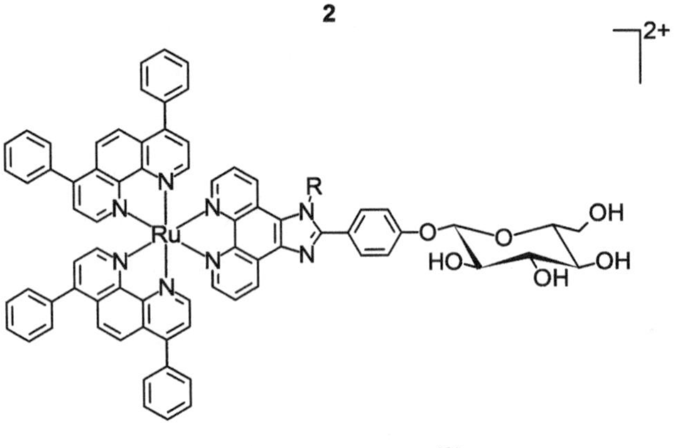

R = H **(3)**

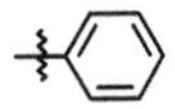

(4)

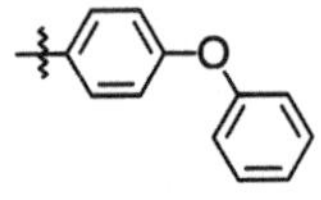

(5)

The substrates of GLUTs are not limited to glucose; for example, GLUT5 is known to display selective uptake of fructose.[17] This transporter is overexpressed in breast cancer tissues, but its expression in other cancer cells and normal breast tissues is limited.[18] A rhenium(I) complex carrying a

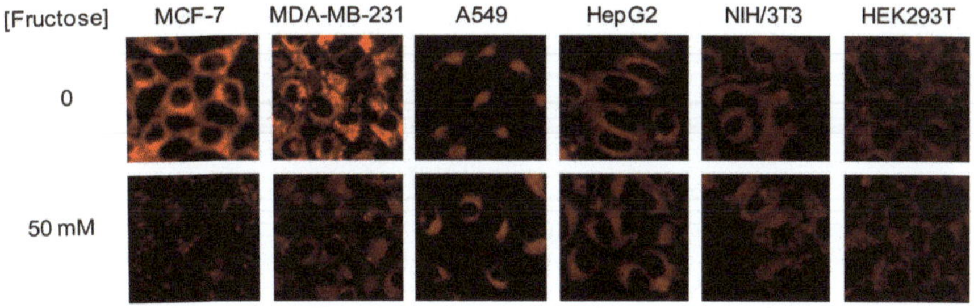

Fig. 1 LSCM images of six different types of cells incubated with complex **6** (50 μM, 37 °C, 1 h) in the absence (top) or presence (bottom) of fructose (50 mM). *Reproduced from Zhang, K. Y.; Tso, K. K.-S.; Louie, M.-W.; Liu, H.-W.; Lo, K. K.-W.* Organometallics ***2013,*** 32, *5098–5102, with permission from American Chemical Society. Copyright 2013.*

D-fructose moiety (**6**) exhibits more efficient uptake by human breast adenocarcinoma cell lines (MCF-7 and MDA-MB-231) over non-breast cancer (A549 and HepG2) and non-transformed cell lines (NIH/3T3 and HEK293T) (Fig. 1).[19] The addition of exogenous fructose significantly inhibits the uptake of the complex by breast cancer cells; however, treatment of the cells with glucose or glucose-uptake inhibitors exerts negligible effects on the uptake. Thus, the cellular uptake of the fructose complex is mediated by fructose transporters but independent of glucose-specific transporters. Additionally, complex **6** shows higher cytotoxic activity toward breast cancer cells ($IC_{50,dark}=9.6$ and 4.9 μM for MCF-7 and MDA-MB-231 cells, respectively) than non-breast cancer cells ($IC_{50,dark}=26.8$ and 33.9 μM for A549 and HepG2 cells, respectively), and becomes more cytotoxic toward MCF-7 cells ($IC_{50,light}=2.0$ μM) upon irradiation due to efficient $^1O_2$ photosensitization. Iridium(III) complexes containing a D-fructose pendant (**7** and **8**) are also taken up more efficiently by MCF-7 cells than HEK293T cells.[20] These complexes are essentially non-cytotoxic toward MCF-7 cells in the dark ($IC_{50,dark}>498.4$ and 128.4 μM, respectively), but display high photocytotoxicity ($IC_{50,light}=5.0$ and 11.2 μM, respectively) due to photoinduced $^1O_2$ generation. Interestingly, a related ruthenium(II) D-fructose complex (**9**) is accumulated in the plasma membrane of cancerous MCF-7 and HeLa cells but not in non-cancerous Madin-Darby canine kidney and 3T3 cells.[21] The exclusive cell surface staining is associated with the binding to the overexpressed GLUT5, which is supported by the reduced cellular uptake of the complex in the presence of exogenous fructose.

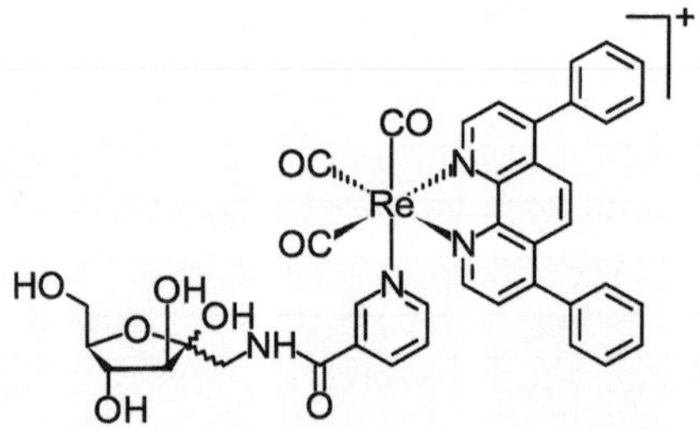

6

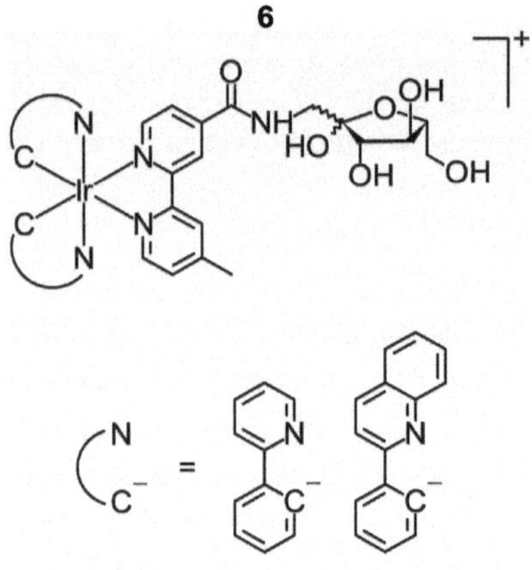

7 **8**

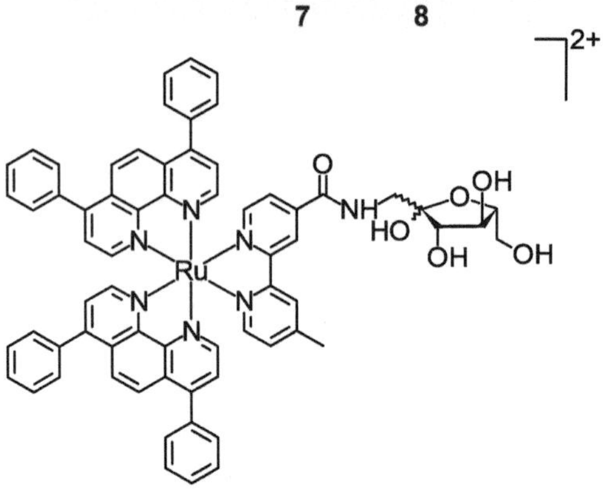

9

2.2 Biotin receptors

Biotin (vitamin B_7) is required for cell growth, proliferation, and differentiation. Cancer cells express high levels of biotin receptors (BRs) for this nutrient to support their rapid cell proliferation.[22] The uptake of a rhenium(I) complex appended with a biotin unit (**10**) by HeLa cells that possess a high level of BR expression is significantly inhibited upon incubation at 4 °C, suggesting that the biotin complex is internalized into the cells via an energy-dependent pathway.[23] Iridium(III) complexes bearing a biotin moiety (**11–13**) exhibit distinct intracellular localization properties depending on their lipophilicity; the less lipophilic complexes **11** and **12** are enriched in the surface of HeLa cells, whereas the more lipophilic complex **13** is localized in the perinuclear region of the cells.[24] A ruthenium(II) complex has been functionalized with a biotin pendant (**14**) for cancer-targeted, two-photon-induced PDT.[25] The complex shows strong TPA ($\delta_{820nm} = 140.26$ GM) and efficient $^1O_2$ photosensitization ($\Phi_\Delta = 0.87$). Results of ICP-MS measurements reveal that the uptake of the complex by BR-overexpressing A549R cells is 13.9 fold higher than that by BR-negative human lymphatic fibroblast (HLF)-a cells. The involvement of BRs in the cellular uptake of the complex is further supported by the reduced uptake in the presence of exogenous biotin. The BR-mediated cellular uptake allows the complex to display higher photocytotoxic activity toward cancer cells (A549, A549R, and HeLa) ($IC_{50,light} = 3.3$–3.9 μM) than non-cancerous HLF-a cells ($IC_{50,light} = 36.4$ μM) through the induction of apoptosis upon two-photon excitation ($\lambda_{ex} = 820$ nm). The high cancer selectivity and therapeutic efficacy of the complex are retained in 3D multicellular spheroids.

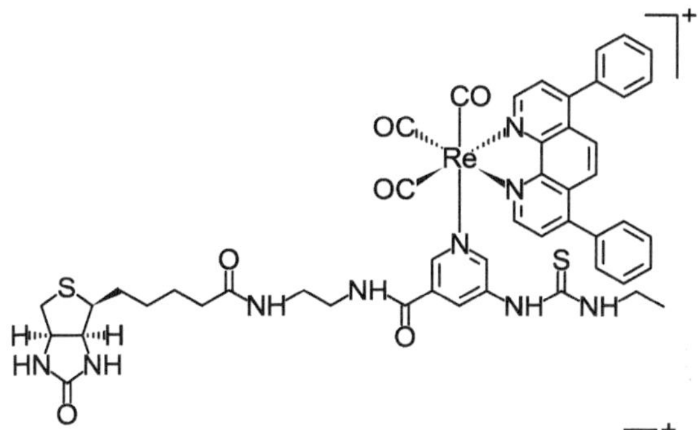

10

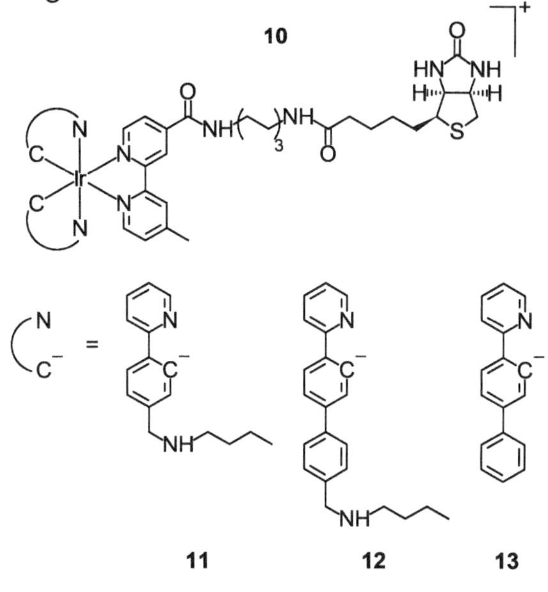

11 **12** **13**

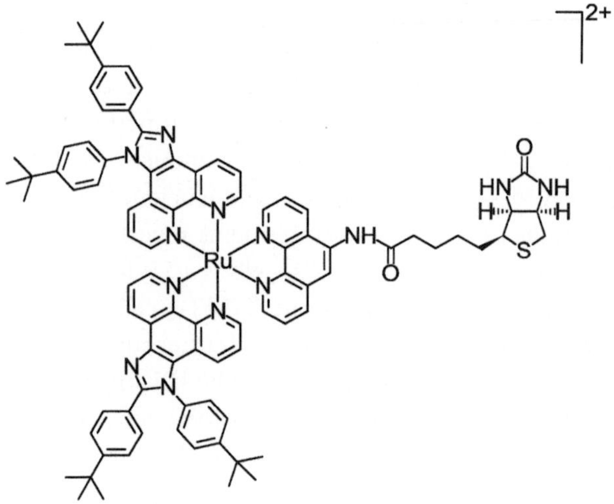

14

2.3 Integrins

Integrins are a family of transmembrane receptors that mediate cell–cell and cell–extracellular matrix adhesion.[26] Integrin $\alpha_v\beta_3$ recognizes ligands containing the tripeptide RGD sequence, and it is overexpressed in cancer cells for promoting tumor growth, invasion, and metastasis. A RGD peptide featuring two additional histidine residues has been cyclized with an iridium(III) aqua complex based on iridium(III)–histidine coordination to afford a phosphorescent peptide conjugate (**15**).[27] The conjugate exhibits much more efficient uptake by $\alpha_v\beta_3$-positive A549 cells than its linear counterpart. Similar uptake is not observed in MCF-7 cells that express $\alpha_v\beta_3$ at low levels. The modification of conjugate **15** with an apoptosis-inducing peptide KLA ($(KLAKLAK)_2$) at the C-terminus gives a conjugate that shows higher cytotoxicity toward A549 cells ($IC_{50,dark} = 4.5$ μM) over MCF-7 cells ($IC_{50,dark} = 37.7$ μM) due to α_v-mediated selective uptake. A two-photon-active ruthenium(II) complex has been modified with a cyclic RGD peptide (**16**) for cancer-targeted PDT.[28] The complex is preferentially taken up by $\alpha_v\beta_3$-overexpressing human glioblastoma U87MG cells over $\alpha_v\beta_3$-negative MCF-7 cells, and enriched in the mitochondria of the cells. The complex is essentially non-cytotoxic toward MCF-7 cells under both dark and light conditions ($IC_{50} > 1$ mM), but displays high photocytotoxic activity toward U87MG cells and multicellular tumor spheroids (MCTSs) ($IC_{50,light} =$ 9–34 μM) due to efficient photogeneration of $^1O_2$ ($\Phi_\Delta = 0.52$) that triggers apoptosis via mitochondria-dependent signaling pathways. Remarkably, the complex efficiently inhibits tumor growth by 87% in U87MG tumor-bearing mice upon two-photon excitation ($\lambda_{ex} = 820$ nm), which is much higher than its peptide-free counterpart (15%) due to more selective tumor accumulation.

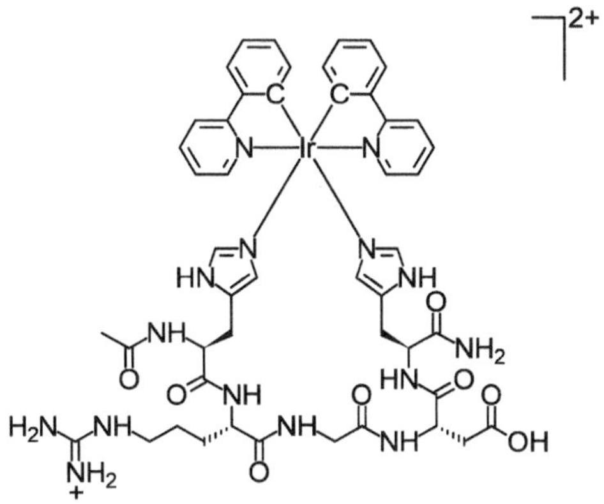

15

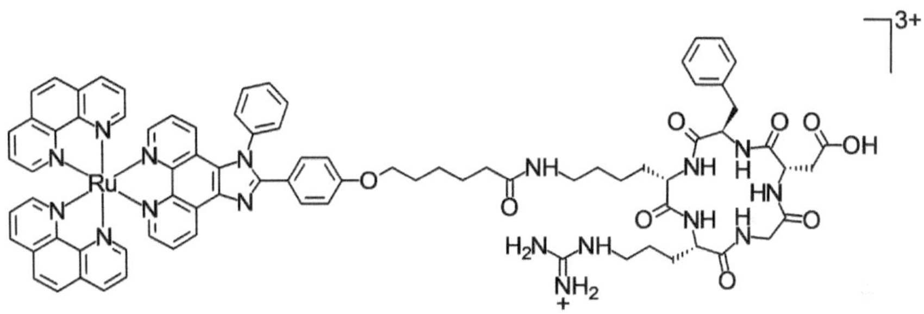

16

2.4 G-Protein-coupled receptors

G-Protein-coupled receptors constitute the largest family of membrane proteins that mediate cellular response to neurotransmitters, hormones, and environmental stimuli in eukaryotic cells, playing important roles in tumor growth, invasion, and metastasis.[29]

Dopamine receptors are upregulated in many types of cancer;[30,31] for example, the overexpression of dopamine D1 receptors in breast cancer tissues is associated with a higher tumor stage and grade and a higher number of nodal metastasis.[30] Iridium(III) complexes carrying a dopamine agonist (**17** and **18**) have been developed to image intracellular dopamine receptors.[32] Incubation of dopamine receptor-expressing A549 cells with the complexes gives rise to intense intracellular emission due to the interaction with intracellular dopamine receptors. This is supported by the reduced emission intensity upon pretreatment of the cells with exogenous dopamine or small interfering RNA (siRNA) that suppresses the dopamine receptor gene expression. Additionally, complex **18** has been applied to track the internalization of dopamine receptors in live cells.

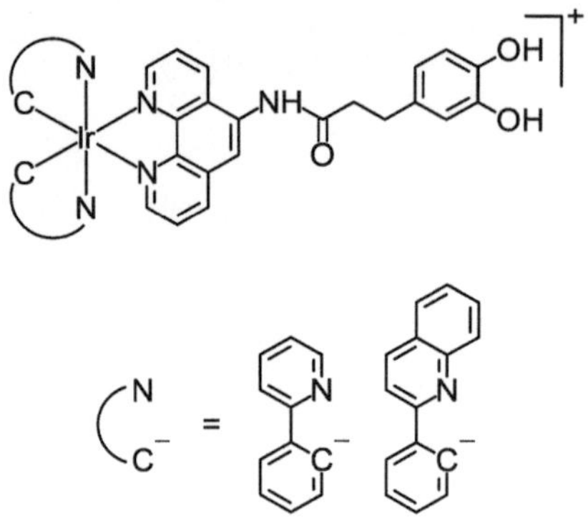

17 **18**

Gastrin-releasing peptide has been recognized as a growth factor for cancer cells and gastrin-releasing peptide receptor (GRPR) is overexpressed in various types of cancer tissues but has limited expression in normal tissues.[33] An iridium(III) complex has been conjugated with the GRPR antagonist JMV594 (**19**) for imaging GRPR in GRPR-positive A549 cells.[34] The specific binding of the complex to GRPR on the plasma membrane is confirmed by the GRPR antagonist (e.g., JMV594 and RC-3095) competition and GRPR gene knockdown experiments. Thus, the complex can discriminate cancerous A549 cells from normal LO2 cells due to different GRPR expression levels.

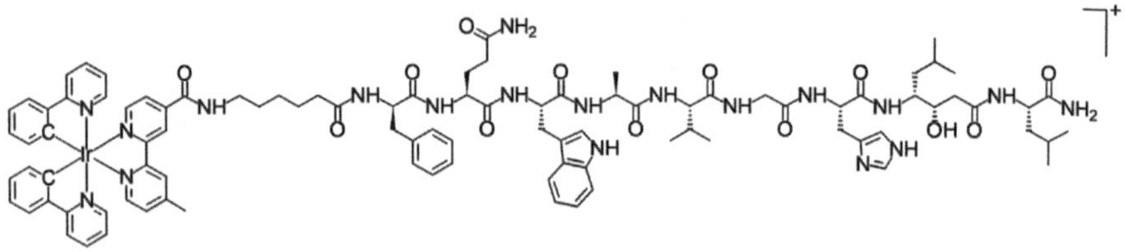

19

Somatostatin (SST) is a neuropeptide that regulates cell proliferation and hormone secretion through interaction with the somatostatin receptors (SSTRs) located on the surface of the target cells.[35] The expression of SSTRs is known to be higher in cancer tissues than in normal tissues.

An alkyne-containing ruthenium(II) complex has been conjugated with an azide-functionalized SST (**20**) via copper(I)-catalyzed azide–alkyne cycloaddition (CuAAC) for cancer-targeted PDT.[36] As shown by LSCM, the uptake of the conjugate by SSTR-overexpressing A549 cells is *ca.* 115 times higher than its peptide-free counterpart. The conjugate exhibits higher photocytotoxicity ($IC_{50,light}=13.2$ μM) toward A549 cells than its peptide-free counterpart ($IC_{50,light}=67.5$ μM). An iridium(III) complex has been conjugated with the peptide-based SST agonist octreotide (**21**) for cancer-targeted PDT.[37] The conjugate shows more efficient uptake by SSTR2-positive HeLa cells than MDA-MB-231 cells that possess a low level of SSTR2 expression via SSTR2-mediated endocytosis. The complex becomes more cytotoxic toward HeLa cells upon irradiation with an IC_{50} value decreasing from 30.9 to 15.5 μM due to photoinduced ROS generation, but displays similar cytotoxic activity toward MDA-MB-231 cells under both dark and light conditions ($IC_{50}=49.2$ and 41.9 μM, respectively).

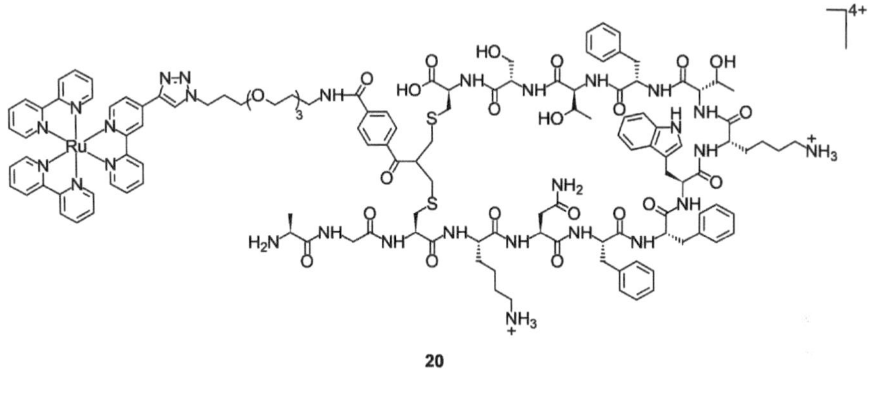

20

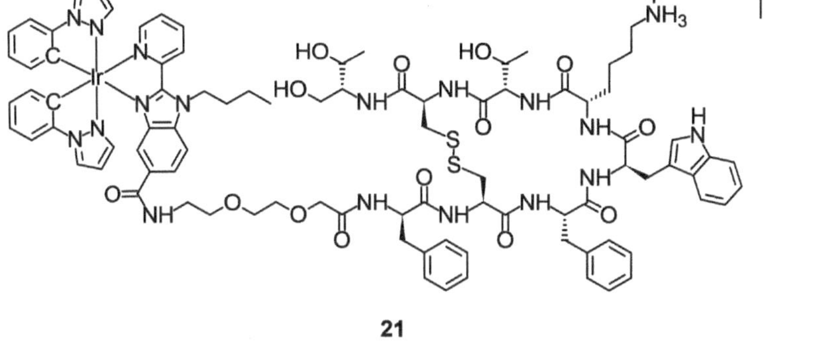

21

2.5 Epidermal growth factor receptor

Epidermal growth factor receptor (EGFR) belongs to the ErbB family of receptor tyrosine kinases that are overexpressed on the plasma membrane of many types of cancer cells, which contribute to increased signaling and promote cancer cell proliferation, invasion, and metastasis.[38] An iridium(III) complex that is modified with the EGFR kinase inhibitor tyrphostin 1 (**22**) has been identified to be capable of simultaneously visualizing and inhibiting EGFR in live cells.[39] Incubation of EGFR-overexpressing human epidermoid carcinoma A431 cells with the complex leads to strong intracellular emission, which is significantly reduced in intensity when the cells have been pretreated with tyrphostin 1 or siRNA to inhibit EGFR activity and suppress EGFR gene expression, respectively. Immunofluorescence staining confirms the specific binding of the complex to EGFR in the cells. The higher intracellular emission in A431 cells than in LO2 cells that express less EGFR indicates that the complex can discriminate EGFR-positive cancer cells from normal cells. Additionally, the binding of the complex to EGFR inhibits EGFR activity, which downregulates the phosphorylation of downstream proteins ERK and MEK and the activity of downstream transcription factor AP1.

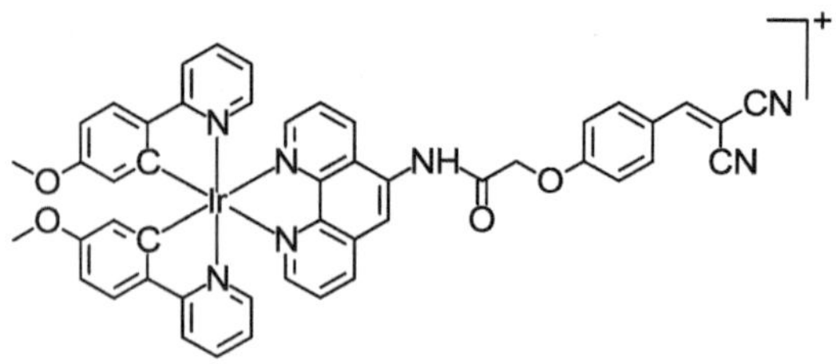

22

2.6 Carbonic anhydrase

Carbonic anhydrase (CA) is a family of enzymes that catalyze the reversible hydration of carbon dioxide. The membrane-bound isoforms CA-IX and CA-XII are overexpressed in cancer cells in response to hypoxia to promote tumor growth by counteracting acidosis through the regulation of intracellular pH.[40] A rhenium(I) complex has been functionalized with the CA inhibitor benzenesulfonamide via a cleavable *N*-acyl imidazole linker (**23**) to label endogenous CA-IX and CA-XII in A549 cells through ligand-directed acyl imidazole chemistry.[41] Upon binding of the complex to the specific protein via substrate recognition, the cleavable linker reacts with the nucleophilic residues that are in close proximity to the substrate-binding site of the protein, resulting in the covalent attachment of the complex to the protein. Synchrotron-based X-ray fluorescence microspectroscopy shows that upon incubation of the CA-overexpressing A549 cells with the complex, the rhenium signal is detected over the whole cell with a stronger signal in the nucleus, which is not observed in the cells that are treated with the competitive CA inhibitor ethoxyzolamide. A rhenium(I) complex featuring an α-oligothiophene moiety and a benzenesulfonamide unit (**24**) has been designed for cancer-targeted PDT.[42] The α-oligothiophene pendant imparts a long-lived and low-lying triplet intraligand charge transfer (^{3}ILCT) state to facilitate the generation of ROS via both Type I and Type II photoreactions. This enables the complex to exhibit very high photocytotoxicity toward MDA-MB-231 and murine mammary carcinoma 4T1 cells ($IC_{50,light}$ = 0.0073–0.22 μM) with photocytotoxicity index (PI) values >13,700 and >5000 under normoxia and hypoxia, respectively, which is much higher than that toward non-cancerous human breast epithelial MCF-10A cells ($IC_{50,light}$ = 0.12–0.39 μM). Studies on the mechanisms of action reveal that the complex causes plasma membrane rupture and induces gasdermin D-mediated pyroptosis that stimulates tumor immunogenicity upon irradiation. This promotes the maturation of dendritic cells and activates T cell-dependent adaptive immune response to eliminate both primary and distant tumors *in vivo*.

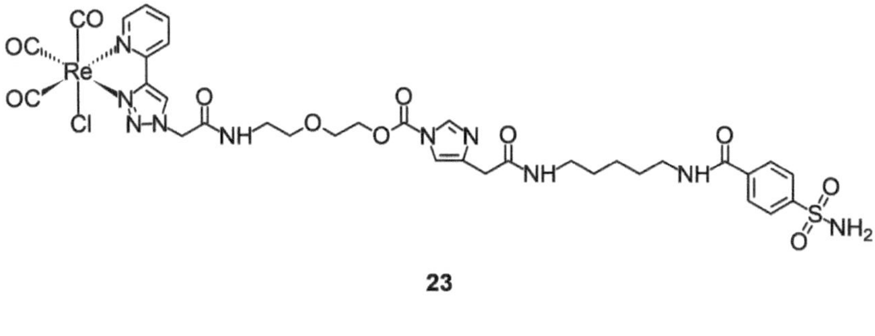

23

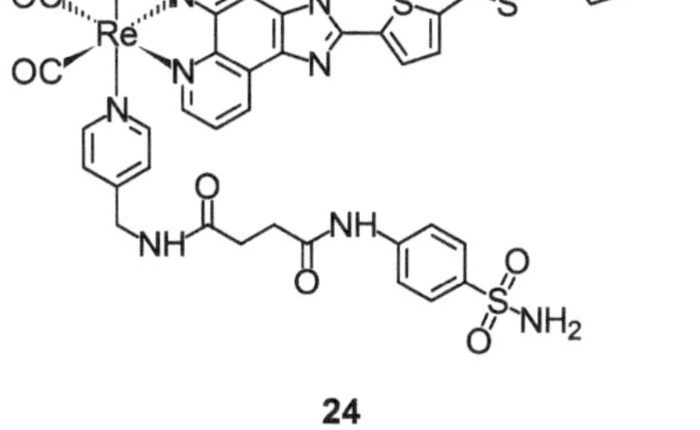

24

2.7 Cyclooxygenase

Cyclooxygenase (COX) is a membrane-embedded, heme-containing enzyme that is involved in the conversion of arachidonic acid to proinflammatory prostaglandins. The inducible isoform COX-2 is usually upregulated in inflamed and neoplastic tissues.[43] An iridium(III) complex appended with a structural analog of the COX-2 inhibitor indomethacin (**25**) has been developed to monitor the COX-2 expression in live cells.[44] Incubation of COX-2-overexpressing HeLa cells with the complex gives rise to intense intracellular emission, which is not observed when HeLa cells have been pretreated with curcumin to inhibit the expression of COX-2 or when normal LO2 cells are used. A rhodamine-bearing iridium(III) complex has been modified with indomethacin (**26**) for cancer-targeted PDT.[45] The heavy iridium(III) center facilitates the population of the long-lived triplet excited state of rhodamine for ROS photosensitization via two processes: (1) intersystem crossing from the singlet excited state of rhodamine; and (2) triplet–triplet energy transfer from the triplet excited state of the iridium(III) polypyridine unit. The uptake of the complex by COX-2-overexpressing cancer cells (HeLa and MCF-7) is much higher than that by COX-2-negative non-cancerous cells (HEK293T and LO2), which allows the complex to cause more significant cell death in MCF-7 cells than in LO2 cells upon irradiation. The high $^1O_2$-photosensitization capability ($\Phi_\Delta = 0.70$) and selective tumor accumulation of the complex enable a substantial decrease in tumor size from *ca.* 1000 to 200 mm^3 in MCF-7 tumor-bearing mice upon irradiation.

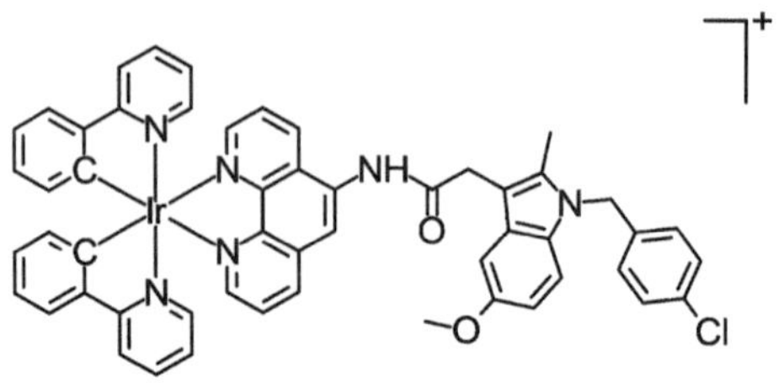

25

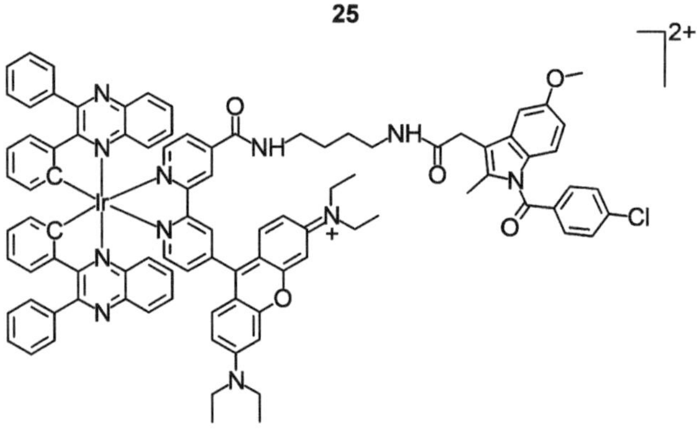

26

2.8 Sialylated glycoproteins and glycolipids

Sialic acids are a family of negatively-charged, nine-carbon monosaccharides derived from neuraminic acid, with the most common member being *N*-acetylneuraminic acid (Neu5Ac).[46] The covalent attachment of sialic acid to the terminal end of the glycan chains of membrane glycoproteins and glycolipids (sialylation) is an important modification involved in mediating cell–cell interactions, signaling, and immune response. Increased cell-surface sialylation in cancer cells is associated with cancer progression, invasion, and metastasis.[47]

Three iridium(III) complexes carrying a phenylboronic acid (PBA) moiety (**27–29**) have been designed to label sialic acids on the plasma membrane.[48] Spectrophotometric titrations indicate that these complexes show higher binding affinity to Neu5Ac (log K_b = 3.61–3.85) over other sugars such as D-glucose, D-galactose, and D-mannose (log K_b = 1.68–3.34) that are commonly found on glycoproteins. This results from the cooperative two-site binding through: (1) PBA ester formation with the vicinal diol of Neu5Ac, and (2) electrostatic interaction between the ammonium group of the complexes and carboxylate group of Neu5Ac under physiological conditions. Incubation of HepG2 cells with complex **29** results in intense emission on the cell surface with much weaker emission in the cytoplasm (Fig. 2), which is not observed in the cases of complexes **27** and **28** due to their much more efficient cellular uptake. The plasma membrane staining disappears when the cells are co-treated with exogenous Neu5Ac or pretreated with neuraminidase to catalyze the hydrolysis of α(2,3) sialic acid linkages (Fig. 2). The specific recognition of cell-surface sialic acid residues allows complex **29** to distinguish cancerous HepG2 cells from non-cancerous HEK293T cells.

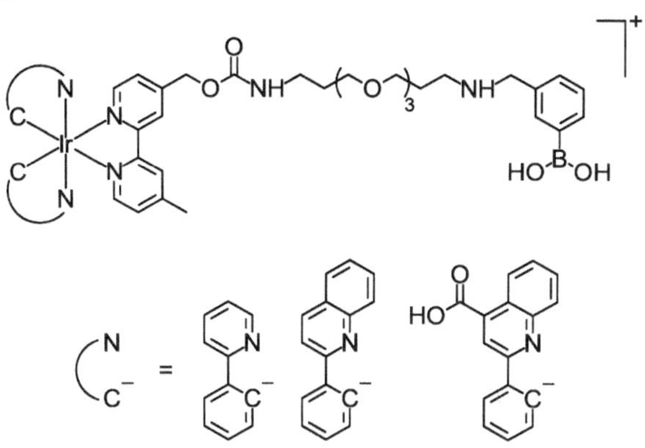

27 **28** **29**

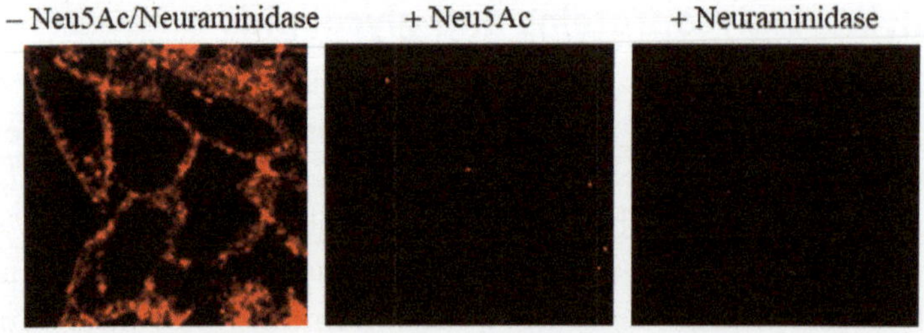

Fig. 2 LSCM images of HepG2 cells incubated with complex **29** (25 μM, 37 °C, 1 h) without any treatment (left), with co-incubation with Neu5Ac (500 μM, 37 °C, 1 h) (middle), or with pretreatment with neuraminidase (0.1 U mL^{-1}, 37 °C, 1 h) (right). *Adapted from Liu, H.-W.; Law, W. H.-T.; Lee, L. C.-C.; Lau, J. C.-W.; Lo, K. K.-W.* Chem. Asian J. ***2017,*** 12, *1545–1556, with permission from Wiley-VCH. Copyright 2017.*

Bioorthogonal chemistry, which refers to the chemoselective reactions that can occur between two abiotic functionalities under physiological conditions without interfering with the native biological processes, has offered a versatile platform for the detection and visualization of various biomolecules including glycans, lipids, nucleic acids, proteins, and other small metabolites in their native environments.[49]

Azide is a promising bioorthogonal chemical reporter due to its small size, non-native nature, and inertness to biomolecules. It undergoes CuAAC[50] and strain-promoted azide–alkyne cycloaddition (SPAAC) reaction[51] with terminal alkynes and cyclooctynes, respectively, to give substituted triazoles under physiological conditions. The unnatural azide-containing monosaccharide 1,3,4,6-tetra-*O*-acetyl-*N*-azidoacetyl-D-mannosamine ($Ac_4ManNAz$) has been developed for metabolic oligosaccharide engineering; it is metabolically converted by the cells to *N*-azidoacetyl sialic acids and incorporated into cell-surface glycans.[52] Thus, pretreatment of cells with $Ac_4ManNAz$ is expected to accumulate the alkyne-functionalized labeling reagents on the plasma membrane through a bioorthogonal CuAAC or SPAAC reaction. The first phosphorescent bioorthogonal probes are derived from iridium(III) complexes featuring a dibenzocyclooctyne (DIBO) pendant (**30–32**).[53] Complexes **30** and **31** are enriched in the mitochondria of both $Ac_4ManNAz$-pretreated and untreated Chinse hamster ovary (CHO) cells (Fig. 3) due to cationic charge and high lipophilicity. However, incubation of $Ac_4ManNAz$-pretreated cells with complex **32** leads to intense membrane and granular cytoplasmic staining (Fig. 3), as the ionized carboxylate groups suppress the internalization of the complex and assist in the SPAAC reaction

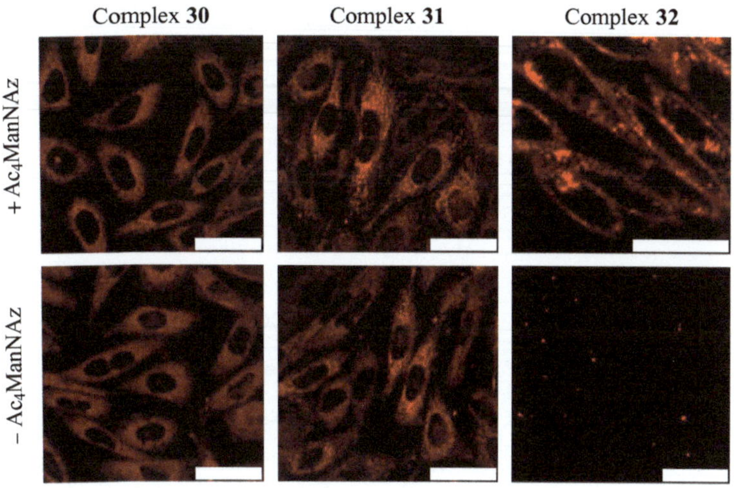

Fig. 3 LSCM images of CHO cells incubated with complexes **30**, **31** (30 μM, 37 °C, 20 min), and **32** (30 μM, 37 °C, 5 h) with (top) or without (bottom) pretreatment with $Ac_4ManNAz$ (50 μM, 37 °C, 72 h). Scale bar = 25 μm. *Adapted from Lo, K. K.-W.; Chan, B. T.-N.; Liu, H.-W.; Zhang, K. Y.; Li, S. P.-Y.; Tang, T. S.-M.* Chem. Commun. ***2013,*** *49, 4271–4273, with permission from Royal Society of Chemistry. Copyright 2013.*

with membrane-bound azide units. Similar results have been observed for related rhenium(I) (**33**–**35**)[54] and ruthenium(II) DIBO complexes (**36** and **37**).[55] The less lipophilic ruthenium(II) complex **36** has been employed to monitor the dynamic behavior of glycans by LSCM.[55] The results show the active involvement of the endosomes, Golgi apparatus, and lysosomes in the intracellular trafficking of the membrane-bound glycans.

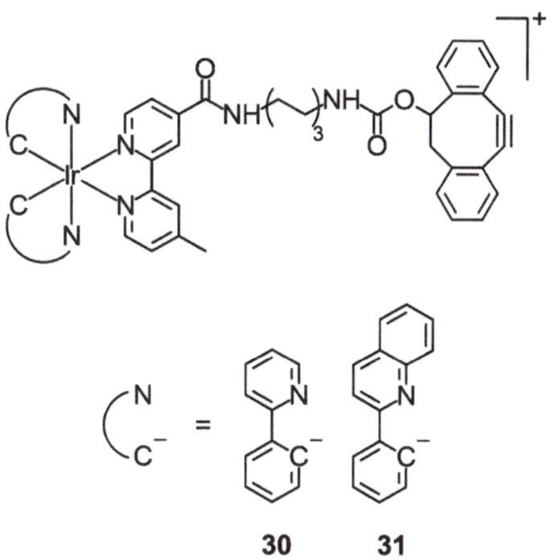

30 **31**

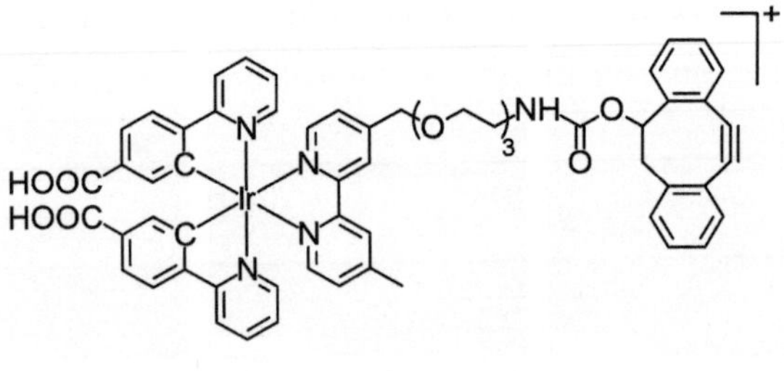

32

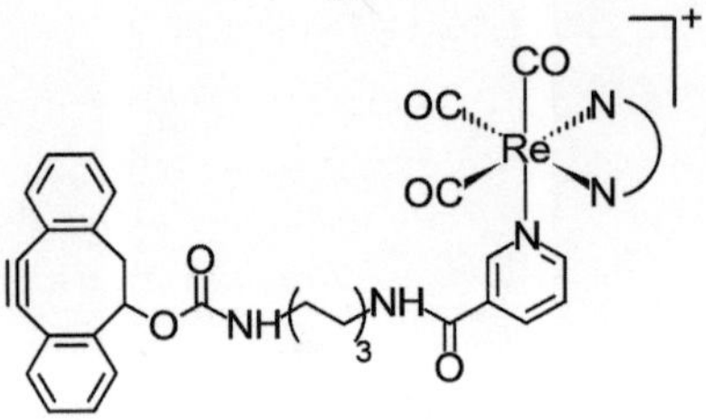

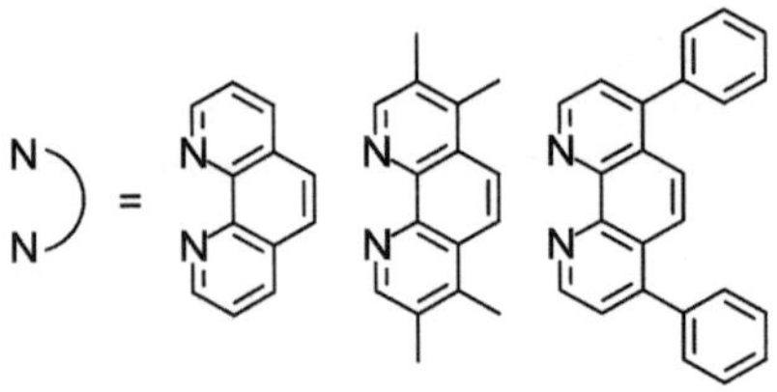

33 **34** **35**

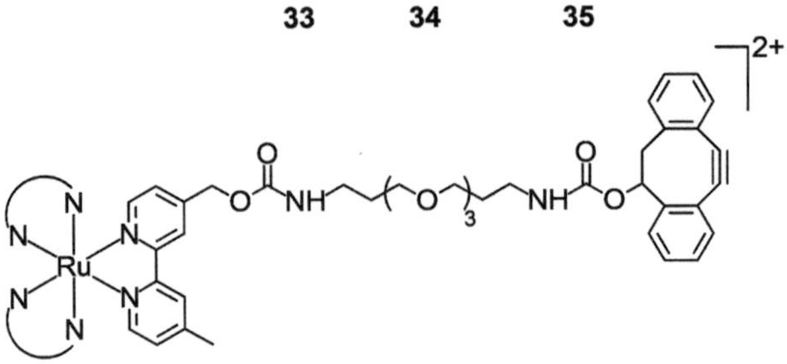

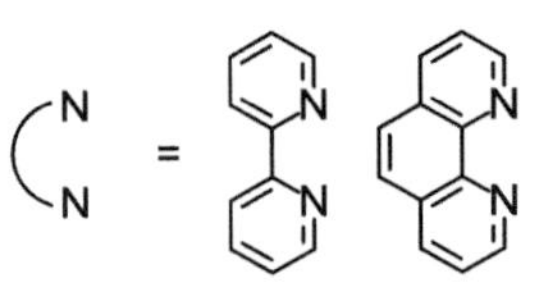

36 **37**

The bioorthogonal chemical reporter strategy has been expanded to the PDT treatment of malignant triple-negative breast cancer (TNBC) that is more aggressive and difficult to treat than other types of invasive breast cancer due to the lack of estrogen receptor, progesterone receptor, and human epidermal growth factor receptor-2 for targeted cancer therapy.[56]

The CuAAC reaction of the azide-modified cell-surface glycans with a membrane-impermeable, alkyne-appended ruthenium(II) complex (**38**) generates a triazole derivative with large TPA cross-section (δ_{810nm} = 212 GM) and efficient $^1O_2$ photosensitization (Φ_Δ = 0.80) on the plasma membrane for two-photon-induced PDT. The complex displays low cytotoxic activity toward TNBC MDA-MB-231 cells under both dark and light conditions (IC_{50} > 40 μM), but becomes more cytotoxic ($IC_{50,light}$ = 10.6 μM) upon two-photon excitation (λ_{ex} = 810 nm) when the cells have been pretreated with Ac_4ManNAz. The complex is anchored on the cell surface and the photogenerated ROS cause strong membrane disruption to induce cell death. The high photocytotoxicity of the complex is retained in Ac_4ManNAz-pretreated, MDA-MB-231-derived MCTSs. Remarkably, the complex does not exhibit any significant (photo)cytotoxic activity toward both Ac_4ManNAz-pretreated and untreated non-cancerous MCF-10A cells (IC_{50} > 40 μM), suggesting that metabolic oligosaccharide engineering is an effective means to improve the cancer selectivity of transition metal complexes.

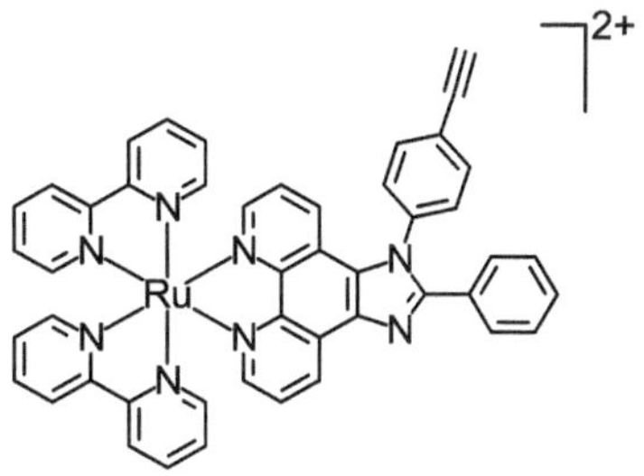

38

3. Activatable transition metal complexes for cancer diagnosis and therapy

Although targeting specific cell-surface proteins allows transition metal complexes to be selectively taken up by cancer cells over normal cells, the expression of these proteins varies among cancer cells within the same tumor.[57] Additionally, the possible non-specific interactions of the complexes with other biomolecules may generate false positives in imaging and weaken the therapeutic efficacy. Thus, developing transition metal complexes using an activity-based approach, which leverages the unique reactivity of the target to induce an increase in the emission intensity and ROS-photosensitization efficiency of the complexes, provides a versatile paradigm to realize selective cancer imaging and therapy.

3.1 Hypoxia

Tumor hypoxia, which refers to a condition where the cancer cells are deprived of an adequate oxygen supply due to uncontrollable cell proliferation and abnormal tumor vasculature, is one of the major characteristics shared by most solid tumors.[58] This phenomenon is associated with poor prognosis and resistance to chemo- and radiotherapy. Phosphorescent transition metal complexes are ideal candidates for detecting tumor hypoxia in a non-destructive, real-time, and reversible manner.[59] These complexes display more intense and longer-lived emission under hypoxic conditions, as emission quenching caused by the diffusion-controlled collisional interaction between the triplet excited state of the complex and the triplet ground state of O_2 is inhibited at low O_2 concentrations.

A representative example is the 2-(2-pyridyl)benzothiophene-based iridium(III) acetylacetonate complex (**39**), which has been applied to image hypoxic tumors in live mice.[60] The complex exhibits intense emission in cells that are incubated at low O_2 levels (5%) but not in those under normoxia (20%). Functionalization of the complex with a protonatable dimethylamino group (**40**) significantly enhances the cellular uptake efficiency of the complex and shifts the intracellular localization of the complex from the endoplasmic reticulum (ER) to the lysosomes.[61] Additionally, ICP-MS measurements reveal that complexes **39** and **40** show distinct distributions *in vivo*; complex **39** is mainly accumulated in the spleen and lung of tumor-bearing mice, whereas complex **40** is predominantly enriched in the liver and kidney. Remarkably, intense emission is readily observed in the tumor region upon intravenous injection of the complexes.[60,61] The longer emission lifetimes detected in the tumor region over the normal tissues further support that the observed emission is due to tumor hypoxia-induced emission enhancement instead of tumor accumulation of the complexes. Increasing the π-conjugation of the 2-(2-pyridyl)benzothiophene ligand reduces the highest occupied molecular orbital (HOMO)–lowest unoccupied molecular orbital (LUMO) gap, affording an near-infrared (NIR)-emitting complex (**41**) ($\lambda_{em} = 720$ nm) that can image tumors that are transplanted 6–7 mm below the abdominal skin surface as NIR emission offers deeper penetration.[60] The complex has been modified with a poly(*N*-vinylpyrrolidone) (PVP) pendant (**42**) to increase tumor retention by the enhanced permeability and retention (EPR) effect for the specific detection of tumors and cancer cells *in vivo*.[62] It has been further functionalized with an O_2-insensitive rhodamine (**43**) or cyanine derivative (**44**) as an internal standard for assessing the local probe concentration and ratiometric O_2 measurement.

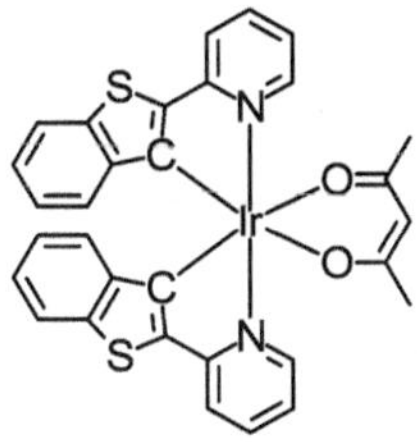
S
C
N
Ir
O
O
C
S
N

39

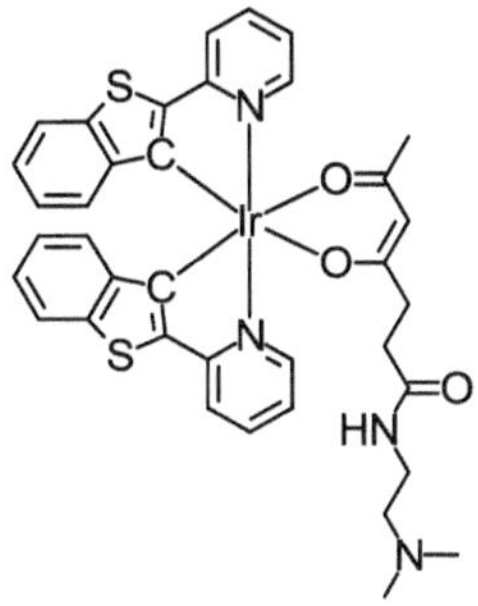
S
C
N
Ir
O
O
C
S
N
O
HN
N

40

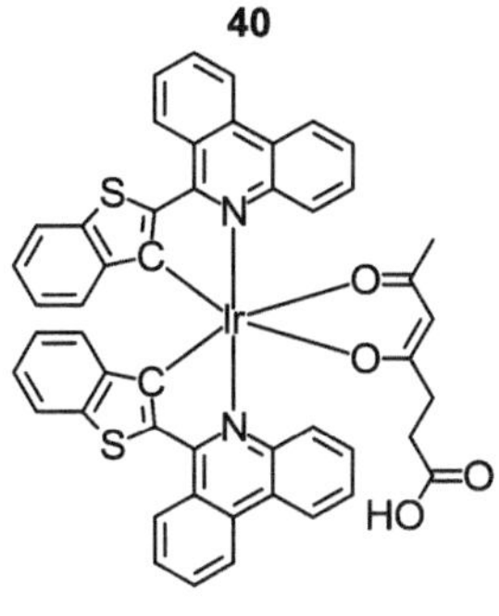
S
C
N
Ir
O
O
C
S
N
HO
O

41

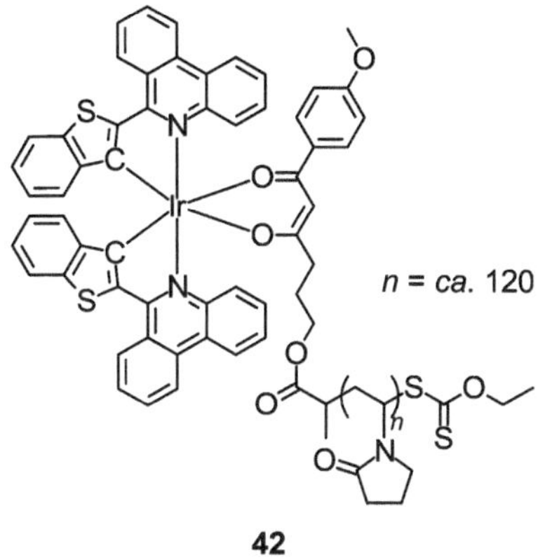
O
S
C
N
Ir
O
O
C
S
N
n = ca. 120
O
O
n
S
O
S
O
N

42

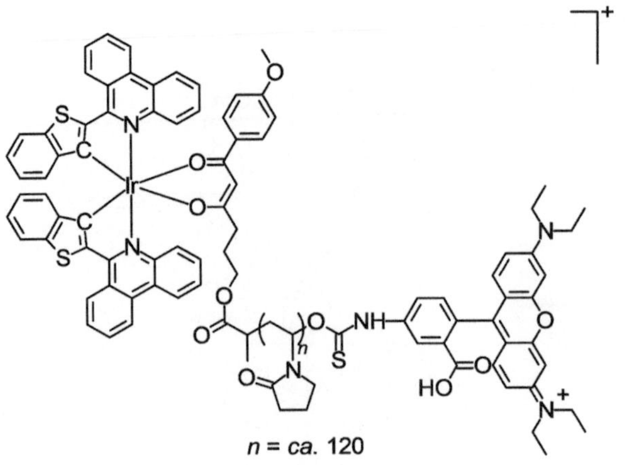

n = *ca.* 120

43

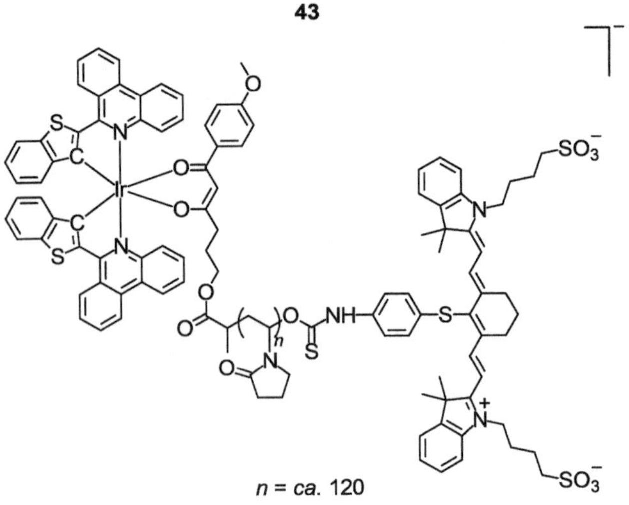

n = *ca.* 120

44

Iridium(III) complexes incorporated with non-conjugated amino groups display dual phosphorescence from two independent emissive states due to the interruption of the internal conversion from the higher-lying triplet intraligand (^{3}IL) state to the lowest-lying triplet charge transfer (^{3}CT) state by an additional triplet amine-to-ligand charge transfer (^{3}NLCT) state.[24,63] Rational tuning of the efficiency of internal conversion between the two emissive states through structural manipulation has identified the first molecular probe (**45**) that can differentiate hypoxia, normoxia, and hyperoxia in living systems.[64] The complex exhibits dual emission bands at *ca.* 520 and 620 nm with comparable intensities in aqueous solution under ambient conditions (Fig. 4a). The intensity of the high-energy (HE) emission decreases

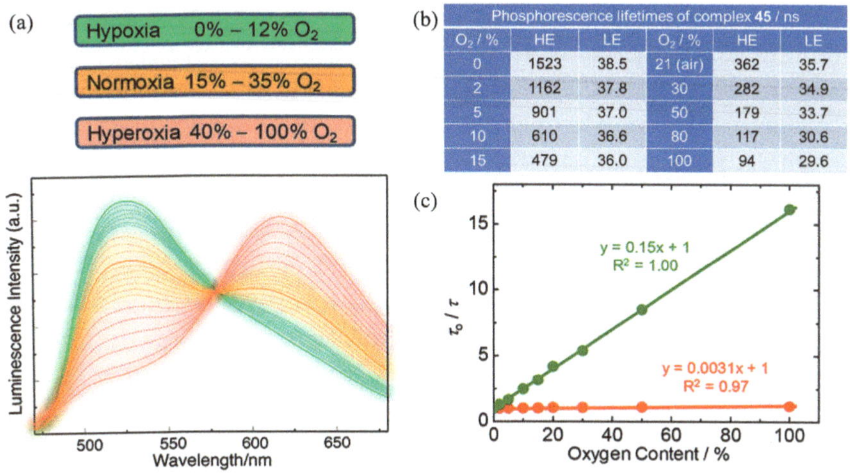

Phosphorescence lifetimes of complex 45 / ns					
O_2 / %	HE	LE	O_2 / %	HE	LE
0	1523	38.5	21 (air)	362	35.7
2	1162	37.8	30	282	34.9
5	901	37.0	50	179	33.7
10	610	36.6	80	117	30.6
15	479	36.0	100	94	29.6

Fig. 4 (A) Emission spectra and (B) phosphorescence lifetimes of complex **45** in phosphate-buffered saline/CH_3OH (9,1, *v/v*) under an atmosphere of 0–100% O_2. (C) Stern–Volmer plots of the HE (green) and LE (red) emission toward O_2 quenching. *Adapted from Zhang, K. Y.; Gao, P.; Sun, G.; Zhang, T.; Li, X.; Liu, S.; Zhao, Q.; Lo, K. K.-W.; Huang, W.* J. Am. Chem. Soc. ***2018,*** 140, *7827–7834, with permission from American Chemical Society. Copyright 2018.*

while that of the low-energy (LE) emission increases upon increasing the O_2 concentration, and an opposite phenomenon is observed upon decreasing the O_2 level. The sensitive spectral response gives rise to naked-eye distinguishable green, orange, and red emission under an atmosphere of 0%, 21%, and 100% O_2, respectively. The ratiometric response of the complex has been ascribed to the different quenching efficiencies of the two emissive states by O_2. This is supported by the much more significant reduction in the lifetime of the HE emission (from 1523 to 94 ns) than the LE emission (from 38.5 to 29.6 ns) upon increasing the O_2 concentration from 0% to 100% (Fig. 4b,c), which indicates that the HE emission is more sensitive to O_2 quenching.

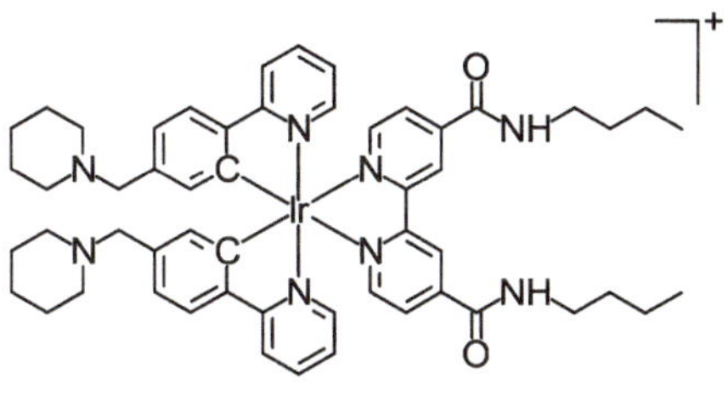

45

Phosphorescent platinum(II) porphyrins are attractive for imaging tumor hypoxia due to their intense red to NIR emission with long emission lifetimes (in the range of 10–1000 μs). The introduction of platinum(II) tetraphenylporphyrin into the backbone of an O_2-insensitive, blue-emitting polyfluorene-based conjugated polyelectrolyte gives an amphiphilic dual-emissive polymer (**46**) that self-assembles into ultrasmall conjugated polymer dots (size = *ca.* 5 nm) in aqueous solution for ratiometric and lifetime imaging of intracellular O_2 levels.[65] Upon decreasing the O_2 concentration, there is a remarkable increase in the intensity and lifetime of the red phosphorescence, whereas those of the blue fluorescence remain almost the same. The polymer dots have been used to image hypoxic tumors in live mice. Modification of platinum(II) tetraphenylporphyrin with a PVP pendant affords a monochromophoric ratiometric O_2 probe (**47**).[66] The probe self-assembles to form micelle-like nanoparticles with an average diameter of *ca.* 118 nm in aqueous solution due to its amphiphilic nature. In addition to the long-lived phosphorescence at 760 nm ($\tau_o = ca.$ 40 μs) from the platinum(II) tetraphenylporphyrin moiety, the probe shows a short-lived fluorescence band at 660 nm ($\tau_o = ca.$ 7 ns) that originates from the assembled clusters of the PVP chain. The probe is enriched in the tumor region of tumor-bearing mice due to the EPR effect, which enables the visualization of the hypoxic tumors with high spatiotemporal resolution. Platinum(II) tetraphenylporphyrins appended with four cationic ammonium-containing oligofluorene pendants (**48–50**) have been developed for tumor imaging and PDT.[67] The cationic fluorene arms are employed to increase the aqueous solubility of the complexes to prevent aggregation of the platinum(II) tetraphenylporphyrin unit, which in turns enhances the O_2 sensitivity and $^1O_2$-generation efficiency of the complexes. Thus, among the three complexes, complex **50** displays the largest photophysical changes upon changing the O_2 levels and the most efficient $^1O_2$ photosensitization ($\Phi_\Delta = 0.92$). These features allow the complex to exhibit more intense and longer-lived emission in the tumor region of a mouse model due to the hypoxic nature of solid tumors, and achieve almost complete tumor ablation upon irradiation due to efficient generation of $^1O_2$.

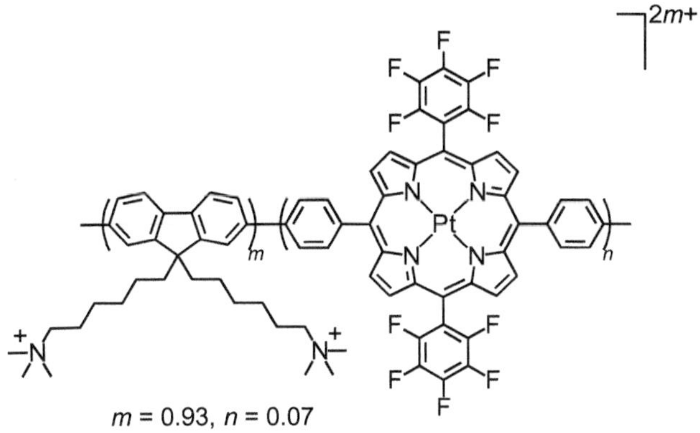

m = 0.93, *n* = 0.07

46

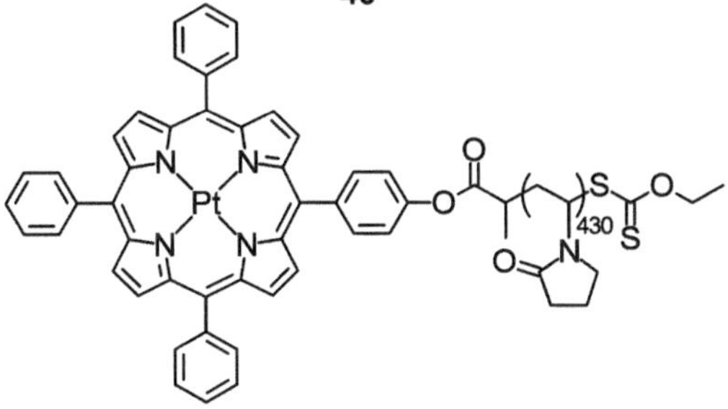

47

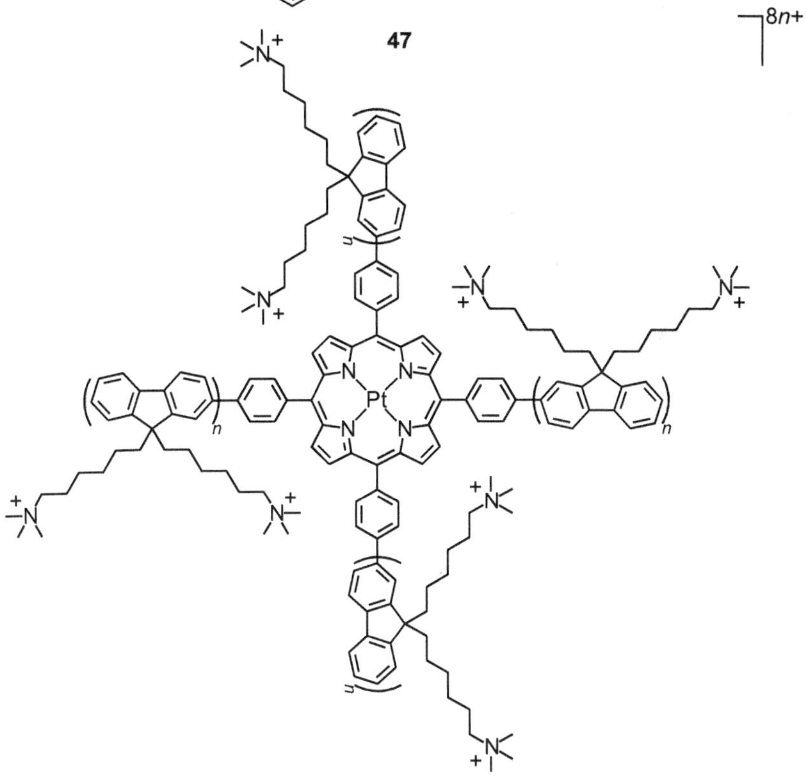

n = 1 (**48**)
2 (**49**)
3 (**50**)

3.2 pH

Intracellular pH homeostasis is crucial to the regulation of cell proliferation and cell death[68] and metabolic activities.[69] Cancer cells undergo glycolysis as the major metabolic pathway for energy production, which generates large amounts of lactate and protons.[70] To avoid intracellular acidification, cancer cells upregulate various pH-regulating proteins on the plasma membrane to transport lactate and protons out of the cells, resulting in a lower extracellular pH (6.7–7.1) than normal cells (~7.4).[71] The acidic extracellular microenvironment promotes migration, invasion and metastasis of cancer cells and increases the resistance to anticancer drugs.

An iridium(III) complex conjugated with a poly(ethylene glycol) (PEG) pendant via an imine linkage (**51**) has been designed to be responsive to both tumor acidosis and hypoxia, the two hallmarks of solid tumor microenvironment.[72] Upon exposure to an acidic environment, the imine bond is hydrolyzed to a more electron-withdrawing aldehyde moiety, inducing a large bathochromic shift in the absorption (from 450 to 580 nm) and emission bands (from 610 to 705 nm). Thus, only the emission at 705 nm is detected upon photoexcitation at 580 nm under acidic conditions, and the emission intensity significantly increases upon decreasing the O_2 concentration from 20% to 0%. The pH- and O_2-sensitive emission of the complex enables *in vivo* imaging of primary tumors with a high tumor-to-normal tissue ratio. Additionally, the complex can detect metastatic liver lesions as small as 1 mm in mice bearing orthotopic breast tumors and monitor cancer cell metabolism *in vivo*. Another iridium(III) complex has been functionalized with a phenol group via a π-conjugated ethynyl linker (**52**) for the synergistic sensing of acidity and O_2 levels in tumor microenvironment.[73] The phenol group renders the emission of the complex to be pH-sensitive, as the deprotonation of the phenolic proton under alkaline conditions allows efficient emission quenching via photoinduced electron transfer (PET). The emission quantum yield and lifetime of the complex increase from 0.002 to 0.015 and 85 to 161 ns, respectively, upon decreasing the pH from 7.4 to 6.4 and O_2 concentration from 21% to 0%. Thus, the intracellular emission intensity and lifetime of the complex-loaded MCF-7 cells that are incubated under an acidic (pH 6.4) and hypoxic (0% O_2) environment are higher and longer than those under normal physiological conditions (pH 7.4 and 21% O_2). The complex has been utilized to visualize the microenvironment of MCF-7-derived MCTSs. As shown by PLIM, the complex-treated MCTSs show a longer emission lifetime at the center (*ca.* 120 ns) than the edge of the spheroids (*ca.* 90 ns), and the emission lifetime of the whole spheroids is extended to *ca.* 120–140 ns upon decreasing the pH of the culture medium from 7.4 to 6.4. These findings highlight the potential of the complex in imaging solid tumors *in vivo* based on its synergistic response toward both tumor acidosis and hypoxia.

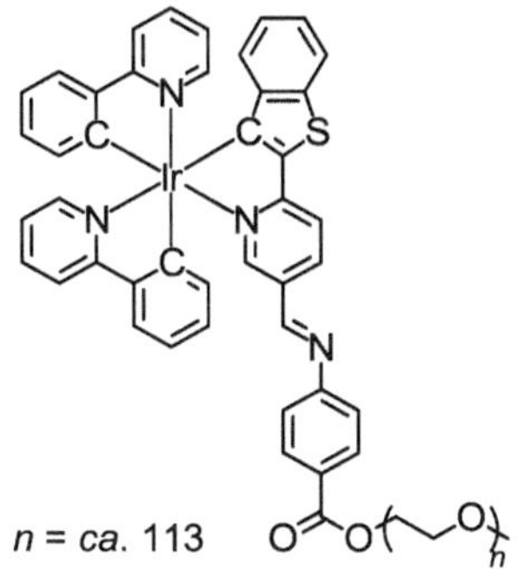

51

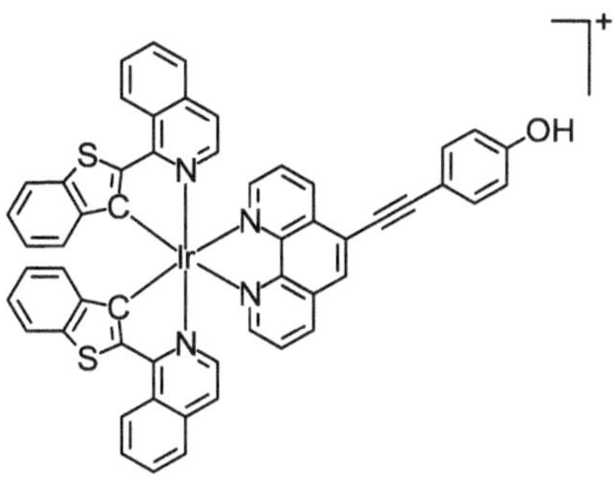

52

Iridium(III) complexes coordinated with a pH-sensitive imidazolyl- or benzimidazolyl-β-carboline ligand (**53–56**) have been designed for cancer- and lysosome-targeted imaging and PDT.[74] Among these complexes, complex **56** displays the largest changes in emission intensity and $^1O_2$-generation efficiency in response to pH; the emission and $^1O_2$-generation quantum yields of this complex increase from 0.019 and 0.05 to 0.092 and 0.51, respectively, upon decreasing the pH from 7.4 to 3.0. This complex is taken up by A549 cells via an energy-dependent mechanism and accumulated in the lysosomes of the cells. Remarkably, complex **56** is essentially non-cytotoxic in the dark ($IC_{50,dark} > 100$ μM), but exhibits high photocytotoxicity toward cancerous A549 and A549R cells ($IC_{50,light} = 0.12$ and 0.16 μM, respectively) with PI values >833 and >625, which is much higher than that toward normal

LO2 cells ($IC_{50,light}$ = 4.4 μM). This results in selective killing of A549 cells over LO2 cells in a co-culture environment. Studies on the mechanism of action reveal that the complex induces caspase- and ROS-dependent apoptosis through causing lysosomal damage and cathepsin B release upon irradiation. Additionally, the complex has been applied to track the changes in lysosomal integrity during PDT in a real-time manner, which provides a convenient method for *in situ* monitoring of the phototherapeutic effect. Selective killing of cancer cells over normal cells has also been observed for a related iridium(III) pyridyl-β-carboline complex (**57**).[75] The complex shows enhanced emission and $^1O_2$ photosensitization under acidic conditions, leading to higher cytotoxic activity toward HeLa cells ($IC_{50,dark}$ = 31.3 μM; $IC_{50,light}$ = 3.8 μM) over non-cancerous human embryonic lung fibroblast MRC-5 cells ($IC_{50,dark}$ > 100 μM; $IC_{50,light}$ = 50.7 μM) under both dark and light conditions.

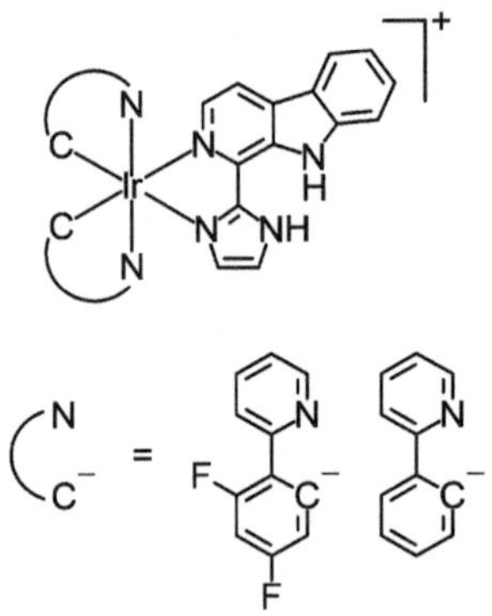

53 **54**

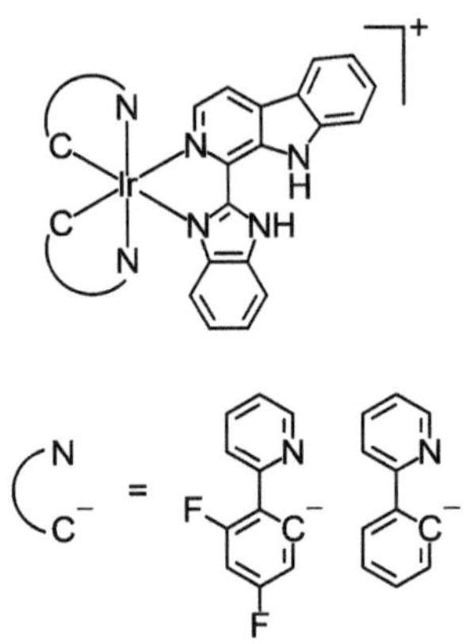

55 **56**

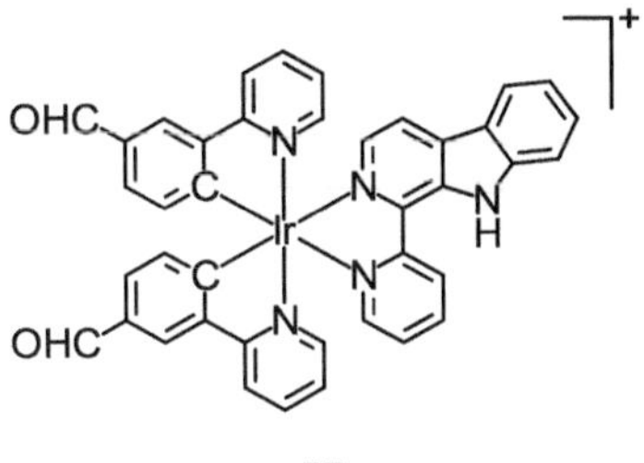

57

3.3 Polarity

Cellular polarity is an important environmental factor that controls various cellular processes associated with protein activities and membrane structures. Each organelle possesses an optimal polarity for its specific roles and functions. Aberrant polarity is implicated in organelle dysfunction and various pathological conditions including cancer.[76] An iridium(III) complex bearing an *o*-carborane unit (**58**) has been employed to track the changes in mitochondrial polarity using PLIM.[77] The introduction of an electron-deficient *o*-carborane moiety to the diimine ligand enhances the transition dipole moment of the complex. Thus, the complex displays a large bathochromic shift in the emission maximum from 535 to 577 nm upon changing the solvent from less polar toluene (0.36 D) to more polar DMSO (3.96 D). This was accompanied by a substantial decrease in the emission quantum yield and lifetime from 0.80 to 0.21 and from 794.5 to 185.2 ns, respectively. The complex is enriched in the mitochondria of both cancerous HepG2 and non-cancerous human liver HL-7702 cells. The more intense and longer-lived intracellular emission in the complex-loaded HepG2 cells with respect to HL-7702 cells suggests a less polar intracellular environment in cancer cells. Additionally, the complex can differentiate live, apoptotic, and dead cells based on mitochondrial polarity. As revealed by PLIM, the emission lifetime of the complex in live cells is shorter than that in apoptotic cells and dead cells, suggestive of decreasing mitochondrial polarity during apoptosis.

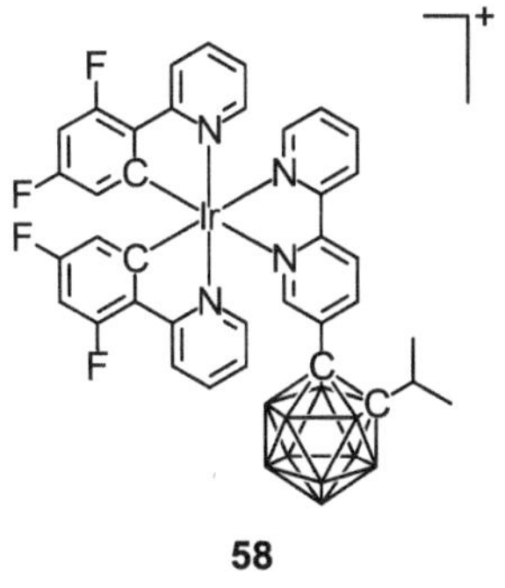

58

3.4 Viscosity

Intracellular viscosity determines various diffusion-mediated cellular processes such as macromolecular interactions, metabolite transport, and signal transduction.[78] Abnormal intracellular viscosity is associated with various disorders and diseases. There has been emerging evidence that the viscosity in cancer cells is higher than normal cells.[79] A dual-emissive binuclear iridium(III) complex (**59**) has been developed for ratiometric monitoring of changes in intracellular viscosity.[80] The complex exhibits two emission bands at 521 and 711 nm, and their intensities increase by 16.9 and 2.5 fold, respectively, upon increasing the viscosity of the solution from 1.1 (water) to 950 cP (99% glycerol). The emission enhancement is a consequence of inhibited motion of the two luminophores in highly viscous media. Treatment of co-cultured HepG2 and HL-7702 cells with the complex leads to more intense emission in HepG2 cells, suggestive of a higher intracellular viscosity in cancer cells. The complex has been used to track the viscosity changes in MCF-7 cells in real time, which indicates that the intracellular viscosity increases during apoptosis. A viscosity-sensitive iridium(III) complex featuring a rotatable pyridine ring (**60**) has been reported to distinguish cancer stem cells from cancer cells at the tissue level.[81] The complex shows increased emission intensity ($I/I_o = ca.$ 8) and extended lifetime (from 12.5 to 63.2 ns) upon increasing the glycerol concentration in a water–glycerol mixture due to the restriction of the free rotation of the pyridine unit. The complex is taken up by HepG2 cells via an energy-dependent, non-endocytic pathway and accumulated in the mitochondria of the cells. Upon the addition of the ionophore monensin to increase the viscosity of mitochondria through disrupting the ion balance, the average emission lifetime of the complex increases from 9.9 to 11.6 ns, indicating that it can monitor intracellular viscosity in real time using PLIM. Remarkably, as revealed by the co-staining experiments with the cell-surface glycoprotein CD44 that serves as a biomarker for cancer stem cells, the complex displays a much longer emission lifetime in cancer stem cells than the surrounding cancer cells in the same tumor tissue, highlighting the potential of the complex in image-guided PDT for treating cancer stem cells.

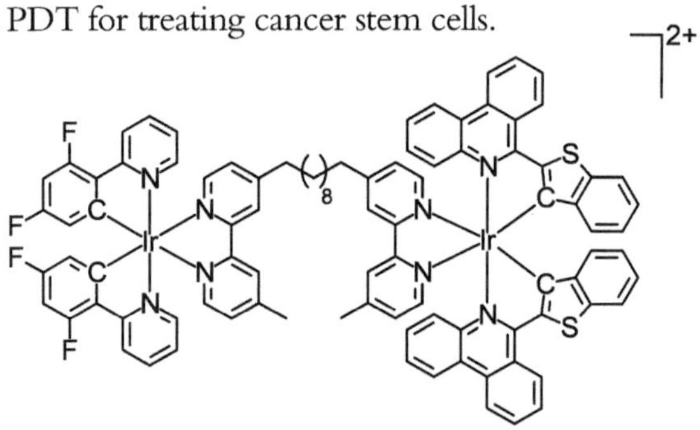

59

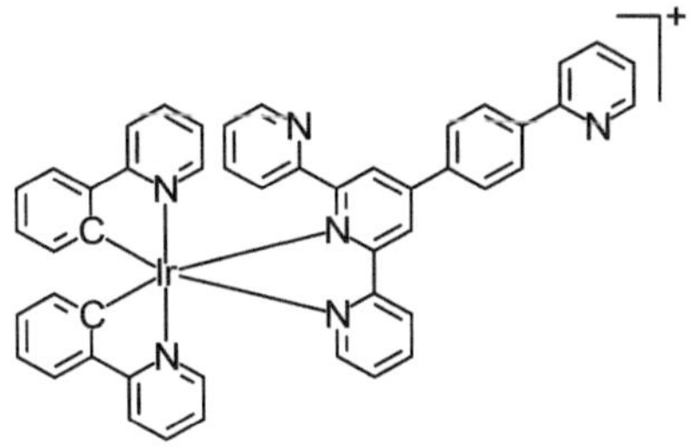

60

3.5 Redox status

The balance of ROS generation and elimination is crucial to the maintenance of intracellular redox homeostasis for normal cell function and survival.[82] Excessive production of ROS disrupts redox homeostasis, leading to oxidative stress that causes severe cellular damage and induces cell death. To accommodate high levels of ROS to activate pro-tumorigenic signaling, enhance cell survival and proliferation, and induce DNA damage and genomic instability, cancer cells express elevated levels of antioxidants to scavenge excess ROS to prevent oxidative stress-induced cell death.[83]

Glutathione (GSH) is the most abundant non-protein thiol in mammalian cells (*ca.* 1–10 mM). It serves as an antioxidant and plays an important role in maintenance of redox homeostasis, metabolism of xenobiotics, and modulation of immune response.[84] Increased cellular GSH levels are implicated in cancer development and growth and resistance to chemotherapeutic drugs. A rhenium(I) complex has been conjugated with a perylene diimide (PDI) derivative via a thiol-sensitive disulfide linkage (**61**) for imaging GSH in HeLa cells.[85] The weak emission of the complex in aqueous solution is due to: (1) efficient Förster resonance energy transfer (FRET) from the rhenium(I) polypyridine unit to the PDI moiety and (2) aggregation-induced quenching of PDI. Reaction with GSH triggers the reductive cleavage of the disulfide bond and thus separates the rhenium(I) polypyridine unit from the quenching PDI moiety, resulting in an increase in the intensity and lifetime of the rhenium(I) emission. Rhenium(I) complexes have also been functionalized with a thiol-responsive 2,4-dinitrophenylsulfonamide (DNPS) moiety (**62–65**) for bioimaging and PDT.[86] The complexes exhibit weak emission ($\Phi_{em} = 0.003$–0.043) and $^1O_2$-generation quantum yields ($\Phi_\Delta = 0.06$–0.15) due to efficient PET quenching from the rhenium(I) polypyridine unit to the electron-withdrawing DNPS moiety. Upon reaction with

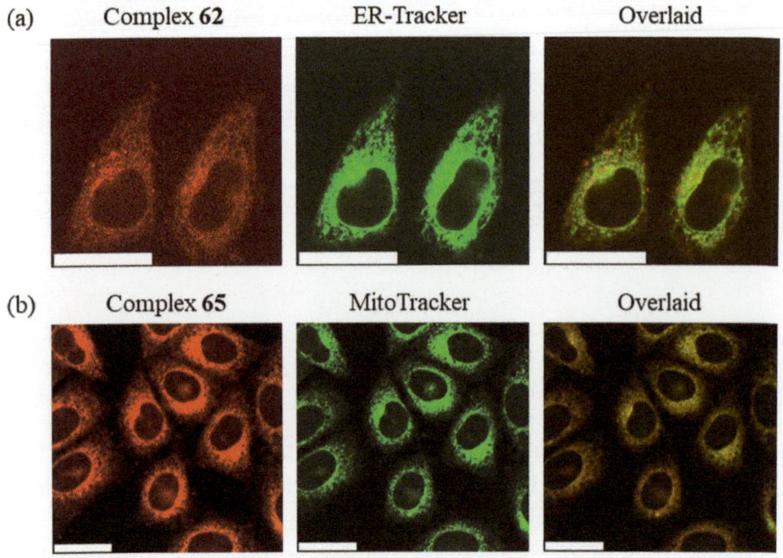

Fig. 5 LSCM images of HeLa cells incubated with (A) complexes **62** and (B) **65** (10 μM, 37 °C, 2 h; λ_{ex} = 405 nm) and then ER-Tracker Green (1 μM, 37 °C, 20 min; λ_{ex} = 488 nm, λ_{em} = 500–505 nm) or MitoTracker Deep Red FM (100 nM, 37 °C, 20 min; λ_{ex} = 635 nm, λ_{em} = 654–674 nm). Pearson's correlation coefficient = (A) 0.89 and (B) 0.93. Scale bar = 25 μm. *Adapted from Xu, G.-X.; Lee, L. C.-C.; Kwok, C. W.-C.; Leung, P. K.-K.; Zhu, J.-H.; Lo, K. K.-W.* Eur. J. Inorg. Chem. ***2021,*** 2021, *3432–3442, with permission from Wiley-VCH. Copyright 2021.*

intracellular GSH, the quenching DNPS unit is departed; thus, the complexes show more intense and longer-lived emission and more efficient $^1O_2$ photosensitization, which enable GSH-responsive imaging and PDT, respectively. Complex **62** is enriched in the ER of HeLa cells due to the tosylamide moiety (Fig. 5a), while complex **65** is accumulated in the mitochondria due to the cationic charge and high lipophilicity (Fig. 5b).

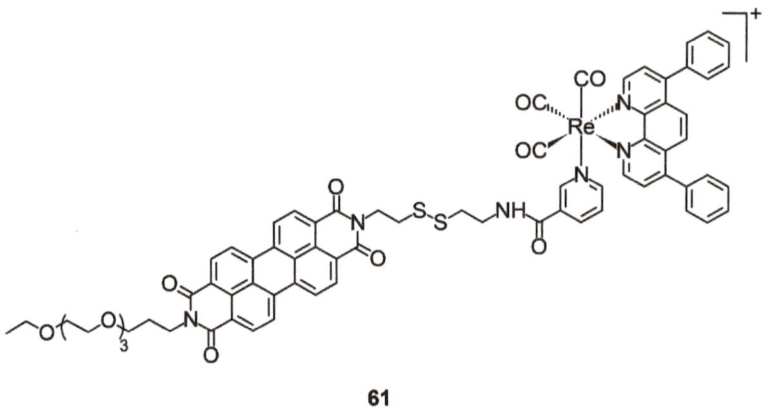

61

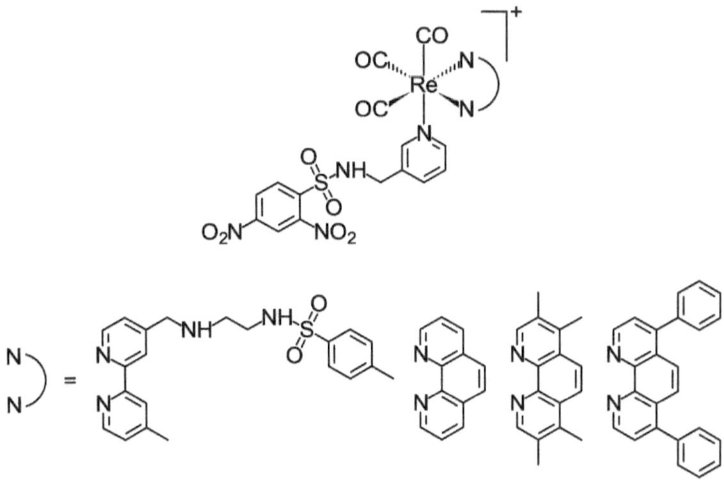

62 **63** **64** **65**

A binuclear ruthenium(II) complex featuring an electron-withdrawing azo group as both a bridge and an emission quencher (**66**) has been developed as a GSH-activatable photosensitizer for two-photon-induced PDT.[87] Upon reduction of the azo group by GSH to a non-quenching hydrazo derivative, the complex displays increased emission intensity ($I/I_o = 50$), TPA cross-section ($\delta_{810nm} = 210$ GM), and $^1O_2$-generation efficiency ($\Phi_\Delta = 0.40$). Thus, the complex exhibits much higher cytotoxicity toward HeLa-derived MCTSs ($IC_{50,dark} > 100$ μM; $IC_{50,light} = 5.71$ μM) upon two-photon excitation ($\lambda_{ex} = 810$ nm). Iridium(III) complexes functionalized with an electron-withdrawing 2,4-dinitrophenyl (DNP) unit (**67–69**) have been utilized for GSH sensing in HeLa cells.[88] These complexes show enhanced emission upon the nucleophilic aromatic substitution reaction with GSH, which leads to the departure of the DNP moiety from the iridium(III) polypyridine unit and thus inhibits PET quenching. Iridium(III) complexes have been modified with a PEG pendant via a thiol-sensitive disulfide linker (**70** and **71**) as GSH-activatable cytotoxic agents.[89] The complexes are essentially non-cytotoxic toward HeLa cells in the dark, but display increased cytotoxicity when the cells have been pretreated with GSH due to the release of the biocompatible PEG unit upon the reductive cleavage of the disulfide bond. An iridium(III) complex carrying two *N*-alkylpyridinium moieties (**72**) has been developed for GSH-selective imaging and PDT.[90] The electron transfer from GSH to the electron-deficient pyridinium unit enhances its susceptibility toward nucleophilic attack and promotes the substitution reaction by hydroxide, resulting in a substantial increase in the emission intensity ($I/I_o = 12.5$) and lifetime (from 84.7 to 690.5 ns) and $^1O_2$-photosensitization quantum yield (from 0.12 to 0.68). The complex can differentiate cancerous (HeLa, A549, and HepG2) and non-cancerous (3T3-L1 and HL-7702) cells using LSCM and PLIM, and exhibit higher photocytotoxic activity toward cancer cells than normal cells due to different intracellular GSH levels.

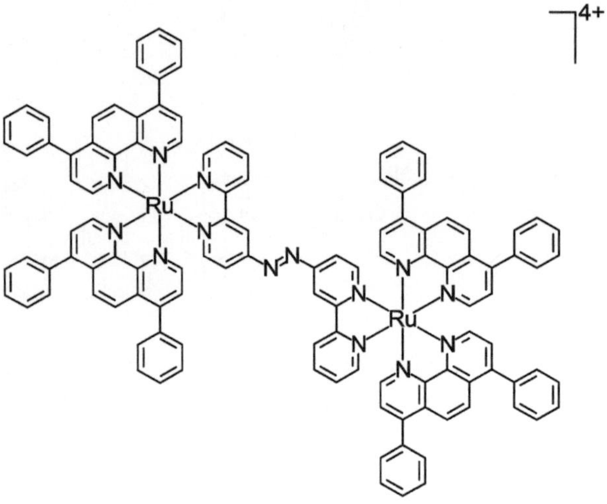

66

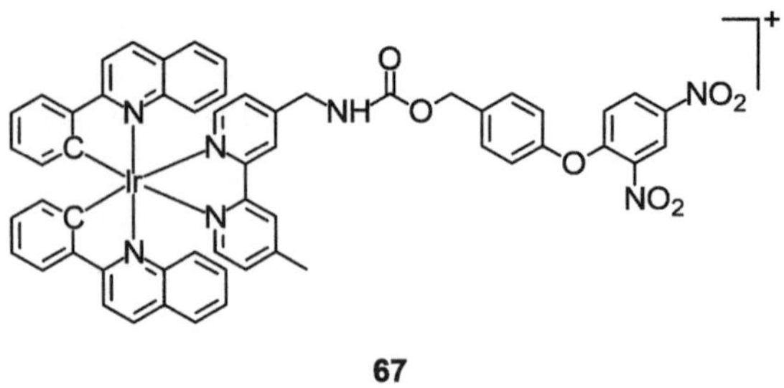

67

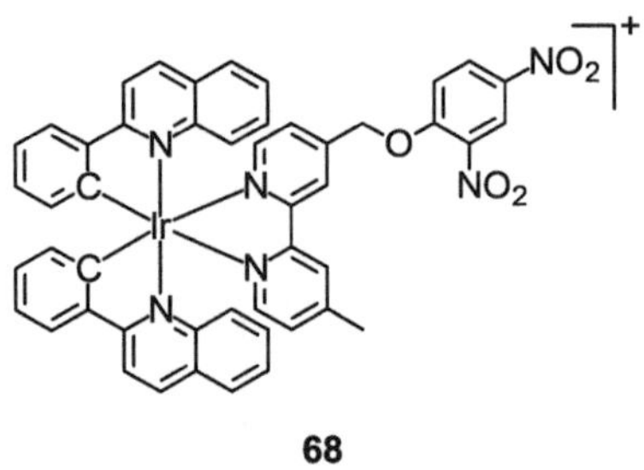

68

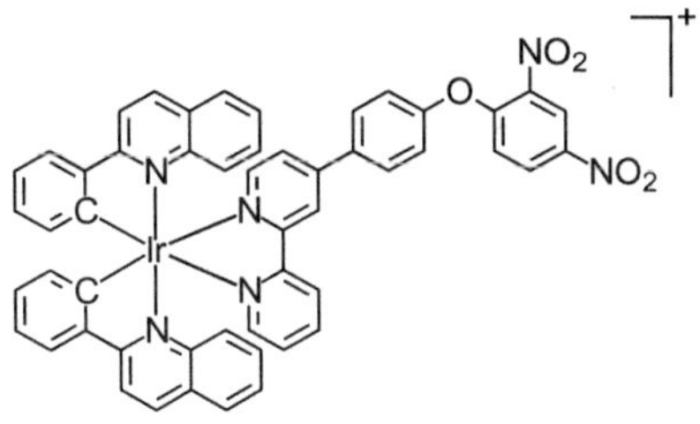

69

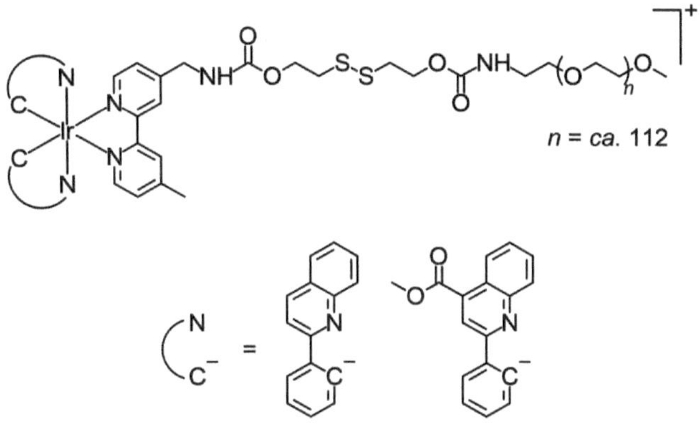

70 **71**

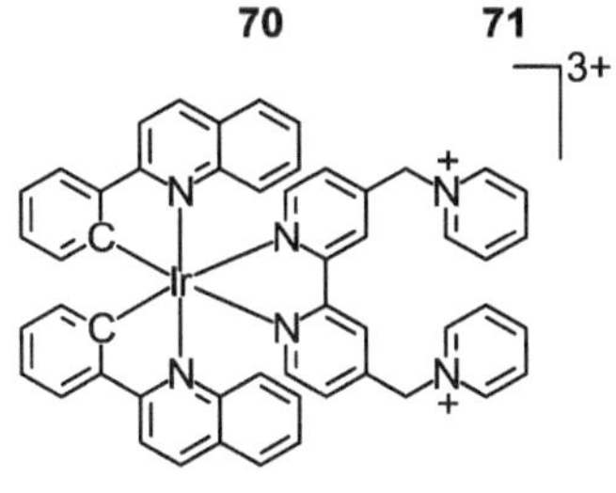

72

Reduced nicotinamide adenine dinucleotide (NADH) and its phosphorylated derivative reduced nicotinamide adenine dinucleotide phosphate (NADPH) are essential electron donors that are involved in various redox reactions associated with calcium homeostasis, energy metabolism, and redox homeostasis.[91] Elevated levels of NAD(P)H in cancer cells are critical to overcoming oxidative stress to support cell proliferation and survival.[92] An iridium(III) complex featuring two *N*-methylquinolinium moieties (**73**) has been designed as an NADH-activatable phototheranostic agent.[93] Reduction of the electron-deficient quinolinium moiety by NADH through hydride transfer not only increases the emission intensity and $^1O_2$-generation efficiency due to inhibited PET quenching, but also reduces the surface-enhanced Raman scattering signal due to weakened electrostatic interaction between the complex and gold nanoparticles, offering a possibility of bimodal imaging of NADH. Remarkably, complex **73** functions as an NADH scavenger in the dark, and it generates $^1O_2$ and $NAD^{\bullet}$ upon irradiation in the presence of NADH, inducing redox imbalance through NADH depletion to trigger apoptosis.

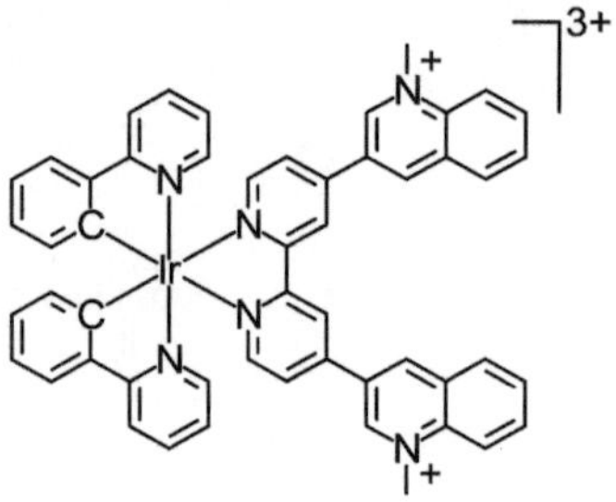

73

3.6 Enzymes

Enzymes usually exhibit high substrate specificity and catalytic efficiency, and are essential to the maintenance of cellular homeostasis and regulation of metabolic activities. Aberrant enzyme activity is often implicated in the occurrence and development of diseases including cancer.[94] Thus, cancer cells show enzyme activity profiles that are different to normal cells as a result of the requirement for activated cell proliferation, invasion, and metastasis.[95]

Reductases catalyze reduction reactions in living systems and play a crucial role in regulating cellular redox homeostasis. Specific reductases such as azoreductases, nitroreductases, and quinone reductases are overexpressed in hypoxic cancer cells, serving as useful biomarkers for cancer diagnosis[96] and therapy.[97] Azoreductases catalyze the reductive cleavage of azo compounds into their corresponding amines using NAD(P)H as an electron donor. Iridium(III) complexes have been modified with an azo group (**74–81**) to image hypoxic cancer cells and MCTSs.[98] These complexes are non-emissive due to efficient quenching associated with photoinduced isomerization of the N═N group, but display substantial emission enhancement ($I/I_o = 11.4$–58.8) upon reduction of the quenching azo group by reductases to a non-quenching amine derivative in the presence of NADPH under hypoxia. Incubation of HeLa cells with the complexes under hypoxic conditions results in intense intracellular emission, which disappears when the cells have been pretreated with the flavoprotein inhibitor diphenyliodonium chloride or when the cells are incubated under normoxic conditions. Additionally, more intense emission is observed in the complex-treated HeLa-derived MCTSs at a depth of *ca.* 40–80 μm from the outer edge of the spheroids, indicative of a hypoxic core in MCTSs.

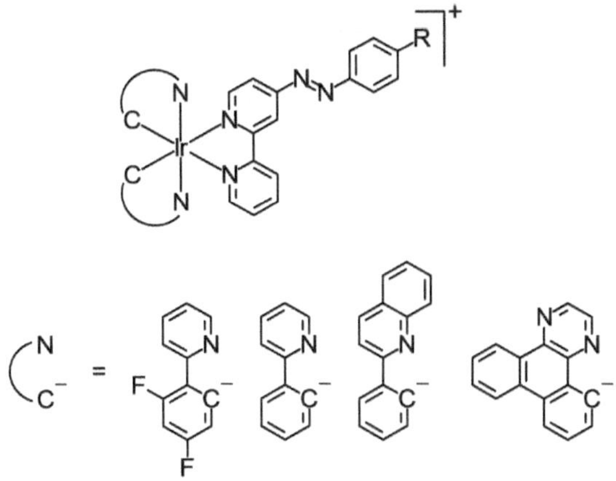

R = H	**74**	**75**	**76**	**77**
$N(CH_3)_2$	**78**	**79**	**80**	**81**

Nitroreductases catalyze the NAD(P)H-dependent reduction of nitroaromatics to their respective amines. An iridium(III) complex coordinated with a 5-nitro-1,10-phenanthroline ligand (**82**) has been applied to hypoxia imaging.[99] The complex is non-emissive due to efficient PET quenching by the nitro unit, but exhibits increased emission intensity upon reduction of the quenching nitro group by nitroreductases to a non-quenching amine derivative in the presence of NADH. Incubation of HeLa cells with the complex under normoxic conditions gives negligible intracellular emission, and the emission intensity gradually increases upon decreasing the O_2 level from 20% to 1% due to enhanced nitroreductase activity under hypoxia. A ruthenium(II) complex has been functionalized with a nitroimidazole moiety (**83**) to monitor O_2 fluctuation in tumor tissues.[100] The introduction of the nitroimidazole unit has weak emission quenching effect on the complex due to the long alkyl spacer; however, it increases the retention of the complex in hypoxic cancer cells as the nitroimidazole moiety is reduced by the overexpressed nitroreductases to generate a reactive hydroxylamine intermediate that irreversibly reacts with biomolecules inside the cells. This is supported by the ICP-MS data which show that the ruthenium content in hypoxic A549 cells is 1.8-fold higher than that in normoxic cells. Thus, the complex is enriched in the tumor region of a mouse model, which allows the real-time monitoring of the changes in the O_2 concentrations in the tumor tissues.

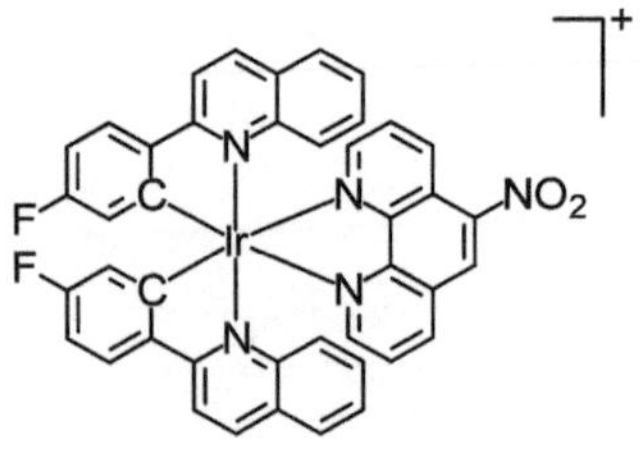

82

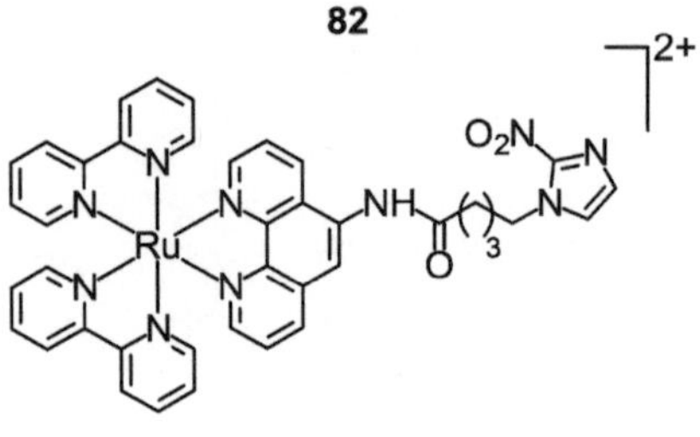

83

Quinone reductases catalyze the reduction of quinones to their hydroquinone derivatives using NAD(P)H as an electron source. Ruthenium(II) complexes have been modified with a redox-active anthraquinone moiety (**84**–**86**) for imaging cycling hypoxia.[101] The complexes are non-emissive due to efficient PET from the ruthenium(II) polypyridine unit to the electron-withdrawing anthraquinone moiety. Reduction of the complexes by reductases in the presence of NADPH under hypoxic conditions affords the dihydroanthraquinone derivatives that show strong TPA (δ_{800nm} = 147–196 GM) and intense emission (Φ_{em} = 0.0040–0.0092). Remarkably, the emission intensities of the complexes rapidly decrease upon exposure to air, and the changes are reversible. The reversible emission response enables the imaging of cycling hypoxia (i.e., cycles of hypoxia followed by reoxygenation) in live cells and zebrafish upon two-photon excitation (λ_{ex} = 800 nm). Iridium(III) complexes appended with two anthraquinone units (**87**–**90**) have been developed to image the mitochondria in hypoxic A549 cells and monitor the morphological changes in the mitochondria induced by the membrane potential uncoupler carbonyl cyanide *m*-chlorophenyl hydrazone.[102]

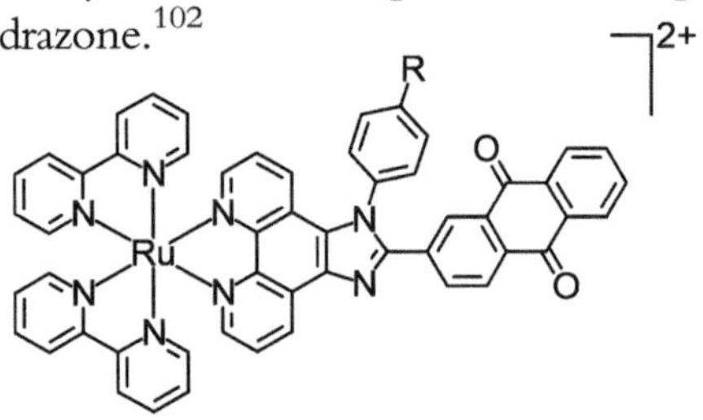

R = H (**84**)
CH_3 (**85**)
$C(CH_3)_3$ (**86**)

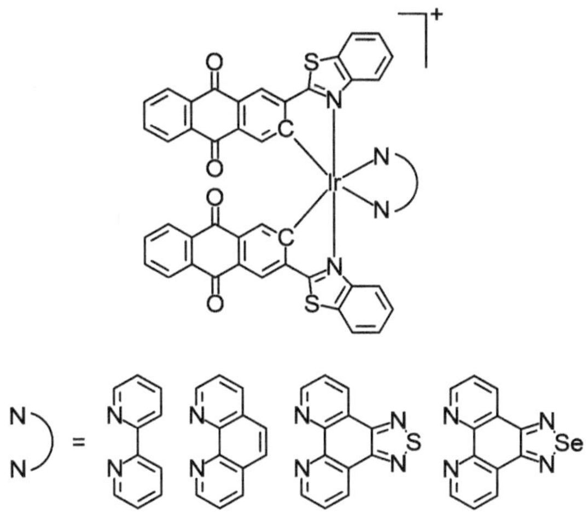

87 **88** **89** **90**

β-Galactosidase is a glycoside hydrolase that catalyzes the hydrolysis of β-galactosides into monosaccharides through cleavage of the glycosidic bond. Enhanced β-galactosidase activity is an important biomarker for primary ovarian cancer.[103] An iridium(III) complex bearing a galactose moiety (**91**) has been employed to visualize β-galactosidase in ovarian cancer cells.[104] The complex displays increased emission intensity in the presence of β-galactosidase, which is ascribed to the binding of the corresponding cleavage product to the enzyme. Remarkably, intense intracellular emission is observed in the complex-loaded human ovarian adenocarcinoma cell lines (OVCAR3 and SKOV3) but not in non-cancerous cell lines (HEK293T and human umbilical vein endothelial cells). The emission intensity is reduced when the ovarian cancer cells have been pretreated with exogenous galactose, suggesting that the complex can distinguish ovarian cancer cells from normal cells based on different β-galactosidase activity.

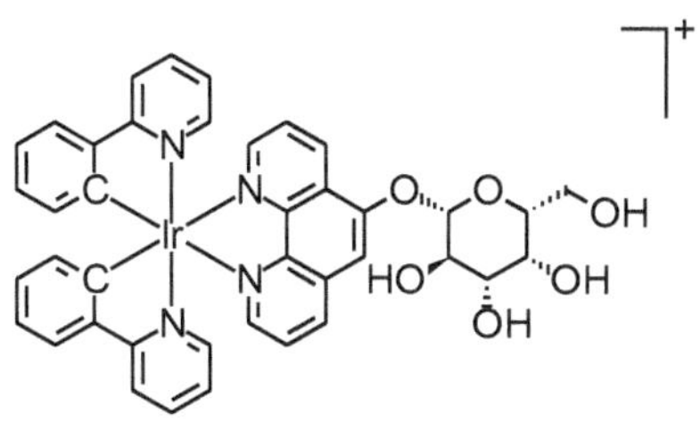

91

3.7 Mismatched DNA molecules

The DNA mismatch repair (MMR) system is responsible for correcting spontaneous base–base mismatches and small insertions or deletions that arise during DNA replication.[105] Inactivation of the MMR system leads to genomic instability and is associated with the development of cancer such as hereditary nonpolyposis colorectal cancer. A binuclear platinum(II) complex bridged with a diphosphine ligand (**92**) has been developed as a selective probe for mismatched and abasic DNA.[106] The complex exhibits up to 13.5-fold higher emission enhancement upon binding to mismatched DNA over matched DNA, as the larger pockets of the mismatched sites provide a better shielding of the complex from the aqueous environment and/or organize the complex molecule into a more ordered structure. The complex also shows larger increases in emission intensity ($I/I_o=$ 17.0–29.3) in the presence of abasic DNA than normal double-stranded DNA ($I/I_o=6.1$). Incubation of MMR-deficient human colorectal carcinoma HCT116 cells with the complex results in intense intracellular emission, which is much higher in intensity than the cells that are transfected to restore the MMR activity. Additionally, the complex displays much enhanced emission (I/I_o up to 5.5) upon treatment with the DNA samples extracted from the colorectal adenocarcinoma tissues than those from adjacent normal tissues of the same patient (I/I_o up to 1.3), highlighting the capability of the complex in differentiating tumor tissues from normal tissues based on different levels of DNA mismatches.

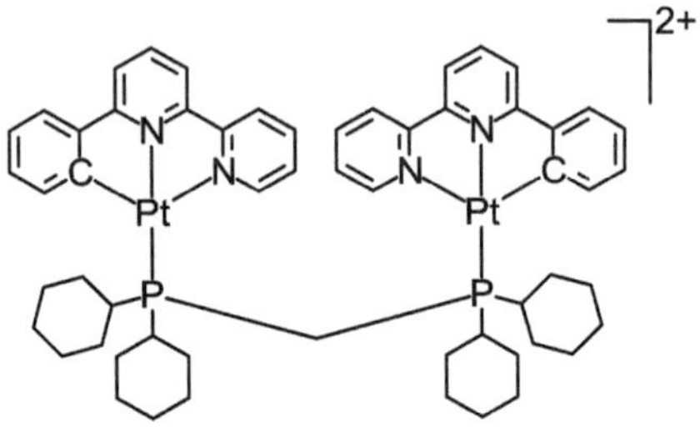

92

3.8 Bioorthogonal chemical reporters

The bioorthogonal chemical reporter strategy has been developed as a promising pretargeting approach for cancer diagnosis and therapy.[107] The introduction of unnatural functional groups into the cells can overcome the variations in the microenvironment such as pH, redox status, and enzyme activity among different cancer cells due to tumor heterogeneity, allowing the activation of the complexes at the targeted site in a controllable manner.

Nitrone, which undergoes strain-promoted alkyne–nitrone cycloaddition (SPANC) reaction with cyclooctynes to give isoxazoline derivatives,[108] has been exploited as both a bioorthogonal handle and an emission quencher for iridium(III) complexes (**93–95**).[109] The very weak emission (Φ_{em} = 0.0002–0.021) is a consequence of efficient quenching associated with photoinduced isomerization of the C=N group in the nitrone unit. However, the complexes exhibit substantial emission enhancement (I/I_o = 28.8–135.0) (Fig. 6) upon the SPANC reaction with (1*R*,8*S*,9*s*)-bicyclo[6.1.0]non-4-yn-9-ylmethanol (BCN-OH) that converts the quenching nitrone moiety to a non-quenching isoxazoline derivative, which enables the selective imaging of cells that have been pretreated with a BCN-modified decane molecule (BCN-C10). Regioselective cellular labeling has been observed in related phosphorogenic ruthenium(II) nitrone complexes (**96–100**).[110] Incubation of BCN-C10-pretreated HeLa cells with the complexes leads to different subcellular staining: punctate cell surface staining with complexes **96** and **98**, exclusive plasma membrane staining with complexes **97** and **100**, and cytoplasmic staining with complex **99** (Fig. 7). The regioselective accumulation of the complexes in the membrane or cytoplasm of the treated cells is not only associated with the subcellular distribution of BCN-C10, but also the lipophilicity and bioorthogonal reactivity of the complexes. Additionally, the most lipophilic complex **99** shows increased cytotoxic activity toward BCN-C10-pretreated HeLa cells, as the SPANC reaction with BCN-C10 in the cytoplasm generates a more lipophilic adduct that displays enhanced non-specific interactions with intracellular species to induce more extensive cell death.

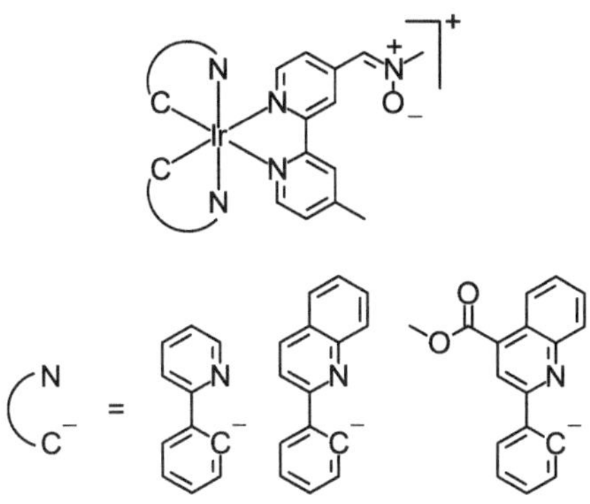

93 **94** **95**

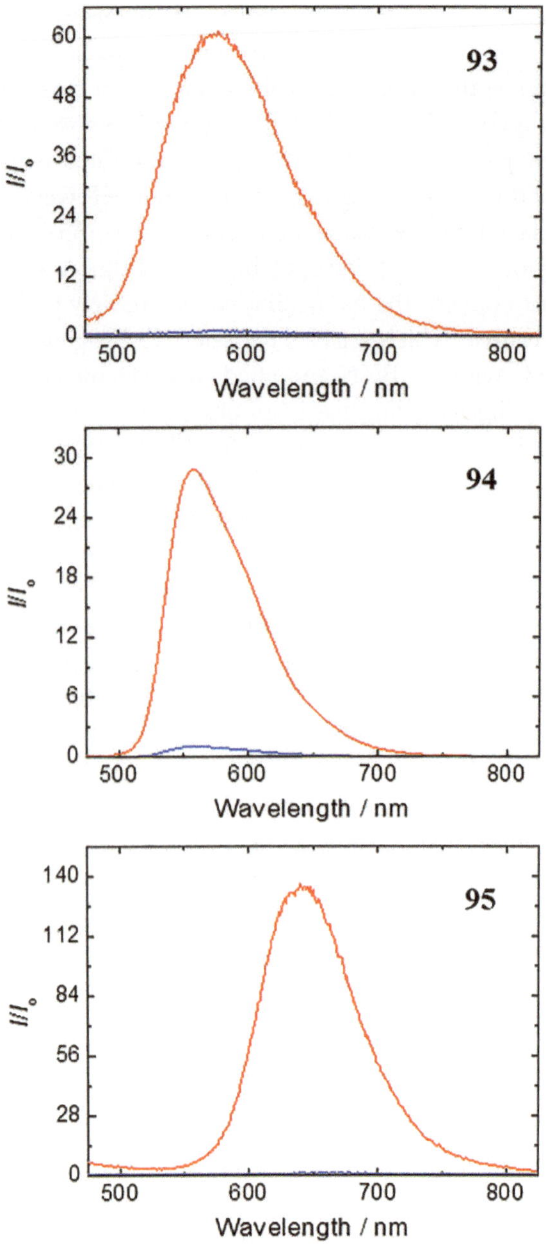

Fig. 6 Emission spectra of complexes **93–95** (10 μM) in the absence (blue) and presence (red) of BCN-OH (250 μM) in aerated potassium phosphate buffer (50 mM, pH 7.4)/CH_3OH (9:1, *v/v*) at 298 K for 24 h. *Adapted from Lee, L. C.-C.; Lau, J. C.-W.; Liu, H.-W.; Lo, K. K.-W.* Angew. Chem. Int. Ed. ***2016,*** 55, *1046–1049, with permission from Wiley-VCH. Copyright 2016.*

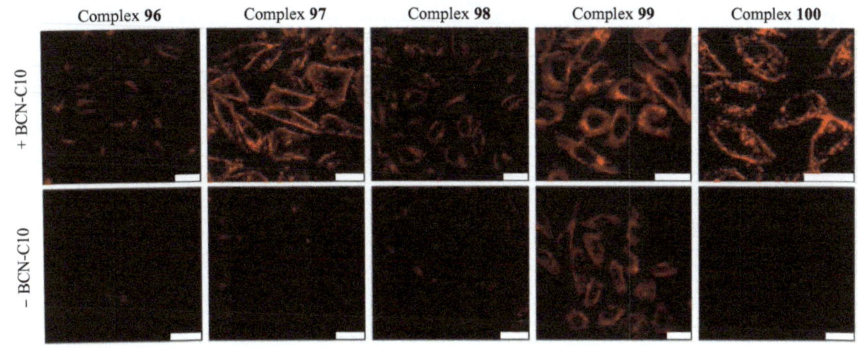

Fig. 7 LSCM images of HeLa cells incubated with complexes **96–100** (40 μM, 37 °C, 12 h) with (top) or without (bottom) pretreatment with BCN-C10 (200 μM, 37 °C, 1 h). Scale bar = 25 μm. *Adapted from Tang, T. S.-M.; Liu, H.-W.; Lo, K. K.-W.* Chem. Eur. J. ***2016,*** 22, *9649–9659, with permission from Wiley-VCH. Copyright 2016.*

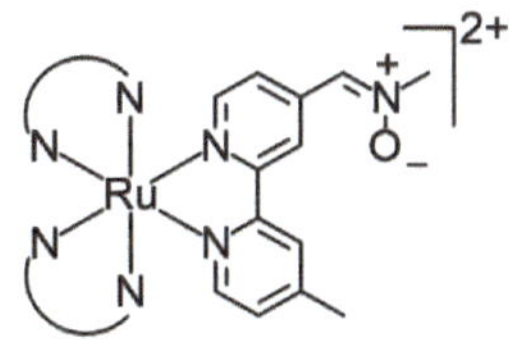

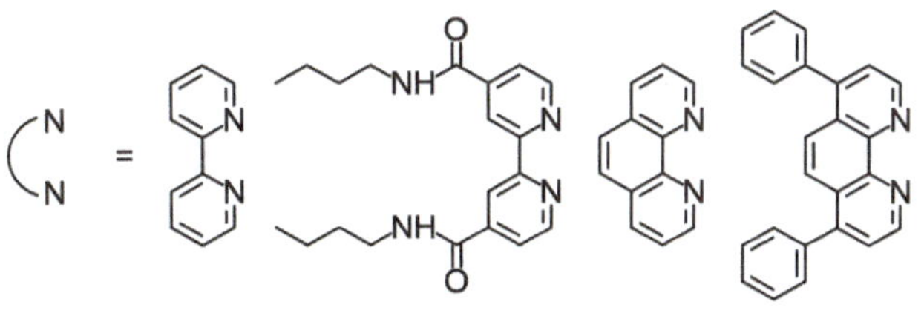

96 **97** **98** **99**

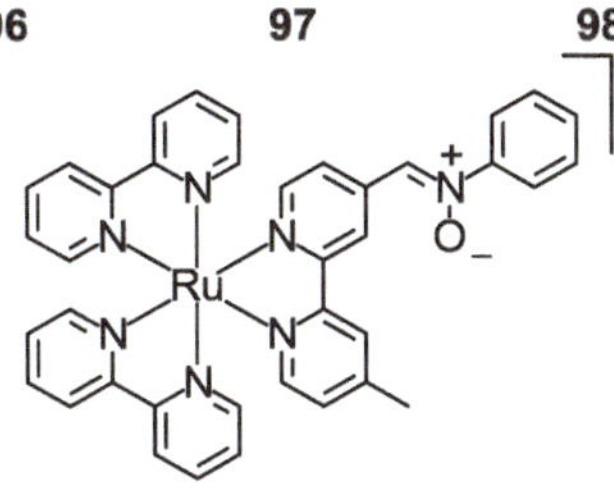

100

Remarkably, the introduction of a nitrone unit not only modulates the photophysical properties but also the $^1O_2$-photosensitization efficiency of the complexes. Iridium(III) nitrone complexes (**101**–**104**) exhibit a large increase in the emission intensity ($I/I_o = 7.2$–47.1) and $^1O_2$-generation quantum yield (from 0.19–0.26 to 0.42–0.91) upon the SPANC reaction with BCN-OH, which enables selective imaging and photoactivated killing of CHO-K1 cells that have been transfected to express SNAP-tag proteins to accommodate a larger amount of BCN units inside the cells using BCN-modified O^6-benzylguanine as a substrate.[111]

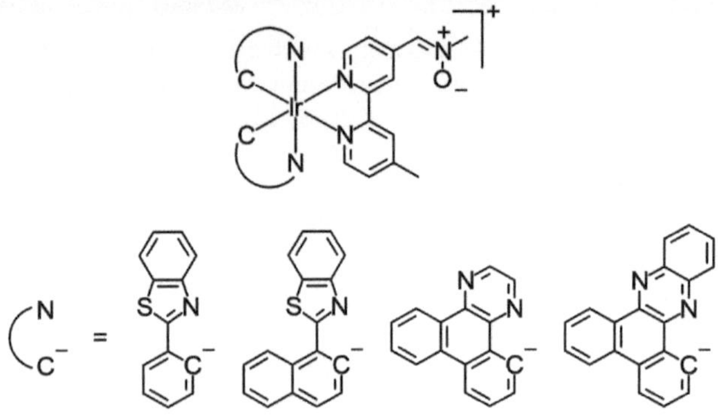

101 **102** **103** **104**

Sydnones are a class of mesoionic compounds that undergo strain-promoted sydnone–alkyne cycloaddition (SPSAC) reaction with cyclooctynes to give substituted pyrazoles.[112] Since the polarity of sydnone and pyrazole is drastically different, iridium(III) complexes with environment-sensitive emission have been functionalized with a sydnone moiety (**105**–**107**) to report the SPSAC reaction by photophysical changes.[113] Upon the SPSAC reaction with BCN-OH, the complexes show enhanced emission intensities ($I/I_o = 1.1$–7.9) and extended lifetimes ($\tau = 0.10$–0.25 μs) with a hypsochromic shift in their emission maxima, which is due to the conversion of the dipolar mesoionic sydnone ring to a less polar pyrazole derivative that results in a more hydrophobic pendant. Complex **106**, which displays the largest photophysical changes, has been used to image HeLa cells that have been pretreated with a BCN-modified ceramide derivative (BCN-Cer) (Fig. 8A,B). The iridium(III)–ceramide adduct is enriched in the mitochondria of the cells due to its cationic and lipophilic character (Fig. 8C). Additionally, the complex exhibits higher photocytotoxicity toward BCN-Cer-pretreated HeLa cells than the untreated cells due to more efficient photosensitization of $^1O_2$ that induces cell death through causing mitochondrial dysfunction.

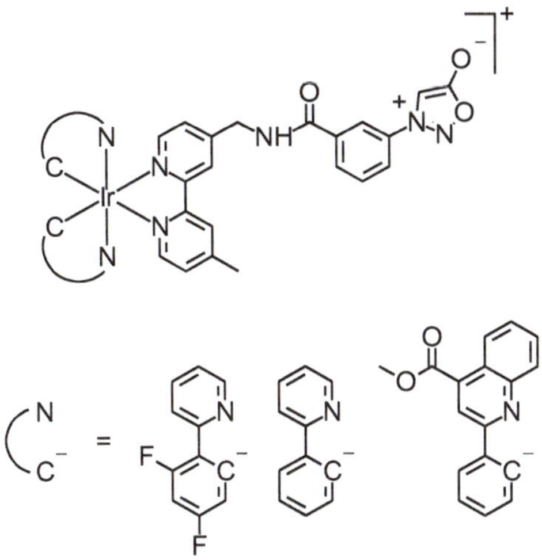

105 **106** **107**

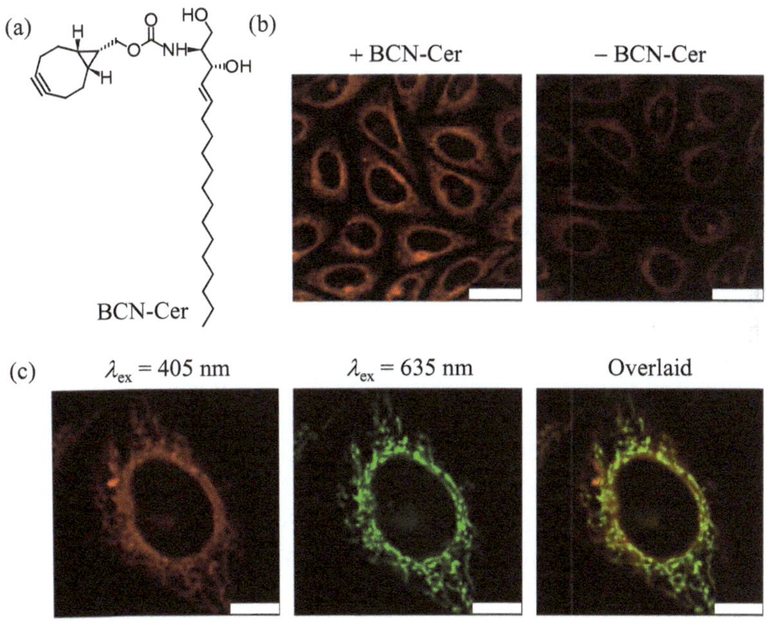

Fig. 8 (A) Structure of BCN-Cer. (B) LSCM images of HeLa cells incubated with complex **106** (5 μM, 37 °C, 4 h) with (left) or without (right) pretreatment with BCN-Cer (125 μM, 37 °C, 1 h). Scale bar = 25 μm. (C) LSCM images of BCN-Cer-pretreated (125 μM, 37 °C, 1 h) HeLa cells incubated with complex **106** (5 μM, 37 °C, 4 h; λ_{ex} = 405 nm) and then MitoTracker Deep Red FM (100 nM, 37 °C, 20 min; λ_{ex} = 635 nm, λ_{em} = 655–675 nm). Scale bar = 10 μm. Pearson's correlation coefficient = 0.93. *Adapted from Lee, L. C.-C.; Cheung, H. M.-H.; Liu, H.-W.; Lo, K. K.-W.* Chem. Eur. J. ***2018,*** 24, *14064–14068, with permission from Wiley-VCH. Copyright 2018.*

Rhenium(I) complexes coordinated with a 4-(4-pyridyl)sydnone ligand (**108–110**) have been developed as phosphorogenic bioorthogonal reagents.[114] These complexes show very weak emission ($\Phi_{em}=0.0005–0.0017$) as the π-conjugation between the pyridine ring and the sydnone moiety allows more efficient emission quenching, which leads to larger emission enhancement ($I/I_o=8.8–17.3$) upon the SPSAC reaction with BCN-OH. Complex **108** has been coupled with a BCN-modified morpholine derivative to target the lysosomes of HeLa cells with emission turn-on.

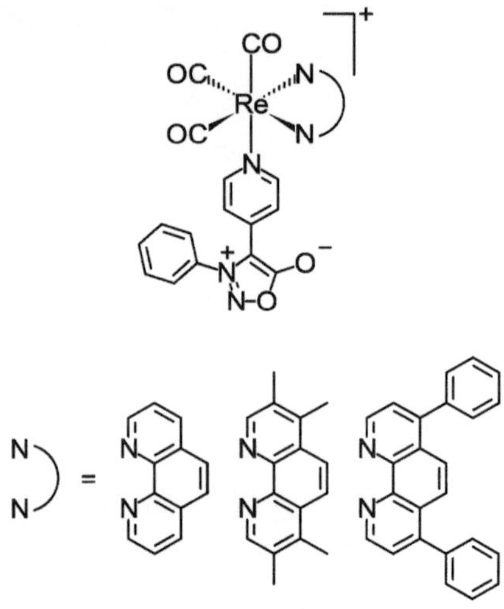

108 **109** **110**

The inverse electron-demand Diels–Alder (IEDDA) reaction of 1,2,4,5-tetrazine with dienophiles such as strained alkenes and alkynes has been developed as a promising approach for *in vivo* pretargeting due to its high reaction selectivity, rapid reaction kinetics, and remarkable biocompatibility.[115] Modification of transition metal complexes with a tetrazine moiety quenches the emission of the complexes via efficient FRET and/or PET from the metal polypyridine unit to the tetrazine moiety. Thus, rhenium(I) (**111–116**)[116] and iridium(III) tetrazine complexes (**117–124**)[117,118] display substantially enhanced emission upon the IEDDA reaction with BCN derivatives due to the conversion of the quenching tetrazine unit to a non-quenching pyridazine derivative. This enables the selective imaging of different cellular compartments (e.g., lysosomes and ER) and biomolecules (e.g., proteins and glycans) in live cells using various BCN-modified substrates. Although tetrazines react with *trans*-cyclooct-4-en-1-ol (TCO-OH) much more rapidly than with BCN-OH,[115] it is possible for emission quenching to occur in the resultant dihydropyridazine products due to PET from the dihydropyridazine moiety to the metal polypyridine unit. Thus, the tetrazine complexes **121–124** exhibit higher emission intensities and longer emission lifetimes ($I/I_o = 51.7–105.3$; $\tau = 0.05–0.33\ \mu s$) upon the IEDDA reaction with BCN-OH than with TCO-OH ($I/I_o = 1.8–13.5$; $\tau = 0.04–0.28\ \mu s$).[118] Additionally, the $^1O_2$-generation efficiencies of the resultant pyridazine derivatives ($\Phi_\Delta =$ 0.31–0.76) are higher than those of the dihydropyridazine derivatives ($\Phi_\Delta = 0.12–0.36$). Thus, complex **122** shows higher photocytotoxic activity toward HeLa cells containing BCN-modified HaloTag protein ($IC_{50,light} = 0.26\ \mu M$) than those containing unmodified and TCO-modified HaloTag protein ($IC_{50,light} = 0.68$ and $1.00\ \mu M$, respectively).

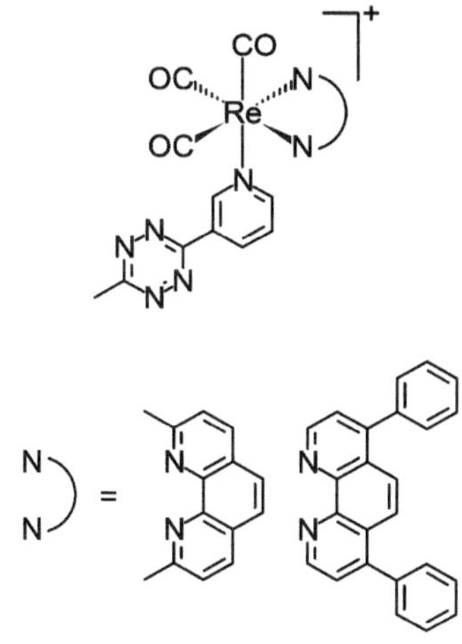

111 **112**

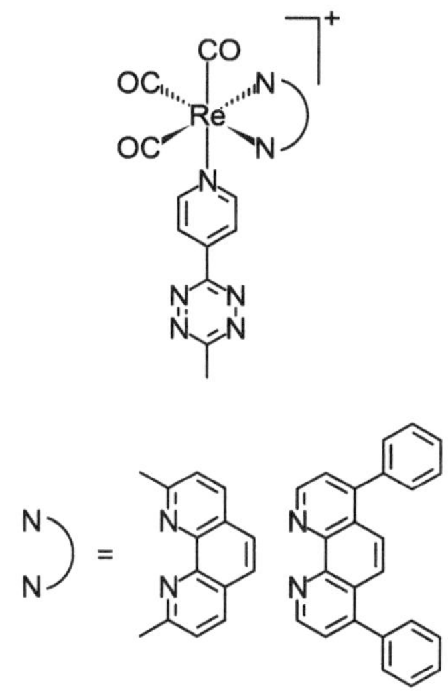

113 **114**

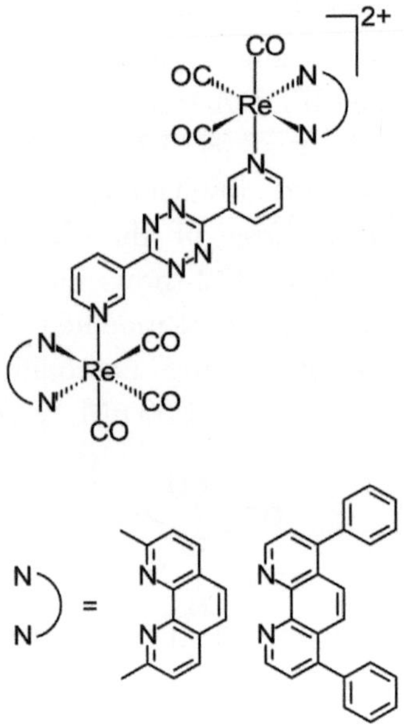

115 **116**

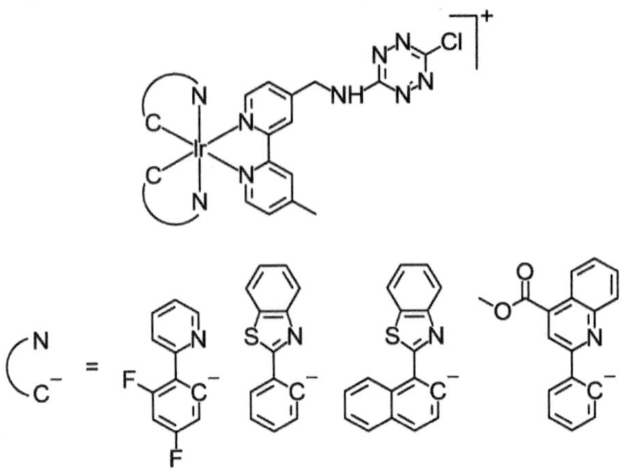

117 **118** **119** **120**

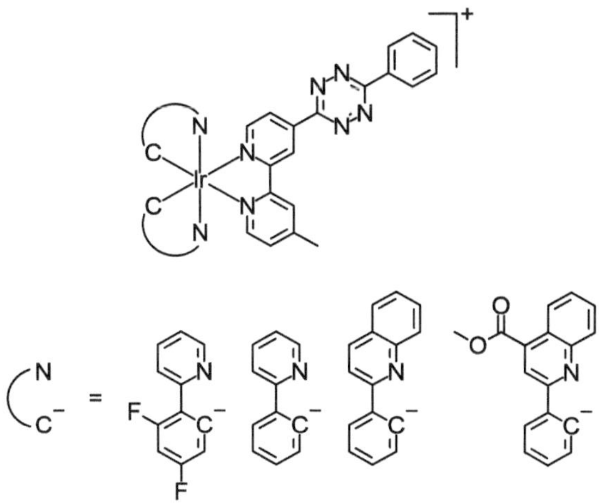

121 **122** **123** **124**

Notably, incorporation of the electron-deficient tetrazine moiety into iridium(III) complexes as part of the coordinating ligand (**125**–**127**) not only allows the complexes to display substantial emission enhancement upon the IEDDA reaction with BCN derivatives by perturbing the π-conjugation of the ligand, but also increases the reactivity of the tetrazine unit due to metal-coordination effects.[119] Remarkably, iridium(III) complexes featuring two coordinated tetrazine moieties (**128**–**130**) exhibit extremely weak emission ($\Phi_{em}=0.0001$–0.0009) and $^1O_2$ photosensitization ($\Phi_{\Delta}=0.02$–0.07), but react very rapidly with 1,13-*bis*-((1*R*,8*S*,9*s*)-bicyclo [6.1.0]non-4-yn-9-ylmethyloxycarbonylamino)-4,7,10-trioxatridecane (*bis*-BCN) ($k_2=62.7$–$14{,}713.6$ M^{-1} s^{-1}) to give *bis*-pyridazine conjugates that show significantly enhanced emission ($I/I_o=$ 1008.2–3885.7) (Fig. 9) and $^1O_2$-generation efficiencies ($\Phi_{\Delta}=0.62$–0.77).[120] Incubation of *bis*-BCN-pretreated HeLa cells with the complexes results in intense emission in the lysosomes (Fig. 10), indicative of the strong targeting capability of the *bis*-pyridazine conjugates. The complexes display similar cytotoxicity toward both *bis*-BCN-pretreated ($IC_{50,dark}=7.76$–16.03 μM) and untreated HeLa cells ($IC_{50,dark}=6.38$–15.64 μM) in the dark, but exhibit higher photocytotoxic activity toward *bis*-BCN-pretreated cells ($IC_{50,light}=$ 0.45–1.60 μM) than untreated cells ($IC_{50,light}=6.94$–13.25 μM) as the IEDDA reaction facilitates the photosensitization of $^1O_2$ to induce lysosomal cell death.

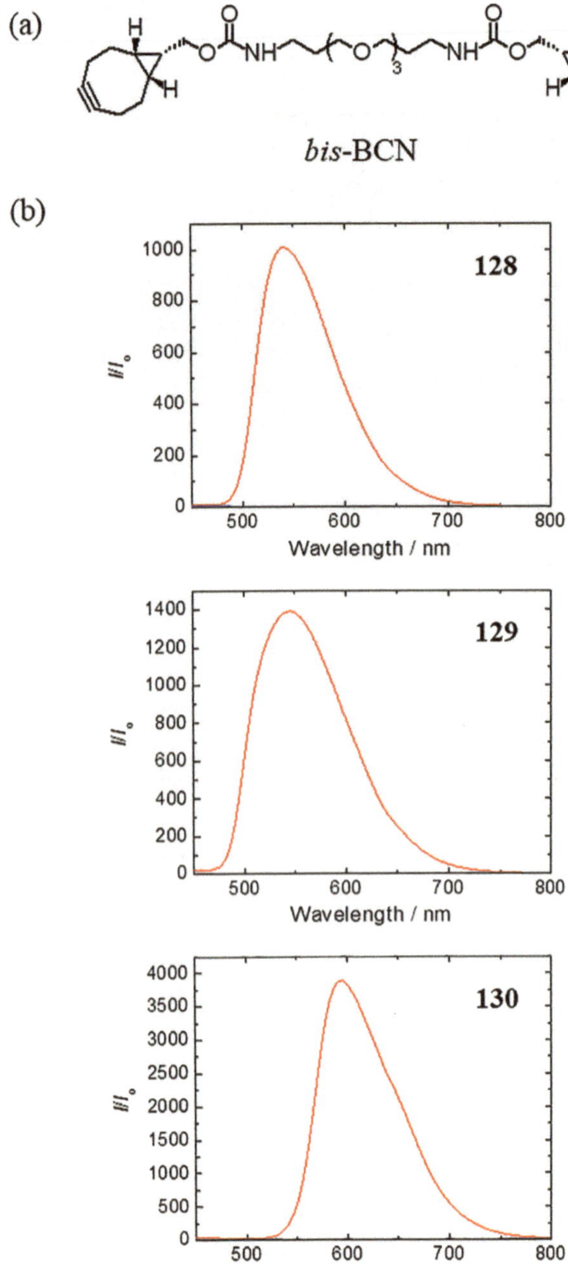

Fig. 9 (A) Structure of *bis*-BCN. (B) Emission spectra of complexes **128–130** (5 μM) in the absence (blue) and presence (red) of *bis*-BCN (250 μM) in H_2O/DMSO (4:1, *v/v*) at 298 K for 2 h. *Adapted from Yip, A. M.-H.; Lai, C. K.-H.; Yiu, K. S.-M.; Lo, K. K.-W.* Angew. Chem. Int. Ed. ***2022,*** 61, *e202116078, with permission from Wiley-VCH. Copyright 2022.*

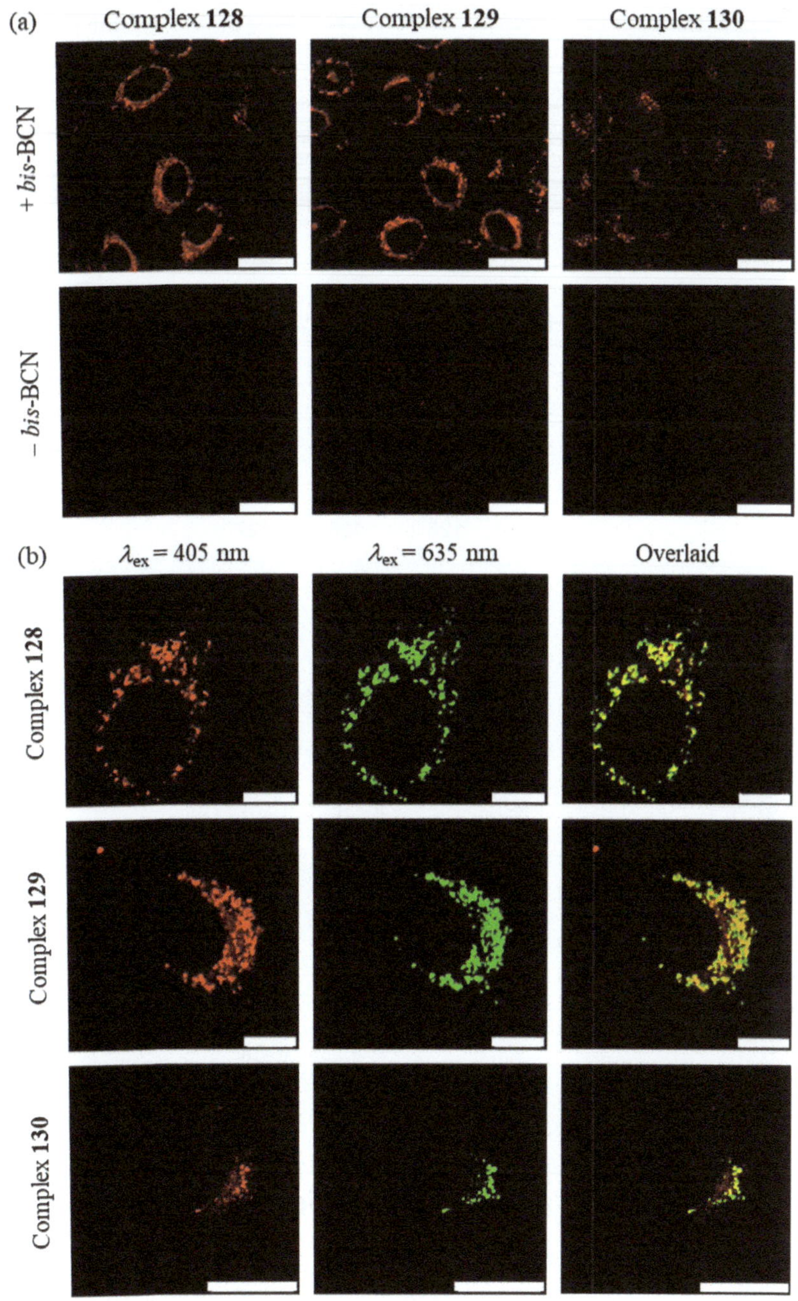

Fig. 10 (A) LSCM images of HeLa cells incubated with complexes **128–130** (5 μM, 37 °C, 2 h) with (top) or without (bottom) pretreatment with *bis*-BCN (100 μM, 37 °C, 2 h). Scale bar = 25 μm. (B) LSCM images of *bis*-BCN-pretreated (100 μM, 37 °C, 2 h) HeLa cells incubated with complexes **128–130** (5 μM, 37 °C, 2 h; λ_{ex} = 405 nm) and then LysoTracker Deep Red (100 nM, 37 °C, 30 min; λ_{ex} = 635 nm, λ_{em} = 650–700 nm). Scale bar = 10 (top and middle) and 25 μm (bottom). Pearson's correlation coefficient = 0.88 (top), 0.85 (middle), and 0.79 (bottom). *Adapted from Yip, A. M.-H.; Lai, C. K.-H.; Yiu, K. S.-M.; Lo, K. K.-W.* Angew. Chem. Int. Ed. ***2022,*** 61, *e202116078, with permission from Wiley-VCH. Copyright 2022.*

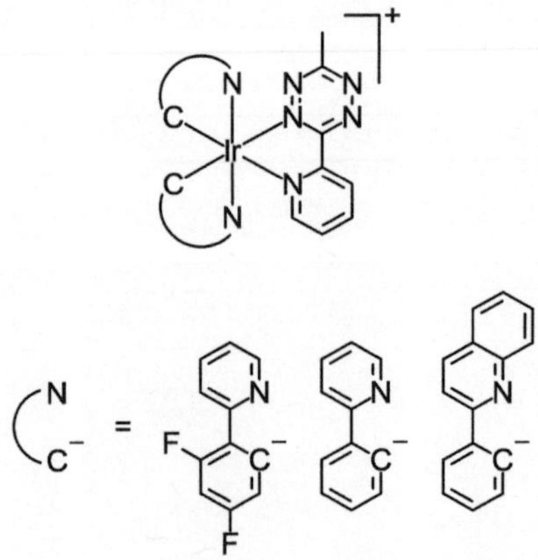

125 **126** **127**

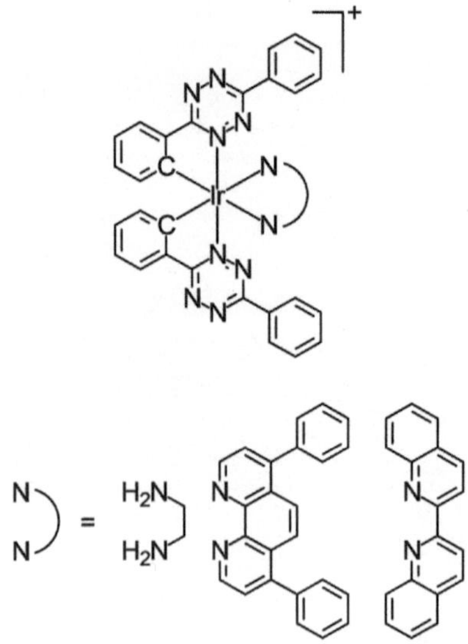

128 **129** **130**

4. Conclusions and future perspective

In this chapter, we summarize the recent advances in the development of photofunctional transition metal complexes for cancer diagnosis and therapy. The strategic molecular designs have led to the successful applications of these complexes in selective cancer imaging and therapy both *in vitro* and *in vivo*. To translate these exciting results into a clinical setting, it is imperative to overcome some major challenges. One of them is the relatively weak absorption and emission of transition metal complexes in the far-red to NIR region, which constrains the utilization of these complexes for imaging and

PDT treatment of deep-seated or large solid tumors due to limited light penetration. Long-wavelength light offers several advantages for biomedical applications including deeper tissue penetration, lower phototoxicity, and minimal autofluorescence background. To this end, one of the future research directions is to shift the absorption and emission of the complexes to longer wavelengths, particularly the far-red to NIR region. This can be achieved by reducing the HOMO–LUMO gap of the complexes through manipulating the electronic properties or increasing the π-conjugation of the ligands. Increasing the TPA cross-sections of the complexes can also render them compatible with two-photon NIR excitation. Another problem is the strong O_2-dependency of the Type II photoreaction involved in $^1O_2$ photosensitization, which poses an obstacle toward the effective treatment of solid tumors that are characterized by a hypoxic microenvironment. Additionally, the significant O_2 consumption for $^1O_2$ generation during PDT further increases tumor hypoxia and thus weakens the therapeutic outcome. Thus, more effort should be devoted to the development of transition metal complexes that undergo a Type I photoreaction or other types of O_2-independent photochemical reactions such as photosubstitution and photocleavage reactions to exert cytotoxic effects that can combat hypoxia-induced therapeutic resistance. Biomedical applications of transition metal complexes are also confronted by the generally low aqueous solubility and high dark cytotoxicity of the complexes. Modification of the complexes with PEG pendants is a useful approach to enhance their solubility and biocompatibility, ensuring that the complexes can be administrated to patients without causing undesirable local or systemic effects.

Most importantly, although the activity-based design approach allows transition metal complexes to display enhanced emission intensity and photocytotoxic activity in cancer cells in response to a specific cancer-associated stimulus, the current one-to-one molecular designs can still generate false-positive signals when the target exists in normal tissues in moderate concentration. For this reason, future research should focus on the development of double-locked complexes, which remains relatively unexplored. The introduction of two different responsive moieties to the complexes enables the emission and ROS-photosensitization capability of the complexes to be activated only in the co-presence of the two distinct stimuli, precluding the possibility of non-specific activation in normal tissues. Also, the attachment of a cancer-targeting motif to the complexes for binding to specific cell-surface proteins can further improve the precision and specificity of cancer imaging and therapy. Remarkably, the bioorthogonal chemical reporter strategy not only allows the incorporation of an abiotic functional group into cancer cells for pretargeting, but also offers opportunities to modulate the photophysical and photochemical properties of the complexes through a specific bioorthogonal reaction, enabling the simultaneous accumulation and activation of the complexes at the tumor site in a highly controllable manner. The continued development of bioorthogonal chemistry will provide an expanded set of targets for selection in the design of activatable complexes for more accurate detection and effective treatment of cancer in complex biological milieu. In conclusion, we believe that photofunctional transition metal complexes will play an important role in the development of innovative diagnostic and therapeutic agents for biomedical applications.

Acknowledgments

We thank the Hong Kong Research Grants Council (Project No. CityU 11300019, CityU 11302820, CityU 11301121, T42-103/16-N, C6014-20 W, C7075-21GF, and N_CityU104/21) for financial support. We also acknowledge the funding support from "Laboratory for Synthetic Chemistry and Chemical Biology" under the Health@InnoHK Program launched by Innovation and Technology Commission, The Government of Hong Kong SAR, PR China.

References

1. Siegel, R. L.; Miller, K. D.; Fuchs, H. E.; Jemal, A. *CA Cancer J. Clin.* **2022**, *72*, 7–33.
2. Lo, K. K.-W. *Acc. Chem. Res.* **2015**, *48*, 2985–2995.
3. Chen, Y.; Guan, R.; Zhang, C.; Huang, J.; Ji, L.; Chao, H. *Coord. Chem. Rev.* **2016**, *310*, 16–40.
4. Ko, C.-N.; Li, G.; Leung, C.-H.; Ma, D.-L. *Coord. Chem. Rev.* **2019**, *381*, 79–103.
5. Tan, C.-P.; Zhong, Y.-M.; Ji, L.-N.; Mao, Z.-W. *Chem. Sci.* **2021**, *12*, 2357–2367.
6. Xu, G.-X.; Mak, E. C.-L.; Lo, K. K.-W. *Inorg. Chem. Front.* **2021**, *8*, 4553–4579.
7. Lo, K. K.-W.; Choi, A. W.-T.; Law, W. H.-T. *Dalton Trans.* **2012**, *41*, 6021–6047.
8. Zhang, K. Y.; Yu, Q.; Wei, H.; Liu, S.; Zhao, Q.; Huang, W. *Chem. Rev.* **2018**, *118*, 1770–1839.
9. McKenzie, L. K.; Bryant, H. E.; Weinstein, J. A. *Coord. Chem. Rev.* **2019**, *379*, 2–29.
10. Puckett, C. A.; Ernst, R. J.; Barton, J. K. *Dalton Trans.* **2010**, *39*, 1159–1170.
11. Grimm, D.; Bauer, J.; Pietsch, J.; Infanger, M.; Eucker, J.; Eilles, C.; Schoenberger, J. *Curr. Med. Chem.* **2011**, *18*, 176–190.
12. Mueckler, M. *Eur. J. Biochem.* **1994**, *219*, 713–725.
13. Macheda, M. L.; Rogers, S.; Best, J. D. *J. Cell. Physiol.* **2005**, *202*, 654–662.
14. Louie, M.-W.; Liu, H.-W.; Lam, M. H.-C.; Lam, Y.-W.; Lo, K. K.-W. *Chem. Eur. J.* **2011**, *17*, 8304–8308.
15. Law, W. H.-T.; Lee, L. C.-C.; Louie, M.-W.; Liu, H.-W.; Ang, T. W.-H.; Lo, K. K.-W. *Inorg. Chem.* **2013**, *52*, 13029–13041.
16. Liu, J.; Liao, X.; Xiong, K.; Kuang, S.; Jin, C.; Ji, L.; Chao, H. *Chem. Commun.* **2020**, *56*, 5839–5842.

17. Douard, V.; Ferraris, R. P. *Am. J. Physiol. Endocrinol. Metab.* **2008**, *295*, E227–E237.
18. Zamora-León, S. P.; Golde, D. W.; Concha, I. I.; Rivas, C. I.; Delgado-López, F.; Baselga, J.; Nualart, F.; Vera, J. C. *Proc. Natl. Acad. Sci. U. S. A.* **1996**, *93*, 1847–1852.
19. Zhang, K. Y.; Tso, K. K.-S.; Louie, M.-W.; Liu, H.-W.; Lo, K. K.-W. *Organometallics* **2013**, *32*, 5098–5102.
20. Lo, K. K.-W.; Law, W. H.-T.; Chan, J. C.-Y.; Liu, H.-W.; Zhang, K. Y. *Metallomics* **2013**, *5*, 808–812.
21. Lau, C. T.-S.; Chan, C.; Zhang, K. Y.; Roy, V. A. L.; Lo, K. K.-W. *Eur. J. Inorg. Chem.* **2017**, *2017*, 5288–5294.
22. Russell-Jones, G.; McTavish, K.; McEwan, J.; Rice, J.; Nowotnik, D. *J. Inorg. Biochem.* **2004**, *98*, 1625–1633.
23. Lo, K. K.-W.; Louie, M.-W.; Sze, K.-S.; Lau, J. S.-Y. *Inorg. Chem.* **2008**, *47*, 602–611.
24. Zhang, K. Y.; Liu, H.-W.; Tang, M.-C.; Choi, A. W.-T.; Zhu, N.; Wei, X.-G.; Lau, K.-C.; Lo, K. K.-W. *Inorg. Chem.* **2015**, *54*, 6582–6593.
25. Li, J.; Zeng, L.; Xiong, K.; Rees, T. W.; Jin, C.; Wu, W.; Chen, Y.; Ji, L.; Chao, H. *Chem. Commun.* **2019**, *55*, 10972–10975.
26. Cox, D.; Brennan, M.; Moran, N. *Nat. Rev. Drug Discov.* **2010**, *9*, 804–820.
27. Ma, X.; Jia, J.; Cao, R.; Wang, X.; Fei, H. *J. Am. Chem. Soc.* **2014**, *136*, 17734–17737.
28. Zhao, Z.; Qiu, K.; Liu, J.; Hao, X.; Wang, J. *Chem. Commun.* **2020**, *56*, 12542–12545.
29. Dorsam, R. T.; Gutkind, J. S. *Nat. Rev. Cancer* **2007**, *7*, 79–94.
30. Sobczuk, P.; Łomiak, M.; Cudnoch-Jędrzejewska, A. *Cancers* **2020**, *12*, 3232.
31. Weissenrieder, J. S.; Neighbors, J. D.; Mailman, R. B.; Hohl, R. J. *J. Pharmacol. Exp. Ther.* **2019**, *370*, 111–126.
32. Vellaisamy, K.; Li, G.; Ko, C.-N.; Zhong, H.-J.; Fatima, S.; Kwan, H.-Y.; Wong, C.-Y.; Kwong, W.-J.; Tan, W.; Leung, C.-H.; Ma, D.-L. *Chem. Sci.* **2018**, *9*, 1119–1125.
33. Patel, O.; Shulkes, A.; Baldwin, G. S. *Biochim. Biophys. Acta* **2006**, *1766*, 23–41.
34. Wang, W.; Wu, K.-J.; Vellaisamy, K.; Leung, C.-H.; Ma, D.-L. *Angew. Chem. Int. Ed.* **2020**, *59*, 17897–17902.
35. Patel, Y. C. *Front. Neuroendocrinol.* **1999**, *20*, 157–198.
36. Wang, T.; Zabarska, N.; Wu, Y.; Lamla, M.; Fischer, S.; Monczak, K.; Ng, D. Y. W.; Rau, S.; Weil, T. *Chem. Commun.* **2015**, *51*, 12552–12555.
37. Novohradsky, V.; Zamora, A.; Gandioso, A.; Brabec, V.; Ruiz, J.; Marchán, V. *Chem. Commun.* **2017**, *53*, 5523–5526.
38. Hynes, N. E.; Lane, H. A. *Nat. Rev. Cancer* **2005**, *5*, 341–354.
39. Wu, C.; Wu, K.-J.; Liu, J.-B.; Wang, W.; Leung, C.-H.; Ma, D.-L. *Chem. Sci.* **2020**, *11*, 11404–11412.
40. Chiche, J.; Ilc, K.; Laferrière, J.; Trottier, E.; Dayan, F.; Mazure, N. M.; Brahimi-Horn, M. C.; Pouysségur, J. *Cancer Res.* **2009**, *69*, 358–368.
41. Hostachy, S.; Masuda, M.; Miki, T.; Hamachi, I.; Sagan, S.; Lequin, O.; Medjoubi, K.; Somogyi, A.; Delsuc, N.; Policar, C. *Chem. Sci.* **2018**, *9*, 4483–4487.
42. Su, X.; Wang, W.-J.; Cao, Q.; Zhang, H.; Liu, B.; Ling, Y.; Zhou, X.; Mao, Z.-W. *Angew. Chem. Int. Ed.* **2022**, *61*, e202115800.
43. Subbaramaiah, K.; Dannenberg, A. J. *Trends Pharmacol. Sci.* **2003**, *24*, 96–102.
44. Liu, C.; Yang, C.; Lu, L.; Wang, W.; Tan, W.; Leung, C.-H.; Ma, D.-L. *Chem. Commun.* **2017**, *53*, 2822–2825.
45. Liu, C.; Xiang, J.; Xiang, C.; Li, H. *Bioorg. Chem.* **2021**, *114*, 105142.
46. Kelm, S.; Schauer, R. *Int. Rev. Cytol.* **1997**, *175*, 137–240.
47. Fuster, M. M.; Esko, J. D. *Nat. Rev. Cancer* **2005**, *5*, 526–542.
48. Liu, H.-W.; Law, W. H.-T.; Lee, L. C.-C.; Lau, J. C.-W.; Lo, K. K.-W. *Chem. Asian J.* **2017**, *12*, 1545–1556.
49. Sletten, E. M.; Bertozzi, C. R. *Angew. Chem. Int. Ed.* **2009**, *48*, 6974–6998.

50. Li, L.; Zhang, Z. *Molecules* **2016**, *21*, 1393.
51. Sletten, E. M.; Bertozzi, C. R. *Acc. Chem. Res.* **2011**, *44*, 666–676.
52. Saxon, E.; Bertozzi, C. R. *Science* **2000**, *287*, 2007–2010.
53. Lo, K. K.-W.; Chan, B. T.-N.; Liu, H.-W.; Zhang, K. Y.; Li, S. P.-Y.; Tang, T. S.-M. *Chem. Commun.* **2013**, *49*, 4271–4273.
54. Choi, A. W.-T.; Liu, H.-W.; Lo, K. K.-W. *J. Inorg. Biochem.* **2015**, *148*, 2–10.
55. Tang, T. S.-M.; Yip, A. M.-H.; Zhang, K. Y.; Liu, H.-W.; Wu, P. L.; Li, K. F.; Cheah, K. W.; Lo, K. K.-W. *Chem. Eur. J.* **2015**, *21*, 10729–10740.
56. Lin, M.; Zou, S.; Liao, X.; Chen, Y.; Luo, D.; Ji, L.; Chao, H. *Chem. Commun.* **2021**, *57*, 4408–4411.
57. Marusyk, A.; Polyak, K. *Biochim. Biophys. Acta* **2010**, *1805*, 105–117.
58. Harris, A. L. *Nat. Rev. Cancer* **2002**, *2*, 38–47.
59. Yoshihara, T.; Hirakawa, Y.; Hosaka, M.; Nangaku, M.; Tobita, S. *J. Photochem. Photobiol. C: Photochem. Rev.* **2017**, *30*, 71–95.
60. Zhang, S.; Hosaka, M.; Yoshihara, T.; Negishi, K.; Iida, Y.; Tobita, S.; Takeuchi, T. *Cancer Res.* **2010**, *70*, 4490–4498.
61. Yoshihara, T.; Hosaka, M.; Terata, M.; Ichikawa, K.; Murayama, S.; Tanaka, A.; Mori, M.; Itabashi, H.; Takeuchi, T.; Tobita, S. *Anal. Chem.* **2015**, *87*, 2710–2717.
62. Zheng, X.; Wang, X.; Mao, H.; Wu, W.; Liu, B.; Jiang, X. *Nat. Commun.* **2015**, *6*, 5834.
63. Lo, K. K.-W.; Zhang, K. Y.; Leung, S.-K.; Tang, M.-C. *Angew. Chem. Int. Ed.* **2008**, *47*, 2213–2216.
64. Zhang, K. Y.; Gao, P.; Sun, G.; Zhang, T.; Li, X.; Liu, S.; Zhao, Q.; Lo, K. K.-W.; Huang, W. *J. Am. Chem. Soc.* **2018**, *140*, 7827–7834.
65. Zhao, Q.; Zhou, X.; Cao, T.; Zhang, K. Y.; Yang, L.; Liu, S.; Liang, H.; Yang, H.; Li, F.; Huang, W. *Chem. Sci.* **2015**, *6*, 1825–1831.
66. Wang, S.; Gu, K.; Guo, Z.; Yan, C.; Yang, T.; Chen, Z.; Tian, H.; Zhu, W.-H. *Adv. Mater.* **2019**, *31*, 1805735.
67. Lv, Z.; Zou, L.; Wei, H.; Liu, S.; Huang, W.; Zhao, Q. *ACS Appl. Mater. Interfaces* **2018**, *10*, 19523–19533.
68. Lagadic-Gossmann, D.; Huc, L.; Lecureur, V. *Cell Death Differ.* **2004**, *11*, 953–961.
69. Busa, W. B.; Nuccitelli, R. *Am. J. Physiol.* **1984**, *246*, R409–R438.
70. Gatenby, R. A.; Gillies, R. J. *Nat. Rev. Cancer* **2004**, *4*, 891–899.
71. Webb, B. A.; Chimenti, M.; Jacobson, M. P.; Barber, D. L. *Nat. Rev. Cancer* **2011**, *11*, 671–677.
72. Zheng, X.; Mao, H.; Huo, D.; Wu, W.; Liu, B.; Jiang, X. *Nat. Biomed. Eng.* **2017**, *1*, 0057.
73. Han, Z.; Wang, Y.; Chen, Y.; Fang, H.; Yuan, H.; Shi, X.; Yang, B.; Chen, Z.; He, W.; Guo, Z. *Chem. Commun.* **2020**, *56*, 8055–8058.
74. He, L.; Li, Y.; Tan, C.-P.; Ye, R.-R.; Chen, M.-H.; Cao, J.-J.; Ji, L.-N.; Mao, Z.-W. *Chem. Sci.* **2015**, *6*, 5409–5418.
75. Yang, P.; Zhang, S.; Wang, K.; Qi, H. *Dalton Trans.* **2021**, *50*, 17338–17345.
76. Xiao, H.; Li, P.; Tang, B. *Coord. Chem. Rev.* **2021**, *427*, 213582.
77. Li, X.; Tong, X.; Yin, Y.; Yan, H.; Lu, C.; Huang, W.; Zhao, Q. *Chem. Sci.* **2017**, *8*, 5930–5940.
78. Luby-Phelps, K. *Int. Rev. Cytol.* **1999**, *192*, 189–221.
79. Xiao, H.; Li, P.; Tang, B. *Chem. Eur. J.* **2021**, *27*, 6880–6898.
80. Liu, F.; Wen, J.; Chen, S.-S.; Sun, S. *Chem. Commun.* **2018**, *54*, 1371–1374.
81. Xiang, P.; Huang, R.; Shao, T.; Shu, Y.; Li, D.; Zhang, Q.; Tian, Y.; Tian, X. *Sens. Actuators B Chem.* **2022**, *361*, 131677.
82. Valko, M.; Leibfritz, D.; Moncol, J.; Cronin, M. T. D.; Mazur, M.; Telser, J. *Int. J. Biochem. Cell Biol.* **2007**, *39*, 44–84.

83. Liou, G.-Y.; Storz, P. *Free Radic. Res.* **2010**, *44*, 479–496.
84. Estrela, J. M.; Ortega, A.; Obrador, E. *Crit. Rev. Clin. Lab. Sci.* **2006**, *43*, 143–181.
85. Yip, A. M.-H.; Shum, J.; Liu, H.-W.; Zhou, H.; Jia, M.; Niu, N.; Li, Y.; Yu, C.; Lo, K. K.-W. *Chem. Eur. J.* **2019**, *25*, 8970–8974.
86. Xu, G.-X.; Lee, L. C.-C.; Kwok, C. W.-C.; Leung, P. K.-K.; Zhu, J.-H.; Lo, K. K.-W. *Eur. J. Inorg. Chem.* **2021**, *2021*, 3432–3442.
87. Zeng, L.; Kuang, S.; Li, G.; Jin, C.; Ji, L.; Chao, H. *Chem. Commun.* **2017**, *53*, 1977–1980.
88. Tso, K. K.-S.; Liu, H.-W.; Lo, K. K.-W. *J. Inorg. Biochem.* **2017**, *177*, 412–422.
89. Li, S. P.-Y.; Yim, V. M.-W.; Shum, J.; Lo, K. K.-W. *Dalton Trans.* **2019**, *48*, 9692–9702.
90. Huang, T.; Yu, Q.; Liu, S.; Zhang, K. Y.; Huang, W.; Zhao, Q. *ChemBioChem* **2019**, *20*, 576–586.
91. Ying, W. *Antioxid. Redox Signal.* **2008**, *10*, 179–206.
92. Xiao, W.; Wang, R.-S.; Handy, D. E.; Loscalzo, J. *Antioxid. Redox Signal.* **2018**, *28*, 251–272.
93. Shamjith, S.; Joseph, M. M.; Murali, V. P.; Remya, G. S.; Nair, J. B.; Suresh, C. H.; Maiti, K. K. *Biosens. Bioelectron.* **2022**, *204*, 114087.
94. Zhang, J.; Chai, X.; He, X.-P.; Kim, H.-J.; Yoon, J.; Tian, H. *Chem. Soc. Rev.* **2019**, *48*, 683–722.
95. Liu, H.-W.; Chen, L.; Xu, C.; Li, Z.; Zhang, H.; Zhang, X.-B.; Tan, W. *Chem. Soc. Rev.* **2018**, *47*, 7140–7180.
96. Elmes, R. B. P. *Chem. Commun.* **2016**, *52*, 8935–8956.
97. Guise, C. P.; Mowday, A. M.; Ashoorzadeh, A.; Yuan, R.; Lin, W.-H.; Wu, D.-H.; Smaill, J. B.; Patterson, A. V.; Ding, K. *Chin. J. Cancer* **2014**, *33*, 80–86.
98. Sun, L.; Li, G.; Chen, X.; Chen, Y.; Jin, C.; Ji, L.; Chao, H. *Sci. Rep.* **2015**, *5*, 14837.
99. Wang, W.; Wu, C.; Yang, C.; Li, G.; Han, Q.-B.; Li, S.; Lee, S. M.-Y.; Leung, C.-H.; Ma, D.-L. *Sens. Actuators B Chem.* **2018**, *255*, 1953–1959.
100. Son, A.; Kawasaki, A.; Hara, D.; Ito, T.; Tanabe, K. *Chem. Eur. J.* **2015**, *21*, 2527–2536.
101. Zhang, P.; Huang, H.; Chen, Y.; Wang, J.; Ji, L.; Chao, H. *Biomaterials* **2015**, *53*, 522–531.
102. Sun, L.; Chen, Y.; Kuang, S.; Li, G.; Guan, R.; Liu, J.; Ji, L.; Chao, H. *Chem. Eur. J.* **2016**, *22*, 8955–8965.
103. Chatterjee, S. K.; Bhattacharya, M.; Barlow, J. J. *Cancer Res.* **1979**, *39*, 1943–1951.
104. Wang, W.; Vellaisamy, K.; Li, G.; Wu, C.; Ko, C.-N.; Leung, C.-H.; Ma, D.-L. *Anal. Chem.* **2017**, *89*, 11679–11684.
105. Li, G.-M. *Cell Res.* **2008**, *18*, 85–98.
106. Fung, S. K.; Zou, T.; Cao, B.; Chen, T.; To, W.-P.; Yang, C.; Lok, C.-N.; Che, C.-M. *Nat. Commun.* **2016**, 7, 10655.
107. Patra, M.; Zarschler, K.; Pietzsch, H.-J.; Stephan, H.; Gasser, G. *Chem. Soc. Rev.* **2016**, *45*, 6415–6431.
108. Bilodeau, D. A.; Margison, K. D.; Serhan, M.; Pezacki, J. P. *Chem. Rev.* **2021**, *121*, 6699–6717.
109. Lee, L. C.-C.; Lau, J. C.-W.; Liu, H.-W.; Lo, K. K.-W. *Angew. Chem. Int. Ed.* **2016**, *55*, 1046–1049.
110. Tang, T. S.-M.; Liu, H.-W.; Lo, K. K.-W. *Chem. Eur. J.* **2016**, *22*, 9649–9659.
111. Leung, P. K.-K.; Lo, K. K.-W. *Chem. Commun.* **2020**, *56*, 6074–6077.
112. Porte, K.; Riomet, M.; Figliola, C.; Audisio, D.; Taran, F. *Chem. Rev.* **2021**, *121*, 6718–6743.
113. Lee, L. C.-C.; Cheung, H. M.-H.; Liu, H.-W.; Lo, K. K.-W. *Chem. Eur. J.* **2018**, *24*, 14064–14068.

114. Shum, J.; Zhang, P.-Z.; Lee, L. C.-C.; Lo, K. K.-W. *ChemPlusChem* **2020**, *85*, 1374–1378.
115. Oliveira, B. L.; Guo, Z.; Bernardes, G. J. L. *Chem. Soc. Rev.* **2017**, *46*, 4895–4950.
116. Choi, A. W.-T.; Tso, K. K.-S.; Yim, V. M.-W.; Liu, H.-W.; Lo, K. K.-W. *Chem. Commun.* **2015**, *51*, 3442–3445.
117. Li, S. P.-Y.; Yip, A. M.-H.; Liu, H.-W.; Lo, K. K.-W. *Biomaterials* **2016**, *103*, 305–313.
118. Leung, P. K.-K.; Lee, L. C.-C.; Yeung, H. H.-Y.; Io, K.-W.; Lo, K. K.-W. *Chem. Commun.* **2021**, *57*, 4914–4917.
119. Tang, T. S.-M.; Liu, H.-W.; Lo, K. K.-W. *Chem. Commun.* **2017**, *53*, 3299–3302.
120. Yip, A. M.-H.; Lai, C. K.-H.; Yiu, K. S.-M.; Lo, K. K.-W. *Angew. Chem. Int. Ed.* **2022**, *61*, e202116078.

CHAPTER THREE

Advances in the design of photoactivated platinum anticancer complexes

Huayun Shi and Peter J. Sadler*
Department of Chemistry, University of Warwick, Coventry, United Kingdom
*Corresponding author: e-mail address: p.j.sadler@warwick.ac.uk

Contents

1. Introduction 96
2. Discovery of photoactive Pt(IV) anticancer complexes 99
3. Mechanism of action 103
   3.1 Radical and ROS generation 104
   3.2 DNA as a target 105
   3.3 Peptide and protein targets 107
   3.4 Induction of immunogenic cell death 108
4. Axial derivatives of Pt(IV) complexes 108
   4.1 Targeting vectors 108
   4.2 Multi-action derivatives 110
5. Pt(IV) azide–acetylene derivatives via "click" reactions 112
6. Nanocarriers 114
   6.1 Proteins 114
   6.2 Polysaccharides 115
   6.3 Hydrogels 116
   6.4 Block copolymers 117
   6.5 Polymetallodrugs 118
   6.6 Carbon dots 119
   6.7 Up-converting nanoparticles (UCNPs) 120
   6.8 Summary and prospects 122
Acknowledgments 124
References 124

Abstract

Platinum(II) complexes are currently the most widely used drugs in cancer chemotherapy. One strategy to increase their selectivity for cancer cells vs normal cells, so reducing side effects, and introduce new mechanisms of action which might overcome resistance and broaden the spectrum of treatable cancers, is the use of non-toxic platinum prodrugs which can be activated by spatially-directed photoirradiation. We describe the

Advances in Inorganic Chemistry, Volume 80
ISSN 0898-8838
https://doi.org/10.1016/bs.adioch.2022.07.001

design of such complexes, focusing especially on octahedral diazido Pt(IV) complexes, as well as their photochemistry and photobiology. The family of diazido dihydroxido diam(*m*)ine complexes can be further targeted through conjugation to the axial hydroxide ligands, by variations in the amines, and their stereochemistry. A wide range of targeting vectors and drug delivery methods is attracting attention. Interestingly, photoirradiated *trans*-diamine complexes often have higher activity than the *cis* isomers. The use of methods such as synchrotron X-ray imaging and spectroscopy allow their distribution and speciation in intact cells to be investigated. The potential for clinical use of photoactivated chemotherapy (PACT) is discussed.

1. Introduction

As a safe and non-invasive external trigger for prodrug activation, the application of spatially-directed light can lead to high efficiency and accurate localization, and reduce side effects and systemic toxicity.[1,2] Thus, phototherapy has been widely used to treat various diseases, especially skin diseases and tumors. Clinical use of light to activate prodrugs began in about 1974 with the combination of psoralen, a naturally-occurring member of the coumarin family of organic compounds, and UVA.[3] Later hematoporphyrin derivatives were introduced as photosensitizers, and the mixture called Photofrin® activated by red light, is still in use today in so-called Photodynamic Therapy (PDT).[4–6] Typically, current clinical photosensitizers such as Photofrin®, convert ground-state triplet oxygen ($^3O_2$), into highly reactive excited-state singlet oxygen ($^1O_2$), on irradiation with light.[1,7,8] Singlet oxygen then causes oxidation of proteins, lipids and other biomolecules, which leads to cell death.[9] PDT photosensitizers are therefore most effective in tissues with a high oxygen content.

The depth of penetration of the light is determined by its wavelength, with longer wavelength light penetrating more deeply into tissues[10,11] (Fig. 1). Typically, red light activation is used for porphyrin photosensitizers.[4] However, although the depth of activity within tumors is increased using longer wavelength light, and hypoxic regions of tumors can be reached, too deep a penetration can also lead to damage to underlying normal tissue and limits the clinical application of PDT photosensitisers.[12] As described here, new photoactive platinum complexes have the potential for activity in hypoxic tumors with a different mechanism of action.

A wide range of low-spin d^6 transition metal complexes are photoactive, including Mn(I), Re(I), Fe(II), Ru(II), Os(II), Rh(III), Ir(III), and of particular interest here, Pt(IV).[13] Depending on the ligand types, some complexes can have long-lived excited states and undergo phosphorescence decay from

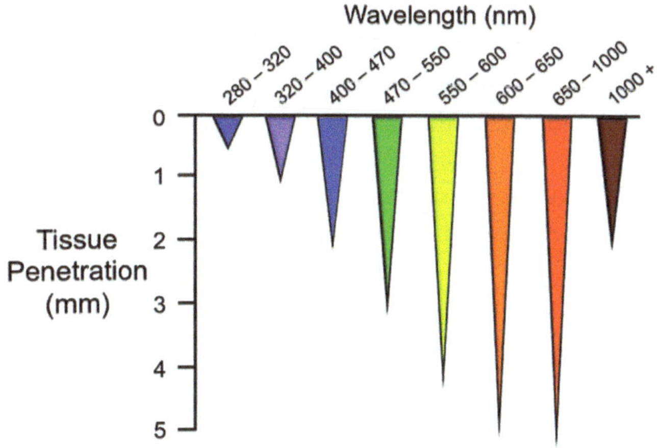

Fig. 1 Tissue penetration depth of irradiance. *Reprinted from Ruggiero, E.; Alonso-de Castro, S.; Habtemariam, A.; Salassa, L.* Dalton Trans. ***2016***, 45, *13012–13020 with permission.*

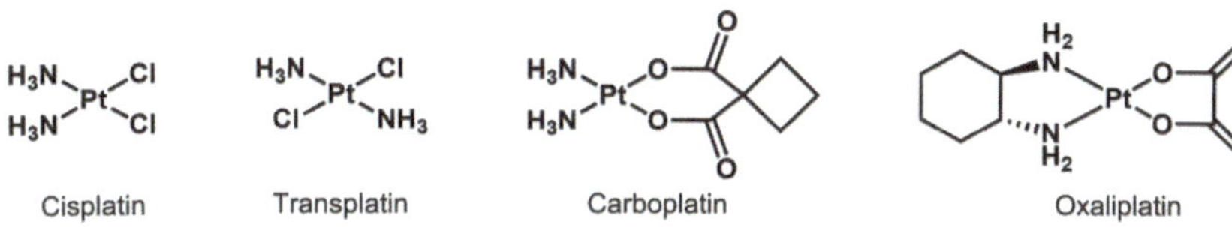

Fig. 2 Clinically used platinum drug cisplatin, its inactive *trans* isomer transplatin, and clinical analogues carboplatin and oxaliplatin.

a triplet excited state. Others can undergo metal-ligand bond weakening in singlet or triplet excited states, thereby inducing ligand dissociation, or experience redox reactions arising from photo-induced metal and/or ligand electron transfer.[12] The $4d^6$ Ru(II) complex TLD1433, which is currently in clinical trials for treatment of bladder cancer, is activated by green light, generates $^1O_2$ and is also luminescent.[12] Different from the PDT effects of TLD1433, $5d^6$ Pt(IV) complexes often decompose directly to generate cytotoxic species independent of the presence of oxygen, with potential applications in Photoactivated Chemotherapy (PACT), including phototherapy toward hypoxic tumors.[14]

The serendipitous discovery of the anticancer activity of platinum coordination complexes by Barnett Rosenberg and co-workers at Michigan State University in the 1960s led to the clinical use of cisplatin from about 1979, and to later generation carboplatin and oxaliplatin drugs for the treatment of solid tumors (Fig. 2).[15–17] Today about 50% of chemotherapies for cancer use a platinum drug.[18,19]

Our own lab contributed to the development of these platinum drugs through studies of their chemistry and biochemistry. We established the structure of carboplatin for example,[20] and used multinuclear NMR methods for elucidating interactions of these platinum drugs with nucleotides, oligonucleotides, amino acids and proteins.[21] In particular these studies highlighted the importance of ligand substitution reactions in their pharmacology, including the strong *trans* effect of sulfur ligands such as methionine.[22]

Despite the clinical success of cisplatin, carboplatin and oxaliplatin for the treatment of solid tumors, there is a need to introduce new therapies which combat resistance to cisplatin, increase the range of treatable cancers, and reduce the side effects. The prospect of designing non-toxic platinum drugs which are selectively activated in cancer cells, and not in normal tissues, is attractive for avoiding unwanted side-effects from platinum therapy.[23,24] One such strategy involves the use of relatively non-toxic platinum prodrugs which are activated selectively in the region of a tumor using spatially-directed light.[14,25] This approach also has the possibility of destroying cancer cells by new mechanisms of action and therefore of combatting resistance.[14]

Photoactive platinum complexes have long been known.[13] Cyclometalated Pt(II) complexes can generate $^1O_2$ upon irradiation, while non-luminescent Pt(II) complexes can be conjugated to highly fluorescent photosensitizers to exert photocytotoxicity.[25,26] In contrast, kinetically-inert Pt(IV) complexes have potential as PACT metallodrugs due to their high dark stability.[14,27] Interestingly the photoactivity of platinum complexes played a role in the discovery of the clinical drug cisplatin in Rosenberg's laboratory over 50 years ago.[28] Rosenberg and colleagues suspected that Pt(IV) complexes were generated in the *E. coli* culture medium from electrolysis using Pt electrodes. However, simple salts such as $[Pt^{IV}Cl_6]^{2-}$ were poorly active in causing filamentous growth of the bacteria until the solutions had been exposed to light for several days. Light was able to catalyze ligand substitution reactions with formation of ammine adducts such as *cis*-$[PtCl_4(NH_3)_2]$ involving the ammonium salts in the medium. It was the *cis*-$\{Pt(NH_3)_2\}$ configuration which turned out to be critical to the eventual discovery of cisplatin as a new anticancer drug.

Here we give a brief account of research that has taken place in our labs over the past 20 years or so on the design of azido-Pt(IV) complexes as phototherapeutic agents together with their photochemical, photobiological and photopharmacological properties, including their mechanism of action. As is often the case with drug discovery and development pipelines, progress has been relatively slow. For example, the clinical platinum drug oxaliplatin was approved for anticancer use 20 years after it was first patented. Although passage along the drug development pipeline has been slow for photoactivatable platinum complexes, this has been offset by the novel chemistry which has emerged in this relatively new area of research.

2. Discovery of photoactive Pt(IV) anticancer complexes

Bednarski and co-workers discovered in the early 1990s at the University of Regensburg that relatively inert octahedral *cis*-diam(*m*)ine di-iodido Pt(IV) complexes (e.g., complexes **1–5**, Fig. 3) have anticancer activity when irradiated with visible light.[29–33] They viewed these as prodrugs for cisplatin, undergoing photoreduction on irradiation with electron donation from the axial iodido ligands (via iodide-to-Pt(IV) charge transfer) and formation of a more reactive square-planar Pt(II) center.

However, these complexes did not show sufficient selectivity toward cancer cells in the light compared to the dark.[31] Kratchowil from Regensburg joined the Sadler lab in Edinburgh as a Marie Curie Fellow in 1997 to gain a deeper insight into their chemistry and biochemistry. We discovered that the complex *trans,cis*-[Pt(en)$(OH)_2I_2$] (**1**, en = ethylenediamine) formed a bis-guanine Pt(II) adduct on photoactivation with blue light,[32] but was readily reduced in the dark by the thiol tripeptide glutathione (γ-L-Glu-L-Cys-Gly),

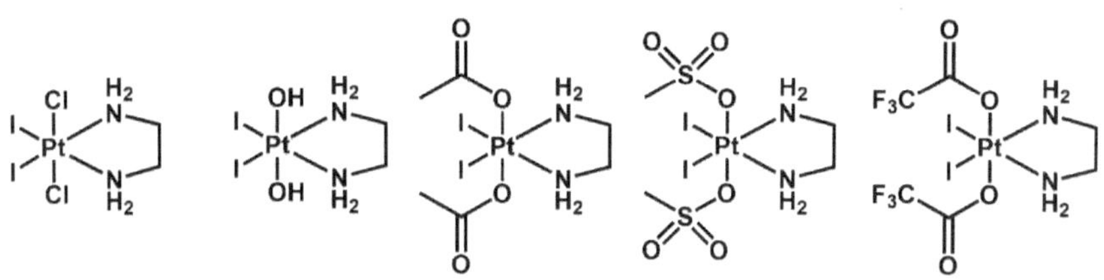

1 2 3 4 5

Fig. 3 Photoactive diiodido-Pt(IV) complexes.

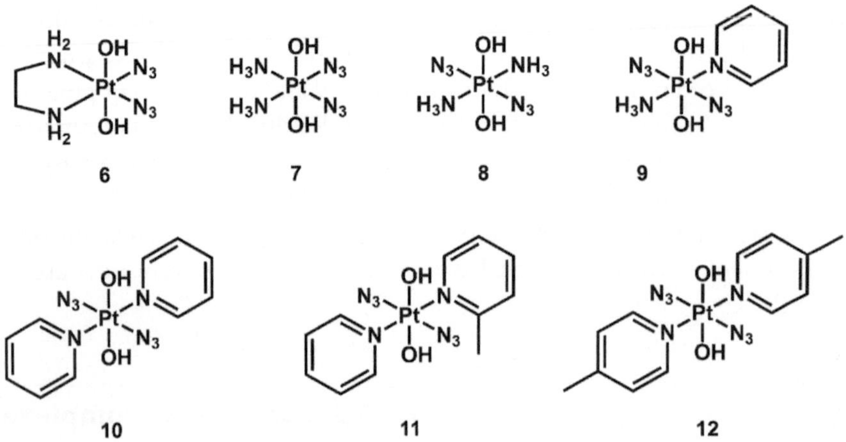

Fig. 4 Photoactive diazido-Pt(IV) diamine complexes.

by a mechanism involving attack of the thiolate S on a coordinated iodide.[31] This was accompanied by en ring-opening, and generation of more-reactive Pt(II) species which could bind to DNA.

Müller then joined the Sadler group in Edinburgh as a PhD student from Switzerland, and we replaced the iodide ligands with azides in an attempt to retain photoactivity, but reduce reactivity toward thiols such as GSH.[34] This was successful. We synthesized the Pt(IV) complexes *cis,trans*-[Pt(en)$(N_3)_2(OH)_2$] (**6**) and *cis,trans,cis*-[Pt$(N_3)_2(OH)_2(NH_3)_2$] (**7**) (Fig. 4), which had intense LMCT (N_3 to Pt^{IV}) bands in the UV region, and were potential photoactive prodrugs for active Pt(II) complexes.[34] Importantly they were relatively inert toward reactions with GSH in the dark. Upon UVA exposure, these complexes exhibited significantly enhanced cytotoxicity in both wild-type 5637 and cisplatin-resistant 5637 bladder cancer cells, producing dramatic cellular morphological changes, despite their low dark cytotoxicity.[35]

Density functional theory (DFT) and time-dependent DFT (TD-DFT) calculations on complex **7** allowed assignments of ^{1}LMCT transitions in the visible region of the UV–vis spectrum. Release of one N_3 ligand is facilitated when these low-energy LMCT states are excited. The dissociation of both NH_3 and N_3 groups *trans* to each other is promoted on triplet formation with no apparent reduction of the Pt(IV) centre.[36] Compared to the singlet-state geometry, an elongated Pt—NH_3 bond and an elongated

Pt—N_3 bond *trans* to it were observed in the optimized triplet-state structure, which increase in length from 2.097 to 2.538 Å, and from 2.027 to 2.221 Å, respectively.

Ronconi used direct NMR observation of the quadrupolar $I=1$ nucleus ^{14}N in the NH_3 ligands (99.6% natural abundance), as well as 2D HSQC ^{1}H—^{15}N NMR studies of >90% ^{15}N-enriched complexes to elucidate the photodecomposition pathways of *cis* and *trans* diamine complexes.[37] They exhibited a strong dependence on the solution conditions including the presence of biomolecules. Notably, Ronconi detected photo-induced release of both N_2 and generation of O_2. He also trapped nitrene intermediates in reactions of complex **7** with the thioether Me_2S, which might have implications for reactions with the thioether side-chain of the amino acid methionine in proteins.[38] ^{1}H and ^{15}N NMR studies of diazido dihydroxido ^{15}N-am(*m*)ine complexes were used to determine a pK_a of ca. 3.4, assignable to the protonation of an axial hydroxido ligand.[39] Since metal—OH_2 bonds are typically more labile than metal—OH bonds, there may be acidic biological compartments in which the complexes are more reactive (e.g., stomach).

Mackay followed up this work as a PhD student in Edinburgh, and discovered that the azide-to-Pt(IV) LMCT band of *trans,trans,trans*-[Pt$(N_3)_2$$(OH)_2$$(NH_3)_2$] (**8**, Fig. 4) was more intense, and shifted to longer wavelength compared to that of the *cis*-diamine, *cis*-diazido isomer.[40] Also, the *trans* complex had a much higher aqueous solubility (>20 mM), and rapidly formed DNA lesions of a different type to those formed by the transplatin isomer of cisplatin when photoactivated (Fig. 2). The *trans* Pt(IV) complex was as cytotoxic as cisplatin on UVA irradiation for a short time (50 min).

Next, we replaced an NH_3 ligand by pyridine. The complex *trans,trans,trans*-[Pt$(N_3)_2$$(OH)_2$$(NH_3)$(py)] (**9**, **FM165**, Fig. 4) was also stable in the dark and stable toward reaction with GSH. Moreover, it was more potent than cisplatin, retained activity toward cisplatin-resistant cancer cells, and rapidly formed interstrand DNA GG crosslinks on irradiation.[41] Compared with cisplatin, complex **9** exhibits 13–80 × higher photocytotoxicity, and is 15 × more effective toward cisplatin-resistant human ovarian A2780cis cancer cells. Notably, in vivo experiments carried out in nude mice bearing OE19 esophageal cancer xenografts showed that two of seven mice treated with complex **9** at low doses with short irradiation times (420 nm blue light,

100 J cm^{-2}, 2 × 30 min) survived at 35 days, while none of the control mice survived.[42] The higher photocytotoxicity of *trans*-ammine/amine Pt(IV) complexes compared to *cis* isomers seemed to be a general feature of this class of complexes.[43]

Deeth et al.[44] demonstrated by TD-DFT that such calculations can help to optimize the photochemical properties of this class of diazido Pt(IV) complexes, e.g., by introducing intramolecular H-bonds to the axial OH ligands and electron-withdrawing substituents on pyridine ligands. Interestingly, solvent can have a significant effect on the prediction of the intensities and energies of the lower energy bands.

However, the synthetic yields of these diazido Pt(IV) ammine complexes were very low. After the Sadler group moved to the University of Warwick in 2007, Farrer and Shaili devised efficient gram-scale syntheses of these *trans*-di-azido, *trans* di-amine complexes, including di-pyridine complex *trans,trans,trans*-[$Pt(N_3)_2(OH)_2(py)_2$] (**10**, **FM190**, Fig. 4). This complex was efficiently photoactivated by visible light as well as by UVA.[45] We were aware of the need to handle azido complexes very carefully to minimize the possibility of detonation. That meant avoiding the use of Ag(I) to remove chloride from chloride Pt(II) precursors in order to introduce azide ligands since Ag(I) azide is known (since 1890) to be highly explosive by impact or UV exposure.[46] We also focused on higher MW complexes to minimize the risk for the platinum products. Reducing the contribution of azide to the MW, usually lessens to likelihood of explosive properties, and we developed more efficient synthetic methods. These complexes killed cancer cells most efficiently on irradiation with UVA or blue light, which is suitable for surface cancers, e.g., bladder or esophageal, whereas longer wavelengths are needed to penetrate tissues and reach deeper cancers.

Interestingly, the ^{14}N NMR peaks for N_α of bound azide (coordinated to Pt) of these *cis*- and *trans*-diazido Pt(IV) complexes have linewidths at half-height of ca. 1–2 kHz and are too broad to observe unless anti-ringing NMR pulse sequences are used. As illustrated for complex **10** in Fig. 5, this peak sharpens with decrease in viscosity of the solvent.[47]

Methyl substituents on the pyridine ligands affect the dark stability and cytotoxicity of derivatives of **10**. For example, a methyl substituent at the 2-position of pyridine causes steric hindrance and leads to the rapid photodecomposition and dark cytotoxicity of complex **11**; while a methyl substituent at the 4-position of both pyridines in **12** results in the highest photocytotoxicity in this series (Fig. 4).[48]

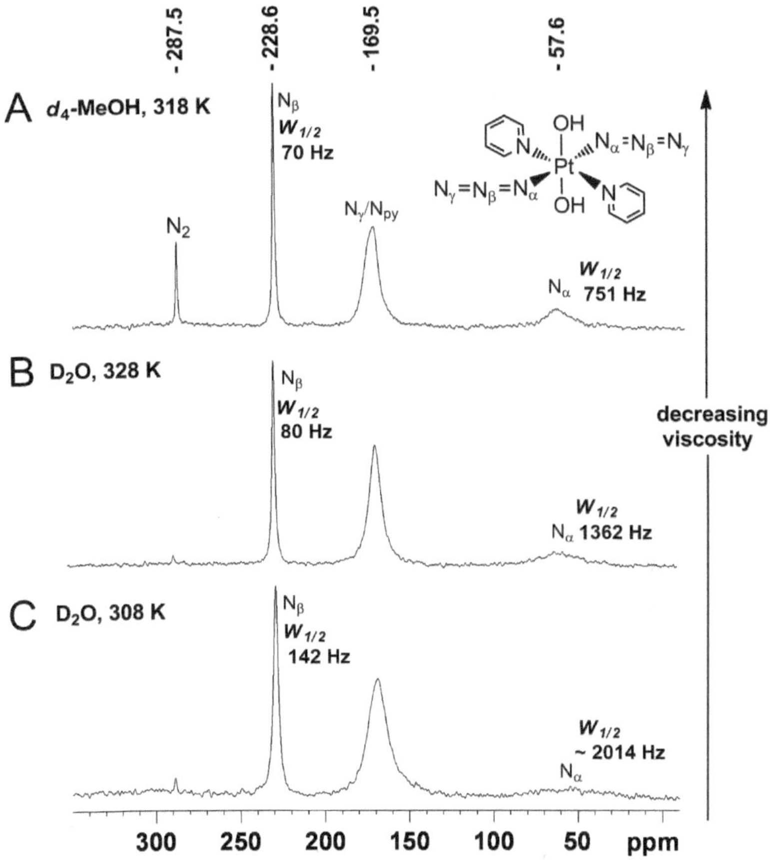

Fig. 5 ^{14}N NMR spectra of *trans,trans,trans*-[Pt(N_3)$_2$(OH)$_2$(py)$_2$] (**10**) (A) 10 mM in [D_4] MeOH, 318 K, $\eta \approx 0.42$ cP; (B) 20 mM in D_2O, 328 K, $\eta \approx 0.50$ cP; (C) 20 mM in D_2O, 308 K, $\eta \approx 0.72$ cP, showing the sharpening of the N_α peak with decreasing viscosity. *Reproduced from Farrer, N. J.; Gierth, P.; Sadler, P. J.* Chem. Eur. J. ***2011***, 17, *12059–12066 with permission.*

3. Mechanism of action

Photoactivation of the above diazido Pt(IV) complexes generates Pt(II) species which form unusual lesions on DNA together with radicals which attack proteins in cancer cells, both of which can contribute to the cytotoxicity. This combination of different mechanisms of action probably allows photoactive Pt(IV) complexes to overcome resistance to cisplatin.

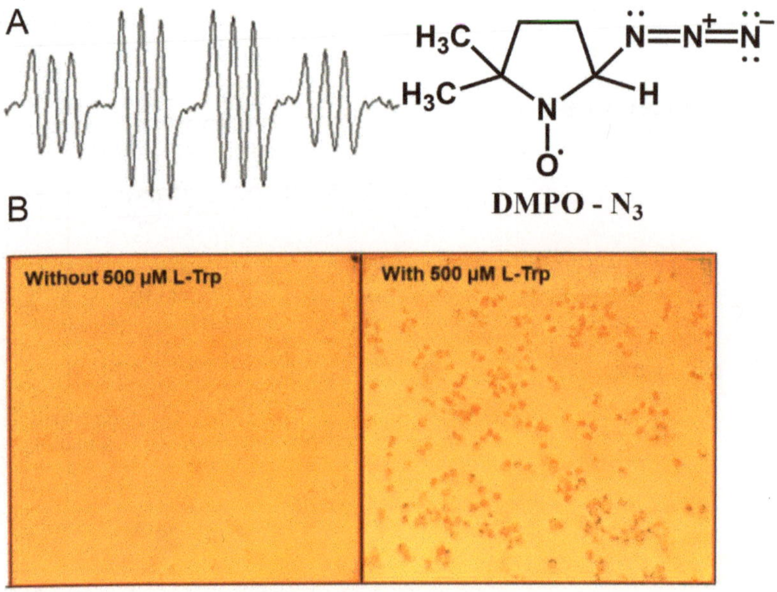

Fig. 6 (A) X-band EPR spectra of azidyl radicals, products of photoactivation with blue light of aqueous solutions of complex **10** trapped by the spin trap DMPO; (B) A2780 human cancer cells treated with **10** (left) in the absence, and (right) in the presence of L-Trp after blue light irradiation. This behavior is consistent with the trapping of damaging azidyl radicals by L-Trp. *Reproduced from Butler, J. S.; Woods, J. A.; Farrer, N. J.; Newton, M. E.; Sadler, P. J.* J. Am. Chem. Soc. ***2012****,* 134*, 16508–16511 with permission.*

3.1 Radical and ROS generation

Azido ligands in diazido Pt(IV) complexes can be released upon irradiation in the form of azidyl radicals, which can contribute to cell death by oxidative attack.[49] These azidyl radicals can be detected by EPR, using 5,5-dimethyl-1-pyrroline N-oxide (DMPO) as a spin trap (Fig. 6A). Besides transfer of an electron from azide ligands onto Pt(IV), the axial hydroxide ligand can also transfer an electron onto Pt(IV), generating hydroxyl radicals. These hydroxyl ligands can also be trapped, and contribute to the increased ROS level of cancer cells treated with Pt(IV) complexes and irradiated.[50]

The amino acid L-tryptophan (Trp), antioxidant melatonin (MLT) and the Trp-peptide pentagastrin all quench the formation of azidyl radicals arising from visible-light excited complex **10**, giving rise to C3-centered Trp and MLT radicals that can be trapped by 2-methyl-2-nitrosopropane (MNP) and detected using EPR.[51] Thus, the acute photocytotoxicity can

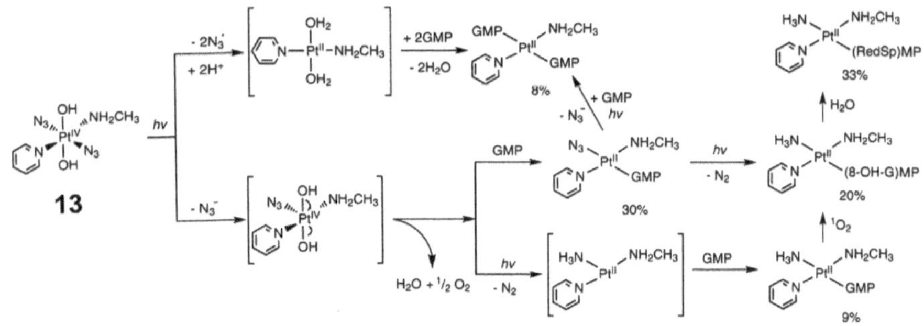

Fig. 7 Possible pathways for photodecomposition and photoreactions of complex **13** with guanosine monophosphate (5′-GMP) upon irradiation with UVA. *Reproduced from Zhao, Y.; Farrer, N. J.; Li, H.; Butler, J. S.; McQuitty, R. J.; Habtemariam, A.; Wang, F.; Sadler, P. J.* Angew. Chem., Int. Ed. ***2013***, 52, *13633–13637 with permission.*

be switched off by low doses (500 μM) of Trp (Fig. 6B). EPR and NMR spectroscopic experiments using spin traps suggest that L-Trp quenches the formation of azidyl radicals, by acting as an electron donor.[49]

The photodecomposition of complex **13** involves Pt-nitrene intermediates and formation of singlet oxygen, free azide, azidyl radicals, and nitrogen gas (Fig. 7). The unexpected photooxidation of guanine is likely to arise from reactions of singlet oxygen, which can be detected by the fluorescence probe Singlet Oxygen Sensor Green (SOSG).[52] Argon- or nitrogen-saturated samples gave strong SOSG fluorescence upon irradiation, which suggests that the $^1O_2$ was not generated from the dissolved oxygen through energy transfer from the photosensitizer. Oxidative damage to guanine may contribute to the potent photocytotoxic effects of this complex (Fig. 7).

As a result, cancer cells treated with diazido Pt(IV) complexes exhibited intense dichlorodihydrofluorescein (DCF) green fluorescence after irradiation, indicating the high level of induced ROS, while no fluorescence was detected in the dark.[50]

3.2 DNA as a target

DNA is a target for Pt anticancer drugs such as cisplatin, and the RNA monomer nucleotide guanosine monophosphate (5′-GMP) is frequently employed as a model for G in oligonicleotides.[53,54] Different from Pt(II) complexes such as cisplatin, diazido Pt(IV) complexes are stable toward 5′-GMP in the absence of light, but form mono- and/or bis-GMP adducts only upon irradiation.

Complexes **6–8** form *cis*-bis-GMP adducts on irradiation with blue light and are therefore likely to be able to bind to DNA.[34,40] Among them, *trans* complex **8** forms bis(5′-GMP) adducts at a faster rate than transplatin under UVA irradiation.[40] *Cis* diazido complex **6** preferentially platinates DNA by forming bifunctional GG adducts at GG sequences, similarly but more rapidly than cisplatin.[55] DNA fragments treated with the *trans* diazido complex **9** and irradiation give rise to similar major stop sites as transplatin, as revealed by transcription mapping. However, the DNA adducts induced by irradiated **9** were significantly more abundant than those formed by cisplatin and transplatin.[41] DNA interstrand cross-links and DNA–protein cross-links were also induced by irradiated **9**. Remarkably, damage-induced plasmid DNA repair synthesis of lesions generated by UVA-irradiated **9** was considerably lower than that from cisplatin treatment. Complex **13** (Fig. 7) displays a preference for nucleobases G and C, which is similar to transplatin, but with remarkably larger unwinding angles and a higher percentage of interstrand crosslinks.[56] Different from cisplatin-type lesions, DNA lesions induced by **13** on a 50 bp duplex were not recognized by HMGB1 protein, indicative of a different type of structural perturbation of the DNA structure induced by platinum binding.

Photoactivated **10** induced distinctly different conformational alterations in DNA adducts, compared to those produced by cisplatin or transplatin.[45] Also, the ability of their DNA adducts to inhibit DNA transcription by stalling RNA pol II, a multiprotein enzyme that transcribes DNA into RNA precursors, is critically different for photoactivated **10** compared to cisplatin.[57] Photoactivated **10** exhibits a significantly more potent inhibition of RNA synthesis compared with cisplatin.

The amount of DNA platination by UVA-photoactivated **10** was ca. 16× higher than that for cisplatin in the dark at the same dose and for the same time.[57] Notably, when the cells were treated with **10** in the dark, no platinum-bound DNA was observed. The DNA binding mode for photoactivated **10** involves combined coordination of platinum to a base residue and interaction of the pyridine ligand(s) in photoactivated **10** with the duplex. Thus, photoactivated **10** can form DNA adducts, and inducing conformational distortions in DNA which extend over more base pairs than DNA adducts formed by cisplatin or transplatin. The modeled structural deviations in DNA induced by platination by photoactivated complex **10** show a bend toward the minor groove, a global bend of ca. 67° and an unwinding of ca. 20°, as indicated by the minor groove average LFMD structure of *trans*-$\{Pt(py)_2\}^{2+}$ – DNA simulated by computational studies (Fig. 8).[58]

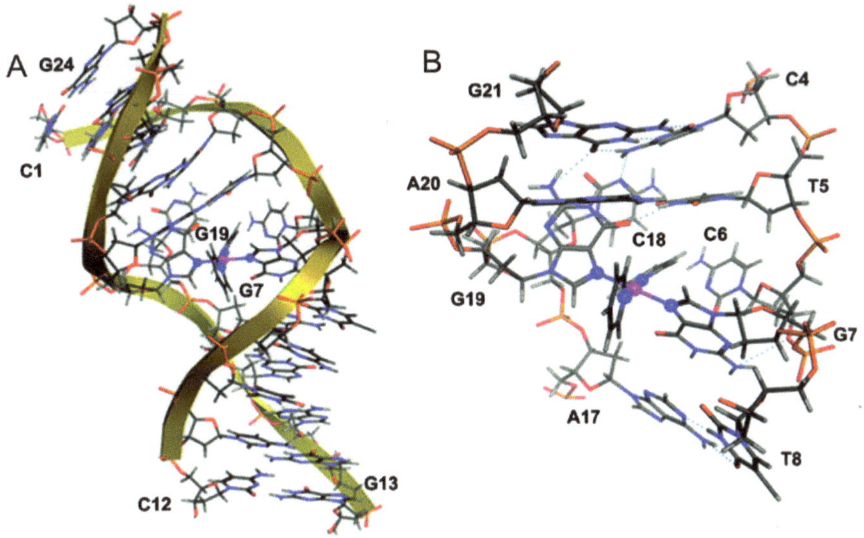

Fig. 8 (A) The average ligand field molecular dynamics (LFMD) structure of *trans*-$\{Pt(py)_2\}^{2+}$–DNA derived from simulations; (B) an expanded view of the platinum-binding site showing the G7–G19 interstrand cross-link within the DNA duplex dodecamer, d(5′-C1C2T3C4T5C6G7T8C9T10C11C12-3′)·d(5′-G13G14A15G16A17C18G19A20G21A22G23G24–3′). *Reproduced from Tai, H. C.; Brodbeck, R.; Kasparkova, J.; Farrer, N. J.; Brabec, V.; Sadler, P. J.; Deeth, R. J.* Inorg. Chem. ***2012***, 51, *6830–6841 with permission.*

3.3 Peptide and protein targets

Diazido Pt(IV) complexes attack peptides via a combined mechanism of sequence-dependent platination and radical generation. After treatment with complex **10** and irradiation with blue light, only monoplatinated adducts formed with the various amino acids in the undecapeptide Substance P (SubP): RPKPQQFFGLM-NH_2, while both mono- and di-platinated adducts with a preference for His, were observed for the tetradecapeptide K^3-Bombasin (K^3-Bom): pEQKLGNQWAVGHLM-NH_2.[59] In contrast, oxidation occurred only at Met of SubP, but both Met and Trp of K^3-Bom were oxidized.

Upon the irradiation with blue light, complex **10** platinates His, Glu, and Gln residues in the enzyme thioredoxin (Trx) and induces oxidation of Met, Trp, and catalytic Cys residues to form disulfide bonds. These reactions inhibit the activity of Trx, and increase the cellular ROS level, which may play a major role in the potent activity of complex **10**.[60]

3.4 Induction of immunogenic cell death

Calreticulin (CRT) expression was induced at the surface of dying cancer cells after treatment with all *trans*-Pt(IV) complex **10** followed by visible light irradiation.[61] The release of high mobility group protein-1B (HMGB1) and secretion of ATP were also observed. In this chemotherapeutically-stimulated ICD, autophagy plays an important role. The engulfment of CT26 colorectal cancer cells treated with irradiated **10** by J774.A1 macrophages confirmed that the uneven distribution of ecto-CRT promotes phagocytosis.

4. Axial derivatives of Pt(IV) complexes

4.1 Targeting vectors

The spatial and temporal control of the activation of prodrugs provides a major advantage for phototherapy over traditional chemotherapy, and can reduce systemic cytotoxicity.[1] Efficient accumulation of prodrug in target tumor sites can significantly improve the utilization of the prodrug and circumvent unwanted side effects. Modification of the axial ligands in octahedral Pt(IV) complexes is a feasible method for improving their cancer targeting ability and photocytotoxicity.[14] These axial ligands are usually photoreleasable and greatly affect the reduction potential of Pt(IV).

The RGD sequence (-Arg-Gly-Asp-) in peptides is a widely used ligand that can bind selectively to overexpressed $\alpha_v\beta_3$ and $\alpha_v\beta_5$ integrins on tumor cell surfaces.[62] Despite its lower photocytotoxicity than the parent succinylated complex, conjugate **14** (Fig. 9) with a cyclic RGD-containing peptide c(RGDfK), displays remarkably enhanced selectivity and increased cellular accumulation in cancer cells with overexpression of $\alpha_v\beta_3$ and $\alpha_v\beta_5$ integrins.[63] Conjugate **15** (Fig. 9) with the cyclic peptide c(CRWYDENAC) containing the RWY sequence (Arg-Trp-Tyr), targets the integrin $\alpha6$ receptor, and exhibits higher photocytotoxicity toward cancer cells upon blue light irradiation compared with its parent compound **10**.[64] Notably, the cellular accumulation of **15** improves by ca. 47 × after irradiation, which results in 3 × higher Pt accumulation than the parent compound **10**, suggesting a role for the cyclic peptide in delivery of the complex.

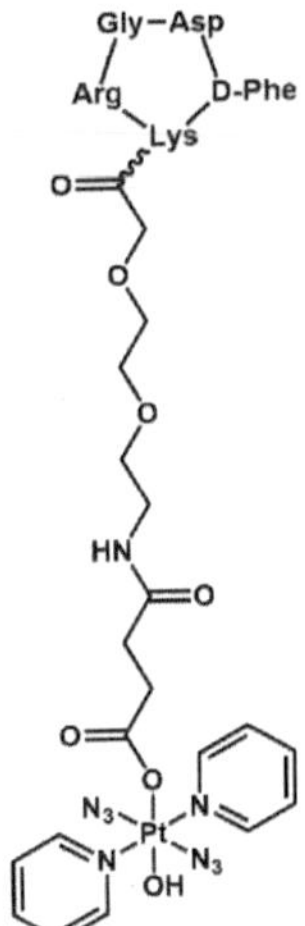

14

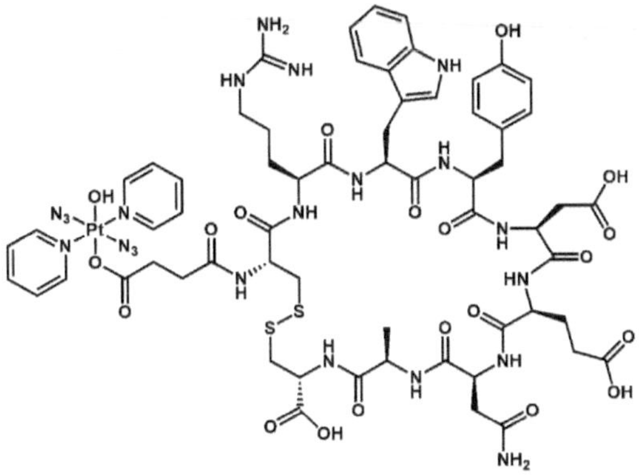

15

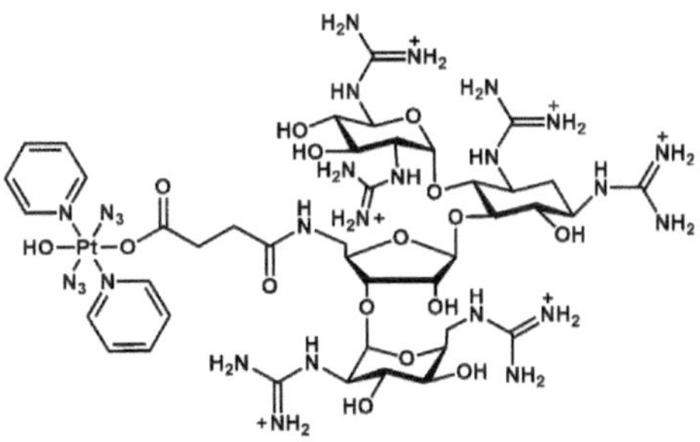

16

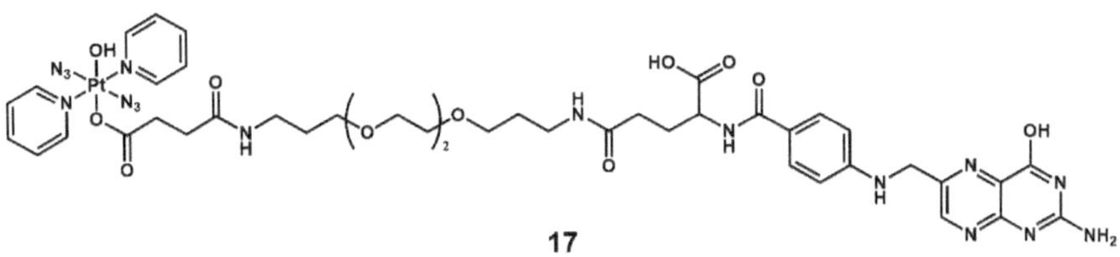

17

Fig. 9 Diazido-Pt(IV) complexes with cancer cell-targeting vectors conjugated to axial hydroxido ligands.

RNA-binding guanidinoneomycin allows the conjugate **16** (Fig. 9) to be taken up in a selective proteoglycan-dependent manner by cancer cells. Due to the overexpression of negatively-charged cell-surface proteoglycans in the malignant melanoma cell line, conjugate **16** displays an enhanced cellular uptake with a preference for SK-MEL-28 malignant melanoma melanocytes.[65]

The Pt–folate conjugate **17** (Fig. 9) shows ca. 2 × higher photocytotoxicity toward the MCF-7 breast cancer cell line compared to its parent complex **10** with a photoselectivity index (PI) > 6.9, which results from its higher cellular accumulation due to the overexpressed folate receptors in MCF-7.[66]

4.2 Multi-action derivatives

Other than cancer-targeting vectors, anticancer agents can also be attached to the axial position of photoactive Pt(IV) complexes.[14] These cytotoxic agents can be released from prodrugs upon irradiation, accompanied by reactive Pt(II) species and azidyl radicals. The anticancer efficacy of these drugs can therefore be enhanced owing to their different cellular targets and mechanisms of action. The combination of these cytotoxic species can result in greater cytotoxicity than the simple combinations of the two agents via a synergistic effect.

The most straightforward example is the dinuclear complex **18** (Fig. 10) bridged by an axial bisamide dicarboxylato linker. This complex displays

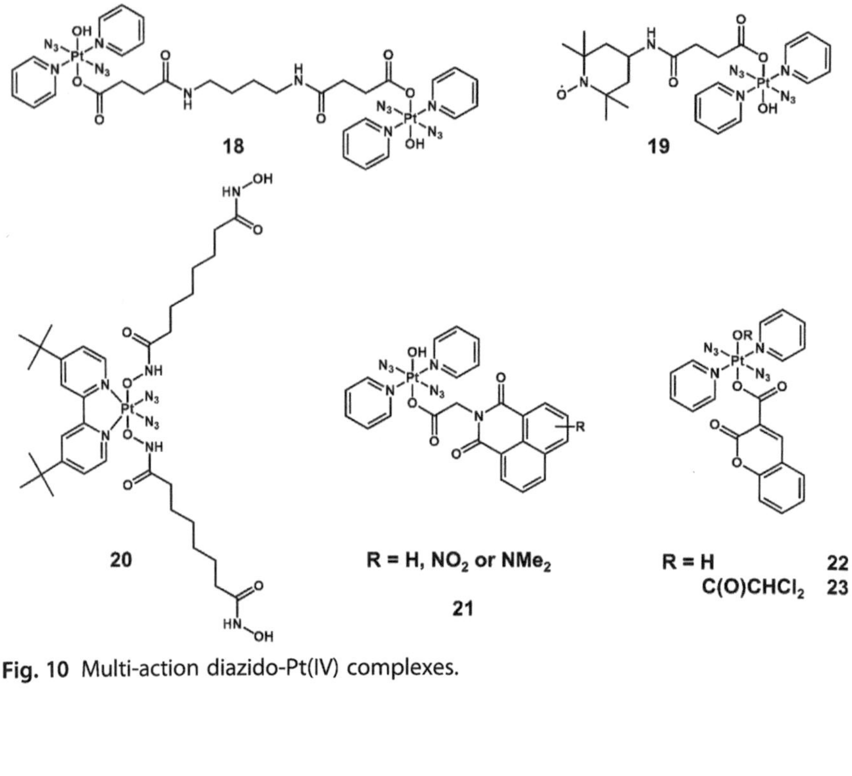

Fig. 10 Multi-action diazido-Pt(IV) complexes.

photocytotoxicity toward both wild-type and cisplatin-resistant ovarian A2780 cancer cells with a relatively low cytotoxicity toward normal MRC5 lung cells.[67]

The nitroxyl radical TEMPO retains its EPR signal when conjugated to the Pt(IV) complex **19** (Fig. 10), which releases azidyl and TEMPO radicals upon blue light irradiation, and exhibits enhanced photocytotoxicity.[68]

Complex **20** (Fig. 10) contains two suberoyl-bis-hydroxamic acid (SubH) ligands, a histone deacetylase (HDAC) inhibitor, that can be released upon UVA irradiation accompanied by Pt(II) species.[69] The released SubH allows Pt(II) species to access chromatin DNA via HDAC inhibition, introduces effective steric blockage of RNA polymerase II, and forms DNA adducts (e.g., interstrand cross-links). Thus, complex **20** has a low resistance factor compared with its Pt(IV) analogue containing inactive axial ligands and cisplatin, but with significant cytotoxicity.

The presence of a 1,8-naphthalimide derivative not only red-shifts the absorbance of Pt(IV) complexes (e.g., complex **21**, Fig. 10), but also allows pre-intercalation into DNA, which results in enhanced photo-induced DNA crosslinking and photocytotoxicity with light of a longer wavelength (520 nm).[70]

Coumarins contain fused benzene and α-pyrone rings and occur naturally in plants. Their anticancer properties have been widely investigated. *Trans* diazido Pt(IV) complex **22** (Fig. 10) with an axial coumarin ligand induces remarkable cellular damage upon blue light irradiation with extreme alterations to multiple cellular components, including formation of vacuoles, and significantly increased Pt accumulation in human PC3 prostate cancer cells compared to treatment in the dark. Only partial reduction of the prodrug occurs in PC3 cells upon blue light irradiation as determined by X-ray absorption near-edge structure (XANES) analysis, indicating that photoexcited Pt(IV) species might also exert phototoxicity as well as the Pt(II) photoproducts (Fig. 11).[71] Derivatization of the second axial hydroxide in complex **22** leads to the dichloroacetate complex **23** (Fig. 10) and increases its reduction potential to -0.886 V from -1.699 V for the unsubstituted complex **10**.[50] In addition, increases of $568\times$ and $26\times$ in cellular accumulation and photocytotoxicity to A2780 cells, respectively, were observed for complex **22** compared with the parental complex **10**, respectively.

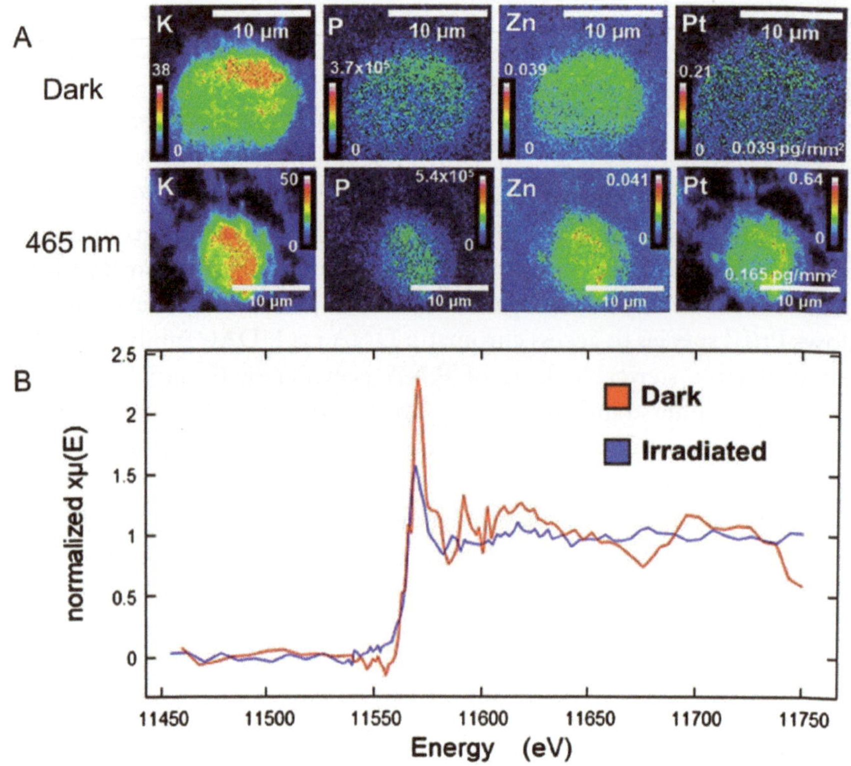

Fig. 11 (A) Comparison of the distribution of K, P, Zn and Pt in cryofixed whole PC3 human prostate cancer cells after treatment with complex **22** without or with irradiation of blue light; (B) Comparison of X-ray absorption near-edge structure (XANES) features for cryofixed whole PC3 human prostate cancer cells treated with **22** in the dark or irradiated with blue light. A shift in the Pt-edge peak to lower energies is observed upon irradiation with blue light, in addition to a reduction in peak height, suggesting that more Pt(II) is present upon irradiation. *Reproduced from Bolitho, E. M.; Sanchez-Cano, C.; Shi, H.; Quinn, P. D.; Harkiolaki, M.; Imberti, C.; Sadler, P. J.* J. Am. Chem. Soc. ***2021***, 143, *20224–20240 with permission.*

5. Pt(IV) azide–acetylene derivatives via "click" reactions

Azide-alkyne cycloaddition represents an important type of click reaction, which has been widely employed in synthetic chemistry and biology.[72] Innovative electron- or strain-promoted azide-alkyne cycloaddition-based approaches have been used in the synthesis of novel diazido Pt(IV) triazolato complexes.[73] Pt(IV) azides are less reactive with electron-deficient alkynes

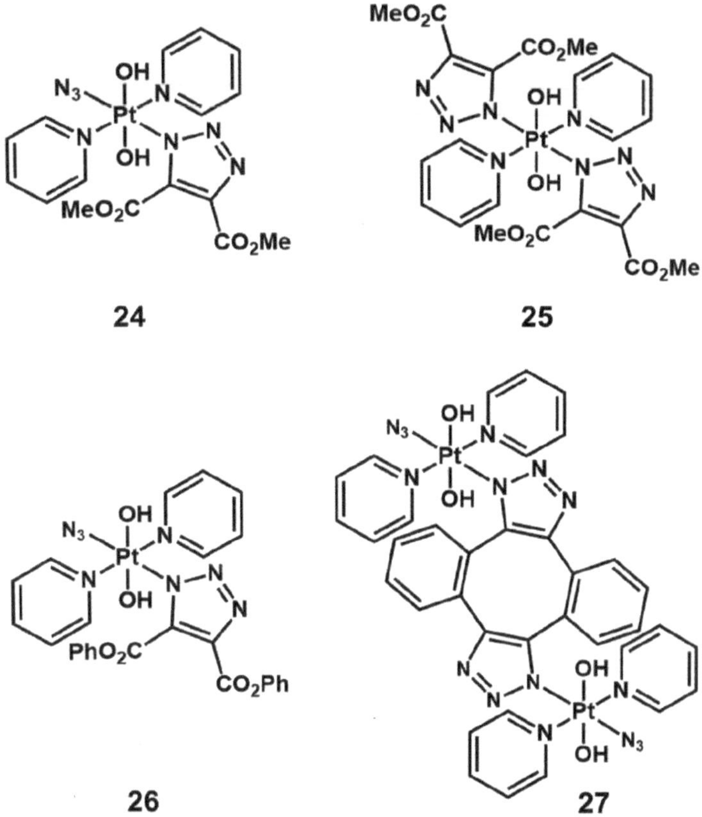

Fig. 12 Pt(IV) triazolato complexes formed by azide-alkyne cycloaddition.

than their Pt(II) counterparts. However, due to the reduced electron density on the azide ligand induced by the higher oxidation state of Pt(IV), both *mono* triazolato (**24**) and *bis* triazolato (**25**) Pt(IV) complexes can be synthesized when **10** reacts with dimethyl acetylenedicarboxylate (DMAD) in MeOH (308 K, Fig. 12).[74] The corresponding cyclometallated derivatives were also detected after attack of the axial hydroxide on the ester group and elimination of MeOH.

Platinum(IV) *mono* azido *mono* triazolato complex **26** (Fig. 12) has been prepared under mild, catalyst-free conditions by cycloaddition of **10** with 1,4-diphenyl-2-butyne-1,4-dione.[75] Upon visible light irradiation (440–480 nm), **26** released azidyl radicals, but no hydroxyl radicals, as demonstrated by EPR spin-trapping with DMPO. The major photoproduct was assigned as a [**26**-N_3]$^+$ species, as confirmed by HPLC and ^1H—^{195}Pt HMBC NMR. No Pt 5′-GMP adduct was observed, which indicates its significantly simplified biological activity compared with **10**, as no DNA binding is likely to occur.

Dinuclear Pt(IV) triazolato complex **27** (Fig. 12) was also synthesized via a strain-promoted double-click reaction between complex **10** and Sondheimer diyne.[76] Both Pt(IV) and Pt(II) 5′-GMP species were detected from **27** as the result of visible light irradiation (452 nm). The photo-generation of both azidyl and hydroxyl radicals was confirmed by EPR spectroscopy, which was significantly reduced in the presence of 5′-GMP.

6. Nanocarriers

Drug delivery nanocarriers, including organic (proteins, hydrogels, micelles, vesicles and other polymers) and inorganic (carbon-based nanomaterials, silica nanoparticles, gold nanoparticles, magnetic nanoparticles, metal-organic frameworks and upconverting nanoparticles (UCNPs)) materials, can improve drug delivery efficiency and diminish unwanted side effects.[77] Elaborate nanocarriers for photoactive Pt(IV) complexes can introduce not only accurate drug delivery by site-specific targeting, red-shifted activation wavelengths and novel mechanisms of action.[14]

6.1 Proteins

Loading Pt(IV) complexes in proteins that possess upregulated transmembrane transporters in tumors is a well-known drug delivery strategy to improve their cancer-selective accumulation.[78]

An apoptosis-sensitive probe that contains a caspase-3 activatable peptide-bridged fluorescence resonance energy transfer (FRET) pair, a far-red fluorescence donor Cy5 and a NIR quencher Qsy21, has been conjugated to photoactive Pt(IV)-loaded HSA nanocarriers **28** (Fig. 13A).[79] Upon UVA irradiation, **28** exerts cytotoxicity and triggers apoptosis, thus activating caspase-3, a key enzyme in programmed cell death, that cleaves the peptide, switches on the Cy5 emission and allows imaging of cellular apoptosis.

Biotinylated **29** (Fig. 13B) shows retained affinity toward avidin with a high binding constant.[80] Although the compound alone induces dramatic cellular morphological changes in A2780 ovarian cancer cells upon irradiation, enhanced photocytotoxicity and cellular accumulation were observed when it was co-incubated with avidin.

Myoglobin is a small protein with accessible Lys side chains, which can form conjugate **30** (Fig. 13C) with Pt(IV) complexes via thiourea linkages.[81] The average Pt/myoglobin ratio was determined to be 5.6. The conjugate is

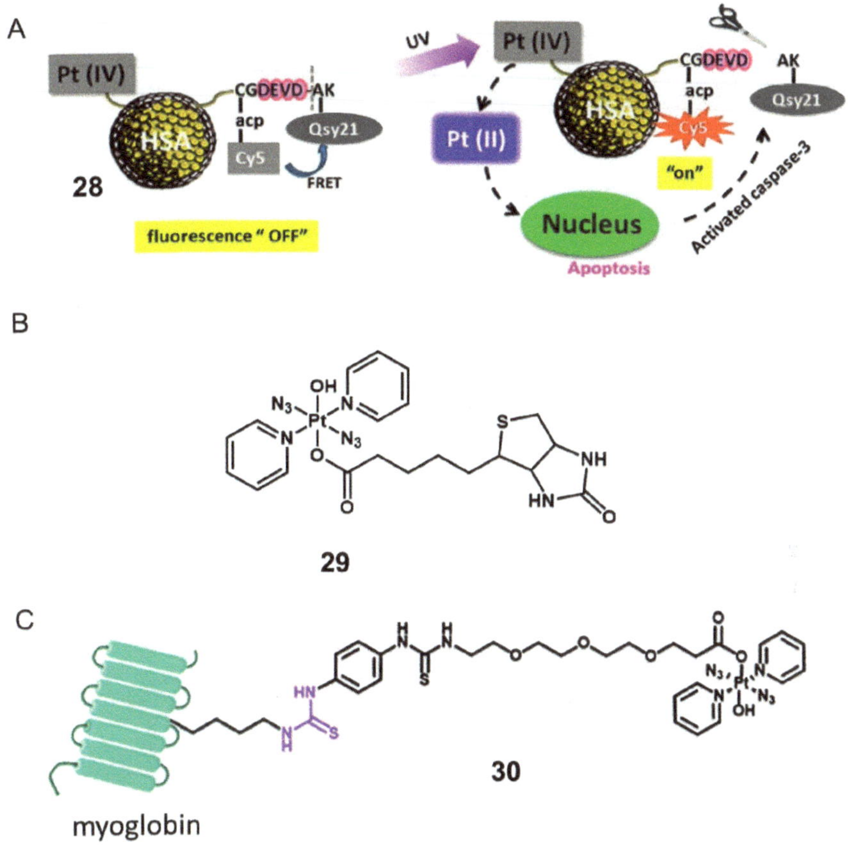

Fig. 13 Photoactive Pt(IV) complexes delivered by proteins.

stable in the dark and decomposes with blue and green light. Using the same strategy, Pt(IV) complexes have been conjugated to the Her2-targeting monoclonal antibody trastuzumab. The mixture that formed contained ca. 55% unmodified trastuzumab, with 30 and 15% of adducts having 1 and 2 Pt per antibody, respectively, resulting in an average Pt/trastuzumab ratio of ca. 0.6. Such a conjugation strategy for diazido Pt(IV) complexes and antibodies is promising for selective delivery of these photoactivatable prodrugs to cancer cells.

6.2 Polysaccharides

Self-assembled amphiphilic nanoparticles **31** (Fig. 14A), ca. 200 nm in diameter, containing *cis,cis,trans*-[$Pt^{IV}(N_3)_2(NH_3)_2(OH)$(succinate)] conjugated to dextran via the terminal carboxyl group in the axial succinate ligand have

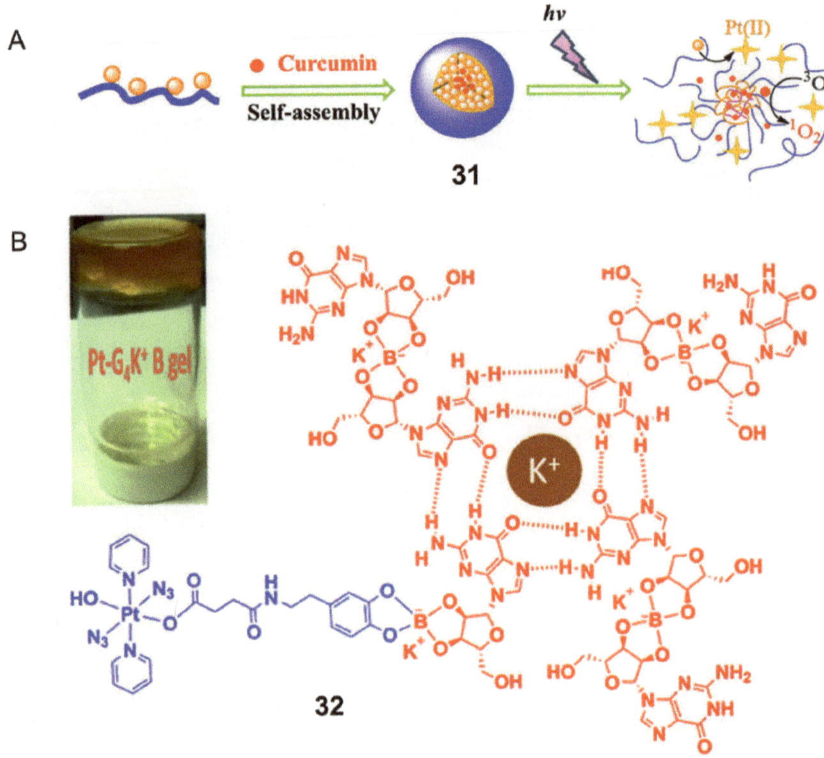

Fig. 14 (A) Preparation and photodecomposition of curcumin-loaded dextran-Pt(N_3) nanoparticles **31**; (B) self-assembled photoactive boronyl G-quadruplex Pt-G_4K^+B anticancer hydrogel **32**. *Reproduced from He, S.; Qi, Y.; Kuang, G.; Zhou, D.; Li, J.; Xie, Z.; Chen, X.; Jing, X.; Huang, Y.* Biomacromolecules ***2016****, 17, 2120–2127 and Venkatesh, V.; Mishra, N. K.; Romero-Canelón, I.; Vernooij, R. R.; Shi, H.; Coverdale, J. P. C.; Habtemariam, A.; Verma, S.; Sadler, P. J.* J. Am. Chem. Soc. ***2017****,* 139*, 5656–5659, respectively.*

been synthesized, and then loaded with curcumin.[82] This loading increases the solubility and half-life of curcumin. Photoactivation with UVA led to rapid production of ROS by curcumin, release of Pt(II) species, and gave rise to 17 × enhanced cytotoxicity in HeLa cells compared to dark treatment, and in vivo antitumor efficacy in KM mice bearing subcutaneous H22 murine hepatocarcinoma tumor, with low systemic toxicity.[82]

6.3 Hydrogels

G-quadruplex G_4K^+ borate ester hydrogels with *trans,trans,trans*-[Pt(N_3)$_2$(py)$_2$(OH)(succinate)] conjugated via the terminal carboxyl of the axial succinate ligand have promising biocompatibility due to their natural tissue similarity, and thus may be suitable for topical medication to allow precise localization in superficial tumors. Self-assembled Pt-G_4K^+B hydrogels **32** (Fig. 14B) are photoactive with blue light and show potent photocytotoxicity toward cisplatin-resistant A2780cis ovarian cancer cells ($IC_{50}=3$ μM). They are also selective for cancer cells versus normal MRC-5 cells.[83]

6.4 Block copolymers

Amphiphilic block copolymers (BCP) can self-assemble into micelles with insoluble blocks forming the core, to which Pt(IV) complexes can be conjugated. The resulting micelles exhibit enhanced accumulation in tumor tissue through the enhanced permeation and retention (EPR) effect, which is a result of the leaky nature of angiogenic blood vessels of solid tumors.[77] The triblock copolymer methoxy-poly(ethylene glycol)-*block*-poly (ε-caprolactone)-*block*-poly-L-lysine, mPEG$_{114}$-*b*-PCL$_{20}$-PLL$_{10}$ that comprised of mPEG ($n=114$), polycaprolactone ($p=20$) and poly-L-lysine ($q=10$) self-assembles in aqueous solution into micelles with polycaprolactone forming the core and Pt(IV) complexes **33–35** (Fig. 15) covalently encapsulated inside the hydrophobic core via amide linkages.[84–86] Compared with polymer-free prodrug analogues, micelles containing the

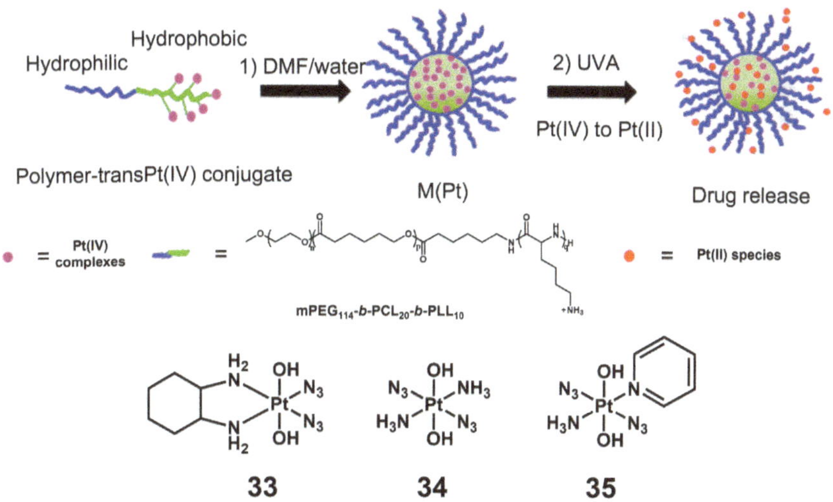

Fig. 15 Preparation and photodecomposition of micelles containing complexes **33–35** assembled from mPEG$_{114}$-*b*-PCL$_{20}$-PLL$_{10}$ polymer and the diazido-Pt(IV) complexes.

oxaliplatin-related *cis*-diazido, *trans*-dihydroxido Pt(IV) complex **33** show greatly enhanced cellular accumulation and photocytotoxicity to ovarian cancer SKOV-3 cells.[84] Importantly, in vivo studies revealed that the blood circulation half-life of the Pt(IV) complex was enhanced by 10-fold when conjugated to micelles. Also, inhibition efficacy against H22 murine hepatocarcinoma was improved and systemic toxicity decreased due to the formation of micelles. Micelles containing *trans*-diammine, *trans*-dihydroxido *trans*-diazido complex **34** also exhibited impaired deactivation in blood circulation for Pt(IV) prodrugs.[85] Photo IC_{50} values comparable to cisplatin and a resistant factor toward cisplatin-resistant ovarian cancer cells (A2780CDDP/A2780) 5 times lower than cisplatin were observed for **34**. Upon UVA irradiation, micelles **35** containing the more sterically hindered Pt(IV) complex (one NH_3 ligand changed to pyridine), were > 100 times more cytotoxic than cisplatin toward liver cancer HepG-2 cells.[86]

6.5 Polymetallodrugs

Different from metal complex-loading of micelles, polymers assembled from photoactive metal-containing monomers can give rise to a higher drug content.[87] Platinum(IV) prodrug-backboned block copolymer **36** (Fig. 16) self-assembles into photosensitive micelles that decompose with wide-range of light-exposure wavelengths to release cytotoxic Pt(II) species.[88] The irradiated micelle (430 nm) shows 3–4 × higher efficacy in reducing A549

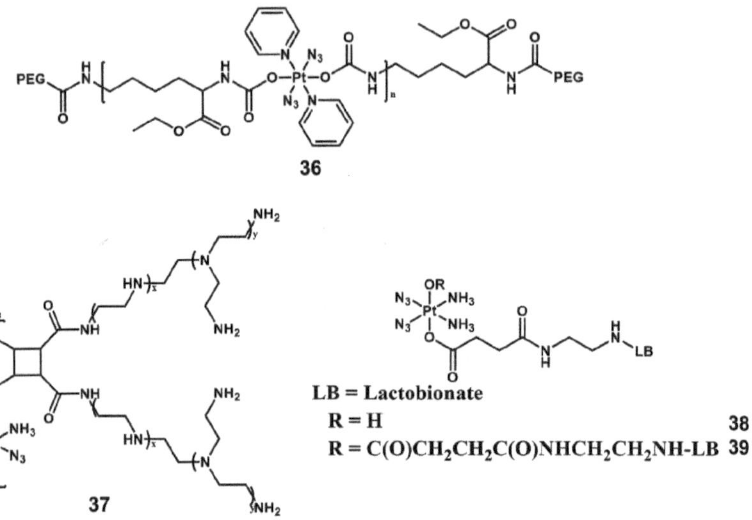

Fig. 16 Photoactive diazido Pt(IV) polymetallodrugs **36–39**.

xenografts in BALB/c nude mice than the prodrug and cisplatin, with low weight loss on day 27. Poly-Pt(IV) prodrug **37** (Fig. 16) provides synergistic cancer therapy via light-controllable codelivery of Pt(IV) prodrug and siRNA.[89] Upon blue light irradiation, the released azidyl radicals from irradiated Pt(IV) prodrug facilitate endosomal/lysosomal escape of active species. The concurrent Pt(II) release and siRNA unpacking result in synergistic therapeutic efficacy in cisplatin-resistant ovarian cancer cells.

Pt(IV) prodrug monomers can also self-assemble into nanoparticles via non-covalent interactions. Complex **38** with one lactose in an axial hydrophilic ligand self-assembles into micelles, but its analogue **39** with two lactose ligands forms vesicles instead (Fig. 16).[90] These nanoparticles display photocytotoxicity with liver cancer-targeting ability in vivo. The platinum distribution in mice can be determined by fluorescence, CT and ICP-MS. Notably, vesicles formed by **35** behave not only as a self-delivery system, but also as a nanocarrier of curcumin, showing synergistic photocytotoxicity in vitro.

6.6 Carbon dots

By coupling with folic acid (FA, vitamin B9) molecules, carbon dots (small carbon nanoparticles, <10 nm), decorated with diazido Pt(IV) prodrugs (both of which are photoactive) **40** (Fig. 17) show selective accumulation via the folate receptor of FR-positive HeLa cancer cells, which can be traced using their inherent fluorescence.[91] Despite its high dark stability, **40** displays photocytotoxicity comparable with that of cisplatin.

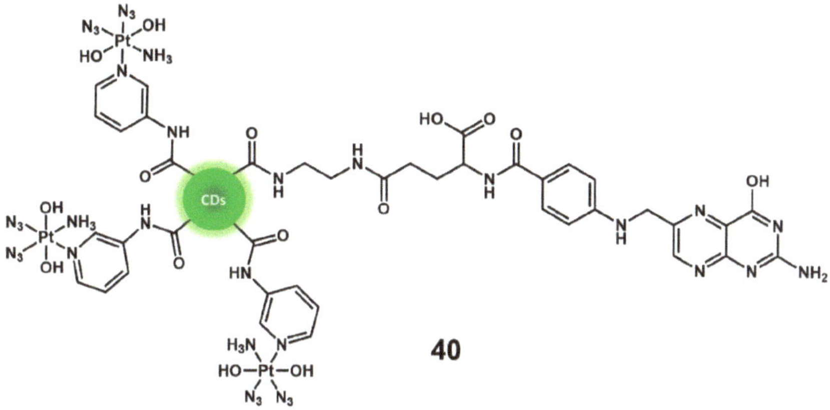

40

Fig. 17 Carbon dots decorated with folate and photoactive diazido Pt(IV) derivative.

6.7 Up-converting nanoparticles (UCNPs)

Light of longer wavelength penetrates more deeply into tissues than shorter wavelength radiation such as UVA or blue light (Fig. 1), and thus has a potentially wider range of applications in the clinic.[92] However, most reported Pt(IV) complexes have strong absorption bands in the UVA region of the spectrum.[14] To achieve activation of these prodrugs by irradiation within the "phototherapeutic window" between 620 and 850 nm, UCNPs can be used to upconvert red or NIR light into UVA/blue radiation, which can decompose Pt(IV) complexes and kill cancer cells.[11] Typically UCNPs are yttrium fluoride ($NaYF_4$) doped with donor (Yb^{3+}, ca. 20 mol%) or acceptor (Er^{3+}, <5 mol%) lanthanide ions: $NaYF_4$: Yb,Tm and $NaYF_4$: Yb,Er.

UCNPs **41** (Fig. 18) provide another system containing the caspase-3 activatable FRET pair and photoactive Pt(IV) complexes, similar to the HSA conjugate **28**.[93] Due to the presence of UCNPs, Pt(IV) complexes can be activated by 980 nm NIR irradiation, and exert cytotoxicity toward A2780 ovarian cancer cells.

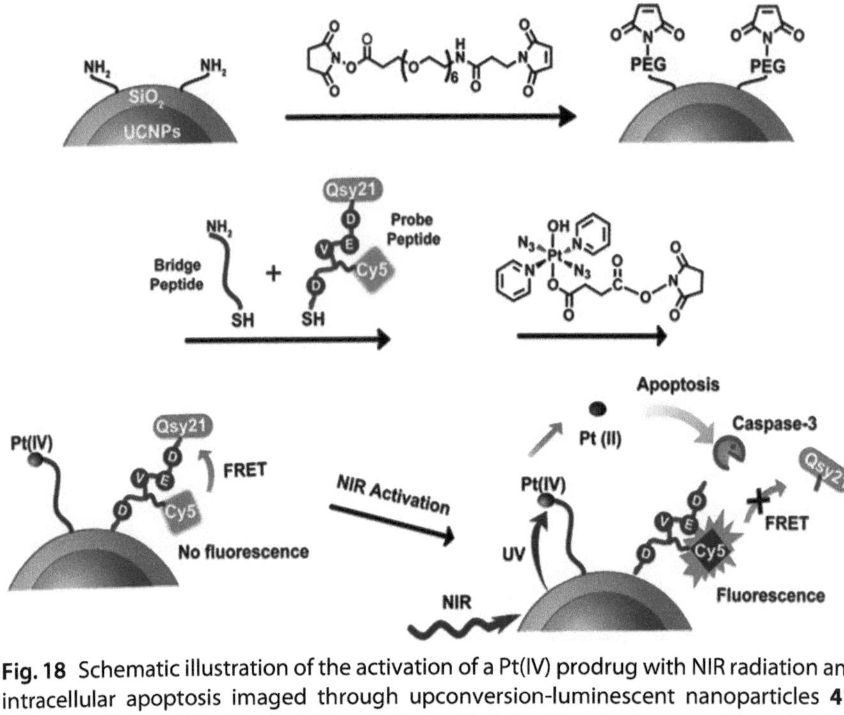

Fig. 18 Schematic illustration of the activation of a Pt(IV) prodrug with NIR radiation and intracellular apoptosis imaged through upconversion-luminescent nanoparticles **41**. *Reproduced form Min, Y.; Li, J.; Liu, F.; Yeow, E. K. L.; Xing, B.* Angew. Chem., Int. Ed. ***2014***, 53, *1012–1016 with permission.*

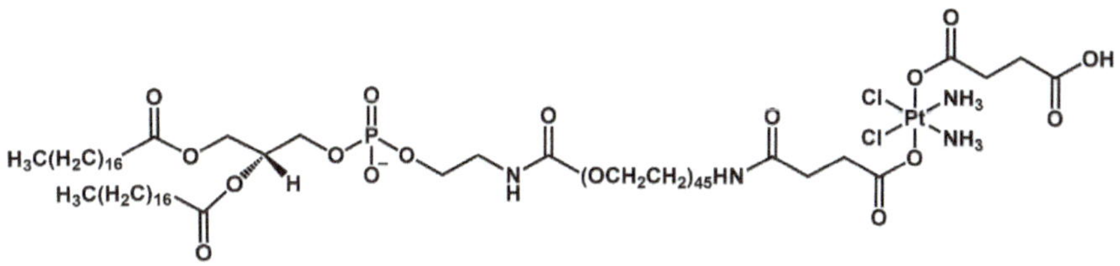

42

Fig. 19 Photoactive complex **42** containing long alkyl side chains which bind to hydrophobic coatings on UCNPs.

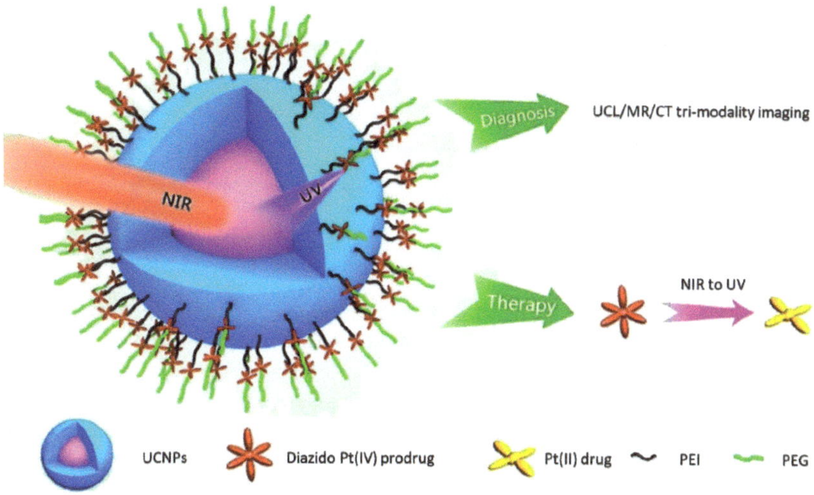

Fig. 20 Schematic illustration of the photodecomposition of conjugate **43**, Pt-Yb/Tm-codoped UCNPs system which delivers Pt(IV) prodrugs to cancer cells by endocytosis, and exhibits higher tumor inhibition with NIR irradiation. *Reproduced from Dai, Y.; Xiao, H.; Liu, J.; Yuan, Q.; Ma, P.; Yang, D.; Li, C.; Cheng, Z.; Hou, Z.; Yang, P.; Lin, J.* J. Am. Chem. Soc. ***2013***, 135, *18920–18929 with permission.*

PEGylated phospholipid-functionalized *cis*-dichorido *cis*-diamine Pt(IV) complex **42** (Fig. 19) alone decomposed with UVA irradiation. However when it decorated the surface of $NaYF_4$:Yb^{3+},Tm^{3+}@$NaYF_4$ core-shell UCNPs, irradiation with 980 nm NIR light decomposed it and released cytotoxic cisplatin.[94]

The Pt-Yb/Tm-codoped UCNPs system **43** (Fig. 20) delivers Pt(IV) prodrugs to cancer cells effectively by endocytosis (formation of intracellular membrane-bound vesicles at the cell surface) and exhibits higher tumor inhibition with NIR irradiation than direct UV irradiation.[95] In addition, this nanosystem is a theranostic agent, whose therapeutic action can be informed and guided by various imaging modalities, such as upconversion luminescence (UCL), magnetic resonance (MR) and computerized tomography (CT).

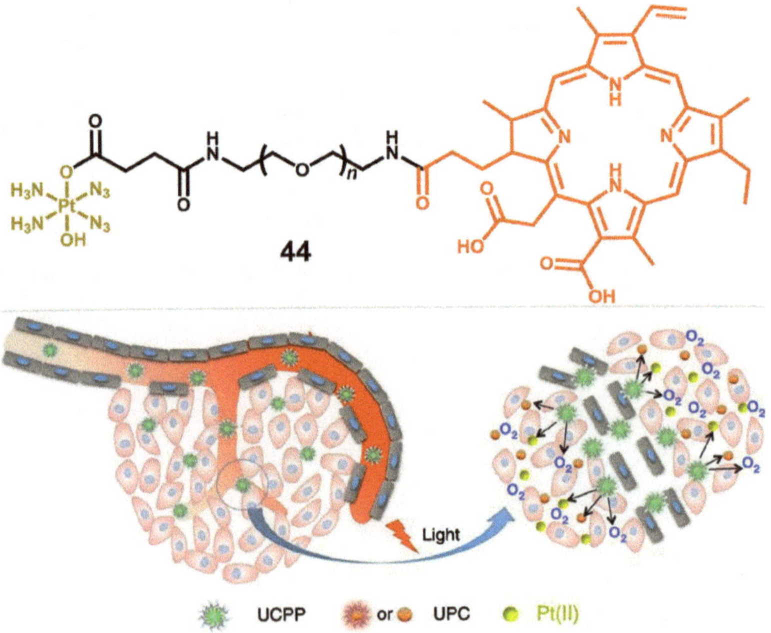

Fig. 21 The amphiphilic oligomer **44** and the photodecomposition of UCPP in a tumor microenvironment. The nanocomposite contains Pt(IV) conjugated to chlorin e6, and upconversion nanoparticles convert 980nm near-infrared light into 365 and 660 nm emissions, generating $O_2/^1O_2$ for PDT processes and active Pt(II) species. UCPP = $NaYbF_4$: Tm@CaF_2 co-assembled as nanoparticles with the Pt(IV) prodrug **44**. *Reproduced from Xu, S.; Zhu, X.; Zhang, C.; Huang, W.; Zhou, Y.; Yan, D.* Nat. Commun. ***2018****, 9, 2053 with permission.*

Amphiphilic oligomer **44** (Fig. 21) with a terminal diazido-Pt(IV) and conjugated chlorin chromophores, self-assembles into micelles, and co-assembles with $NaYbF_4$: Tm@CaF_2 UCNPs, forming nanoparticles which allow the photoactivation of **44** with NIR radiation.[96] Diazido Pt(IV) complexes can generate O_2 independently (via the formation of H_2O_2),[14] and this can be utilized by the photosensitizer Chlorin e6. With these two parts, **44** becomes an O_2-self generating PACT-PDT agent, which exhibits promising photocytotoxicity even in hypoxic environments.

6.8 Summary and prospects

The remarkable success of Pt(II) anticancer drugs, especially cisplatin, carboplatin and oxaliplatin, currently in widespread clinical use for the treatment of solid tumors, has led to interest in the design of new platinum agents which might expand the range of treatable cancers with fewer off-target side effects and with novel mechanisms of action to combat resistance to current platinum drugs. We have described one approach in our laboratory to providing such new agents, the use of inert Pt(IV) prodrugs which are relatively non-toxic in the dark, but can be activated selectively in tumor tissue by the use of spatially-directed light, photo-activated chemotherapy (PACT).

Diazido Pt(IV) complexes are promising PACT agents on account of their intense ligand (azide)-to-metal (Pt^{IV}) charge-transfer bands, their high stability and inertness in the dark, and formation of DNA lesions different from cisplatin. Moreover, in contrast to cisplatin and its analogues, *trans*-diamine *trans*-diazido photoactivatable Pt(IV) complexes have more favorable properties than their *cis*-diamine isomers. Axial hydroxide ligands are important for inertness, e.g., toward abundant bioreductants such as glutathione, and can be derivatized so as to introduce vectors for targeting cellular receptors.

Although they are efficiently photoactivated by UVA, blue or green light, activation by longer wavelength light that penetrates more deeply into tissues can be achieved, for example by appropriate choice of large conjugated ligands or by attachment to upconverting nanoparticles. In addition, a variety of other formulation strategies have been reported, including formation of hydrogels, incorporation of protein inhibitors, conjugation to proteins (including antibodies) and to polymers. Such materials introduce potential for multimodal and multitargeting mechanisms of action, and combating resistance.

Photodecomposition pathways not only include the generation of Pt(II) photoproducts which readily bind to DNA, but also azidyl radicals which can extract an electron from the indole ring of the amino acid tryptophan giving Trp radicals, and hydroxyl radicals. In addition, nitrenes and singlet oxygen can be formed under some conditions.

An important next step is to investigate the contributions of these photochemical pathways to the mechanism of anticancer activity in cells. Cells contain a range of compartments and organelles, and are inherently dynamic non-equilibrium systems undergoing constant transport, synthesis and decomposition processes.

Photoactivatable octahedral Pt(IV) offers a wide and versatile platform for design of novel prodrugs and offer promise for treatment of cancer, and perhaps also infections, as agents with low side-effects and novel mechanisms of action which combat resistance.

Acknowledgments

We acknowledge financial support for our recent research on photoactive metal complexes from the EPSRC (EP/G006792, EP/F034210/1), Anglo American Platinum and University of Warwick. We are also grateful for stimulating discussions on this topic with our research group, collaborators, and colleagues.

References

1. Shi, H.; Sadler, P. J. *Br. J. Cancer* **2020**, *123*, 871–873.
2. Torres, A. E.; Lyons, A. B.; Hamzavi, I. H.; Lim, H. W. *J. Am. Acad. Dermatol.* **2021**, *84*, 479–485.
3. Pathak, M. A.; Fitzpatrick, T. B. *J. Photochem. Photobiol. B Biol.* **1992**, *14*, 3–22.
4. Hamblin, M. R. *Photochem. Photobiol.* **2020**, *96*, 506–516.
5. Shi, X.; Zhang, C. Y.; Gao, J.; Wang, Z. *Wiley Interdiscip. Rev. Nanomed. Nanobiotechnol.* **2019**, *11*, e1560.
6. Abdel-Kader, M. H., Ed. *Photodynamic Therapy From Theory to Application*; Springer-Verlag: Berlin Heidelberg, 2014.
7. Correia, J. H.; Rodrigues, J. A.; Pimenta, S.; Dong, T.; Yang, Z.; Wojcik, M.; Preis, E.; Litscher, G. *Pharmaceutics* **2021**, *13*, 1332.
8. Karges, J. *Angew. Chem. Int. Ed.* **2022**, *61*, e202112236.
9. DeRosa, M. C.; Crutchley, R. J. *Coord. Chem. Rev.* **2002**, *233–234*, 351–371.
10. Sandell, J. L.; Zhu, T. C. *J. Biophotonics* **2011**, *4*, 773–787.
11. Ruggiero, E.; Alonso-de Castro, S.; Habtemariam, A.; Salassa, L. *Dalton Trans.* **2016**, *45*, 13012–13020.
12. Monro, S.; Colón, K. L.; Yin, H.; Roque, J.; Konda, P.; Gujar, S.; Thummel, R. P.; Lilge, L.; Cameron, C. G.; McFarland, S. A. *Chem. Rev.* **2019**, *119*, 797–828.
13. Shi, H.; Sadler, P. J. In *Comprehensive Inorganic Chemistry III: CH 00037*; Reedijk, J., Poeppelmeier, K., Eds.; Elsevier, 2022. https://doi.org/10.1016/B978-0-12-823144-9.00037-6.
14. Shi, H.; Imberti, C.; Sadler, P. J. *Inorg. Chem. Front.* **2019**, *6*, 1623–1638.
15. Rosenberg, B.; Vancamp, L.; Krigas, T. *Nature* **1965**, *205*, 698–699.
16. Rosenberg, B.; Renshaw, E.; Vancamp, L.; Hartwick, J.; Drobnik, J. *J. Bacteriol.* **1967**, *93*, 716–721.
17. Rosenberg, B.; Vancamp, L.; Trosko, J. E.; Mansour, V. H. *Nature* **1969**, *222*, 385–386.
18. Aldossary, S. A. *Biomed. Pharmacol. J.* **2019**, *12*, 7–15.
19. Kelland, L. *Nat. Rev. Cancer* **2007**, 7, 573–584.
20. Neidle, S.; Ismail, I. M.; Sadler, P. J. *J. Inorg. Biochem.* **1980**, *13*, 205–212.
21. Berners-Price, S. J.; Sadler, P. J. *Coord. Chem. Rev.* **1996**, *151*, 1–40.
22. Barnham, K. J.; Djuran, M. I.; Murdoch, P. S.; Ranfordb, J. D.; Sadler, P. J. *J. Chem. Soc. Dalton Trans.* **1995**, *372*, 3721–3726.
23. Wang, X.; Wang, X.; Jin, S.; Muhammad, N.; Guo, Z. *Chem. Rev.* **2019**, *119*, 1138–1192.
24. Deo, K. M.; Ang, D. L.; McGhie, B.; Rajamanickam, A.; Dhiman, A.; Khoury, A.; Holland, J.; Bjelosevic, A.; Pages, B.; Gordon, C.; Aldrich-Wright, J. R. *Coord. Chem. Rev.* **2018**, *375*, 148–163.
25. Mitra, K. *Dalton Trans.* **2016**, *45*, 19157–19171.
26. Gurruchaga-Pereda, J.; Martínez, Á.; Terenzi, A.; Salassa, L. *Inorg. Chim. Acta* **2019**, *495*, 118981.
27. Mu, M.; Zhan, J.; Dai, X.; Gao, H. *Eur. J. Med. Chem.* **2022**, *227*, 113927.
28. Rosenberg, B.; Vancamp, L.; Grimley, E. B.; Thomson, A. J. *J. Biol. Chem.* **1967**, *242*, 1347–1352.

29. Kratochwil, N. A.; Zabel, M.; Range, K. J.; Bednarski, P. J. *J. Med. Chem.* **1996**, *39*, 2499–2507.
30. Kratochwil, N. A.; Bednarski, P. J.; Mrozek, H.; Vogler, A. *Anti-Cancer Drug Des.* **1996**, *11*, 155–171.
31. Kratochwil, N. A.; Guo, Z.; Murdoch, P. S.; Parkinson, J. A.; Bednarski, P. J.; Sadler, P. J. *J. Am. Chem. Soc.* **1998**, *120*, 8253–8254.
32. Kratochwil, N. A.; Parkinson, J. A.; Bednarski, P. J.; Sadler, P. J. *Angew. Chem., Int. Ed.* **1999**, *38*, 1460–1463.
33. Kratochwil, N. A.; Bednarski, P. J. *Arch. Pharm. Pharm. Med. Chem.* **1999**, *332*, 279–285.
34. Philippe, M.; Schröder, B.; Parkinson, J. A.; Kratochwil, N. A.; Coxall, R. A.; Parkin, A.; Parsons, S.; Sadler, P. J. *Angew. Chem., Int. Ed.* **2003**, *42*, 335–339.
35. Bednarski, P. J.; Grünert, R.; Zielzki, M.; Wellner, A.; Mackay, F. S.; Sadler, P. J. *Chem. Biol.* **2006**, *13*, 61–67.
36. Salassa, L.; Phillips, H. I. A.; Sadler, P. J. *Phys. Chem. Chem. Phys.* **2009**, *11*, 10311–10316.
37. Ronconi, L.; Sadler, P. J. *Dalton Trans.* **2011**, *40*, 262–268.
38. Ronconi, L.; Sadler, P. J. *Chem. Commun.* **2008**, 235–237.
39. Ronconi, L.; Pizarro, A. M.; McQuitty, R. J.; Sadler, P. J. *Chem. Eur. J.* **2011**, *17*, 12051–12058.
40. Mackay, F. S.; Woods, J. A.; Moseley, H.; Ferguson, J.; Dawson, A.; Parsons, S.; Sadler, P. J. *Chem. Eur. J.* **2006**, *12*, 3155–3161.
41. Mackay, F. S.; Woods, J. A.; Heringova, P.; Kašpárková, J.; Pizarro, A. M.; Moggach, S. A.; Parsons, S.; Brabec, V.; Sadler, P. J. *Proc. Natl. Acad. Sci. U. S. A.* **2007**, *104*, 20743–20748.
42. Westendorf, A. F.; Woods, J. A.; Korpis, K.; Farrer, N. J.; Salassa, L.; Robinson, K.; Appleyard, V.; Murray, K.; Grünert, R.; Thompson, A. M.; Sadler, P. J.; Bednarski, P. J. *Mol. Cancer Ther.* **2012**, *11*, 1894–1904.
43. Farrer, N. J.; Woods, J. A.; Munk, V. P.; MacKay, F. S.; Sadler, P. J. *Chem. Res. Toxicol.* **2010**, *23*, 413–421.
44. Tai, H. C.; Zhao, Y.; Farrer, N. J.; Anastasi, A. E.; Clarkson, G.; Sadler, P. J.; Deeth, R. J. *Chem. Eur. J.* **2012**, *18*, 10630–10642.
45. Farrer, N. J.; Woods, J. A.; Salassa, L.; Zhao, Y.; Robinson, K. S.; Clarkson, G.; MacKay, F. S.; Sadler, P. J. *Angew. Chem., Int. Ed.* **2010**, *49*, 8905–8908.
46. https://en.wikipedia.org/wiki/Silver_azide.
47. Farrer, N. J.; Gierth, P.; Sadler, P. J. *Chem. Eur. J.* **2011**, *17*, 12059–12066.
48. Shaili, E.; Salassa, L.; Woods, J. A.; Clarkson, G.; Sadler, P. J.; Farrer, N. J. *Chem. Sci.* **2019**, *10*, 8610–8617.
49. Butler, J. S.; Woods, J. A.; Farrer, N. J.; Newton, M. E.; Sadler, P. J. *J. Am. Chem. Soc.* **2012**, *134*, 16508–16511.
50. Shi, H.; Imberti, C.; Clarkson, G. J.; Sadler, P. J. *Inorg. Chem. Front.* **2020**, 7, 3533–3540.
51. Vallotto, C.; Shaili, E.; Shi, H.; Butler, J. S.; Wedge, C. J.; Newton, M. E.; Sadler, P. J. *Chem. Commun.* **2018**, *54*, 13845–13848.
52. Zhao, Y.; Farrer, N. J.; Li, H.; Butler, J. S.; McQuitty, R. J.; Habtemariam, A.; Wang, F.; Sadler, P. J. *Angew. Chem., Int. Ed.* **2013**, *52*, 13633–13637.
53. Bancroft, D. P.; Lepre, C. A.; Lippard, S. J. *J. Am. Chem. Soc.* **1990**, *112*, 6860–6871.
54. Crul, M.; van Waardenburg, R. C. A. M.; Beijnen, J. H.; Schellens, J. H. M. *Cancer Treat. Rev.* **2002**, *28*, 291–303.
55. Kašpárková, J.; Mackay, F. S.; Brabec, V.; Sadler, P. J. *J. Biol. Inorg. Chem.* **2003**, *8*, 741–745.
56. Zhao, Y.; Woods, J. A.; Farrer, N. J.; Robinson, K. S.; Pracharova, J.; Kasparkova, J.; Novakova, O.; Li, H.; Salassa, L.; Pizarro, A. M.; Clarkson, G. J.; Song, L.; Brabec, V.; Sadler, P. J. *Chem. Eur. J.* **2013**, *19*, 9578–9591.

57. Pracharova, J.; Zerzankova, L.; Stepankova, J.; Novakova, O.; Farrer, N. J.; Sadler, P. J.; Brabec, V.; Kasparkova, J. *Chem. Res. Toxicol.* **2012**, *25*, 1099–1111.
58. Tai, H. C.; Brodbeck, R.; Kasparkova, J.; Farrer, N. J.; Brabec, V.; Sadler, P. J.; Deeth, R. J. *Inorg. Chem.* **2012**, *51*, 6830–6841.
59. Wootton, C. A.; Sanchez-Cano, C.; Lopez-Clavijo, A. F.; Shaili, E.; Barrow, M. P.; Sadler, P. J.; O'Connor, P. B. *Chem. Sci.* **2018**, *9*, 2733–2739.
60. Du, J.; Wei, Y.; Zhao, Y.; Xu, F.; Wang, Y.; Zheng, W.; Luo, Q.; Wang, M.; Wang, F. *Inorg. Chem.* **2018**, *57*, 5575–5584.
61. Novohradsky, V.; Pracharova, J.; Kasparkova, J.; Imberti, C.; Bridgewater, H. E.; Sadler, P. J.; Brabec, V. *Inorg. Chem. Front.* **2020**, 7, 4150–4159.
62. Desgrosellier, J. S.; Cheresh, D. A. *Nat. Rev. Cancer* **2010**, *10*, 9–22.
63. Gandioso, A.; Shaili, E.; Massaguer, A.; Artigas, G.; González-Cantó, A.; Woods, J. A.; Sadler, P. J.; Marchán, V. *Chem. Commun.* **2015**, *51*, 9169–9172.
64. Shi, H.; Wang, Q.; Venkatesh, V.; Feng, G.; Young, L. S.; Romero-Canelón, I.; Zeng, M.; Sadler, P. J. *Dalton Trans.* **2019**, *48*, 8560–8564.
65. Shaili, E.; Fernández-Giménez, M.; Rodríguez-Astor, S.; Gandioso, A.; Sandín, L.; García-Vélez, C.; Massaguer, A.; Clarkson, G. J.; Woods, J. A.; Sadler, P. J.; Marchán, V. *Chem. Eur. J.* **2015**, *21*, 18474–18486.
66. Gandioso, A.; Rovira, A.; Shi, H.; Sadler, P. J.; Marchán, V. *Dalton Trans.* **2020**, *49*, 11828–11834.
67. Shi, H.; Romero-Canelón, I.; Hreusova, M.; Novakova, O.; Venkatesh, V.; Habtemariam, A.; Clarkson, G. J.; Song, J.-I.; Brabec, V.; Sadler, P. J. *Inorg. Chem.* **2018**, *57*, 14409–14420.
68. Venkatesh, V.; Wedge, C. J.; Romero-Canelón, I.; Habtemariam, A.; Sadler, P. J. *Dalton Trans.* **2016**, *45*, 13034–13037.
69. Kasparkova, J.; Kostrhunova, H.; Novakova, O.; Křikavová, R.; Vančo, J.; Trávníček, Z.; Brabec, V. *Angew. Chem., Int. Ed.* **2015**, *54*, 14478–14482.
70. Shi, H.; Kasparkova, J.; Soulié, C.; Clarkson, G. J.; Imberti, C.; Novakova, O.; Paterson, M. J.; Brabec, V.; Sadler, P. J. *Chem. Eur. J.* **2021**, *27*, 10711–10716.
71. Bolitho, E. M.; Sanchez-Cano, C.; Shi, H.; Quinn, P. D.; Harkiolaki, M.; Imberti, C.; Sadler, P. J. *J. Am. Chem. Soc.* **2021**, *143*, 20224–20240.
72. Hein, C. D.; Liu, X. M.; Wang, D. *Pharm. Res.* **2008**, *25*, 2216–2230.
73. Farrer, N. J.; Griffith, D. M. *Curr. Opin. Chem. Biol.* **2020**, *55*, 59–68.
74. Farrer, N. J.; Sharma, G.; Sayers, R.; Shaili, E.; Sadler, P. J. *Dalton Trans.* **2018**, *47*, 10553–10560.
75. Yao, K.; Bertran, A.; Morgan, J.; Hare, S. M.; Rees, N. H.; Kenwright, A. M.; Edkins, K.; Bowen, A. M.; Farrer, N. J. *Dalton Trans.* **2019**, *48*, 6416–6420.
76. Yao, K.; Bertran, A.; Howarth, A.; Goicoechea, J. M.; Hare, S. M.; Rees, N. H.; Foroozandeh, M.; Bowen, A. M.; Farrer, N. J. *Chem. Commun.* **2019**, *55*, 11287–11290.
77. Xiao, H.; Yan, L.; Dempsey, E. M.; Song, W.; Qi, R.; Li, W.; Huang, Y.; Jing, X.; Zhou, D.; Ding, J.; Chen, X. *Prog. Polym. Sci.* **2018**, *87*, 70–106.
78. Imberti, C.; Zhang, P.; Huang, H.; Sadler, P. J. *Angew. Chem., Int. Ed.* **2020**, *59*, 61–73.
79. Li, X.; Mu, J.; Liu, F.; Tan, E. W. P.; Khezri, B.; Webster, R. D.; Yeow, E. K. L.; Xing, B. *Bioconjugate Chem.* **2015**, *26*, 955–961.
80. Shi, H.; Imberti, C.; Huang, H.; Hands-Portman, I.; Sadler, P. J. *Chem. Commun.* **2020**, *56*, 2320–2323.
81. Imberti, C.; Lermyte, F.; Friar, E. P.; O'Connor, P. B.; Sadler, P. J. *Chem. Commun.* **2021**, *57*, 7645–7648.
82. He, S.; Qi, Y.; Kuang, G.; Zhou, D.; Li, J.; Xie, Z.; Chen, X.; Jing, X.; Huang, Y. *Biomacromolecules* **2016**, *17*, 2120–2127.
83. Venkatesh, V.; Mishra, N. K.; Romero-Canelón, I.; Vernooij, R. R.; Shi, H.; Coverdale, J. P. C.; Habtemariam, A.; Verma, S.; Sadler, P. J. *J. Am. Chem. Soc.* **2017**, *139*, 5656–5659.

84. Xiao, H.; Noble, G. T.; Stefanick, J. F.; Qi, R.; Kiziltepe, T.; Jing, X.; Bilgicer, B. *J. Controlled Release* **2014**, *173*, 11–17.
85. Song, H.; Li, W.; Qi, R.; Yan, L.; Jing, X.; Zheng, M.; Xiao, H. *Chem. Commun.* **2015**, *51*, 11493–11495.
86. Song, H.; Kang, X.; Sun, J.; Jing, X.; Wang, Z.; Yan, L.; Qi, R.; Zheng, M. *Chem. Commun.* **2016**, *52*, 2281–2283.
87. Sun, W.; Li, S.; Häupler, B.; Liu, J.; Jin, S.; Steffen, W.; Schubert, U. S.; Butt, H. J.; Liang, X. J.; Wu, S. *Adv. Mater.* **2017**, *29*, 1603702.
88. Zhou, D.; Guo, J.; Kim, G. B.; Li, J.; Chen, X.; Yang, J.; Huang, Y. *Adv. Healthcare Mater.* **2016**, *5*, 2493–2499.
89. Zhang, Q.; Kuang, G.; He, S.; Lu, H.; Cheng, Y.; Zhou, D.; Huang, Y. *Nano Lett.* **2020**, *20*, 3039–3049.
90. He, S.; Li, C.; Zhang, Q.; Ding, J.; Liang, X. J.; Chen, X.; Xiao, H.; Chen, X.; Zhou, D.; Huang, Y. *ACS Nano* **2018**, *12*, 7272–7281.
91. Yang, X. D.; Xiang, H. J.; An, L.; Yang, S. P.; Liu, J. G. *New J. Chem.* **2015**, *39*, 800–804.
92. Szaciłowski, K.; Macyk, W.; Drzewiecka-Matuszek, A.; Brindell, M.; Stochel, G. *Chem. Rev.* **2005**, *105*, 2647–2694.
93. Min, Y.; Li, J.; Liu, F.; Yeow, E. K. L.; Xing, B. *Angew. Chem., Int. Ed.* **2014**, *53*, 1012–1016.
94. Ruggiero, E.; Hernández-Gil, J.; Mareque-Rivas, J. C.; Salassa, L. *Chem. Commun.* **2015**, *51*, 2091–2094.
95. Dai, Y.; Xiao, H.; Liu, J.; Yuan, Q.; Ma, P.; Yang, D.; Li, C.; Cheng, Z.; Hou, Z.; Yang, P.; Lin, J. *J. Am. Chem. Soc.* **2013**, *135*, 18920–18929.
96. Xu, S.; Zhu, X.; Zhang, C.; Huang, W.; Zhou, Y.; Yan, D. *Nat. Commun.* **2018**, *9*, 2053.

CHAPTER FOUR

Adventures in the photo-uncaging of small molecule bioregulators

Peter C. Ford*
Department of Chemistry and Biochemistry, University of California, Santa Barbara, CA, United States
*Corresponding author: e-mail address: pcford@ucsb.edu

Contents

1. Introduction 131
2. PhotoNORMs and PhotoCORMs 133
   2.1 Uncaging nitric oxide 133
   2.2 Uncaging carbon monoxide 139
3. Enhancing photoreaction SMB release rates with antennas 143
   3.1 Organic chromophore antennas 143
   3.2 Semiconductor quantum dot antennas 146
4. Moving toward the phototherapeutic window 148
   4.1 Single photon excitation strategies 149
   4.2 Simultaneous two photon excitation 150
   4.3 Upconverting nanoparticles 153
   4.4 Nano-carriers 157
5. Biological targeting 159
   5.1 Liposomes 159
   5.2 Hollow gold nanoshells 160
   5.3 Macrophage Trojan Horses 161
6. Summary 163
Acknowledgments 164
References 164

Abstract

The spatial-temporal controlled delivery of bioactive substances at physiological sites is a highly desirable for therapeutic and diagnostic applications. Photochemical release (photo-uncaging) of the bioactive compound from a biologically benign precursor offers exquisite control of timing, location and dosage. Two challenging issues need to be addressed in order to facilitate this methodology. The first is the poor transmission of visible and UV light through mammalian tissue; the second is the need to deliver the appropriate precursors to the targeted sites. In this chapter we discuss research in our laboratory concerned with these two issues for the photo-uncaging of the small molecule bioregulators nitric oxide and carbon monoxide.

Advances in Inorganic Chemistry, Volume 80
ISSN 0898-8838
https://doi.org/10.1016/bs.adioch.2022.04.002

Abbreviations

bpy	2,2'-bipyridine
CORM	carbon monoxide releasing moiety
CrONO	trans-Cr(cyclam)$(ONO)_2^+$
CW	continuous wave
cyclam	1,4,8,11 tetraazacyclotetradecane
DAF-2	4,5-diaminofluorescein
DAPTA	1,4-diacetyl-1,3,7-triaza-5-phosphabicylco[3.3.1]nonane
DFT	density functional theory
DSPE-PEG	phospholipid-functionalized poly(ethylene glycol)
ES	excited state
Fluor	fluorescein
FRET	Förster resonance energy transfer
GM	Goeppert–Mayer unit for two photon absorption
HGN	hollow gold nanosphere
I_0	incident light intensity
I_{abs}	intensity of light absorbed
LF	ligand field (excited state)
Mb	myoglobin
MLCT	metal to ligand charge transfer (excited state)
NIR	near infrared
NOA	Sievers nitric oxide analyzer
NP	nanoparticle
PDMS	polydimethylsiloxane
PEG	polyethylene glycol
PetA	5,12-dimethyl-7,14-diphenyl-1,4,8,11-tetra-aza-cyclotetradecane
phen	1,10-phenanthroline
photoCORM	photo-activated CO releasing moiety
photoNORM	photo-activated NO releasing moiety
PL	photoluminescence
PLGA	poly(D,L-lactic-co-glycolic) acid
Por	porphyrin
PPIX	protoporphyrin-IX
PTA	1,3,5-triaza-7-phosphaadamantane
py	pyridine
QD	quantum dot
R_i	photochemical rate
RNS	reactive nitrogen species
RBS	Roussin's black salt
RRS	Roussin's red salt
RSE	Roussin's red ester
salen	*N,N'*-ethylenebis(salicylideneiminato)dianion
salophen	*N,N'*-1,2-phenylenebis(salicylideneiminato)dianion
SMB	small molecule bioregulator
SPE	single photon excitation
TCF	thiol functionalized derivative of cupferron
THP	tris(hydroxymethyl)phosphine

TPYOH	4'-p-*N,N*-bis(2-hydroxyethyl)benzyl-2,2':6',2"-terpyridine
TPE	two-photon excitation
TPEG	thiolated polyethylene glycol
TPPTS	tris(sulfonatophenyl)phosphine trianion
UCNP	upconverting nanoparticle
UV	ultraviolet
Φ	quantum yield
λ_{irr}	irradiation wavelength
β	two-photon absorption cross section

1. Introduction

This laboratory's long-term research into the bioinorganic chemistry of nitric oxide (NO) was first stimulated by a 1990 telephone call from former graduate student Dr. David A. Wink. David had recently been appointed to a research position in the Laboratory of Comparative Carcinogenesis at the National Cancer Institute in Frederick, MD, where he was working with Dr. Larry Keefer on compounds that release NO thermally under physiological conditions.[1] Discussions with David, and subsequently with David and Larry, regarding the exciting, recent discoveries of the important roles of NO in mammalian cardiovascular, immune response and neurological biology,[2,3] inspired our laboratory to initiate (still continuing) investigations into the fundamental chemistry of NO with biologically relevant metal centers.[4–11] Photochemistry was always a key component, since many of our studies utilized flash photolysis methodologies to probe reaction dynamics and mechanisms.[12–14] In addition, we initiated a parallel, complementary effort to develop compounds and strategies for the photochemical delivery of NO to physiological targets. These and related studies are the subject of the current chapter. While we will describe some other relevant research in this area, this article will not be a comprehensive review.

The unexpected discoveries that NO has important signaling pathways in mammalian biology also triggered interest in the chemical biology of other reactive nitrogen species (RNS) such as nitrite (NO_2^-)[15] and nitroxyl (HNO).[16] More recently discovered small molecule bioregulators (SMBs) such as hydrogen sulfide (H_2S) and carbon monoxide (CO) are often grouped with NO as "gasotransmitters."[17] For example, although it has long been known that CO is one product of heme degradation by heme oxygenase,[18] the discovery of CO's potentially therapeutic roles is a much more recent discovery.[19] There is a growing understanding that there is a vast scope of biological mechanisms and potential therapies in which such SMBs participate.

For investigative and therapeutic applications, photochemistry can provide precise spatial-temporal control of SMB delivery. When a SMB precursor is benign, the gasotransmitter is "caged." Release of the bioactive form can be termed "photo-uncaging." The use of light as the uncaging signal determines the *location* and *timing* of release, and in principle controls the *dosage*. In addition, if the precursor itself and/or a photoproduct formed after uncaging is luminescent, this property can be used to image how the caged compound is distributed and/or where the bioactive materials are uncaged.

When one envisions materials to be used to photo-uncage a SMB or other bioactive molecule in an organism, there are key questions that should be addressed.

(a) What is the target? A subcutaneous site would be less challenging than deep tissue site such as a tumor. The poor transmission of UV and higher energy visible light through mammalian tissue requires excitation with deep red or near-infrared (NIR) wavelengths.

(b) How biocompatible are the photochemical precursor and the photoproduct residue after uncaging? Do either display acute or long-term toxicities?

(c) How can such precursors be delivered to the desired physiological target

Another issue of importance is the rate (R_i) of photochemical release. This is defined in Eq. (1) in terms of the uncaging quantum yield (Φ_i) and the intensity of light *absorbed* by the precursor (I_{abs}) in the units of Einsteins/unit time/unit volume. Both Φ_I and I_{abs} are generally dependent on the irradiation wavelength (λ_{irr}), while the latter is also a function of the excitation light intensity I_0 at that wavelength and the concentration and absorbance of precursor. Integration of R_i over the excitation time determines the total amount SMB delivered, while the competition between R_i and depletion mechanisms such as reaction or diffusion away from the target defines the steady state concentrations.

$$R_i = \Phi_i \, x \, I_{abs} \quad (1)$$

where according to Beers law:

$$I_{\mathrm{abs}} = I_0 \left(1 - 10^{-\mathrm{Abs}(\lambda)}\right) \quad (2)$$

As discussed later in this chapter, in some cases multiphoton excitation may prove to be the more effective approach to uncage a SMB rather than the more typical single photon excitation (SPE) methods for which Eq. (1) is relevant.[20] Such multiphoton excitation may occur either simultaneously or sequentially. With simultaneous two-photon excitation (TPE), the excited state initially formed equals the summed energy of the two photons. Often the origin of both photons is the intense laser pulse from an ultra-fast laser, but two-color experiments are also known. The selection rules for TPE are different than those for SPE. For example, the rate of excited state (ES) formation via TPE is proportional to the square of the incident intensity (I_0^2) rather being the linearly dependent on I_0 as indicated by Eq. (2). A focused beam will have the highest intensity at the focal point, so multiphoton processes may provide 3D spatial resolution in the excitation profile.

Sequential multi-electron excitation involves first populating one excited state by SPE, then a second, higher energy, state by photons of the same or different frequencies. This is how lanthanide ion doped upconverting nanoparticles (UCNPs) generate ultraviolet (UV) or visible photoluminescence (PL) upon NIR excitation function.[21] The rates by which the higher energy ES responsible for these emissions are formed are typically non-linear with respect to I_0, with the exponent n of the I_0^n relationship often between 1 and 2. Examples will be discussed in greater detail below.

2. PhotoNORMs and PhotoCORMs

Studies of the potentially therapeutic roles of carbon monoxide has led to the development and study of various metal complex and organic precursors that can release CO under biological conditions. These CO releasing moieties were termed CORMs at an early stage of these studies.[19a] Subsequently, the term photoCORM was coined for photo-activated CO releasing moieties.[22] For consistency, we will use the term photoNORM for photo-activated NO releasing moieties.

2.1 Uncaging nitric oxide

Our early adventures in the photo-uncaging focused on several types of metal nitrosyl complexes including ruthenium porphyrins Ru(Por)(X)(NO) ($X=Cl^-$ or ONO^-; Por=a porphyrinato dianion),[23] the Roussin's red and black salts (RRS and RBS, respectively)[24] and sodium nitroprusside[25] (Fig. 1). Metallo-porphyrins were of interest because of their strong absorptions through the visible region as well as the analogy to the heme protein targets of NO in mammalian signaling mechanisms. Roussin's salt anions had qualitatively been shown by Mathews et al. to trigger NO signaling pathways under photoexcitation[26] and nitroprusside was used in the clinic as a vasodilator during hypertensive events.[27] A notable study from our

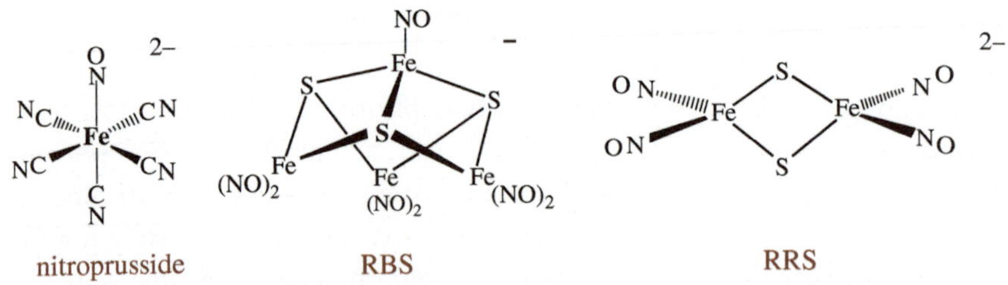

Fig. 1 Formulas for photoNORMs: nitroprusside, Roussin's black salt anion (RBS) and Roussin's red salt anion (RRS) anions.

laboratory was in collaboration with James Mitchell and David Wink of the Radiation Biology Branch of the National Cancer Institute, where it was shown that NO released by white light irradiation of RRS that had been infused into hypoxic Chinese hamster V79 lung cells strongly sensitized gamma-radiation killing of those cells.[24]

Flash photolysis studies showed the photo-labilization of NO from both RRS and RBS to be reversible, especially in the presence of excess NO (Scheme 1).[28,29] As a consequence, oxygen trapping of the iron sulfur

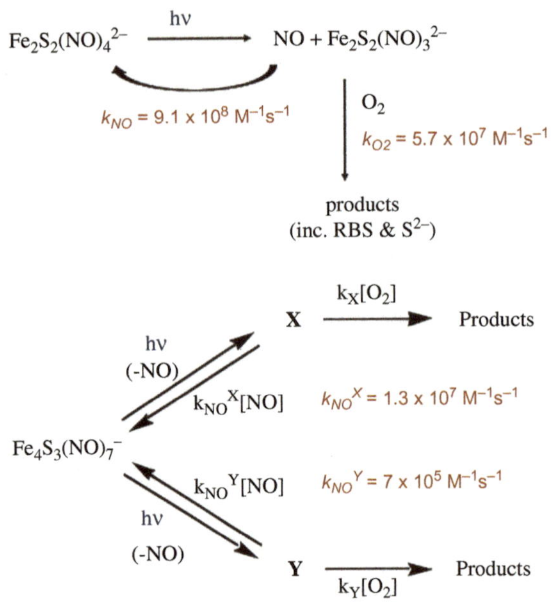

Scheme 1 Flash photolysis kinetics of RRS and RBS in ambient temperature aqueous solutions. Intermediates X and Y formed upon photolysis of RBS are thought to result from photodissociation of NO from the $Fe(NO)_2$ and the apical Fe(NO) units, respectively (Refs. 28, 29).

nitrosyl intermediates formed, proved necessary to ensure the net release of NO at least in solution. In the case of RBS, as many as 6 NO's are photo-uncaged in aerobic media (Eq. 3).

$$Fe_4S_3(NO)_7^- \xrightarrow[O_2]{h\nu} 3.9Fe^{2+} + 5.9NO + 3S^{2-} \underset{+\text{other products}}{} \quad (3)$$

Quantifying photochemical NO release is challenging, especially *in vivo*. In our earlier solution phase studies, NO release was measured and quantified using NO specific electrodes.[25] This method is very sensitive and has been used very successfully by others;[30] however, in our hands, reproducibility proved to be challenging. Our later studies mostly quantified nitric oxide release by bubbling a gas (usually He, but sometimes medical grade air) through the photolysis system and measuring the uncaged NO entrained with a Sievers Nitric Oxide Analyzer (NOA). (Fig. 2).[31] This proved to be a very sensitive, reliably quantitative and reproducible analytical method for photoNORM reactions in solution. Several detection methods are available for NO release within *in vitro* cell cultures, an example being the reaction of 4,5-diaminofluorescein (DAF-2) to give a fluorescent triazole.[32]

As will be seen in subsequent sections, the Roussin's iron/sulfur/nitrosyl clusters have proved valuable photoNORMs in several different configurations. A key modification of the RRS motif is reflected in the Roussin's red

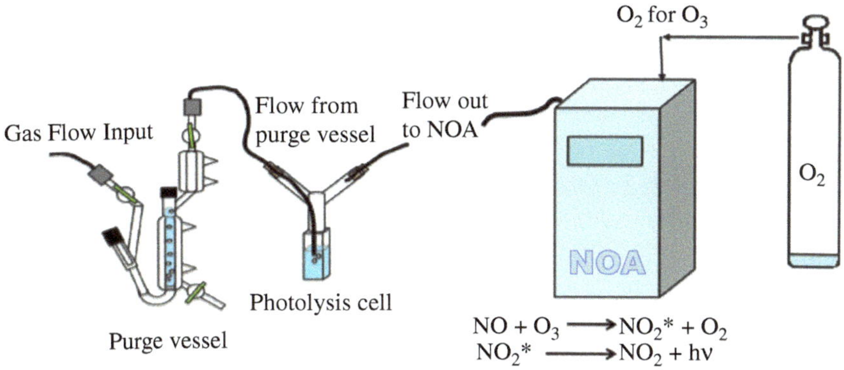

Fig. 2 Apparatus for measuring NO release quantum yields in solution where the entraining gas passes first through a solvent purge vessel to reduce evaporative losses then through the photolysis "Y" cell. NO from the photolysis cell is transferred to the Sievers Nitric Oxide Analyzer (NOA, GE model NOA-280i), where it is detected via a very sensitive, calibrated, chemiluminescence technique. *From the Supporting information of Huang, P.-J.; Garcia, J.V.; Fenwick, A.; Wu, G.; Ford, P.C.* ACS Omega ***2019***, 4, *9181–9187, Copyright 2019 American Chemical Society.*

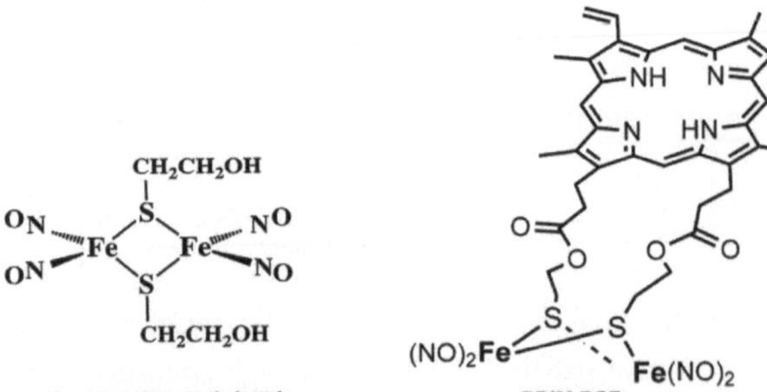

$Fe(\mu\text{-}SCH_2CH_2OH)_2(NO)_4$ PPIX-RSE

Fig. 3 Representative Roussin's red salt esters $Fe_2(\mu\text{-}SCH_2CH_2OH)_2(NO)_4$ (Ref. 33) and PPIX-RSE (Ref. 34), where PPIX is the bis(2-thiolatoethyl) diester of protoporphyrin IX.

salt esters $Fe_2(\mu\text{-}SR)_2(NO)_4$ (RSEs) for which two examples are displayed in Fig. 3.[33,34] These are prepared by derivatizing the bridging sulfides of the red salt. Doing so provides opportunities for tuning the physical and chemical properties using different RS- derivatives. The esters also offer better thermal stability in aerobic solution and higher quantum yields for NO release than RRS. For example, laser flash photolysis studies of the red salt ester $Fe_2(\mu\text{-}SCH_2CH_2OH)_2(NO)_4$ in methanol solution gave behavior similar to that described above for RRS in Scheme 1. Photodissociation of NO proved to be reversible with $k_{NO} = 1.1 \times 10^9\ M^{-1}\ s^{-1}$ for the back reaction (Eq. 4).[29] In aerobic solution, however, the initial Fe-containing intermediate is trapped by O_2 with an estimated k_{O2} of $\sim 10^7\ M^{-1}\ s^{-1}$. The net photolysis result in aerobic media is the release of all four NO's.

$$Fe_2(\mu\text{-}SCH_2CH_2OH)_2(NO)_4 \underset{k_{NO}}{\overset{h\nu}{\rightleftharpoons}} Fe_2(\mu\text{-}SCH_2CH_2OH)_2(NO)_3 + NO \qquad (4)$$

The quantitative photo-dynamics of the RSEs were recently revisited by Manho and coworkers who used ultrafast laser flash photolysis to probe the initial stages of the photolysis mechanism over the time frame 0.3 ps to 10 μs.[35,36] In one study, in which the bridging RS^- thiolate was derived from cysteine, i.e., $Fe_2(\mu\text{-cysteinate})_2(NO)_4$ (Cys-RSE) the primary quantum yields of the NO dissociation from the photo-excited Cys-RSE were found to be 0.24 − 0.54 depending on the excitation wavelength. However, recombination of the initially formed geminate pair occurs within a few ps, thereby reducing the net quantum yield of the NO released into the median that would be biologically available.

Subsequent to the early photochemical studies in this laboratory of nitrosyl ruthenium porphyrins by Katrina Miranda, Ivan Lorkovic and coworkers,[23] other ruthenium photoNORMs have been introduced in a number of laboratories.[37–40] A key advantage of the ruthenium nitrosyls is the high thermal stability of the Ru-NO bonding in the ground state. At UCSB, Carmen Works et al. initiated studies of the ruthenium nitrosyl platform based on the salen complexes Ru(salen)(X)(NO) (salen= *N,N′*-ethylenebis(salicylideneiminato) dianion, X=Cl^-, H_2O, ONO^-) and analogous salophen complexes (salophen= *N,N′*-1,2-phenylene-bis (salicylideneiminato) dianion).[41,42] Photolysis of these precursors in solution releases NO with modest quantum yields to give the corresponding solvento species Ru^{III}(salen)(X)(Sol) and Ru(salophen)(X)(Sol) (e.g., Eq. 5). These reactions are reversible, but the back reaction is very slow in coordinating solvents, thus, NO photo-uncaging is effectively irreversible under those conditions. These complexes proved to be thermally quite stable. For examples, the water soluble complex [Ru(salen)(H_2O)(NO)]NO_3 maintains its integrity in the dark for weeks even in aerated solution, as does the analogous complex where the salen aromatic rings were modified by carboxylate functionalities to improve aqueous solubility.[43] Notably, the salophen complexes show stronger absorptions in the mid-visible range than do the salen analogs.[41b]

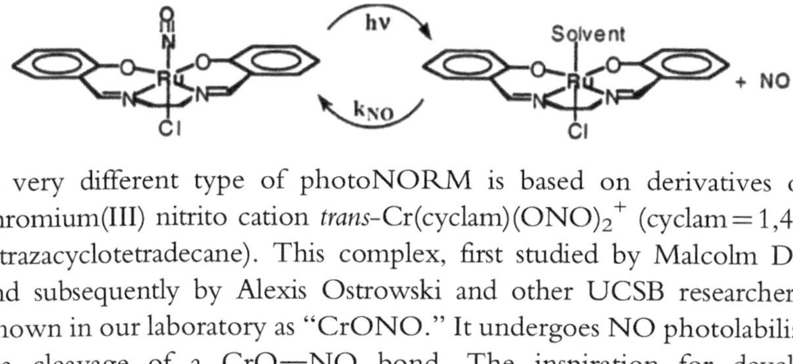

(5)

A very different type of photoNORM is based on derivatives of the chromium(III) nitrito cation *trans*-Cr(cyclam)$(ONO)_2^+$ (cyclam=1,4,8,11-tetrazacyclotetradecane). This complex, first studied by Malcolm Deleo[44] and subsequently by Alexis Ostrowski and other UCSB researchers,[45] is known in our laboratory as "CrONO." It undergoes NO photolabilization via cleavage of a CrO—NO bond. The inspiration for developing CrONO was drawn from an earlier study by Mikio Hoshino and coworkers who showed irreversible formation of Cr^{IV}(TPP)(O) plus NO when Cr^{III}(TPP)(ONO) (TPP=tetraphenyl porphyrin) was irradiated in benzene.[46] Our premise was that the CrO—NO cleavage was due to the oxophilic character of Cr(III), and we confirmed this idea by demonstrating that photolysis of the analogous nitrate complex Cr^{III}(TPP)(ONO_2) leads to formation of Cr^{IV}(TPP)(O) plus NO_2 (Eq. 6).[47]

$$Cr^{III}(TPP)(ONO_2) \xrightarrow{h\nu(366\ nm)} Cr^{IV}(TPP)(O) + NO_2 \quad (6)$$

CrONO is photoactive with a quantum yield of NO release (Φ_{NO}) of ~0.25 under visible wavelength excitation out to at least 546 nm.[45b] The NO photolabilization is reversible (Eq. 3) with a k_{NO} of $3.1 \times 10^6\ M^{-1}\ s^{-1}$ in aq. buffer. Therefore, net NO release depends on trapping the metal containing intermediate, presumably a Cr^{IV} oxo species, by an oxidant such as O_2 or by a reductant such as glutathione (Scheme 2).[45a] CrONO is stable in aerated aq. buffers even at physiological temperatures (37 °C) and shows no acute toxicity. Thus, in many respects, CrONO would appear to be well suited for NO photo-uncaging. However, its photochemically active, visible range absorption bands are metal-centered ligand field (LF) d-d transitions, which for this centrosymmetric complex, are quite weak. For photoNORM biological applications, the key issue is the rate of NO uncaging R_{NO} as defined by $\Phi_{NO}\ x\ I_{abs}$ (Eq. 1). A moderately high value of Φ_{NO} is one important part contribution, but the rate is also determined by I_{abs}. As defined by Eq. 2, I_{abs} is dependent both on the initial intensity I_0 of the exciting light and on the absorbance of the sample at the wavelength of irradiation (λ_{irr}). Thus, the effectiveness of a photoNORM or photochemical precursor of other bioactive or therapeutic compounds will depend on the absorptivity at the λ_{irr} that can be delivered to the physiological target. Section 3 summarizes efforts in our laboratory to address this issue by

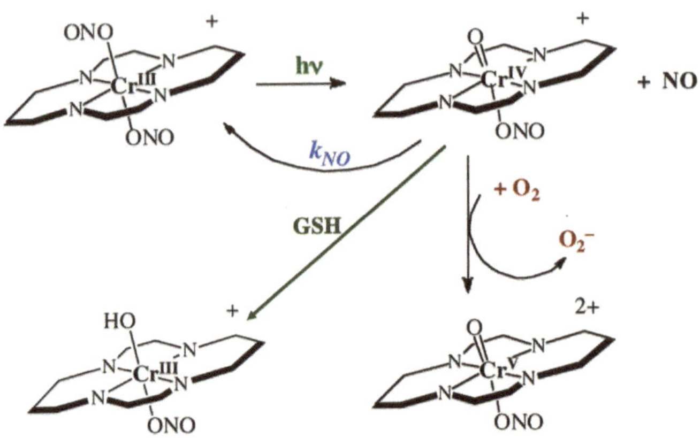

Scheme 2 The photolysis of *trans*-Cr(cyclam) $(ONO)_2^+$ showing that the initially formed Cr(IV) intermediate can undergo the reverse reaction with NO or be trapped by reaction with O_2 or a reductant such as glutathione (Ref. 45a).

coupling photoNORMs to chromophores that can act as antennas to enhance the absorptivity of CrONO and related systems.

2.2 Uncaging carbon monoxide

As noted above, there is considerable interest in developing CO releasing moieties (CORMs) as potential therapeutics for disease states including vascular- and immuno-related dysfunctions,[48] antibiotic resistant microbial infections[49] and cancer.[50] Various triggers for CO release from CORMs have been discussed; for example, reactions of metal carbonyls with H_2O_2 and other reactive oxidative species (ROS) may serve this purpose.[51] Photo-uncaging is another approach that offers temporal and spatial resolution. The photo-lability of metal carbonyls has long been known although largely under conditions (non-aqueous, anaerobic media) incompatible with therapeutic tasks.[52] In this context, design goals for photoCORMs are similar to those for other caged bioactive substances, namely:

Photoactivation at a therapeutically suitable wavelength.

Stability in the dark and in aerobic, aqueous media.

Biocompatibility of the original photoCORM and of other products of CO release

Specificity of delivery methods

The first photoCORM examples in a biological setting were qualitative studies with $Mn_2(CO)_{10}$ and $Fe(CO)_5$, which have poor aqueous solubility.[53] A subsequent report[54] described metal carbonyl, *fac*-$[Mn(CO)_3(tpm)]PF_6$ (tpm = tris(pyrazolyl)methane). Dimethyl sulfoxide solutions of this compound were delivered to HT29 colon cancer cells. Photoinduced toxicity was noted upon 365 nm excitation.

Our laboratory has largely focused on characterizing transition metal photoCORMs that are soluble and stable in aerobic aqueous media and/or can be activated by irradiation with longer visible or NIR wavelengths. These involve relatively low valent, electron-rich metals bound to CO plus other ligands L that impart desirable properties. As noted above, one key property would be the absorption spectrum, since the photoCORM needs to absorb light at a therapeutically suitable wavelength. The characteristic spectra of such complexes often display metal-to-CO charge transfer bands, but these generally occur at short wavelengths. At lower energies will be found LF and metal to ligand charge transfer (MLCT) bands, especially if L is a diimine ligand with relatively low-lying empty π^*-orbitals. Both of these excited state (ES) types would have weakened M—CO bonds relative to the ground state (GS). In the former, electron density is transferred from orbitals that participate in M—CO π-backbonding to ones that are σ* with respect to the M-CO bond. For the MLCT ES, the metal center is formally oxidized, and the π-backbonding is depleted. Based on such simplistic descriptions, both of these ES types should have increased CO lability, but the LF ES should be the more reactive,[55] as has been noted experimentally with the tungsten(0) complexes $W(CO)_5$py-X (where py-X is a substituted pyridine ligand).[56] MLCT transitions typically display the much higher extinction coefficients, so net photoreactivity will likely depend on rapid internal conversion/intersystem crossing processes that lead from the ES initially populated by excitation to lower energy ES. The nature and relative energies of the latter often define the net photochemical/physical properties, and their roles must be considered in the photoCORM design.

Another important issue is how the CO uncaged from a photoCORM is detected and quantified. The myoglobin (Mb) assay[57] takes advantage of spectral changes of the heme chromophore upon coordination of CO to Mb from which one is able to directly quantify the amount of CO released. However, this assay suffers from the disadvantage that it functions only in an anaerobic environment. In our laboratory several methods have been utilized, but we have found that the most effective and quantitative approach is to carry out the photoreaction in a closed vessel such as shown in Fig. 4.[58] The gas phase can be periodically sampled by syringe and analyzed using a calibrated gas chromatograph with a thermal conductivity detector (GC-TCD).[22] For detection of carbon monoxide release in living cells, several "turn-on" sensors that become photoluminescent upon reacting with CO have been developed.[59]

The first photoCORM studied in our laboratory[22] was the air and aqueous stable and water soluble complex $W(CO)_5(TPPTS)^{3-}$ (TPPTS = tris(sulfonatophenyl)-phosphine trianion).[60] Dale Rimmer showed that this complex anion released one CO upon photolysis in anaerobic pH 7.4 buffered aq. solution with quantum yields of CO release (Φ_{CO}) of 0.90 for λ_{irr} 313 nm and 0.6 for λ_{irr} 405 nm. Although, photolysis in deaerated solution led to release of only one CO equivalent, in aerated solutions subsequent reactions led to release of 1.2–1.6 additional equivalents of CO concomitant with oxidation of the initially formed $W(CO)_4(H_2O)(TPPTS)^{3-}$ (Scheme 3). The key lesson is that the primary photoactivation may be followed by secondary reactions relevant (negatively or positively) to the initial goal. In this publication, we coined the term "photoCORM" for a species that releases CO directly upon photoactivation,[22] a designation that

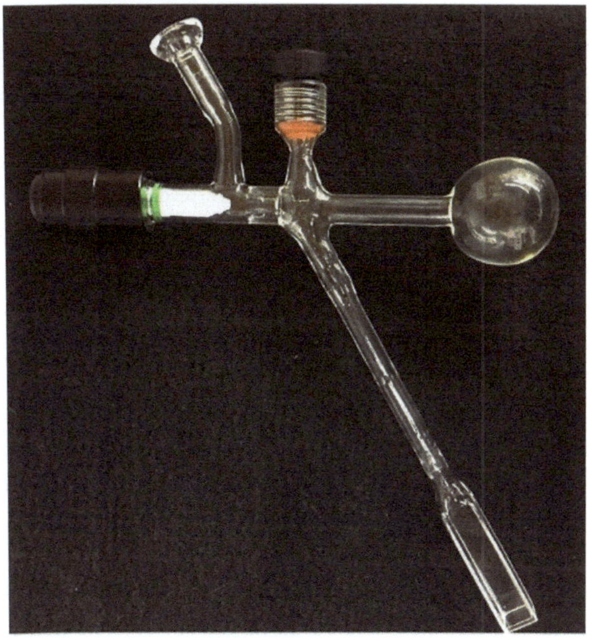

Fig. 4 Schlenk cell for anaerobic photolysis experiments. The sample is located in the quartz fluorimeter cell. Gas samples are withdrawn through the syringe port at the top. *Photo by Zhi Li. With permission from Li, Z. Toward Long Wavelength Absorption: Dinuclear Photo-activated CO Releasing Moieties, Ph.D. Dissertation. University of California, Santa Barbara,* ***2017***.

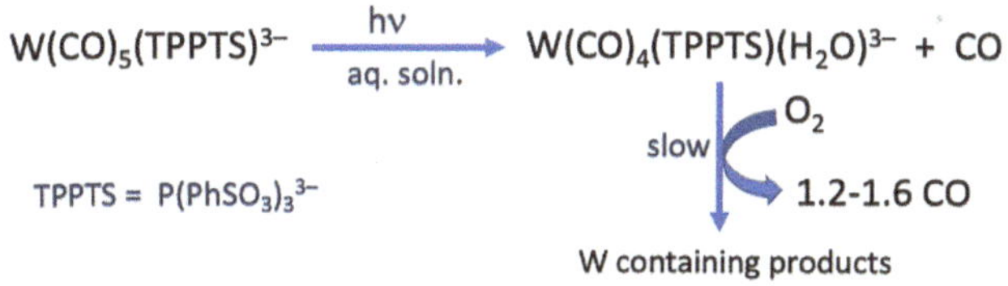

Scheme 3 Photoreaction of $W(CO)_5(TPPTS)^{3-}$ showing that initial photos-uncaging of one CO is followed by slow oxidation of the W(0) center to an unidentified product and the release of additional equivalents of CO (Ref. 22).

has found use with other researchers. Since the primary photoproduct $W(CO)_4(H_2O)(TPPTS)^{3-}$ is stable in deaerated aqueous buffer solution but undergoes oxidation and CO loss in aerobic solution, we designated this as a "proCORM," but that term has not caught on.

As noted above, an attractive feature for any photochemical uncaging precursor would be the opportunity to track the location of that compound in the physiological medium where it is employed. The term "theranostic" has been applied to such systems. If a photoCORM or the resulting photoproducts are luminescent, such tracking can be accomplished by sensitive optical methods. For examples, several mononuclear rhenium(I) carbonyl complexes of the type $ReL(CO)_3(NN)^{0/1+}$ (where L is a halide, phosphine or other uncharged ligand and NN is a bidentate diimine ligand such as 1,10 phenanthroline (phen) or 2,2'-bipyridine (bpy)) have been shown to demonstrate both CO photolability and photoluminescence (PL) in solution and in cellular media.[61–63]

From our laboratory, Agustin Pierre et al.[61] reported the biocompatible, water-soluble photoCORM $[Re(bpy)(CO)_3(THP)](CF_3SO_3)$ (THP = tris (hydroxymethyl)phosphine). This compound proved to be stable in aerated solution, strongly luminescent and photolabile toward CO release when irradiated with 405 nm light. It was readily incorporated into human prostatic carcinoma PPC-1 cells, where it showed no apparent dark cytotoxicity and the PL properties served to indicate the location of photo-induced CO release. Subsequently, Mascharak and coworkers[62] reported several luminescent $ReL(CO)_3(phen)^{0/1+}$ complexes that are photolabile under UV excitation and that can be internalized into MDA-MB-231 human breast adenocarcinoma cells.

In this context, Wilson and coworkers,[63] published an in-depth study of 15 members of the series $Re(PR_3)(CO)_3(NN)^+$ (PR_3 = THP, 1,3,5-triaza-7-phosphaadamantane (PTA), or 1,4-diacetyl-1,3,7-triaza-5-phosphabicylco[3.3.1]nonane (DAPTA), and NN is one of five different diimine ligands) prepared according to Scheme 4. The photochemical and PL properties in aerated and deaerated solutions were evaluated as were the phototoxicities with human cervical (HeLa), ovarian (A2780), and cisplatin-resistant ovarian (A2780CP70) cancer cell lines. When irradiated with 365 nm light in aqueous solutions, the THP and DAPTA complexes are luminescent and undergo CO photo-uncaging. In aerobic media, the PL is partially quenched with the concomitant formation of singlet oxygen ($^1O_2$). These complexes displayed little dark toxicity when incorporated into the cancer cell lines but each was phototoxic, although the least phototoxic was the $Re(bpy)(CO)_3(THP)^+$ complex described above. In contrast, the PTA complexes proved to be photo-inert. These in-depth studies suggested that the observed cellular phototoxicity may result, not only from the CO release, but also from the rhenium photoproduct and from $^1O_2$ production.[63]

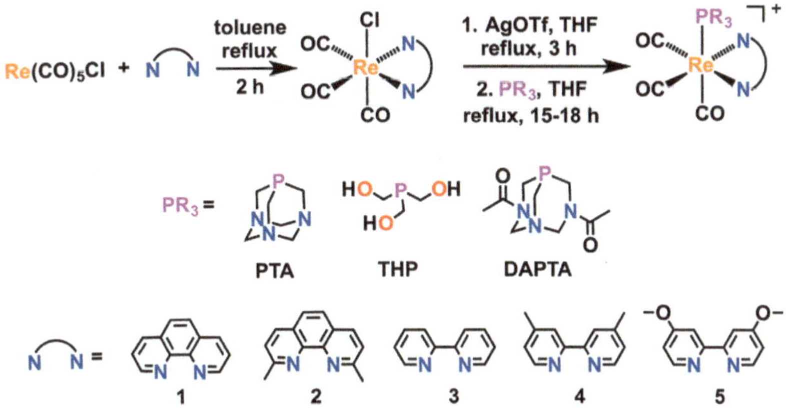

Scheme 4 Synthesis scheme for the $Re(PR_3)(CO)_3(NN)^+$ complexes described by Wilson and coworkers. *Reprinted with permission from Marker, S.C.; MacMillan, S.N.; Zipfel, W.R.; Li, Z.; Ford, P.C.; Wilson, J. J.* Inorg. Chem. ***2018***, 57, *1311–1331, Copyright 2018 American Chemical Society.*

3. Enhancing photoreaction SMB release rates with antennas

There are multiple reasons that simply being photoactive toward NO or CO release does not necessarily make a particular photochemical precursor a valuable photoNORM or photoCORM. As noted above, the rate of SMB release R_i (Eq. 1) is a key parameter as is the irradiation wavelength λ_{irr} necessary to carry out such excitation. One way to increase R_i is to increase I_{abs}, the intensity of the light absorbed by attaching chromophores to the caged photoreaction precursor serve as antennas at the desired λ_{irr} as illustrated by Scheme 5. Excitation of a strongly absorbing chromophore can be followed by energy transfer to the site responsible for NO or CO release, if acceptor states on the latter are at appropriate energies to stimulate the desired photoreactions.

3.1 Organic chromophore antennas

The complex PPIX-RSE prepared by Christa Conrado et al.[34] (Fig. 3) illustrates this approach. Spectra of Roussin's red salt esters such as $Fe_2(\mu\text{-}SCH_2CH_2OH)_2(NO)_4$ or $Fe_2(\mu\text{-}SCH_2CH_3)_2(NO)_4$ display only weak absorption at longer visible wavelengths ($<10^3\ M^{-1}\ cm^{-1}$ at 546 nm). However, the visible spectrum of the conjugate PPIX-RSE is dominated by strong Q-bands characteristic of the PPIX porphyrin functionality with

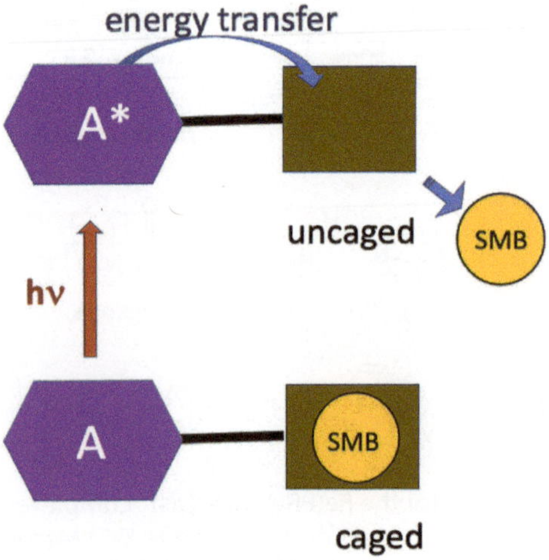

Scheme 5 Attaching a strongly absorbing chromophore A to a photochemical precursor carrying a caged SMB can enhance the rate of uncaging by photosensitizing the reaction at the precursor.

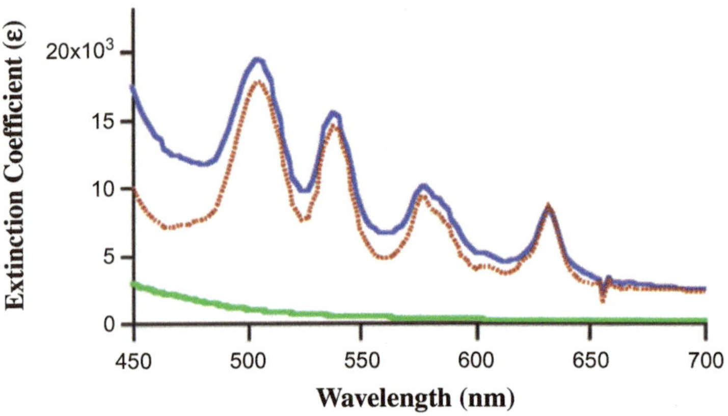

Fig. 5 Quantitative spectra of PPIX-RSE (blue), PPIX (red), and $Fe_2(\mu\text{-}SCH_2CH_3)_2(NO)_4$ (green) in THF in the Q-band region of the visible spectrum. Note that the spectrum of PPIX-RSE is approximately the sum of the spectra of the PPIX and RSE chromophores. *With permission from Conrado, C.L.; Weckler, S.; Egler C.; Magde, D.; Ford, P.C.* Inorg. Chem. ***2004***, 43, *5543–5549, Copyright 2018 American Chemical Society.*

extinction coefficients of $\sim 10^4\ M^{-1}\ cm^{-1}$ (Fig. 5). When excited with 404 nm light, the PL at 640 nm characteristic of PPIX is about 90% quenched in PPIX-RSE, indicating that energy transfer is occurring from the porphyrin chromophore to the RSE core. This is further demonstrated by comparing the photoreaction quantum yields for 546 nm excitation of PPIX-RSE and $Fe_2(\mu\text{-}SCH_2CH_3)_2(NO)_4$ in aerated solutions. The Φ_{NO} values (2.5×10^{-4} and 1.7×10^{-4}, respectively) were quite similar. However, given that its absorbance at 546 nm was more than an order of magnitude larger, it is clear that PPIX-RSE is much more efficient at absorbing red light and therefore gave a greater rate of net photochemistry at this longer excitation wavelength. For example, the rate of NO photogeneration by PPIX-RSE at an initial concentration of 10 μM would be about 19-fold faster than for $Fe_2(\mu\text{-}SCH_2CH_3)_2(NO)_4$ at the same concentration. (Incidentally, the residual luminescence properties of PPIX-RSE suggest that it might have theranostic value, although this possibility was not further studied in this laboratory.)

As noted above CrONO displays very attractive characteristics as a photoNORM, but its ligand field absorption bands in the visible are quite weak. As a result, the rate of photochemical NO production in a dilute environment or with low intensity light (both likely in a physiological environment) would be low. To address this issue, Frank DeRosa in our laboratory prepared derivatives of CrONO with pendant organic dyes such as anthracene and pyrene, Anth-CrONO and Pyr-CrONO, respectively (Fig. 6).[64] These antennas strongly enhanced the molecular absorbances in the 300–400 nm region of the spectrum, but did not extend this effect to longer visible wavelengths. Photoexcitation in aqueous solutions at wavelengths corresponding to the pendant chromophore absorptions led to enhanced NO release. Furthermore, the expected fluorescence from the pendant chromophores is largely, but not entirely, quenched in the complexes, a result interpreted in terms of fast energy transfer from ligand-centered

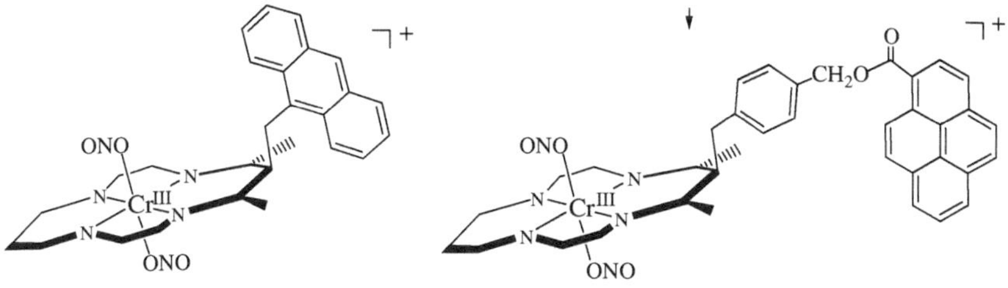

Anth-CrONO **Pyr-CrONO**

Fig. 6 Anth-CrONO (*trans*-[Cr(5,7-dimethyl-6-anthracylcyclam)(ONO)$_2$] and Pyr-CrONO (*trans*-[Cr(5,7-dimethyl-6-(1′-methyl-4′-(1″-carboxymethylpyrene)]benzyl)-cyclam)(ONO)$_2$]) (Ref. 64).

π–π* states to the metal-centered LF states responsible for the net photochemistry. Thus, these tethered chromophores can serve as light-gathering antennae for intramolecular sensitization of CrONO.

Not surprisingly, ruthenium(II) diimine platforms, which are widely used for other photo-uncaging applications,[65] have drawn attention as photoCORMs.[66] These include complexes of the general formulas $Ru^{II}(NN)(CO)_2X_2$, $Ru^{II}(NNN)(CO)_2X$, and $Ru^{II}(NN)_2(CO)Y$ (NN = a bidentate aromatic heterocycle; NNN = tridentate aromatic heterocycle; X = halide, Y = X or other ligand). However, CO coordination shifts the characteristic Ru^{II}(diimine) MLCT band(s) to the blue such that their strong absorptions do not extend very far into the visible. This property can be addressed by attaching an antenna such as the pendant boron-dipyrromethene (BODIPY) dye shown in Fig. 7.[67] This chromophore displays very strong visible absorbance (λ_{max} 509 nm) and PL, allowing one to image uptake and location in biological structures. However, the CO photo-uncaging studies were carried out in the near-UV, so it is unclear whether energy transfer from the BODIPY chromophore to the metal center is effective in this case.

3.2 Semiconductor quantum dot antennas

Nanocrystal semiconductor core/shell quantum dots (QDs) represent different types of antennas. The absorption and PL spectra of such QDs across the visible spectral range are size dependent,[68] thus QDs may serve as continuously tunable photosensitizers.[69,70] In this context, Danial Neuman, Alexis Ostrowski and coworkers at UCSB[71,72] showed that the emission from

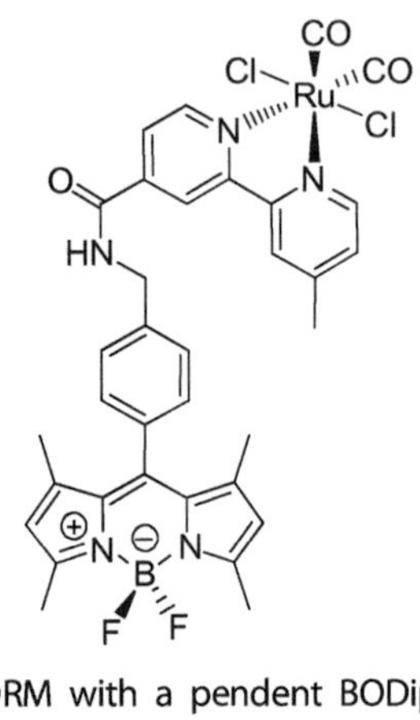

Fig. 7 A ruthenium photoCORM with a pendent BODipy dye antenna. *Adapted from Geri, S.; Krunclova, T.; Janouskova, O.; Panek, J.; Hruby, M.; Hernández-Valdés, D.; Probst, B.; Alberto, R.A.; Mamat, C.; Kubeil, M.; Stephan, H.* Chem. Eur. J. ***2020***, 26, *10992–11006.*

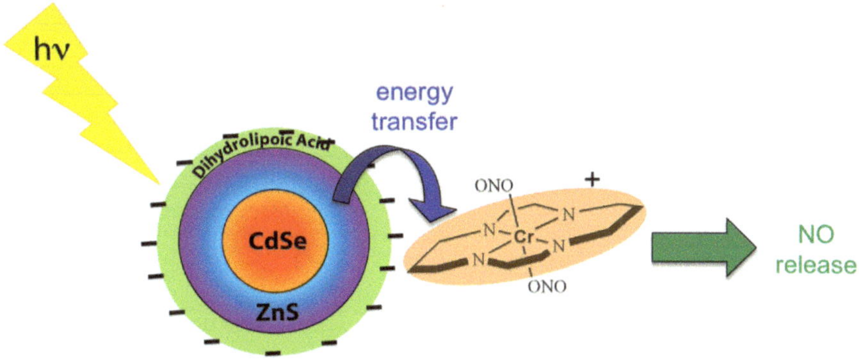

Fig. 8 Representation of NO release from the *trans*-Cr(cyclam)$(ONO)_2^+$ cation using CdSe/ZnS core/shell quantum dots, surface modified with dihydrolipoate to induce aqueous solubility, as a photosensitizer. The association between the CrONO cation and the anionic surface of the QD is electrostatic rather than covalent. *With permission from Burks, P.T., Ostrowski, A.D.; Mikhailovsky, A.A.; Chan, E.M.; Wagenknecht, P.S.; Ford, P.C.* J. Am. Chem. Soc. ***2012***, 134, *13266–13275 Copyright 2012 American Chemical Society.*

3.8 nm (est.) diameter core/shell CdSe/ZnS QDs, decorated on the surface with dihydrolipoic acid ligands to facilitate solubility, is quenched by CrONO in aqueous solution. Furthermore, this quenching was accompanied by photosensitized NO release. The surfaces of these aqueous soluble QDs are anionic, so the ensemble with the CrONO cation(s) in solution is electrostatic rather than covalent (Fig. 8). It is likely that multiple CrONO cations are associated with a single QD, but the number will be concentration dependent.[73]

One key question is by what mechanism(s) does this photosensitization occur? This was probed by quantitative studies of the quenching of the PL from water-soluble CdSe/ZnS and CdSeS/ZnS core/shell QDs by several *trans*-$Cr(N_4)(X)_2^+$ complexes (N_4 represents different tetraazamacrocycle ligands and $X^- = CN^-$, Cl^-, or ONO^-) in aqueous solution.[73] These studies by Peter Burks and coworkers demonstrated a clear correlation between quenching efficiency and the emission/absorption overlap integral in a manner consistent with a Förster resonance energy transfer (FRET) mechanism.

QD photosensitizers are not limited to energy transfer pathways. For example, Lilian Franco, a visiting scholar in our laboratory from Brazil, concluded that the CdTe QD photosensitized NO uncaging from *cis*-$[Ru(bpy)_2(4\text{-ampy})(NO)]^{3+}$ (4-ampy is 4-aminopyridine)[74] occurs through charge transfer mechanism. Furthermore, in our initial foray into the uncaging of the potential SMB carbon disulfide, Chris Bernt et al.[75]

demonstrated that aqueous CdSe QDs photocatalyze the aerobic oxidation of 1,1-dithioloxalate (DTO) via a charge transfer mechanism to release CS_2 plus CO_2 (Eq. 7) with quantum yields of ~0.03 under 498 nm irradiation. Notably the DTO anions serve to solubilize these QDs in aqueous media but quench the PL, so, as this ligand is oxidized, these QDs eventually become insoluble and precipitate as luminescent colloids from the solution.

$$[O_2C - CS_2]^{2-} \xrightarrow[\text{oxidant}]{h\nu,\,\text{CdSe QD}} CO_2 + CS_2 \quad (7)$$

Cadmium is toxic, and this provides a strong incentive to develop conjugates with more benign nanomaterials.[76] For example, Lui et al. described several luminescent nanoplatforms consisting of N-doped graphene QDs (GQDs, diameters 7–10 nm) functionalized with a ruthenium photoNORM.[77] On-going studies in our laboratory involving GQDs are described below.

4. Moving toward the phototherapeutic window

It is generally understood that the ideal wavelengths for penetrating mammalian tissue fall in the range 650–1100 nm, that is, deep red and NIR wavelengths with the maximum penetration occurring at ~800 nm (Fig. 9).[78] This is often termed the "phototherapeutic window." This property was exploited in our laboratory by Burks et al.,[79] who demonstrated that

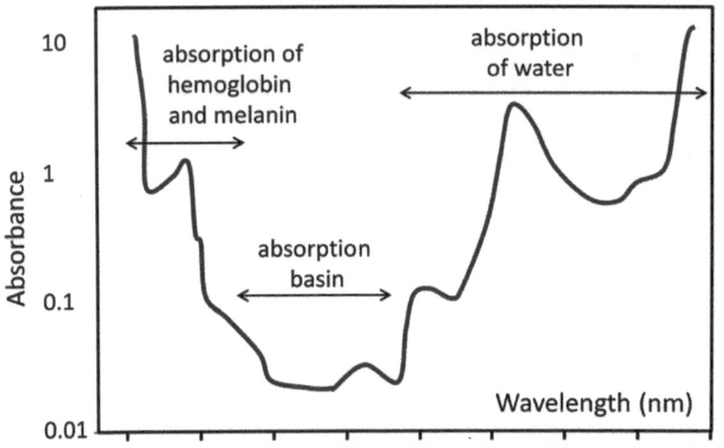

Fig. 9 Illustration of the phototherapeutic window for mammalian tissue. *With permission from Polesskaya, O., Baranova, A., Bui, S., et al. Optogenetic regulation of transcription.* BMC Neurosci ***2018***, 19, *12.*

a photochemical process could be activated by the light from a NIR diode laser passed first through a tissue filter (see below). For this reason, our more recent uncaging efforts have largely focused on methodologies to extend effective photoNORM and photoCORM excitation to these longer wavelengths.

4.1 Single photon excitation strategies

One such approach is to design compounds whose photochemistry can be triggered at such wavelengths with single photon excitation (SPE), examples being several manganese nitrosyl derivatives reported to have strong absorptions at longer visible wavelength. For example, aqueous solutions of $Mn(PaPy_2Q)(NO)^+$ ($PaPy_2QH$ = *N*,*N*-bis(2-pyridylmethyl)amine-*N*-ethyl-2-quinoline-2-carbox-amide) (Fig. 10) gave high quantum yields for NO release at λ_{irr} 550 nm.[80] The Φ_{NO} decreased at longer wavelengths but some photoactivity was observed upon 810 nm excitation. Similarly, Hitomi and coworkers[81] described manganese nitrosyl complexes of the pentacoordinate anionic ligand $dpaq^-$ (2-[*N*,*N*-bis(pyridine-2-ylmethyl)]-amino-*N*′-quinoline-8-yl-acetamido) modified by various substituents. In acetonitrile solution, the photoNORM of the nitro derivative $Mn(dpaq^{NO2})(NO)^+$ (Fig. 10) impressively demonstrated Φ_{NO} values of 0.78 and 0.18 upon direct irradiation at 650 nm and at 794 nm, respectively.

With regard to photoCORMs, there are numerous examples of direct CO photolabilization from metal carbonyl complexes at visible excitation wavelengths, but pushing SPE into the deep red or NIR has been challenging. In our laboratory, Li et al.[82] addressed this question using a different strategy, namely to use long wavelength excitation to generate an intermediate (a proCORM) that reacts further with the oxygen in an aerobic medium to trigger CO release. The photochemical precursors were the dinuclear Re-Mn complexes $(CO)_5ReMn(CO)_3(NN)$ (NN = phen, bpy, biquinoline (biq) or phenanthroline-carboxaldehyde (phen-CHO)). These metal carbonyls display strong metal-metal bond to ligand ($\sigma_{MM} \rightarrow \pi_L^*$)

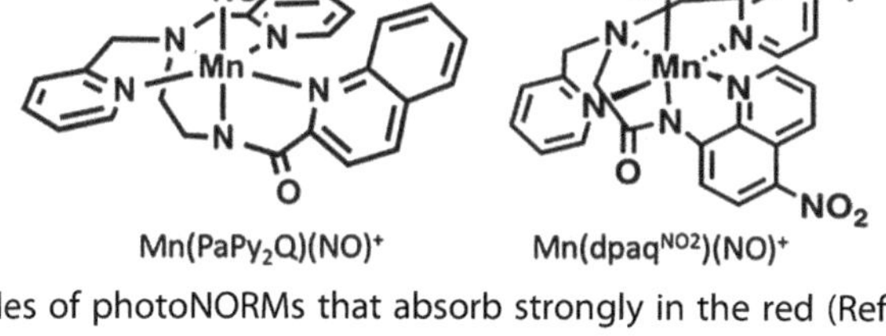

Fig. 10 Examples of photoNORMs that absorb strongly in the red (Refs. 80, 81).

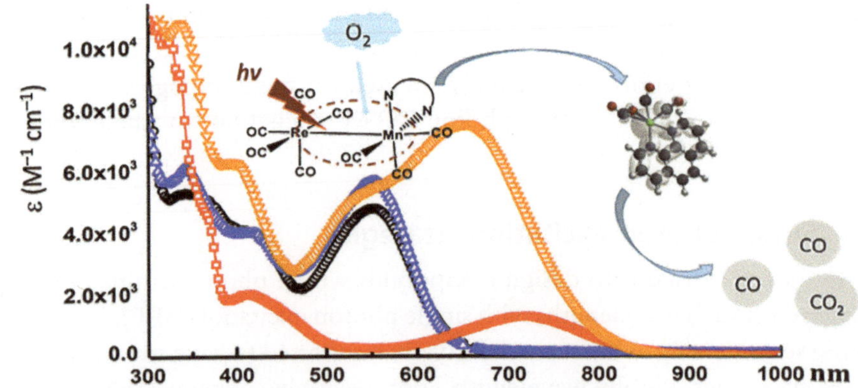

Fig. 11 Absorption spectra of $(CO)_5ReMn(CO)_3(phen)$ (blue, $\lambda_{max}=550$ nm), $(CO)_5ReMn(CO)_3(bpy)$ (black, 550 nm), $(CO)_5ReMn(CO)_3(biq)$ (red, 719 nm) and $(CO)_5ReMn(CO)_3(phen\text{-}CHO)$ orange, 652 nm) in acetonitrile showing the strong MMLCT absorption bands. *From Li, Z.; Pierri, A.E.; Huang, P.-J.; Wu, G.; Iretskii, A.V.; Ford, P.C.* Inorg. Chem. ***2017***, 56, *6094–6104 Copyright 2017 American Chemical Society.*

charge transfer (MMLCT) absorptions at longer visible wavelengths that extend into the NIR for NN=biq or phen-CHO (Fig. 11). SPE of these MMLCT bands in the deep red (659 nm) or NIR (794 nm for NN=biq or phen-CHO) leads to homolytic Re—Mn bond cleavage to give metal radicals (Eq. 8). These radicals are trapped by dioxygen to form intermediates that are labile toward CO release via secondary reactions. Density functional theory (DFT) computations explained the observation that the CO's are derived from the manganese fragment while the rhenium carbonyl fragment stays intact. Unpublished studies[58] indicate that if the rhenium fragment includes a diimine ligand such as a phenanthroline, the resulting rhenium carbonyl species is luminescent. This property suggests that such systems can be designed to be theranostic. Our ongoing studies are directed toward modifying these complexes to be compatible with biological media.

$$(CO)_5Re\text{-}Mn(CO)_3(phen) \xrightleftharpoons{h\nu} (CO)_5Re\cdot + \cdot Mn(CO)_3(phen) \tag{8}$$

4.2 Simultaneous two photon excitation

The reactive excited states of many otherwise attractive photoNORMs and photoCORMs are not low enough in energy to be accessed by SPE of an antenna with strong absorption bands in the phototherapeutic window. Our laboratory addressed this problem with multi-photon excitation, where the

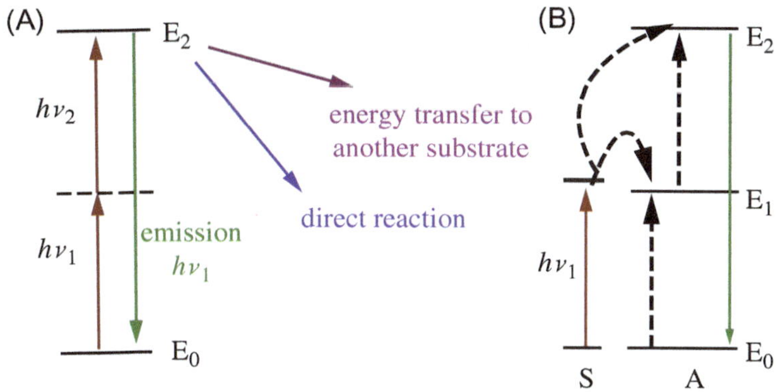

Scheme 6 Simplified models for multi-photon excitation. (A) TPE involves the simultaneous absorption of two photons, usually, but not necessarily, of the same energy. The ES formed may undergo emission, direct reaction to product or energy transfer to another substrate. (B) In this model, the NIR SPE of the sensitizer component (S) gives an ES that can sensitize an acceptor (or "activator," if emissive) component (A) having a long-lived ES (the shorter curved dashed line). Further excitation of S results in additional energy transfer to A* to give a higher energy ES E_2 (the longer curved dashed line) from which PL or (possibly) energy transfer occurs.

additive energy of two or more NIR photons would be sufficient to generate the desired precursor state for uncaging.[20] One approach is simultaneous two photon excitation (TPE) using very high intensity laser pulses owing to the intensity squared (I_0^2) relationship noted above (Scheme 6A). The other is sequential two (or more) photon absorption which is illustrated in Scheme 6B[21] and which will be the subject of the next section.

In 2004, Steve Wecksler in collaboration with Alexander Mikhailovsky of UCSB[83] demonstrated, for the first time, the activation of a photoNORM by two photon excitation (TPE) with NIR light from an ultra-fast pulsed Ti-sapphire laser operating at 800 nm. Photolysis of the Roussin's red ester PPIX-RSE (Fig. 3) led both to weak fluorescence at ~630 nm from the PPIX chromophore and to NO release detected using a nitric oxide specific electrode. Given that PPIX-RSE shows no apparent absorption at this NIR wavelength, both the PL and the NO release must result from multiphoton absorption.

Wecksler and coworkers extended these studies to two other antenna-RSE conjugates $Fe_2(\mu\text{-}RS)_2(NO)_4$, Fluor-RSE (RS = 2-thioethyl ester of fluorescein)[84] and AFX-RSE (RS = the 2-thioethyl ester of N-phenyl-N-(3-(2-ethoxy)phenyl)-7-(benzothiazol-2-yl)-9,9-diethyl-fluoren-2-yl-amine)[85] (Fig. 12). Both ligands display strong PL under SPE, and in both

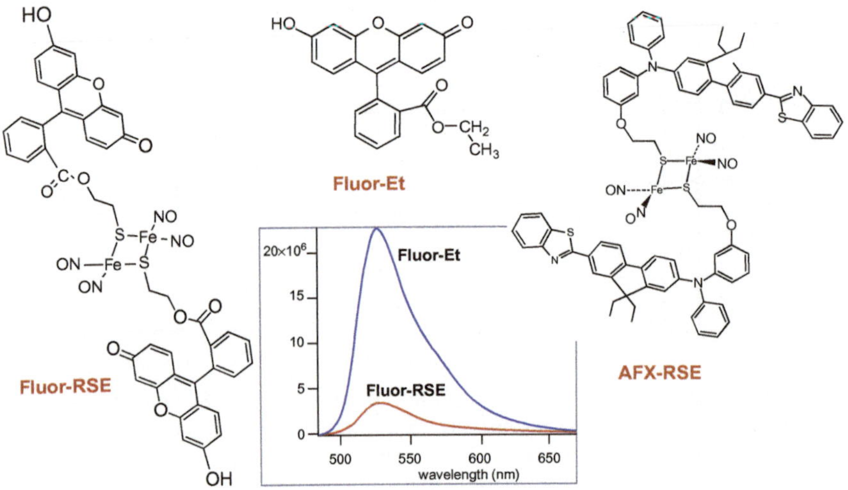

Fig. 12 Structures of the -Roussin's red ester conjugates Fluor-RSE and AFX-RSE with SPE and TPE antennas. The PL spectra shown compare the photoluminescence from the model compound Fluor-Et to that of Fluor-RSE, under SPE excitation at 480 nm.

cases the PL from the RSE conjugate was quenched ~90%, indicating energy transfer from the antenna ligand to the RSE center (see insert of Fig. 12). In this context, Fluor-RSE[84] and AFX-RSE[85] displayed PL and moderate quantum yields for NO release under SPE of the strong chromophore visible absorbance bands. Furthermore, under 800 nm excitation using the high intensity pulses from an ultra-fast Ti:sapphire laser, both complexes displayed visible PL (at ~530 nm for Fluor-RSE) with the respective two-photon absorption cross-sections δ of 63±7 GM and 246±8 GM (1 GM $=10^{-50}$ cm^4 s $photon^{-1}$ $molecule^{-1}$). The δ value at 800 nm for Fluor-RSE was nearly twice that of the model compound Fluor-Et, a not surprising observation given that Fluor-RSE includes two fluorescein chromophores. Both conjugates released NO under such TPE, and for Fluor-RSE, it was shown that that NO release was a function of the excitation intensity squared I_0^2, as expected for a two-photon process (see above).[84a]

A subsequent publication by Prasad and coworkers[86] described a RSE incorporating a two-photon absorbing chromophore with tetra(ethylene glycol) units to increase aqueous solubility. These were incubated into HeLa cells, and TPE was shown to impact the viabilities of these *in vitro* cultures. More recent developments concerned with photo-uncaging of NO via simultaneous two-photon excitation are summarized in several recent review articles.[87–89]

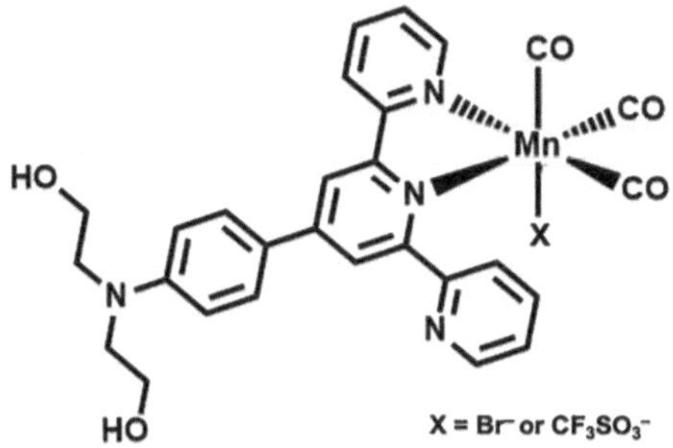

Fig. 13 $Mn(CO)_3(\eta^2\text{-TPYOH})X$ (TPYOH = 4′-p-*N,N*-bis(2-hydroxyethyl)benzyl-2,2′:6′,2″-terpyridine). Shown is only one enantiomer of the racemic mix that was studied (Ref. 90).

Photo-uncaging of CO with TPE is a more recent development with studies by Qin Jiang et al. from our laboratory[90] and by Ramu et al.[91] appearing in 2019. In both cases, the complexes were based on the *fac*-Mn^I tricarbonyl platform that has been used for a number of photoCORMs,[92] but with antenna ligands that have known two-photon absorption cross sections. For both examples, SPE in the visible led to photo-uncaging of CO, while TPE with an ultrafast pulsed laser operating at 800 nm similarly led to CO release. With $Mn(CO)_3(\eta^2\text{-TPYOH})X$, ($X = Br^-$ or $CF_3SO_3^-$, Fig. 13) the triflate complex proved to be moderately soluble in water, while the bromo complex was soluble in mixed ethanol/water (v:v 2:1) solutions. The former proved to be quite stable in aerobic aqueous solution at 37 °C, indicating its likely stability under physiological conditions. However, while the Jiang et al. and Ramu et al. studies both provide valuable proof-of-principle examples of photoCORM activation by TPE with tissue transmitting NIR light, the TPE cross sections were rather small. Practical application of NIR TPE to uncage CO under physiological conditions will require compounds with stronger cross sections.

4.3 Upconverting nanoparticles

Sequential, as opposed to simultaneous, multiphoton excitation is an alternative approach for photo-uncaging with NIR light. This can be achieved by employing NaYF4 or NaGdF4 nanoparticles that are doped with a sensitizer ion (S in Scheme 6) such as Yb^{3+} which has a transition that can be pumped with simple, continuous wave (CW) diode laser operating at 980 nm. Energy transfer from S* to an acceptor (A in Scheme 6B) with a long-live excited state such as Er^{3+} or Tm^{3+} is followed by another single photon excitation of S. Energy transfer from S* to already excited A leads

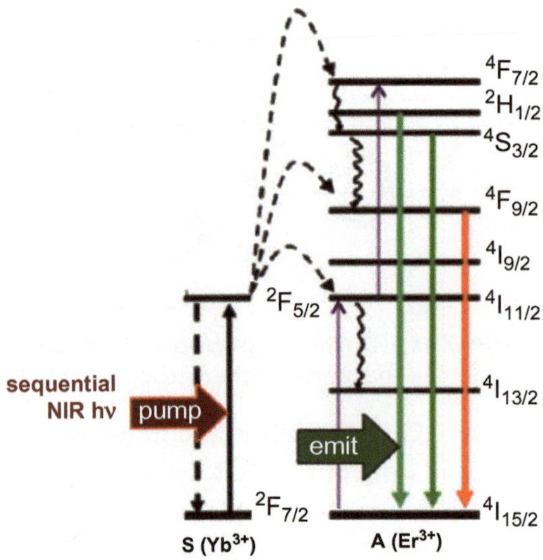

Fig. 14 Diagram showing the excitation of an unconverting material with Yb^{3+} as the sensitizer pumped with NIR light (940 nm and Er^{3+} as one type of acceptor/emitter leading to emission at several visible wavelengths).

to even higher energy excited states of A.[21] The latter can decay by emission of visible and (sometimes) UV light. Fig. 14 illustrates this process for an upconverting material where the sensitizer is Yb^{3+} (usually present at a fairly high concentration) and the acceptor/emitter is Er^{3+}.[93] Different PL spectra are obtained when other acceptor lanthanide ions are used. The net NIR-to-visible (or UV) up-conversion by *sequential* multi-photon excitation is much more efficient than simultaneous TPE, thereby allowing one to use CW NIR diode lasers for this purpose. The upconversion efficiency remains a function of I^n with $n > 1$, so spatial resolution is a potential feature.

The unique optical properties and low toxicities of Ln^{3+}-doped upconverting nanoparticles (UCNPs)[94,95] provide new opportunities for biological imaging and photosensitization. Such UCNPs can act as NIR antennas to activate a photoCORM or photoNORM to uncage its payload, either by upconverted PL followed by reabsorption (trivial energy transfer) or perhaps by direct energy transfer. Fig. 15 illustrates the visible PL spectrum from ($NaYF_4$:Yb/Er (20/2%)@$NaYF_4$) core:shell UCNPs with a 20 nm average diameter excited at 980 nm. John Garcia et al.[96] coated these UCNPs with a 10 nm thick layer of dense silica, then treated the outer silica shell with (3-aminopropyl)triethoxysilane to incorporate primary amines on the surface. Treatment with acid gave surface —NH_3^+ groups making the

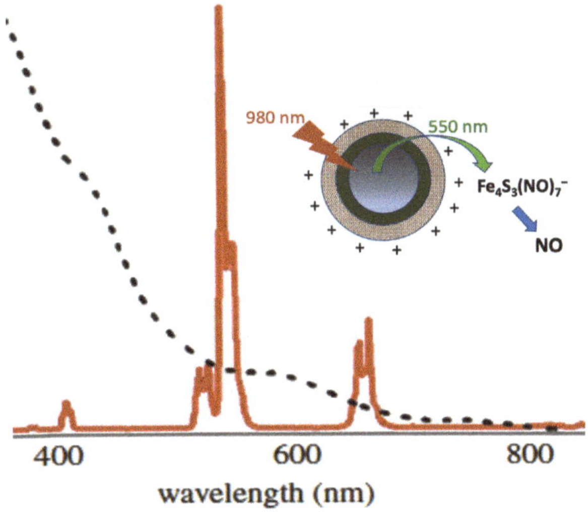

Fig. 15 An overlay of PL spectrum from an aq. suspension of β-(NaYF4:Yb/Er (20/2%)@NaYF) core:shell UCNPs excited at 980 nm and optical spectrum of Roussin's black salt $Na[Fe_4S_3(NO)_7]$ (42 μM) in aq. solution (dashed line). Inset: cartoon representation of the NIR excitation of a UCNP modified with surface $-NH^{3+}$ groups and the resulting visible PL activating NO release from a RBS anion.

resulting UCNPs@SiO_2 nanoparticles aqueous soluble. These were added to a solution of the Roussin's black salt whose absorption spectrum overlaps with the UCNP PL (Fig. 15). When the resulting mixture was irradiated with 980 nm light, NO was generated as analyzed using the NOA (Fig. 2). In contrast, NO release was not observed when the RBS solution alone was irradiated at this wavelength, thereby demonstrating that the UCNP antennas were able to sensitize NO photo-uncaging using NIR light as illustrated by the cartoon inset in the figure.

As an extension of this study, Burks et al.[79] suspended NIR to visible UCNPs in poly(dimethylsiloxane) (PDMS), and these composites were cast into biocompatible polymer disks (PDs). These PDs were then impregnated with RBS to give UCNP/RBS/PD devices that generate physiologically relevant concentrations of NO when subjected to 980 nm irradiation. In addition, these workers demonstrated that such NO uncaging could be facilitated by NIR irradiation through procine tissue filters. In terms of delivery, one can view these disks as potential implants from which a bioactive agent can be delivered upon triggering to enable on-demand timing and dosage control.[97] For such devices, energy transfer from UCNP to RBS was likely the result of emission followed by reabsorption, a requirement being that the photoNORM or photoCORM absorbs strongly at the wavelengths of the UCNP PL. In order to use a focused laser beam, Burks et al. designed a photolysis train where the laser beam was scanned over the PD surface to induce photo-uncaging while minimizing damage from localized heating.[79]

The Ostrowski group at Bowling Green University[98] carried out a related study using several different polymers to prepare PDs: poly(methyl methacrylate) (PMMA), polyurethane (PU) and PDMS. The photoNORM was the hydrophobic CrONO salt *trans*-[Cr(cyclam)(ONO)$_2$]BPh$_4$. This was dissolved in 1:1 acetone:tetrahydrofuran and the resulting solution was mixed with the polymer before gelling to encapsulate CrONO in the solid matrix. PDs were built in this manner from the three polymers with the photoNORM alone and (for the PDMS disks) with the photoNORM and a core:shell UCNP (NaYF$_4$):Yb/Er(20/2%)@NaYF$_4$ or (NaYF$_4$):Yb/Tm (20/1%)@NaYF$_4$. When irradiated with visible light (405 nm), NO release was noted for the PDMS samples but none was seen for PDs prepared from the other two polymers. The latter behavior was explained in terms of the poor gas permeability of those two materials. NIR excitation of the PDMS samples containing the UCNPs gave weak, but detectible, NO release.[98]

Po Ju Huang et al.[31] extended such studies of silicone polymer disks by examining the rates of NO release from disks prepared from six different biocompatible commercial formulations. The photoNORM in this case was the newly synthesized O-nitrito Cr(III) complex salt *trans*-[Cr(PetA)(ONO)$_2$](BF$_4$) (PetA = 5,14-dimethyl-7,12-diphenyl-1,4,8,11-tetraazacyclotetra-decane) (Fig. 16). The hydrophobic PetACrONO salt is photoactive toward NO release in isobutyronitrile solution when photolyzed

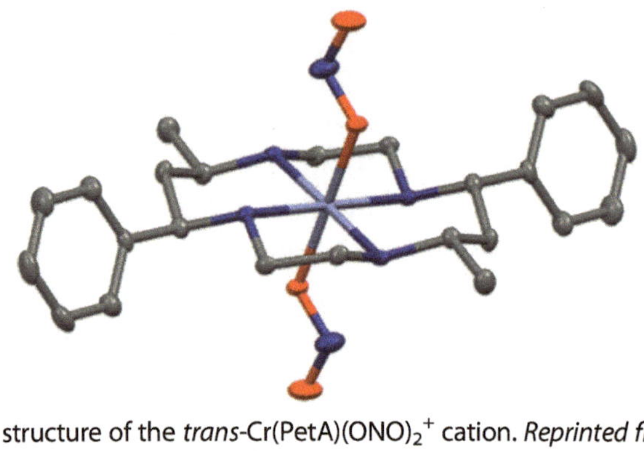

Fig. 16 Crystal structure of the *trans*-Cr(PetA)(ONO)$_2^+$ cation. *Reprinted from Huang, P.-J.; Garcia, J.V.; Fenwick, A.; Wu, G.; Ford, P.C.* ACS Omega ***2019****,* 4*, 9181 copyright 2019, American Chemical Society.*

with 451 nm light, giving a Φ_{NO} value of 0.32 ± 0.01, very similar to that seen for CrONO in aqueous solution. Rather than encapsulating PetACrONO, the photoNORM was infused into the PDs by soaking in a concentrated solution of the PetACrONO salt in tetrahydrofuran. The resulting devices were active for NO release from the 451 nm light from a blue light-emitting diode (LED) with the NO excaping over a period of a few 10's of seconds after a short (1–5 s) excitation pulse. Analogous PDs were prepared by encapsulating Nd^{3+}-doped UCNPs (β-$NaYF_4$:Tm/Nd/Yb/Gd@$NaYF_4$:Nd) into the polymer mixture before gelling and infusing the resulting PDs with $[Cr(PetA)(ONO)_2]$ (BF_4). The resulting devices could be activated with NIR light from a fiber-coupled diode laser operating at 794 nm, a wavelength that avoids the weak absorbance from water at 980 nm.

4.4 Nano-carriers

The studies described above that incorporate both UCNPs and photoNORMs in a solution or polymer disk, provide valuable proof of principle that NIR excitation using a simple CW diode laser can be used to uncage NO. However, although there are possible applications for the PDs as implants, one needs to combine together the photosensitizer and the photoNORM or photoCORM in some type of carrier for any application involving the circulatory system. In this context, John Garcia in our laboratory in collaboration with Fan Zhang at Fudan U[96] prepared core:shell $NaYF_4$:Yb/Er@$NaYF_4$ UCNPs with ~10 nm diameter cores and deposited onto these a layer of silica, then one of mesoporous silica. Roussin's black salt was then infused into the mesoporous silica, and the pores were sealed by coating the device with an amphiphilic polymer (poly(allylamine hydrochloride, PAH, Mw=56,000) to give a water-soluble nano-carrier (Fig. 17). Upon irradiation at 980 nm with a CW diode laser, these UCNPs displayed bright visible PL that overlapped the RBS absorption spectrum (Fig. 15) and photoactivated NO uncaging (Eq. 3). The result was a water-soluble nano-carrier that releases biologically relevant amounts of NO upon NIR excitation.

In a similar vein, Agustin Pierri in our laboratory in collaboration with Nanfeng Zheng of Xiamen University[99] designed nano-carriers with $NaGdF_4$:Tm,Yb@$NaGdF_4$ core:shell UCNPs terminated with hydrophobic oleate surface ligands. Coating with an amphiphilic phospholipid-functionalized PEG (DSPE-PEG2000) made these constructs water-soluble.

Fig. 17 The formation of a water soluble nanocarrier $NaYF_4$:Yb/Er@SiO_2@mpSiO_2 UCNP. The mesoporous silica shell was loaded with the photoNORM RBS and capped with an amphiphilic polymer PAH. NIR photolysis at 980 nm resulted in NO uncaging (Ref. 96).

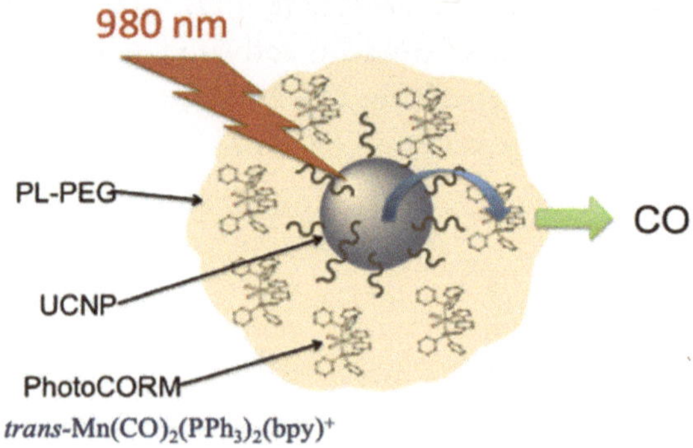

Fig. 18 Water-soluble photoCORM nano-carriers with a UCNP core. *Adapted with permission from Pierri, A.E.; Huang, P.-J.; Garcia, J.V.; Stanfill, J.G.; Chui, M.; Wu, G.; Zheng, N.; Ford, P.C.* Chem. Commun. ***2015****,* 51*, 2072, Pierri et al.* Chem. Commun. ***2015****,* 51*, 2072–2074, published by The Royal Society of Chemistry.*

The hydrophobic photoCORM *trans*-[$Mn(CO)_2(PPh_3)_2$(bpy)](CF_3SO_3), which has an absorption spectrum well-matched to the UCNP PL, was then infused into the lipid-like interior of these nano-carriers. NIR excitation (980 nm) of these stable, water-soluble nano-carrier ensembles led to the CO release as confirmed by gas chromatography of the head space (Fig. 18). A similar photoCORM nano-carrier activated by NIR-to-visible upconversion was subsequently reported by Liu and coworkers.[100]

Po-Ju Huang of this laboratory has developed a nano-carrier with N-doped graphene QD cores.[101] Amide coupling of surface carboxylates with oleylamine gave lipophilic surface domains that were assembled with the amphiphilic biotinylated phospholipid-functionalized poly(ethylene glycol) DSPE-PEG to give the nano-carrier GQD@ DSPE-PEG. The resulting lipid domains were loaded with PetACrONO, giving fully loaded nano-carriers in the 20–70 nm size range. NO uncaging was observed upon 794 nm excitation with a CW diode laser of an aqueous suspension of these loaded nano-carriers in pH 7.4 PBS solution. It is noteworthy the amount of

NO uncaged was approximately linear in laser power and in irradiation time in contrast to the expectation for a multi-photon process. Thus, the preliminary conclusion of this incomplete study was that a photothermal rather than an upconversion process may be responsible for triggering NO release.

5. Biological targeting

Physiological targeting can be accomplished by direct injection of photocages into the desired site or by implanting a device such as the polymer disks described above. Indeed, if such an implant were attached to an optical fiber, there would be less need to make the specific photoNORM or photoCORM responsive to tissue-transmitting wavelengths. Such implants would require that the bioactive compound uncaged can diffuse out of the PD. This should not be a serious problem for NO or CO assuming that the polymer is sufficiently gas permeable,[98] but might not be as effective with larger molecules.

A less invasive approach would be multi-function nanoparticles (NPs) that could be introduced to sites of therapeutic interest via the circulatory system. Quantum dots (QDs), upconverting nanoparticles (UCNPs) and metallic NPs[102,103] can serve as antennas to enhance light absorption at desired wavelengths or simply as photochemically benign carriers to transport a caged compound to targeted sites.[104] Described here are several approaches utilized for photouncaging NO and CO, but these methods potentially have broader applications.

5.1 Liposomes

Liposomes are also possible vehicles for delivering lipophilic photoNORMs to specific targets owing to the enhanced permeation and retention (EPR) effect which accumulates particles in tumors.[105] Such carriers provide the means to maintain a localized high concentration of a photocage and can be modified for in vivo targeting. In this context, Ostrowski and coworkers[106] encapsulated the BF_4^- salts of CrONO and of the anthracene tethered complex Anth-CrONO (Fig. 6) in phosphatidylcholine liposomes (Fig. 19). These constructs allowed for a controlled and steady release of NO after photolysis. In addition, the residual fluorescence from Anth-CrONO allows tracking these vessels simultaneous with therapeutic NO release. Subsequently, Nakanishi et al.[107] reported the loading of liposomes with a photoNORM based on the Ru salen nitrosyl platform described above[41] modified by attaching highly lipophilic, pendant cholesterol groups.

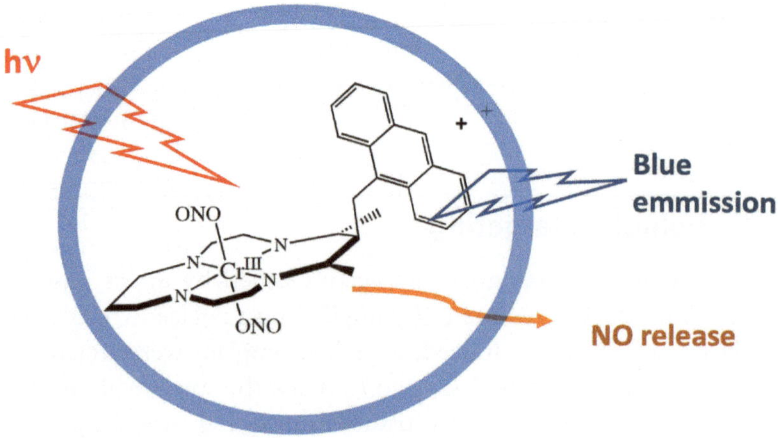

Fig. 19 Illustration of a liposome encapsulating the luminescent salt [Anth-CrONO]BF_4. Photolysis leads both to NO release and detectable emission. *Reprinted with permission from Ostrowski, A.D.; Lin, B.F.; Tirrell, M.V.; Ford, P.C.* Mol. Pharm. ***2012****, 9, 2950–2955 Copyright 2012 American Chemical Society.*

5.2 Hollow gold nanoshells

The affinity of gold or platinum for thiols has been used to facilitate the assembly of different functionalities on nanocarriers composed of these metals.[102,108] When light is absorbed by the size dependent localized surface plasmon resonance of noble metal nanoparticles the heat generated can trigger photothermal processes.[102a] In a collaborative study with UCSB colleague Norbert Reich, we have demonstrated that these properties could be combined to deliver NO to specific cell types. Liz Levy et al.[103] assembled nano-carriers based hollow gold nanoshells (HGNs, ~65 nm diameters) with various functionalities (Fig. 20). The ensemble was built from a HGN core, the thiolated NO precursor ammonium N-nitroso (4-mercaptomethylphenyl)-hydroxylamine (TCF), and a thiolated poly-ethylene glycol (TPEG) to provide aqueous solubility. The HGN core displayed a broad, strong plasmon absorption band centered at ~750 nm. NIR excitation (800 nm) of these TCF–HGN–TPEG constructs in pH 7.4 buffered solution using the output from an ultrafast Ti:sapphire laser uncage the NO from the TCF. Notably, NO release could also be effected by excitation with a CW diode laser operating at 800 nm, although the process was considerably less efficient. This behavior is consistent with non-linear absorption of light by the HGN surface plasmon leading to localized heating and consequently to thermochemical NO release from the TCF.

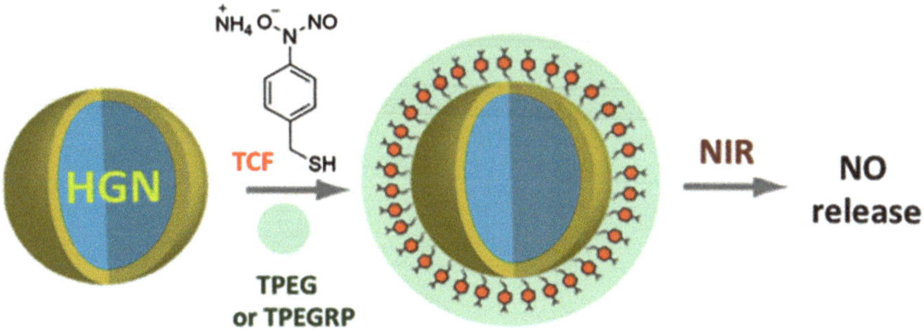

Fig. 20 HGN surfaces are coated by co-absorption of TCF and TPEG or TPEGRP including a cell-targeting peptide. Laser excitation (800 nm) of these conjugates results in NO release with spatial-temporal control. *With permission from Levy, E.S.; Morales, D.P.; Garcia, J.V.; Reich, N.O.; Ford, P.C.* Chem. Commun. ***2015***, 51, *17692–17695 The Royal Society of Chemistry.*

Analogous water HGN constructs were prepared with a thiolated polyethylene glycol modified to include the terminally bound targeting peptide RPARPAR (TCF–HGN–TPEGRP).[103] The TPEGRP peptide mediates cellular endocytosis for cells that overexpress the Neuropilin-1 receptor.[109] Thus, these conjugates were readily internalized by PPC-1 or 22Rv1 prostate cancer cells, but not by HeLa cells which lack this receptor. NIR photolysis at 800 nm of nano-carriers internalized in 22Rv1cell clearly demonstrated intracellular NO-release as shown with the DAF-2A indicator. The cell-specific targeting peptide in the TCF–HGN–TPEGRP conjugates and the NIR-mediated control of NO release provides an excellent basis for investigating the spatio-temporal response to this important SMB.

5.3 Macrophage Trojan Horses

Another very promising strategy would be to recruit the host's immune cells to serve as "Trojan Horses" to carry payloads such as photocages to the physiological targets (Fig. 21).[110] Macrophages and monocytes can cross the blood brain barrier and will accumulate in tumors and other sites in response to inflammatory signals.[111] Our premise in the study described here was that these immune cells could be loaded in some manner with a photoNORM and that once these cells localize in the targeted tissue, NO can be photo-uncaged by NIR excitation.

In a collaboration with bioengineering colleague Samir Mitagotri, Michael Evans, Po-Ju Huang et al.[112] prepared micron-sized, poly(lactic-co-glycolic acid) (PLGA) biodegradable microparticles into which was

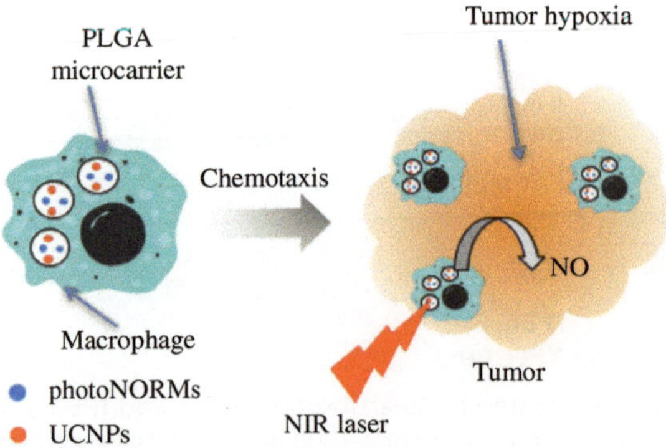

Fig. 21 Trojan Horse mechanism using macrophages to transport PLGA microcarriers loaded with UCNPs and a photoNORM to tumors or other sites of inflammation. *Adapted from Evans, M.A.; Huang, P.-J.; Iwamoto, Y.; Ibsen, K.N.; Chan, E.M.; Hitomi, Y.; Ford, P.C.; Mitragotri, S.* Chem. Sci. ***2018****, 9, 3729–3741 with permission from The Royal Society of Chemistry.*

incorporated a theranostic payload. The encapsulated payload here included Yutaka Hitomi's photoNORM $[Mn(NO)dpaq^{NO2}]BPh_4$ (Fig. 10)[81] together with Nd^{3+} doped upconverting nanoparticles (Nd-UCNPs) to provide imaging capabilities. These micro-carriers were essentially miniaturized versions of the biocompatible PDs described above. Confocal microscopy showed that the macrophages readily ingested such micron-sized particles as illustrated in Fig. 22, when IgG antibodies were incorporated in the micro-carrier preparation. The microcarrier loaded macrophages were then shown to invade laboratory grown tumor spheroids as seen in the microscope image shown in Fig. 22. The presence of the PLGA microcarriers is evidenced by the green emission from the incorporated Nd-UCNPs when irradiated with a CW diode laser operating at 794 nm. Notably, the PLGA microcarriers alone were ineffective in penetrating the spheroids.

NIR Irradiation (λ_{irr} 794 nm) of these tumor spheroids invaded by Trojan Horse macrophages carrying PLGA microcarriers loaded with UCNPs and the photoNORM led to quantifiable NO release as illustrated by the cartoons in Fig. 21. Flow cytometry experiments demonstrated that with low NIR light doses, the result of photoexcitation was a reduction in hypoxia inducible factor-1α (HIF-1α) levels in the tumor spheroids. In contrast, high NO doses promoted cell death, presumably by nitrosative stress induced apoptosis triggered by the higher amounts of uncaged NO.[113] Thus, these macrophage Trojan Horses were shown to carry a NIR-activated theranostic payload into the tumor spheroid model, and this

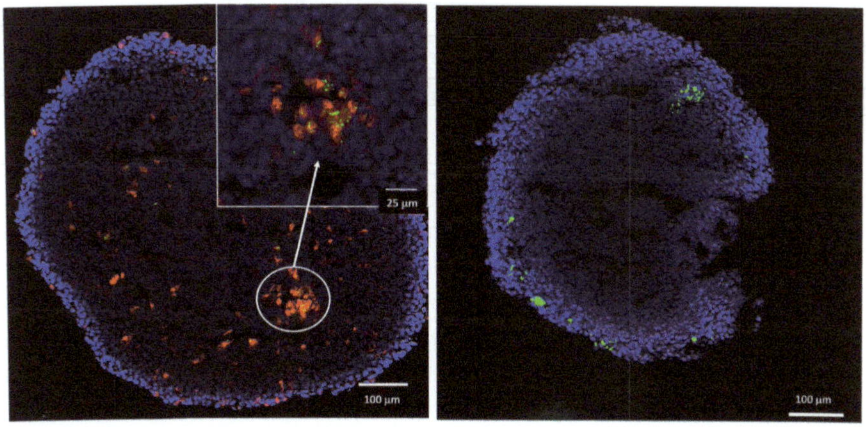

Fig. 22 Left: Slices of 4T1/NIH-3T3 cocultured breast tumor spheroids incubated with microparticles-laden macrophages and imaged by confocal microscopy. Blue is 40,6-diamidino-2-phenylindole (DAPI) and red represents macrophages stained with CellTracker™ deep red. The green spots (inset) are Cy-3–IgG labeled microparticles. Right: Confocal image of tumor spheroids incubated with Cy-3–IgG labled microparticles only. Note that the microparticles alone do not penetrate into the spheroid. *Adapted from Evans, M.A.; Huang, P.-J.; Iwamoto, Y.; Ibsen, K.N.; Chan, E.M.; Hitomi, Y.; Ford, P.C.; Mitragotri, S.* Chem. Sci. ***2018****, 9, 3729–3741 with permission from The Royal Society of Chemistry.*

allowed us to control biological response by varying the timing and intensity of the irradiation. This study[111] provides a proof-of-principle model illustrating the potential of immune cells to transport a payload to targeted tissue, where photo-uncaging of NO (or some other therapeutic drug) can be effected by tissue penetrating NIR light. However, while this strategy appears to be very promising, it will have to be extended to animal models to demonstrate that the host's immune cells themselves can be used to serve in this manner.

6. Summary

This chapter has described the evolution of research in our laboratories at the University of California, Santa Barbara largely oriented toward applications of transition metal complexes photo-cages of small molecule bioregulators for potential therapeutic applications. Although the physiological applications of such delivery may differ, the fundamental principles remain the same. The caged therapeutic should be biocompatible and relatively benign before activation with light and needs to be responsive to the photolysis wavelengths that can be delivered to the desired target. For deep tissue such as a tumor, these wavelengths will be in the deep red or near infrared, the most penetrating wavelengths. In addition, it is highly desirable to deliver the photochemical precursor to the target site selectively. While delivery might be accomplished by injection or implantation, biological targeting mechanisms would be superior. With regard to the latter issue, the use of cell selective targeting peptides or of the macrophage "Trojan Horse" strategy would appear to be especially attractive.

There are many directions this type of research may take, but at UCSB we have recently become intrigued by the idea that the radiation that is used in the radiotherapy of tumors and other deep tissue sights might also be used to trigger the release of a chemotherapeutic drug or a radiation sensitizer such as NO. In this context we are now exploring radiation induced scintillators that might be used to trigger the release of an SMB from a photoNORM or photoCORM.

Acknowledgments

The research at UC Santa Barbara described in this article was largely supported by a series of grants from the Chemistry Division of the US National Science Foundation. Many of the nanoparticles used in these studies were prepared by collaborators at the Molecular Foundry of the Lawrence Berkeley National Laboratory, supported by the Office of Science, Office of Basic Energy Sciences, of the U. S. Department of Energy. A large number of outstanding undergraduate, graduate and postdoctoral research scholars, as well as visiting scholars and other collaborators, contributed to the experimental studies and intellectual concepts described in this review.

References

1. Maragos, C. M.; Morley, D.; Wink, D. A.; Dunams, T. M.; Saaverda, J. E.; Hoffman, A.; Bove, A. A.; Isaac, L.; Hrabie, J. A.; Keefer, L. K. *J. Med. Chem.* **1991**, *34*, 3242–3247.
2. Moncada, S.; Palmer, R. M. J.; Higgs, E. A. *Pharmacol. Rev.* **1991**, *43*, 109–142.
3. Ignarro, L. J.; Freeman, B., Eds. *Nitric Oxide: Biology and Pathobiology*; 3rd ed.; Academic Press: Burlington, MA, 2017. and references therein.
4. Ford, P. C.; Wink, D. A.; Stanbury, D. M. *FEBS Lett.* **1993**, *326*, 1–3.
5. Tran, D.; Shelton, B. W.; White, A. H.; Laverman, L. E.; Ford, P. C. *Inorg. Chem.* **1998**, *37*, 2505–2511.
6. Fernandez, B. O.; Ford, P. C. *J. Am. Chem. Soc.* **2003**, *125*, 10510–10511.
7. Tsuge, K.; DeRosa, F.; Lim, M. D.; Ford, P. C. *J. Am. Chem. Soc.* **2004**, *126*, 6564–6565.
8. Heinecke, J. L.; Khin, C.; Pereira, J. C. M.; Suárez, S. A.; Iretskii, A. V.; Doctorovich, F.; Ford, P. C. *J. Am. Chem. Soc.* **2013**, *135*, 4007–4017.
9. Ford, P. C.; Pereira, J. C. M.; Miranda, K. M. *Struct. Bond.* **2014**, *154*, 99–136.
10. Truzzi, D. R.; Augusto, O.; Iretskii, A. V.; Ford, P. C. *Inorg. Chem.* **2019**, *58*, 13446–13456.

11. Kurtikyan, T. S.; Hayrapetyan, V. A.; Hovhannisyan, A. A.; Martirosyan, G. G.; Hovhannisyan, G. S.; Iretski, A. V.; Ford, P. C. *Inorg. Chem.* **2020**, *59*, 17224–17233.
12. Hoshino, M.; Ozawa, K.; Seki, H.; Ford, P. C. *J. Am. Chem. Soc.* **1993**, *115*, 9568–9575.
13. (a) Laverman, L. E.; Hoshino, M.; Ford, P. C. *J. Am. Chem. Soc.* **1997**, *119*, 12663–12664; (b) Laverman, L. E.; Ford, P. C. *J. Am. Chem. Soc.* **2001**, *123*, 11614–11622.
14. Pereira, J. C. M.; Iretskii, A. V.; Han, R.-M.; Ford, P. C. *J. Am. Chem. Soc.* **2015**, *137*, 328–336.
15. (a) Lundberg, J. O.; Weitzberg, E.; Gladwin, M. T. *Nat. Rev. Drug Discov.* **2008**, *7*, 156–167; (b) DeMartino, A. W.; Kim-Shapiro, D. B.; Patel, R. P.; Gladwin, M. T. *Brit. J. Pharm.* **2019**, *176*, 228–245; (c) Feelisch, M.; Fernandez, B. O.; Bryan, N. S.; Garcia-Saura, M. F.; Bauer, S.; Whitlock, D. R.; Ford, P. C.; Janero, D. R.; Rodriguez, J.; Ashrafian, H. *J. Biol. Chem.* **2008**, *283*, 33927–33934; (d) Zweier, J. L.; Li, H.; Samouilov, A. *Nitric Oxide-Biol. Chem.* **2010**, *22*, 83–90.
16. (a) Miranda, K. M. *Coord. Chem. Rev.* **2005**, *249*, 433–455; (b) Miao, Z.; King, S. B. *Nitric Oxide-Biol. Chem.* **2016**, *57*, 1–14; (c) Fukuto, J. M. *Brit. J. Pharm.* **2019**, *176*, 135–146.
17. (a) Szabo, C. *Nat. Rev. Drug Discov.* **2016**, *15*, 185–203; (b) Fukuto, J. M.; Carrington, S. J.; Tantillo, D. J.; Harrison, J. G.; Ignarro, L. J.; Freeman, B. A.; Chen, A.; Wink, D. A. *Chem. Res. Toxicol.* **2012**, *25* (4), 769–793.
18. Sjostrand, T. *Scand. J. Clin. Lab. Invest.* **1949**, *1*, 201. Also Nature 1949, 164, 580–581.
19. (a) Motterlini, R.; Mann, B. E.; Foresti, R. *Exp. Opin. Invest. Drugs* **2005**, *14*, 1305–1318; (b) Siracusa, R.; Schau, A.; Calabrese, V.; Fuller, P. M.; Otterbein, L. E. *Trends Pharmacol. Sci.* **2021**, *42*, 329–339.
20. Ford, P. C. *Accts. Chem. Res.* **2008**, *41*, 190–200.
21. Garcia, J. V.; Zhang, F.; Ford, P. C. *Philos. Trans. R. Soc. A* **2013**, *371* (1995), 20120129/1–20120129/25.
22. Rimmer, R. D.; Richter, H.; Ford, P. C. *Inorg. Chem.* **2010**, *49*, 1180–1185.
23. (a) Ford, P. C.; Bourassa, J.; Miranda, K. M.; Lee, B.; Lorkovic, I.; Boggs, S.; Kudo, S.; Laverman, L. *Coord. Chem. Rev.* **1998**, *171*, 185–202; (b) Miranda, K. M.; Bu, X.; Lorkovic, I.; Ford, P. C. *Inorg. Chem.* **1997**, *36*, 4838–4848; (c) Lorkovic, I. M.; Miranda, K. M.; Lee, B.; Bernhard, S.; Schoonover, J. R.; Ford, P. C. *J. Am. Chem. Soc.* **1998**, *120*, 11674–11683.
24. Bourassa, J.; DeGraff, W.; Kudo, S.; Wink, D. A.; Mitchell, J. B.; Ford, P. C. *J. Am. Chem. Soc.* **1997**, *119*, 2853–2860.
25. (a) Kudo, S.; Bourassa, J. L.; Boggs, S. E.; Sato, Y.; Ford, P. C. *Analyt. Biochem* **1997**, *247*, 193–202; (b) Christodoulou, D.; Kudo, S.; Cook, J. A.; Krishna, M. C.; Miles, A.; Grisham, M. B.; Ford, P. C.; Wink, D. A. *Meth. Enzymol.* **1996**, *268*, 69–83.
26. Matthews, E. K.; Seaton, E. D.; Forsyth, M. J.; Humphrey, P. P. A. *Br. J. Pharmacol.* **1994**, *113*, 87–94.
27. Bates, J. N.; Baker, M. T.; Guerra, R., Jr.; Harrison, D. G. *Biochem. Pharmacol.* **1991**, *42*, S157–S165.
28. Bourassa, J.; Lee, B.; Bernhard, S.; Schoonover, J.; Ford, P. C. *Inorg. Chem.* **1999**, *38*, 2947–2952.
29. Bourassa, J.; Ford, P. C. *Coord. Chem. Rev.* **2000**, *200–202*, 887–900.
30. Rodrigues, F. P.; Macedo, L. J. A.; Maximo, L. N. C.; Sales, F. C. P. F.; da Silva, R. S.; Crespilho, F. N. *Nitric Oxide Biol. Chem.* **2020**, *96*, 29–34.
31. Huang, P.-J.; Garcia, J. V.; Fenwick, A.; Wu, G.; Ford, P. C. *ACS Omega* **2019**, *4*, 9181–9187.
32. Kojima, H.; Sakurai, K.; Kikuchi, K.; Kawahara, S.; Kirino, Y.; Nagoshi, H.; Hirata, Y.; Nagano, T. *Chem. Pharm. Bull.* **1998**, *46*, 373–375.

33. Conrado, C. L.; Bourassa, J. L.; Egler, C.; Wecksler, S.; Ford, P. C. *Inorg. Chem.* **2003**, *42*, 2288–2293.
34. Conrado, C. L.; Wecksler, S.; Egler, C.; Magde, D.; Ford, P. C. *Inorg. Chem* **2004**, *43*, 5543–5549.
35. Yoon, H.; Park, S.; Lim, M. *J. Phys. Chem. Lett.* **2020**, *11*, 3198–3202.
36. Yoon, H.; Park, S.; Lim, M. *ACS Omega* **2021**, *6*, 27158–27169.
37. (a) Rose, M. J.; Mascharak, P. K. *Coord. Chem. Rev.* **2008**, *252*, 2093–2114; (b) Tfouni, E.; Doro, F. G.; Gomes, A. J.; Silva, R. S.d.; Metzker, G.; Benini, P. G. A. Z.; Franco, D. W. *Coord. Chem. Rev.* **2010**, *254*, 355–371.
38. Merkle, A. C.; McQuarters, A. B.; Lehnert, N. *Dalton Trans.* **2012**, *41*, 8047–8059.
39. Gomes, A. J.; Espreafico, E. M.; Tfouni, E. *Mol. Pharm.* **2013**, *10*, 3544–3554.
40. Ramos, L. C. B.; Pereira, M. S. P.; Callejon, D.; Baruffi, M.; Lunardi, C. N.; Slep, L. D.; Bendhack, L. M.; da Silva, R. S. *Eur. J. Inorg. Chem.* **2016**, *22*, 3592–3597.
41. (a) Works, C. F.; Ford, P. C. *J. Am. Chem. Soc.* **2000**, *122*, 7592–7593; (b) Works, C. F.; Jocher, C. J.; Bart, G. D.; Bu, X.; Ford, P. C. *Inorg. Chem.* **2002**, *41*, 3728–3739.
42. (a) Bordini, J.; Hughes, D. L.; Da Motta Neto, J. D.; Jorge da Cunha, C. *Inorg. Chem.* **2002**, *41*, 5410–5416; (b) Bordini, J.; Ford, P. C.; Tfouni, E. *Chem. Commun.* **2005**, 4169–4171; (c) Bordini, J.; Novaes, D.; Borissevitch, I.; Owens, B.; Ford, P. C.; Tfouni, E. *Inorg. Chim. Acta* **2008**, *361*, 2252–2258.
43. Crisalli, M. A.; Franco, L. P.; Silva, B. R.; Holanda, A. K. M.; Bendhack, L. M.; da Silva, R. S.; Ford, P. C. *J. Coord. Chem.* **2018**, *71*, 1690–1703.
44. (a) De Leo, M. A.; Ford, P. C. *J. Am. Chem. Soc.* **1999**, *121*, 1980–1981; (b) De Leo, M. A.; Ford, P. C. *Coord. Chem. Rev.* **2000**, *208*, 47–59.
45. (a) Ostrowski, A. D.; Deakin, S. J.; Azhar, B.; Miller, T. W.; Franco, N.; Cherney, M. M.; Lee, A. J.; Burstyn, J. N.; Fukuto, J. M.; Megson, I. L.; Ford, P. C. *J. Med. Chem.* **2010**, *53*, 715–722; (b) Ostrowski, A. D.; Absalonson, R. O.; DeLeo, M. A.; Wu, G.; Pavlovich, J. G.; Adamson, J.; Azhar, B.; Iretskii, A. V.; Megson, I. L.; Ford, P. C. *Inorg. Chem.* **2011**, *50*, 4453–4462. Correction: *Inorg. Chem.* **2011**. *50*, 5848.
46. Yamaji, M.; Hama, Y.; Miyazake, M.; Hoshino, M. *Inorg. Chem.* **1992**, *31*, 932–934.
47. Radford, R. J.; Lim, M. D.; da Silva, R. S.; Ford, P. C. *J. Coord. Chem.* **2010**, *63*, 2743–2749.
48. (a) Motterlini, R.; Otterbein, L. E. *Nat. Rev. Drug Discov.* **2010**, *9*, 728–743; (b) Romao, C. C.; Blaettler, W. A.; Seixas, J. D.; Bernardes, G. J. L. *Chem. Soc. Rev.* **2012**, *41*, 3571–3583; (c) García-Gallego, S.; Bernardes, G. J. L. *Angew. Chem. Int. Ed.* **2014**, *53*, 9712–9721; (d) Schatzschneider, U. *Brit. J. Pharmacol.* **2015**, *172*, 1638–1650; (e) Ling, K.; Men, F.; Wang, W.-C.; Zhou, Y.-Q.; Zhang, H.-W.; Ye, D. W. *J. Med. Chem.* **2018**, *61*, 2611–2635.
49. Wareham, L. K.; Poole, R. K.; Tinajero-Trejo, M. *J. Biol. Chem.* **2015**, *290*, 18999–19007.
50. (a) Pinto, M. N.; Mascharak, P. K. *J. PhotoChem. Photobiol. C* **2020**, *42*, 100341; (b) Marhenke, J.; Travino, K.; Works, C. *Coord. Chem. Rev.* **2016**, *306*, 533–543.
51. (a) Fayad-Kobeissi, S.; Ratovonantenaina, J.; Dabiré, H.; Wilson, L. J.; Rodriguez, A. M.; Berdeaux, A.; Dubois-Randé, J. L.; Mann, B. E.; Motterini, R.; Foresti, R. *Biochem. Pharmacol.* **2016**, *102*, 64–77; (b) Jin, Z.; Wen, Y.; Xiong, L.; Yang, T.; Zhao, P.; Tan, L.; Wang, T.; Qian, Z.; Su, B. L.; He, Q. *Chem. Commun.* **2017**, *53*, 5557–5560; (c) Barrett, J. A.; Li, Z.; Garcia, J. V.; Wein, E.; Zheng, D.; Hunt, C.; Ngo, L.; Sepunaru, L.; Iretskii, A. V.; Ford, P. C. *R. Soc. Open Sci.* **2021**, *8*, 211022.
52. Wrighton, M. S. *Chem. Rev.* **1974**, *74*, 401–430.
53. Motterlini, R.; Clark, J. E.; Foresti, R.; Sarathchandra, P.; Mann, B. E.; Green, C. *Circ. Res.* **2002**, *90*, e17–e24.

54. Niesel, J.; Pinto, A.; Dongo, H. W. P. N.'.; Merz, A. K.; Ott, I.; Gust, B. R.; Schatzschneider, U. *Chem. Commun.* **2008**, *44*, 1798–1800.
55. Wagenknecht, P. S.; Ford, P. C. *Coord. Chem. Rev.* **2011**, *255*, 591–616.
56. (a) Wrighton, M. S.; Abrahamson, H. B.; Morses, D. L. *J. Am. Chem. Soc.* **1976**, *98*, 4105; (b) Lees, A. J.; Adamson, A. W. *J. Am. Chem. Soc.* **1980**, *102*, 6874; (c) Wieland, S.; van Eldik, R.; Crane, D. R.; Ford, P. C. *Inorg. Chem.* **1989**, *28*, 3663–3666.
57. Atkin, A. J.; Lynam, J. M.; Moulton, B. E.; Sawle, P.; Motterlini, R.; Boyle, N. M.; Pryce, M. T.; Fairlamb, I. J. S. *Dalton Trans.* **2011**, *40*, 5755.
58. Li, Z. *Toward Long Wavelength Absorption: Dinuclear Photo-activated CO Releasing Moieties*; Ph.D. Dissertation, University of California: Santa Barbara, 2017.
59. (a) Mukhopadhyay, S.; Sarkar, A.; Chattopadhyay, P.; Dhara, K. *Chem. Asian J.* **2020**, *15*, 3162–3179; (b) Ohata, J.; Bruemmer, K. J.; Chang, C. J. *Acc. Chem. Res.* **2019**, *52*, 2841–2848.
60. Darensbourg, D. J.; Bischoff, C. J. *Inorg. Chem.* **1993**, *32*, 47–53.
61. Pierri, A. E.; Pallaoro, A.; Wu, G.; Ford, P. C. *J. Am. Chem. Soc.* **2012**, *134*, 18197–18200. correction: J. Am. Chem. Soc. **2018**, 140, 525.
62. Chakraborty, I.; Carrington, S. J.; Roseman, G.; Mascharak, P. K. *Inorg. Chem.* **2017**, *56*, 1534–1545.
63. Marker, S. C.; MacMillan, S. N.; Zipfel, W. R.; Li, Z.; Ford, P. C.; Wilson, J. *J. Inorg. Chem.* **2018**, *57*, 1311–1331.
64. DeRosa, F.; Bu, X.; Ford, P. C. *Inorg. Chem.* **2005**, *44*, 4157–4165.
65. (a) Filevich, O.; Zayat, L.; Baraldo, L. M.; Etchenique, R. *Struct. Bond.* **2015**, *165*, 47–68; (b) Respondek, T.; Garner, R. N.; Herroon, M. K.; Podgorski, I.; Turro, C.; Kodanko, J. J. *J. Am. Chem. Soc.* **2011**, *133*, 17164–17167; (c) Knoll, J. D.; Albani, B. A.; Turro, C. *Acc. Chem. Res.* **2015**, *48*, 2280–2287; (d) Toupin, N. P.; Nadella, S.; Steinke, S. J.; Turro, C.; Kodanko, J. *J. Inorg. Chem.* **2020**, *59*, 3919–3933; (e) Heinemann, F.; Karges, J.; Gasser, G. *Accts Chem. Res.* **2017**, *50*, 2727–2736; (f) Wachter, E.; Heidary, D. K.; Howerton, B. S.; Parkin, S.; Glazer, E. C. *Chem. Commun.* **2012**, *48*, 9649–9651; (g) Zhao, J.; Sun, S.; Gou, S.; Wang, X.; Wang, Z.; Li, X.; Zhang, W. *J. Inorg. Biochem.* **2019**, *196*, 110684–110693.
66. (a) Bischof, C.; Joshi, T.; Dimri, A.; Spiccia, L.; Schatzschneider, U. *Inorg. Chem.* **2013**, *52*, 9297–9308; (b) Kubeil, M.; Vernooij, R. R.; Kubeil, C.; Wood, B. R.; Graham, B.; Stephan, H.; Spiccia, L. *Inorg. Chem.* **2017**, *56*, 5941–5952; (c) de Sousa, A. P.; Carvalho, E. M.; Ellena, J.; Sousa, E. H. S.; de Sousa, J. R.; Lopes, L. G. F.; Ford, P. C.; Holanda, A. K. M. *J. Inorg. Biochem.* **2017**, *173*, 144–151; (d) Mansour, A. M. *Eur. J. Inorg. Chem.* **2018**, 852–860; (e) Gonzalez, M. A.; Carrington, S. J.; Chakraborty, I.; Olmstead, M. M.; Mascharak, P. K. *Inorg. Chem.* **2013**, *52*, 11320–11331.
67. Geri, S.; Krunclova, T.; Janouskova, O.; Panek, J.; Hruby, M.; Hernández-Valdés, D.; Probst, B.; Alberto, R. A.; Mamat, C.; Kubeil, M.; Stephan, H. *Chem. Eur. J.* **2020**, *26*, 10992–11006.
68. Alivisatos, A. P. *J. Phys. Chem.* **1996**, *100*, 13226–13239.
69. Samia, A. C. S.; Chen, X.; Burda, C. *J. Am. Chem. Soc.* **2003**, *125*, 15736–15737.
70. Burks, P. T.; Ford, P. C. *Dalton Trans.* **2012**, *41*, 13030–13042.
71. Neuman, D.; Ostrowski, A. D.; Absalonson, R. O.; Strouse, G. F.; Ford, P. C. *J. Am. Chem. Soc.* **2007**, *129*, 4146–4147.
72. Neuman, D.; Ostrowski, A. D.; Mikhailovsky, A. A.; Absalonson, R. O.; Strouse, G. F.; Ford, P. C. *J. Am. Chem. Soc.* **2008**, *130*, 168–175.
73. Burks, P. T.; Ostrowski, A. D.; Mikhailovsky, A. A.; Chan, E. M.; Wagenknecht, P. S.; Ford, P. C. *J. Am. Chem. Soc.* **2012**, *134*, 13266–13275.

74. Franco, L. P.; Cicillini, S. A.; Biazzotto, J. C.; Schiavon, M. A.; Mikhailovsky, A.; Burks, P.; Garcia, J.; Ford, P. C.; da Silva, R. S. *J. Phys. Chem. A* **2014**, *118*, 12184–12191.
75. (a) Bernt, C. M.; Burks, P. T.; DeMartino, A. W.; Pierri, A. E.; Levy, E. S.; Zigler, D. F.; Ford, P. C. *J. Am. Chem. Soc.* **2014**, *136*, 2192–2195; (b) DeMartino, A. W.; Zigler, D. F.; Fukuto, J. M.; Ford, P. C. *Chem. Soc. Rev.* **2017**, *46*, 21–39.
76. (a) Tan, L.; Wan, A.; Li, H. *Langmuir* **2013**, *29*, 15032–15042; (b) Tan, L.; Wan, A.; Li, H. *ACS Appl. Mater. Interfaces* **2013**, *5*, 11163–11171.
77. (a) Guo, M.; Xiang, H. J.; Wang, Y.; Zhang, Q. L.; An, L.; Yang, S. P.; Ma, Y.; Wang, Y.; Liu, J. G. *Chem. Commun.* **2017**, *53*, 3253–3256; (b) Deng, Q.; Xiang, H. J.; Tang, W. W.; An, L.; Yang, S. P.; Zhang, Q. L.; Liu, J. G. *J. Inorg. Biochem.* **2016**, *165*, 152–158; (c) Li, Y.-H.; Guo, M.; Shi, S.-W.; Zhang, Q.-L.; Yang, S.-P.; Liu, J.-G. *J. Mater. Chem. B* **2017**, *5*, 7831–7838.
78. (a) Konig, K. *J. Microsc.* **2000**, *200*, 83–104; (b) Polesskaya, O.; Baranova, A.; Bui, S.; et al. Optogenetic regulation of transcription. *BMC Neurosci.* **2018**, *19*, 12.
79. Burks, P. T.; Garcia, J. V.; GonzalezIrias, R.; Tillman, J. T.; Niu, M.; Mikhailovsky, A. A.; Zhang, J.; Zhang, F.; Ford, P. C. *J. Am. Chem. Soc.* **2013**, *135*, 18145–18152.
80. Eroy-Reveles, A. A.; Leung, Y.; Beavers, C. M.; Olmstead, M. M.; Mascharak, P. K. *J. Am. Chem. Soc.* **2008**, *130*, 4447–4458.
81. Hitomi, Y.; Iwamoto, Y.; Kodera, M. *Dalton Trans.* **2013**, *43*, 2161–2167.
82. Li, Z.; Pierri, A. E.; Huang, P.-J.; Wu, G.; Iretskii, A. V.; Ford, P. C. *Inorg. Chem.* **2017**, *56*, 6094–6104.
83. Wecksler, S. R.; Mikhailovsky, A.; Ford, P. C. *J. Am. Chem. Soc.* **2004**, *126*, 13566–13567.
84. (a) Wecksler, S. R.; Mikhailovsky, A.; Korystov, D.; Ford, P. C. *J. Am. Chem. Soc.* **2006**, *128*, 3831–3837; (b) Wecksler, S. R.; Hutchinson, J.; Ford, P. C. *Inorg. Chem.* **2006**, *45*, 1192–1200.
85. Wecksler, S. R.; Mikhailovsky, A.; Korystov, D.; Buller, F.; Kannan, R.; Tan, L.-S.; Ford, P. C. *Inorg. Chem.* **2007**, *46*, 395–402.
86. Zheng, Q.; Bonoiu, A.; Ohulchansky, T. Y.; He, G. S.; Prasad, P. N. *Mol. Pharm.* **2008**, *5*, 389–398.
87. Lehnert, N.; Kim, E.; Dong, H. T.; Harland, J. B.; Hunt, A. P.; Manickas, E. C.; Oakley, K. M.; Pham, J.; Reed, G. C.; Alfaro, V. S. *Chem. Rev.* **2021**, *121*, 14682–14905.
88. Gong, W.; Xia, C.; He, Q. *Wiley Interdiscip. Rev. Nanomed. Nanobiotech.* **2021**, e1744.
89. Weinstain, R.; Slanina, T.; Kand, D.; Klan, P. *Chem Rev.* **2020**, *120*, 13135–13272.
90. Jiang, Q.; Xia, Y.; Barrett, J.; Mikhailovsky, A.; Wu, G.; Wang, D.; Shi, P.; Ford, P. C. *Inorg. Chem.* **2019**, *58*, 11066–11075.
91. Ramu, V.; Reddy, G. U.; Liu, J.; Hoffmann, P.; Sollapur, R.; Wyrwa, R.; Kupfer, S.; Spielmann, C.; Bonnet, S.; Neugebauer, U.; Schiller, A. *Chem. Eur. J.* **2019**, *25*, 8453–8458.
92. Ford, P. C. *Coord. Chem. Rev.* **2018**, *376*, 548–564.
93. Zhang, F.; Wan, Y.; Yu, T.; Zhang, F.; Shi, Y.; Xie, S.; Li, Y.; Xu, L.; Tu, B.; Zhao, D. *Angew. Chem. Int. Ed.* **2007**, *46*, 7976–7979.
94. (a) Wang, F.; Banerjee, D.; Liu, Y.; Chen, X. Y.; Liu, X. G. *Analyst* **2010**, *135*, 1839–1854; (b) Zhou, B.; Shi, B.; Jin, D.; Liu, X. G. *Nat. Nanotech.* **2015**, *10*, 924–936.
95. (a) Park, Y.; Lee, K. T.; Suh, Y. D.; Hyeon, T., Chem. Soc. Rev., 2015, 44, 1302-1317; (b) Haase, M.; Schaefer, H. Angew. Chem. Int. Ed. 2011, 50, 5808-5829.
96. Garcia, J. V.; Yang, J.; Shen, D.; Yao, C.; Li, X.; Wang, R.; Stucky, G. D.; Zhao, D.; Ford, P. C.; Zhang, F. *Small* **2012**, *8*, 3800–3805.

97. Timko, B. P.; Arruebo, M.; Shankarappa, S. A.; McAlvin, J. B.; Okonkwo, O. S.; Mizrahi, B.; Stefanescu, C. F.; Gomez, L.; Zhu, J.; Zhu, A.; Santamaria, J.; Langer, R.; Kohane, D. S. *Proc. Natl. Acad. Sci. U. S. A.* **2014**, *111*, 1349–1354.
98. Mase, J. D.; Razgoniaev, A. O.; Tschirhart, M. K.; Ostrowski, A. D. *PhotoChem. Photobiol. Sci.* **2015**, *14*, 777–785.
99. Pierri, A. E.; Huang, P.-J.; Garcia, J. V.; Stanfill, J. G.; Chui, M.; Wu, G.; Zheng, N.; Ford, P. C. *Chem. Commun.* **2015**, *51*, 2072.
100. Ou, J.; Zheng, W.; Xiao, Z.; Yan, Y.; Jiang, X.; Dou, Y.; Jiang, R.; Liu, X. *J. Mater. Chem. B* **2017**, *5*, 8161–8168.
101. Huang, P.-J. 2017. Ph.D. Dissertation, University of California, Santa Barbara.
102. (a) Jain, P. K.; Huang, X.; El-Sayed, I. H.; El-Sayed, M. A. *Accts. Chem. Res.* **2008**, *41*, 1578–1586; (b) Taladriz-Blanco, P.; Pérez-Juste, J.; Kandoth, N.; Hervés, P.; Sortino, S. J. *Colloid Interfac. Sci.* **2013**, *407*, 524–528.
103. Levy, E. S.; Morales, D. P.; Garcia, J. V.; Reich, N. O.; Ford, P. C. *Chem. Commun.* **2015**, *51*, 17692–17695.
104. (a) Carpenter, A. W.; Johnson, J. A.; Schoenfisch, M. H. *Physicochem. Eng. Asp.* **2014**, *454*, 144–151; (b) Quinn, J. F.; Whittaker, M. R.; Davis, T. P. *J. Controlled Release* **2015**, 190–205; (c) Afonso, D.; Valetti, S.; Fraix, A.; Bascetta, C.; Petralia, S.; Conoci, S.; Feiler, A.; Sortino, S. *Nanoscale* **2017**, *9*, 13404–13408.
105. Maeda, H.; Nakamura, H.; Fang, J. *Adv. Drug Deliv. Rev.* **2013**, *65*, 71–79.
106. Ostrowski, A. D.; Lin, B. F.; Tirrell, M. V.; Ford, P. C. *Mol. Pharm.* **2012**, *9*, 2950–2955.
107. Nakanishi, K.; Koshiyama, T.; Iba, S.; Ohba, M. *Dalton Trans* **2015**, *44*, 14200–14203.
108. Barone, M.; Sciortino, M. T.; Zaccaria, D.; Mazzaglia, A.; Sortino, S. *J. Mater. Chem.* **2008**, *18*, 5531–5536.
109. Sugahara, K. N.; Teesalu, T.; Karmali, P. P.; Kotamraju, V. R.; Agemy, L.; Girard, O. M.; Hanahan, D.; Mattrey, R. F.; Ruoslahti, E. *Cancer Cell* **2009**, *16*, 510–520.
110. (a) Shi, C.; Pamer, E. G. *Nat. Rev. Immunol.* **2011**, *11*, 762–774; (b) Anselmo, A. C.; Mitragotri, S. *J. Controlled Release* **2014**, *190* (531-541), 112.
111. Murdoch, C.; Giannoudis, A.; Lewis, C. E. *Blood* **2004**, *104*, 2224–2234.
112. Evans, M. A.; Huang, P.-J.; Iwamoto, Y.; Ibsen, K. N.; Chan, E. M.; Hitomi, Y.; Ford, P. C.; Mitragotri, S. *Chem. Sci.* **2018**, *9*, 3729–3741.
113. (a) Ridnour, L. A.; Thomas, D. D.; Switzer, C.; Flores-Santana, W.; Isenberg, J. S.; Ambs, S.; Roberts, D. D.; Wink, D. A. *Nitric Oxide* **2008**, *19*, 73–76; (b) Mocellin, S.; Bronte, V.; Nitti, D. *Med. Res. Rev.* **2007**, *27*, 317–352.

CHAPTER FIVE

Light-triggered unconventional therapies with engineered inorganic nanoparticles

Aurore Fraix and Salvatore Sortino*
PhotoChemLab, Department of Drug and Health Sciences, University of Catania, Catania, Italy
*Corresponding author: e-mail address: ssortino@unict.it

Contents

1. Introduction 172
2. Unconventional phototherapy: Basic concepts 175
   2.1 PDT 175
   2.2 NO-PDT 176
   2.3 PTT 177
3. Nonmetallic inorganic NPs 179
4. Noble metal-based NPs 189
5. Concluding remarks and future perspectives 199
Acknowledgments 200
References 200

Abstract

The utilization of nanomaterials to release therapeutic agents under the control of light stimuli to enhance the therapeutic efficacy is one of the major challenges in nanomedicine. Because of their wide variety, good biocompatibility and facile functionalization, many inorganic nanoparticles (NPs) offer great potential in this respect. Their engineering with suitable photoactivatable components can open up exciting avenues in the field of the most intriguing unconventional therapeutic approaches for fighting cancer and bacterial diseases, such as photodynamic therapy (PDT), nitric oxide (NO) photodynamic therapy (NO-PDT) and photothermal therapy (PTT). These emerging treatment modalities are based on the localized cytotoxic action induced by a "burst" of singlet oxygen, nitric oxide or heat, generated with superb spatiotemporal control by appropriate photoprecursors, and in principle, they can be alternatives to, or used in combination with, conventional chemotherapeutic and antibiotic drugs. This account illustrates an overview of the work carried out in our research group over the last years and specifically focuses on inorganic NPs based on non-metallic- and noble metal scaffolds, highlighting the logical design and their potential applications in PDT, NO-PDT and PTT.

Advances in Inorganic Chemistry, Volume 80
ISSN 0898-8838
https://doi.org/10.1016/bs.adioch.2022.04.004

Abbreviations

Φ_Δ	singlet oxygen quantum yield
α-CD	α-cyclodextrin
β-CD	β-cyclodextrin
$^1O_2$	singlet oxygen
AFM	atomic force microscopy
APES	3-aminopropyltriethoxysilane
BODIPY	difluoro-4-bora-3a,4a-diaza-s-indacene
BP	benzophenone
CDs	carbon dots
Chl	chlorophyll
DOX	doxorubicin
FA	folic acid
FQs	fluoroquinolones
FRET	Förster resonance energy transfer
GO	graphene oxide
ISC	intersystem crossing
LbL	layer-by-layer
LS	Langmuir-Schaefer
LSPR	localized surface plasmon resonance
MSNPs	mesoporous silica nanoparticles
NF	norfloxacin
NIR	near infrared
NO	nitric oxide
NO-PDT	nitric oxide photodynamic therapy
NPs	nanoparticles
O_2^-	superoxide anion
Ox-S^4TdR	oxidized 4-thiothymidine
PDT	photodynamic therapy
PS	photosensitizer
PTT	photothermal therapy
RF	rufloxacin
S^4TdR	4-thiothymidine
SAAD	selected area electron diffraction
TCPP	meso-tetrakis(4-carboxyphenyl)porphyrin
TEOS	tetraethoxysilane
Vis	visible

1. Introduction

Inorganic nanoparticles (NPs), including nonporous and mesoporous silica NPs, carbon-based NPs, quantum dots, and noble metal-based plasmonic NPs, display a wealth of structural and optical properties.

These materials provide intriguing building blocks for the realization of functional nanomaterials with great prospects in diagnostic, therapeutic and theranostic applications.[1–7] In this frame, the achievement of NPs addressed to the release of therapeutics with precise spatiotemporal control is one of the most active areas.[8–11] Site, timing and dosage of such release play key roles in determining the efficacy of the therapeutic outcome.

The fast rates of the photochemical processes, many of which occur in the ns and ps time regime, and the easy manipulation of photons in terms of energy, intensity, location, and duration, make light, a powerful and minimally invasive trigger for the fine regulation of the release process with exquisite spatiotemporal control.[12] Moreover, light does not affect physiological values of pH and temperature, essential criteria for biomedical applications.

The convergence of consolidated knowledges in molecular photochemistry and the significant breakthroughs in materials chemistry have led in recent years to a successful marriage between photo-science and nanomaterials science resulting in creative examples of photoactivatable NPs for therapeutic release.[8–12] Two-photon excitation, by using femtosecond lasers,[13] allows one to photoactivate many chromogenic centers in the so-called therapeutic window (650–1300 nm)[14] with high penetrating near infrared (NIR) light. Besides, the parallel technological advances in fiber optic light-guides are considerably pushing photoactivatable nanomaterials towards less utopian applications, opening great prospects for an entirely new category of clinical solutions.

In the frame of NPs for light-activated therapy, those devoted to "unconventional" therapeutic approaches are very appealing. Photodynamic therapy (PDT),[15] nitric oxide (NO) photodynamic therapy (NO-PDT)[16] and photothermal therapy (PTT)[17] play dominant roles in this context. These emerging treatment modalities exploit localized cytotoxic action induced by a "burst" of harmful species such as singlet oxygen ($^1O_2$), nitric oxide (NO) or heat, generated with superb spatiotemporal control by appropriate photoprecursors. Such strategic approaches hold a high degree of innovation to be used either as alternative to, or in combination with, conventional therapies based on chemotherapeutic and antibiotic drugs.

Fabrication of these photoresponsive NPs requires an appropriate design phase to obtain a specific response of the engineered NPs to light stimuli and interdisciplinary efforts, because of the synthetic procedures necessary to integrate the photoactive components into nanoscaffolds characterized by specific shapes, sizes and coordination environment. In this regard, inorganic scaffolds can be engineered with photoresponsive units, which need to preserve, amplify or suppress their photochemical properties, depending on the final goal. In the case of multiple photoresponsive components integrated in the same nanoscaffold, the final material can be designed to allow each individual unit either to work in tandem or to communicate with each other (via energy/electron transfer) upon light excitation. Whatever the final goal, once engineered, these NPs possess the great advantage to concentrate a large number of photoactivatable units in a small volume and thus allow producing a "burst" of cytotoxicity in a precisely confined region of space.

This contribution starts with a brief introductory description of the main "unconventional" light-driven therapeutic approaches, which can be useful for the readers not familiar with photochemistry. Afterward, we shall illustrate an overview of the work carried out in our research group over the last years specifically focused on inorganic NPs based on non-metallic- and noble metal scaffolds and highlight the logical design and their potential applications in PDT, NO-PDT and PTT (Fig. 1). In the last part, we also report some examples emphasizing the role of light as convenient reactant in the fabrication of metallic NPs for therapeutic uses, through simple and eco-friendly synthetic procedures.

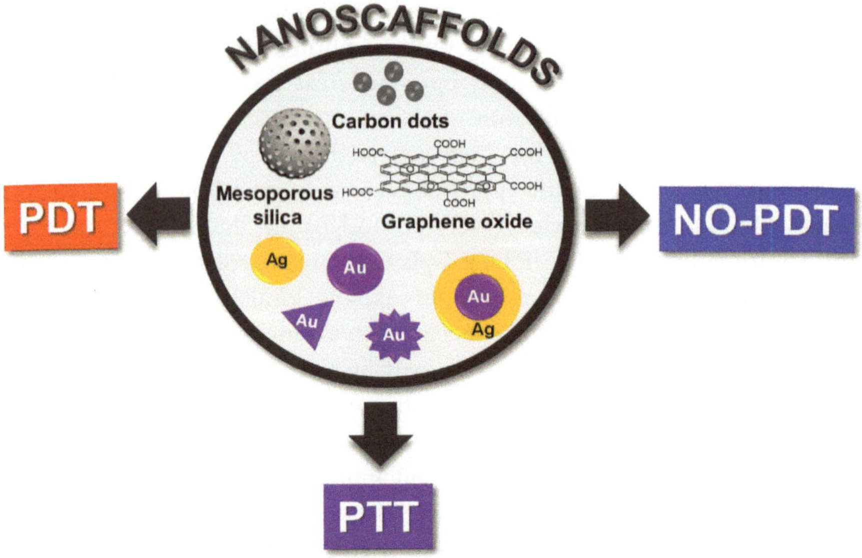

Fig. 1 Schematic illustration of the non-metal and metal-based nanoscaffolds described herein and main potential applications after their engineering.

2. Unconventional phototherapy: Basic concepts

PDT, NO-PDT and PTT represent major emerging and minimally invasive unconventional therapeutic modalities for cancer, infectious and other malignant diseases. As anticipated above, these approaches exploit appropriate photoprecursors for the generation of $^1O_2$, NO or heat whose cytotoxic effects can be confined to the desired area with great accuracy by directing the incident light. These species share common advantages with respect to conventional drugs such as (i) capability to be multi target cytotoxic agents able to damage all biological components, (ii) absence of multidrug resistance, (iii) confinement of their action to short distances from the production site inside the cells, due to their transient nature. Furthermore, since NO and heat production are independent of O_2 availability, NO-PDT and PTT may successfully complement PDT at the onset of hypoxic conditions, typical for some tumors and anaerobic bacteria, where PDT may fail due to the poor oxygen concentration.

2.1 PDT

This promising treatment modality, which finds already clinical applications, takes advantage of the effects originated by the appropriate combination of three main ingredients: Vis/NIR light, a photosensitizer (PS) and molecular oxygen. Upon light excitation and subsequent intersystem crossing (ISC), the PS reaches its lowest-energy excited triplet state. Due to its long lifetime (μs-ms time regime), this excited state can be effectively quenched by nearby molecular oxygen via either collisional energy transfer process (Type II reaction) producing the highly oxidizing $^1O_2$ or electron transfer (Type I reaction) producing superoxide anion O_2^- (Fig. 2).[18,19] The former, having a lifetime of ca. 3 μs and a diffusion radius <20 nm, is accepted to be the main mediator of cytotoxic reactions in PDT treatments. Most importantly, its production is a catalytic process, since after the Type II reaction, the PS reverts to the ground state. Porphyrinoid derivatives are the most common PSs used in PDT by virtue of their high absorption coefficient in the VIS/NIR region, high ISC quantum yields, very long triplet lifetimes and effective generation of $^1O_2$.[20] However, in recent years derivatives of difluoro-4-bora-3a,4a-diaza-s-indacene (BODIPY) are emerging as a novel class of PSs, because associated with the above properties, a remarkable photostability is reached.[21]

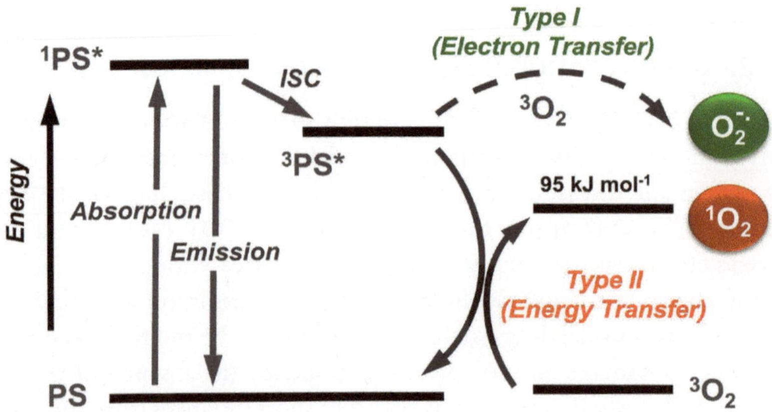

Fig. 2 Simplified Jablonski diagram showing the two main quenching processes of the excited triplet state of the PS by molecular oxygen, leading to the formation of $^1O_2$ and $O_2^{\cdot-}$.

2.2 NO-PDT

Differently to PDT, recently NO-PDT has come to the limelight and, although still confined to the research area, holds very promising features for anticancer and antibacterial applications.[16,22] Besides its multiple roles in the bioregulation of a broad array of physiological processes,[23,24] NO, having a lifetime of ca 5 s and a diffusion radius <200 μm in tissues, has in fact proven to be an effective anticancer and antibacterial agent.[25,26] However, NO may have conflicting roles especially in cancer biology, as it can be implicated in both tumor growth and suppression, depending whether the dosage falls in the nM or μM range, respectively.[27] Because of this, NO-PDT depends on production of NO with the concentration accurately regulated by light stimuli through appropriate NO photocages. Such photocages are chromogenic organic or inorganic metal complex-based molecular scaffolds that covalent integrate NO and exploit the excitation light to liberate NO and, concomitantly, a co-product (Fig. 3).[28–31] Therefore, in contrast to traditional PDT, NO-PDT is not a photocatalytic process, since it implies a net photochemical reaction to generate NO.

This, of course, on one hand limits the reservoir of NO to the concentration of the photocages but, on the other hand, does not require O_2 availability to function, thereby offering a great advantage under hypoxia. In principle, light triggering permits a control of NO concentration on the basis of the knowledge of the light intensity and the NO quantum

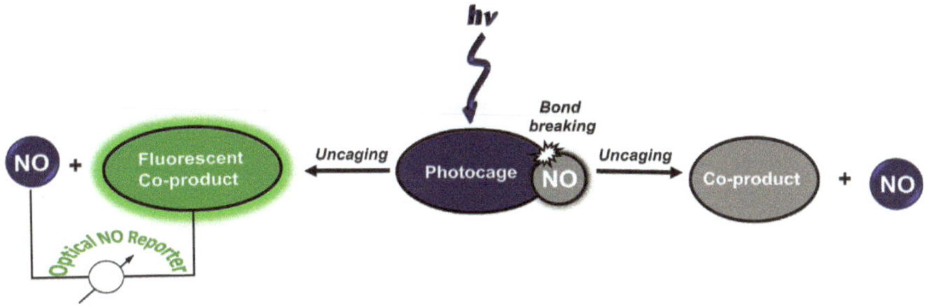

Fig. 3 The NO photo-uncaging process: light excitation of the photocage is exploited to break the covalent bond with the caged and unactive NO molecule, liberating it in its active form. In the case of NO uncaging with a fluorescent reporter, the photocage is designed to be nonfluorescent and to generate a fluorescent co-product concomitant to NO photorelease.

yield of the photocage. However, some uncertainties due to the cell environment and related to unknown medium characteristics such as optical non-homogeneity, viscosity, volume, etc, make the calibration of light intensity difficult. Therefore, additional strategies for a better control over the amount of NO liberated in biological environment are necessary. Accurate control of NO dosage can be achieved by designing NO photocages that possess fluorescent reporting functionalities.[32] In these cases, NO photocages are designed to incorporate a latent fluorogenic unit whose fluorescence is restored after the NO release. The highly fluorescent co-product released concomitantly to NO acts as an optical counter for this species, offering the possibility for its monitoring and quantification via non-invasive fluorescence techniques even in a cellular environment (Fig. 3).[32]

2.3 PTT

The basic and simple principle of this treatment modality takes advantage of Vis/NIR absorbing components able to convert a large part of the excitation energy into heat to increase the temperature of the surrounding region in the range 41–47 °C for tens of minutes. This light-triggered hyperthermia results in irreversible cell damage by loosening cell membranes, denaturating proteins and, eventually, leads to cellular death with great efficiency and selectivity.[33] The impressive breakthrough of nanomaterial science has had a tremendous impact in PTT.[34] In particular, Au nanostructures are ideal for PTT applications due to the combination of their unique spectroscopic and photophysical properties, such as: (i) localized surface plasmon resonance (LSPR) in the VIS/NIR region, which is associated with a

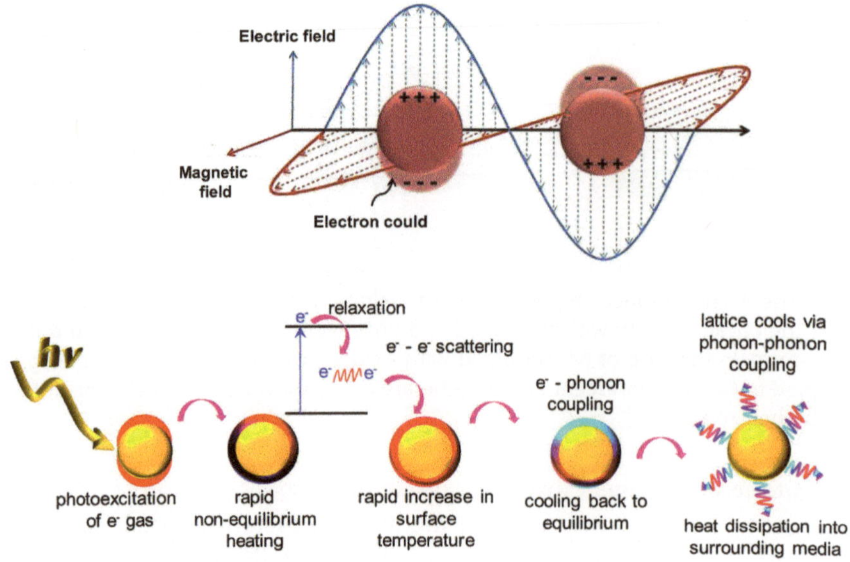

Fig. 4 Schematic description of localized surface plasmon resonance in metal NPs (top) and schematic describing the principle of photothermal light to heat conversion by plasmonic nanostructures (bottom). *Adapted from Webb, J.A.; Bardhan, R.* Nanoscale ***2014,*** *6, 2502–2530 with permission from the Royal Society of Chemistry.*

collective excitation of conduction electrons (Fig. 4), (ii) absorption coefficients several orders of magnitude greater than those of other photo-absorbing dyes, (iii) high light-to-heat conversion (Fig. 4); (iv) superb photostability and (v) good biocompatibility.[35] Non-spherical Au nanomaterials are better suited to PTT applications because their LSPR absorption falls in the therapeutic window (650–1300 nm) where tissues, hemoglobin and water display high optical transmission. A wide range of Au nanorods,[36] nanoshells,[37] nanocrosses,[38] nanostars,[39] nanotriangles[40] and nanoflowers[41] has been achieved over the last years, all displaying the above optical features.

In recent years, the use of PTT in combination with PDT[42] and the latter with NO-PDT,[43] have been suggested to be highly desirable. PDT, PTT and NO-PDT have also been combined with conventional drugs.[44–46] These combined approaches aim to exploit either additive or synergistic effects due to the simultaneous generation of more active species in the same region of space with the final goal to amplify the therapeutic action. Such strategies may definitely open new doors in the emerging field of multimodal therapies for both cancer and infective diseases.[47,48]

3. Nonmetallic inorganic NPs

Because of their mesoporous structure and the good biocompatibility, mesoporous silica nanoparticles (MSNPs) are very suitable scaffolds for the integration of functional phototherapeutic components through non-covalent and covalent methodologies. The pores of MSNPs can be fabricated with designed sizes to be accessible to appropriate guest molecules and their inner and outer walls can be readily tailored with a large variety of molecules through well-established synthetic protocols.[49] From the photochemical viewpoint, MSNPs offer the great advantage of optical transparency allowing the chromogenic photoresponsive species integrated within their structure to absorb the incident light and thereby permitting both spectroscopic monitoring and photoactivation.

MSNPs have proven among the most promising nanocarriers enabling the incorporation of PSs for PDT applications.[50] To this purpose, the porous nature of these nanoscaffolds can strongly facilitate not only the solubilization of hydrophobic PS, but can also avoid typical aggregation phenomena, which may preclude the photochemical response due to detrimental self-quenching processes. Recently, we have demonstrated the suitability of highly ordered amino-grafted MSNPs ca. 100 nm in diameter and with pore of ca. 4 nm as valuable nanocarriers for chlorophyll *a* (Chl).[51] Chl is the main light-absorbing pigment in the light harvesting complex and in the reaction centers, and its use as PSs has been already proposed by virtue of its long triplet lifetime and capability to photogenerate $^1O_2$.[52] However, its low water solubility and formation of non-photoresponsive aggregates represent major limitations for its use in biomedical applications. In this case, the presence of the amino groups both at the external surface and inside the pores of the nanoscaffolds significantly enhances the absorption of the Chl from an ethanolic solution, compared with the same MSNPs without the amino termination (Fig. 5). The zeta potential values of the modified MSNPs dramatically changes from positive to negative after the Chl absorption in accord with a key role of electrostatic interactions between the PS and the inorganic scaffold. Absorption and fluorescence emission spectroscopy, as well as time-resolved fluorescence, provide clear evidence that a large portion of Chl is embedded in the MSNPs as the photoactivatable monomeric form. The red emission of the encapsulated Chl allows one to investigate the nanoconstruct by fluorescence microscopy which shows a quite homogeneous distribution of the PS inside the MSNPs. Photogeneration of $^1O_2$ from

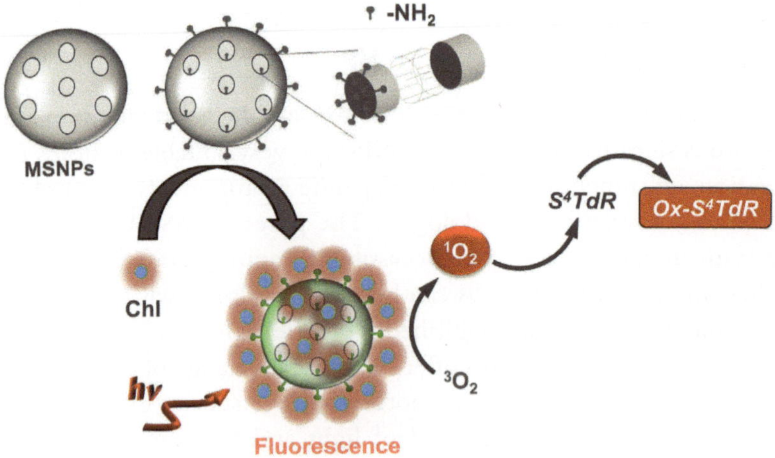

Fig. 5 Schematic of the as prepared MSNPs and after the grafting with amino groups which encourage the absorption of the Chl mainly as photoactive monomeric form able to retain fluorescence and $^1O_2$ photogeneration properties. *Adapted from Rizzi, V.; Gubitosa, J.; Fini, P.; Fanelli, F.; Fraix, A.; Sortino, S.; Agostiano, A.; De Cola, L.; Nacci, A.; Cosma, P.* J. Photochem. Photobiol. A Chem. ***2019,*** 377, *149–158, Copyright (2019), with permission from Elsevier.*

the nanoconstruct is demonstrated by using suitable chemical probes such as 4-thiothymidine (S^4TdR). The formation of its oxidation product (Ox-S^4TdR) observed upon visible light excitation of the Chl incorporating MSNPs provides a clear cut evidence that the $^1O_2$ is able to diffuse outside the nanoscaffolds inducing damage to the biomolecule, making the present system in principle suited for PDT applications.[51]

Although light excitation ensures a high level of targeting to PDT, the partial localization of PSs also in healthy tissues may result in adverse side effects.[53] A widely explored strategy to increase the accumulation of the PS at the target site and, as a consequence, to maximize the PDT efficacy and reduce side effects, is the targeted delivery of the PS.[54] Combination of PS with ligands able to target for specific receptors overexpressed by many cancer cells is very suited to this purpose.[54] Folic acid (FA) has been widely used for targeted PDT[55] because of its non-toxicity and the high binding affinity for folate receptors typically overexpressed by many types of cancer cells.[56]

Covalent molecular hybrids between FA and PSs have already demonstrated the improvement of photodynamic activity.[57] In this frame, we have recently shown an original strategy to obtain photoresponsive non-covalent supramolecular assemblies of FA and the meso-tetrakis(4-carboxyphenyl) porphyrin (TCPP) for targeted PDT, starting from a mesoporous silica material.[58] This work was inspired by the mesoporous materials prepared by exploiting the ability of FA molecules to form columnar hexagonal liquid crystalline phases by Hoogesteen-bonded tetrads and π–π stacking interactions with the addition of a co-structure directing agent.[59] On these grounds, TCPP can be in-situ encapsulated in the silica material exploiting the well-known properties of the porphyrins to intercalate in the stacks of tetrads by the π–π interaction (Fig. 6). The material is obtained using FA as template, 3-aminopropyltriethoxysilane (APES) as co-structure agent and tetraethoxysilane (TEOS) as silica source. XRD analysis confirm the stacking of the porphyrin between the FA tetramers within the silica channels. Interestingly, after stirring overnight in phosphate buffer and filtering the suspension of the mesoporous material, absorption and emission spectroscopic evidences demonstrate that FA and the PS, are released in the form of a stable supramolecular assembly with the PS still stacked between the FA tetramers. The assembling process does not occur in aqueous solution, as proven by the simple mixing of FA and TCPP, but it takes place exclusively in the interior of the mesoporous silica material entrapping both components. $^1O_2$ photorelease from the nanoassembly is demonstrated by the direct monitoring of its typical phosphorescence in the near-IR spectral window with maximum at ca. 1270 nm,[60] which represents the most unambiguous tool to prove the formation of the key species involved in PDT.

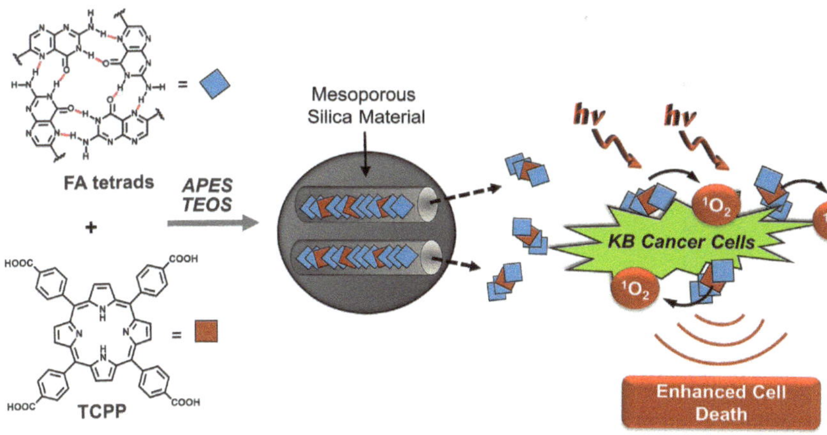

Fig. 6 Schematic for the targeted PDT with FA/TCPP supramolecular assembly released from the mesoporous silica material. *Adapted from Zhou, C.; Afonso, D.; Valetti, S.; Feiler, A.; Cardile, V.; Graziano, A.C.E.; Conoci, S.; Sortino, S.* Chem. Eur. J. ***2017,*** 23, *7672–7676 with permission from John Wiley and Sons.*

Interestingly, the quantum yield for production of $^1O_2$ ($\Phi_\Delta = 0.32$) is even slightly higher than that of the free TCPP ($\Phi_\Delta = 0.25$). This is due to a higher efficiency of ISC in the nanoassembly, leading to a larger population of the excited triple state, the precursor for the $^1O_2$ photogeneration of the PS, as demonstrated by the direct time-resolved spectroscopy. Cell cytotoxicity experiments performed in the dark and after irradiation with visible light, demonstrate that the nanoassembly exhibits photoinactivation of KB cancer cells that overexpress the folate receptors, compared with the free TCPP, confirming the FA targeting effect.[58]

MSNPs are also very versatile scaffolds for the integration of multiple therapeutic components in view of multimodal treatment which aim to attack cancer cells on different sides by acting either on a single oncogenic pathway or across parallel pathways without amplification of side effects.[61–63] MSNPs entrapping PDT agents in combinations with chemotherapeutics have been already reported.[50,64] With the goal to go further, we have recently reported the first example of engineered MSNPs able to deliver "three bullets" for potential anticancer therapy (Fig. 7).[65] These photoactivatable MSNPs covalently integrate erythrosine as PS and a nitroaniline derivative as an NO photocage,[66] respectively, and encapsulate

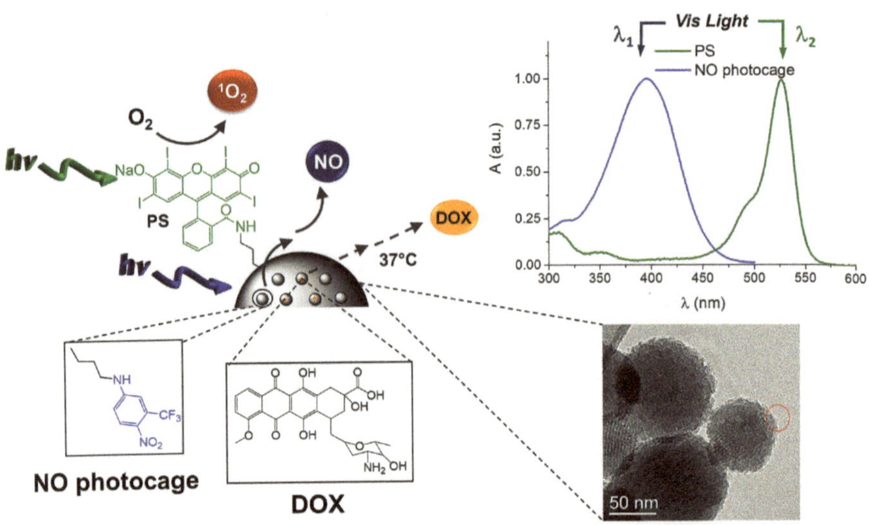

Fig. 7 Schematic for the "three-bullets" MSNPs and their working principle. The normalized absorption spectra of the photoactivatable components and a representative TEM image of the engineered MSNPs are also shown. *Adapted from Tessaro, A.L.; Fraix, A.; Pedrozo da Silva, A.C.; Gazzano, E.; Riganti, C.; Sortino, S.* Nanomaterials ***2019,*** 9 *(6), pii: E823.*

the chemotherapeutic doxorubicin (DOX) in a non-covalent fashion. Incorporation of multiple components within the same nanoscaffolds implies, of course, several aspects to be taken into consideration in the design phase. The components need to be appropriately chosen in order to ensure the absence of any communication between them in the excited state. Only in such a way, they can be operated in tandem without reciprocal interferences. Besides, based on the different diffusion radii of $^1O_2$ (<20 nm) and NO (<200 μm), MSNPs need to be designed to integrate the PS mainly in the outer part of the scaffold and the NO photocage in the inner pores. This strategy avoids $^1O_2$ decay before it escapes from the scaffold due to its short diffusion radius. On the other hand, despite being photogenerated in the core of the scaffold, NO can diffuse out due to its long diffusion radius. Another issue to be addressed is in regard to the mutual concentrations of $^1O_2$ and NO generated upon light excitation. As outlined above, $^1O_2$ is produced through a photocatalytic process that, in principle, does not consume the PS, whereas photogeneration of NO occurs through a neat photochemical reaction with consequent consumption of the NO photocage. Therefore, the regulation of the reservoir of NO with respect to $^1O_2$ is a critical point to observe effective bimodal photodynamic action. This can be accomplished in different ways: (i) by using a larger amount of the NO photocage with respect to the PS, (ii) by choosing PS and NO photocage with absorption spectra falling in different spectral windows permitting the tuning of the photogenerated species by regulating the intensity of the excitation light or (iii) by combining both conditions. The appropriate synthetic protocol for the anchoring of the PS and the NO photocage allows the regulation of the relative concentrations of distinct subsets of chromogenic units integrated within the very same scaffold. For such a reason, MSNPs are engineered with the amount of the NO photocage more than an order of magnitude larger than the PS. Erythrosine and the nitroaniline-based NO photocage have been chosen by virtue of their absorption in distinct regions of the visible spectral range. These spectral features permit their selective or simultaneous excitation by using blue light, green light or both. The covalent binding of the NO photocage mainly in the inner part of the MSNPs does not preclude the capability to encapsulate additional therapeutic compounds in a non-covalent fashion, due to the large number of pores available in the nanoscaffold. As a proof of principle, we selected DOX as a suitable chemotherapeutic. This drug has a wide spectrum of activity in clinical application but it is limited by the development of MDR in cancer cells.[67] The validity of the "three bullets" MSNPs is demonstrated by the detection of $^1O_2$ and NO upon visible light excitation and the DOX release under physiological conditions. Biological results performed on A375 cells expressing ABC transporters show absence of toxicity exerted by the engineered MSNPs in the absence of irradiation but potentiated activity of DOX due to the effect of NO photorelease. The reduction of the cell viability observed under light excitation is effective in overcoming DOX-resistance in cells expressing ABC transporters, making the nanoconstruct an appealing system to overcome the severe limitations of currently used chemotherapy.

As outlined in the NO-PDT section, due to the double sword activity of NO, especially in cancer, the real time quantification of the NO release is a challenging issue to be faced, and the release and report strategy is the most appropriate at this regard.[32] We have recently shown the use of carbon dots (CDs) as alternative to conventional organic fluorophores as fluorescent reporters for the NO photorelease.[68] CDs are widely employed in biomedicine and display intense emission bands in the visible region whose position is strongly dependent on the excitation wavelength.[69] This characteristic allows tuning the excitation wavelength at an appropriate value to obtain an emission spectrum that considerably overlaps the absorption of the nitroaniline-based NO photocage shown above. On this basis, amino-terminated MSNPs embedding CDs formed by calcination were functionalized with the NO photocage to give the robust MSNPs-1 (Fig. 8).[68] The spectral overlap, the close spatial proximity between the CDs and the NO photocage and the extensive functionalization of the MSNPs with the photocage lead to quenching of the CDs emission by Förster Resonance Energy Transfer (FRET). NO photorelease can be monitored by direct amperometric techniques and strictly depends on the light excitation. In addition, photodecomposition of the engineered MSNPs-1 into MSNPs-2 is accompanied by the complete revival of the characteristic emission of the CDs, which behave as optical reporters for the NO release. The restoring of the fluorescence agrees very well with the disappearance of the absorption by the nitroaniline fragment in the visible region, which does not allow any spectral overlap with the emission of the CDs and thus precludes the FRET process. The restoration of the emission can be detected even with the naked eye (Fig. 8) providing readable information about the NO photogenerated. Another interesting aspect of this nanoconstruct is its capability to encapsulate a highly hydrophobic PS extensively used for PDT applications such as Temoporfin. The encapsulation process occurs with high efficiency and the optical transparency of the MSNPs allows the easy

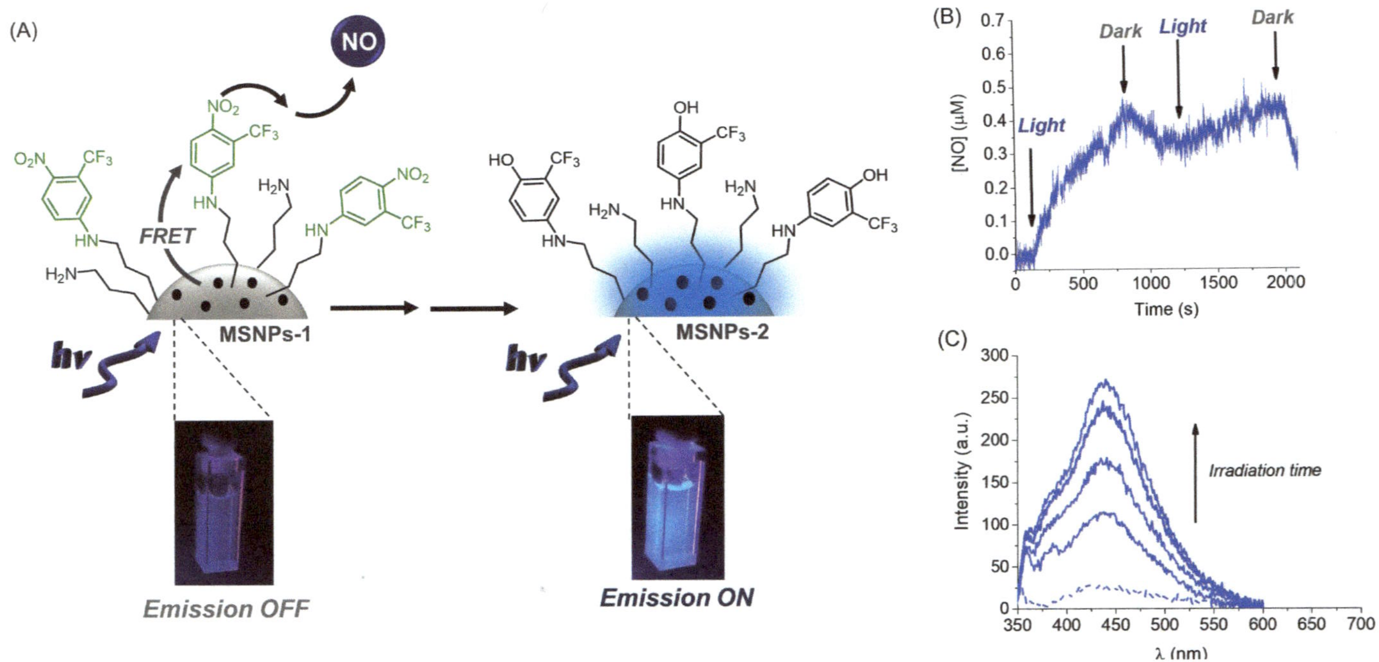

Fig. 8 (A) Schematic of the NO photoreleasing MSNPs-1 and their working principle. The typical blue fluorescence from the carbon dots embedded within the MSNPs-1 is quenched by FRET due to the presence of the nitroaniline-based NO photocage. The phenol derivatives formed after NO release in MSNPs-2 are not a good energy acceptors thus restoring of the emission of the carbon dots. The inset shows the actual images of the suspensions acquired before and after the photolysis. (B) NO release profile observed for aqueous suspensions (0.1 mg mL^{-1}) of MSNPs-1 and (C) evolution of the emission spectra of the same suspension recorded at $\lambda_{exc} = 340$ nm. *Adapted from Afonso, D.; Valetti, S.; Fraix, A.; Bascetta, C.; Petralia, S.; Conoci, S.; Feiler, A; Sortino, S.* Nanoscale ***2017,*** *9, 13404–13408 with permission from the Royal Society of Chemistry.*

detection of the encapsulation by absorption and emission spectroscopy. Interestingly, the emission of the CDs and that of the PS are not affected after encapsulation of the latter, suggesting the lack of spatial proximity between these fluorogenic components within the silica scaffold. Such a feature is remarkable since it offers the great advantage to preserve the fluorescent reporting functionality of the CDs in the blue region and the imaging properties of the porphyrinoid guest in the red region.

One of the most interesting properties of the CDs in the perspective of bioapplications is their huge two-photon cross section in the NIR region, which is about three orders of magnitude larger than that of organic small molecules.[70] This has stimulated their use as suitable two-photon antenna to encourage NO release from the NO photocage exploiting the above FRET conditions. The nanoconstruct illustrated in Fig. 9 is able to photogenerate the NO radical via FRET mechanism activated by two-photon excitation with the highly biocompatible 800 nm NIR light, overcoming the limitation of this organic photocage due to its absorption below 500 nm.[71] These engineered CDs are well tolerated in the dark by cancer cells in vitro, induce significant cell mortality upon NIR light and demonstrate successful in vivo applications. In fact, they are able to significantly reduce the size of human pancreatic BxPc-3 mice xenograft, which is relatively hypoxic, using highly biocompatible 800 nm light, thus demonstrating the potential of NO as therapeutic alternative to PDT under hypoxic conditions.[71]

Graphene oxide (GO) represent another intriguing nanomaterial very suitable for engineering with photoactivatable components. It is low cost, synthesized from graphite, and in recent years has also come to the limelight as a promising nanomaterial in drug delivery applications.[72] The sheet-like structure makes GO an ideal platform with loading capacity considerably higher than those observed for NPs thanks to the two external surfaces, which can be derivatized through both covalent or non-covalent conjugation. GO-based nanoplatforms exhibiting light-regulated NO delivery are very limited, suggesting a critical need to develop more photoresponsive nanomedicine releasing NO based on GO scaffolds. The nitroaniline-based NO photocage offers the ideal requisites for its facile integration into GO nanosheets. Carboxylic groups located at the periphery of GO sheets are utilized to covalently graft the amino-terminated NO photocage via amide bonds (Fig. 10).[73]

The engineered GO hybrid material is well-dispersible in water and stable in the dark. The TEM image reported in Fig. 10 indicates that

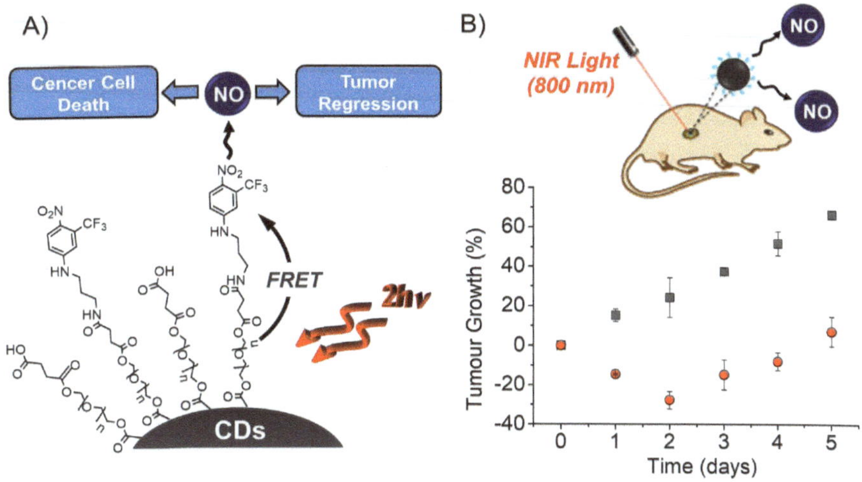

Fig. 9 (A) Schematic of the CDs engineered to release NO under two-photon excitation with 800 nm NIR light. (B) Change in tumor volume for Balb/c SCID mice treated with the engineered CDs in the dark (■) and after 800 nm two-photon excitation (●). *Adapted from Fowley, C.; McHale, A.P.; McCaughan, B.; Fraix, A.; Sortino, S.; Callan, J. F.* Chem. Commun. ***2015,*** 51, *81–84 with permission from the Royal Society of Chemistry.*

engineered nanosheets are corrugated, typical for GO sheet and due to the disruption of the planar sheet by the introduction of sp^3-hybridized carbon upon the oxidation process. The selected area electron diffraction (SAED) pattern (see inset) indicates a crystalline structure. The Atomic Force Microscopy (AFM) analysis reveals the presence of sheets with heights ranging between 0.9 and 1.5 nm and lateral dimensions ranging between 35 and 150 nm, in excellent agreement with the predominance of single layers of exfoliated engineered GO nanosheets.

The choice of this specific NO photoreleaser for the preparation of the GO-NO hybrid is not random. It finds the rationale in the potential absence of quenching effects by the GO platform via photoinduced energy and/or electron transfer processes, which would preclude NO photorelease. In fact, the lack of any significant emission of the NO photocage makes quenching by FRET mechanism not feasible. Besides, the values of the reduction and oxidation potentials of GO (-0.75 V and >1.4 V vs NHE, respectively),[74,75] and the energy of the lowest excited state (ca 2.4 eV) of the NO photocage, make quenching by photoinduced electron transfer only slightly exergonic. This view is experimentally confirmed by the excellent preservation of the photochemical properties after integration of the NO photocage into the GO nanoplatform. This is demonstrated by the

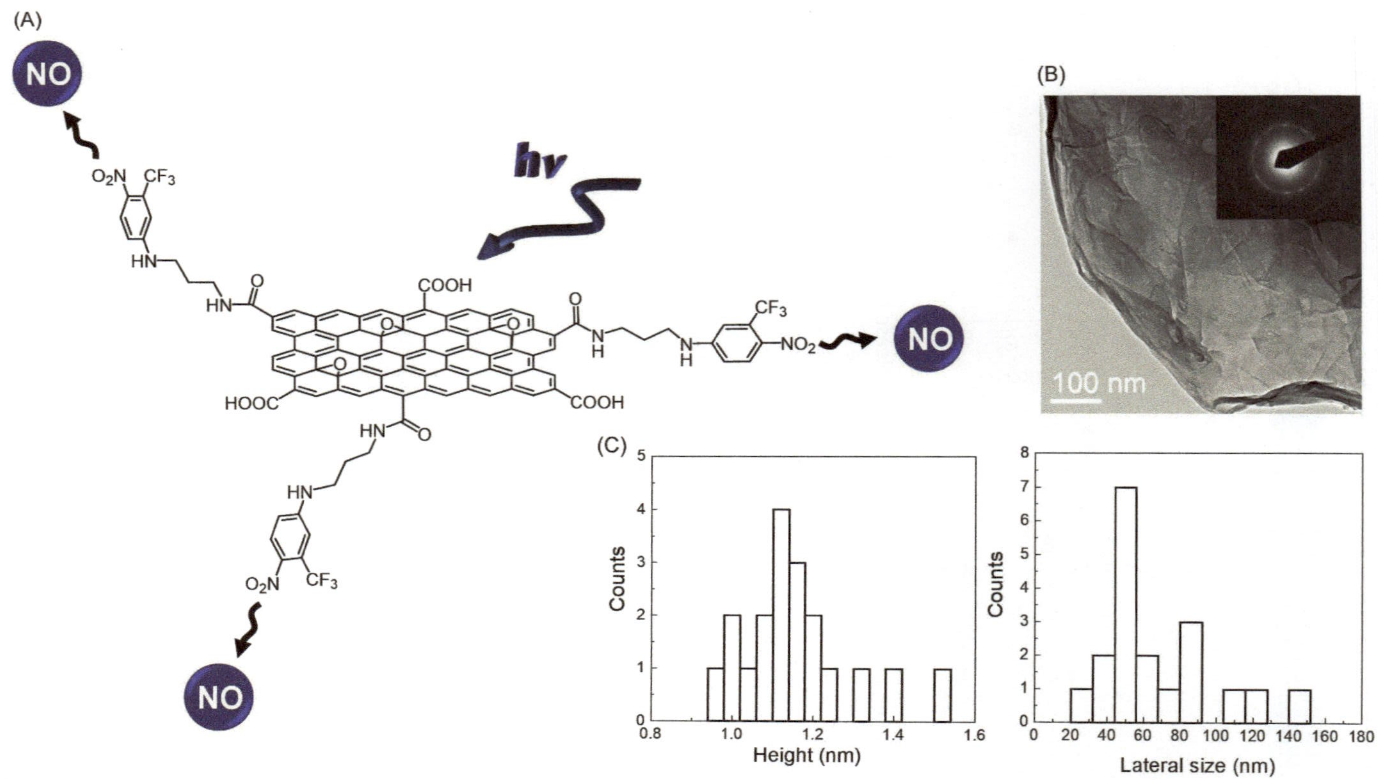

Fig. 10 (A) Schematic of the photoactivatable engineered GO for the light-regulated release of NO. (B) Representative TEM image (the inset shows the SAED pattern) and (C) height and lateral dimension distribution, obtained by AFM, of the engineered GO. *Adapted from Marino, N.; Petralia, S.; Perez-Lloret, M.; Mosinger. J.; Conoci, S.; Sortino, S.* J. Mater. Chem. B. ***2016,*** 4, *5825–5830 with permission from the Royal Society of Chemistry.*

photoregulated NO release shown by both direct and indirect detection methodologies. Thus, these graphene-based hybrid materials are intriguing platforms for further studies in biomedical research where light-triggered NO delivery is required.

4. Noble metal-based NPs

Surface chemistry based on thiol terminated units makes the engineering of noble metal NPs quite easy. In the case of photoactivatable compounds, the three-dimensional shape of NPs offers some advantages compared to the bidimensional analogues, such a reduced quenching effects of the excited states of the photoactive units and, in some cases, even enhanced photochemical response due to plasmonic effects.[76]

Decoration of AuNPs with NO photocages generates nanohybrids of great potential in bimodal phototherapy. Such systems may combine, in principle, the excellent LSPR absorption of the Au core for PTT with a photoactivatable shell for NO release. Combination of supramolecular chemistry with nanomaterials once again reveals a suitable approach. A water insoluble thiol-terminated nitroaniline-based NO photocage (the guest) can be promptly dissolved in a water solution of α-cyclodextrin (α-CD) (the host), exploiting the instantaneous formation of a host-guest inclusion complex. The self-assembling of this complex at the surface of AuNPs through the thiol-termination results into a rotaxane-like nanoconstruct. It is well soluble in water medium due to the presence of the α-CD rings, and it releases NO under the exclusive control of blue light stimuli (Fig. 11).[77] The distinct absorption spectral regions of the Au core at ca 530 nm and the NO photocage at ca 400 nm, represent ideal conditions for the selective excitation of the metal core to induce photothermia, the shell to induce NO release, or both to trigger bimodal effects by the appropriate tuning of the visible light energy in the range 400–550 nm.

Thiol-based chemistry for decoration of AuNPs leads mainly to the formation of a monolayer of functional molecules at the NPs surface. This, of course, not only limits the reservoir of NO to that available in a monolayer but, in the case of chromophores with modest extinction coefficient, implies prolonged irradiation time to observe a satisfactory release of NO, due to the reduced fraction of absorbed photons. The layer-by-layer (LbL) approach would be an alternative strategy to achieve stable AuNPs with enhanced payload of NO photocages at the NPs periphery. This is based on the deposition of oppositely charged polyelectrolytes and is extensively employed for

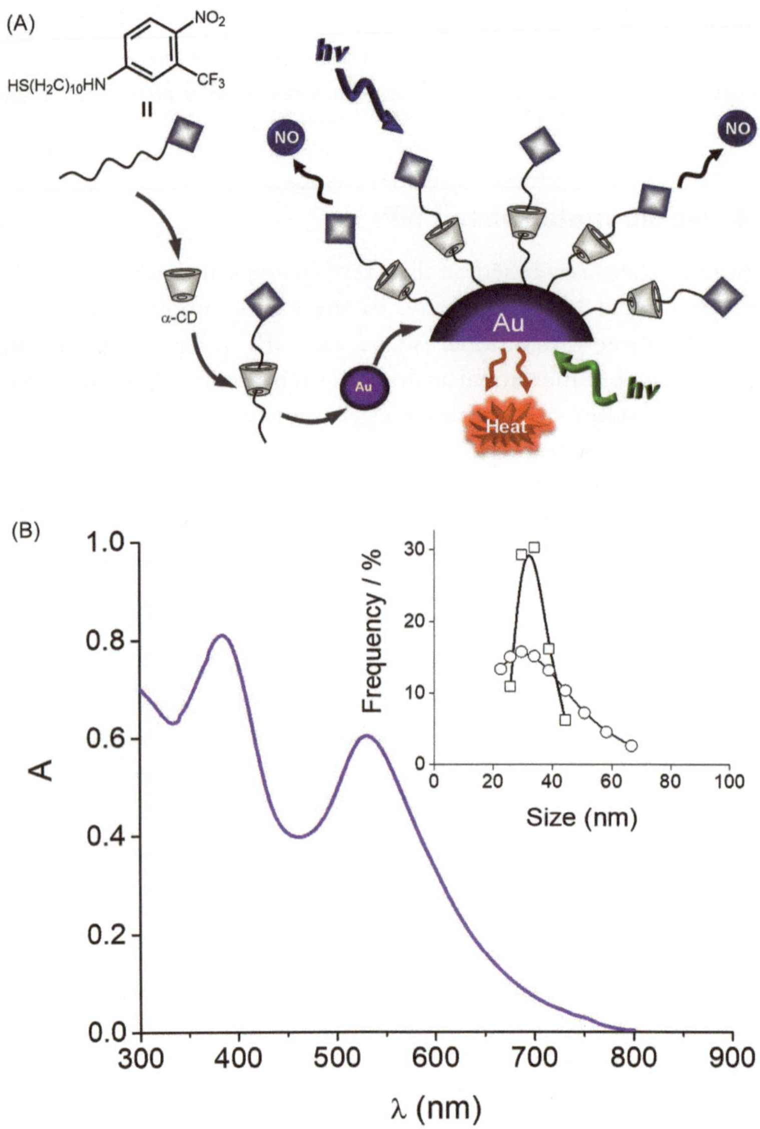

Fig. 11 (A) Schematic of the preparation of the dual-function AuNPs. (B) Absorption spectrum of the engineered rotaxane-like AuNPs. The inset shows the size distribution of the AuNPs before (□) and after (×) the assembling of the host-guest complex at their surface. *Adapted from Kandoth, N.; Vittorino, E.; Sortino, S.* New J. Chem. ***2011,*** 35, *52–56 with permission from the Centre National de la Recherche Scientifique (CNRS) and the Royal Society of Chemistry.*

the fabrication of two and three-dimensional nanoconstructs.[78,79] In this framework, water soluble, colloidal platforms for NO photorelease with increasing payloads have been obtained by alternate coating of the citrate stabilized AuNPs template with oppositely charged polyelectrolytes, one of which integrates the NO photocage (Fig. 12).[80] The LbL coating can be easily monitored by the shift in the plasmon absorption band as a function of the number of layers deposited, which is due to the influence of the dielectric constant of the surrounding medium, and by the parallel changing of the zeta potential values. The layered AuNPs do not show significant clustering under physiological conditions and remain stable for weeks preserving their nanodimensional character. Interestingly, despite the close packing of the photoresponsive units, the engineered AuNPs does not exhibit significant self-quenching phenomena. The result is photoregulated NO release detected via amperometric techniques. Comparative experiments performed with samples with different loading reveals that the amount of the NO photoreleased, is ca twofold higher in the case of the AuNPs coated with two layers of polyelectrolyte that integrate the NO photocage.[80]

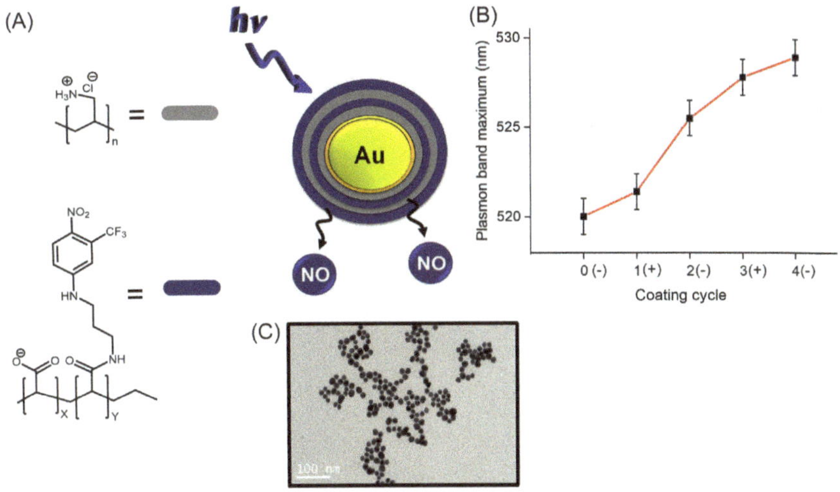

Fig. 12 (A) Schematic for the NO photoreleasing AuNPs prepared by alternate LbL strategy. (B) Absorption maximum of the plasmon absorption band as a function of the coating cycles. The sign in brackets is related to the zeta potential of the AuNPs after each coating cycle. (C) Representative TEM image of AuNPs coated with two layers of the photoactive polyelectrolyte. *Adapted from Taladriz-Blanco, P.; Pérez-Juste, J.; Kandoth, N.; Hervés, P.; Sortino, S.* J. Coll. Interface Sci. ***2013,*** 407, *524–528, Copyright (2013), with permission from Elsevier.*

A similar approach based on the surface non-covalent interaction between an NO photocage and metallic nanoparticles has been used to prepare bidimensional nanostructures based on silver nanoparticles (AgNPs), and an amino-terminated amphiphilic NO photocage (Fig. 13).[81] In this case, the well-known antibacterial properties of AgNPs[82] can be, in principle, combined with the bactericidal action of the NO photoreleased. The assembling of these hybrids onto solid support in the form of molecular thin films is of interest in view, for example, of topical applications and the achievement of bifunctional bioactive coatings for medical devices.

The Langmuir-Schaefer (LS) technique is one of the most simple and elegant methods to fabricate multilayers on almost any type of substrate, and offers the advantage to control thickness, packing and molecular orientation by exploiting the properties of amphiphilic compounds arranged at the air-water interface.[83] Hybrid multilayer films incorporating the tailored NO photocage and AgNPs can easily be achieved by LS approach by dissolving naked and water soluble AgNPs, prepared with a green procedure,[84] in the water subphase, and spreading the amphiphilic NO photocage on the water surface.[81] The floating monolayers at the water-air interface interact with the AgNPs through the amino functionality, encouraging their transfer onto quartz slides by LS deposition to give hybrid multilayer films containing two potential antibacterial agents, AgNPs and the caged NO. The approach used does not alter the nanodimensional character

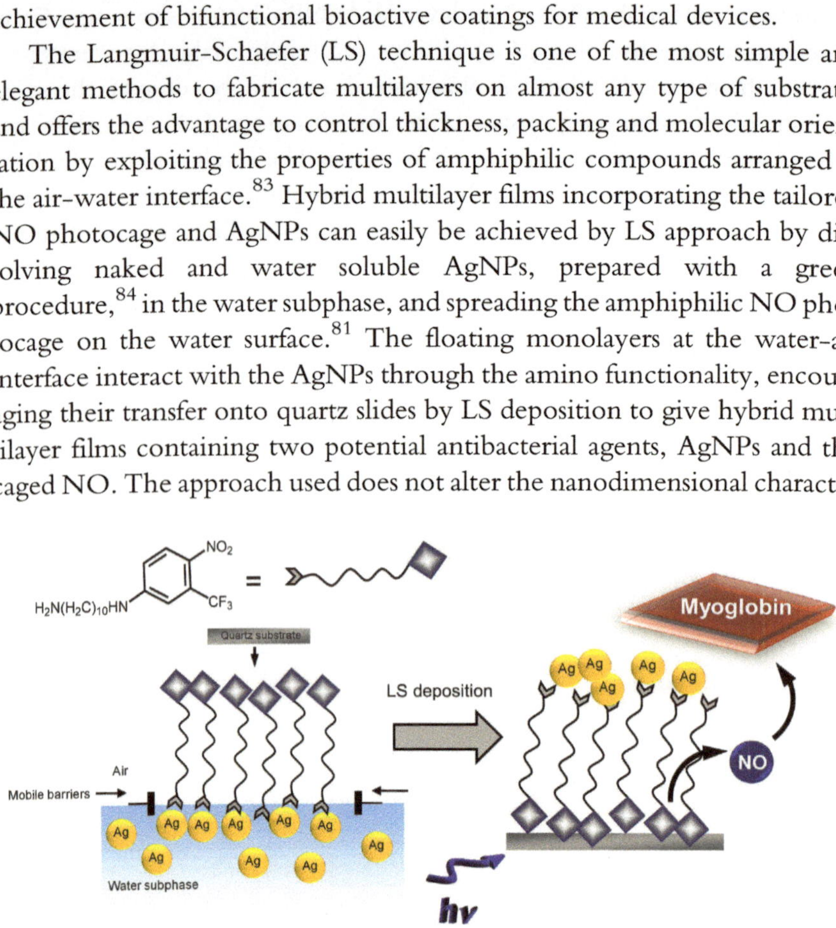

Fig. 13 Schematic view of the arrangement of an amphiphilic NO photocage and the AgNPs at the air/water interface and after their transfer on a quartz substrate by LS deposition. Irradiation of the thin film leads to release of NO from the solid surface and its transfer to myoglobin. *Adapted from Vittorino, E.; Giancane, G.; Manno, D.; Serra, A.; Valli, L.; Sortino, S. J.* Coll. Interface Sci. ***2012,*** 368, *191–196, Copyright (2012), with permission from Elsevier.*

of the AgNPs, which is an indispensable requisite for the antibacterial action. In addition, the NO photodonor retains its photochemical properties even after embedding within the hybrid nanostructure. This is confirmed by the effective transfer of the photo-liberated NO to a protein such as myoglobin. This strategy allows, with a relative simplicity, to regulate the reservoir of AgNPs and NO available in the films by changing the number of layers, making these multilayers promising bifunctional model systems to be tested in biomedical research studies.[81]

One of the main effects that noble metal NPs may exert on chromophores intimate to their surface is the quenching of excited states via energy/electron transfer mechanisms. This certainly represents a serious limitation in the case of molecules devised to perform specific functions under light inputs, precluding their photoactivation. However, in some cases, such quenching phenomena can be successfully exploited for the stabilization of therapeutic molecules which suffer of undesired photolability that limits their pharmacological properties. Fluoroquinolones (FQs), for example, are widely used broad-spectrum antimicrobial agents which are indeed characterized by uncommonly high photolability under environmental light. Their photodecomposition mechanisms have been the object of interest from others and from our group in the past.[85,86] In particular FQs photodegradation is mediated by radical and ionic pathways involving transient species which are responsible for effective side effects such as phototoxicity and photocarcinogenicity.[86] Due to the ionization of the carboxylic group and the 4′-amino of the piperazinyl ring common to their structures, many FQs are amphoteric substrates with the zwitterionic form (Fig. 14) being the dominant as well as the most photolabile at physiological pH. On these grounds, the achievement of AgNPs capped with FQs may lead, in principle, to potential bifunctional antibacterial nanodrugs with an antibacterial "core" able to photostabilize the antibacterial "shell". A proof of principle of this concept has been demonstrated with the FQs norfloxacin (NF) and rufloxacin (RF). These two compounds were chosen as prototype examples of FQs whose photodecomposition is mediated by type I (electron transfer)[85,86] and type II pathways (energy transfer), respectively. Naked AgNPs were used as metallic scaffolds because of their excellent water solubility, exceptional stability, largely unprotected surface and negatively zeta potential due to the adsorption of hydroxide ions at their exterior.[84] Therefore, they present ideal requisites for the decoration of their shell with FQs, exploiting electrostatic interaction with the piperazinyl ring of the FQs, which is positively charged at physiological pH values (Fig. 14).[87]

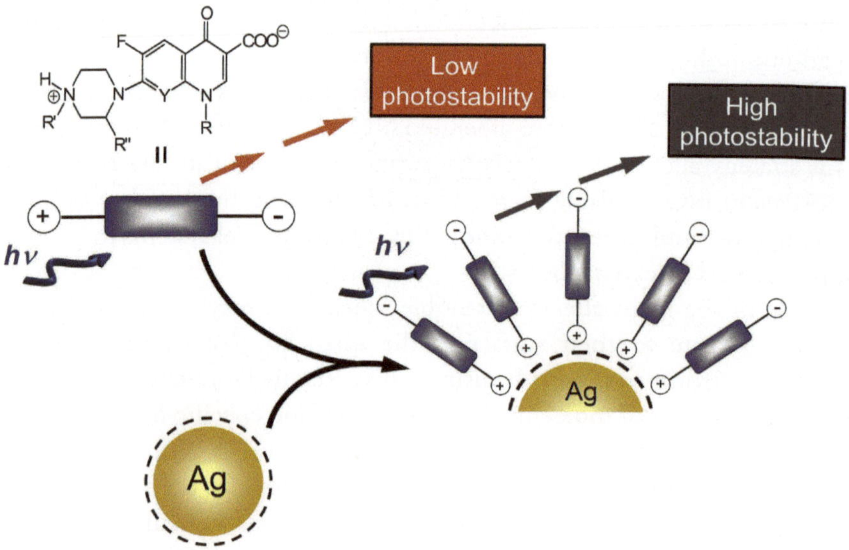

Fig. 14 The zwitterionic form of the FQs and idealized view of the FQs-capped AgNPs. *Adapted from Giuffrida, S.; Sortino, S.* Adv. Biomater. ***2011,*** 6, *B356–B359 with permission from John Wiley and Sons.*

Despite the formation of some inter-particle aggregates, the dimensions of the resulting Ag@FQs NPs remained in the nanometer size regime with diameters ca. 85 nm. These nanoclusters exhibit a remarkable photostability if compared with the free FQs and retain their fluorescent properties. In addition, in view of the noncovalent nature of the core-shell binding, the FQs could be released without affecting their chemical structure.[87] The results of this study anticipate that the behavior observed for NF and RF when absorbed on AgNPs surface could be general, at least for structurally related FQs, paving the way for straightforward strategies aimed at enhancing the FQs photostability with consequent prevention of their undesired photosensitization effects.

One of the most challenging goals in the fascinating realm of plasmonic NPs, is the development of "green" synthetic procedures not involving toxic reducing agents and that are less reliant on the chemisorption of conventional stabilizers containing S, N or P. From an eco-friendly perspective, the three main criteria that should be evaluated in the preparation of metal NPs have been outlined in a seminal paper by Raveendran et al.[88]: (i) appropriate choice of the solvent medium used for the synthesis, (ii) utilization of environmentally benign reducing agents, and (iii) choice of nontoxic materials for the stabilization of the NPs. Light, once again, emerges as a very convenient tool for this purpose. Photons can be considered easy to handle, low cost "green reactants" that offer the peculiar advantage to produce highly performing reducing agents by choosing suitable photoprecursors. Scaiano's group have given a decisive input in this field, providing an elegant demonstration of how consolidated concepts of organic molecular photochemistry translated to the modern material nanoscience can result in the ingenuous achievement of a variety of metallic NPs[89–92]

In the field of nanoplasmonic photochemistry, bimetallic Au—Ag core-shell NPs (Au@AgNPs) are very appealing nanostructures that are opening additional prospects in the biomedical field by combining the properties of the individual components in a single scaffold. For example, the antibacterial properties typical of the Ag shell combined with the heat photodelivery of the Au core, may result in an amplified antibacterial action otherwise precluded with the individual NP elements. Inspired by the cornerstone studies from Scaiano's group,[89–92] we have recently reported facile one-step green synthetic protocols for the preparation of AgNPs, AuNPs and core-shell Au@AgNPs by irradiation of a water soluble, amphiphilic, tailored polymer in the presence of $AgNO_3$, $HAuCl_4$ or both.[93] This polymer consists of the highly biocompatible dextran covalently integrating benzophenone (BP) as a suitable photoprecursor (Fig. 15). The presence of the hydrophobic BP in the skeleton of the water soluble dextran, makes the resulting polymer amphiphilic and able to form micelles in aqueous solution. The high number of easily abstractable H-atoms from the dextran scaffold and close to BP are expected to make the typical H-abstraction by the excited triplet state of BP[94] very effective and faster than the potential triplet quenching by the metal ions.[89–92] This H-abstraction process leads to the formation of ketyl radicals which are well-known reducing agents ($E^0 = -1.80$ V vs NHE) towards both Au^{3+} and Ag^+ (M^{n+}) ($E^0 = +1.50$ V and $+0.70$ vs NHE, respectively). Finally, the dextran is a well-known stabilizing agent for metal NPs that does not present S, N or P centers.[95]

Irradiation of the polymer leads to the formation of the highly reducing ketyl radicals in less than 50 ns and, in a few minutes, produces AgNPs, AuNPs and Au@AgNPs whose formation can be promptly followed through their typical plasmonic absorption at ca 414, and 528 nm and the double plasmon band associated to the Au core and the Ag shell, respectively. Thanks to the role of the dextran as capping agent, the different types

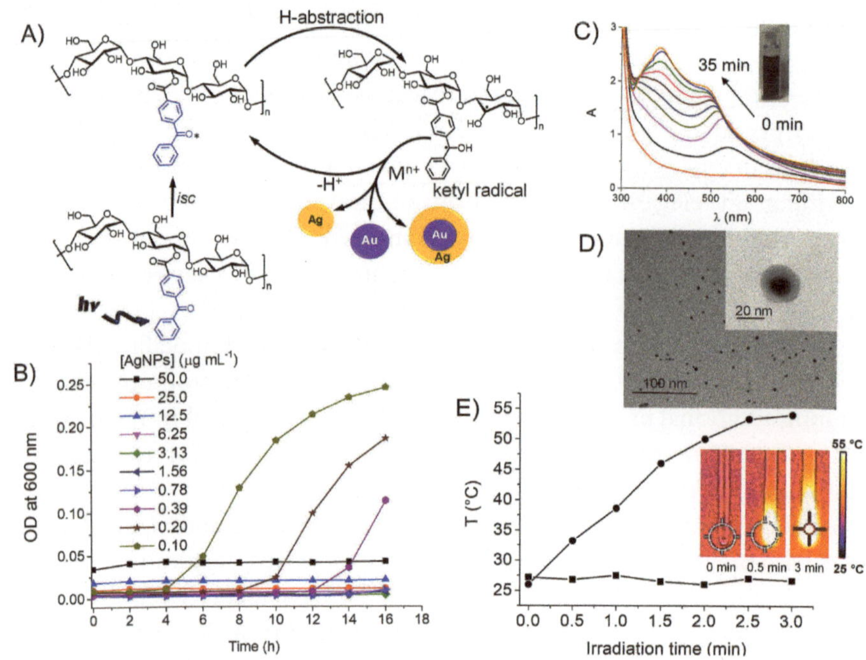

Fig. 15 (A) Phototriggered generation of the metal NPs. Photoexcitation of the BP unit, followed by ISC, generates its excited triplet state which produces the highly reducing ketyl radicals upon H-abstraction from the dextran scaffold. $[M^{n+}] = Ag^{+}$ or Au^{3+}. (B) Antibacterial activity of the photochemically synthesized AgNPs observed against *S. aureus* bacterial strain. (C) Absorption spectral changes observed at regular intervals of 5 min upon 350 nm light irradiation a N_2-saturated aqueous solution of the BP-modified dextran containing $HAuCl_4$ and $AgNO_3$ both at 0.3 mM. The inset shows the actual image of the solution at the end of the photolysis. (D) Representative TEM image of the above sample at the end of the photolysis. The inset shows a magnified image which better show the core-shell nature of the Au@AgNPs. (E) Photothermal effect of the core-shell Au@AgNPs; temperature changes of 0.1 mL of water in the presence (●) and in the absence (■) of Au@AgNPs upon 532 nm laser irradiation (6 W cm^{-2}). The inset shows representative thermographic images of the sample containing the Au@AgNPs recorded at 3 different irradiation times. *Adapted from Perez-Lloret, M.; Fraix, A.; Petralia, S.; Conoci, S.; Tafani, V.; Cutrone, G.; Vargas-Berenguel, A.; Gref, R.; Sortino, S.* Chem. Eur. J. ***2019,*** 25, *14638–14643 with permission from John Wiley and Sons.*

of metallic NPs are soluble in water and remain stable for weeks. Furthermore, AgNPs and Au@AgNPs exhibit encouraging antibacterial activity and excellent photothermal action, respectively (Fig. 15). The combined features are beneficial in terms of environmental safety and ready integration in biologically relevant systems.[93]

As anticipated in the PPT section, the LSPR band of Au is strictly dependent on the shape of the nanostructure. In this regard, the development of green synthetic protocols for the fabrication of non-spherical Au nano-objects with LSPR falling in the NIR region is very challenging for applications in nanomedicine due to the possibility to activated photothermia by light excitation in the therapeutic window (λ_{max} 650–1300 nm). Taking this need into account and stimulated by continuing interest in NO photoreleasing systems, we recently showed an unprecedented, simple and "green" methodology for preparing water dispersible Au nanotriangles and nanoflowers using NIR LSPR (λ_{max} 850–1250 nm) and exploiting NO as reducing agent.[96] These intriguing nanostructures can be obtained in a single-step at room temperature and in few minutes by visible light irradiation of an NO photoreleasing tailored β-CD branched polymer, without the need of preformed seeds, external reducing and sacrificial agents, conventional surfactants and stabilizing ligands (Fig. 16).

The redox process between NO and Au(III) is favored by the well-known complexation capability of the β-CD branched polymer towards Au metal ions ensuring that the bimolecular reaction between the redox partners occurs mainly within the polymeric network and thus more effectively due to the proximity effect.[97] The resulting Au nanostructures are excellent photothermal agents as demonstrated by the fast increase of the temperature to a plateau value above 40 °C after few minutes of irradiation with an 808 nm laser. Photothermal applications of these metallic nanostructures are demonstrated by a photothermally activated enzymatic reaction and photothermal-induced sarcoma cancer cell mortality using 808 nm light. Effective catalytic conversion of L-phenylalanine into trans-cinnamic acid in the presence of the enzyme phenylalanine Ammonia Lyase, which typically occurs at 35 °C, was observed upon irradiation of the reactant solution at 808 nm in the presence of the Au nanotriangles. Au nanoflowers, are well tolerated in the dark by human Ewing Sarcoma cell lines (A673). In contrast, they induce a decrease of cell viability upon NIR excitation. The evaluation of the apoptotic process 24 h after Au and/or light treatments displayed that the number of cleaved caspase-3 (Casp3) cells increased considerably in the combined Au + Light treatment with respect to the control samples.

Besides satisfying the main important green chemistry criteria that a synthetic approach for metal nanostructures should meet (vide supra), this approach can open up an exciting research direction for novel synthetic methodologies for the preparation of plasmonic Au nanostructures for bio-applications based on the underestimated role of NO as reducing agent.[98]

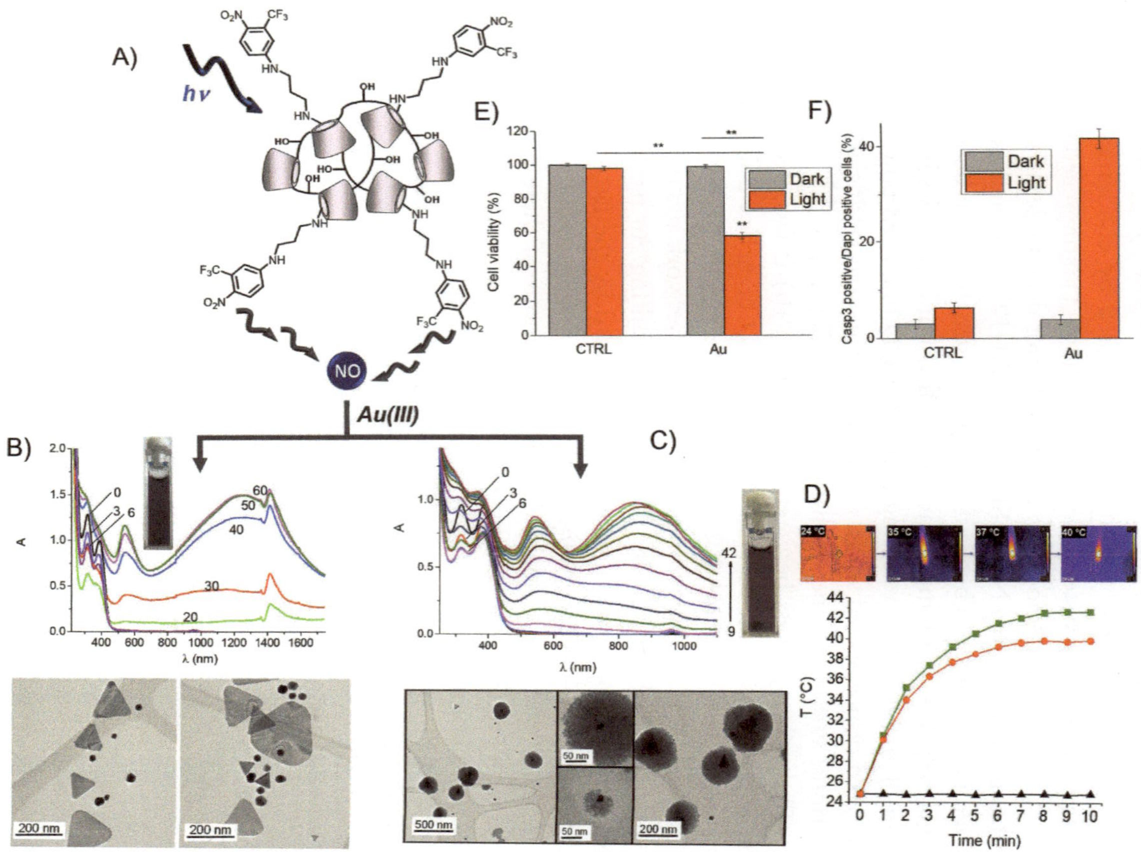

Fig. 16 See figure legend on opposite page.

5. Concluding remarks and future perspectives

In this contribution we have illustrated a panel of inorganic NPs made of different materials, ad-hoc engineered for potential use in "unconventional" therapies. The work described herein demonstrates the central role light plays, not only in the activation of specific therapeutic functionalities with precise spatiotemporal control, but also as a convenient reactant in the fabrication of NPs through simple and eco-friendly synthetic procedures. The examples described show that the creation of these photoactivatable nanoconstructs is the result of merging photochemistry, synthesis and materials science. This blending is fundamental for permitting the photoactive components to retain, amplify or suppress their photochemical behavior once integrated within a specific nanoscaffold. Hopefully the readers may find in this paper fruitful inspirations for further developments in this exciting research field. Inspired by the cornerstone sentences by Richard P. Feynman's about nanotechnology "*There is Plenty of Room at the Bottom*," we are convinced that at this bottom there is plenty of room for photochemists, whose creativity in the service of biomedicine foresees a "bright" future for innovative research in this field.

Fig. 16 (A) Sketch of the molecular structures of the NO photoreleasing β-CD branched polymer. (B, C) Absorption spectral changes observed upon 420 nm light irradiation of an aqueous solution of the NO photoreleasing polymer (1 mg mL^{-1}) containing 0.5 mM $HAuCl_4$ (B) or 0.25 mM $HAuCl_4$ (C). The numbers indicate the irradiation times in minutes. The insets show the actual images of the solutions at the end of the photolysis. Below, representative TEM images of the above sample at the end of the photolysis are shown. (D) Representative thermographic images of the sample containing the Au nanoflowers recorded at 4 different irradiation times (top) and temperature changes observed upon 808 nm CW irradiation (6 W cm^{-2}) of the Au nanotriangles (●) and nanoflowers (■) solutions (100 μL) (bottom) and net water (▲) is also shown, for sake of comparison. (E) Cell viability of A673 cells observed 24 h after incubation without and with Au nanoflowers in the dark and after 30 min 808 nm CW irradiation (6 W cm^{-2}). Tukey HSD post hoc test (**$P<0.01$) indicates significant differences between groups. (F) Quantification of the number of cleaved caspase-3 positive cells. Data are expressed as number of cleaved caspase-3 and DAPI double positive cells over total DAPI positive cells. *Adapted with permission from Nocito, G.; Petralia, S.; Malanga, M.; Beni, S.; Calabrese, G.; Parenti, R.; Conoci, S; Sortino, S.* ACS Appl. Nano Mater. ***2019,*** 2, *7916–7923. Copyright {2019} American Chemical Society.*

Acknowledgments

S.S. is very grateful to his former students and the collaborators, whose names are listed in the references, for their inestimable contribution in the work described herein. We thank AIRC (Project IG-12834) for financial support.

References

1. Liu, Q.; Kim, Y.-J.; Im, G.-B.; Zhu, J.; Wu, Y.; Liu, Y.; Bhang, S. H. *Adv. Funct. Mater.* **2021**, *31*, 2008171–2008194.
2. Giner-Casares, J. J.; Henriksen-Lacey, M.; Coronado-Puchau, M.; Liz-Marzan, L. M. *Mater. Today* **2016**, *19*, 19–28.
3. Kim, D.; Kim, J.; Park, Y. I.; Lee, N.; Hyeon, T. *ACS Cent. Sci.* **2018**, *4*, 324–336.
4. Nasirzadeh, K.; Nazarian, S.; Hayat, S. M. G. *J. App. Biotech. Rep.* **2016**, *3*, 395–402.
5. Pandey, P.; Dahiya, M. *J. Crit. Rev.* **2016**, *3*, 18–26.
6. Urie, R.; Ghosh, D.; Ridha, I.; Rege, K. *Annu. Rev. Biomed. Eng.* **2018**, *20*, 353–374.
7. Jaleel, J. A.; Pramod, K. *J. Contr. Rel.* **2018**, *269*, 302–321.
8. Zhang, L.; Li, Y.; Yu, J. C. *J. Mater. Chem. B* **2014**, *2*, 452–470.
9. Khan, M. A.; Singh, D.; Ahmad, A.; Siddique, H. R. *Eur. J. Pharm. Sci.* **2021**, *164*, 105892–105903.
10. Dai, Y.; Xu, C.; Sun, X.; Chen, X. *Chem. Soc. Rev.* **2017**, *46*, 3830–3852.
11. Chen, J.; Fan, T.; Xie, Z.; Zeng, Q.; Xue, P.; Zheng, T.; Chen, Y.; Luo, X.; Zhang, H. *Biomaterials* **2020**, *237*, 119827–119854.
12. Sortino, S. *J. Mater. Chem.* **2012**, *22*, 301–318.
13. Zipfel, W. R.; Williams, R. M.; Webb, W. W. *Nat. Biotech.* **2003**, *21*, 1369–1377.
14. Weissleder, R. A. *Nat. Biotechnol.* **2001**, *19*, 316–317.
15. Dolmans, D. E. J. G. J.; Fukumura, D.; Jain, R. K. *Nat. Rev. Cancer* **2003**, *3*, 380–387.
16. Ostrowski, A. D.; Ford, P. C. *Dalton Trans.* **2009**, *48*, 10660–10669.
17. Li, X.; Lovell, J. F.; Yoon, J.; Chen, X. *Nat. Rev. Clin. Oncol.* **2020**, *17*, 657–674.
18. Celli, J. P.; Spring, B. Q.; Rizvi, I.; Evans, C. L.; Samkoe, K. S.; Verma, S.; Pogue, B. W.; Hasan, T. *Chem. Rev.* **2010**, *12*, 2795–2838.
19. Castano, A. P.; Mroz, P.; Hamblin, M. R. *Nat. Rev. Cancer* **2006**, *6*, 535–545.
20. Wainmwright, M. *Photosensitizers in Biomedicine*; Wiley-Blackwell, 2009.
21. Kamkaew, A.; Lim, S. H.; Lee, H. B.; Kiew, L. V.; Chung, L. Y.; Burgess, K. *Chem. Soc. Rev.* **2013**, *42*, 77–88.
22. Halpenny, G. M.; Mascharak, P. K. *Anti-Infect Agents Med. Chem.* **2010**, *9*, 187–197.
23. Ignarro, L. J., Ed. *Nitric Oxide: Biology and Pathobiology*; Elsevier Inc., 2010.
24. Ignarro, L. J., Ed. *Special Journal Issue on Nitric Oxide Chemistry and Biology*; Archives of Pharmacal Research, 2009.
25. Fukumura, D.; Kashiwagi, S.; Jain, R. K. *Nat. Rev. Cancer* **2006**, *6*, 521–534.
26. Fang, F. C., Ed. *Nitric Oxide and Infection*; Kluwer Academic/Plenum Publishers: New York, 1999.
27. Huang, Z.; Fu, J.; Zhang, Y. *J. Med. Chem.* **2017**, *60*, 7617–7635.
28. Sortino, S. *Chem. Soc. Rev.* **2010**, *39*, 2903–2913.
29. Ford, P. C. *Nitric Oxide* **2013**, *34*, 56–64.
30. Fry, N. L.; Mascharak, P. K. *Acc. Chem. Res.* **2011**, *44*, 289–298.
31. Ieda, N.; Oka, Y.; Yoshihara, T.; Tobita, S.; Sasamori, T.; Kawiguchi, M.; Nakagawa, H. *Sci. Rep.* **2019**, *9*, 1430–1437.
32. Fraix, A.; Parisi, C.; Seggio, M.; Sortino, S. *Chem. Eur. J.* **2021**, *27*, 12714–12725.
33. Camerin, M.; Rello, S.; Villanueva, A.; Ping, X.; Kenney, M. E.; Rodgers, M. A. J.; Jori, G. *Eur. J. Cancer* **2005**, *41*, 1203–1212.

34. Hu, M.; Chen, J.; Li, Z.-Y.; Au, L.; Hartland, G. V.; Li, X.; Marquez, M.; Xia, Y. *Chem. Soc. Rev.* **2006**, *35*, 1084–1094.
35. Webb, J. A.; Bardhan, R. *Nanoscale* **2014**, *6*, 2502–2530.
36. Murphy, C. J.; Gole, A.; Stone, J. W.; Sisco, P. N.; Alkilany, A. M.; Goldsmith, E. C.; Baxter, S. C. *Acc. Chem. Res.* **2008**, *41*, 1721–1730.
37. Song, G.; Li, C.; Hu, J.; Zou, R.; Xu, K.; Han, L.; Yang, J.; Chen, Z.; Qin, Z.; Ran, K.; Hu, R. C.; Qian, W. *J. Mater. Chem.* **2012**, *22*, 17011–17018.
38. Ye, E.; Win, K. Y.; Tan, H. R.; Lin, M.; Teng, C. P.; Mlayah, A.; Han, M. Y. *J. Am. Chem. Soc.* **2011**, *133*, 8506–8509.
39. He, R.; Wang, Y. C.; Wang, X.; Wang, Z.; Liu, G.; Zhou, W.; Wen, L.; Li, Q.; Wang, X.; Chen, X.; Zeng, J.; Hou, J. G. *Nat. Commun.* **2014**, *5*, 4327–4338.
40. Scarabelli, L.; Coronado-Puchau, M.; Giner-Casares, J. J.; Langer, J.; Liz-Marzan, L. M. *ACS Nano* **2014**, *8*, 5833–5842.
41. Hao, E.; Bailey, R. C.; Schatz, G. C.; Hupp, J. T.; Li, S. *Nano Lett.* **2004**, *42*, 323–327.
42. Wang, Y.; Gong, N.; Li, Y.; Lu, Q.; Wang, X.; Li, J. *J. Am. Chem. Soc.* **2020**, *142*, 1735–1739.
43. Fraix, A.; Sortino, S. *Photochem. Photobiol. Sci.* **2018**, *17*, 1709–1727.
44. Maiolino, S.; Moret, F.; Conte, C.; Fraix, A.; Tirino, P.; Ungaro, F.; Sortino, S.; Reddi, E.; Quaglia, F. *Nanoscale* **2015**, 7, 5643–5653.
45. Nam, J.; Son, S.; Ochyl, L. J.; Kuai, R.; Schwendeman, A.; Moon, J. J. *Nat. Commun.* **2018**, *9*, 1074–1086.
46. Chegaev, K.; Fraix, A.; Gazzano, E.; Abd-Ellatef, G. E. F.; Blangetti, M.; Rolando, B.; Conoci, S.; Riganti, C.; Fruttero, R.; Gasco, A.; Sortino, S. *ACS Med. Chem. Lett.* **2017**, *8*, 361–365.
47. Kremsner, P. G.; Krishna, S. *Lancet* **2004**, *364*, 285–294.
48. Sengupta, S.; Eavarone, D.; Capila, I.; Zhao, G.; Watson, N.; Kiziltepe, T.; Sasisekharan, R. *Nature* **2005**, *436*, 568–572.
49. Narayan, R.; Nayak, U. Y.; Raichur, A. M.; Garg, S. *Pharmaceutics* **2018**, *10*, 118–166.
50. Brevet, D.; Gary-Bobo, M.; Raehm, L.; Richeter, S.; Hocine, O.; Amro, K.; Loock, B.; Couleaud, P.; Frochot, C.; MorHre, A.; Maillard, P.; Garcia, M.; Durand, J. O. *Chem. Commun.* **2009**, 1475–1477.
51. Rizzi, V.; Gubitosa, J.; Fini, P.; Fanelli, F.; Fraix, A.; Sortino, S.; Agostiano, A.; De Cola, L.; Nacci, A.; Cosma, P. *J. Photochem. Photobiol. A Chem.* **2019**, *377*, 149–158.
52. Krieger-Liszkay, A. *J. Exp. Bot.* **2004**, *56*, 337–346.
53. Jori, G. J. *J. Photochem. Photobiol. B Biol.* **1996**, *36*, 87–93.
54. Bugaj, A. M. *Photochem. Photobiol. Sci.* **2011**, *10*, 1097–1109.
55. Leamon, C. P.; Reddy, J. A. *Adv. Drug Delivery Rev.* **2004**, *56*, 1127–1141.
56. Allen, T. M. *Nat. Rev. Cancer* **2002**, *2*, 750–763.
57. Schneider, R.; Schmitt, F.; Frochot, C.; Fort, Y.; Lourette, N.; Guillemin, F.; Muller, J.-F.; Barberi-Heyob, M. *Biorg. Med. Chem.* **2005**, *13*, 2799–2808.
58. Zhou, C.; Afonso, D.; Valetti, S.; Feiler, A.; Cardile, V.; Graziano, A. C. E.; Conoci, S.; Sortino, S. *Chem. Eur. J.* **2017**, *23*, 7672–7676.
59. Atluri, R.; Hedin, N.; Garcia-Bennett, A. E. *J. Am. Chem. Soc.* **2009**, *131*, 3189–3191.
60. Wilkinson, F.; Helman, W. P.; Ross, A. B. *J. Phys. Chem. Ref. Data* **1993**, *22*, 113.
61. Kemp, J. A.; Shim, M. S.; Heo, C. Y.; Kwon, Y. J. *Adv. Drug Deliv. Rev.* **2016**, *98*, 3–18.
62. Li, Y.; Atkinson, K.; Zhang, T. *Cancer Lett.* **2017**, *396*, 103–109.
63. Dawson, M. A. *Science* **2017**, *355*, 1147–1152.
64. Wong, R. C. H.; Ng, D. K. P.; Fong, W.-P.; Lo, P.-C. *Chem. Eur. J.* **2017**, *23*, 16505–16511.
65. Tessaro, A. L.; Fraix, A.; Pedrozo da Silva, A. C.; Gazzano, E.; Riganti, C.; Sortino, S. *Nanomaterials* **2019**, *9* (6). pii: E823.

66. Caruso, E. B.; Petralia, S.; Conoci, S.; Giuffrida, S.; Sortino, S. *J. Am. Chem. Soc.* **2007**, *129*, 480–481.
67. Krishna, R.; Mayer, L. D. *Eur. J. Pharm. Sci.* **2000**, *11*, 265–283.
68. Afonso, D.; Valetti, S.; Fraix, A.; Bascetta, C.; Petralia, S.; Conoci, S.; Feiler, A.; Sortino, S. *Nanoscale* **2017**, *9*, 13404–13408.
69. Liu, W.; Li, C.; Ren, Y.; Sun, X.; Pan, W.; Li, Y.; Wang, J.; Wang, W. *J. Mater. Chem. B* **2016**, *4*, 5772–5788.
70. Cao, L.; Wang, X.; Meziani, M. J.; Lu, F.; Wang, H.; Luo, P. G.; Lin, Y.; Harruff, B. A.; Veca, L. M.; Murray, D.; Xie, S.-Y.; Sun, Y.-P. *J. Am. Chem. Soc.* **2007**, *129*, 11318–11319.
71. Fowley, C.; McHale, A. P.; McCaughan, B.; Fraix, A.; Sortino, S.; Callan, J. F. *Chem. Commun.* **2015**, *51*, 81–84.
72. Mao, H. Y.; Laurent, S.; Chen, W.; Akhavan, O.; Imani, M.; Ashkarran, A.; Mahmoudi, M. *Chem. Rev.* **2013**, *113*, 3407–3424.
73. Marino, N.; Petralia, S.; Perez-Lloret, M.; Mosinger, J.; Conoci, S.; Sortino, S. *J. Mater. Chem. B.* **2016**, *4*, 5825–5830.
74. Shao, Y.; Wang, J.; Engelhard, M.; Wang, C.; Lin, Y. *J. Mater. Chem.* **2010**, *20*, 743–748.
75. Jonsson, M.; Lind, J.; Eriksen, T. E.; Merenyi, C. *J. Am. Chem. Soc.* **1994**, *116*, 1423–1427.
76. Pacioni, N. L.; González-Béjar, M.; Alarcón, E.; McGilvray, K. L.; Scaiano, J. C. *J. Am. Chem. Soc.* **2010**, *132*, 6298–6299.
77. Kandoth, N.; Vittorino, E.; Sortino, S. *New J. Chem.* **2011**, *35*, 52–56.
78. Decher, G.; Schlenoff, J. *Multilayer Thin Films*; Wiley-VCH: Weinheim, 2002.
79. Tang, A.; Wang, Y.; Podsiadlo, P.; Kotov, N. A. *Adv. Mater.* **2006**, *18*, 3203–3224.
80. Taladriz-Blanco, P.; Pérez-Juste, J.; Kandoth, N.; Hervés, P.; Sortino, S. *J. Coll. Interface Sci.* **2013**, *407*, 524–528.
81. Vittorino, E.; Giancane, G.; Manno, D.; Serra, A.; Valli, L.; Sortino, S. *J. Coll. Interface Sci.* **2012**, *368*, 191–196.
82. Pal, S.; Tak, Y. K.; Song, J. M. *Appl. Environ. Microbiol.* **2007**, *73*, 1712–1720.
83. Ulman, A., Ed. *An Introduction to Ultrathin Organic Films: From Langmuir-Blodgett to Self-Assembly*; Academic Press Inc.: New York, 1991.
84. Giuffrida, S.; Ventimiglia, G.; Sortino, S. *Chem. Commun.* **2009**, *48*, 4055–4057.
85. Monti, S.; Sortino, S.; Fasani, E.; Albini, A. *Chem. Eur. J.* **2001**, *7*, 2185–2196.
86. Albini, A.; Monti, S. *Chem. Soc. Rev.* **2003**, *32*, 238–250.
87. Giuffrida, S.; Sortino, S. *Adv. Biomater.* **2011**, *6*, B356–B359.
88. Raveendran, P.; Fu, J.; Wallen, S. L. *J. Am. Chem. Soc.* **2003**, *125*, 13940–13941.
89. Scaiano, J. C.; Billone, P.; Gonzalez, C. M.; Maretti, L.; Marin, M. L.; McGilvray, K. L.; Yuan, N. *Pure Appl. Chem.* **2009**, *81*, 635–647.
90. Scaiano, J. C.; Aliaga, C.; Maguire, S.; Wang, D. *J. Phys. Chem. B* **2006**, *110*, 12856–12859.
91. Marin, M. L.; McGilvray, K. L.; Scaiano, J. C. *J. Am. Chem. Soc.* **2008**, *130*, 16572–16584.
92. McGilvray, K. L.; Decan, M. R.; Wang, D.; Scaiano, J. C. *J. Am. Chem. Soc.* **2006**, *128*, 1598–1604.
93. Perez-Lloret, M.; Fraix, A.; Petralia, S.; Conoci, S.; Tafani, V.; Cutrone, G.; Vargas-Berenguel, A.; Gref, R.; Sortino, S. *Chem. Eur. J.* **2019**, *25*, 14638–14643.
94. Turro, N. J.; Ramamurthy, V.; Scaiano, J. C. *Modern Molecular Photochemistry of Organic Molecules*; University Science Books: Sausalito, CA, 2010; p. 1110.
95. Bankura, K. P.; Maity, D.; Mollick, M. M. R.; Mondal, D.; Bhowmick, B.; Bain, M. K.; Chakraborty, A.; Sarkar, J.; Acharya, K.; Chattopadhyay, D. *Carbohydr. Polym.* **2012**, *89*, 1159–1165.

96. Nocito, G.; Petralia, S.; Malanga, M.; Beni, S.; Calabrese, G.; Parenti, R.; Conoci, S.; Sortino, S. *ACS Appl. Nano Mater.* **2019**, *2*, 7916–7923.
97. Martin-Trasanco, R.; Cao, R.; Esparza-Ponce, H. E. M.; Montero-Cabrera, E.; Arratia-Perez, R. *Carbohydr. Polym.* **2017**, *175*, 530–537.
98. Tran, D.; Ford, P. C. *Inorg. Chem.* **1996**, *35*, 2411–2412.

CHAPTER SIX

Photoactive manganese carbonyl complexes with *fac*-{$Mn(CO)_3$} moiety: Design, application, and potential as prodrugs in CO therapy

Indranil Chakraborty and Pradip K. Mascharak*
Department of Chemistry and Biochemistry, University of California, Santa Cruz, CA, United States
*Corresponding author: e-mail address: pradip@ucsc.edu

Contents

1. Introduction 206
2. Chemical and photochemical properties of the *fac*-{$Mn(CO)_3$} moiety 207
3. Moving photosensitivity to visible region of light; Smart design of photoCORMs 208
4. Water-soluble photoCORMs for chemotherapeutic applications 216
5. Trackable photoCORMs that allow visualization of the CO donor within cellular matrix 217
6. Turn-ON photoCORMs that confirm CO delivery at the target site 220
7. PhotoCORMs with various coligands that confer specific drug actions 222
8. CO-releasing materials and platform design 224
9. Conclusion and future perspectives 227
Acknowledgment 229
References 229

Abstract

Application of low doses of carbon monoxide selectively at biological targets has been shown to elicit salutary effects. Metal carbonyl complexes have been shown to be very convenient as CO delivery agents in hospital settings. Photoactive carbonyl complexes have additional advantage in this regard since they could be triggered with light on demand. Research in our and others' laboratory has shown that manganese carbonyl complexes bearing the *fac*-{$Mn(CO)_3$} moiety can be synthesized on the basis of smart design to systematically produce biocompatible CO donors that exhibit sensitivity to visible/NIR light, solubility in water, trackability in cellular matrices, easy permeability through cellular membranes, and ease with which they can be incorporated in various delivery platforms suitable for specific drug use. Results of extensive *in vitro* experiments with these photoactive prodrugs (CO is the actual drug) provide convincing examples

Advances in Inorganic Chemistry, Volume 80
ISSN 0898-8838
https://doi.org/10.1016/bs.adioch.2022.04.003

of biomedical applications of inorganic photochemistry. In addition, these complexes could significantly expand the potential of photopharmacology, an emerging field in which light is used to remotely control the activity of pharmaceuticals.

1. Introduction

During the past two decades, carbon monoxide (CO) has been identified as a signaling molecule (gasotransmitter) that participates in multiple pathways in mammalian pathophysiology.[1] Further research has indicated its therapeutic promise in inflammation, metabolic disorders, cardiovascular diseases and cancer. This surprising discovery of the therapeutic effects of CO and the potential of CO-derived therapy[2–5] have prompted intense interest and research activity in development of CO delivery platforms to administer low (50–200 ppm) doses of CO to biological targets to elicit salutary outcomes. Despite its "silent killer" moniker, multiple clinical studies have demonstrated that low dose inhaled CO (100–200 ppm) can be safely administered to patients in a controlled environment and CO(g) has recently been recommended for the treatment of a phase I trial of low dose of inhaled CO in acute respiratory distress syndrome (ARDS). Overall, use of gaseous CO poses serious challenge in hospital settings and therefore researchers have looked into different CO-releasing molecules (CORMs) as alternatives and metal carbonyl complexes have drawn attention in this regard.[6–11] Indeed the therapeutic effects of CO were originally documented with the use of CORMs such as CORM-1 and CORM-3 (Fig. 1).

Although metal carbonyls like CORM-3 releases CO in contact with water and biological fluids, the loss of CO (the actual drug) from the "prodrug" cannot be controlled and thus proper dosing of CO at a certain site can hardly be attained. On the contrary, metal carbonyls such as CORM-1 are photosensitive and provide a means for control of the amount of CO release. For this extra advantage of "on demand CO delivery" the

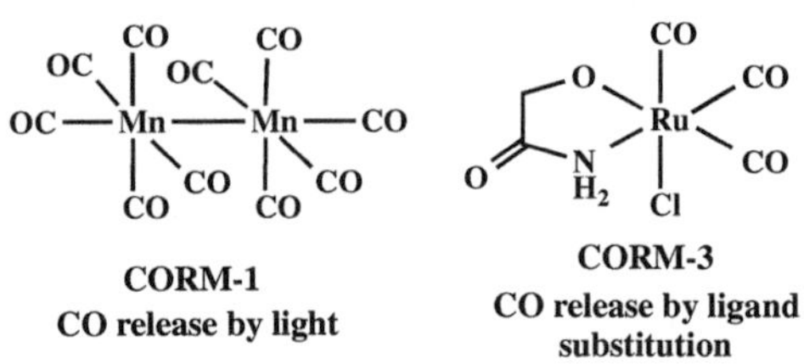

Fig. 1 CORMs used in previous biological studies.

photoactive metal carbonyl complexes or photoCORMs[6–11] have taken the center stage for CO phototherapy. Because simple metal carbonyls like CORM-1 requires UV light and UV light is detrimental to tissues, photoCORMs that could be triggered by visible and NIR light (500–900 nm) are more desirable as CO prodrugs.[12–15] This range of light also travels longer in human tissues and hence facilitates release of CO from photoCORMs within biological milieu. Research during the past decade has demonstrated that the photosensitivity of the photoCORMs can be tuned by including various coligand(s) L in the coordination sphere of metal carbonyl complexes of the type *fac*-$[M(CO)_n(L)_m]X$ (*fac* = facial isomer). Among the photoCORMs reported so far, manganese photoCORMs with *fac*-{Mn(CO)$_3$} moiety have shown a wide range of photosensitivity in addition to their ease of incorporation into various delivery platforms for controlled CO delivery. This account is focused on such CO donors developed by our research group and others *to highlight biomedical applications of inorganic photochemistry*. Excellent reviews on CORMs and photoCORMs in general have already been published.[8,10–15]

2. Chemical and photochemical properties of the *fac*-{Mn(CO)$_3$} moiety

The *fac*-{Mn(CO)$_3$} unit appears to have some inherent chemical stability as evident from a variety of syntheses of manganese carbonyl complexes with coligand(s) L starting from $[MnBr(CO)_5]$. A close look at the structure of the $[MnBr(CO)_5]$ reveals that out of the five CO ligands two CO occupy positions *trans* to other CO (shown in red in Fig. 2) and hence are easily displaced by incoming ligand(s) L. The combination of the Mn(I) center with three CO ligands in *facial* disposition is presumably a good balance in terms of π backbonding between the metal center and three COs giving rise to a stable moiety. This is further supported by the fact that with strong-field tridentate ligands like terpy (terpy = 2,2′:6′,2″-terpyridine), reaction with $[MnBr(CO)_5]$ affords the carbonyl complex

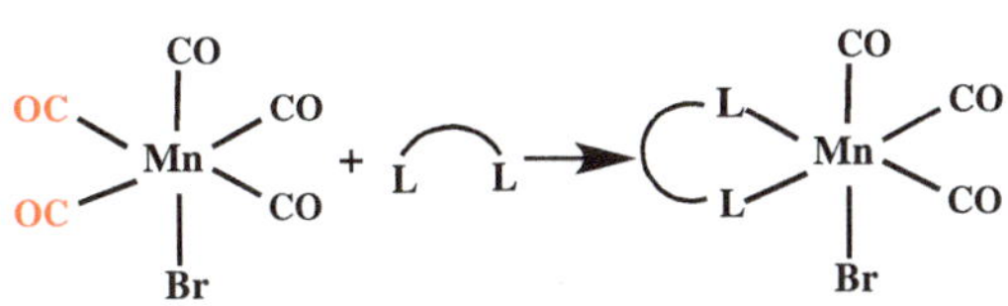

Fig. 2 In ligand substitution reactions, the *fac*-{Mn(CO)$_3$} moiety remains intact (reaction with a bidentate ligand is shown).

fac-[MnBr(CO)$_3$(terpy-κ^2*N*^{1},*N*^{2})] (yield 90%) in which only two of the three N donor centers of terpy are ligated to the Mn(I) center. The *fac*-{Mn(CO)$_3$} unit remains intact in this reaction. The expected *fac*-[Mn(MeCN)(CO)$_2$(terpy-κ^3*N*^{1},*N*^{2},*N*^{3})] complex, in which the terpy ligand is tri-coordinate, can only be prepared from the original complex by selective photolysis at 372 nm UV light in acetonitrile (when one CO is photoreleased and allows binding of the third N donor center of terpy).[16] However, despite this chemical stability, most photoCORMs with the *fac*-{Mn(CO)$_3$} moiety releases all three CO molecules upon illumination under aerobic conditions. This unique property boosts the chemotherapeutic potential of this unit as part of a CO donor because even a small dose of such a photoCORM (pro-drug) is able to deliver three times its equivalent of CO (the actual drug) to the intended site. Careful spectroscopic studies reveal that photoCORMs with *fac*-{Mn(CO)$_3$] moiety such as *fac*-[Mn(CO)$_3$(tpm]$^+$ (where tpm = tris(pyrazolyl)methane) first loses one CO upon irradiation, and then the last two COs are lost in a reaction accompanied by oxidation of the Mn(I) center by O_2.[17]

3. Moving photosensitivity to visible region of light; Smart design of photoCORMs

The initial photoCORMs like $Mn_2(CO)_{10}$ (CORM-1) undergo CO photorelease only when irradiated with UV light in the 300–350 nm region because the metal-CO bonds typically require high energies for dissociation owing to strong backbonding interactions.[18] In order to move the photosensitivity to visible and NIR region (lower energy light), we argued that if the metal carbonyl complex contains coligand(s) with significant π conjugation in their structures, *photo-triggered excitation of electron(s) from the low-valent metal center to the low-lying MOs of such coligand frames will effectively weaken the M-CO backbonding and cause CO loss.* With that in mind we synthesized three manganese carbonyl complexes **1–3** (Fig. 3) with increasing

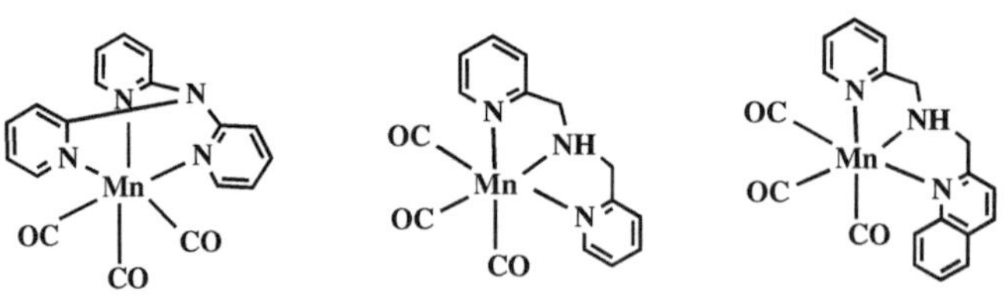

fac-[Mn(CO)$_3$(tpa)]$^+$ (1) *fac*-[Mn(CO)$_3$(dpa)]$^+$ (2) *fac*-[Mn(CO)$_3$(pqa)]$^+$ (3)

Fig. 3 Designed photoCORMs with moderate sensitivity to visible light.

number of pyridine/quinoline rings in the ligand frame and the *fac*-{Mn(CO)$_3$} moiety.[19] Complexes of this type are readily synthesized starting from [MnBr(CO)$_5$] or *fac*-[Mn(MeCN)$_3$(CO)$_3$]ClO$_4$. These carbonyl complexes exhibited increasing metal-to-ligand charge transfer (MLCT) absorption in the 400–450 nm region in the order **3** > **1** > **2** as well as increasing CO photolability upon illumination with low power visible light (with cut-off UV filter). This result demonstrates that coligands with extended π-conjugation do increase the amount of light absorbed by the designed carbonyl complexes and moves sensitivity to longer wavelengths of light. The loss of CO from **3** upon such illumination was confirmed by reduced myoglobin assay and vasorelaxation of mouse aorta muscle ring experiments in tissue bath.[19] Interestingly, Schatzschneider and coworkers have reported formation of *fac*-[Mn(CO)$_3$(tpa-κ^3N)]$^+$ (with one –CH$_2$Py arm not coordinated) when [MnBr(CO)$_5$] was heated to reflux in diethyl ether under *anaerobic* conditions.[20] This photoCORM releases CO under UV irradiation and has been shown to exhibit strong antibacterial activity. Replacement of the pyridine-N donors with pyrazol-N, as in the photoCORM *fac*-[Mn(CO)$_3$(tpr)]$^+$ (tpr = tris(pyrazolyl)methane),[21] leads to shift of photosensitivity toward the UV region. Photorelease of CO from this photoCORM upon exposure to UV light has been shown to cause damage to HT29 colon cancer cell.

In order to further examine the role of the extended π-conjugation of the coligand on energy levels of the MLCT band of Mn-based photoCORMs bearing the *fac*-{Mn(CO)$_3$} moiety, we focused on bidentate coligands along with one σ-donor or π-acceptor ancillary ligand. A number of photoCORMs in this category has been reported in the literature. In our research, we synthesized a bidentate ligand pimq and two derivatives pmtpm and qmtpm with SMe appendage (additional electron density) on the ligand frame (Fig. 4).

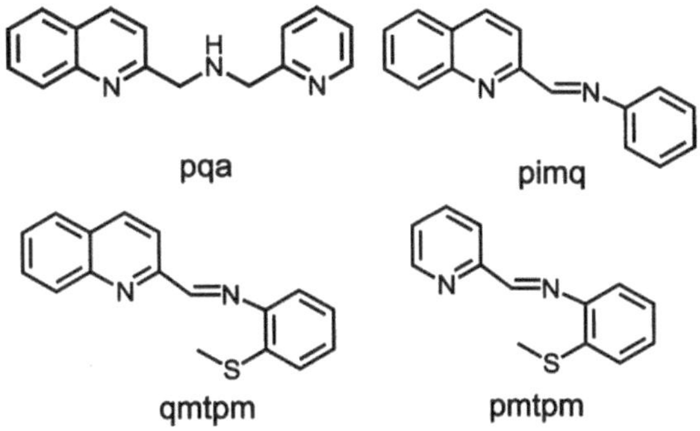

Fig. 4 Coligands for photoCORM to alter the energy of the MLCT band.

Replacement of the amine nitrogen of pqa with an imine function in pimq was expected to stabilize the lowest unoccupied MO (LUMO) associated with the MLCT transition and thus could lead to a red shift of the photoactive band toward the visible region. Similarly, more π-conjugation in qmtpm compared to pmtpm was introduced to confirm the validity of our design concept. Quite interestingly, the four manganese complexes namely, *fac*-$[MnBr(CO)_3(pmtpm)]$ (**4**), *fac*-$[Mn(MeCN)(CO)_3(pmtpm)]^+$ (**5**), *fac*-$[MnBr(qmtpm)(CO)_3]^+$ (**6**) and *fac*-$[Mn(MeCN)(CO)_3(qmtpm)]^+$ (**7**) exhibited electronic absorption spectral data that provided excellent clue regarding the MLCT transitions in this series of designed photoCORMs with similar structures.[22] As shown in Fig. 5, the effect of increased conjugation in the ligand frame is readily indicated by the red shift of the absorption maximum of **7** (435 nm) compared to that of **5** (390 nm). Replacement of the MeCN ligand with the σ-donating Br^- ligand results in further red shift of **6** and **4** to 535 and 500 nm, respectively. All these photoCORMs with strong MLCT band in the visible region exhibited CO release upon exposure to visible light as confirmed by myoglobin assay.

Density functional theory (DFT) and time-dependent DFT (TDDFT) results afforded more insight into molecular orbital electron densities and calculated electronic transitions for these photoCORMs. As shown in Fig. 6, these studies showed that for each photoCORM, the lowest energy band corresponds to transitions that promote electron density from HOMO-2 (predominantly Mn-CO bonding character) to the LUMO

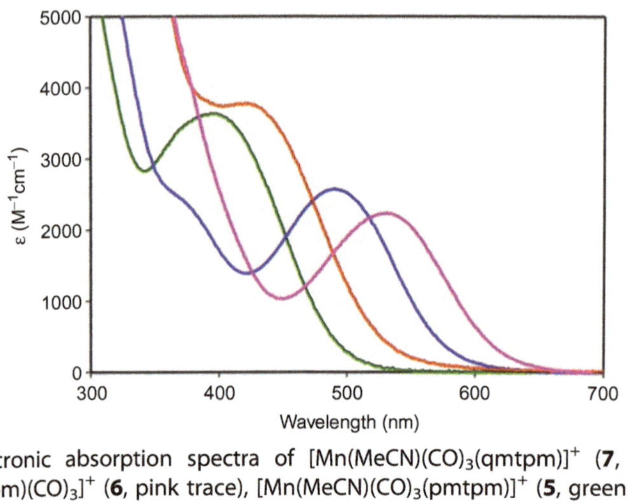

Fig. 5 Electronic absorption spectra of $[Mn(MeCN)(CO)_3(qmtpm)]^+$ (**7**, red trace), $[MnBr(qmtpm)(CO)_3]^+$ (**6**, pink trace), $[Mn(MeCN)(CO)_3(pmtpm)]^+$ (**5**, green trace), and $[MnBr(CO)_3(pmtpm)]$ (**4**, blue trace) in $CHCl_3$.

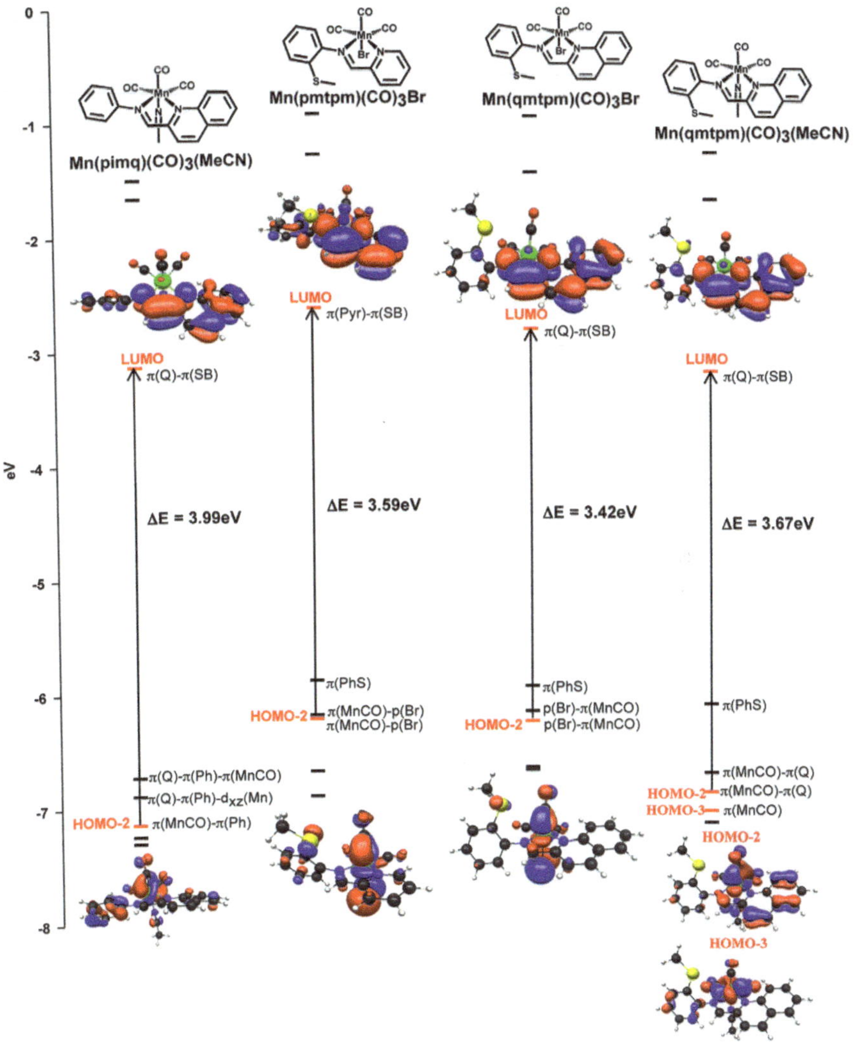

Fig. 6 Calculated HOMO/LUMO energy diagram of photoCORMs *fac*-[Mn(MeCN)(CO)$_3$(pimq)] (**8**), **4**, **6**, and **7** (from left to right). The most prominent MOs involved with transitions under the low energy band and their diagrams are shown. Transitions discussed in the text are shown in red (all other orbitals in TDDFT calculations are labeled in black).

mostly composed of the imine functionality and pyridine or quinoline ring. These MLCT transitions therefore shift electron density from the metal center to the ligand π frame, an event that curbs the affinity of the Mn center toward CO (because of loss of backbonding from the metal). The fact that photoCORMs **4** and **6** rapidly release CO upon exposure to 500 nm visible light confirms that the MLCT transitions are responsible for the photolability of these photoCORMs and these "photobands" could be moved to longer wavelengths of light through proper design and choice of the coligand(s).

To further establish the role of the ligand frame structures on the shift of the photobands to the longer wavelengths of light, we selected 2-phenylazopyridine (azpy) as the coligand and synthesized the photo CORM *fac*-[MnBr(CO)$_3$(azpy)] (**9**, Fig. 7).[23] This photoCORM exhibited a strong photoband with λ_{max} at 586 nm and rapidly release CO (quantum yield $\phi_{550} = 0.48 \pm 0.01$) upon illumination with low power (10–15 mW/cm^2) visible light of wavelengths $\lambda > 520$ nm (as evident by cut-off filter). DFT and TDDFT studies provided helpful insights into the origin of the strong MLCT band of **9** at 586 nm and the associated photorelease of CO under illumination. As shown in the right panel of Fig. 7, the release of CO from **9** has its origin in a transition from an occupied combination orbital of π(Mn-CO) (41%) and p-Br (37%) character to the LUMO primarily made up of the π*-orbital of the 2-phenylazopyridine ligand. Reduction in the Mn-CO backbonding during this transition weakens the Mn-CO bonds. It is important to note that the absorption maximum of **9** is 160 nm red-shifted compared to the absorption maximum of the structurally related [MnBr(CO)$_3$(bpy)] (bpy = 2,2′-bipyridine), a fact demonstrating that the low-lying (compared to bpy) π*-orbital of the azpy ligand is mostly responsible for the red-shift of the MLCT band maximum of **9**. The photoCORM **9** has been shown to effectively induce dose-dependent apoptotic cell death of both HeLA (human cervical cancer cell) and human breast cancer cell MDA-MB-231. The extent of cell kill was comparable when the cells were exposed to 10 μM of 5-fluorouracil. Later work by Zobi and coworkers demonstrated that the photosensitivity of **9** could be further red shifted to

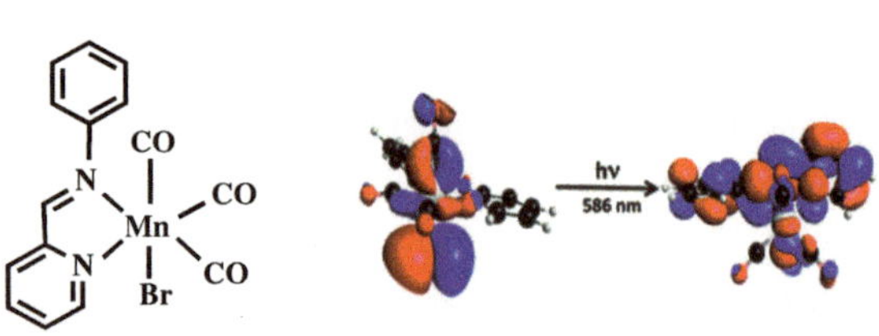

fac-[MnBr(CO)$_3$(azpy)] (**9**)

Fig. 7 Left panel: Structure of photoCORM **9**; Right panel: MLCT-XLCT transition in **9** leading to the release of CO.

625–693 nm by utilizing a derivative of the azpy ligand frame with various substituents (with groups such as CH_3, Br, CF_3).[24] One of the derivative photoCORMs showed photorelease of CO upon exposure to 810 nm. Theoretical studies revealed that the presence of electron-withdrawing substituents induces a decrease of the HOMO-LUMO gap with a consequent bathochromic shift of the visible band (photoband) in these photoCORMs.

As described above, the coligand azpy was chosen due to its known superior π-acceptor properties by virtue of its vacant low-lying π* antibonding orbital(s) that can accept electron density from the low-valent metal centers. Indeed examination of the metric parameters of **9** revealed noticeable lengthening of N═N (azo group) bond indicating significant metal-to-azo π-backbonding.[1] Consistent with all these observations, **9** revealed a strong MLCT band in the visible region centered at 586 nm. Interestingly, when we synthesized the structurally analogous rhenium complex, namely *fac*-[$ReBr(CO)_3$(azpy)], it also exhibited a prominent MLCT band in the visible region centered at 530 nm.[25] However, despite the presence of a strong MLCT band in the visible region, this Re-based photoCORM showed *no CO release under visible light illumination*. Ishitani and co-workers have delineated the excited state photo behavior of a series of Re(I) carbonyl complexes derived from some traditional N-donor ligands (such as bpy and phen) as well as their derivatives.[26] In presence of coordinating solvents like acetonitrile or THF, these Re(I) carbonyl complexes exhibit photochemical ligand substitution reactions (PLS) through uncaging of CO *strictly under UV irradiation*.[27] With these Re(I) carbonyl complexes, irradiation at ≥400 nm populates the ^{1}MLCT and/or ^{1}LLCT (ligand-to-ligand charge transfer) and eventually through intersystem crossing (ISC) to the ^{3}MLCT or ^{3}LLCT excited states. The latter states are found to be nonproductive in terms of CO release but are responsible for the emission properties of the complexes. Strong spin-orbit coupling (prominent in heavy metals) appears to promote this ISC. Contrary to this, irradiation at UV region (λ =300–350 nm) populates the ^{1}LF (ligand field) and subsequently the ^{3}LF state. The ^{3}LF is a non-emissive state but leads to metal-CO bond dissociation and thus is the active state for PLS reactions. The ^{3}LF state resides in relatively higher energy range compared to the ^{3}MLCT state. These two states are separated by a thermal activation barrier (ΔG^*), which explains the temperature dependence of the emission quantum yields and lifetimes (Scheme 1). Taken together, it is evident that while selecting the transition metal from group 7, the manganese carbonyl complexes are

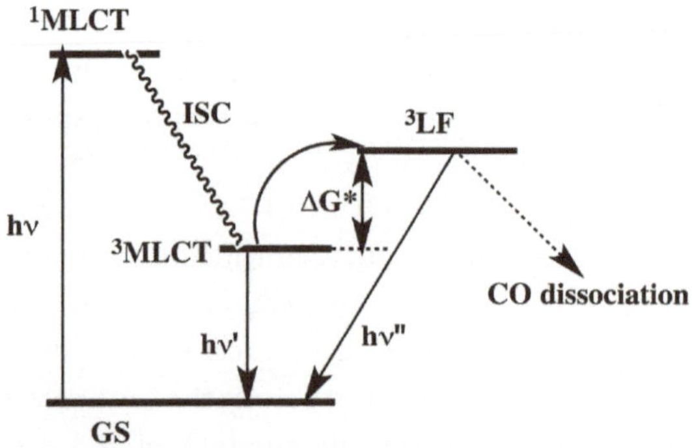

Scheme 1 Simplified Jablonski diagram illustrating the active excited state for CO dissociation upon illumination of the Re-based photoCORMs.

the only choice for developing photoCORMs that could be activated with the aid of visible/NIR light.

Our photoCORM design strategy appeared to indicate that a combination of a coligand with low-lying extended π^* molecular orbital and a strong σ-donating ancillary ligand is required to red-shift the photosensitivity of the resulting carbonyl complex with the *fac*-{$Mn(CO)_3$} moiety to the intended visible-NIR region. A close scrutiny of the literature reveals that a variety of metal carbonyl complexes have been derived from α,α′-diimine ligands such as 2,2′-bipyridine.[28,29] For example [$MnBr(CO)_3$(bpy)] releases CO when exposed to 420 nm light. In order to utilize this functionality in our coligand we selected BIAN (bis(4-chlorophenyl-imino)acenaphthene) and MIAN (2-[(4-chlorophenyl)imino]acenaphthylen-1-one), two α,α′-diimine ligand frame with extended conjugation (Fig. 8). Both *fac*-[$MnBr(CO)_3$(BIAN)] (**10**) and *fac*-[$MnBr(CO)_3$(MIAN)] (**11**, Fig. 9) exhibited strong MLCT absorption with λ_{max} at 570 and 630 nm, respectively.[30] Although both these photoCORMs rapidly release CO upon exposure to low power

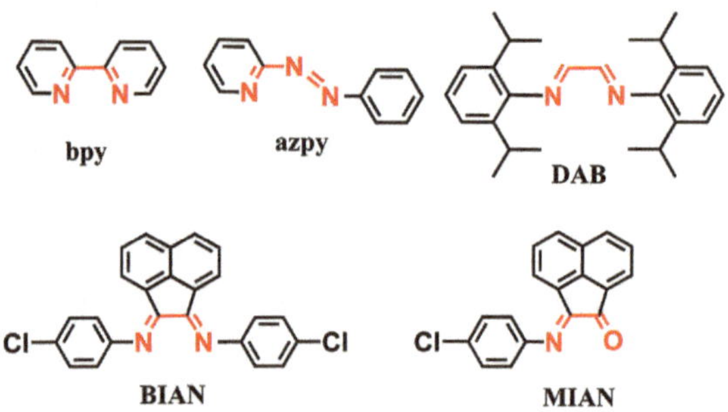

Fig. 8 Structures of coligands with α,α′-diimine functionality.

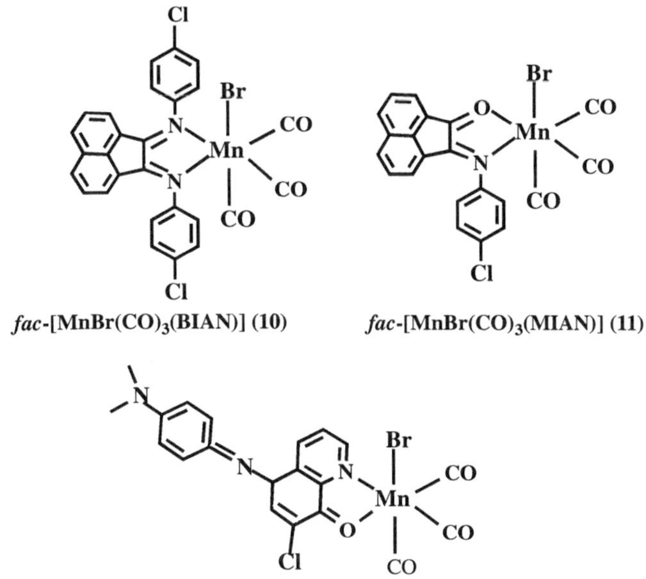

fac-[MnBr(CO)$_3$(L1)] (12)

Fig. 9 Examples of photoCORMs that release CO upon exposure to light in the 600–800 nm range.

(5–10 mW/cm^2) broadband visible light (**11** is sensitive to 800 nm light), the photosensitivity of **10** is rather exceptional. It releases CO even in the crystalline state upon illumination.

In acetonitrile solution, **10** exhibits a quantum yield $\phi_{545} = 0.70 \pm 0.2$. DFT and TDDFT results confirm that for both complexes, the MLCT transitions originate from the occupied orbitals primarily composed of π(Mn-CO) and p-Br character and terminate in the LUMO primarily made up of π*-imine and acenaphthene moiety. It is important to note that as one goes from bpy to BIAN to azpy to MIAN, the photoband maximum red-shifts from 420 nm to 570 nm to 590 nm to 630 nm (in dichloromethane).

As mentioned before, poor transmission of UV and visible light through biological fluids and tissues has led to considerable interest in developing photoactivatable systems that could be triggered by tissue penetrating near-infrared (NIR) irradiations for photo-uncaging of CO. As continuation of our effort to shift the MLCT band(s) of the designed photoCORMs to the therapeutic window (700–1100 nm), we have recently synthesized and structurally characterized a manganese carbonyl complex *fac*-[MnBr(CO)$_3$(L1)]

(**12**, Fig. 9, L1 = (5-(4′-dimethylamino-phenylimino)-7-chloroquinolin-8-one) that exhibits a strong absorption in the 750–900 nm range with λ_{max} at 785 nm.[31] Results of DFT studies indicate that the HOMO-LUMO gap of the ligand L1 (1.11 eV) drops to 0.97 eV in the complex and as a consequence the absorption maximum of the complex moves to longer wavelength compared to the ligand. Also, the HOMO of the complex has significant Mn-CO bonding character while the LUMO is devoid of any such interaction. Thus the HOMO→LUMO excitation with 730 nm LED light lowers the M-CO bond order and causes rapid CO release (quantum yield $\phi_{730} = 0.090 \pm 0.004$). The CO release was dependent on the power of the LED light source and IR spectroscopic study indicated loss of all 3 COs upon illumination. Collectively these results demonstrate that proper design of the coligand in Mn-based photoCORMs with the *fac*-$\{Mn(CO)_3\}$ moiety can dramatically alter their ability to deliver CO upon exposure to a wide range of lights in the visible/NIR region.

4. Water-soluble photoCORMs for chemotherapeutic applications

The chemotherapeutic applications of the photoCORMs for delivery of CO to biological targets require that the photoCORM under consideration is soluble in aqueous media. Because we intended to isolate the designed photoCORMs for that ultimate purpose, we also focused on water-soluble photoCORMs that could be triggered by visible or NIR light. A number of photoCORM bearing the *fac*-$\{Mn(CO)_3\}$ moiety have been synthesized by several groups with coligands such as 1,10-phenanthroline (phen) and 2,2′-bipyridine (bpy). Both these ligands are amenable to substitution of various kinds and hence offer the opportunity of synthesizing a variety of photoCORMs with different chemical and photochemical properties. We utilized the cage-like monodentate phosphane, 1,3,5-phosphaadamantane (PTA) to confer water solubility in the synthesis of *fac*-[$Mn(CO)_3$(phen)(PTA)] (**13**, Fig. 10).[1] PTA is an amphiphilic, air-stable, neutral ligand with high solubility in water. This photoCORM releases CO upon exposure to visible light and is readily internalized by cancer cells. Light-triggered CO delivery from **13** has been shown to induce apoptotic cell death of human breast cancer cell MDA-MB-231.[32] In addition this photoCORM has proven to be a *convenient tool to explore the salutary effects of CO delivery* in combination chemotherapy. For example, we have shown that sensitization of human breast

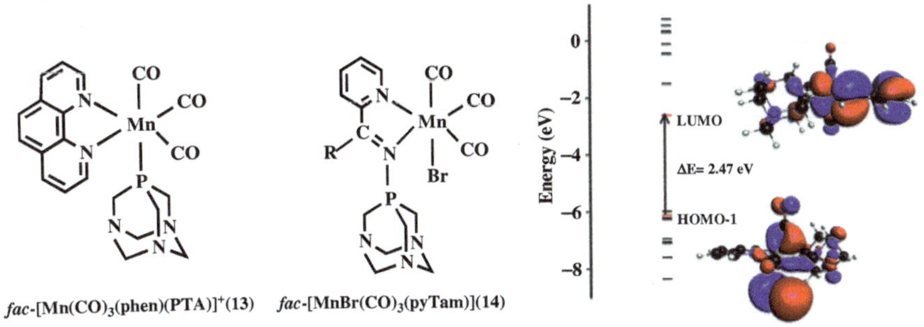

Fig. 10 Left panel: Water soluble photoCORMs with the *fac*-{$Mn(CO)_3$} moiety. Right panel: Calculated energy diagram of **14**; the most prominent MOs involved with transitions under the photoband associated with the CO release.

cancer cells to paclitaxel can be enhanced with delivery of CO (from the photoCORM **13**) through inhibition of cytochrome P450 enzymes (CYP) specifically CYP3A4 and CYP2C8.[33] Sensitization of ovarian cancer cells to conventional chemotherapeutics such as cisplatin could also be achieved by site-specific delivery of CO and this benefit arises from modulation of cellular glutathione and metallotheionein.[34] In addition, using photoCORM **13**, we demonstrated that CO inhibits cystathionine beta-synthase, an important enzyme that maintains cellular redox homeostasis and attenuates the overall antioxidant capacity of cancer cells, leading to more drug sensitivity.[35]

Solubility in water could be conferred to photoCORMs by using α-diimine ligands derived from triazaadmantane amine. For example we have reported the photoCORM *fac*-[$MnBr(CO)_3$(pyTam)] (**14**, Fig. 10, pyTam=2-(pyridyl)iminotriazaadamantane), which is readily soluble in PBS buffer and releases CO upon exposure to visible light.[36] The calculated energy diagram of **14** indicates that the 450 nm MLCT band originates from the HOMO-1 mostly comprised off π(Mn-CO) and p-Br and ends in LUMO primarily made up of π* orbitals of the imine and pyridine moieties.

5. Trackable photoCORMs that allow visualization of the CO donor within cellular matrix

Direct ligation of fluorescent ligands such as phen to the manganese center of the photoCORMs leads to quenching of the luminescence and it is not possible to trace the internalization of a prodrug like **13** into the targeted cell by confocal microscopy. However, ligation of a fluorophore to the metal center through a spacer often provides a fluorescent photoCORM that could be tracked within the target. For example, when the dansyl fluorophore is tethered to the metal center of the photoCORM *fac*-[Mn(CO)$_3$(Imdansyl)(phen)]$^+$ (**15**) through an imidazole spacer, the overall luminescence is only partly quenched and luminescent photoCORM exhibits a broad emission band with maximum at 600 nm in acetonitrile solution.[37] However, when the dansyl unit is tethered to the metal center through an unconjugated piperidine linker as in **16**, the fluorescence of the axially bound fluorophore is not quenched at all. Both these visible light-active photoCORMs (Fig. 11) have been tracked within human colorectal adenocarcinoma HT-29 cells following incubation of the cells with the photoCORMs in the dark. Exposure of the prodrug-loaded cells to visible light led to dose-dependent apoptotic death of the cancer cells. We have further fine-tuned this type of "trackable" photoCORM by using more lipophilic coligands instead of phen to achieve

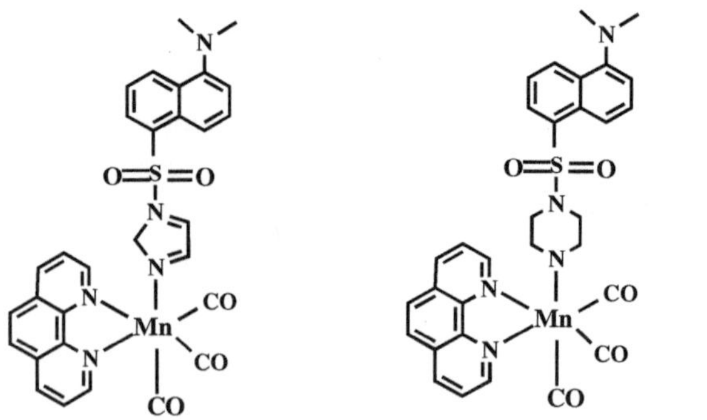

fac-[Mn(CO)$_3$(Imdansyl)(phen)]$^+$ (15) *fac*-[Mn(CO)$_3$(Imdansyl)(phen)]$^+$ (16)

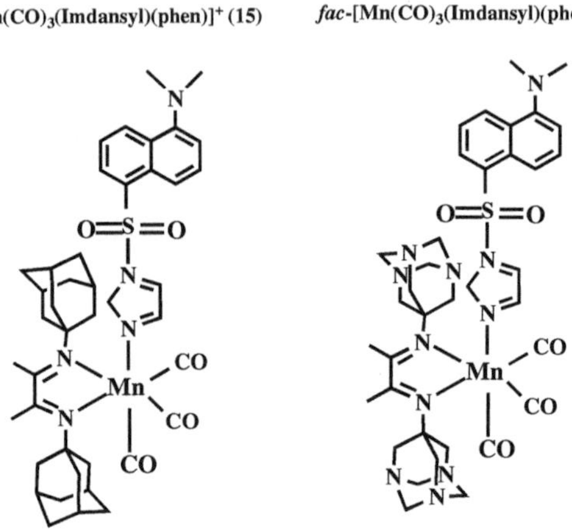

fac-[Mn(CO)$_3$(Imdansyl)L1]$^+$ (17) *fac*-[Mn(CO)$_3$(Imdansyl)L2]$^+$ (18)

Fig. 11 Examples of fluorescent photoCORMs bearing *fac*-{Mn(CO)$_3$} moiety.

better internalization of the prodrug into the target cells. For example, co-localization experiments using membrane stain indicated different extents of localization of the two photoCORMs **17** and **18** (Fig. 11) within HT-29 cells.[38] As expected, the fluorescent photoCORM **18** bearing triazaadamantyl groups (more water soluble) was more localized within the cytoplasm whereas the lipophilic photoCORM **17** was mostly associated with the cellular membrane. Exposure of the loaded cells to visible light led to CO-induced cellular death.

Zobi and coworkers have exploited the transcobalamine-mediated uptake of photoCORMs in a series of complexes (such as B_{12}-Mn-2, **19**, Fig. 12) by using bpy-based ligands tethered to Vitamin B_{12} frame.[39] This essential vitamin is internalized in cells following a receptor-mediated endocytosis. Because vitamin B12 is also a tumor-targeting agent, these VitB_{12}-conjugates are readily internalized in human breast cancer MCF-7, lung carcinoma A549 and colon adenocarcinoma HT20 cells and cause apoptotic cell death upon photo-irradiation with visible light (420 nm). Although vitamin B12 itself is luminescent (emission at 337 nm), modification of the sugar moiety of the cobalamine (a modification that is well-tolerated in cell internalization mechanisms) resulted in fluorescent photoCORMs bearing the *fac*-{Mn(CO)$_3$} unit that can be tracked in cellular matrices by its bright green fluorescence.

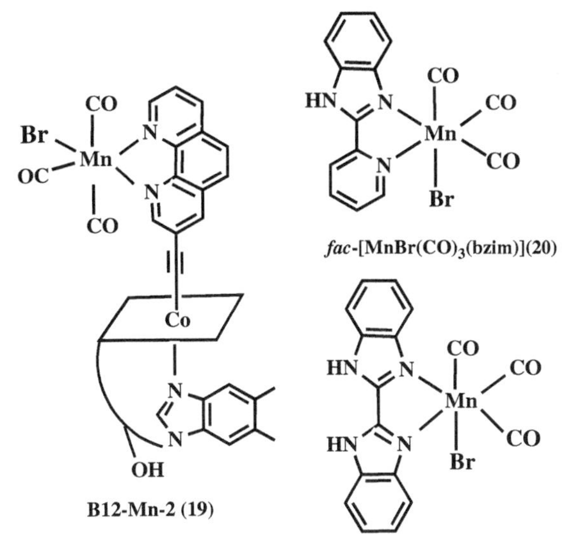

B12-Mn-2 (19)

***fac*-[MnBr(CO)$_3$(bzim)](20)**

***fac*-[MnBr(CO)$_3$(bisbzim)](21)**

Fig. 12 Trackable photoCORMs derived from therapeutic coligands.

Luminescent photoCORMs derived from 2-(2-pyridyl)benzimidazole and its derivatives such as *fac*-[MnBr$(CO)_3$(bzim)] (**20**) and *fac*-{MnBr$(CO)_3$(bis-bzim)] (**21**, Fig. 12) also exhibit cytotoxicity as shown with liver cancer SK-Hep1 cells.[40] The benzimidazole ligand itself possesses antibacterial and antiparasitic properties. In addition, the anthelmintic mebendazole, which consists of a benzimidazole ring, has been shown to possess anticancer properties in a broad range of preclinical studies across various cancer types. The strong luminescence of **21** in the 500–600 nm region makes it a suitable candidate for optical imaging. Cell viability studies indicate that the lipophilic photoCORM **21** rapidly releases CO when exposed to light in the 300–450 nm range and causes CO-induced cell death.

6. Turn-ON photoCORMs that confirm CO delivery at the target site

A close scrutiny of literature reveals that photoCORMs bearing the *fac*-{Mn$(CO)_3$} moiety release all three CO molecules upon photolysis. Careful FTIR studies of the exhaustively photolyzed products exhibit no IR peak around 2000 cm^{-1}.[32,36] Also, CO loss from these photoCORMs under aerobic conditions leads to oxidation of the Mn(I) center to Mn(II) (as shown by EPR studies).[28,32,36] Interestingly, in many cases deligation of the coligand is also observed. This release of coligand has been exploited in synthesizing "turn-ON" photoCORMs that indicates delivery of CO within target cells. For example, we have reported the water soluble photoCORM *fac*-[MnBr$(CO)_3$(pbt)] (**22**, Fig. 13, pbt=2-(2-pyridyl)-benzothiazole) that acts as a "turn-ON" CO donor.[41] This non-luminescent complex is readily internalized by the breast cancer cell MDA-MB-231. When the prodrug-loaded cells were exposed to visible light ($\lambda > 450$ nm), CO release from the photoCORM caused dose-dependent apoptotic death of the cancer cells (as evidenced by cell viability assay) along with emergence of deep blue fluorescence of the free pbt ligand confirming delivery of the drug (CO) within the target cells.

Schiller and coworkers have also attached various light-absorbing dye appendage and isolated photoCORMs bearing the *fac*-{Mn$(CO)_3$} unit. These photoCORMs exhibit both colorimetric and fluorometric dual-response upon releasing of CO upon exposure to visible light. For example visible light-induced CO release from the nitrobenzoxadiazole (NBD) fluorophore containing photoCORM NBD-CORM (**23**, Fig. 13) brings about

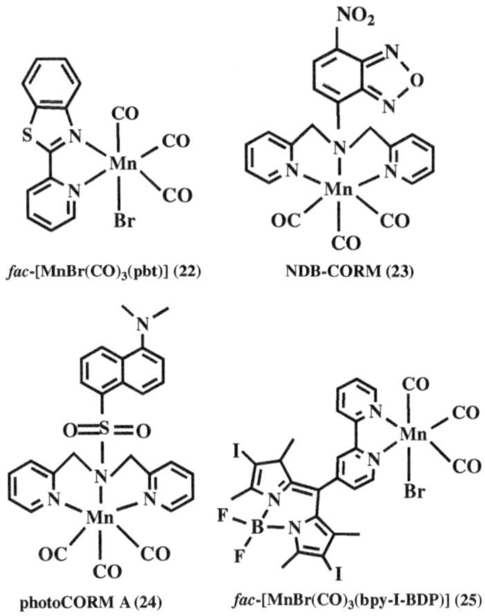

Fig. 13 Turn-ON photoCORMs reported by various groups (see text).

a color change (due to deligation of NBD from the Mn center following CO loss) and a strong emergence of fluorescence ($\lambda_{em} = 553$ nm).[42]

The same group has previously reported the photoCORM **24** (referred to as photoCORM **A**, Fig. 13) derived from dipicolylamine tethered with a dansyl fluorophore. CO release from this photoCORM (upon exposure to 405 nm light) is chaperoned by "turn-ON" fluorescence (from deligation of the ligand) in aqueous phosphate buffer. Addition of H_2O_2 also leads to the same results. The authors have designated such behavior as a logic "OR" gate *via* co-registering the inputs of irradiation (light) and peroxide (matter) into the concomitant outputs fluorescence (light) and matter (CO).[43] This claim is somewhat superfluous since many photoCORMs do respond to both light and H_2O_2 in this regard. CO release from photoCORM **A** has been employed to cause apoptotic death of HCT16 colon cancer cells where the CO delivery within the target cells is confirmed by the "turn-ON" phenomenon.

Very recently, Mn-based photoCORMs with the *fac*-{Mn(CO)3} moiety has been designed to release both CO and singlet oxygen ($^1O_2$) upon illumination with visible light. For example, the photoCORM *fac*-[MnBr(CO)3(bpy-I-BDP)] (**25**, Fig. 13, bpy-BDP = bpy tethered to the emissive BODIPY unit with iodide substituents in the 2,6 positions) releases

CO and generates $^1O_2$ when exposed to 590 nm visible light.[44] DFT and TDDFT results indicate that excitation of the BDP core with low energy light results in the photodissociation of CO. Also, introduction of two heavy atom substituents on the 2,6 positions of BDP core in **25** enhances ISC leading to $^1O_2$ production. Thus upon illumination, the BDP excites to its singlet excited state ($^1\pi\pi^*$) and due to the heavy atom effect, undergoes ISC to populate its log-lived $^3\pi\pi^*$ state which allows for reaction with molecular oxygen to produce $^1O_2$ through an energy transfer mechanism. Although the BODIPY green luminescence is 90% quenched when bound to the manganese center, photodissociation of CO leads to deligation of the bpy-I-BDP ligand and restoration of the full emission intensity much like **22**. The potential of delivery of both CO and $^1O_2$ to a target site makes **25** a very desirable chemotherapeutic for Photodynamic Therapy (PDT) of malignant locations.

7. PhotoCORMs with various coligands that confer specific drug actions

It is getting more and more evident that photoCORMs with designed coligand(s) could deliver a "one-two punch" effect at targeted site. In addition to delivering a cargo of CO, photoCORMs could selectively accumulate at desired site and deliver additional drug effect at the site arising from the entire coligand or one part of it. This is possible because under aerobic conditions, photorelease of CO from Mn(I)-based photoCORMs is often accompanied by release of the coligand(s) from the oxidized Mn(II) center. Since the phenanthroline (phen) ligand frame is amenable to functionalization, several photoCORMs derived from derivatives of phen have been synthesized, some of which exhibit novel drug actions. For example, the phen-based photoCORM *fac*-[MnBr(CO)$_3$(aminophen)] (**26**, Fig. 14) exhibits strong antibacterial activity against multidrug-resistant *Klebsiella pneumoniae* and *Acetobacter baumannii* when photoactivated by blue LED light.[45] The antibacterial effect is only observed under illumination. Conjugation of the lipophilic fatty acid group in *fac*-[MnBr(CO)$_3$(phen-oleo)] (**27**, Fig. 14) allows it to better permeate through bacterial membranes thus improving the antibacterial efficacy. The emergence of drug-resistant bacteria, already a global threat, has prompted intense search for novel antibacterial drugs and photoCORM-conjugates of this kind could serve as effective chemotherapeutics to thwart chronic infections by drug-resistant bacteria.

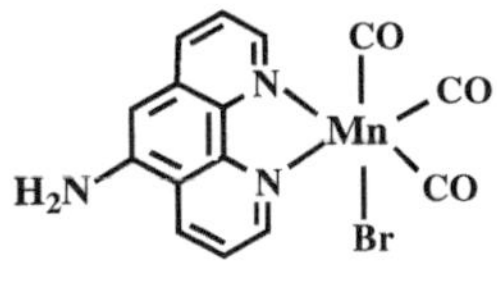

fac-[MnBr(CO)$_3$(aminophen)] (26)

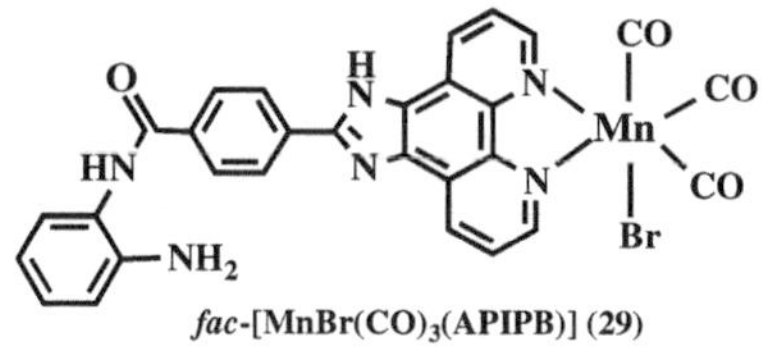

fac-[MnBr(CO)$_3$(APIPB)] (29)

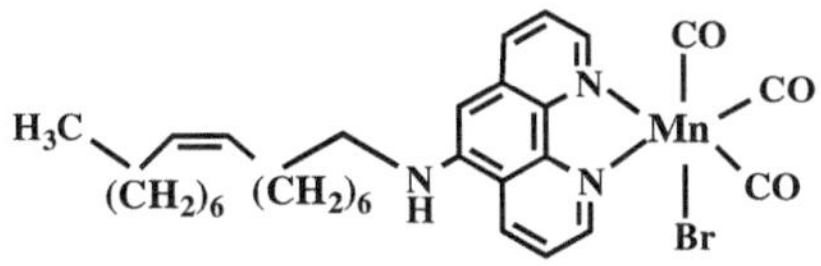

fac-[MnBr(CO)$_3$(phen-oleo)] (27)

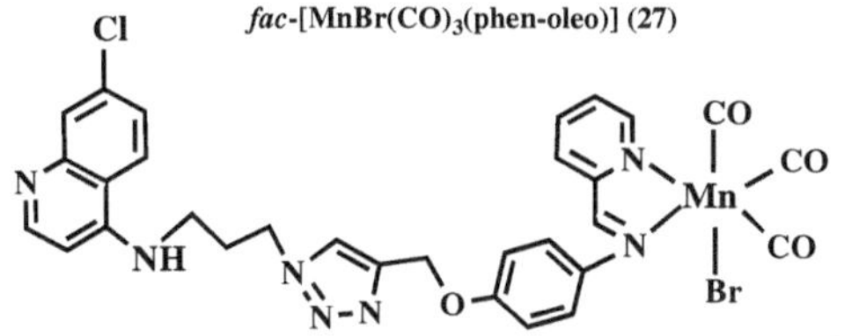

fac-[MnBr(CO)$_3$(am-qui/triz-im)] (28)

Fig. 14 PhotoCORM conjugates that exhibit novel drug actions.

The *fac*-{Mn(CO)$_3$} moiety has also found use in the quest for effective drugs against public health crisis arising out of parasitic infections. For example, Malaria remains an infectious disease of global health concern and emergence of drug-resistant strains of *Plasmodium falciparum* has further exacerbated the threat in under-developed nations. Recently, a series of manganese carbonyl complexes derived from a quinoline-triazole hybrid scaffold has been reported by Smith and coworkers.[46] Among them, the photoCORM *fac*-[MnBr(CO)$_3$(am-qui/triz-im)] (**28**, Fig. 14) exhibit strong inhibitory activities against *Plasmodium falciparum* when applied under irradiation at 365 nm. Results of *in silico* studies indicate that the quinolinyl, triazole, phenyl and pyridyl rings of **28** play a significant role in binding to the scaffold of hemozoin (π-stacked polymer of heam units) and CO release from the photoCORM could be vital for blocking the hemozoin formation pathway.[47] Strategies of this kind could be a valuable and novel route to design photoactive antimalarial agents with high efficiency.

Zhang and coworkers have reported a bifunctional photoCORM *fac*-[MnBr(CO)$_3$(APIPB)] (**29**, APIPB = N-(2-aminophenyl)-4-(1H-imidazo[4,5-f](1,10-phenanthrolin-2-yl)-benzamide, Fig. 14) in which a nuclear histone deacetylase inhibitor (HDACI) analogue has been conjugated to the phenanthroline ligand.[48] Because excessive HDAC activity

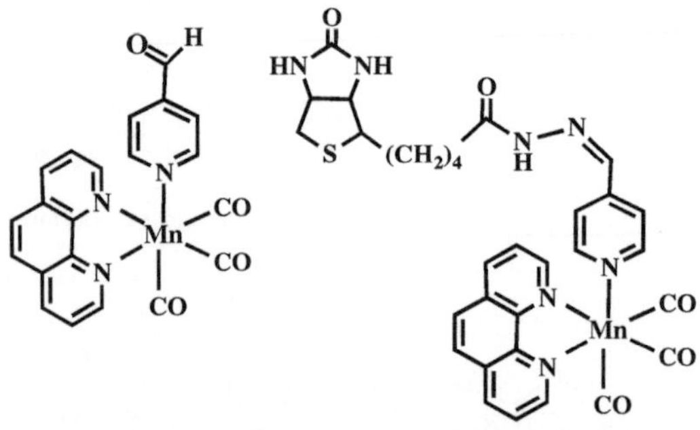

fac-$[Mn(CO)_3(phen)(4\text{-aldPy})]^+$ (**29**) biotinylated photoCORM (**30**)

Fig. 15 Biotinylated photoCORM for conjugation to antibodies.

causes aberrant expression of various genes leading to chromatin condensation and transcriptional repression associated with cancer cell proliferation, the ADACI moiety in the ligand frame was expected to ameliorate cancer cell proliferation. Indeed the cytotoxic effects of **29** on cervical carcinoma HeLa cells was greater than that of the commercially available HDAC inhibitor MS-275 under visible light illumination, an observation confirming that the combined salutary effects of the HDACI and CO do increase the efficacy of the bifunctional photoCORM compared to the chemotherapeutic (MS-275) alone.

Precise spatial and temporal control over release of CO at a target site has been further extended in our recent design of *antibody-tethered* photoCORMs.[49] In this work, the photoCORM *fac*-$[Mn(CO)_3(phen)(4\text{-aldPy})]^+$(**29**) (4-aldPy = pyridine-4-aldehyde, Fig. 15) was first conjugated to biotin and then the biotinylated photoCORM (**30**) was conjugated to mouse monoclonal antibodies through a biotin-streptavidin system. Different monoclonal antibodies raised against different surface-expressed antigens that are implicated in ovarian cancer afforded a family of antibody-photoCORM conjugates (Ab-photoCORMs) that were tested in an immunosorbent/cell viability assay. These Ab-photoCORMs accumulated onto ovarian cancer cells expressing the target antigens and delivered cytotoxic doses of CO upon illumination. It is therefore evident that "immunoCORM" of this type could deliver a gaseous drug CO to a cancer site quite precisely depending on the grade and type of the malignancy.

8. CO-releasing materials and platform design

The need for site-specific delivery of controlled doses of CO to elicit salutary effects has prompted development of various CO-delivery platforms by different groups. These attempts have often been beset with problems related to biocompatibility and to clearance of the CO-spent platforms from the patient's body. Despite such impediments, progress has been made in the area of CO delivery platforms with a variety of CO donors. In addition to other CO donors, manganese carbonyls with {Mn(CO)$_3$} moiety have either been incorporated within or grafted (or conjugated) on the surface of nanoparticles,[50,51] and porous materials[52] of various kind including metal-organic frameworks (MOF).[53]

Among porous materials, the non-toxic zeolites have drawn much attention as drug carriers.[54] Several groups have incorporated CO gas as well as metal carbonyls within the mesopores of the silicate-based nanoparticles for CO delivery. We have incorporated the cationic *fac*-[Mn(pqa)(CO)]$^+$ (pqa=(2-pyridylmethyl)(2-quinolylmethyl)amine) carbonyl within the negatively-charged pores of the aluminosilicate Al-MCM-41 nanoparticles. CO released from the particles upon visible light illumination has been shown to cause vasorelaxation of rat aorta muscle rings (Fig. 16).[55] Barea and coworkers have also incorporated *fac*-[Mn(CO)$_3$(tacn)]$^+$ (tacn=1,4,7-triazacyclononane) into the negatively charged pores of Al-MCM-41 nanoparticles and within the framework of bio-MOF-1 (a MOF derived from zinc and 4,4′-biphenyldicarboxylate). The visible light-induced CO release properties of these materials showed that the anionic silica matrix of Al-MCM-41 exhibits more controlled release of CO upon illumination by visible light.[56]

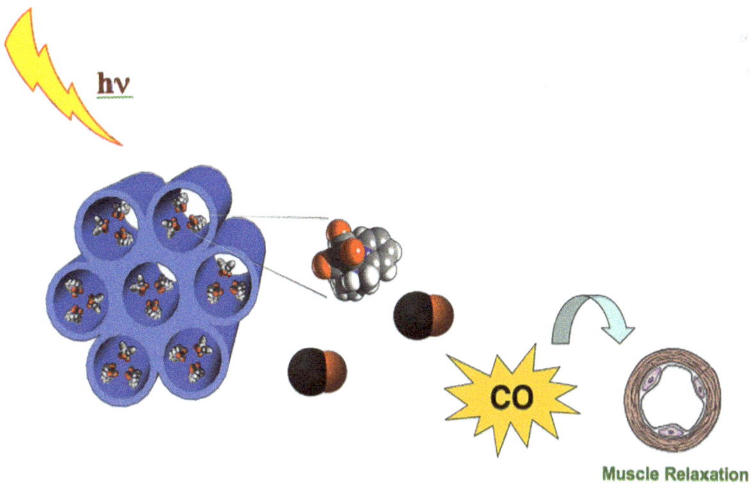

Fig. 16 CO released from aluminosilicate mesoporous particles upon visible light illumination causes vasorelaxation of rat aorta muscle rings (Ref. 55).

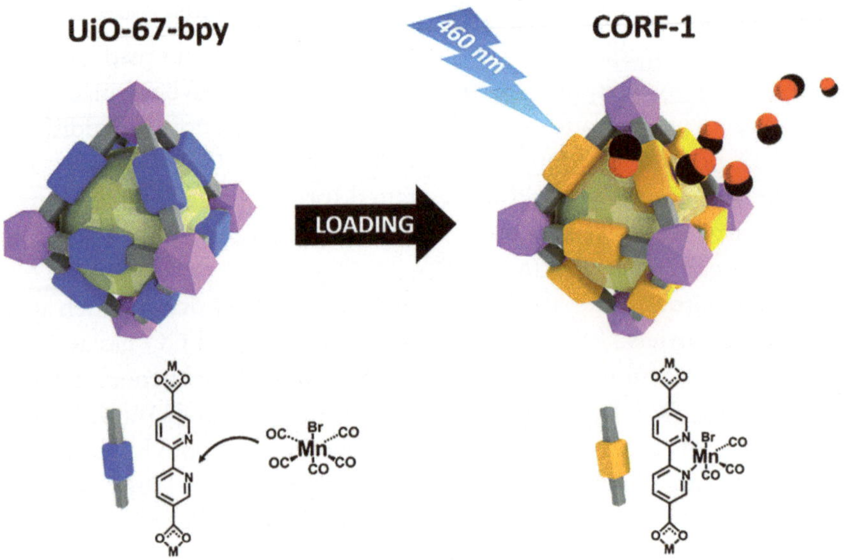

Fig. 17 Schematic drawing of the photosensitive CO-releasing (Ref. 57).

Furukawa and coworkers have loaded [MnBr(bpydc)$(CO)_3$] (where bpydc = 5,5′-dicarboxylate-2,2′-bipyridine) onto the surface of a MOF called UiO-67-bpy (a Zr-acetate-bipyridine derived MOF) through covalent coordination resulting in a photosensitive CO-releasing MOF framework (called CORF-1) which rapidly release CO upon exposure to 460 nm radiation (Fig. 17).[57] The macromolecular scaffold based CO donor releases CO faster than the manganese carbonyl itself in solid state. Instability of CORF-1 in aqueous PBS solution and related media however restricts use of this CO-donating platform mostly in colloidal form and their integration into extracellular devices.

Because of their high degree of biocompatibility peptide-based porous materials have found applications as CO delivery platforms. Recently, Cai and coworkers reported a CO-releasing material based on a self-assembly of fluorinated amphiphilic dendritic peptide (FADP) with a hydrophobic core, co-encapsulating chlorin e6 (as photosensitizer) and the CO-releasing molecule [$Mn(CO)_4(S_2CNME)(CH_2COOH)$] (CORM-401).[58] When this porous material is exposed to 665 nm light, excitation of Chlorin e6 generates $^1O_2$ and H_2O_2. As CORM-401 is a H_2O_2-sensitive CORM, the generation of H_2O_2 triggers the CO release. While the fluorinated surface of this material enhances binding affinity toward bacterial membranes, the $^1O_2$ and H_2O_2 cocktail exerts strong bactericidal effects

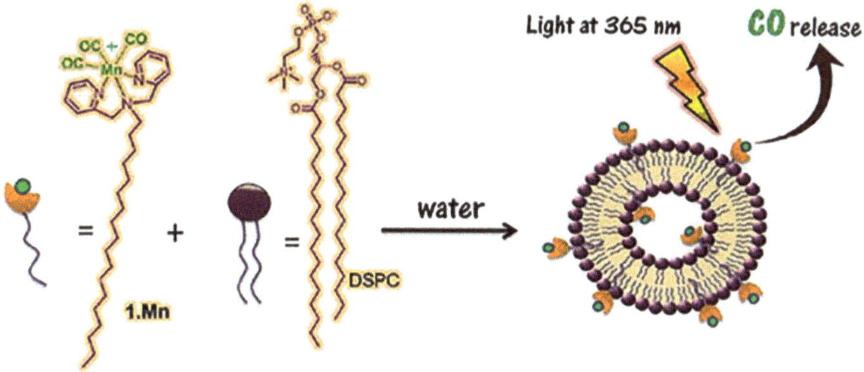

Fig. 18 Biocompatible glycero-phosphocholine lipid vesicle with incorporated *fac*-{$Mn(CO)_3$} moiety as nanocarrier for CO delivery (Ref. 60).

and causes ablation of biofilms. Smith and coworkers have modified polypyridyl dendritic scaffolds with *fac*-[$MnBr(CO)_3(bpy)$] moiety to obtain CO-releasing metallodendrimers.[59] The CO-releasing rate of the first- and second-generation metallodendrimers upon irradiation with 415 nm light were found to depend on the molecular weights of the dendritic scaffolds. These composites might be suitable for sustained release of large amounts of CO (~15 equiv. of CO per dendrimer).

In an alternative approach, Sakala and Jose have incorporated an amphiphilic manganese carbonyl complex into the 1,2-distearoyl-sn-glycero-3-phosphocholine lipid vesicles (Fig. 18) and employed the liposomes for photoinduced release of CO under UV light (365 nm). Clearly, these rational approaches could be applied to exploit water-insoluble photoactive manganese carbonyls in aqueous media by using vesicles as nanocarrier for CO release.[60]

Collectively these examples demonstrate that manganese carbonyl complexes with the {$Mn(CO)_3$} moiety could be incorporated in a wide variety of materials that could find applications in medicine[61–63] and constitute a new field of CO phototherapy that has its origin in inorganic photochemistry.

9. Conclusion and future perspectives

Research during the past three decades has brought the field of inorganic photochemistry into the drug design arena and synthetic chemists have isolated a trove of photoactive carbonyl complexes (photoCORMs) as CO donors that deliver this physiologically critical gasotransmitter site-specifically to biological targets. Research from our laboratory and others' has confirmed that these photoCORMs can be synthesized on the basis of smart design to systematically improve their sensitivity to visible/NIR light, solubility in water, permeability through cellular membranes, partitioning preferentially to specific regions within the cellular matrix, trackability within cellular matrix, and convenient incorporations into biocompatible delivery platforms for site-specific CO delivery. Among these CO donors, the manganese-base photoCORMs bearing the *fac*-$\{Mn(CO)_3\}$ moiety have drawn special attention because of their sensitivity to visible/NIR light and the ability to donate 3 equivalents of CO per molecular unit. In addition, various biocompatible delivery platforms with incorporated *fac*-$\{Mn(CO)_3\}$ moiety have opened up the possibility of administering CO (a noxious gas) in hospital settings to achieve its salutary effects in human pathophysiology.

The use of light for remote control over the activity of pharmaceuticals, a concept known as photopharmacology, is a recent trend in the area of drug delivery.[64–66] Successful clinical use of visible light to control drug activity in time and space to regulate biological processes is already well-documented.[67,68] Photopharmacological treatment is also naturally suited to localized and exposed targets; optic fibers inserted through small and minimally invasive incisions allow for illumination of most body organs to be illuminated with intense, visible, broad spectrum light from non-laser sources.[69] Indeed, photopharmacological approaches have even been successfully applied for metastatic ovarian cancer, where intraoperative and laparoscopic light sources were successfully used in photodynamic therapy, resulting in substantial benefits for patients in clinical trials.[70] Visible light offers unparalleled therapeutic benefits as an external control element for pharmacological activity, which allows for the delivery and activity of photoactivatable pharmaceuticals with very high spatiotemporal precision. Furthermore, unlike chemicals, light exhibits high orthogonality toward biological systems with minimal contamination of the study subject and low to negligible toxicity. Collectively, results of studies on the photoCORMs with the *fac*-$\{Mn(CO)_3\}$ moiety and biocompatible materials bearing such CO-donors, as described above, unequivocally demonstrate the utility of inorganic photochemistry in clinical use of CO therapy under the control of light as part of the burgeoning field of photopharmacology and photodynamic therapy.

Acknowledgment

Research in authors' laboratory was sponsored by National Science Foundation.

References

1. Wang, R., Ed. *Signal Transduction and the Gasotransmitters: NO, CO and H_2S in Biology and Medicine*; Humana Press: Totowa, NJ, 2004.
2. Goebel, U.; Wolborn, J. *Int. Care Med. Exp.* **2020**. https://doi.org/10.1186/s40635-020-0292-8.
3. Rochette, L.; Cottin, Y.; Zeller, M.; Vergely, C. *Pharmacol. Therapeut.* **2013**, *137*, 133–152.
4. Motterlini, R.; Otterbein, L. E. *Nat. Rev. Drug Discov.* **2010**, *9*, 728–743.
5. Kaczorowski, D. J.; Zuckerbraun, B. S. *Curr. Med. Chem.* **2007**, *14*, 2720–2725.
6. Kourti, M.; Jiang, W. G.; Cai, J. *Oxid. Med. Cell. Longev.* **2017**. https://doi.org/10.1155/2017/9326454.
7. Heinemann, S.; Hoshi, T.; Westerhausen, M.; Schiller, A. *Chem. Commun.* **2014**, *50*, 3644–3660.
8. Garcia-Gallego, S.; Bernardes, G. J. L. *Angew. Chem., Int. Ed.* **2014**, *53*, 9712–9721.
9. Wareham, L. K.; Poole, R. K.; Tinajero-Trejo, M. *J. Biol. Chem.* **2015**, *290* (31), 18999–19007.
10. Zobi, F. *Fut. Med. Chem.* **2013**, *5*, 175–188.
11. Romao, C. C.; Blattler, W. A.; Seixas, J. D.; Bernardes, G. J. L. *Chem. Soc. Rev.* **2012**, *41*, 3571–3583.
12. Kottelat, E.; Zobi, F. *Inorganics* **2017**, 5. https://doi.org/10.3390/inorganics5020024.
13. Wright, M. A.; Wright, J. A. *Dalton Trans.* **2016**, *45*, 6801–6811.
14. Chakraborty, I.; Carrington, S. J.; Mascharak, P. K. *Acc. Chem. Res.* **2014**, *47*, 2603–2611.
15. Heilman, B. J.; Gonzalez, M. A.; Mascharak, P. K. *Prog. Inorg. Chem.* **2014**, *58*.
16. Compain, J.-D.; Bourrez, M.; Haukka, M.; Deronzier, A.; Chardon-Noblat, S. *Chem. Commun.* **2014**, *50*, 2539–2542.
17. Berends, H.-M.; Kurz, P. *Inorg. Chim. Acta* **2012**, *380*, 141–147.
18. Yesaka, H.; Kobayashi, T.; Yasufuku, K.; Nagakura, S. *J. Am. Chem. Soc.* **1983**, *105*, 6249–6252.
19. Gonzalez, M. A.; Yim, M. A.; Cheng, S.; Moyes, A.; Hobbs, A. J.; Mascharak, P. K. *Inorg. Chem.* **2012**, *51* (1), 601–608.
20. Nagel, C.; McLean, S.; Poole, R. K.; Braunschweig, H.; Kramer, T.; Schatzschneider, U. *Dalton Trans.* **2014**, *43*, 9986–9997.
21. Niesel, J.; Ainto, A.; N'Dongo, H. W. P.; Merz, K.; Ott, I.; Gust, R.; Schatzschneider, U. *Chem. Commun.* **2008**, 1798–1800.
22. Gonzalez, M. A.; Carrington, S. J.; Fry, N. L.; Martinez, J. L.; Mascharak, P. K. *Inorg. Chem.* **2012**, *51*, 11930–11940.
23. Carrington, S. J.; Chakraborty, I.; Mascharak, P. K. *Chem. Commun.* **2013**, *49*, 11254–11256.
24. Kottelat, E.; Ruggi, A.; Zobi, F. *Dalton Trans.* **2016**, *45*, 6920–6927.
25. Chakraborty, I.; Carrington, S. J.; Mascharak, P. K. *ChemMedChem* **2014**, *9*, 1266–1274.
26. Sato, S.; Ishitani, O. *Chem. Rev.* **2015**, *50-59*, 282–283.
27. Saita, K.; Harabuchi, T.; Taketsugu, T.; Ishitani, O.; Maeda, S. *Phys. Chem. Chem. Phys.* **2016**, *18*, 17557–17564.
28. Jimenez, J.; Chakraborty, I.; Mascharak, P. K. *Eur. J. Inorg. Chem.* **2015**, *30*, 5021–5026.

29. Pordel, S.; White, J. K. *Inorg. Chim. Acta* **2020**, *500*, 119206.
30. Carrington, S. J.; Chakraborty, I.; Mascharak, P. K. *Dalton Trans.* **2015**, *44*, 13828–13834.
31. Stenger-Smith, J.; Chakraborty, I.; Ouattara, R.; Sameera, W. M. C.; Rue, K.; Mascharak, P. *Chem. Commun.* **2021**, *57*, 1101–1104.
32. Chakraborty, I.; Carrington, S. J.; Roseman, G.; Mascharak, P. K. *Inorg. Chem.* **2017**, *56*, 1534–1545.
33. Kawahara, B.; Faull, K. F.; Janzen, C.; Mascharak, P. K. *J. Med. Chem.* **2021**, *64*, 8437–8446.
34. Kawahara, B.; Ramadoss, S.; Chaudhuri, G.; Janzen, C.; Sen, S.; Mascharak, P. K. *J. Inorg. Biochem.* **2019**, *191*, 29–39.
35. Kawahara, B.; Moller, T.; Hu-Moore, K.; Rodriguez, E.; Carrington, S. J.; Faull, K.; Sen, S.; Mascharak, P. K. *J. Med. Chem.* **2017**, *60*, 8000–8010.
36. Jimenez, J.; Chakraborty, I.; Carrington, S. J.; Mascharak, P. K. *Dalton Trans.* **2016**, *45*, 13204–13213.
37. Jimenez, J.; Chakraborty, I.; Dominguez, A.; Martinez-Gonzalez, J.; Sameera, W. M. C.; Mascharak, P. K. *Inorg. Chem.* **2018**, *57*, 1766–1773.
38. Pinto, M. N.; Chakraborty, I.; Jimenez, J.; Murphy, K.; Wenger, J.; Mascharak, P. K. *Inorg. Chem.* **2019**, *58*, 14522–14531.
39. Rossier, J.; Delasoie, J.; Haeni, L.; Hauser, D.; Rothen-Rutishauser, B.; Zobi, F. *J. Inorg. Biochem.* **2020**, *209*, 111122.
40. Hu, M.; Yan, Y.; Zhu, B.; Chang, F.; Yu, S.; Alatan, G. *RSC Adv.* **2019**, *9*, 20505–20512.
41. Carrington, S. J.; Chakraborty, I.; Bernard, J. M. L.; Mascharak, P. K. *ACS Med. Chem. Lett.* **2014**, *5*, 1324–1328.
42. Liu, J.; Hoffmann, P.; Steinmetzer, J.; Askes, S. H. C.; Kupfer, S.; Gorls, H.; Grafe, S.; Neugebauer, U.; Gandra, U. R.; Schiller, A. *Polyhedron* **2019**, *172*, 175–181.
43. Reddy, U. G.; Axthelm, J.; Hoffmann, P.; Taye, N.; Gläser, S.; Görls, H.; Hopkins, S.; Plass, W.; Neugebauer, U.; Bonnet, S.; Schiller, A. *J. Am. Chem. Soc.* **2017**, *139*, 4991–4994.
44. Pordel, S.; Pickens, R. N.; White, J. K. *Organometallics* **2021**, *40*, 2983–2994.
45. Toscani, A.; Hind, C.; Clifford, M.; Kim, S.-H.; Gucic, A.; Woolley, C.; Saeed, N.; Rahman, K. M.; Sutton, J. M. *Eur. J. Med. Chem.* **2021**, *213*, 113172.
46. Ishmail, F.-Z.; Melis, D. R.; Mbaba, M.; Smith, G. S. *J. Inorg. Biochem.* **2021**, *215*, 111328.
47. Pamplona, A.; Ferreira, A.; Balla, J.; Jeney, V.; Balla, G.; Epiphanio, S.; Chora, A.; Rodrigues, C. D.; Gregoire, I. P.; Cunha-Rodrigues, M.; Portugal, S.; Soares, M. P.; Mota, M. M. *Nat. Med.* **2007**, *13*, 703–710.
48. Zhang, H.-L.; Yu, Y.-T.; Wang, Y.; Tang, Q.; Yang, S.-P.; Liu, J.-G. *J. Inorg. Biochem.* **2021**, *216*, 111354.
49. Kawahara, B.; Goa, L.; Cohn, W.; Whitelegge, J. P.; Sen, S.; Janzen, C.; Mascharak, P. K. *Chem. Sci.* **2020**, *11*, 467–473.
50. Yan, H.; Du, J.; Zhu, S.; Nie, G.; Zhang, H.; Gu, Z.; Zhao, Y. *Small* **2019**, *15* (49), e1904382.
51. Musib, D.; Raza, M. K.; Pal, M.; Roy, M. *Appl. Organomet. Chem.* **2021**, *35*, e6110.
52. Carne-Sanchez, A.; Carmona, F. J.; Kim, C.; Furukawa, S. *Chem. Comm.* **2020**, *56*, 9750–9766.
53. Zhang, M.; Quio, R.; Hu, J. *Nanomaterials* **2020**, *10*, 1134.
54. Chakraborty, I.; Mascharak, P. K. *Micropor. Mesopor. Mater.* **2016**, *234*, 409–419.
55. Gonzales, M. A.; Han, H.; Moyes, A.; Radinos, A.; Hobbs, A. J.; Coombs, N.; Oliver, S. R. J.; Mascharak, P. K. *J. Mater. Chem. B* **2014**, *2*, 2107–2113.

56. Carmona, F. J.; Rojas, S.; Sanchez, P.; Jeremias, H.; Marques, A. R.; Romão, C. C.; Choquesillo-Lazarte, D.; Navarro, J. A. R.; Maldonado, C. R.; Barea, E. *Inorg. Chem.* **2016**, *55*, 6525–6531.
57. Diring, S.; Carne-Sanchez, A.; Zhang, J.; Ikemura, S.; Kim, C.; Inaba, H.; Kitagawa, S.; Furukawa, S. *Chem. Sci.* **2017**, *8*, 2381–2386.
58. Ma, W.; Chen, X.; Fu, L.; Zhu, J.; Fan, M.; Chen, J.; Yang, C.; Yang, G.; Wu, L.; Mao, G.; Yang, X.; Mou, X.; Gu, Z.; Cai, X. *ACS Appl. Mater. Interfaces* **2020**, *12*, 22479–22491.
59. Govender, P.; Pai, S.; Schatzschneider, U.; Smith, G. S. *Inorg. Chem.* **2013**, *52*, 5470–5478.
60. Sakla, R.; Jose, D. A. *ACS Appl. Mater. Interfaces* **2018**, *10* (16), 14214–14220.
61. Yan, H.; Du, J.; Zhu, S.; Nie, G.; Zhang, H.; Gu, Z.; Zhao, Y. *Small* **2019**, *15* (49), e1904382.
62. Zhou, Y.; Yu, W.; Cao, J.; Gao, H. *Biomaterials* **2020**, *255*, 120193.
63. Inaba, H.; Fujita, K.; Ueno, T. *Biomater. Sci.* **2015**, *3*, 1423–1438.
64. Vickerman, B. M.; Zywot, E. M.; Tarrant, T. K.; Lawrence, D. S. *Nat. Rev.* **2021**. https://doi.org/10.1038/s41570-021.00326.w.
65. Hull, K.; Morstein, J.; Trauner, D. *Chem. Rev.* **2018**, *118*, 10710–10747.
66. Velema, W. A.; Szymanski, W.; Feringa, B. L. *J. Am. Chem. Soc.* **2014**, *136*, 2178–2191.
67. Mehta, Z. B.; Johnston, N. R.; Nguyen-Tu, M. S.; Broichhagen, J.; Schultz, P.; Larner, D. P.; Leclerc, I.; Trauner, D.; Rutter, G. A.; Hodson, D. J. *Sci. Rep.* **2017**, 7, 291.
68. Broichhagen, J.; Schonberger, M.; Cork, S. C.; Frank, J. A.; Marchetti, P.; Bugliani, M.; Shapiro, A. M. J.; Trapp, S.; Rutter, G. A.; Hodson, D. J.; Trauner, D. *Nat. Comm.* **2014**, *5*, 5116.
69. Allison, R. R.; Cuenca, R.; Downie, G. H.; Randall, M. E.; Bagnato, V. S.; Sibata, C. H. *Photodiagn. Phododyn. Ther.* **2005**, *2*, 51–63.
70. Wierrani, F.; Fiedler, D.; Grin, W.; Henry, M.; Dienes, E.; Gharehbaghi, K.; Krammer, B.; Grunberger, W. *BJOG* **2005**, *104*, 376–378.

CHAPTER SEVEN

Mechanistic insight into photoactivation of small inorganic molecules from the biomedical applications perspectives

Agnieszka Kyzioł*, Łukasz Orzeł, Ilona Gurgul, Olga Mazuryk, Przemysław Łabuz, and Grażyna Stochel*

Faculty of Chemistry, Jagiellonian University, Kraków, Poland

*Corresponding authors: e-mail address: kyziol@chemia.uj.edu.pl; stochel@chemia.uj.edu.pl

Contents

1. Introduction 234
2. Reactive small inorganic molecules (RSIM) 234
3. RSIM photogeneration in biological systems - Pathways and mechanisms 236
   3.1 Light management strategies 237
   3.2 Photobiostimulation 238
   3.3 Optogenetics 238
   3.4 Photocleavage 240
   3.5 Photosensitization 240
   3.6 Photocatalysis 244
4. Photoresponsive materials - Examples and development trends 246
   4.1 Molecular systems 246
   4.2 Nanostructured systems 258
   4.3 Multimodal systems 269
5. Conclusions and future directions 275
Acknowledgments 277
References 277

Abstract

Photogeneration of reactive small inorganic molecules (RSIM) by molecular and nanostructured systems with an emphasis on the mechanisms involved in the process are reviewed. RSIM, including reactive oxygen species (ROS), nitrogen species (RNS), sulfur species (RSS), chlorine species (RClS), and carbon species (RCS), play a pivotal role in physiological and pathological processes. Photodissociation, photoisomerization, photosubstitution, or photoredox reactions of RSIM precursors constituted pathways of their direct generation and have been studied mostly in molecular systems, whereas

Advances in Inorganic Chemistry, Volume 80
ISSN 0898-8838
https://doi.org/10.1016/bs.adioch.2022.06.004

research interest in photosensitized RSIM generation processes is focused on both molecular and nanostructured systems. *In vivo* light management strategies and mechanisms of light-stimulated processes leading to spatially and temporally controlled RSIM generation such as optogenetics, photocleavage, photosensitization, and homogeneous and heterogeneous photocatalysis are discussed, and representative examples are presented. A variety of photoactive molecular, macromolecular, nanostructured, and hybrid materials are discussed considering their RSIM photogeneration properties and potent biological and medical applications.

1. Introduction

The biochemistry of small inorganic molecules such as O_2, N_2, H_2O, NO, H_2S, CO, *etc.* and their derivatives is extremely rich and complex.[1–3] These molecules can perform various biological functions by participating in the networks of matter, energy, and information flow, ranging from participation in the processes of synthesis and degradation of many macromolecules and ending with signaling, regulatory, and control functions. The fulfillment of such different activities is possible thanks to the systems developed and optimized in the course of evolution of compounds with an appropriate electronic and geometric structure connected by networks of redox and acid-base reactions. One of the key stages of the operation of these types of systems is the ability to generate reactive forms of the considered small inorganic molecules (RSIM) appropriate for a given process in the appropriate concentration, place, and time. Uncontrolled disturbance of reactive small inorganic molecules homeostasis can lead to pathological states and a variety of diseases. Understanding the mechanisms of the generation and functioning of reactive small inorganic molecules in biological systems inspires the design of synthetic systems for their use in catalysis, biomedicine, or environmental protection. This paper will present mechanistic insight into the light-induced generation of reactive small inorganic molecules from biomedical application perspectives.

2. Reactive small inorganic molecules (RSIM)

Small inorganic molecules, because of their fundamental roles in biology and the environment, have been intensively studied in recent decades. Among them, small redox active inorganic molecules such as reactive oxygen (ROS), nitrogen (RNS), sulfur (RSS), chlorine (RClS),

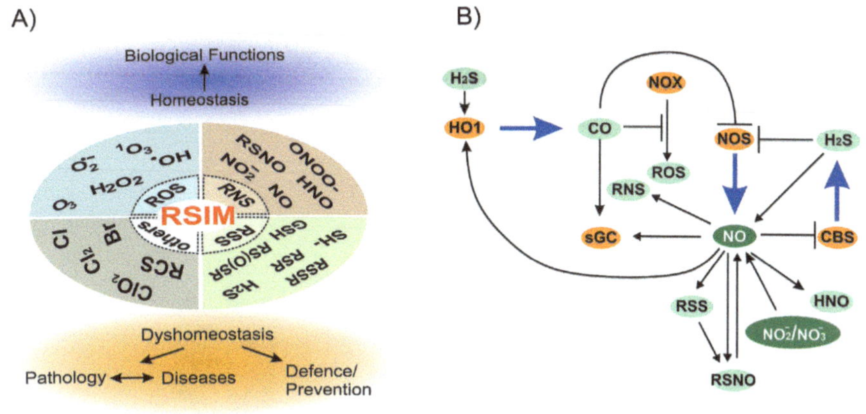

Fig. 1 Reactive small inorganic molecules (RSIM) (A) and crosstalk of signaling inorganic molecules (adapted from Ref. 1) (B). Nitric oxide synthase (NOS), NADPH oxidases (NOX), soluble guanylyl cyclase (sGC), cystathionine β-synthase (CBS), heme oxygenase (HO1).

or carbonyl (RCS) species play an important role in various physiological and pathological processes[1,2,4] (Fig. 1).

These reactive small molecules react independently or in a concentrated manner to mediate physiological responses to various inner and outer stimuli. The major ROS/RNS/RSS species include hydroxyl radicals, hydrogen peroxide, superoxide radical, nitric oxide, peroxynitrite, hydrogen sulfide, thiols, thiyl radicals, and nitrosothiols. Antioxidant enzymes, including superoxide dismutase, catalase, glutathione peroxidase, glutathione S transferase, glutaredoxin, and S-nitrosoglutathione reductase, are the main macromolecular redox regulating systems. The redox balance of cells provides a stable microenvironment for many biomolecules and their physiological functions. RSIM dyshomeostasis leads to pathological states and various diseases such as aging, cancer, diabetes, cardiovascular, and neurodegenerative diseases. Restoring SRIM homeostasis and physiological redox signaling processes can be achieved by RSIM regulating biomedical strategies focused on appropriate cells, time, type of chemical species, and their physiological concentration ranges. On the other hand, a high excess of appropriate RSIM may be used for destructions of specific biological targets such as biomolecules or cells in various pathological states or for inactivation of pathological microorganisms[5,6] (Fig. 2). Among various exogenous stimuli suggested for the control and regulation of reactive small inorganic molecules, light offers many interesting possibilities.[7–20]

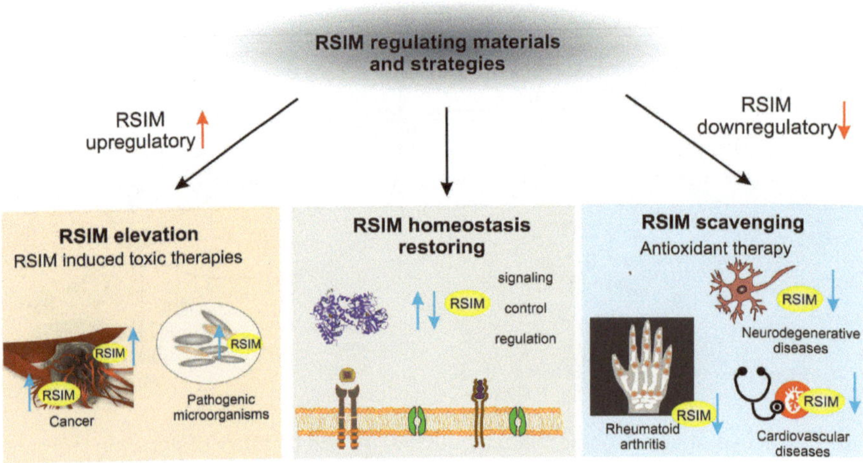

Fig. 2 RSIM regulating scenarios for biomedical applications.

3. RSIM photogeneration in biological systems - Pathways and mechanisms

Light is necessary for the proper functioning of our body, and at the same time solar radiation and radiation obtained by means of artificial sources can be used in a planned manner for prophylactic, diagnostic and therapeutic purposes (Fig. 3). The effects of light can be considered at different levels of biomaterial organization, from biomolecules, cells, and tissues to whole organisms and biological systems. In the case of the interaction of laser radiation with the tissue, three main types of effects can be distinguished depending on the energy of the radiation and the exposure time, namely: electromechanical (photodestruction), thermal (coagulation, ablation) and photochemical. For low light intensities, the processes of light absorption by endogenous or exogenous chromophores and the deactivation processes of electronic excited states are important.

The deactivation of excited electronic states can take place by physical and chemical processes. In the first case, light may be reemitted (used in sensing, imaging, and diagnostics), converted into thermal energy (used for example in photothermal therapy), or undergo reactions constituting an important path of deactivation processes. Photochemical reactions can be used in therapy, prophylaxis, delivery of bioactive substances, biostimulation, and other biomedical applications. Molecules in electronic excited states in relation to molecules in ground states are characterized

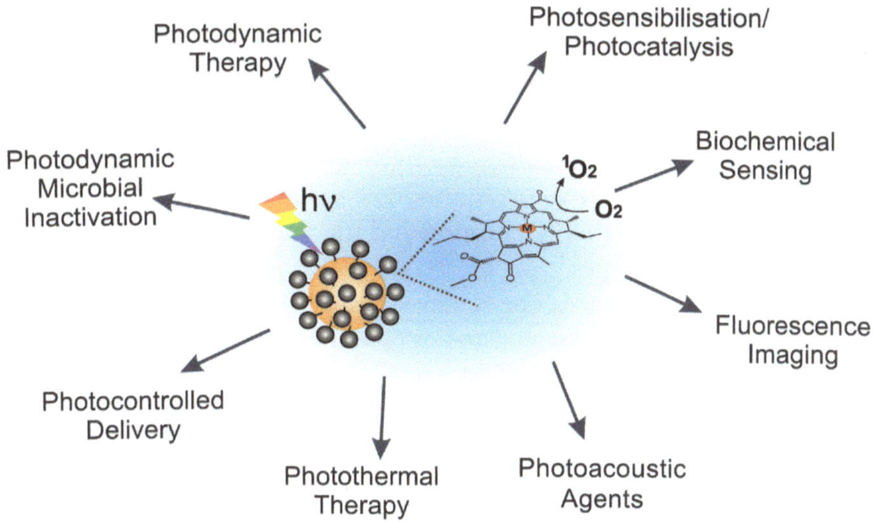

Fig. 3 Biomedical and environmental applications of light-controlled processes.

not only by much greater energy, but also by a different electronic configuration, electronic density distribution, charge distribution, molecule geometry, lengths and energies of individual bonds, better oxidizing and reducing properties, and alkalinity or acidity. Thus, a molecule in the electronically excited state is characterized by a different and often greater reactivity than this molecule in the ground state, and the photochemical reactions have much lower activation energy barriers and offer greater selectivity in relation to thermal reactions. Additionally, in the biological system, both direct and indirect reactions can occur. Photocatalysis and photosensitization are phenomena that play important roles in both natural and human-designed processes. Light-induced RSIM generation can proceed through direct photoreactions of endogenous or exogenous chromophores or by photosensitized processes with a variety of chemical and genetic photosensitizers.[7–26]

3.1 Light management strategies

Tissue is a heterogeneous material. The depth of penetration of light in tissue depends on the energy of photons. Due to the absorption of endogenous chromophores, the best radiation range for excitation of the applied exogenous chromophores is the range between 650 and 900 nm, called the phototherapeutic window. Compounds for potential applications in photomedicine are designed to have high absorption in this area.

Often, the light suitable for activating a given compound is of a higher energy range, which is not optimal for biological reasons. Lasers that can deliver high intensity pulses may allow the use of longer wavelength radiation to activate higher-energy transitions through population of intermediate virtual excited electron states (*e.g.*, two- and three-photon absorption) (Fig. 4). It is also possible to use appropriate spectral sensitizers that allow the absorption of low-energy radiation and the use of a population of real intermediate excited states.

In vivo photoreactions can be achieved using 1P or 2P excitation of the photoresponsive target through tissue or by using transducing materials that mediate in situ conversion of low energy incident photons into higher energy secondary photons which then excite the nearby target. Most of the photoresponsive biomaterials require UV light for photoactivation, which hampers transfer to targets deep in the body. Beyond phototoxicity associated with UV light, absorption and scattering by biological tissues rule out the possibility to activate photochemical processes even just below the skin by such materials. Upconverting nanoparticles (UCNPs) can activate UV/Vis-responsive reactions using NIR light. This process is significantly more efficient than simultaneous 2P excitation and does not require high energy femtosecond pulsed lasers.[7,15,24]

3.2 Photobiostimulation

Photobiostimulation (or laser biostimulation) with the use of low-power near-infrared or visible lasers is now considered as a part of light therapy (phototherapy). Various mechanisms of the biological action of light at the cellular or tissue level are proposed (Fig. 5). They are obviously dependent on endogenous chromophores (*e.g.* porphyrins, melanins, some amino acids, *etc.*) and the deactivation pathways of their electronic excited states. Among the proposed mechanisms are also those related to the generation of reactive oxygen species or nitric oxide. Due to the very low concentrations of endogenous biomolecules involved in the processes, photobiostimulated generation of reactive small molecules has very limited therapeutic possibilities.[15]

3.3 Optogenetics

Optogenetics, which uses light to control the activity of genetically encoded effector proteins, is one of the pathways to photogenerate RSIM. Photoinducible genetically encoded ROS and RNS generating proteins (RGPs) were originally employed for their phototoxic effects and cell

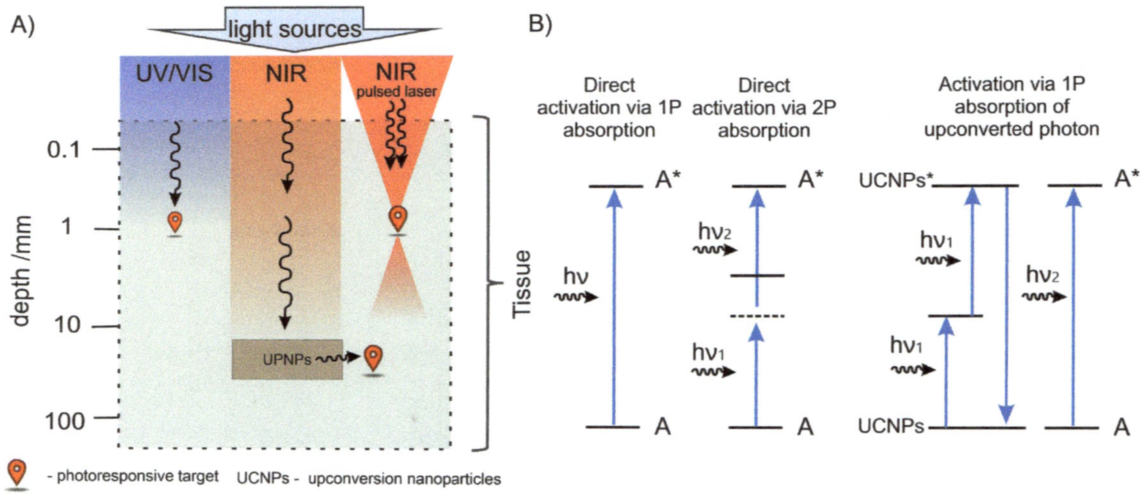

Fig. 4 Light management strategies (A) simplified energy level diagrams for one (1P) or two (2P) photon photoactivation (B).

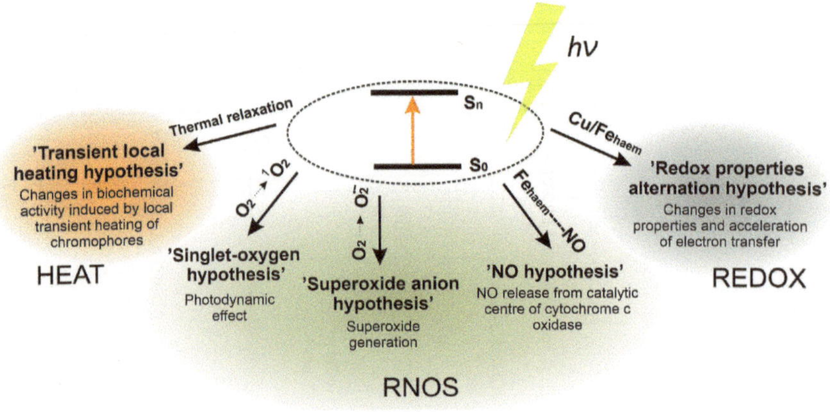

Fig. 5 Possible pathways of photobiostimulation.[15,20]

destruction. By reducing irradiance and/or fluency, sublethal levels of ROS and RNS can be achieved that may mediate subtle signaling effects. In fact, temporal and spatial control of ROS production is possible using RGPs, since they can be applied to tag proteins of interest and expressed at physiological levels.[26]

3.4 Photocleavage

Light is a very useful tool among exogenous stimuli for controlling the delivery of reactive small inorganic molecules to appropriate cellular targets in a safe manner. This result may be achieved by using exogenous molecular or nanostructured systems that are inactive until they are irradiated with the light of appropriate energy. In these cage systems, a bioactive small molecule is covalently bound to a light absorbing photolabile compound (phototrigger), yielding biologically inactive delivery systems (Fig. 6). Photoreactions such as photodissociation, photosubstitution, photoredox, or photoisomerization, could be involved in the precise delivery of reactive small inorganic molecules.[7,15,16]

3.5 Photosensitization

Photosensitization reactions are one of the most widely used methods for the generation of reactive forms of small inorganic molecules. Photosensitized reactions are indirect photochemical reactions in which molecules of one chemical species (photosensitizer) absorb light and induce chemical changes in molecules of another chemical species. The excitation of photosensitizer

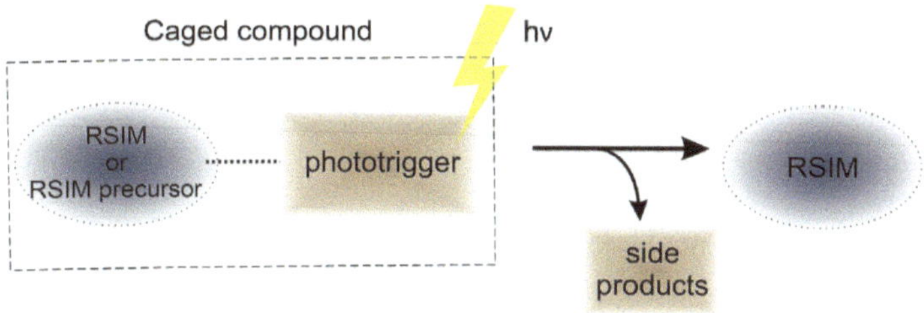

Fig. 6 Photoactivation of RSIM precursors.

may be followed by several physical processes (heat dissipation, light emission, or energy transfer), as well as by chemical reactions.

Photosensitizers in biological systems can be of endogenous or exogenous origin. In the former group, many natural tetracyclic compounds can act as photosensitizers, such as porphyrins, flavins, and pterins. Although endogenous photosensitizers are usually present at very low concentrations and their photoactivity is limited, they explain part of the deleterious effects of UVA and visible solar radiation. Also, they can accumulate under certain pathological conditions, increasing photodamage. Among the latter group, a large number of xenobiotics can be found, mainly pharmaceutical drugs or pollutants. The main biological targets of photosensitization reactions include some amino acids (tryptophan, tyrosine, histidine, methionine, and cysteine), nucleobases (guanine, adenine, thymine, cytosine) and unsaturated fatty acids.

Apart from the harmful effects induced in biological systems, photosensitization reactions can be beneficial for medical and environmental applications such as photodynamic therapy (PDT) of cancer cells, photodynamic inactivation of microorganisms (PDI) or pollutants photodegradation.[7–9,14–16,18]

3.5.1 Photosensitized ROS generation—Molecular systems

Upon light absorption by many organic photosensitizers, the resulting singlet excited state undergoes intersystem crossing to yield a longer-lived triplet state. The first bimolecular process is the reaction of the singlet or triplet excited state of the photosensitizer with the substrate or with dissolved molecular oxygen. Photosensitized reactions that apply to biomolecules may generally be classified as oxygen-dependent or independent. In photosensitized oxidation, molecular oxygen (O_2) can react with the electronically excited photosensitizer or participate in the secondary step, such as the reaction with radicals obtained from the photosensitizer or substrate molecules (Fig. 7).

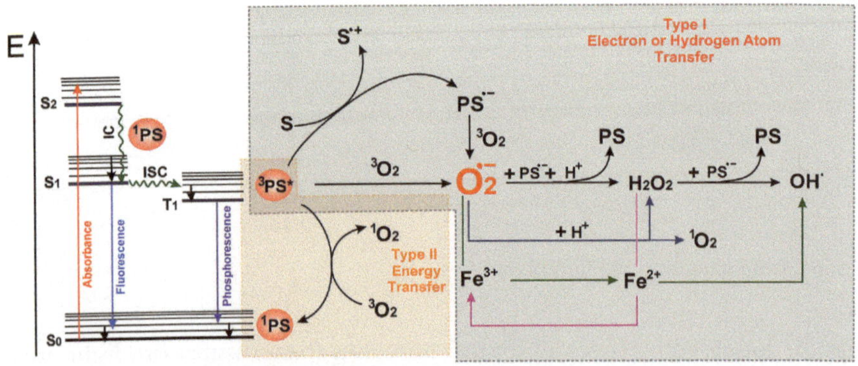

Fig. 7 Mechanisms of molecular oxygen photosensitization. The Type I mechanism involves electron or hydrogen atom transfer, whereas Type II involves energy transfer. Redox metal ions such as Fe^{2+}, Cu^{+}, or ascorbate ions may catalyze the generation of hydroxyl radicals.

Singlet oxygen is generated by energy transfer from the photosensitizer triplet state to molecular oxygen (type II reaction), but the triplet state of the photosensitizer may also participate in electron or hydrogen atom transfer reactions (type I reactions) that eventually lead to oxygen-centered radicals. Interestingly, charge transfer interactions between the triplet state of the photosensitizer and molecular oxygen facilitate the generation of the superoxide ion and hydroxyl radical, in addition to singlet oxygen, and increase phototoxicity (Fig. 8A). Photodynamic efficacy intimately depends on the dynamics of photosensitizer-oxygen interactions: charge transfer to molecular oxygen with generation of both singlet oxygen and superoxide ion (high k_D) must be tempered by photostability (low k_d). These properties depend on the oxidation potential of the photosensitizer. The dynamics of the interaction between the triplet state of the photosensitizer and oxygen determine both the nature of the ROS generated and the stability of the photosensitizer toward such ROS. The ratios between the triplet photosensitizer (^{3}PS)—O_2 interaction rate constant (k_D) and the photosensitizer decomposition rate constant (k_d), *e.g.* k_D/k_d determine the relative photodynamic activities against various cancer cells (Fig. 8B).

According to the mechanism proposed by Wilkinson and slightly modified by Schmidt, the diffusion controlled encounter complex of the excited triplet photosensitizer and the triplet ground state O_2 leads to the reversible formation of excited encounter complexes of singlet, triplet, and quintet multiplicities. Statistically, there will be one singlet, three triplets, and five quintet complexes. The quintet pathway has no spin- or energy-allowed

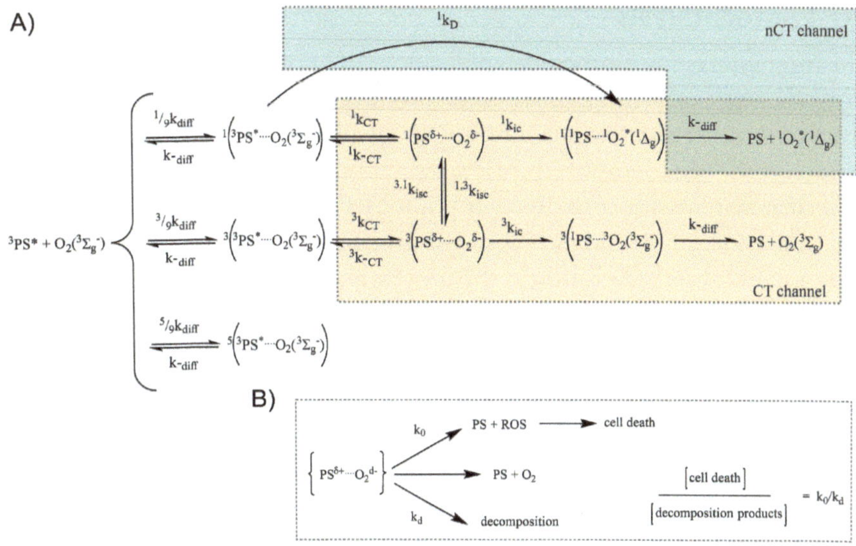

Fig. 8 Wilkinson mechanism for photosensitizer triplet quenching (A) and competition between ROS generation and PS degradation in CT channels that affect PDT efficacy (B).[27,28]

products, and quenching occurs only through internal conversion processes in the singlet and triplet pathways. In the absence of charge transfer interactions, the triplet pathway is unproductive; the rate of formation of singlet oxygen is given by Eq. (1):

$$k_\Delta = \frac{k_{diff}\,{}^1k_\Delta}{9(k_{-diff} + {}^1k_\Delta)} \leq \frac{k_{diff}}{9} \tag{1}$$

and the quantum yield of singlet oxygen generation may reach $\Phi_\Delta = 1$. In the presence of charge transfer, CT channel, and for small energetic differences between the singlet and triplet states of the intermediates, the rate determining step is the parallel formation of singlet and triplet exciplexes; the limiting rate increases to $k_\Delta < 4/9\ k_{diff}$ and the limiting quantum yield decreases to $\Phi_\Delta = 0.25$.

The type I reaction involves an electron transfer from the excited PS, which is a good electron donor in its triplet state, to O_2 in the triplet state, resulting in the formation of a superoxide ion radical that can be further transformed to H_2O_2 and finally to the OH radical. The lower the redox potential of PS in the excited state, the greater the tendency of PS to undergo the reaction of type I. On the other hand, lower values of redox potentials result in reduced photostability of photosensitizers.[8,9,15,27,28]

3.6 Photocatalysis

The most interesting biomedical applications of photogenerated reactive small inorganic molecules (RSIM) are those produced by photocatalytic and nanostructured molecular systems. Homogenous photocatalysis is generally connected with photosensitized reactions, although not all photosensitized molecular systems are considered as photocatalytic. According to the definition of photocatalysis and photocatalyst given in the IUPAC "Glossary," the photocatalyst absorbs radiation in the process and always regenerates itself after each cycle of interaction with the reaction partners (Fig. 9). The photosensitizer always fulfills the first condition, while the second does not. The differentiation between both terms can be controversial.[7,8,15]

3.6.1 Heterogenous photocatalysis

Even more complicated are heterogenous photocatalytic systems. Reactive oxygen species (ROS) and other reactive species generation processes occurring in the presence of heterogeneous photocatalysts differ significantly from those observed in homogeneous systems. The most active nanomaterials for ROS generation are semiconductors, which are mostly based on d-block metal compounds, such as oxides and chalcogenides (Fig. 10).

The excitation of such semiconductors, because of their electronic band structure, results in the generation of electrons (e^-) and holes (h^+). Photogenerated charges may recombine or be spatially separated and long-lived, if the recombination is suppressed. Such long-lived electrons and holes may migrate to the surface, and these charges are mainly responsible for the activity of heterogeneous photocatalysts. Therefore, in comparison to homogeneous photocatalysis, ROS photogeneration processes in heterogeneous photocatalysis are based mainly on type I reactions. The reduction pathways proceed according to the same mechanisms as in the case of homogenous

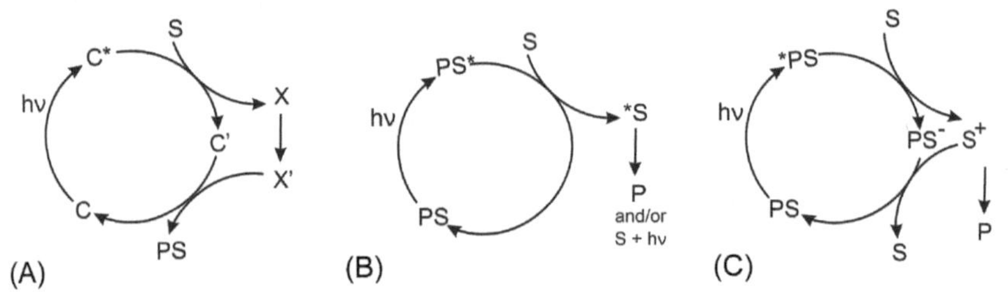

Fig. 9 Photocatalysis and photosensitization mechanisms in homogeneous systems. Photocatalysis (A), energy-transfer photosensitization (B), electron-transfer photosensitization (C) in homogenous systems C, catalyst; PS, photosensitizer; R, reagent; P, product; X, intermediate species.

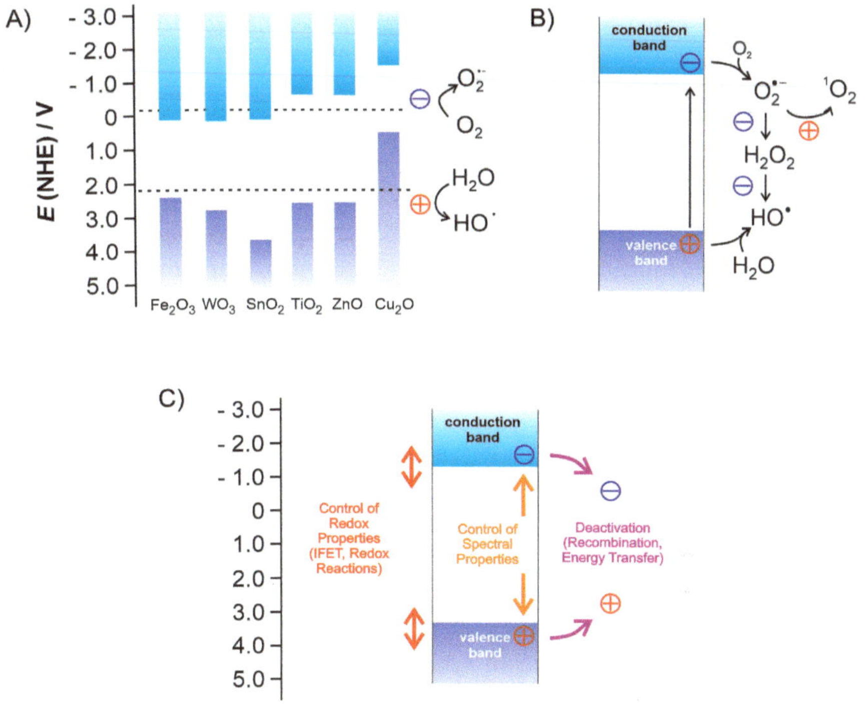

Fig. 10 Heterogeneous photocatalysis principles. Diversity of band-edge positions energy levels of conduction and valence bands in relation to fundamental redox couples involved in photocatalysis (A) Exemplary reductive and oxidative pathway of ROS generation at the surface of TiO_2 (B) Crucial factors affecting semiconductor properties (C).

systems, that is, through the consecutive one-electron reduction of an oxygen molecule. The oxidation pathways are based on the oxidation reactions of water molecules or hydroxyl ions adsorbed at the surface of the photocatalyst, leading to the formation of hydroxyl radicals. The possibility of the reduction and oxidation pathways depends on the potentials of the CB and VB edges, respectively, as well as on the redox potentials of the charge traps within the band gap.

Their activity is primarily dependent on redox processes, and unlike molecular systems, they are capable of very efficient, multi-electron and concurrent oxidation and reduction processes, both of which result in the formation of highly reactive radicals. Furthermore, the use of well-developed materials engineering techniques allows for the modification of the potentials of CB and VB edges, which affect the selectivity of the photocatalysts. The greater photostability of the chosen semiconductor nanoparticles over molecular systems is another benefit over molecular systems (which is characteristic for metallic and f-block metal-based nanostructures). The fundamental limitation that prevents these chemicals from being used in photoinduced deep tissue treatment is their relatively high excitation energy. Because most active materials absorb only UV light, they cannot be used in PDT as direct photosensitizers. As a result, the development of semiconducting particulate systems should focus on combining photosensitizing agents with semiconductors that are efficient in ROS formation. All key properties of semiconductors can be improved by conjugation with well-known molecular and macrocyclic photosensitizers,[29] nanosized metal particles,[30] or NIR active upconverting nanoparticles. The strategy for future advancements is to focus on the creation of systems designed to microorganism inactivation, activated with an energy of NIR light.[31]

Unlike photocatalytically active semiconductors, metallic and f-block metal doped/based nanoparticles may be excited within the phototherapeutic window because of their optical characteristics. Despite their high absorption coefficient, the yields of ROS production on bare noble metal nanoparticles are substantially lower than those of molecular photosensitizers. When this is taken into account, metallic NPs seem to be more viable for future development of systems including metallic NPs coupled with molecular or nanostructured photosensitizers. Such devices might be used to generate ROS through PS plasmons or plasmon-amplified ROS. Furthermore, they may be a component of multimodal therapeutic drugs that combine PDT and PTT capabilities.

The photoactivity of f-block metal-based/doped nanostructures, in turn, is confined to singlet oxygen production and is constrained to materials containing certain lanthanide ions. The capacity of such devices to convert NIR radiation to visible and ultraviolet light is their primary benefit. In terms of effective ROS producers, there are several experiments underway for the creation of hybrid materials for PDT, in which molecular or nanoparticular PS is indirectly activated through an NIR-triggered upconversion process.

4. Photoresponsive materials - Examples and development trends

4.1 Molecular systems

4.1.1 Molecular nitric oxide generating systems

The wide range of potential biomedical applications offered by nitric oxide, resulting from both its primary biological functions (regulatory, signaling, *etc.*) and cytostatic activity, has aroused increasing interest in recent years and stimulated the design of new therapeutic and diagnostic strategies. The high activity of NO paradoxically poses a key problem in its direct medical application. Taking into account the multitude of its nonspecific interactions with biological molecules, sensitivity to even fine changes in physiological conditions, and above all, short lifetime, the efficiency of applications must depend on the precision of NO delivery and release in close proximity to biological targets (<200 μm). The combination of appropriately stable carrier systems with high affinity for specific cellular structures and their physiological states and the availability of a light source with sufficient energy to initiate photochemical release of NO seem to ensure successful circumvention of the aforementioned drawbacks.

The structural diversity of NO delivery and photo-release systems ranges from single molecules of organic compounds and metal complexes *via* dendrimers to MOFs and functionalized nanoparticles. This diversification is motivated by a number of practical aspects related, among others, to the possibility of operating *via* direct or indirect (*i.e.* sensitized) photochemical reaction, the intended increase in affinity to the cellular environment and permeability through biological membranes, or the intended use for applications requiring different dynamics of NO release. Low molecular weight organic NO-releasing motifs are relatively diverse and include, among others, N-nitrosamines, S-nitrosothiols, organic nitrites, nitrates, or N-diazeniumdiolates. One of their important advantages is the possibility of extending the structure with fluorophore residues to facilitate the tracking of their activity.[32,33]

When molecular organic photodonors are considered, particular attention should be paid to S-nitrosothiols (RSNO), which serve as an endogenous reservoir of NO. Its gradual release, dependent on physiological states, ensures maintenance of homeostasis and proper functioning of numerous 'mild' biological processes, including cell signaling pathways.[34] The decomposition of RSNO in cells is induced, among other pathways, by acidic conditions, ascorbic acid or Cu^{2+} ions. Nevertheless, it is still worth using light as a trigger, which, in cases where a fast and efficient release of NO is required, will provide precise spatiotemporal control over it. The illumination of nitrosoglutathione (Fig. 11A) at around 340 and 545 nm proves to be particularly productive in this regard.[35] Multiplication of donor groups within one molecule is one of the ways to significantly increase the concentration of released NO. Such a possibility is offered, among others, by dendrimers. As shown by Schoenfisch et al.[36] light-sensitive 4-generation dendrimers with polyamidoamine arms terminated by nitrosothiol groups (Fig. 11B) allow storage of up to 2 μmol of NO per mg of compound.

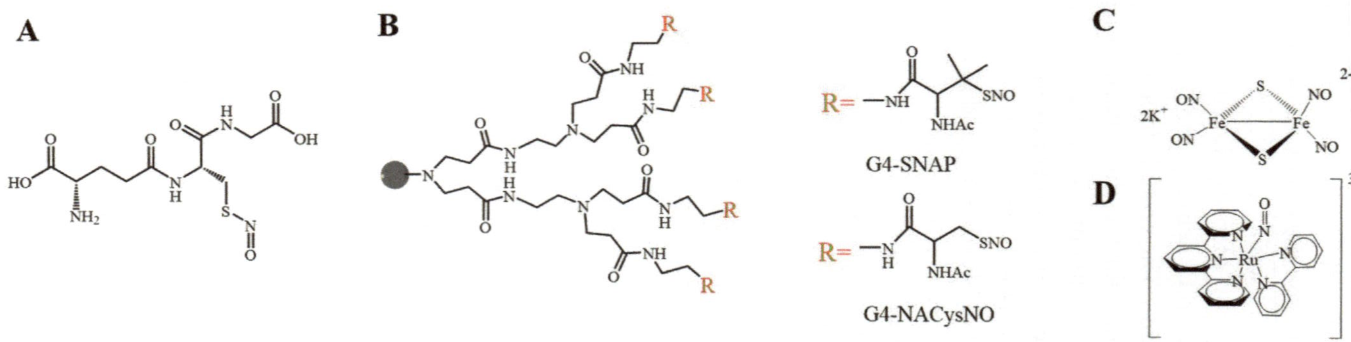

Fig. 11 Structures of selected NO photodonors: S-nitrosoglutathione (A), polyamidoamine dendrimer with S-nitrosothiol end groups (B), the Roussin's red salt (C) and $[Ru(bpy)(terpy)(NO)]^{3+}$ (D).

Among metal compounds capable of NO storage and light-triggered release, nitrosyl Ru(II) complexes rank particularly high.[37] This is due to the high efficiency of NO photodissociation under irradiation in the near UV and visible light range, in addition to their thermal stability and low unintentional cytotoxicity.[38] As a source of NO obtained by direct or sensitized photodissociation, Cr(III) and Fe(III) complexes have also been reported,[39,40] including probably the most popular Roussin's red salt (Fig. 11C), in research to which our group also contributed.[41] However, to provide full functionality of these systems in therapeutic applications, it is necessary to shift the excitation range to the therapeutic window. In recent years, ruthenium-based systems activated by near-infrared photon absorption have been presented, both in the form of single molecule complexes with strongly chromophoric aromatic ligands (Fig. 11D)[42] as well as various types of Ru-NO functionalized nanoparticles.[43] The use of low-energy light has become possible due to the development of two-photon excitation techniques. Strategies in development include up-conversion using lanthanide-based sensitizers[44] or macrocyclic π-conjugated molecules.[39]

Under aerobic conditions, NO forms various derivatives with O_2, such as nitrite, nitrate, and nitrogen dioxide. Their action in biological systems is of interest; however, due to spontaneous generation, the ways of their direct delivery into cells are not extensively studied. For example, peroxynitrite was shown to be an alternative product to NO for photosensitized decomposition of (3-(hydroxynitrosoamino)-D,L-alanine).[45] The photogeneration of other reactive NO species (RNOS) has gained some more attention. A growing interest in nitroxyl (HNO) is evident due to the increased recognition of its role in the treatment of heart failure. Among the proposed potential donors, compounds based on the hydrophilic hetero-Diels-Alder cycloadducts, deserve attention due to the possibility of obtaining different photoreaction products (NO or HNO) depending on the polarity of the environment (Fig. 12A).[46] Recently, (hydroxynaphthalen-2-yl)methyl-photocaged

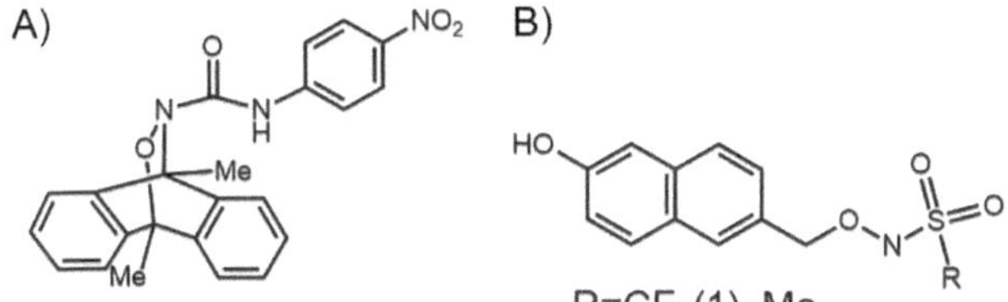

Fig. 12 Examples of nitroxyl photodonors proposed by Miyata et al. (A)[46] and Brasch et al. (B).[47]

N-hydroxysulfonamides have been introduced as a promising class of novel HNO photodonators (Fig. 12B). Mechanistic studies revealed that the key step for photogeneration of nitroxyl is protonation of the donor compound in its ground state. Such protonation provides a much higher probability of heterolytic cleavage in the C—O/N—S system with respect to the undesired breakdown of the N—O bond.[47]

The development of ever more specialized systems capable of controlled release of specific reactive NO species brings closer the perspective of effective use of various therapeutic strategies. Still, increasing the accuracy of delivery and the efficiency of release remains a real challenge.

4.1.2 Phototriggered NO release as a tool in the investigation of bioinorganic reaction mechanisms

The high chemical reactivity of NO that results from its radical nature is manifested in the diversity of implemented reaction pathways. This poses a particular challenge in mechanistic studies. The variety of available substrates in such complex systems as the cellular environment causes that reliable results can be obtained almost exclusively in sterile, highly simplified model systems. A certain difficulty of these studies is due to the very high rates of the reaction involving NO, making them inaccessible by standard time-resolved spectroscopic techniques. Hence, in many cases the only possible approach is to observe restoration of the equilibrium in a system disturbed by a strong and rapid pulse. The photochemical reactivity of nitrosyl metal complexes makes it possible to use light as a trigger for the release of NO, which reconstitutes the product in a reaction that is already fully controlled by the observer. The combination of laser flash photolysis with high-pressure techniques provides unique opportunities to determine the type of mechanism in ultrafast ligand exchange reactions. Such a strategy has yielded excellent results in elucidating the reaction mechanisms involving biologically relevant macrocyclic iron complexes. Hem, the reactive center of many essential enzymes, has been the primary subject of this research for many years. Particular attention has been paid to the mechanism of NO binding and release from cytochrome P450. By using flash photolysis, it was shown that the binding of a substrate, such as camphor, enhances the binding of NO to the iron complex.[18] By expanding the study to synthetic Fe(III) porphyrins and sulfonate ligands, it became possible to identify the structural elements of the macrocyclic system and the reaction medium as factors influencing the mechanism of axial ligand substitution.[48]

These results stimulated further development of the research on the catalytic activity of tetrapyrrolic iron complexes in the activation of peroxides for the oxidation and hydroxylation of biologically relevant substrates. Recently, on the basis of kinetic parameters obtained by photoinduced NO release, a mechanism was demonstrated in which the variation of the reactivity of ferric microperoxidase in the formation of nitrosyl complexes is controlled by tautomerism associated with proton transfer between coordinated histidine and water molecules.[49]

4.1.3 Molecular photosensitizers

Various molecular photosensitizers such as organic, inorganic, coordination or metalloorganic compounds have been studied for many years.[7–9,11–13,15,16,18,19,50,51] Nowadays, nanostructured systems are developing rapidly as a new promising group of photosensitizers.[5–7,10,11,14–16,22,29,30,51,52] Organic dyes (*e.g.* rose bengal, methylene blue, indocyanine) or aromatic hydrocarbons (naphthalenes, anthracenes or quinones) are very effective molecular photosensitizers, as they possess triplet states of appropriate energies for sensitization of molecular oxygen. But the most interesting photosensitizers on the biomedical applications side occurred to be porphyrins, phthalocyanines, texaphyrins and related tetrapyrroles.[7,9,11–13,15,16,21,23,25,27,53,54] Naturally occurring macrocyclic photosensitizers such as heme cofactors, chlorophylls, bacteriochlorophylls and their derivatives especially pheophorbides and their metal derivatives are interesting due to the very low dark cytotoxicity. In biomedical applications, such as photodynamic therapy of cancers or photodynamic microorganism inactivation, synthetic porphyrins, chlorins and bacteriochlorins are among the most studied groups of photosensitizers. Various structural and electronic configuration modifications of the tetrapyrrolic ring, its peripheral substituents, and coordination of various metal ions allowed for the receiving of many interesting photosensitizers. Modifications of the porphyrin ring via reduction of pyrroles results in the largest differences in the ROS generation mechanisms. Porphyrins generate essentially more singlet oxygen than chlorins or bacteriochlorins. In the case of bacteriochlorins, both types of ROS generation mechanisms are possible. Modifications applied to tetraphenylporphyrin (TPP) to obtained required spectroscopic, physical, chemical or biological properties of the halogenated bacteriochlorins, chlorins and porphyrins studied by us are shown in Fig. 13.

The generally accepted properties of a good photosensitizer are: simple formulation, low dark toxicity, amphiphilicity, selectivity toward tumor

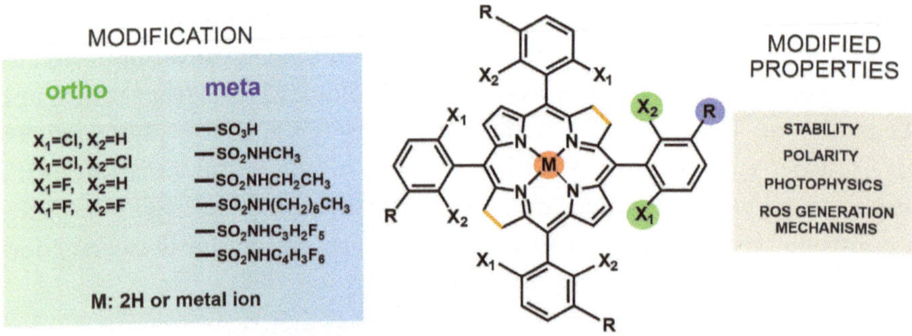

Fig. 13 Chemical and structural modifications of the tetraphenylporphyrin (TPP).[21]

tissue, long shelf-life, rapid clearance from the body, facile synthesis, feasible scale-up, $700\ \text{nm} < \lambda_{max} < 850\ \text{nm}$, $\varepsilon_{infrared} > 10^5\ \text{M}^{-1}\ \text{cm}^{-1}$, $\Phi_F \geq 0.2$, $\Phi_T \geq 0.7$, triplet lifetime $\tau_T \geq 100\ \mu s$, $\Phi_\Delta \geq 0.5$, and $\Phi_{pd} < 10^{-5}$, in which Φ refers to quantum yields of fluorescence (F), triplet (T), singlet oxygen (Δ), and photodecomposition (pd). Halogenated and sulfonated synthetic bacteriochlorins meet most of these expectations[11,12,15,16,21,23,27,53,55] (Fig. 14).

4.1.4 Molecular metal complexes

The generation of reactive oxygen species (ROS) upon light activation of metal complexes has been used efficiently in photodynamic therapy (PDT) for the treatment of various types of cancers. Nowadays, the most widely studied photosensitizers (PSs) are tetrapyrrole macrocycles, including porphyrins, chlorins, bacteriochlorins, and phtalocyanines.[56] Their popularity is explained by their photophysical, photochemical, and biological properties, including long-wavelength absorption peaks, triplet quantum yields and lifetimes, and low toxicities in the dark. Most porphyrin-based PSs act predominantly through the energy transfer mechanism and generate singlet oxygen.[57] Therefore, the efficacy of such PDT systems is strictly related to tissue oxygenation. To overcome the restricted therapeutic effect of PDT under hypoxic conditions PSs should be able to operate *via* both type I and type II mechanisms.[58] Modifications of the structures of porphyrins are widely used to modulate obtained photophysical properties, including the efficiency of electron/energy transfer processes. One of the possible ways to affect the properties of PSs is to include various metal ions in the coordination center of the macrocycle.[59] The tuning of the properties of the tetrapyrroles depends almost entirely on the electronic configuration of the chosen metal center.[21] Diamagnetic, closed shell divalent metal ions such

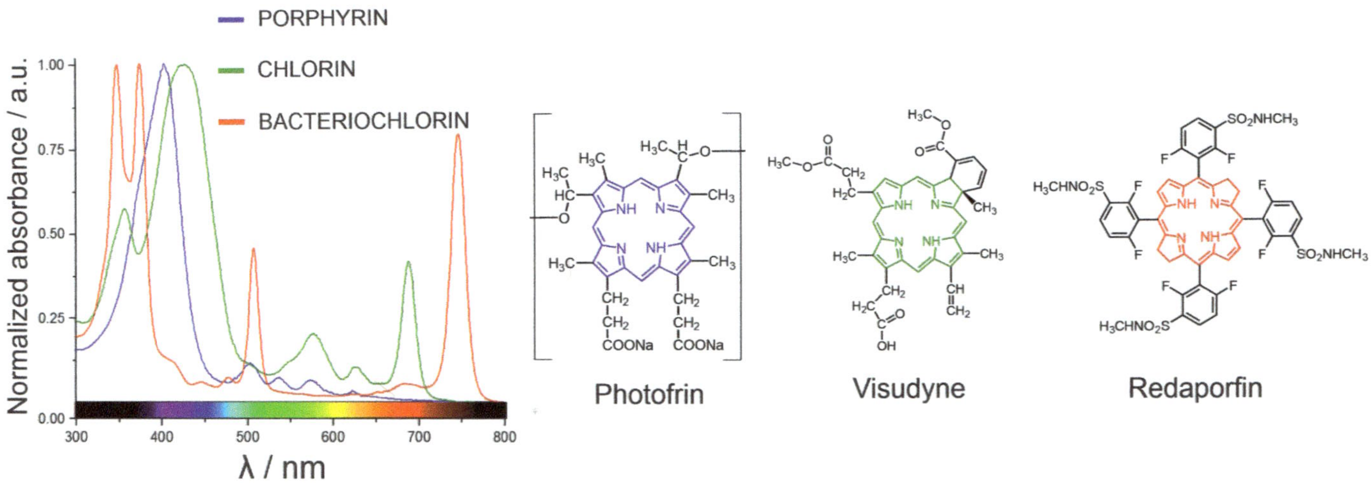

Fig. 14 Electronic absorption spectra and chemical structures of three clinically used tetrapyrrolic photosensitizers: porphyrin (photofrin), chlorin (visudyne) and bacteriochlorin (redaporfin).[21]

as Mg^{2+}, Zn^{2+} or Pd^{2+} increase the intersystem crossing quantum yield, which is crucial for subsequent ROS generation. On the other hand, paramagnetic metal ions (Co^{2+}, Mn^{2+}, Fe^{2+}) in the structure of porphyrins increase the rate of internal conversion and decrease the excited state lifetime, strongly restricting the use of such derivatives as PDT agents.[21] The coordinated metal ion can also influence the cellular uptake and localization of the PSs. PSs that are localized in mitochondria are known to be much more phototoxic than PSs that are localized in lysosomes.[60] Among metallotetracycles, palladium derivatives attract significant attention, as they are more photoactive than free-based or even analogue Zn tetrapyrroles. Mroz et al.[61] showed that Pd-porphyrins produce ROS mainly *via* the type I mechanism, which explains their high photodynamic reactivity. However, in consequence, they are more prone to photodegradation and are less photostable than their Zn-analogues. For *in vivo* studies, these compounds should be further modified to shift the absorption spectra closer to the phototherapeutic window. The efficacy of Pd-tetrapyrroles is being confirmed by Padeliporfin/WST11 (TOOKAD soluble). This palladium-coordinated bacteriochlorophyll α is clinically approved in Mexico, Israel, and 31 countries in the EU for the treatment of prostate cancer.[62] Interestingly, although most clinically used PSs operate *via* the type II mechanism, TOOKAD soluble functions almost exclusively *via* the generation of superoxide and hydroxyl radicals.[12]

Tetrapyrrolic structures are also applied in photodynamic inactivation of bacteria (PDI). While the transfer of knowledge from anticancer PDT to antimicrobial PDI is tempting, several differences in the structure of PSs should be highlighted. PDI agents do not have the requirement to absorb in a phototherapeutic window, as the application of blue light for infectious diseases is well-documented. It is also well-recognized that incorporation of halogen atoms (fluorine or chlorine) into the macrocycle structure improves photophysical parameters by promoting intersystem crossing and increases $^1O_2$ quantum yields.[63] Then, the introduction of sulfonic or sulfonamide groups modulates the PS water-solubility and can influence their interaction with biological membranes. As an additional benefit, these modifications also prevent aggregation and increase the photostability of PSs. To increase cellular uptake, intracellular ROS generation and overall photodynamic effect, the incorporation of tetrapyrroles into polymeric micelles of triblock copolymers (L121) has recently been proposed by Pucelik et al.[63] The charge of tetrapyrrole structures influences their PDI activities: positively charged PSs efficiently accumulate in Gram-negative bacteria, while negative charged PSs are useful in PDI treatment of Gram-positive bacteria.[64] It is noteworthy that the use of efflux pump inhibitors as adjuvant molecules of PDI could additionally improve the antibacterial effect of therapy due to increased accumulation of PSs. The antimicrobial activity of tetrapyrroles is based on the generation of singlet oxygen and hydroxyl radicals and can be further enhanced with the use of KI.[64] Furthermore, as PSs are indiscriminate in their actions, it is important to ensure a high selectivity of PDI agents toward bacterial cells rather than healthy mammalian cells. Recently, Aroso et al.[65] reported the activity of a Zn(II) tetra-ethyl cationic phthalocyanine (Fig. 15A), that efficiently reduce colonies of Gram-negative bacteria (*E. coli*, *P. aeruginosa*) and *C. albicans* by 7 log units while leaving mammalian cells unharmed. The success of this PS results from carefully designed structure: a moderate number of positive charges, a short alkyl chain, a Zn(II) cation that favors intersystem crossing to the triplet state and no low-lying charge-transfer states, and high $^1O_2$ quantum yield.

Metal complexes, as an alternative to tetrapyrroles structures, are also drawing particular attention as potential PDT cancer agents due to their attractive photophysical properties. Ru(II) polypyridyl complexes have attracted the most attention among scientists. The interest in Ru(II) complexes is fueled by the development of TLD-1433 (Fig. 15), a novel Ru(II) compound that just entered phase II of clinical studies for the treatment of non-muscle invasive bladder cancer.[62] The Ru(II) core facilitates an efficient population of triplet excited states, while an organic α-terthienyl group promotes the formation of intraligand (^{3}IL)/intraligand charge transfer (^{3}ILCT) triplet excited states with prolonged lifetimes. These states can generate $^1O_2$ at high quantum yields or interact with the biological environment producing ROS *via* the type I mechanism.[66] The obtained PDT effect can be further modified by formulating TLD1433 with transferrin (Rutherrin), which has been demonstrated using bladder cancer cells.[70] The rational design of ligands in Ru(II) complexes is one of the strategies used to obtain PSs with dual type I/II activity.[68,69,71] Other explored strategies are focused on modifying the absorption spectra of complexes closer to the phototherapeutic window[67–69,72] or using two-photon irradiation.[73]

PDT finds its application not only in the treatments of cancers. Recently, Son et al.[74] have demonstrated photodissociation of β-amyloids into a small, non-toxic fragments through photoactivation of $[Ru(bpy)_3]^{2+}$. The disassembling of pre-formed highly stable Aβ aggregates into less amyloidic fragments occurred due to the generation of $^1O_2$ by the Ru complex *via* the energy transfer mechanism. *In vitro* studies revealed that photoexcited

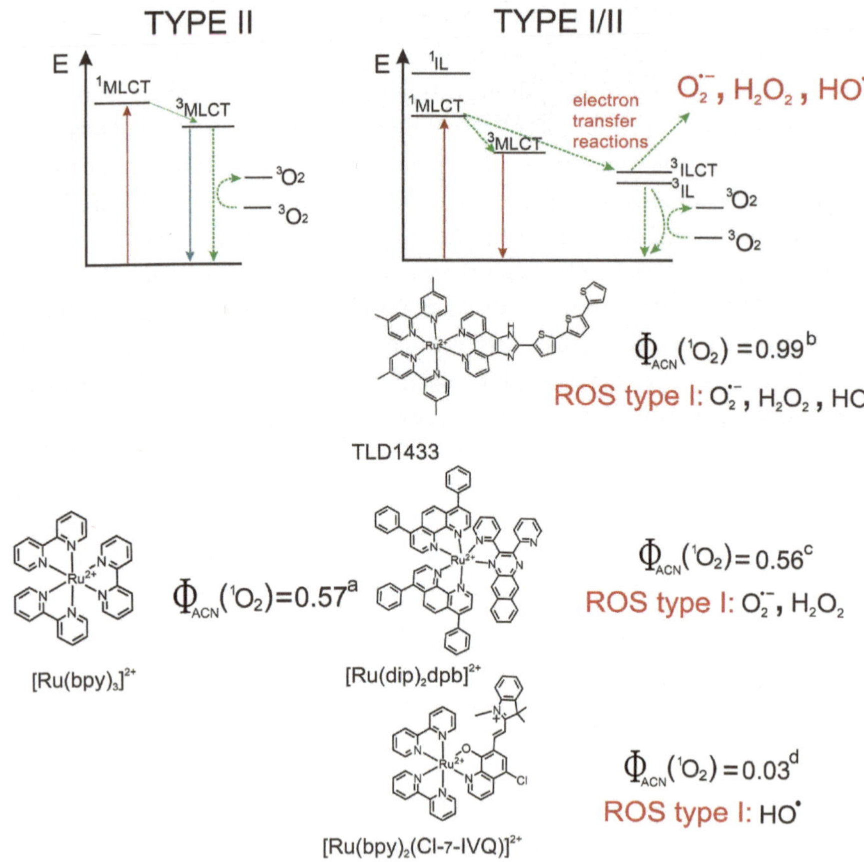

Fig. 15 Excited-state deactivation pathways accessible with visible light activation to Ru(II) polypyridyl complexes with dual type I/II PSs activity (based on Ref. 59, 66). Quantum yields of singlet oxygen production in acetonitrile $\Phi_{ACN}(^1O_2)$ were taken from the references[59] (a),[67] (b),[68] (c),[69] (d). Abbreviations of excited states: IL, intraligand; ILCT, intraligand charge transfer; MLCT, metal-to-ligand charge transfer. Solid lines, radiative transitions; dashed lines, non-radiative transitions.

Ru-treated Aβ aggregates became significantly less cytotoxic toward PC12 cells. Furthermore, light activation of $[Ru(bpy)_3]^{2+}$ inhibited the self-assembly of Aβ monomers into β-sheet-packed aggregates. Photoexcited Ru(II) complexes as potential therapeutic light-responsive anti-Aβ agents were indicated to be further verified by *in vivo* studies.

Among other metal complexes, iridium(III) complexes became increasingly popular due to their high quantum yields for triplet excited state formation and long triplet excited state lifetimes.[75] These properties provide efficient $^1O_2$ photogeneration with high luminescence that can be used for imaging as well. Generally, bis-tridentate Ir(III) compounds are explored for PDT applications.[76,77] One of the strategies to explore the potential of metal complexes as PDT agents involves the conjugation of complexes to biologically active molecules to modulate their intracellular behavior.[76] The conjugation of a cyclometalated Ir(III) complex to a xanthene dye was developed to obtain a mitochondria-targeted photosensitizer.[78] The construct exhibited low dark cytotoxicity, selective mitochondria-targeting uptake, high molar absorptivity, high photostability, and a high yield of $^1O_2$ photogeneration. Its photodynamic antitumor efficiency was proven *in vivo*. Recently, Liu et al.[79] reported mononuclear bis-terpyridine Ir(III) complexes with excellent photodynamic activity. Ligands were designed to increase the intraligand charge transfer (ILCT) character of the excited state and ensure appropriate water solubility. These compounds efficiently generate $^1O_2$ and exhibit potent photobiological activity with a photodynamic index higher than 1657 *in vitro* and *in vivo*. Then Kang et al. reported iridium(III) complexes that are able to modify Aβ peptides through coordination and photooxidation.[80] The simple incorporation of 2-phenylquinoline into the structure of Ir(III) complexes guaranteed the ability of the compounds to efficiently generate ROS such as $^1O_2$ and $O_2^{\cdot-}$ *via* electron or energy transfer processes.

Light activated metal complexes can also be used for imaging. Alyian et al.[81] synthesized the luminescent rhenium(I) compound $[Re(CO)_3(dppz)(Py)]^+$ (dppz = dipyrido[3,2- a: 2′,3′-c]phenazine), Py = pyridine) which exhibits primary and secondary light-switching effects for the detection of Aβ aggregates in real time. The compound is only weakly photoluminescent in phosphate buffer, but when aggregate Aβ is present, an ~18-fold increase in photoluminescence is observed. A further increase (second-light switch effect) in luminescence intensity (105-fold increase in initial value) is observed upon 362 nm irradiation of the Re-Aβ aggregates solution. This effect occurred due to catalytic oxidation of Aβ fibrils by the rhenium complex *via* ROS production. The latter is realized because the oxidized amino acid residues in Aβ fibrils are not able to deactivate the excited state of the metal complex or substitute the ligand with natural or oxidized amino acid during irradiation.

Although metal complexes often provide a superior effect in photobiological activity compared to organic compounds, only a few complexes are used clinically,[59,62] and a limited number of metals (Ru, Ir, Pt, Re) are being explored. Additionally, most of the reported systems operated in the visible range of the spectrum. Therefore, modifications of the compounds toward absorption in the near-infrared part of the spectrum should be an obvious next step.

4.2 Nanostructured systems

4.2.1 NO delivery systems

The main advantages of nanostructured NO-delivery systems are their protective properties against environmental effects, while ensuring high efficiency of intentional light-induced processes. In addition, for some applications, primarily anti-cancer therapies, their ability to penetrate the membranes of diseased cells, which reduces side effects, is of considerable importance. Nanostructured systems are often inspired by molecular nitric oxide donors. Among the latter, NONOates (N-diazeniumdiolates) hold an important place. Here, the cleavage of the N-N bond occurs upon the action of a number of chemical factors and can result from bimolecular reactions, protonation, or oxidation. However, a major problem occurred when spatio-temporal precision is a key factor in maintaining high therapeutic activity. Embedding NONOates into nanoparticular structures provides a way to reduce spontaneous NO release and make it dependent on controllable factors. A very interesting system of this type was presented by Kim et al.[82] N-diazeniumdiolate was incorporated in mesoporous silica nanoparticles and coated with a calcium phosphide layer. The latter undergoes photoinduced deprotonation upon irradiation in the UV range. This in turn induces simultaneous dissociation of NO from the donor and breakdown of the CaP shell (Fig. 16).

A similar approach has been applied to nitrosyl ruthenium complexes. Their combination with nanoparticle carriers usually brings about additional functionalities. Thus, Ru(NO)-modified TiO_2 nanoparticles show selective targeting of folate receptors overexpressed in cancer cells.[83] In turn, immobilization in bilayers of lipid vesicles provides high biocompatibility and

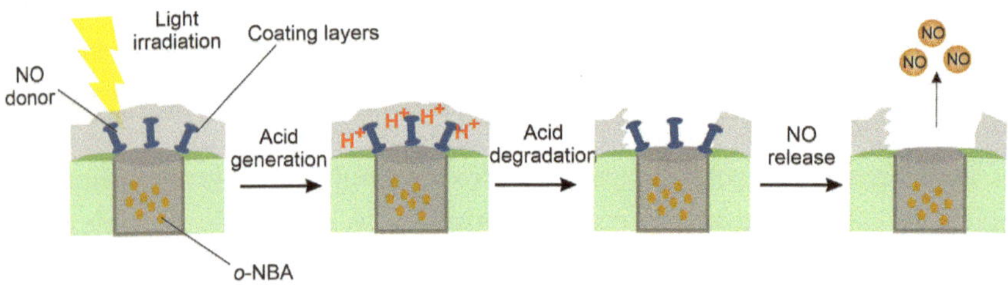

Fig. 16 Mechanism of photoinduced NO release from NONOate entrapped in silica nanoparticles (based on Ref. 82).

water solubility.[84] However, attachment to gold nanoparticles enables synergistic photothermal and NO antibacterial activity.[85] Increasing the functionality of NO photo-releasing systems is directed not only toward achieving high affinity for specific cellular structures and association with supplementary therapeutic strategies. Another strongly developed trend is to enable imaging at accumulation sites and signaling of NO release. In the case of molecular carriers, this function is usually performed by conjugated fluorophores.[33] The variety of mechanisms used in nanostructured systems has expanded considerably. Exceptional imaging performance with high NO accumulation and release ability is provided by the use of mesoporous materials such as MOFs[86] and zeolites.[44]

4.2.2 Nanostructured systems for ROS generation

The mode of action of ROS-(photo)generating nanomaterials enables the disruption of the endogenous redox equilibrium state by intra/extracellular interventions.[8] Elevating ROS levels beyond the cellular tolerability threshold altogether with overwhelmed antioxidant systems can be an attractive and feasible strategy to preferentially destroy cancer cells and/or various pathogens. Noteworthy, such a ROS-mediated therapeutic approach focused on disturbances in the redox biology provided by nanomaterials-guided results in failure of the self-adaptation defense mechanisms of pro- and eukaryotic cells (bacteria, fungi, and cancer cells, respectively).[2c,87] Thus, this has significant optimistic therapeutic implications, since severe homeostatic disorders of many pathways may hold the hope of overcoming multidrug resistance.[88] Moreover, considering the universality of nanomedicine, ROS-driven mechanisms may also be highly competent for antineoplastic therapy. Recently, based on relevant research in the field of nanotechnology balancing between ROS biology, ROS pathology, and ROS oncology, several redox-related therapeutic modalities, including PDT (photodynamic therapy), PDI (photodynamic inactivation of microorganisms), PTT (photothermal therapy), PCT (photochemotherapy), CDT (chemodynamic therapy), SDT (sonodynamic therapy), RT (radiotherapy), CDR (controlled drug release), and synergistic therapies have been proposed in clinics.[6,87,89]

4.2.3 Noble metal nanoparticles

ROS photogeneration, in particular with emphasis on $^1O_2$ generation, by noble metal nanoparticles (NPs), is still an intense and profound subject of investigation in the current literature. Mechanistic approaches that try to explain the photoprocesses that occur have been given less attention.[59,90]

The first study in this area concerned photothermal degradation of lysosome in the vicinity of Au NPs upon the pulsed 532-nm laser irradiation *via* the formation of nanoplasma, described by Takeda et al. in 2006.[91] Then, Sakamoto et al.[92] reported on $^1O_2$ photogeneration by laser irradiated gold clusters and proposed the relevant intersystem crossing (ISC)-type transfer mechanism. Next, the production of reactive oxygen species including $^1O_2$ by irradiation of Au NPs with low-intensity lasers was demonstrated by Krpetic et al.[93] In this study, photoinduced ROS generation was indicated to be the main cell death mechanism, as no significant thermal effects were observed. Thus, the previously postulated photothermal cytotoxicity was then indicated as one of the possible pathways of cytotoxicity. After that, Misawa et al. showed[94] ROS generation, including $^1O_2$, upon X-ray and UV irradiation by naked Au NPs. The study revealed enhanced generations of $^{\cdot}OH$ and $O_2^{\cdot-}$ with dependence on the absorbed energy, quencher concentration, and Au NP concentration. Regarding $^1O_2$ photogeneration, a very significant study was that of Vankayala,[95] who demonstrated the formation of $^1O_2$ through direct excitation of Ag, Pt, and Au NPs without the presence of organic photosensitizers and determined the quantum yields of $^1O_2$ formation as 0.155, 0.085, and 0.037 for Ag, Pt, and Au NPs, respectively. Shortly thereafter, the same research group proved that Au NRs can totally destroy melanoma tumors in mice with NIR light irradiation.[96] Zhang et al.[97] also confirmed ROS photogeneration on naked nanoparticulated metals such as Ag, Au, Ni, and Si. Only $^1O_2$ was concluded to be generated with Au NPs, Ni NPs, and Si NPs, while for Ag NPs, superoxide and hydroxyl radicals were also observed. Recently, Pasparakis[98] has determined the efficiency of $^1O_2$ production from Au NPs irradiated by continuous wave (CW) or pulsed laser sources to be approximately one order of magnitude lower than the reported values for quantum dots.

The mechanistic explanation revealed that ROS photogeneration may occur under two distinct scenarios. First, a plasmon-activated mode is realized by interactions of plasmons and hot-electrons with molecular oxygen resulting in efficient $^1O_2$ production as the reactive species. A second pathway can be an indirect photochemical pathway of $^1O_2$ formation in the case of irradiation with the pulsed laser source. This results in an extreme heat increase, causing particle fragmentation and thermionic electron emission. The latter mechanism is presumably associated with surface defects and various monolayers that protect or functionalize the surface of Au NPs.

The enhanced generation of ROS from metal NPs-based systems can be realized by various surface modifications, for instance photosensitizers, commonly investigated for PDT applications *i.e.*, porphyrins and their derivatives. Such nanostructural hybrid systems based on *e.g.*, Ag, Au NPs, improve PDT effectiveness as the metallic component acts as both an ROS enhancer and a photosensitizer carrier. The resulting ROS-based cytotoxic action can also be associated with $^1O_2$ formation *via* energy transfer from the excited surface modifiers (*e.g.*, indocyanine green,[99] porphyrins,[100] *etc.*) to molecular oxygen apart from the direct contribution of the highly localized plasmonic field of metallic nanostructures. As an example, He et al.[101] described a novel multifunctional system capable of ROS photogeneration, including $^1O_2$, composed of acrylic polymeric nanoparticles decorated with Au NPs functionalized with biotin. Photoresponsive multimodal system exhibited selective cytotoxicity toward C6 glioma cells associated with the enhanced ROS photogeneration upon the UV irradiation. Many other current groundbreaking examples of such hybrid systems are described in-depth in our recent review.[59]

4.2.4 *Semiconductors as nanocatalysts*

The vast majority of current studies on photocatalytic activity focus on TiO_2 due to its advantages, such as chemical stability, high photoactivity, and nontoxicity.[14] Only few reports are devoted to other photoreactive semiconductors, such as simple/mixed oxides (*e.g.*, perovskites, pyrochlores), metal nitrides and sulfides, graphene-based nanocomposites, *etc.*[59] Photodriven ROS generation by semiconductors is possible because of their unique energy band structures, resulting in quantitatively and qualitatively controlled production of various reactive species. Inorganic photocatalysts capable of light-triggered efficient ROS generation are fiercely competitive with most organic PSs that suffer from photoinduced or enzymatic degradation, making them useless in reaching the desired therapeutic outcomes. Furthermore, unlike most organic PSs acting in type II PDT mode, such inorganic materials can participate in type I PDT processes.[102] It is noteworthy that since $^1O_2$ generation, which has a key role in biology, medicine, and overall oxidation processes, cannot be neglected, any investigation focused on its detection in heterogeneous systems of wide-band-gap semiconductors is extremely difficult, nevertheless desirable.

In our group, following Nosaka's pioneering report on $^1O_2$ photogeneration at TiO_2,[103] we demonstrated that modification of titania surface or structure enhances the energy transfer suppressing the interfacial electron transfer. However, complete inhibition of the latter does not take place; therefore, on the surface of unmodified TiO_2, the electron transfer processes prevail, while $^1O_2$ may be generated with lower efficiencies.[50] Continuing this research on ROS photogeneration at TiO_2, the photocatalytic activities of rutile (r-TiO_2), anatase (a-TiO_2), and composites of r-TiO_2/a-TiO_2 were compared. It was concluded that considerably higher oxygen adsorption on the surface of r-TiO_2 and a lower redox potential of the excited electron allow for more efficient $O_2^{\cdot-}$ production in the case of r-TiO_2. Furthermore, the generation of $^1O_2$ (involving O_2 reduction and subsequent superoxide oxidation) and reduction of H_2O_2 to OH^- and $HO^{\cdot}$ are also favored for this titania structure. Efficient $HO^{\cdot}$ generation as the result of water or surface hydroxyl group oxidation altogether with oxidation of H_2O_2 to $O_2^{\cdot-}$ are observed for a-TiO_2. Thus, r-TiO_2 can be considered a good photoactivator of molecular oxygen (reductor), while a-TiO_2 a stronger oxidant (Fig. 17A).[104,107] Any synthetic procedures *e.g.*, sintering, surface coating, and doping can noticeably affect the physicochemical properties of semiconductors resulting in significant influence on ROS photogeneration involving such processes as water reduction, degradation of organic molecules (*e.g.*, herbicides), and overall photoelectrochemical properties.[108]

The photocatalytic degradation of cancer cells or inactivation of microorganisms was first demonstrated on very simple materials. As the first, Matsunaga et al.[109] reported in 1985 photocatalytic inactivation of *Saccharomyces cerevisiae* in the presence of TiO_2 irradiated with UV light. In more recent reports, many different photocatalytic therapeutic, diagnostic, and theragnostic approaches using various semiconductors, and also quantum dots (QDs), have been described in the literature.[105,106,110–113] In 2009, Rozhkova et al.[114] reported that photoactivated TiO_2 NPs (5 nm particles covalently conjugated with the IL13R2R-targeting antibody) caused programmed glioblastoma cell death. Also, heterogeneous photocatalysis has been studied widely as a promising tool allowing detoxification and disinfection. For instance, in our group Mitoraj et al.[111] has demonstrated visible light induced photocatalytic inactivation of bacteria (*Escherichia coli*, *Staphylococcus aureus*, *Enterococcus faecalis*) and fungi (*Candida albicans*, *Aspergillus niger*) by carbon-doped titanium dioxide (C/TiO_2) and TiO_2 modified with platinum(IV) chloride complexes (TiO_2/$PtCl_4$). The formation of hydrogen peroxide, superoxide anion, and hydroxyl radicals was indicated as small redox inorganic agents with a decisive impact on the observed photoinactivation effect (cellular wall damage, cell death, and final mineralization of organic matter). The H_2O_2 is generated on the surface of the irradiated titania (both neat and photosensitized) as a result of a two-step reduction of an adsorbed oxygen molecule with superoxide radical as an intermediate. The highly reactive $^{\cdot}OH$

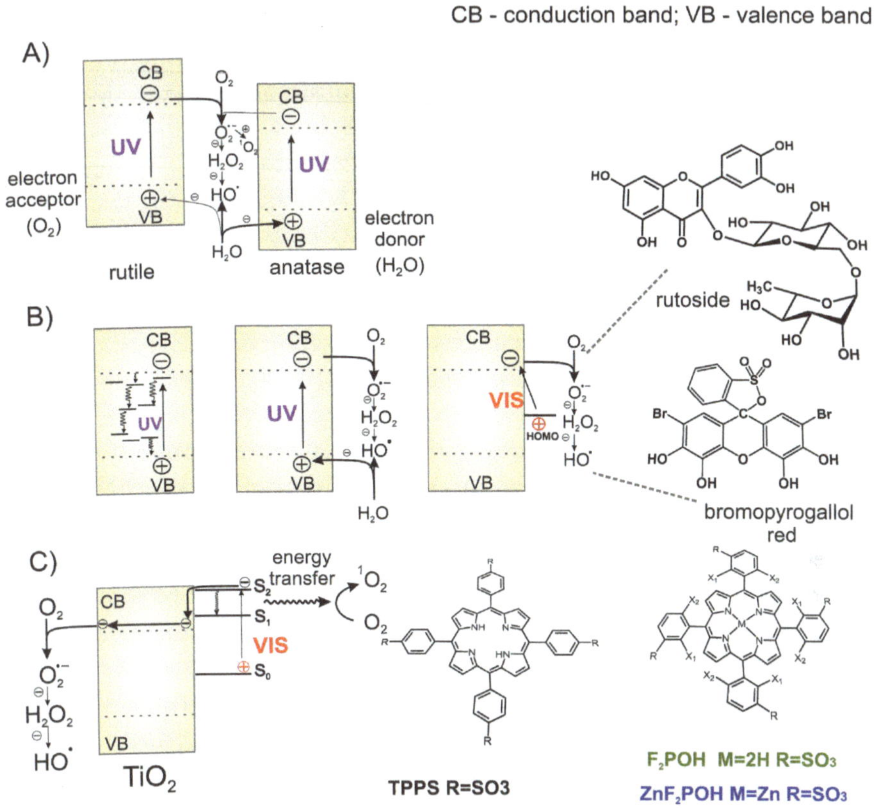

Fig. 17 Main ROS generation pathway in the presence of rutile and anatase heterogeneous photocatalyst based on Ref. 104 (A), in the presence of inactive amorphous nano-TiO_2 (on the left), UV-active crystalline nano-TiO_2 (in the middle), and direct photosensitized TiO_2 toward VIS by rutin or bromopyrogallol red based on Ref. 105 (on the right) (B), and in the presence of indirect photosensitized TiO_2 toward VIS by porphyrin derivatives (including $^1O_2$ generation by porphyrin derivatives through energy transfer mechanism) based on Ref. 106 (C).

is generated either as a result of the described multistep reduction of oxygen or in the process of direct water oxidation by photogenerated holes in the valence band of titania. However, the latter reaction can only take place under direct excitation of TiO_2 with UV light. Titania photosensitized toward visible light with the platinum(IV) complex exhibited a higher inactivation effect when compared to carbon-doped TiO_2. Then, this platinum(IV) chloride surface-modified TiO_2 was tested *in vitro* on mouse melanoma cells (S-91) and murine macrophages (RAW 264.7) with visible light irradiation.[110]

Phototoxicity associated with the photogenerated reactive species can be modulated by pretreatment with antioxidants (*i.e.*, α-tocopherol, β-carotene) indicating, on the one hand, the mechanism of cell death dependent on the concentration of photogenerated ROS, but also pointing to the ease of controlling the toxic effect for these simple photocatalytic systems. This strategy involving photosensibilization of TiO_2, initiated in our team in the early 2000 years, was continued on other systems that were consciously structurally modified to achieve a specific path of ROS photogeneration (electron/energy transfer). For instance, Łabuz et al. studied the structural transformation of amorphous titania by complexation of TiO_2 particles with rutin (Rut@TiO_2)[105] or bromopyrogallol red (Brp@TiO_2).[115] Both functionalizations resulted in photosensitization by visible light with enhanced efficiency of the charge separation and interfacial charge transfer processes. In the case of Rut@TiO_2, the enhancement of the photocatalytic ROS generation of hydroperoxyl ($HOO^{\cdot}$) and hydroxyl ($HO^{\cdot}$) radicals reactions was observed. The process is governed by hydrogen peroxide, which plays the double role of a reactant prone to reduction and generation of hydroxyl radicals or a redox agent destroying the intra-band gap electronic states, thereby suppressing charge recombination. Noteworthy, it was concluded that only stable complexation of rutin to the surface titanium ions promotes spontaneous crystallization of parent amorphous nanoparticles and guaranteed efficient photoactivity. In the case of Brp@TiO_2, ROS-dependent lipids peroxidation by free radicals photogeneration (type I PSO reactions) was indicated as the main mechanism of photosensitized cellular membranes oxidation. In addition, a two-step mechanism involving photoinduced $^1O_2$ generation on the titania surface was considered (Fig. 17B). Even though EPR oximetry and time-resolved detection of $^1O_2$ phosphorescence at 1270 nm did not confirm any measurable formation of $^1O_2$, its formation cannot be ruled out. For the investigated systems, Nosaka's mechanism of $^1O_2$ generation and the Russell reaction involving peroxyl radicals formation were proposed due to visible light absorption by a surface charge transfer complex that results in the generation of electrons in a conduction band and an oxidized modifier.[116]

In addition, the photosensitization of TiO_2 by modification with porphyrins is currently being researched very intensively in our team. Efficient photodynamic inactivation (PDI) of bacteria was demonstrated *in vitro* by Sułek et al.[106] for TiO_2 surface-modified with sulfonated and halogenated phorphyrins. Different mechanisms of bacteria disruption depending on the type of generated oxygen and/or iodide species were suggested in the case of surface-modified TiO_2 with two phorphyrin derivatives: 5,10,15,20-*tetrakis*-(4-sulfonatophenyl)porphyrin (TPPS); 5, 10,15,20-*tetrakis*(2,6-difluoro-3-sulfophenyl)porphyrin (F_2POH), and its metal complexes 5,10,15,20-*tetrakis*(2,6-difluoro-3-sulfophenyl)porphyrin Zn(II) (ZnF_2POH). Visible light-active, nanocolloidal TiO_2-based materials turned out to efficiently photoinactivate *Escherichia coli* and *Staphylococcus aureus* bacteria strains *in vitro*. Noteworthy, in contrast to TPPS in a homogeneous system, the hybrid TPPS@TiO_2 generates not only singlet oxygen, but also oxygen-centered radicals. It was concluded that the photoactivity of phorphyrin-based TiO_2 is derived from multiple mechanisms of ROS generation (Fig. 17C).

In the case of many semiconductors, such as TiO_2, ZnO, *etc.*, serious limitations of their practical photocatalytical applications involving medical therapies (*i.e.*, PDT, PDI) cannot be neglected due to the low quantum yield of ROS generation resulting from the fast electron – hole recombination (50 ± 30 ns). Modification of these semiconductors with noble metals such as Au, Ag, Pt, and Pd can lead to significantly improved quantum yields of ROS generation.[117] The prevention of electron-hole recombination due to trapping the photoexcited electron and the enhancement of absorption spectrum *via* surface plasmon resonance is associated with the photo-induced effective charge separation in the semiconductor shell by transfer of electrons into the metal structure. Importantly, the characteristic metal surface plasmon can be returned, when the electrons are discharged from the metal. Recently, Jiang et al.[118] demonstrated the improved photogeneration of ROS by ZnO modified with noble metals (Ag, Au, Pt) NPs. Simple mixing of the ZnO with metal NPs fine-tunes the yield of photogenerated ROS depending on particles size, composition, and mass ratio of the metal NPs and the ZnO NPs.

One of the novel concepts in cancer treatment can be a photochemotherapy (PCT) strategy of cancer therapy using NIR-assisted tumor-specific Fenton reactions. Hu et al.[119] has recently demonstrated that a serious limitation of PDT consisting of the strong dependence on oxygen level, low UV/vis light penetration depth and the extremely short lifetime of ROS can be overcome by upconverting nanoparticles (UCNPs) catalyzing the intramitochondrial Fenton reaction between the delivered Fe^{2+} and H_2O_2 overexpressed in cancer cell's mitochondria. A photochemotherapeutic nanomaterial composed of three components: (1) a UCNP core based on lanthanide-doped nanocrystals for converting NIR light to UV or visible photons and catalyzing photo-Fenton reaction,

(2) mesoporous silica shell loaded with Fenton reagent (Fe^{2+}), and (3) a Ru(II) complex co-conjugated on the shell to bind mitochondrial DNA is the first PCT agent with potency of O_2-independent but H_2O_2-responsive generation of $^{\cdot}OH$. This highly selective and efficient approach is an excellent example of a controlled apoptotic death scenario driven by mitochondria *via the* assistance of the photo-Fenton reaction. Following this line of research, Wang et al.[120] have proposed ultra-small biomimetic $Cu_{2-x}Se$ NPs acting as nanocatalyst for an efficacious chemodynamic treatment (CDT) of breast cancer cells. A strong NIR localized surface plasmon resonance (600–1000 nm) assures the heat conversion (NIR-I window; 700–900 nm) altogether with electron and energy transfer (NIR-II window; 1000–1700 nm), both to speed up the Fenton reaction. Furthermore, $Cu_{2-x}Se$ NPs (CS) were conjugated with glucose oxidase (GOD) and coated with 4T1 cell membranes (CMs) to enhance biocompatibility, limit cytotoxicity, and improve homologous adhesion. GOD, as a natural aerobic dehydrogenase, can catalyze the oxidation of glucose, abundant in the tumor environment, generating H_2O_2 and delivering it in situ as a Fenton reaction substrate. The resulting CS-GOD@CM NPs can be selectively accumulated inside tumor cells and efficiently increase the content of H_2O_2 through *in situ* oxidization of the glucose within the tumor by GOD. The high enhancement of the NIR-II photo-Fenton reaction resulted in the large amounts of $^{\cdot}OH$ radicals and other ROS in a short time, providing excellent therapeutic efficacy.

The complexity of the intracellular redox environment implies that it is possible to stimulate a specific cellular response at the desired level other than cell toxicity. The photoexcited QDs, tuned through the size-dependent quantum effect, with potency to kill a wide range of multidrug-resistant bacterial clinical isolates *in vitro* (*i.e.*, methicillin-resistant *Staphylococcus aureus* and *Salmonella typhimurium*, carbapenem-resistant *Escherichia coli*, extended-spectrum *β*-lactamase-producing *Klebsiella pneumoniae*) by specific ROS generation have recently been reported by Courtney et al.[121] The dependence of the cellular effect on the quantum dot oxidation and reduction potentials that resulted in the ability to selectively hinder the growth of bacteria. This was demonstrated in co-culture of bacteria and healthy mammalian cells, leaving the latter intact, and even causing a photo-proliferative effect. It was concluded that the killing effect is independent of the type of material but is controlled only by the redox potentials of the photogenerated charge carriers, which selectively alter the cellular redox state. Noteworthy, this very promising research approach was transferred to the animal model (mice) and such engineered LED patches with CdTe QDs (2.4 eV band gap) have been proposed to treat the bacterial infections.[122] A non-toxic effect *in vivo* was demonstrated at a concentration relevant to reducing the bacterial load of multidrug resistant *Escherichia coli* infections in a murine subcutaneous abscess model. A plausible mechanism for the observed antibacterial effect was the generation of superoxide anion. The small size of the superoxide allows it to diffuse easily across bacterial cell walls, eliminating the possibility of resistance associated with adaptations of transport proteins. The dependence of the growth and saturation of triggered oxidative stress on irradiation intensity supports the light-activated redox mechanism. Thus, such light-activated nanotherapeutics that effectively inhibit MDR bacteria through targeted interactions with the cellular redox environment can revolutionize the treatment of burn wounds, cellulitis, and other serious skin infections.

It is also necessary to mention that ROS photogeneration on inorganic particles can be a source of serious human health problems. It has recently been shown that particles present in air pollution can generate singlet oxygen and hydroxyl radical upon UV-Vis-NIR irradiation.[123] Such processes can be harmful to human health, especially when exposure is prolonged. Furthermore, a completely different approach to the therapeutic potency of photocatalysts can be the controlled reduction of photocatalytic reactivity in the case when it is preferable to use deactivated forms of semiconductors, *e.g.*, TiO_2 and ZnO. The latter components of several suntan lotions generate highly oxidizing, cytotoxic, and genotoxic ROS upon UV light absorption. Therefore, efficient quenching of their photocatalytic activity should be introduced to guarantee safe use of semiconductor nanoparticles as ingredients in suntan lotions, cosmetics, and dermatological formulations. Hybridization *via* ionotropic gelation with a natural biopolymers, *e.g.*, chitosan (CS), is one possible way of photocatalytic control.[112] A complete lack of photoelectrochemical response (photocurrent quenching) for nanocomposites in the form of a transparent polymeric coating on ZnO NPs guarantees total inhibition of photoinduced interfacial energy or electron transfer processes. It is noteworthy that such hybrid crossed-linked chitosan-ZnO material not only blocks ROS photogeneration (photoprotection) but also exhibits enhanced antibacterial activity compared to unmodified ZnO NPs.

The explanation of the mechanism of ROS generation on both naked noble metal nanoparticles as well as on nanostructures composed of semiconductors and noble metal NPs is far from clear-cut. However, based on the current state of knowledge, it can be summarized that, in the first case, ROS can be generated by plasmon-produced electrons because of energy and electron-transfer processes. The efficiency of ROS generation by naked metal NPs was shown to be proportional to 1/d (where d is a particle diameter). This indicates the dominant role of the surface area rather than the weight percentage of NPs in this process.[94] Noteworthy, the adequate selection of the source of excitation can cause even a sixfold increase of the efficacy of $^1O_2$ generation (for instance, irradiation of AuNCs with a femtosecond-pulsed Ti:sapphire laser[124]). Unfortunately, most studies have been so far carried out under one-photon irradiation sources rather than with application of two-photon treatment since the latter requires expensive ultrafast pulsed lasers. Plausibly, the two-photon radiation mode will provide a discovery of very little researched and unrecognized nonlinear optical effects by plasmonic metal nanostructures for advanced therapy and diagnosis.

With semiconductor/metal heterojunctions, ROS photogeneration is improved. In the case of hybrid nanostructures such as semiconductors (*e.g.*, TiO_2, ZnO) modified with noble metal nanoparticles, the observed enhanced photocatalytic activity and ROS generation presumably result from the following effects: (1) enhancing light absorption because of the SPR of noble metal, (2) photosensibilization of semiconductor toward visible light (i.e., plasmon-induced resonance energy transfer (PIRET), and hot electron injection, also called as direct electron transfer (DET)), and (3) increasing the efficiency of electron transport and charge carrier separation.[125] These proposed strategies for designing noble metal/semiconductor hybrid nanostructures offer an easy way to tailor ROS and charge carrier production that can be applied in medicine, in particular in PDT and PDI.

Undoubtedly, a further in-depth investigation on plasmonic metal-mediated photogeneration of ROS *via* energy or electron transfer is highly justifiable since the prospects for practical applications of such photoactive nanomaterials are huge in the field of medicine. In particular, great expectations lay in Au NPs with different morphology, such as nonorods, nanocages, nanostars, *etc.*,[126,127] and bimetallic nanostructures (*e.g.*, spiky hollow Ag-Au nanostars,[128] Fe_3O_4@Au core/shell nanostars[129]), since they are outstanding NIR-light active multifunctional photodynamic reagents.[130] For efficient PDT treatment of deep tissue-buried solid tumors, profound tissue penetration is absolutely mandatory, therefore Au NPs are excellent PDT candidates since they are characterized by the ability to absorb NIR light and sensitize formation of singlet oxygen. Furthermore, gold nanostructures can be used simultaneously as fluorescent markers to follow their intracellular accumulation and systemic distribution in the body as they emit excitation wavelength dependent fluorescence in the visible to NIR region.

4.3 Multimodal systems

4.3.1 Molecular systems

Most anticancer therapies currently used are generally not free from drawbacks due to limited efficacy or troublesome side effects. Combining methods that bring together different aspects of drug activity are among the ways to overcome these shortcomings. One direction being developed is to take advantage of the dual cytotoxicity of reactive oxygen and nitrogen species. The design of functional systems to enable the achievement of such a goal needs to meet criteria resulting from the nature of the gaseous agents generated. First, it is necessary to prevent their mutual deactivation by electron or energy transfer. Furthermore, it is particularly advantageous to generate both NO and $^1O_2$ by the action of a single photon. This ensures full cooperation of both cytotoxic agents at a specific location and time. The systems presented to date to provide such a dual therapeutic strategy exhibit varying structural complexity. The simplest approaches are a mixture of compounds, one of which acts as a photosensitizer in the generation of singlet oxygen and the other acts as a photodonor of nitric oxide. Despite their fundamental simplicity in design, such systems must exhibit complex photochemical reactivity due to the different mechanisms of NO and $^1O_2$ generation.

For instance, such a strategy becomes possible when the applied compound has the ability to act simultaneously as a NO donor and a PDT photosensitizer. The use of light as a trigger for both intended actions provides, at least in theory, the opportunity to fully control the activity of the therapeutic agent, resulting in both increased activity and reduced side effects (Fig. 18).[131]

The cooperation of small gaseous molecules, such as nitric oxide and (singlet) oxygen, is all the more promising, as they share common beneficial features that predispose them to multimodal biomedical approaches. Among the physical properties, the small size, lack of charge, and good lipophilicity are of particular importance. Therapeutic efficiency is promoted by cytotoxic activity against a variety of cellular targets and the absence of multidrug resistance. In addition, the very limited time and range of action, resulting

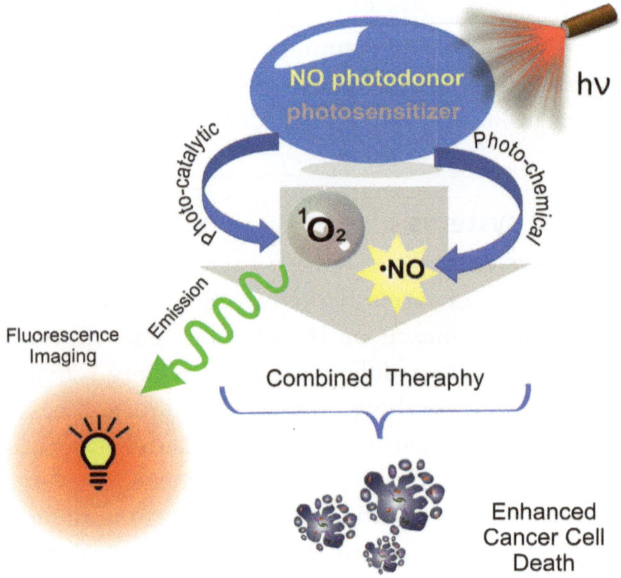

Fig. 18 Simultaneous photo-induced generation of NO and $^1O_2$ (based on 131).

from short lifetimes, and finally the possibility of full light control ensure reduced side effects. A prerequisite for high performance is, first of all the simultaneous release of both cytotoxic agents exactly at the site of their intended action. For this to occur, it is necessary to equip such a system with a strong chromophore that absorbs biocompatible light from the near-infrared or low-energy visible range.[131] The system proposed by Ramos et al.[132] based on the cooperation of two ruthenium complexes is a good example (I and II in Fig. 19A). Absorption within the range of the MLCT band (around 470 nm) by complex I can be employed to generate $^1O_2$ and other ROS. In addition, in the presence of phosphate ions as mediators, photoinduced electron transfer can occur to complex II in a close vicinity, resulting in the release of NO. The therapeutic efficacy of this system was confirmed on selected melanoma cell lines *in vitro*.

Regrettably, the proximity of the donor and acceptor in the step of photoinduced electron transfer does not ensure the generation of both cytotoxic agents in the same small cellular compartment. Such a certainty is only provided by single-molecule bifunctional systems. The phthalocyanine Ru complex having a nitrosyl ligand in the axial position (Fig. 19B) is perhaps the simplest example of structural but not mechanistic coupling.[133] In this case, as a consequence of the separation of the two processes, oxygen

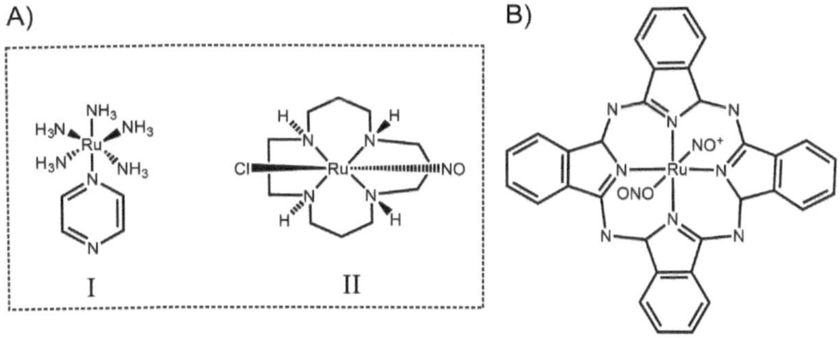

Fig. 19 Ruthenium-based phototherapeutics for NO/PDT combined anticancer therapy.

photosensitization and NO release require irradiation at different wavelengths, 660 and 377 nm, respectively. Regardless of this disadvantage, increasing the efficiency of NO generation requires a higher number of nitrosyl ligands in the complex. In practice, this effect can be achieved by forming a trinuclear conjugate containing phthalocyanine and two NO-releasing complexes.[134] The release of more nitric oxide molecules by bifunctional molecular systems is still a considerable challenge.

A separate issue is the possibility of self-diagnosis of the efficiency of photochemical generation of reactive oxygen and nitrogen species. Such functionality is provided by molecular systems equipped with conjugated chromo- and fluorophoric systems. Switching between emitting and non-emitting states can occur, for example, as a consequence of the photo-induced release of NO. Benkovics et al.[135] proposed a molecular hybrid based on β-cyclodextrin (Fig. 20).

It contains the anthracene group, whose excitation results in the generation of $^1O_2$ that competes with fluorescence emission and the nitroaniline moiety, which plays a role of NO photodonor. Substitution of the nitrosyl group by the hydroxyl group inhibits further intramolecular energy transfer, which results in the prevention of fluorescence.

The fixed stoichiometry of molecular systems considerably limits, if not prevents, the regulation of the generation efficiency of both cystostatic agents. Such situation changes favorably when photoactive molecules are combined with specific nanovehicles, thus forming supramolecular systems. The simplicity of the synthesis of such structures is only the first advantage and not the most important. Of greater importance is the ability to incorporate a large number of chromophores into a single structure, which greatly improves both the light harvesting properties in a specific cellular compartment, and the ability to alter the NO to $^1O_2$ ratio. In assemblies of this type,

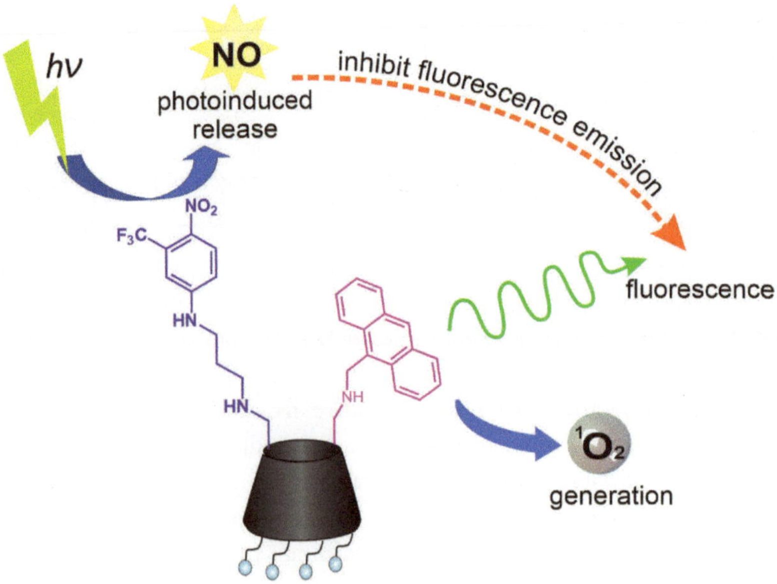

Fig. 20 NO-releasing and $^1O_2$ generating hybrid capable of fluorescent reporting as proposed by Benkovics et al.[135]

presented so far, there is a great variety in terms of both the guest molecules and the structures of the hosts. The latter function can be performed by liposomes, with the added advantage of being able to incorporate with them hydrophobic compounds such as phthalocyanines commonly used as PDT photosensitizers. The embedding of any number of nitrosyl ruthenium complexes in the membrane allows the excitation energy of zinc phthalocyanine to be used for the photoinduced release of a significant amount of NO in the immediate vicinity of the cellular target while maintaining a satisfactory efficiency of $^1O_2$ generation.[136] A different approach in creating bifunctional nanostructured materials is to modify the photoactive components in such a way that they can spontaneously aggregate. This makes the necessity of engaging nanocarriers redundant. An example of such an approach is a nanoassembly formed by incorporation of nitroaniline molecules, purposely tailored by a long side substituent, into the β-cyclodextrin cavity linked to the aggregated porphyrin-based chromophores.[137] Such a system exhibits enhanced activity in inducing cancer cell death compared to monomeric systems based on similar subunits. Some other modes of self-assembly of tetrapyrrolic photosensitizers and NO photodonors were recognized. An interesting example is the formation of micelles by the modified nitroaniline when its long side substituent is terminally equipped with bulky cations, such as the trimethylammonium ion. Hydrophilic porphyrin molecules spontaneously adhere to the surface of such micelles, forming direct contact between the interacting sensitizer and quencher (Fig. 21).[138]

The use of antimicrobial medical devices and various topical applications calls for a different form of spatial organization of the components of coupled systems for photoinduced generation of reactive oxygen and nitrogen species. Thin monolayer and multilayer films containing NO and $^1O_2$ delivering compounds are formed at the liquid-gas interface and then applied to a solid support. The limitations in the applicability of such systems are compensated by the complementary antimicrobial activity of $^1O_2$ and NO, which manifests itself in differences in their diffusion radii.[139]

In addition to the aforementioned types of coupled systems providing the possibility of simultaneous generation of NO and ROS induced by absorption of a single photon, there are several studies in which the release of both cytotoxic factors is conditioned by the action of different triggers. A quite commonly promoted solution assumes photosensitization as a way to generate $^1O_2$, while NO is released by oxidation of its donor. Some systems of this type designed for anticancer therapies may not have the expected

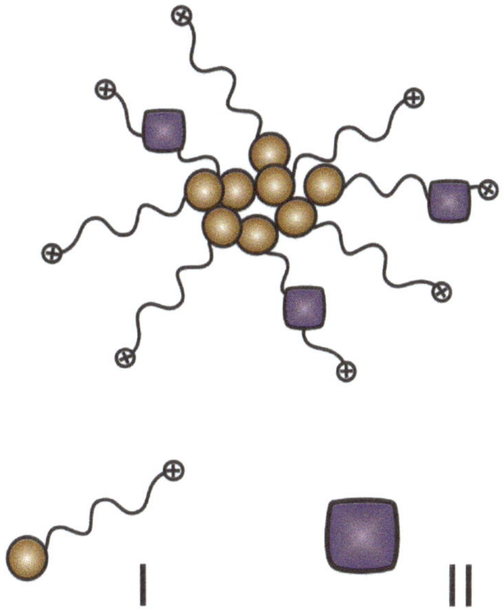

Fig. 21 Hybrid system with increased efficiency of NO photo-release proposed by Caruso et al.[138]

effects, especially when they rely on intracellular oxidant levels. Deficiency of O_2 and H_2O_2 in hypoxic tumors causes insufficient NO release from L-arginine. Attempts are being made to circumvent these problems by photoinduced intracellular oxidation of water molecules. One of the most promising systems operating by this mechanism was based on carbon nitride-carbon dot composite nanoparticles surface-modified with L-arginine polymer. Red-light irradiation triggers charge separation within the nanomaterial. The resulting hole oxidizes H_2O to H_2O_2, which, in turn, by oxidizing arginine residues, causes the release of NO. The yield of this reaction sequence was found to be sufficient to induce cell apoptosis.[140]

4.3.2 Nanostructured systems

Recently, to achieve higher therapeutic efficacy of PDT a novel strategy targeting mitochondria, attractive subcellular organelles, has been proposed that improve photodynamic and photothermic efficacies.[141–143] For instance, Guan et al.[141] developed a ternary polymeric nanosystem involving NIR triggered mitochondria targeting aggregation-induced emission photosensitizer (AIE), triphenylphosphine, and upconverting nanoparticles (UCNPs). The coated with pH-responsive AIE polymer can be photoactivated as a photosensitizer by the upconverted energy of the Tm^{3+} doped UCNPs upon 980 nm laser irradiation. This results in high photogenerated ROS levels at specific locations, as the TPP acts as a mitochondria-targeted group. As a result, mitochondrial collapse causes the induction of cell apoptosis. Thus, such a triple combination of ROS generation by an AIE polymer, subcellular targeted delivery, and NIR-triggered PDT is beneficial for the therapeutic efficacy. In addition, the applied pH-responsive PEG layer, which is easily conjugated to the surface of NPs, shields the surface positive charges, resulting in a stronger protein-resistance ability, while the cleavage of the UCNPs@PAIE-TPP-PEG is possible in an acidic tumor environment. Zhang et al.[123] introduced another innovative example of a mitochondria-targeted NIR-responsive multifunctional nanosystem.[142] Synergistic PDT/PTT by a single NIR irradiation (808 nm), self-sufficient oxygen-augmented PDT, and multiple imaging guidance/monitoring was rationally designed with the efficient ability of targeting mitochondria. Perfluorooctyl bromide (PFOB) was employed for oxygen delivery into tumors in the form of nanoliposomes with a great capacity of O_2 dissolving and excellent medium prolonging the ROS lifetime. These nanoplatforms can efficiently perform the functions of red blood cells ('Nano-RBCs') transporting oxygen and improving the efficacy of PDT by alleviating tumor hypoxia. Lipophilic IR780 iodide was incorporated into the lipid bilayer, while PFOB was encapsulated in the core forming an oxygen reservoir. The resulting PFOB@LIP-IR-780 "Nano-RBCs" enable the following continuous *in vivo* phototherapies (PDT/PTT) and the guidance/monitoring by multimodal imaging, including CT, PA, and FL imaging.

In general, metal-based NPs phototoxicity is affected by many factors, including physicochemical characteristics, such as uniformity and size, morphology, composition, surface functionalization and charge, crystalline structure, and magnetism.[89,126] On the other side, a final activity involving ROS photogeneration in natural media, including intracellular environment *in vivo* (*e.g.*, the complex tumor microenvironment), is governed by various factors such as pH, oxygen concertation, altered redox potentials, *etc.* In addition, the immunosuppressive nature of tumors often limits the clinical benefits of many treatments. With this in mind, Shi et al.[144] have recently developed a novel strategy targeting tumor-associated macrophages (TAMs) as the critical driver of immunosuppressive activity that promotes tumor metastasis. Precise nanoparticle-based ROS photogeneration in TAMs (endosome/lysosome, cytoplasm) can effectively repolarize TAMs, improving antigen presentation *via* direct activation of the M1 signaling pathway, inhibition of the lysosomal proton pump, and induction of proteolytic activity. Such reprogrammed TAMs to the antitumor M1 phenotype efficiently recruit cytotoxic lymphocytes (CTLs) and direct memory T-cells toward tumoricidal responses resulting in metastasis inhibition and tumor eradication. The mechanism of mannose-modified PLGA nanoparticles (TAMs overexpress the mannose receptor) loaded with photosensitizer (indocyanine green), TiO_2 and ammonium bicarbonate (NH_4HCO_3) is associated with the production of CO_2 and NH_3 that disrupt the endosome/lysosome membranes followed by PS release into the cytoplasm upon endosome/lysosome acidification. This skillful combination of cancer immunology with nanotechnology gives hope for an effective breakthrough in photodynamic treatment.

5. Conclusions and future directions

Reactive forms of small inorganic molecules such as reactive oxygen (ROS), nitrogen species (RNS), sulfur species (RSS), and carbon species (RCS) photogenerated by molecular and nanostructured systems in the proper spatial-temporal regimen are central players in the redox biochemistry of cells. Due to the remarkable advances in organic and inorganic chemistry, photochemistry and photophysics, material chemistry, and nanotechnology, a great variety of simple nonmacrocyclic and macrocyclic organic compounds, transition metal complexes with a diversity of ligands, and nanomaterials (nanoparticles, nanoscaffolds, hybrid materials, upconverting nanomaterials) with unique photoresponsive properties have been explored to provide or detect the dynamic behaviors of RSIM in biological systems. Conscious application of light controlled strategies such as optogenetics, photocleavage, photosensitization, and photocatalysis operated by photoactive materials endeavor the control over the RSIM generation and interrelationships under physiological and pathological conditions.

In the near future, successful monitoring of chemical or physical processes at particular locations within a single cell will be one of the common practical methodologies in explaining the modes of action offered by novel materials. Photoactive systems with the possibility of multifunctional modes such as sensing and imaging of dynamic physical/chemical/biological processes in live cells and *in vivo* in the real-time, together with therapeutic potency open up novel teragnostic options. This offers great potential for diagnosing and treatment of many diseases at an early stage, when changes on the tissue level are not yet observable; however, chemical variations inside cells are evident.

Increasing awareness of the chemical reactivity and biological function of various RSIM has led to growing interest in developing ways for their controlled photogeneration. Such an approach makes it possible to take advantage of their complementary properties and combine different therapeutic strategies. The rapid development of research gives hope for a rapid solution to the most pressing problems, such as increasing the RSIM generation efficiency in the immediate vicinity of their predicted site of action. Taking into account the trends of photoactive materials development, the following aspects are considered. As only a few metal complexes are currently the subject of intense research, the range of transition metals used as central ions should be widened. Furthermore, modification of the ligands should focus on shifting the absorption spectra toward a longer wavelength in the visible and near-infrared ranges and increasing the selectivity of the compounds. Finally, metallic nanostructures that do not degrade easily in solution in contrast to molecular compounds and offer a huge diversity of possible functionalization and conjugation with a variety of molecules are outstanding candidates not only as separate materials but also as platforms for the delivery of molecular compounds photogenerating RSIM. Moreover, many strategies to improve the quantum yield of RSIM generation should be developed and investigated, for instance modifications of semiconductors with noble metals NPs such as Au, Ag, Pt, and Pd significantly enhance RSIM photogeneration. It is noteworthy that the intrinsic optical properties of metal-based nanomaterials in the NIR range enable their application in imaging of the whole human body without the use of additional fluorophores. Another promising future direction is the activation of multiple signals using single nanoplatforms. Presumably, multiphoton activation modes will provide unrecognized nonlinear optical effects guaranteed by plasmonic metal nanostructures for advanced diagnosis and therapy.

In the context of research on the mechanisms of RSIM photogeneration, there is still much to be elucidated: (i) secondary "dark" reactions that occur *in vivo* after light exposure, (ii) photosensitized reactions in pathological conditions (*e.g.*, anaerobic environments) requiring molecular contact, (iii) extrapolation of model cellular systems to animal models that should speed of the potential biomedical applications. The diversity, complexity, and heterogeneity of novel multifunctional systems (both molecular and nanostructural) and biosystems *in vivo* make the achievement of these goals seem far away, but the advanced development of research techniques slowly brings us closer to them.

Acknowledgments

The work was supported by the Nacional Science Centre Poland within the OPUS project no 2019/35/B/ST4/04266.

References

1. Basudhar, D.; Ridnour, L. A.; Cheng, R.; Kesarwala, A. H.; Heinecke, J.; Wink, D. A. *Coord. Chem. Rev.* **2016**, *306*, 708–723.
2. (a) Cortese-Krott, M. M.; Koning, A.; Kuhnle, G. G. C.; Nagy, P.; Bianco, C. L.; Pasch, A.; Wink, D. A.; Fukuto, J. M.; Jackson, A. A.; van Goor, H.; et al. *Antioxid. Redox Signal* **2017**, *27* (10), 684–712; (b) Graves, D. B. *J. Phys. D: Appl. Phys* **2012**, *45* (26), 263001. https://doi.org/10.1088/0022-3727/45/26/263001; (c) Meng, J.; Lv, Z.; Zhang, Y.; Wang, Y.; Qiao, X.; Sun, C.; Chen, Y.; Guo, M.; Han, W.; Ye, A.; et al. *Antioxid. Redox Signal* **2021**, *34* (14), 1069–1082.
3. Frausto da Silva, R. J. J.; Williams, R. J. P. *The Biological Chemistry of the Elements: The Inorganic Chemistry of Life*; Clarendon Press, 2001.
4. (a) Shields, H. J.; Traa, A.; Van Raamsdonk, J. M. *Front. Cell Dev. Biol.* **2021**, *9* (628157). Review; (b) Thomas, D. D. *Redox Biol.* **2015**, *5*, 225–233.

5. (a) Canaparo, R.; Foglietta, F.; Limongi, T.; Serpe, L. *Materials* **2020**, *14* (1), 53. https://doi.org/10.3390/ma14010053. PubMed; (b) Cheeseman, S.; Christofferson, A. J.; Kariuki, R.; Cozzolino, D.; Daeneke, T.; Crawford, R. J.; Truong, V. K.; Chapman, J.; Elbourne, A. *Adv. Sci.* **2020**, 7 (10), 1902913. https://doi.org/10.1002/advs.201902913; (c) Yu, Z.; Li, Q.; Wang, J.; Yu, Y.; Wang, Y.; Zhou, Q.; Li, P. *Nanoscale Res. Lett.* **2020**, *15* (1), 115. https://doi.org/10.1186/s11671-020-03344-7; (d) Zhang, L.; Zhu, C.; Huang, R.; Ding, Y.; Ruan, C.; Shen, X.-C. *Front. Chem.* **2021**, *9*. https://doi.org/10.3389/fchem.2021.630969. Mini Review.
6. Zhang, C.; Wang, X.; Du, J.; Gu, Z.; Zhao, Y. *Adv. Sci.* **2021**, *8* (3), 2002797. https://doi.org/10.1002/advs.202002797.
7. Balzani, V.; Ceroni, P.; Juris, A. *Photochemistry and Photophysics: Concepts, Research Applications*; Wiley-VCH, 2014.
8. Baptista, M. S.; Cadet, J.; Greer, A.; Thomas, A. H. *Photochem. Photobiol.* **2021**, *97*, 1456–1483.
9. DeRosa, M. C.; Crutchley, R. J. *Coord. Chem. Rev.* **2002**, *233–234*, 351–371. https://doi.org/10.1016/S0010-8545(02)00034-6.
10. (a) Hong, L.; Li, J.; Luo, Y.; Guo, T.; Zhang, C.; Ou, S.; Long, Y.; Hu, Z. *Biomolecules* **2022**, *12* (1), 81; (b) Li, Y.; Zhang, W.; Niu, J.; Chen, Y. *ACS Nano* **2012**, *6* (6), 5164–5173. https://doi.org/10.1021/nn300934k; (c) Prakash, J.; Sun, S.; Swart, H. C.; Gupta, R. K. *Appl. Mater. Today* **2018**, *11*, 82–135. https://doi.org/10.1016/j.apmt.2018.02.002; (d) Zheng, X.; Jin, Y.; Liu, X.; Liu, T.; Wang, W.; Yu, H. *Bioact. Mater.* **2021**, *6* (12), 4301–4318. https://doi.org/10.1016/j.bioactmat.2021.04.030.
11. Kuncewicz, J.; Dąbrowski, J. M.; Kyzioł, A.; Brindell, M.; Łabuz, P.; Mazuryk, O.; Macyk, W.; Stochel, G. *Coord. Chem. Rev.* **2019**, *398*, 113012. https://doi.org/10.1016/j.ccr.2019.07.009.
12. McFarland, S. A.; Mandel, A.; Dumoulin-White, R.; Gasser, G. *Curr. Opin. Chem. Biol.* **2020**, *56*, 23–27. https://doi.org/10.1016/j.cbpa.2019.10.004.
13. Mfouo-Tynga, I. S.; Dias, L. D.; Inada, N. M.; Kurachi, C. *Photodiagnosis Photodyn. Ther.* **2021**, *34*, 102091. https://doi.org/10.1016/j.pdpdt.2020.102091.
14. Nosaka, Y.; Nosaka, A. Y. *Chem. Rev.* **2017**, *117* (17), 11302–11336. https://doi.org/10.1021/acs.chemrev.7b00161.
15. Stochel, G.; Stasicka, Z.; Brindell, M.; Macyk, W.; Szacilowski, K. *Bioinorganic Photochemistry*; Wiley, 2009.
16. Szaciłowski, K.; Macyk, W.; Drzewiecka-Matuszek, A.; Brindell, M.; Stochel, G. *Chem. Rev.* **2005**, *105* (6), 2647–2694. https://doi.org/10.1021/cr030707e.
17. Wan, Q.; Zhang, R.; Zhuang, Z.; Li, Y. H.; Yuhua; Wang, Z.; Zhang, W.; Hou, J.; Tang, B. Z. *Adv. Funct. Mater.* **2020**, *30* (39), 2002057. https://doi.org/10.1002/adfm.202002057.
18. Zhao, J.; Wu, W.; Sun, J.; Guo, S. *Chem. Soc. Rev.* **2013**, *42* (12), 5323–5351. https://doi.org/10.1039/C3CS35531D. DOI: 10.1039/c3cs35531d.
19. Zheng, Q.; Liu, X.; Zheng, Y.; Yeung, K. W. K.; Cui, Z.; Liang, Y.; Li, Z.; Zhu, S.; Wang, X.; Wu, S. *Chem. Soc. Rev.* **2021**, *50* (8), 5086–5125. https://doi.org/10.1039/D1CS00056J. DOI: 10.1039/d1cs00056j.
20. Karu, T. I. Low-power laser therapy. In *Biomedical Photonics Handbook*; Vo-Dinh, T., Ed.; CRC Press, 2014; p. 1864.
21. Dąbrowski, J. M.; Pucelik, B.; Regiel-Futyra, A.; Brindell, M.; Mazuryk, O.; Kyzioł, A.; Stochel, G.; Macyk, W.; Arnaut, L. G. *Coord. Chem. Rev.* **2016**, *325*, 67–101. Review. https://doi.org/10.1016/j.ccr.2016.06.007.
22. (a) Hiremath, N.; Kumar, R.; Hwang, K. C.; Banerjee, I.; Thangudu, S.; Vankayala, R. *ACS Appl. Nano Mater.* **2022**, *5* (2), 1719–1733. https://doi.org/10.1021/acsanm.2c00170; (b) Leem, J. W.; Kim, S.-R.; Choi, K.-H.; Kim, Y. L. *Nano Converg.* **2018**, *5* (1), 8. https://doi.org/10.1186/s40580-018-0140-7.

23. (a) Oszajca, M.; Brindell, M.; Orzeł, Ł.; Dąbrowski, J. M.; Śpiewak, K.; Łabuz, P.; Pacia, M.; Stochel-Gaudyn, A.; Macyk, W.; van Eldik, R.; et al. *Coord. Chem. Rev.* **2016**, *327–328*, 143–165. https://doi.org/10.1016/j.ccr.2016.05.013; (b) Regiel-Futyra, A.; Dąbrowski, J. M.; Mazuryk, O.; Śpiewak, K.; Kyzioł, A.; Pucelik, B.; Brindell, M.; Stochel, G. *Coord. Chem. Rev.* **2017**, *351*, 76–111.
24. Pearson, S.; Feng, J.; del Campo, A. *Adv. Funct. Mater.* **2021**, *31* (50), 2105989. https://doi.org/10.1002/adfm.202105989.
25. Plekhova, N.; Shevchenko, O.; Korshunova, O.; Stepanyugina, A.; Tananaev, I.; Apanasevich, V. *Bioengineering* **2022**, *9* (2), 82.
26. Wojtovich, A. P.; Foster, T. H. *Redox Biol.* **2014**, *2*, 368–376. https://doi.org/10.1016/j.redox.2014.01.019.
27. (a) Arnaut, L. G.; Pereira, M. M.; Dąbrowski, J. M.; Silva, E. F. F.; Schaberle, F. A.; Abreu, A. R.; Rocha, L. B.; Barsan, M. M.; Urbańska, K.; Stochel, G.; et al. *Chem. Eur. J.* **2014**, *20* (18), 5346–5357. https://doi.org/10.1002/chem.201304202; (b) Silva, E. F. F.; Serpa, C.; Dąbrowski, J. M.; Monteiro, C. J. P.; Formosinho, S. J.; Stochel, G.; Urbanska, K.; Simões, S.; Pereira, M. M.; Arnaut, L. G. *Chem. Eur. J.* **2010**, *16* (30), 9273–9286. https://doi.org/10.1002/chem.201000111.
28. Wilkinson, F.; Abdel-Shafi, A. A. *J. Phys. Chem. A* **1997**, *101* (30), 5509–5516. https://doi.org/10.1021/jp970706o.
29. (a) Bonet-Aleta, J.; Garcia-Peiro, J. I.; Hueso, J. L. *Catalysts* **2022**, *12* (2), 167; (b) Watanabe, M. *Sci. Technol. Adv. Mater.* **2017**, *18* (1), 705–723.
30. Gogoi, N.; Borah, G.; Gogoi, P. K.; Chetia, T. R. *Chem. Phys. Lett.* **2018**, *692*, 224–231.
31. Xu, J.; Liu, N.; Wu, D.; Gao, Z.; Song, Y. Y.; Schmuki, P. *ACS Nano* **2020**, *14* (1), 337–346.
32. Thomsen, H.; Marino, N.; Conoci, S.; Sortino, S.; Ericson, M. B. *Sci. Reports* **2018**, *8*, 9753.
33. Paul, S.; Pan, S.; Mukherjee, A.; De, P. *Mol. Pharmaceutics* **2021**, *18*, 3181–3205.
34. Feldman, P. L.; Griffith, O. W.; Stuehr, D. J. *Chem. Eng. News* **1993**, *71* (51), 26–38.
35. Sexton, D. J.; Muruganandam, A.; McKenney, D. J.; Mutus, B. *Photochem. Photobiol.* **1994**, *59* (4), 463–467.
36. Stasko, N. A.; Fischer, T. H.; Schoenfisch, M. H. *Biomacromolecules* **2008**, *9* (3), 834–841.
37. Xiang, H. J.; Guo, M.; Liu, J. G. *Eur. J. Inorg. Chem.* **2017**, *12*, 1586–1595.
38. Fry, N. L.; Mascharak, P. K. *Acc. Chem. Res.* **2011**, *44* (4), 289–298.
39. Ford, P. C. *Nitric Oxide* **2013**, *1* (34), 59–64.
40. Ford, P. C. *Coord. Chem. Rev.* **2018**, *376*, 548–564.
41. Janczyk, A.; Wolnicka-Glubisz, A.; Chmura, A.; Elas, M.; Matuszak, Z.; Stochel, G.; Urbanska, K. *Nitric Oxide* **2004**, *10*, 42–50.
42. Roose, M.; Sasaki, I.; Bukhanko, V.; Mallet-Ladeira, S.; Barba-Barba, R. M.; Ramos-Ortiz, G.; Enriquez-Cabrera, A.; Farfán, N.; Lacroix, P. G.; Malfant, I. *Polyhedron* **2018**, *151*, 100–111.
43. Guo, M.; Xiang, H. J.; Wang, Y.; Zhang, Q. L.; An, L.; Yang, S. P.; Ma, Y.; Wang, Y.; Liu, J. G. *Chem. Commun.* **2017**, *53*, 3253–3256.
44. Jiang, Y.; Fu, P.; Liu, Y.; Wang, C.; Zhao, P.; Chu, X.; Jiang, X.; Yang, W.; Wu, Y.; Wang, Y.; et al. *Sci. Adv.* **2020**, *6*, eabc3513.
45. Sanchez-Cruz, P.; Alegria, A. E. *J. Photochem. Photobiol. A Chemistry* **2016**, *330*, 79–85.
46. Matsuo, K.; Nakagawa, H.; Adachi, Y.; Kameda, E.; Tsumoto, H.; Suzuki, T.; Miyata, N. *Chem. Commun.* **2010**, *46*, 3788–3790.
47. Cink, R. B.; Zhou, Y.; Du, L.; Rahman, M. S.; Phillips, D. L.; Simpson, M. C.; Seed, A. J.; Sampson, P.; Brasch, N. E. *J. Org. Chem.* **2021**, *86*, 8056–8068.
48. (a) Laverman, L. E.; Wanat, A.; Oszajca, J.; Stochel, G.; Ford, P. C.; van Eldik, R. *J. Am. Chem. Soc.* **2001**, *123* (2), 285–293. https://doi.org/10.1021/ja001696z; (b) Laverman, L. E.; Ford, P. C. *J. Am. Chem. Soc.* **2001**, *123* (47), 11614–11622. https://doi.org/10.1021/ja0113910.

49. Pixton, D. A.; Petersen, C. A.; Franke, A.; van Eldik, R.; Garton, E. M.; Andrew, C. R. *J. Am. Chem. Soc.* **2009**, *131*, 4846–4853.
50. Jańczyk, A.; Krakowska, E.; Stochel, G.; Macyk, W. *J. Am. Chem. Soc.* **2006**, *128* (49), 15574–15575. https://doi.org/10.1021/ja065970m.
51. Macyk, W.; Szaciłowski, K.; Stochel, G.; Buchalska, M.; Kuncewicz, J.; Łabuz, P. *Coord. Chem. Rev.* **2010**, *254* (21), 2687–2701. https://doi.org/10.1016/j.ccr.2009.12.037.
52. Wu, M.; Lu, Z.; Wu, K.; Nam, C.; Zhang, L.; Guo, J. *J. Mater. Chem. B* **2021**, *9*, 7063–7075.
53. (a) Dąbrowski, J. M.; Pereira, M. M.; Arnaut, L. G.; Monteiro, C. J. P.; Peixoto, A. F.; Karocki, A.; Urbańska, K.; Stochel, G. *Photochem. Photobiol.* **2007**, *83* (4), 897–903; (b) Pucelik, B.; Arnaut, L. G.; Stochel, G.; Dąbrowski, J. M. *ACS Appl. Mater. Interfaces* **2016**, *8* (34), 22039–22055. https://doi.org/10.1021/acsami.6b07031.
54. (a) Dąbrowski, J. M.; Pucelik, B.; Pereira, M. M.; Arnaut, L. G.; Macyk, W.; Stochel, G. *RSC Adv.* **2015**, *5* (113), 93252–93261. https://doi.org/10.1039/C5RA19742B; (b) Drzewiecka, A.; Urbańska, K.; Matuszak, Z.; Pineiro, M.; Arnaut, L. G.; Habdas, J.; Ratuszna, A.; Stochel, G. *Acta Biochim. Pol.* **2001**, *48* (1), 277–282; (c) Drzewiecka-Matuszek, A.; Skalna, A.; Karocki, A.; Stochel, G.; Fiedor, L. *J. Biol. Inorg. Chem.* **2005**, *10* (5), 453–462. https://doi.org/10.1007/s00775-005-0652-6; (d) Orzeł, Ł.; Kania, A.; Rutkowska-Żbik, D.; Susz, A.; Stochel, G.; Fiedor, L. *Inorg. Chem.* **2010**, *49* (16), 7362–7371. https://doi.org/10.1021/ic100466s; (e) Szczygieł, M.; Urbańska, K.; Jurecka, P.; Stawoska, I.; Stochel, G.; Fiedor, L. *J. Med. Chem.* **2008**, *51* (15), 4412–4418. https://doi.org/10.1021/jm7016368; (f) Zhu, W.; Gao, Y. H.; Liao, P.-Y.; Chen, D.-Y.; Sun, N.-N.; Thi, P. A. N.; Yan, Y.-J.; Wu, X.-F.; Chen, Z.-L. *Eur. J. Med. Chem.* **2018**, *160*, 146–156.
55. Arnaut, L. G.; Formosinho, S. J.; Simoes, S.; Pereira, M. M.; Stochel, G.; Urbanska, K. *Process for Preparing Chlorins and Their Pharmaceutical Uses*; 2013.
56. Martinez De Pinillos Bayona, A.; Mroz, P.; Thunshelle, C.; Hamblin, M. R. *Chem. Bio. Drug Des.* **2017**, *89* (2), 192–206, Editorial. https://doi.org/10.1111/cbdd.12792.
57. Dabrowski, J. M.; Krzykawska, M.; Arnaut, L. G.; Pereira, M. M.; Monteiro, C. J. P.; Simões, S.; Urbańska, K.; Stochel, G. *ChemMedChem* **2011**, *6* (9), 1715–1726. Article. https://doi.org/10.1002/cmdc.201100186.
58. Wang, Y. Y.; Liu, Y. C.; Sun, H.; Guo, D. S. *Coord. Chem. Rev.* **2019**, *395*, 46–62. Review. https://doi.org/10.1016/j.ccr.2019.05.016.
59. Kuncewicz, J.; Dabrowski, J. M.; Kyzioł, A.; Brindell, M.; Łabuz, P.; Mazuryk, O.; Macyk, W.; Stochel, G. *Coord. Chem. Rev.* **2019**, *398*, 113012.
60. MacDonald, I. J.; Morgan, J.; Bellnier, D. A.; Paszkiewicz, G. M.; Whitaker, J. E.; Litchfield, D. J.; Dougherty, T. J. *Photochem. Photobiol.* **1999**, *70* (5), 789–797. Article. https://doi.org/10.1111/j.1751-1097.1999.tb08284.x.
61. Mroz, P.; Bhaumik, J.; Dogutan, D. K.; Aly, Z.; Kamal, Z.; Khalid, L.; Kee, H. L.; Bocian, D. F.; Holten, D.; Lindsey, J. S.; et al. *Cancer Lett.* **2009**, *282* (1), 63–76. Article. https://doi.org/10.1016/j.canlet.2009.02.054.
62. Karges, J. *Angew. Chem.* **2022**, *61* (5). https://doi.org/10.1002/anie.202112236. Review.
63. Pucelik, B.; Paczyński, R.; Dubin, G.; Pereira, M. M.; Arnaut, L. G.; Dąbrowski, J. M. *PLoS One* **2017**, *12* (10). https://doi.org/10.1371/journal.pone.0185984. Article.
64. Sułek, A.; Pucelik, B.; Kobielusz, M.; Barzowska, A.; Dąbrowski, J. M. *Int. J. Mol. Sci.* **2020**, *21* (22), 1–34. Article. https://doi.org/10.3390/ijms21228716.
65. Aroso, R. T.; Calvete, M. J. F.; Pucelik, B.; Dubin, G.; Arnaut, L. G.; Pereira, M. M.; Dąbrowski, J. M. *Eur. J. Med. Chem.* **2019**, *184*. https://doi.org/10.1016/j.ejmech.2019.111740. Article.

66. Monro, S.; Colon, K. L.; Yin, H.; Roque, J.; Konda, P.; Gujar, S.; Thummel, R. P.; Lilge, L.; Cameron, C. G.; McFarland, S. A. *Chem. Rev.* **2019**, *119*, 797–828.
67. Karges, J.; Blacque, O.; Goldner, P.; Chao, H.; Gasser, G. *Eur. J. Inorg. Chem.* **2019**, *32*, 3704–3712. Article. https://doi.org/10.1002/ejic.201900569.
68. Mazuryk, O.; Janczy-Cempa, E.; Łagosz, J.; Rutkowska-Zbik, D.; Machnicka, A.; Krasowska, A.; Pietrzyk, P.; Stochel, G.; Brindell, M. *Dalton Trans.* **2022**, *51* (5), 1888–1900. https://doi.org/10.1039/D1DT02908H.
69. Zhang, Y.; Zhou, Q.; Tian, N.; Li, C.; Wang, X. *Inorg. Chem.* **2017**, *56* (4), 1865–1873. Article. https://doi.org/10.1021/acs.inorgchem.6b02459.
70. Kaspler, P.; Lazic, S.; Forward, S.; Arenas, Y.; Mandel, A.; Lilge, L. *Photochem. Photobiol. Sci.* **2016**, *15* (4), 481–495. Article. https://doi.org/10.1039/c5pp00450k.
71. (a) Lv, Z.; Wei, H.; Li, Q.; Su, X.; Liu, S.; Zhang, K. Y.; Lv, W.; Zhao, Q.; Li, X.; Huang, W. *Chem. Sci.* **2018**, *9*, 502–512; (b) Monro, S.; Scott, J.; Chouai, A.; Lincoln, R.; Zong, R.; Thummel, R. P.; McFarland, S. A. *Inorg. Chem.* **2010**, *49* (6), 2889–2900. https://doi.org/10.1021/ic902427r.
72. McCain, J.; Colón, K. L.; Barrett, P. C.; Monro, S. M. A.; Sainuddin, T.; Roque, J., III; Pinto, M.; Yin, H.; Cameron, C. G.; McFarland, S. A. *Inorg. Chem.* **2019**, *58* (16), 10778–10790. Article. https://doi.org/10.1021/acs.inorgchem.9b01044.
73. (a) Raza, A.; Archer, S. A.; Fairbanks, S. D.; Smitten, K. L.; Botchway, S. W.; Thomas, J. A.; Macneil, S.; Haycock, J. W. *J. Am. Chem. Soc.* **2020**, *142* (10), 4639–4647. Article https://doi.org/10.1021/jacs.9b11313; (b) Karges, J.; Kuang, S.; Ong, Y. C.; Chao, H.; Gasser, G. *Chem. Eur. J.* **2021**, *27* (1), 362–370. Article https://doi.org/10.1002/chem.202003486; (c) Liu, Y. T.; Yin, X.; Lai, X. Y.; Wang, X. *Dyes Pigment* **2020**, *176*. https://doi.org/10.1016/j.dyepig.2020.108244. Article.
74. Son, G.; Lee, B. I.; Chung, Y. J.; Park, C. B. *Acta Biomater.* **2018**, *67*, 147–155. Article https://doi.org/10.1016/j.actbio.2017.11.048.
75. (a) Lan, M.; Zhao, S.; Liu, W.; Lee, C. S.; Zhang, W.; Wang, P. *Adv. Healthcare Mater.* **2019**, *8* (13). https://doi.org/10.1002/adhm.201900132. Review; (b) Poblocki, K.; Drzezdzon, J.; Kostrzewa, T.; Jacewicz, D. *Int. J. Mol. Sci.* **2021**, *22* (15). https://doi.org/10.3390/ijms22158052. Review; (c) Raza, M. K.; Noor, A.; Samantaray, P. K. *Chembiochem* **2021**, *22* (23), 3270–3272. Review. https://doi.org/10.1002/cbic.202100469.
76. Lee, L. C. C.; Tsang, A. W. Y.; Liu, H. W.; Lo, K. W. *Inorg. Chem.* **2020**, *59* (20), 14796–14806. Article. https://doi.org/10.1021/acs.inorgchem.0c01343.
77. (a) Manav, N.; Lone, M. Y.; Raza, M. K.; Chavda, J.; Mori, S.; Gupta, I. *Dalton Trans.* **2022**, *51* (10), 3849–3863. Article https://doi.org/10.1039/d1dt04218a; (b) Kuang, S.; Wei, F.; Karges, J.; Ke, L.; Xiong, K.; Liao, X.; Gasser, G.; Ji, L.; Chao, H. *J. Am. Chem. Soc.* **2022**, *144* (9), 4091–4101. https://doi.org/10.1021/jacs.1c13137.
78. Wu, Y.; Wu, J.; Wong, W. Y. *Biomater. Sci.* **2021**, *9* (14), 4843–4853. Article. https://doi.org/10.1039/d1bm00128k.
79. Liu, B.; Monro, S.; Li, Z.; Jabed, M. A.; Ramirez, D.; Cameron, C. G.; Colón, K.; Roque, J.; Kilina, S.; Tian, J.; et al. *ACS Appl. Bio Mater.* **2019**, *2* (7), 2964–2977. Article. https://doi.org/10.1021/acsabm.9b00312.
80. Kang, J.; Nam, J. S.; Lee, H. J.; Nam, G.; Rhee, H. W.; Kwon, T. H.; Lim, M. H. *Chem. Sci.* **2019**, *10* (28), 6855–6862. Article. https://doi.org/10.1039/c9sc00931k.
81. Aliyan, A.; Kirby, B.; Pennington, C.; Martí, A. A. *J. Am. Chem. Soc.* **2016**, *138* (28), 8686. Article. https://doi.org/10.1021/jacs.6b04411.
82. Choi, H. W.; Kim, J.; Kim, J.; Kim, Y.; Song, H. B.; Kim, J. H.; Kim, K.; Kim, W. J. *ACS Nano* **2016**, *10*, 4199–4208.
83. Xiang, H. J.; An, L.; Tang, W. W.; Yang, S. P.; Liu, J. G. *Chem. Commun.* **2015**, *51*, 2555–2558.

84. Sharma, N.; Arjunan, P.; Marepally, S.; Jain, N.; Naziruddin, A. R.; Ghosh, A.; Mariappan, C. R.; Jose, D. A. *J. Photochem. Photobiol. A Chem.* **2022**, *425*, 113703.
85. Yang, Y.; Huang, Z.; Li, L. L. *Nanoscale* **2021**, *13*, 444–459.
86. Zhang, M.; Qiao, R.; Hu, J. *Nanomaterials* **2020**, *10*, 1134.
87. Liu, Y.; Shi, J. *Nano Today* **2019**, *27*, 146–177. https://doi.org/10.1016/j.nantod.2019.05.008.
88. Szakacs, G.; Hall, M. D.; Gottesman, M. M.; Boumendjel, A.; Kachadourian, R.; Day, B. J.; Baubichon-Cortay, H.; Di Pietro, A. *Chem. Rev.* **2014**, *114* (11), 5753–5774. https://doi.org/10.1021/cr4006236.
89. Yang, B.; Chen, Y.; Shi, J. *Chem. Rev.* **2019**, *119*, 4881–4985.
90. Ogilby, P. R. *Chem Soc Rev.* **2010**, *39*, 3189–3209.
91. Takeda, Y.; Kondow, T.; Mafuné, F. *J. Phys. Chem. B.* **2006**, *9* (110), 2393–2397.
92. (a) Sakamoto, M.; Tachikawa, T.; Fujitsuka, M.; Majima, T. *J. Am. Chem. Soc.* **2008**, *131*, 6–7; (b) Sakamoto, M.; Tachikawa, T.; Fujitsuka, M.; Majima, T. *Langmuir* **2009**, *25* (24), 13888–13893. https://doi.org/10.1021/la901552f.
93. Krpetić, Z.; Nativo, P.; Sée, V.; Prior, I. A.; Brust, M.; Volk, M. *Nano Lett.* **2010**, *10* (10), 4549–4554.
94. Misawa, M.; Takahashi, J. *Nanomed. Nanotechnol. Biol. Med.* **2011**, *7*, 604–614.
95. Vankayala, R.; Sagadevan, A.; Vijayaraghavan, P.; Kuo, C. L.; Hwang, K. C. *Angew. Chem. Int. Ed.* **2011**, *50* (45), 10640–10644. https://doi.org/10.1002/anie.201105236.
96. Vankayala, R.; Huang, Y. K.; Kalluru, P.; Chiang, C. S.; Hwang, K. C. *Small* **2014**, *10* (8), 1612–1622.
97. Zhang, W.; Li, Y.; Niu, J.; Chen, Y. *Langmuir* **2013**, *29*, 4647–4651. https://doi.org/10.1021/la400500t.
98. Pasparakis, G. *Small* **2013**, *9* (24), 4130–4134. https://doi.org/10.1002/smll.201301365.
99. Li, X.; Kim, J.; Yoon, J.; Chen, X. *Adv. Mater.* **2017**, *29* (23), 1606857. https://doi.org/10.1002/adma.201606857.
100. (a) Khaing, O.; Maung, K.; Yang, Y.; Hu, Y.; Gomez, M.; Du, H.; Wang, H. *ACS Nano* **2012**, *6* (3), 1939–1947. https://doi.org/10.1021/nn300327c; (b) Ferreira, D. C.; Monteiro, C. S.; Chaves, C. R.; Sáfar, G. A. M.; Moreira, R. L.; Pinheiro, M. V. B.; Martins, D. C. S.; Ladeira, L. O.; Krambrock, K. *Colloids Surf. B Biointerfaces* **2017**, *150*, 297–307. https://doi.org/10.1016/j.colsurfb.2016.10.042.
101. He, Y.; Gao, Q.; Lv, C.; Liu, L. *J. Photochem. Photobiol. B Biol.* **2021**, *215*, 112102. https://doi.org/10.1016/j.jphotobiol.2020.112102.
102. Fan, W.; Huang, P.; Chen, X. *Chem. Soc. Rev.* **2016**, *45* (23), 6488–6519. https://doi.org/10.1039/C6CS00616G.
103. Nosaka, Y.; Daimon, T.; Nosaka, A. Y.; Murakami, Y. *Phys. Chem. Chem. Phys.* **2004**, *6* (11), 2917–2918. https://doi.org/10.1039/B405084C.
104. Buchalska, M.; Kobielusz, M.; Matuszek, A.; Pacia, M.; Wojtyła, S.; Macyk, W. *ACS Catal.* **2015**, *5* (12), 7424–7431. https://doi.org/10.1021/acscatal.5b01562.
105. Łabuz, P.; Gryboś, J.; Pietrzyk, P.; Sobańska, K.; Macyk, W.; Sojka, Z. *Res. Chem. Int.* **2019**, *45* (12), 5781–5800. https://doi.org/10.1007/s11164-019-04002-z.
106. (a) Sułek, A.; Pucelik, B.; Kobielusz, M.; Łabuz, P.; Dubin, G.; Dąbrowski, J. M. *Catalysts* **2019**, *9* (10), 821; (b) Sułek, A.; Pucelik, B.; Kuncewicz, J.; Dubin, G.; Dąbrowski, J. M. *Catal. Today* **2019**, *335*, 538–549. https://doi.org/10.1016/j.cattod.2019.02.070.
107. Yaemsunthorn, K.; Kobielusz, M.; Macyk, W. *ACS Appl. Nano Mater.* **2021**, *4* (1), 633–643. https://doi.org/10.1021/acsanm.0c02932.
108. Trochowski, M.; Kobielusz, M.; Mróz, K.; Surówka, M.; Hämäläinen, J.; Iivonen, T.; Leskelä, M.; Macyk, W. *J. Mater. Chem. A* **2019**, 7 (43), 25142–25154. https://doi.org/10.1039/C9TA09400H. DOI: 10.1039/c9ta09400h.

109. Matsunaga, T.; Tomoda, T.; Nakajima, T.; Wake, H. *FEMS Microbiol. Lett.* **1985**, *29*, 211–214.
110. (a) Jańczyk, A.; Wolnicka-Głubisz, A.; Urbanska, K.; Stochel, G.; Macyk, W. *J. Photochem. Photobiol. B Biol.* **2008**, *92* (1), 54–58. https://doi.org/10.1016/j.jphotobiol.2008.05.003; (b) Jańczyk, A.; Wolnicka-Głubisz, A.; Urbanska, K.; Kisch, H.; Stochel, G.; Macyk, W. *Free Radic. Biol. Med.* **2008**, *44* (6), 1120–1130. https://doi.org/10.1016/j.freeradbiomed.2007.12.019.
111. Mitoraj, D.; Jańczyk, A.; Strus, M.; Kisch, H.; Stochel, G.; Heczko, P. B.; Macyk, W. *Photochem. Photobiol. Sci.* **2007**, *6* (6), 642–648. https://doi.org/10.1039/B617043A.
112. Regiel-Futyra, A.; Kus-Liśkiewicz, M.; Wojtyła, S.; Stochel, G.; Macyk, W. *RSC Adv.* **2015**, *5* (97), 80089–80097. https://doi.org/10.1039/C5RA12667C.
113. Sadowski, R.; Strus, M.; Buchalska, M.; Heczko, P. B.; Macyk, W. *Photochem. Photobiol. Sci.* **2015**, *14* (3), 514–519. https://doi.org/10.1039/C4PP00270A.
114. Rozhkova, E. A.; Ulasov, I.; Lai, B.; Dimitrijevic, N. M.; Lesniak, M. S.; Rajh, T. *Nano Lett.* **2009**, *9* (9), 3337–3342. https://doi.org/10.1021/nl901610f.
115. Kozinska, A.; Zadlo, A.; Labuz, P.; Broniec, A.; Pabisz, P.; Sarna, T. *Photochem. Photobiol.* **2019**, *95* (1), 227–236.
116. Buchalska, M.; Łabuz, P.; Bujak, Ł.; Szewczyk, G.; Sarna, T.; Maćkowski, S.; Macyk, W. *Dalton Trans.* **2013**, *42* (26), 9468–9475. https://doi.org/10.1039/C3DT50399B.
117. (a) Qi, H. P.; Wang, H. L.; Zhao, D. Y.; Jiang, W. F. *Appl. Surf. Sci.* **2019**,. https://doi.org/10.1016/j.apsusc.2019.02.194; (b) Bian, Z.; Tachikawa, T.; Kim, W.; Choi, W.; Majima, T. *J. Phys. Chem. C* **2012**, *116* (48), 25444–25453. https://doi.org/10.1021/jp309683f; (c) He, W.; Wu, H.; Wamer, W. G.; Kim, K. H.; Zheng, J.; Jia, H.; Zheng, Z.; Yin, J. J. *ACS Appl. Mater. Interfaces* **2014**, *6*, 15527–15535.
118. Jiang, X.; He, W.; Zhang, X.; Wu, Y.; Zhang, Q.; Cao, G.; Zhang, H.; Zheng, J.; Croley, T. R.; Yin, J. J. *J. Phys. Chem. C* **2018**, *122*, 29414–29425.
119. Hu, P.; Wu, T.; Fan, W.; Chen, L.; Liu, Y.; Ni, D.; Bu, W.; Shi, J. *Biomaterials* **2017**, *141*, 86–95. https://doi.org/10.1016/j.biomaterials.2017.06.035.
120. Wang, T.; Zhang, H.; Liu, H.; Yuan, Q.; Ren, F.; Han, Y.; Sun, Q.; Li, Z.; Gao, M. *Adv. Funct. Mater.* **2020**, *30* (3), 1906128. https://doi.org/10.1002/adfm.201906128.
121. Courtney, C. M.; Goodman, S. M.; McDaniel, J. A.; Madinger, N. E.; Chatterjee, A.; Nagpal, P. *Nat. Mater.* **2016**, *15* (5), 529–534. https://doi.org/10.1038/nmat4542.
122. McCollum, C. R.; Levy, M.; Bertram, J. R.; Nagpal, P.; Chatterjee, A. *ACS Biomater. Sci. Eng.* **2021**, 7 (5), 1863–1875. https://doi.org/10.1021/acsbiomaterials.0c01406.
123. (a) Mikrut, M.; Regiel-Futyra, A.; Samek, L.; Macyk, W.; Stochel, G.; van Eldik, R. *Environ. Pollut.* **2018**, *238*, 638–646. https://doi.org/10.1016/j.envpol.2018.03.068; (b) Mikrut, M.; Mazuryk, O.; Macyk, W.; van Eldik, R.; Stochel, G. *J. Environ. Chem. Eng.* **2021**, *9* (6), 106478. https://doi.org/10.1016/j.jece.2021.106478.
124. Gao, L.; Liu, R.; Gao, F.; Wang, Y.; Jiang, X.; Gao, X. *ACS Nano* **2014**, *8* (7), 7260–7271. https://doi.org/10.1021/nn502325j.
125. Wu, N. *Nanoscale* **2018**, *10*, 2679.
126. Guerrero-Florez, V.; Mendez-Sanchez, S. C.; Patrón-Soberano, O. A.; Rodríguez-González, V.; Blach, D.; Martínez, O. F. *J. Mater. Chem. B* **2020**, *8* (14), 2862–2875. https://doi.org/10.1039/D0TB00240B.
127. Siegel, A. L.; Baker, G. A. *Nanoscale Adv.* **2021**, *3* (14), 3980–4004. https://doi.org/10.1039/D0NA01057J.
128. Ye, Z.; Li, C.; Celentano, M.; Lindley, M.; O'Reilly, T.; Greer, A. J.; Huang, Y.; Hardacre, C.; Haigh, S. J.; Xu, Y.; et al. *JACS Au* **2022**, *2* (1), 178–187. https://doi.org/10.1021/jacsau.1c00462.

129. Hu, Y.; Wang, R.; Wang, S.; Ding, L.; Li, J.; Luo, Y.; Wang, X.; Shen, M.; Shi, X. *Sci. Reports* **2016**, *6* (1), 28325. https://doi.org/10.1038/srep28325.
130. (a) Tee, S. Y.; Ye, E. *Mater. Adv.* **2021**, *2*, 1507–1529; (b) Ma, J.; Liu, X.; Wang, R.; Zhang, J.; Jiang, P.; Wang, Y.; Tu, G. *ACS Appl. Nano Mater.* **2020**, *3* (11), 10885–10894. https://doi.org/10.1021/acsanm.0c02144; (c) Ning, P.; Chen, Y.; Bai, Q.; Xu, C.; Deng, C.; Cheng, Q.; Cheng, Y. ACS Appl. Mater. Interfaces, 14 (6), 7551-7564. DOI: https://doi.org/10.1021/acsami.1c13072.
131. Fraix, A.; Sortino, S. *Photochem. Photobiol. Sci.* **2018**, *17*, 1709–1727.
132. Ramos, L. C. B.; Marchesi, M. S. P.; Callejon, D.; Baruffi, M. D.; Lunardi, L. N.; Slep, L. D.; Bendhack, L. M.; da Silva, R. S. *Eur. J. Inorg. Chem.* **2016**, 3592–3597.
133. Carneiro, Z. A.; de Moraes, J. C. B.; Rodrigues, F. P.; de Lima, R. G.; Curti, C.; da Rocha, Z. N.; Paulo, M.; Bendhack, L. M.; Tedesco, A. C.; Formiga, A. L. B.; et al. *J. Inorg. Biochem.* **2011**, *105*, 1035–1043.
134. Heinrich, T. A.; Tedesco, A. C.; Fukuto, J. M.; da Silva, R. S. *Dalton Trans.* **2014**, *43*, 4021–4025.
135. Benkovics, G.; Perez-Lloret, M.; Afonso, D.; Darcsi, A.; Beni, S.; Fenyvesi, E.; Malanga, M.; Sortino, S. *Int. J. Pharm.* **2017**, 531.
136. Maranho, D. S.; de Lima, R. G.; Primo, F. L.; da Silva, R. S.; Tedesco, A. C. *Photochem. Photobiol.* **2009**, *85*, 705–713.
137. Fraix, A.; Gonçalves, A. R.; Cardile, V.; Graziano, A. C. E.; Theodossiou, T. A.; Yannakopoulou, K.; Sortino, S. *Chem. Asian J.* **2013**, *8*, 2634–2641.
138. Caruso, E. B.; Cicciarella, E.; Sortino, S. *Chem. Commun.* **2007**, 5028–5030.
139. Giancane, G.; Valli, L.; Sortino, S. *ChemPhysChem* **2009**, *10*, 3077–3082.
140. Fang, X.; Cai, S.; Wang, M.; Chen, Z.; Lu, C.; Yang, H. *Angew. Chem. Int. Ed.* **2021**, *60* (13), 7046–7050.
141. Guan, Y.; Lu, H.; Li, W.; Zheng, Y.; Jiang, Z.; Zou, J.; Gao, H. *ACS Appl. Mater. Interfaces* **2017**, *9* (32), 26731–26739. https://doi.org/10.1021/acsami.7b07768.
142. Zhang, L.; Wang, D.; Yang, K.; Sheng, D.; Tan, B.; Wang, Z.; Ran, H.; Yi, H.; Zhong, Y.; Lin, H.; et al. *Adv. Sci.* **2018**, *5* (1800049), 1–15.
143. Jung, H. S.; Lee, J.-H.; Kim, K.; Koo, S.; Verwilst, P.; Sessler, J. L.; Kang, C.; Kim, J. S. *J. Am. Chem. Soc.* **2017**, *139* (29), 9972–9978. https://doi.org/10.1021/jacs.7b04263.
144. Shi, C.; Liu, T.; Guo, Z.; Zhuang, R.; Zhang, X.; Chen, X. *Nano Lett.* **2018**, *18* (11), 7330–7342. https://doi.org/10.1021/acs.nanolett.8b03568.

CHAPTER EIGHT

Ruthenium complexes for photoactivated dual activity: Drug delivery and singlet oxygen generation

Sean J. Steinke[a], Jeremy J. Kodanko[b], and Claudia Turro[a,*]
[a]Department of Chemistry and Biochemistry, The Ohio State University, Columbus, OH, United States
[b]Department of Chemistry, Wayne State University, Detroit, MI, United States
*Corresponding author: e-mail address: turro.1@osu.edu

Contents

1. Introduction 286
2. Photodynamic therapy 289
   2.1 The role of singlet oxygen 290
   2.2 Characteristics affecting singlet oxygen production 291
   2.3 Examples of PDT agents 297
3. Photochemotherapy 298
   3.1 Design strategies for Ru(II) polypyridyl photocages 300
   3.2 Examples of PCT agents 302
4. Dual activity 306
   4.1 Derivatives of the prototypical $[Ru(tpy)(Me_2dppn)(py)]^{2+}$ 309
   4.2 Other dual activity complexes from the Turro group 313
   4.3 Dual activity complexes from other research groups 314
5. Concluding remarks 316
Acknowledgment 317
References 317

Abstract

Light-based cancer treatments offer spatiotemporal control absent in traditional chemotherapies through drug activation only upon irradiation, which prevents the widespread toxicity often associated with current anticancer chemotherapeutics. Photoactivated anticancer drugs containing ruthenium(II) are particularly useful due to their tunable coordination sphere, visible light absorption, and rich excited state photochemistry. The two processes available to these anticancer complexes are photodynamic therapy (PDT), which involves the generation of reactive oxygen species to treat cancer, and photochemotherapy (PCT), whereby a therapeutic agent is released from a ruthenium "cage" upon irradiation. Complexes have typically been designed for either PDT or PCT applications, as they are separate photochemical processes,

Advances in Inorganic Chemistry, Volume 80
ISSN 0898-8838
https://doi.org/10.1016/bs.adioch.2022.05.001

but recent progress has been made with "dual activity" complexes which combine PDT and PCT into a single chromophore. Dual activity complexes have demonstrated cancer toxicity greater than PDT or PCT agents alone and are particularly promising candidates for photoinduced cancer treatment. In this chapter, the design strategies, photochemistry, and cellular toxicity of Ru(II) PDT and PCT anticancer agents are reviewed and compared to dual activity complexes.

1. Introduction

Cancer is currently the second leading cause of death in the United States.[1] Early treatments for cancer involved invasive surgery, beginning with the first reported abdominal surgery in the United States, performed by Ephraim McDowell in 1809, in which an ovarian tumor was removed without anesthesia.[2] Advancements in the field of surgery improved the relative survival rates of cancer patients in the United States, but until the middle of the 20th century surgery was the only treatment option and a minority of patients were cured through surgery alone.[3] The development of radiation therapy provided a new, effective therapy, beginning with the successful treatment of laryngeal cancer with fractionated radiation treatment in the 1930s and advanced with the use of cobalt therapy in 1951.[4,5] However, only around one-third of cancers could be successfully treated with the use of surgery and radiation therapy, alone or together, spurring efforts to discover alternative treatments to improve outcomes.[3]

The use of chemicals in therapy, termed chemotherapy by pioneer of the field Paul Ehrlich, is the basis of the "magic bullet" concept—a treatment affecting only the intended disease target without affecting the healthy areas of the patient in which the ailment resides.[6] At the beginning of the 20th century, Erlich's laboratory discovered the therapeutic use of arsphenamine for the treatment of syphilis, one of the first chemotherapeutic agents, during a screening of hundreds of newly synthesized organic compounds, a method of drug discovery that is still used in much of modern pharmaceutical research.[7] The first use of a chemotherapeutic agent for the treatment of cancer was reported in the 1940s. Following World War II, Goodman and Gilman discovered that nitrogen mustards were effective at lymphoid suppression in rabbits.[8–10] In this work published in 1946 by Goodman et al., clinical trials showed that *bis*-(2-chloroethyl)ethylamine and *bis*-(2-chloroethyl)ethylamine, now known as chlormethine or mechlorethamine, were effective for the treatment of Hodgkin's disease and non-Hodgkin lymphoma.[11] The trials provided optimism for the future of chemotherapeutic treatment of cancer and, in the succeeding decades, were followed by the discovery of and successful trials with various biomolecules, organic compounds, and natural products, including a folic acid antagonist for the treatment of leukemia in children and methotrexate for choriocarcinoma treatment.[12,13]

Despite the early focus on organic and natural product drugs for treating cancer, metals have been used for medicine throughout human history. Gold is known to have been used to treat illness in China as far back as 2500 BC and HgCl and $HgCl_2$ were administered for a variety of ailments since the 16th century, including as an antiseptic, disinfectant, and cathartic.[14,15] Metals possess characteristics that make them attractive for use in medicine, including variable coordination numbers, tunable redox states, visible light absorption, and structural variety that make them unique from organic drugs. The expansion of inorganic chemistry research in the 19th and 20th centuries helped expand the role of metals in medicine to include chrysotherapy, chelation therapy, radiopharmaceuticals, and, pertinent to this chapter, chemotherapy.

The discovery of the platinum complex cisplatin, *cis*–$[PtCl_2(NH_3)_2]$ (Fig. 1), remained unknown as a landmark compound in the history of metal-based chemotherapy when it was first reported in 1844.[16] It was following the finding that cisplatin displayed potent antitumor activity against leukemia and sarcoma cells by Rosenberg et al. in 1969 that its potential for use in treating cancer was realized.[17] Following further testing against other tumor cell lines and completing clinical trials, cisplatin was approved for the treatment of testicular and ovarian cancer in the U.S., U.K., and several European countries in the late 1970s.[18] Cisplatin was found to be an extraordinarily effective chemotherapeutic, particularly in the treatment of prostate cancer where the 5-year survival rates increased from around 50% in the 1960s to greater than 95% in 2016.[19,20] Second and third generation alternatives, carboplatin and oxaliplatin (Fig. 1), are also effective

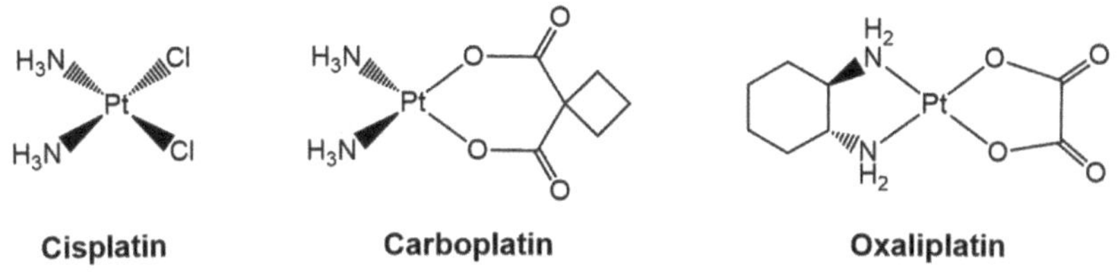

Fig. 1 Molecular structures of cisplatin and its derivatives.

chemotherapeutics, finding success in treating colorectal, ovarian, and head and neck cancers.[21] As of 2005 cisplatin was used in approximately half of all cancer treatments and, combined with later generation Pt-based drugs, metallodrugs are widely used in therapies to this day.[22,23]

Despite the striking increase in the efficacy of treating several types of cancer following the discovery and use of cisplatin, the chemotherapeutic agent has several drawbacks. Ototoxicity, damage to the inner ear and auditory nerve, is a common side effect of cisplatin due to, among other reasons, overproduction of reactive oxygen species (ROS) from the cochlea after interaction with the drug.[24] Due to the potential of permanent hearing damage, clinicians are expected to use well-established ototoxicity monitoring guidelines when administering cisplatin, but a 2021 study found "poor adherence" to guidelines and "minimal data detailing actual ototoxicity monitoring."[25] While auditory damage is common during cisplatin treatment, the primary dose-limiting side effect is nephrotoxicity, damage to the kidneys commonly manifesting in acute kidney injury and hypomagnesemia (abnormally low magnesium blood serum levels).[26] Another drawback to treatments with cisplatin and later generation platinum drugs is intrinsic and acquired tumor resistance. Through mechanisms such as reduced cellular uptake, increased DNA repair processes, and apoptosis resistance in cancerous tumors, patients that initially benefit from Pt-based therapy often become resistant to further treatment.[27] The persistence of side effects, together with tumor resistance to treatments with platinum-based drugs, has led to the search for alternative therapeutic agents.

The approach taken by the Turro group is light activation of chemotherapeutic agents. Activating anticancer drugs that are only operative following light irradiation provides spatial and temporal selectivity absent in traditional cancer treatments and has the potential to reduce negative, systemic side effects. Ruthenium(II) polypyridyl complexes possess strong absorption of visible light, well known excited state photophysical properties, can be easily modified *via* variations in the ancillary ligands, and labile ligands that can be dissociated through the use of light, making them attractive candidates for light-based cancer treatments.

Two methods of light-based treatment with ruthenium(II) polypyridyl complexes are photodynamic therapy (PDT) and photochemotherapy (PCT). Chemotherapeutic Ru(II) drugs that kill cancerous cells through the production of reactive oxygen species (ROS) following irradiation are commonly termed PDT agents, while complexes undergoing PCT utilize light to dissociate a ligand, typically an organic drug or inhibitor.

Complexes that combine PCT and PDT activity into one drug are termed "dual activity" systems and are particularly promising candidates for cancer treatment. The Turro group, in collaboration with the Dunbar and Baptista groups, published the first example of a dual activity complex, *cis*-[Ru(bpy)(dppn)$(CH_3CN)_2]^{2+}$ (dppn = benzo[*i*]dipyrido[3,2-a:2′,3′-c]phenazine, bpy = 2,2′-bipyridine), which showed significant activity against HeLa cells owing to its ability to produce ROS and generate a Ru(II) fragment capable of binding to biomolecules upon irradiation with visible light. The Turro group later developed a pyridine-releasing dual activity model Ru(II) complex, [Ru(tpy)(Me_2dppn)(py)]$^{2+}$ (tpy = 2,2′:6′,2″-terpyridine; Me_2dppn = 3,6-dimethylbenzo[*i*]dipyrido[3,2-a:2′,3′-c]phenazine; py = pyridine), that also produces $^1O_2$ upon irradiation with visible light. This finding became an inspiration for many dual activity complexes that deliver pyridine-containing drugs and inhibitors and will be discussed along with its derivatives and the future outlook for dual activity complexes.

2. Photodynamic therapy

Photodynamic therapy (PDT) involves the use of light to excite a photosensitizer which can subsequently interact with a substrate to kill a cancerous cell, typically through the production of $^1O_2$ or other ROS. For a PDT agent to be effective, it must satisfy several criteria including: low toxicity in the dark, stability in solution, efficient production of ROS or energy/electron transfer, and the ability to selectively accumulate in tumor tissue. Due to the absorption of water and biomolecules, the penetration of light through human tissue is wavelength dependent and results in the optimal absorption range of 600–900 nm, termed the PDT window, in which PDT agents must absorb light for maximum effectiveness.

There are three types of reactions that PDT agents can undergo, each involving an excited photosensitizer but inducing therapeutic effects through different mechanisms. Electron transfer to molecular oxygen or other oxygen-containing molecules to generate cytotoxic ROS is a Type I mechanism. Type II involves energy transfer to $^3O_2$ to generate $^1O_2$ and is the mechanism employed by all clinically approved PDT agents. Unlike the previous two mechanisms, Type III does not require the presence of O_2 nor oxygen-containing molecules but instead involves direct transfer of energy or an electron from the excited photosensitizer to its target.[28] Due to the prevalence of type II PDT drugs and current absence of clinically approved type I or III chemotherapeutics, type II mechanisms will be the focus of the remainder of this section and will be termed PDT agents.

2.1 The role of singlet oxygen

Singlet oxygen, $^1O_2$, is a reactive, electronically excited species commonly employed to induce cell death. Ground state oxygen ($^3O_2$) has the term symbol $^3\Sigma_{g-}$ and following excitation can populate two singlet excited states, the $^1\Sigma_{g+}$ state is 37 kcal above the ground state and rapidly decays to the longer-lived singlet excited state $^1\Delta_g$, which is 22 kcal above the ground state (Fig. 2A and B). The phrase "singlet oxygen" refers to diatomic oxygen in the $^1\Delta_g$ state, which is a reactive species commonly employed to induce cell death. With the exception of the production of $^1O_2$ by eosinophils, a component of the immune system that uses ROS to combat infectious agents, $^1O_2$ is not found in human tissue.[29] Once produced, singlet oxygen can be relatively long-lived within a cell, with lifetimes up to ~3 μs.[30] This relatively long-lived excited state allows for its diffusion throughout the cell, estimated to be capable of traveling ~268 nm in 6 μs in the absence of organelles, which would otherwise be oxidatively damaged, and adds to the reactivity of $^1O_2$; this is a characteristic that may contribute to the tissue damage observed in the eosinophil-related disease Hypereosinophilic Syndrome.[31,32]

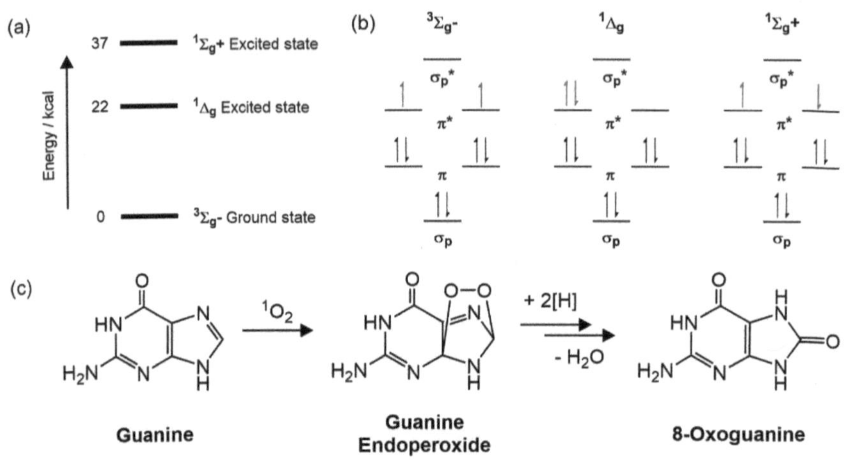

Fig. 2 (A) Simplified energy diagram for the ground and first two excited states of O_2, (B) diagram of the molecular orbitals formed by the p-orbitals of O_2 in the $^3\Sigma_{g-}$, $^1\Delta_g$, and $^1\Sigma_{g+}$ states, and (C) the oxidation of guanine by $^1O_2$ to form 8-oxoguanine.

Singlet oxygen can damage cells by oxidizing lipids, proteins, DNA, and other biomolecules, typically *via* cycloaddition.[33] With DNA in particular, $^1O_2$ selectively oxidizes guanine (G, Fig. 2C), which can lead to single-strand breakages at G bases and eventually lead to cell death.[34] In cancerous cells, this deleterious effect can be used to treat cancerous tumors, a strategy available to type II PDT agents. Due to its high reactivity and short travel length *in vivo*, $^1O_2$ produced in or around cancerous cells reacts within the target tissue, the tumor site, without damaging healthy tissue, achieving the selective and localized treatment desired with PDT.

The location of $^1O_2$ production within a cell influences the potency of its cytotoxicity. A 2020 study using dye-activating proteins targeted to specific subcellular sites in the normal kidney cell lines HEK293 and to the mitochondria in the colorectal tumor cell line CT26 found notable differences in cell death responses based on sub-cellular localization and cell type.[35] In the HEK293 cell line, cells with $^1O_2$ production localized to the mitochondria and nucleus showed the highest resistance to cell death, while a ~threefold increase in cell death was observed in cells where the $^1O_2$ production was localized to the cytosol, endoplasmic reticulum, and cell membrane. The colorectal tumor cell line was remarkably resistant to $^1O_2$ production in the mitochondria, maintaining ~60% cell viability regardless of light dose, while HEK293 cells were reduced to 28% viability under identical conditions. These results are contrary to previous findings, which found ROS generation in the mitochondria or lysosome to be most effective at killing cancerous cells.[36] It is apparent that further research is needed regarding location-dependent ROS production and sub-cellular localization data, and possibly organelle-directing agents, should be utilized in future work to produce the most pronounced therapeutic effect.

2.2 Characteristics affecting singlet oxygen production

In type II PDT agents, singlet oxygen is produced through energy transfer from an excited photosensitizer to ground state $^3O_2$. A Jablonski diagram depicting the production of $^1O_2$ from an organic PDT agent is shown in Fig. 3. Both organic and metal-based PDT agents transfer energy to $^3O_2$ from a long-lived triplet excited state (T_1 or ^{3}ES), which is populated following excitation from the ground state and intersystem crossing (k_{ISC}) from a singlet excited state (S_1 or ^{1}ES). The efficiency of singlet oxygen production from each photon absorbed, termed the singlet oxygen quantum yield (Φ_Δ), is affected by several factors including the local oxygen

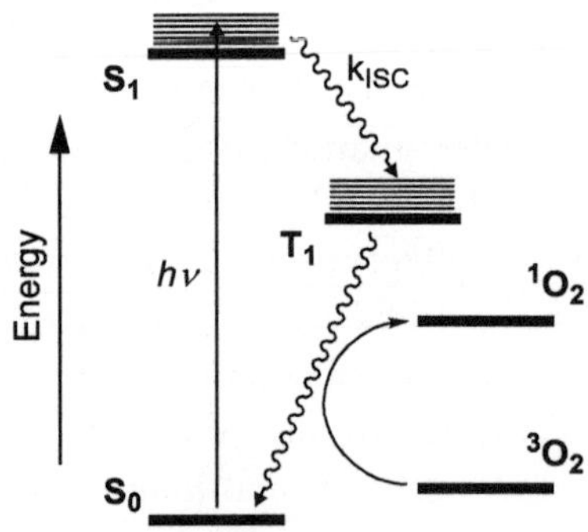

Fig. 3 Jablonski diagram depicting the production of $^1O_2$ from a typical organic PDT agent.

concentration, quantum yield of intersystem crossing (Φ_{ISC}), ^{3}ES lifetime, and the energy of the ^{3}ES. In addition to efficiently producing $^1O_2$, PDT agents must also absorb light in the PDT window to achieve effective photoactivated cancer treatment.

In many small organic molecules lacking heavy atoms, the Φ_{ISC} to the triplet state is relatively low, 0.39 in benzene for example, and the ^{1}ES deactivates through fluorescence in high yield.[37] The addition of heteroatoms, such as nitrogen and sulfur, and heavy atoms, such as iodine and bromine, to aromatic organic molecules results in an increase of Φ_{ISC}. An example of this effect can be found in two pyridyl derivatives of the dye dipyrrometheneboron difluoride (BODIPY), py-BODIPY-H and py-BODIPY-I. Addition of two iodine atoms reduces the fluorescence quantum yield of py-BODIPY-I by nearly half in CH_3CN as compared to py-BODIPY-H and allows only the iodinated dye to populate the long-lived ^{3}ES and produce singlet oxygen with $\Phi_\Delta = 0.75(8)$.[38,39] Two well studied FDA-approved PDT drugs are 5-aminolevulinic acid, marketed as Levulan, and Photofrin, a porphyrin oligomer. Upon entering the body, Levulan is metabolized to protoporphyrin IX, which has $\Phi_\Delta = 0.77$ and while unreported, Φ_{ISC} is likely relatively high due to the conjugation in the molecule.[40,41] Photofrin has been reported to have $\Phi_{ISC} \sim 0.8$ and Φ_Δ from 0.65 to 0.85 and is still one of the most commonly administered PDT drugs since its approval in the 1990s.[42–44] The majority of metal-based PDT drugs being researched are ruthenium(II)-based which, due to the high spin-orbit coupling of the second row transition metal, have $\Phi_{ISC} \sim 1$ and excited state deactivation through fluorescence is not observed, such that it is not a barrier to producing singlet oxygen.

Ru(II) polypyridyl complexes for use in photodynamic therapy are able to produce $^1O_2$ from two different triplet excited states, a triplet metal-to-ligand charge transfer (^{3}MLCT) state and a ligand-centered $^3\pi\pi^*$ state, as shown in Fig. 4. These excited states are accessed following excitation from the ground state, populating a ^{1}MLCT state, followed by rapid intersystem crossing to the ^{3}MLCT state and, if a ligand with a triplet state of appropriate triplet energy is coordinated to ruthenium, internal conversion (IC) to a ligand-centered $^3\pi\pi^*$ state. The triplet state that generates $^1O_2$ is that which is lowest in energy in a given complex. $[Ru(bpy)_3]^{2+}$ (bpy = 2,2′-bipyridine), an archetypal Ru(II) polypyridyl complex, exhibits a ^{3}MLCT lifetime of 1.0 μs in D_2O and produces singlet oxygen with $\Phi_\Delta = 0.22$.[45,46] In contrast, $[Ru(tpy)_2]^{2+}$ (tpy = 2,2′;6′,2″-terpyridine) is another Ru(II) polypyridyl complex with a lowest energy ^{3}ES that is MLCT in nature, but due to steric distortion to the pseudo-octahedral geometry induced by the tridentate tpy ligand, its lifetime is only 250 ps at room temperature in aqueous solution, and it does not produce $^1O_2$, highlighting the importance of a long-lived ^{3}ES.[47] One strategy that has been employed to increase the lifetime of ^{3}MLCT states and subsequently "turn on" singlet oxygen production is to decrease distortion around the metal center, thus reducing the rate of deactivation of the ^{3}MLCT state through thermal population of a low-lying ^{3}LF (ligand field) state. In the case of $[Ru(tpy)_2]^{2+}$, the substitution of the peripheral pyridine units in tpy for quinoline to create 2,6-di(quinolin-8-yl)pyridine (dqp) increased the bite angle from 158° in tpy to 178° in dqp, as seen in Fig. 5, resulting

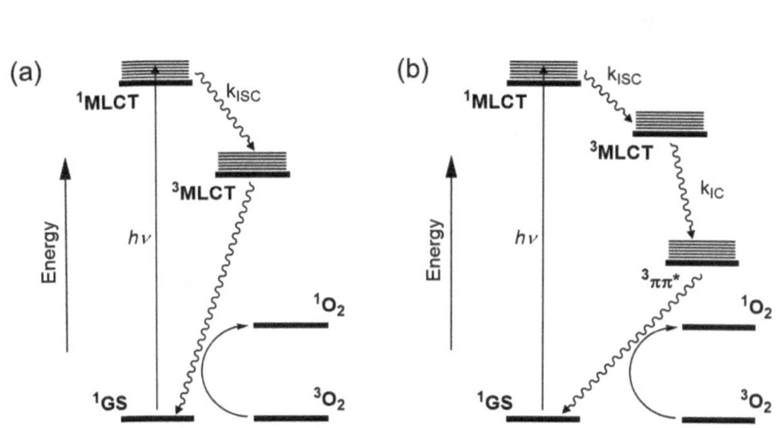

Fig. 4 Simplified Jablonski diagrams depicting the production of $^1O_2$ from (A) ^{3}MLCT and (B) $^3\pi\pi^*$ states of generic Ru(II) polypyridyl complexes.

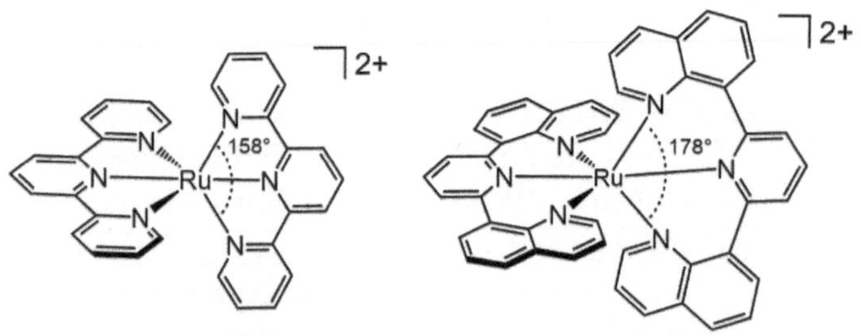

Fig. 5 Schematic representations of $[Ru(tpy)_2]^{2+}$ and $[Ru(dqp)_2]^{2+}$ with the bite angles of the tridentate tpy (left) and dqp (right) ligand coordinated to Ru(II).

in an increase of the room temperature ^{3}MLCT lifetime to 3.0 μs for $[Ru(dqp)_2]^{2+}$ in 4:1 EtOH:MeOH solution.[48]

Ligand-centered $^3\pi\pi^*$ states are the other ^{3}ES in Ru(II) complexes commonly employed to produce singlet oxygen. Extending the conjugation of a ligand can lower the triplet energy of its ligand-centered states such that, upon coordination to ruthenium, the lowest energy triplet excited state of the complex is localized on the aromatic ligand. For example, the "light switch" complex $[Ru(bpy)_2(dppz)]^{2+}$ (dppz = dipyrido-[3,2-*a*:2′,3′-*c*]phenazine, Fig. 6) has lowest energy ^{3}ES that is MLCT in nature with $\tau = 750$ ns in acetonitrile.[49] The addition of another phenyl ring to the dppz ligand to create the dppn ligand, benzo[*i*]dipyrido[3,2-*a*:2′,3′-*c*]-phenazine, results in a dppn-centered lowest energy ^{3}ES that is $^3\pi\pi^*$ in nature. The lifetime of $[Ru(bpy)_2(dppn)]^{2+}$, shown in Fig. 6, in acetonitrile increases to 33 μs, similar to that of the free dppn ligand in solution.[49] The increased lifetimes also impact the efficiency of singlet oxygen production, with $\Phi_\Delta = 0.16(2)$ for $[Ru(bpy)_2(dppz)]^{2+}$ and $\Phi_\Delta = 0.88(5)$ for the longer lived $[Ru(bpy)_2(dppn)]^{2+}$.[49] A similar effect is evident when comparing $[Ru(tpy)_2]^{2+}$ to $[Ru(tpy)(pydppn)]^{2+}$ (pydppn = 3-(pyrid-2′-yl)-4,5,9,16-tetraaza-dibenzo[*a*,*c*]naphthacene, Fig. 6), whereas the short-lived ^{3}MLCT state of the former does not produce $^1O_2$, the substitution of one tpy ligand for the pydppn ligand with an extended π-system results in a lowest energy $^3\pi\pi^*$ state localized on the pydppn ligand with $\tau = 20$ μs and $\Phi_\Delta = 0.92(2)$.[50]

The energy of a ^{3}ES is also important to singlet oxygen production, as sensitization requires ~0.97 eV.[33] Efforts to modify the geometry around the Ru(II) center or increase conjugation in ligands to extend ^{3}ES lifetimes can also impact the energy of the excited states. In the case of $[Ru(dqp)_2]^{2+}$,

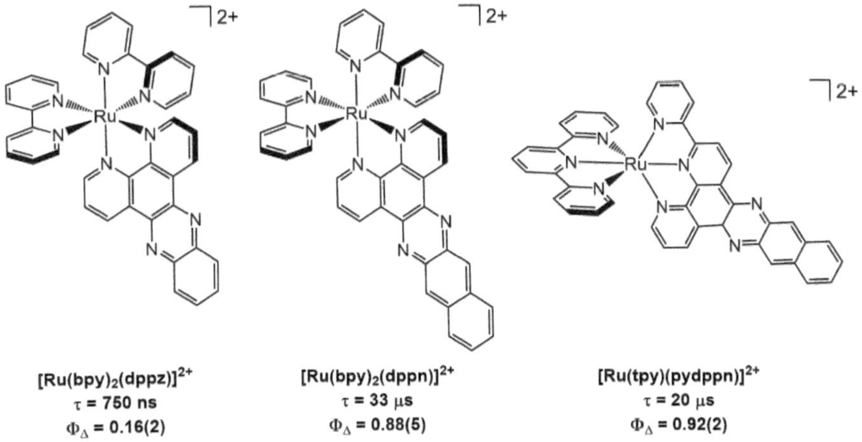

Fig. 6 Schematic representations of the molecular structures of $[Ru(bpy)_2(dppz)]^{2+}$, $[Ru(bpy)_2(dppn)]^{2+}$, and $[Ru(tpy)(pydppn)]^{2+}$ with their corresponding lowest energy ^{3}ES lifetimes and Φ_Δ values.

while increasing the tridentate ligand bite angle successfully increased the 298 K ^{3}MLCT lifetime 12,000-fold as compared to $[Ru(tpy)_2]^{2+}$, it also reduced the energy of the state from 2.07 eV in the latter to 1.84 eV in the former.[48] The change in the nature of the ^{3}ES from MLCT in $[Ru(bpy)_2(dppz)]^{2+}$ to $\pi\pi^*$ in $[Ru(bpy)_2(dppn)]^{2+}$ resulted in a decrease of the energy of the lowest energy excited state that produces $^1O_2$ from ~2.1 to ~1.3 eV, respectively.[50,51] While the lower energies of the aforementioned excited states are still sufficient to produce $^1O_2$, the effect of ligand triplet energies on the ^{3}ES properties of Ru(II) polypyridyl complexes once coordinated must be taken into consideration when designing PDT agents. A recent study on a series of Ru(II) polypyridyl complexes containing an imidazo[4,5-*f*][1,10]phenanthroline (IP) ligand with a number of substituents R (IP-R), demonstrated a direct correlation between the triplet energy of the IP-R ligand and both the resulting ^{3}ES energy and Φ_Δ.[52]

Steady state electronic absorption in the PDT window (600–900 nm) must also be considered when designing PDT agents. To achieve spatiotemporal selectivity in treatment by irradiating drugs through the patient's skin or using an endoscope to access internal organs, such as esophagus, stomach, bladder, and intestines, complexes must be capable of sensitizing $^1O_2$ using light in this range, as other wavelengths are absorbed by biomolecules and water or otherwise scattered by human tissue, thus reducing the depth of light penetration. Ligands with extended π-systems typically have

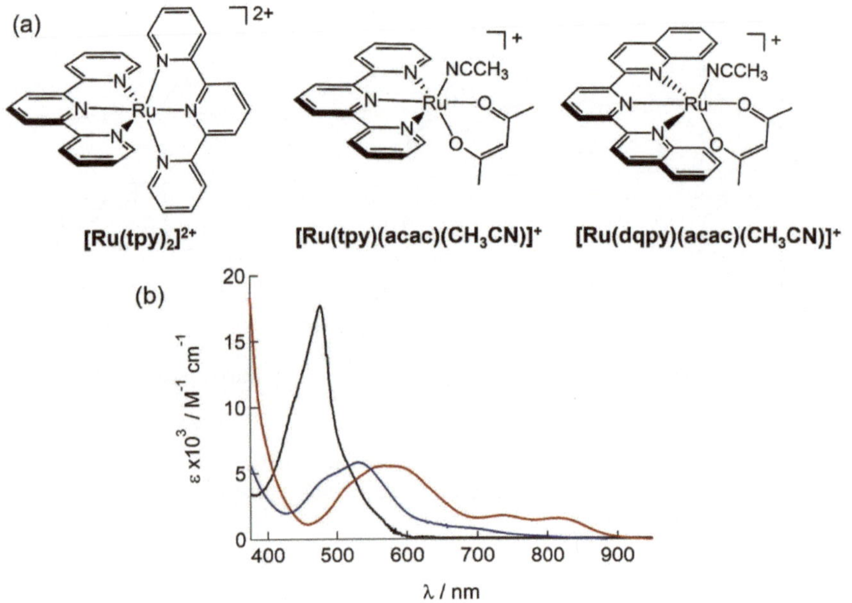

Fig. 7 (A) Molecular structures and (B) electronic absorption spectra of $[Ru(tpy)_2]^{2+}$ (black), $[Ru(tpy)(acac)(CH_3CN)]^+$ (blue), and $[Ru(dqpy)(acac)(CH_3CN)]^+$ (red) in acetone.

red-shifted absorption spectra, fortuitous for PDT complexes designed to have $^3\pi\pi^*$ states capable of producing singlet oxygen, as evident when comparing $[Ru(tpy)_2]^{2+}$ and $[Ru(tpy)\text{-}(pydppn)]^{2+}$ (Fig. 7), which absorb light as far red as ~600 and ~650 nm, respectively.[50,53] Manipulating the geometry around the ruthenium metal center also affects the electronic absorption spectra, although changes are primarily observed in the ligand-centered transitions in the UV region and any changes to the ^{1}MLCT absorptions are nearly imperceptible.[48] Another method of red-shifting absorption in Ru(II) complexes is the modification of the electronic properties of ligands. In a series of $[Ru(tpy)(L)(CH_3CN)]^+$ complexes, where L = acetylacetonate with various substitutions (acac-R), the electron donating nature of L destabilizes highest-occupied Ru(dπ) orbitals and bathochromically shifts the electronic absorption of the complexes, tailing out past 700 nm and well into the PDT window.[54] The effect of conjugation and electron-donating ability together is evident in $[Ru(dqpy)(acac)(CH_3CN)]^+$ (dqpy = 2,6-di (quinolin-2-yl)pyridine), a complex where the peripheral pyridine units of tpy in $[Ru(tpy)(acac)(CH_3CN)]^+$ are substituted for quinoline moieties, resulting in a red-shift of the absorption as far as ~900 nm in acetone.[55]

While $[Ru(dqpy)(acac)(CH_3CN)]^+$ has significant absorption in the PDT window, many of the complexes mentioned herein have relatively low extinction coefficients for absorption in the target range. However, the McFarland group found that, in a series of PDT complexes with >1 μs excited state lifetimes, even those with $\varepsilon < 100\ M^{-1}\ cm^{-1}$ at red wavelengths were toxic toward HL-60 cells with 625 nm irradiation.[56] This is a promising result for Ru(II) PDT agents, as it points to the clinical usefulness of potential PDT drugs with even minimal absorption in the PDT window.

2.3 Examples of PDT agents

There are a number of organic photosensitizers undergoing clinical trials or approved for clinical use, largely split into three classes: porphyrins, chlorins, and other dyes.[57] The aforementioned Photofrin, a mixture of hematoporphyrin and its oligomers, and the metabolic precursor to heme 5-aminolevulinic acid, sold as Levulan, are examples of porphyrin-based PDT drugs that have been approved for clinical use in the United States and their structures are shown in Fig. 8.[58] Talaporfin, sold as Laserphyrin,

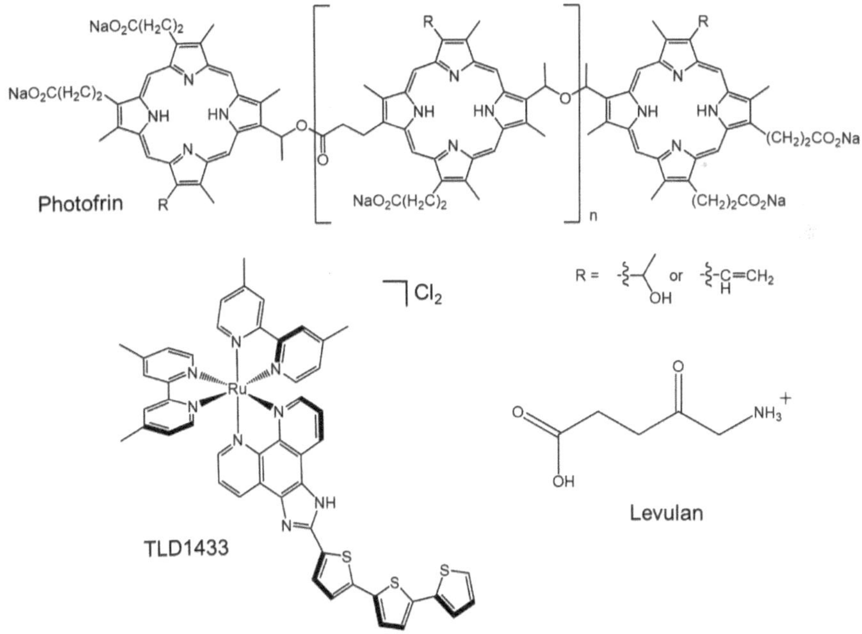

Fig. 8 Molecular structures of Photofrin (top), TLD1433 (bottom left), and Levulan (bottom right).

is a chlorin-based photosensitizer that has been approved for the treatment of lung cancer in Japan and is currently in clinical trials for its use in colorectal neoplasms and liver cancer.[59] There are currently no clinically approved dyes for PDT applications that are not porphyrin- or chlorin-based, but many are undergoing clinical trials or preclinical development, including dyes based on BODIPY which are promising candidates due to their high extinction coefficients and photostability, in addition to sensitizing $^1O_2$.[60]

Several ruthenium based anticancer drugs have undergone clinical trials, including NAMI-A, KP1019, and KP1339, but each encountered issues including dose related side effects and low antitumor activity.[61–63] The only Ru(II) polypyridyl complex capable of sensitizing $^1O_2$, and therefore suitable as a PDT agent, that is undergoing clinical trials in humans is TLD1433 and its structure is shown in Fig. 8. The lowest energy excited state in TLD1433 is ^{3}IL (intraligand) in nature, contains significant $\pi\pi^*$ character, and is localized on the terthienyl group separated from the ruthenium core by linkers. This complex has substantial aqueous solubility, often an issue with Ru(II) complexes containing large aromatic ligands, highly efficient ROS production (Φ_Δ not published), and displays remarkably selective toxicity toward cancerous cells upon irradiation.[64,65] TLD1433 successfully completed Phase I trials (ClinicalTrials.gov Identifier: NCT03053635) in humans for the treatment of non-muscle-invasive bladder cancer in 2018. The PDT agent entered Phase II trials (ClinicalTrials.gov Identifier: NCT03945162) in 2019 with a larger number of patients, is currently ongoing with a projected completion date of May 2023, and bodes well for the future of Ru(II) polypyridyl PDT research.

3. Photochemotherapy

Photochemotherapy (PCT) is the use of light to convert a complex that is non-toxic in the dark into a cytotoxic photoproduct(s) *via* light induced ligand substitution or other photochemical reaction, whereby molecular bonds are made or broken as a primary result of light absorption. Effective PCT agents are stable in solution in the dark, but undergo a photochemical reaction upon visible light irradiation. The earliest recorded use of a PCT agent dates back to ~2000 BC, when psoralens and long wave ultraviolet radiation (UV-A) were used in combination in India and China to treat skin diseases.[66] This therapy, termed "PUVA," is still used in clinical settings today for the treatment of psoriasis, vitiligo, and mycosis fungoides, a type of cutaneous T-cell lymphoma.[67] Transition metal PCT agents, such as those of Ru(II), typically undergo the photoinduced exchange of a ligand, an organic drug or inhibitor, for a solvent water molecule in biological contexts. The drug or inhibitor is usually said to be "photocaged" by ruthenium, bound by a stable bond that is broken only following selective irradiation. Inactivation of drugs by binding them to a chemical scaffold that results in a molecule with little to no biological activity, typically called a prodrug or "caging," that are later activated by bioconversion or photochemical reaction is a technique that has been used in pharmaceutical development since the mid-20th century.[68,69] Organic prodrugs achieve improved solubility, stability, targeting, and pharmacological activity following *in vivo* conversion using the human body's own functions.[69,70] The same conversion is achieved in ruthenium-based PCT agents using irradiation and by combining spatially localized irradiation with the release of drugs or inhibitors that target cancerous tumors, such that significant selectivity in treatment may be achieved without systemic side effects to the patient.

One advantage of photochemotherapy agents is that they do not rely on the presence of O_2 to function, which can be a drawback for PDT agents because of the prevalence of hypoxia in tumors.[71] However, researchers investigating the singlet oxygen sensitization of several porphyrin- and chlorin-based photosensitizers found little variation in Φ_Δ when the concentration of dissolved oxygen in a buffer was varied between 1% and 20%, indicating the impact of hypoxic environments on PDT activity is complex and PDT agents cannot be assumed to be inactive even with low $[O_2]$.[41] Indeed, TLD1433 was found to lose all PDT activity against a glioblastoma cell line (U87) under hypoxic conditions but *increased* its antibacterial activity against *Staphylococcus aureus* and methicillin-resistant *S. aureus* (MRSA).[72,73]

Ruthenium PCT agents are typically thermally stable in solution in the dark and successfully cage drugs/inhibitors as ligands until irradiated. As seen in the Jablonski diagram in Fig. 9, after visible light excitation, ligands in Ru(II) polypyridyl complexes are generally understood to dissociate following thermal population of a higher energy ^{3}LF state from the lowest energy ^{3}MLCT state, where the former possess Ru-ligand σ^* anti-bonding character. The efficiency of the ligand exchange following irradiation at a given wavelength, Φ_λ, is dependent on the energy barrier (E_a) to populating the ^{3}LF state and the nature of the orbitals comprising the excited states. These are characteristics that can be modified by varying the substitutionally inert ancillary ligands coordinated to the ruthenium metal center. Many

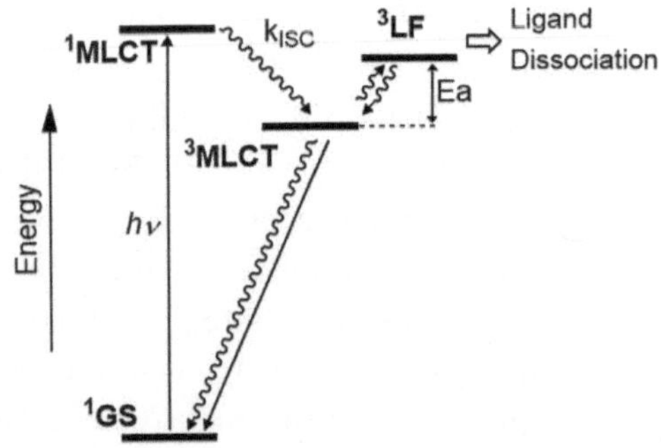

Fig. 9 Simplified Jablonski diagram depicting the photodissociation of a ligand from the ^{3}LF state of a generic Ru(II) polypyridyl complex.

organic drugs and inhibitors contain nitrile motifs or nitrogen-containing heterocycles (imidazole, pyridine, triazole, triazine, etc.), which can coordinate to and undergo photodissociation from ruthenium. The majority of proof of concept complexes are designed using acetonitrile or pyridine as model ligands to later be replaced by clinically relevant molecules.[74,75]

3.1 Design strategies for Ru(II) polypyridyl photocages

Ru(II) polypyridyl complexes are low-spin with a d^6 electron configuration and typically feature a pseudo-octahedral geometry around the metal center. As mentioned previously, distortions away from this octahedral coordination sphere through the introduction of steric strain can reduce the ^{3}MLCT lifetime of Ru(II) complexes by increasing the rate of deactivation through a ^{3}LF state. However, this effect is achieved by decreasing the energy barrier to populating the ^{3}LF state from the ^{3}MLCT state, which generally also increases the efficiency of ligand exchange in Ru(II) complexes and is desirable for a PCT agent. This property has been used as a strategy to increase ligand exchange quantum yields, as in the case of the series of $[Ru(tpy^*)(L)(MeOBN)]^{2+}$, where tpy* = 4′-(3,5-ditertiobutylphenyl)-2,2′:6′,2″-terpyridine, L = phen (1,10-phenanthroline) or dmp (2,9-dimethyl-1,10-phenanthroline), and MeOBN = 2,6-dimethoxybenzonitrile.[76] For the complex with phen, a sterically undemanding ligand, the ligand dissociation quantum yield with 476 nm light, Φ_{476}, in pyridine is 0.0035 (9). The addition of two methyl groups with dmp induces steric strain around ruthenium, reducing the energy barrier to populating the ^{3}LF state, and increasing the value of Φ_{476} 22-fold to 0.079(15). This strategy has also been employed to increase the efficiency of pyridine photosubstitution, where the quantum yield (Φ_{500}) of pyridine exchange in $[Ru(tpy)(bpy)(py)]^{2+}$ in CH_3CN was found to be <0.0001.[77] Substitution of the bpy ligand in this complex for 2,2′-biquinoline (biq) and 4,4′-dimethyl-2,2′-bipyridine (Me_2bpy), more sterically demanding ligands, increased the values of Φ_{500} to 0.033(1) and 0.16(1), respectively. Later experiments aimed at investigating the excited state dynamics of $[Ru(tpy)(Me_2bpy)(py)]^{2+}$ with ultrafast transient absorption spectroscopy, demonstrated that the steric distortion around ruthenium is sufficiently significant that the dissociative ^{3}LF state is lower in energy than the ^{3}MLCT state in the latter, depicted in Fig. 10, which explains its efficient ligand exchange.[78]

The parentage and contributions from the molecular orbitals to the lowest energy excited state of Ru(II) polypyridyl complexes also impacts the efficiency of ligand exchange. In two bis-acetonitrile Ru(II) complexes bearing tetradentate ligands, $[Ru(cyTPA)(CH_3CN)_2]^{2+}$ (cyTPA = 2,2′-((2*R*,6*S*)-1-(pyridin-2-ylmethyl)piperidine-2,6- diyl)dipyridine) and $[Ru(1\text{-isocyTPQA})(CH_3CN)_2]^{2+}$ (1-isocyTPQA = 1-(((2*R*,6*S*)-2,6-di-(pyridin-2-yl)piperidin-1-yl)methyl)isoquinoline), the ^{1}GS HOMOs were found to possess 71% and 62% Ru d-orbital character, respectively.[79] The quantum yield for the exchange of one CH_3CN ligand was reported to be Φ_{400} = 0.0066(3) for $[Ru(cyTPA)(CH_3CN)_2]^{2+}$, whose lowest energy ^{3}ES was calculated to be LF in nature, and Φ_{400} = 0.033(3) for $[Ru(1\text{-isocyTPQA})(CH_3CN)_2]^{2+}$ where a mixed ^{3}MLCT/$^3\pi\pi^*$ state was the lowest energy ^{3}ES. While the former complex possessed a ^{3}LF state as its lowest energy excited state with Ru–$NCCH_3$ anti-bonding character, this state was not observed in ultrafast TA experiments, likely because the energy barrier to its population is too large and cannot be overcome at room temperature. The greater ligand mixing in the ^{1}GS HOMO of the

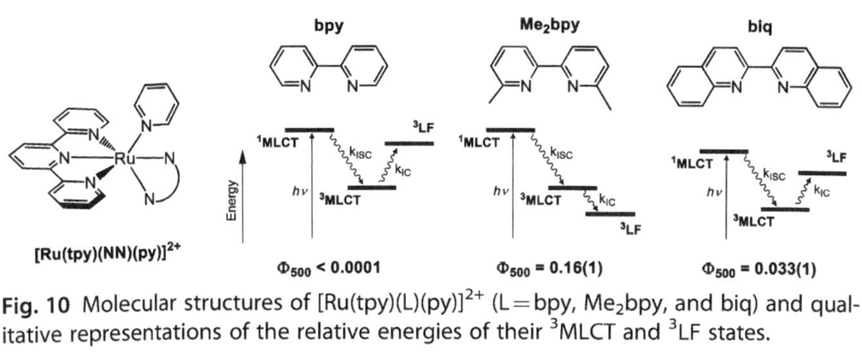

Fig. 10 Molecular structures of $[Ru(tpy)(L)(py)]^{2+}$ (L = bpy, Me_2bpy, and biq) and qualitative representations of the relative energies of their ^{3}MLCT and ^{3}LF states.

1-isocyTPQA complex was determined to lead to a *trans*-type influence being exerted on CH_3CN in the excited state and the origin of greater ligand exchange efficiency.

In a series of $[Ru(tpy)L(CH_3CN)]^{n+}$ complexes, where L is a series of bidentate ligands of varied electron-donating ability, a similar pattern was identified.[80] An inverse relationship between the %Ru d-orbital character in the ^{1}GS HOMO of the complexes and the value of Φ_{450} for CH_3CN exchange was identified - as the ligand mixing increased, so did the efficiency of ligand exchange. In particular, $[Ru(tpy)(acac)(CH_3CN)]^+$ has $\Phi_{450} = 0.014(1)$, 61.3% Ru d-orbital character in the ^{1}GS HOMO, and absorption tailing to ~700 nm. As mentioned previously, the substitution of dqpy for tpy red-shifts the absorption such that it reaches 900 nm and does not notably increase steric strain around ruthenium, but $[Ru(dqpy)(acac)(CH_3CN)]^+$ has $\Phi_{450} = 0.0096(8)$ and 53% Ru d-orbital character in its ^{1}GS HOMO.[55] These results indicate that ligand orbital mixing in the ^{1}GS HOMO can increase the efficiency of ligand exchange, but it must be balanced with red-shifting absorption to reach the PDT window.

3.2 Examples of PCT agents

To our knowledge, there are currently no Ru(II) polypyridyl PCT agents that have undergone clinical trials in humans. However, results from proof-of-concept Ru(II) complexes with caged drugs active against a number of cancer cell lines are promising for future ruthenium-based PCT agents to enter clinical trials. One example is that of imatinib. Chronic myelogenous leukemia (CML), a cancer of the white blood cells, is caused by a specific chromosomal abnormality that produces the abnormal tyrosine kinase protein Bcr-Abl.[81] Prior to 1975 chronic phase CML patients had an 8-year survival rate of ~6%, but studies by Brian Druker and colleagues on the inhibitor that would come to be known as imatinib, depicted in Fig. 11, radically improved the prognosis of CML patients, increasing the 8-year survival rate to ~90%.[82,83] Imatinib functions by inhibiting the tyrosine kinase activity of the Bcr-Abl protein, preventing it from forming or proliferating tumors.[81] Imatinib has also been used in the treatment of gastrointestinal stromal tumors (GISTs) and light-activated delivery of the inhibitor by a ruthenium photocage was theorized to be an effective method of treating GISTs without common side effects.[84,85] The complexes $[Ru(tpy)(Me_2bpy)(imatinib)]^{2+}$ and $[Ru(tpy)(biq)(imatinib)]^{2+}$ were synthesized and found to efficiently exchange imatinib for a solvent

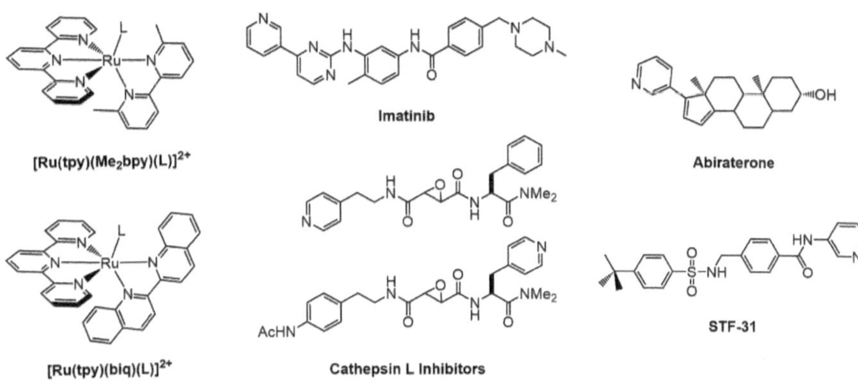

Fig. 11 Molecular structures of the $[Ru(tpy)(Me_2bpy)(L)]^{2+}$ and $[Ru(tpy)(biq)(L)]^{2+}$ scaffolds and the drugs imatinib, Cathepsin L (CTSL) inhibitors, abiraterone, and STF-31.

molecule following irradiation with $\Phi_{500} = 0.25(2)$ and 0.058(1), respectively. After confirming their PCT activity, the tpy ligands of each complex were modified to allow coupling to an antibody that targets a receptor that is overexpressed in GISTs. The resulting antibody-PCT conjugate complexes were found to maintain selectivity for the target receptor and imatinib was successfully released from the photocages with $\geq$590 nm.

The Turro and Kodanko groups have also worked with irreversible inhibition of the cysteine protease class of enzymes. Cathepsin L (CTSL) is a lysosomal cysteine protease overexpressed in a number of cancers, and two CTSL inhibitors based on the well known Katunuma (CLIK) inhibitor were synthesized. These feature an epoxide group that irreversibly binds to a cysteine thiol group in the active site and pyridyl groups available for coordination to a $[Ru(tpy)(Me_2bpy)]^{2+}$ scaffold.[86–88] The resulting complexes were able to inhibit CTSL activity and photodissociate the inhibitors with $\Phi_{500} = 0.08$–0.1. Further experiments with the cathepsin L model papain found that the $[Ru(tpy)(Me_2bpy)\text{-(inhibitor)}]^{2+}$ complexes bound to the cysteine thiol in the dark prior to photoinduced ligand dissociation. Upon irradiation, they released the $[Ru(tpy)(Me_2bpy)]^{2+}$ fragment while the inhibitor remained bound to the active site of the enzyme, similar to the biological targeting and uncaging observed in the ruthenium-imatinib complexes.

Cytochrome P450 (CYP) enzymes are a superfamily of heme-containing proteins that are involved in many important functions, including the synthesis of steroid hormones and drug metabolism.[89] CYP enzymes are known to metabolize and deactivate drugs before they reach their intended target, leading to reduced treatment efficacy. The administration of a given drug alongside CYP inhibitors has been used to perform a controlled inhibition of the enzyme, preventing drug metabolism and improving the therapeutic effect. The 17A1 isoform, CYP17A1, is involved in steroidogenesis, which is upregulated in prostate cancer, and inhibition of CYP17A1 has been shown to be effective in the treatment of this type of cancer.[90] The FDA approved CYP17A1 inhibitor abiraterone (ABI) was coordinated to two Ru(II)-based photocages to create the new complexes $[Ru(tpy)(Me_2bpy)(ABI)]^{2+}$ and $[Ru(tpy)(biq)(ABI)]^{2+}$, which exchange the inhibitor for a solvent CH_3CN molecule when irradiated with $\Phi_{500} = 0.049$ (5) and 0.018(1), respectively.[91] Following irradiation in a buffer solution, $[Ru(tpy)(Me_2bpy)(ABI)]^{2+}$ inhibits CYP17A1 activity at a similar level as free ABI, while no inhibition was observed in the dark, consistent with effective photocaging through coordination of ABI to Ru(II). The complex was then evaluated against the DU-145 human prostate cancer cell line, which shows significant CYP17A1 expression, to determine if effective inhibition was maintained in a cellular environment. In the dark, $[Ru(tpy)(Me_2bpy)(ABI)]^{2+}$ did not display a dose-dependent toxicity to DU-145 cells, only reducing viability to ~70% at the highest tested concentration. Following irradiation, however, the complex mirrored the toxicity of free abiraterone, indicating the complex is itself not toxic, but is an effective photocage of ABI and can provide spatiotemporal control of treatment for prostate cancer.

The Bonnet group has used the $[Ru(tpy)(Me_2bpy)]^{2+}$ and $[Ru(tpy)(biq)]^{2+}$ fragments to photocage STF-31, an inhibitor of the enzyme nicotinamide phosphoribosyltransferase (NAMPT).[92] NAMPT is the rate limiting enzyme in the nicotinamide adenine dinucleotide (NAD) salvage pathway and is known to be upregulated in some cancers. NAMPT inhibition has been shown to be an effective method of treating gliomas.[93,94] $[Ru(tpy)(Me_2bpy)(STF\text{-}31)]^{2+}$ underwent photochemical ligand exchange of the inhibitor at room temperature with $\Phi_{625} = 0.058$ and at biological temperatures with $\Phi_{625} = 0.08$, while $[Ru(tpy)(biq)\text{-}(STF\text{-}31)]^{2+}$ was slightly less efficient with $\Phi_{625} = 0.013$ at 298 K and $\Phi_{625} = 0.019$ at 310 K. The two complexes and free STF-31 were then evaluated under normoxic conditions against the non-cancerous lung cell line MRC-5 and the skin cancer cell line A431, as well as under normoxic and hypoxic conditions against lung cancer (A549) and breast cancer (MCF-7) cell lines. Free STF-31 is toxic to all four cell lines, whereas $[Ru(tpy)(Me_2bpy)\text{-}(STF\text{-}31)]^{2+}$ demonstrated greater toxicity to the three cancerous cell lines, though at roughly the same level in both the dark and following irradiation regardless of O_2. Only $[Ru(tpy)(biq)(STF\text{-}31)]^{2+}$ displayed a notable cytotoxicity difference between light and dark conditions, under both hypoxic and normoxic conditions. The phototoxicity index (PI), the ratio of dark IC_{50} to irradiated IC_{50}, which provides an indication of the effectiveness of the photocage for light-activated treatment, was only as high as 3.6 against cancer cells and 2.4 against the normal MRC-5 cells for $[Ru(tpy)(biq)(STF\text{-}31)]^{2+}$, indicating the complex is not a particularly effective PCT agent for achieving spatiotemporal control selective for cancer. The complex is notable, however, for demonstrating one of the first examples of PCT activity in cancerous cells under hypoxic conditions, activation with 625 nm irradiation at clinically relevant light doses, and serving as the first example of Ru PCT *in vivo*.

Zhao et al. utilized the androgen receptor (AR) antagonist bicalutamide, which is used in multidrug therapies for the treatment of prostate cancer and inhibits the physiological effects of dihydrotestosterone by preventing it from binding.[95,96] Bicalutamide possesses a nitrile group which allows for coordination to ruthenium, resulting in three possible PCT agents of the formula $[Ru(L)_2(bicalutamide)_2]^{2+}$ (Fig. 12, L = bpy, phen, biq). All three complexes exchange the first bicalutamide ligand with $\Phi_{465} = 0.02$ and $[Ru(biq)_2(bicalutamide)_2]^{2+}$ demonstrated the ability to exchange the drug with 660 nm light. Density functional theory (DFT) calculations show that the ^{3}LF state in all three complexes is slightly lower in energy than the ^{3}MLCT state(s), which can explain the similar ligand exchange quantum yields. The ability of $[Ru(biq)_2\text{-}(bicalutamide)_2]^{2+}$ to photodissociate a ligand with 660 nm light is likely due to the extended π-system in the bidentate ligands and red-shifted absorption as compared to the other two complexes. The three complexes and free bicalutamide were then evaluated

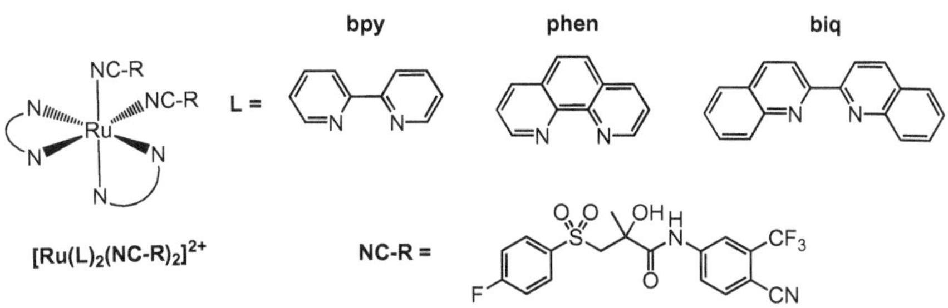

Fig. 12 Schematic representation of $[Ru(L)_2(NC\text{-}R)_2]^{2+}$, where L = bpy, phen, biq and NC-R = bicalutamide.

against AR-positive and AR-negative prostate cancer cell lines, LNCaP and PC-3, respectively. All of the complexes displayed negligible activity toward the AR-negative cells, confirming the selectivity of bicalutamide for the androgen receptor. The three ruthenium complexes were relatively nontoxic in the dark toward LNCaP cells and more toxic than bicalutamide following 465 nm irradiation, with PI values of 9.6, 5.3, and 7.7 for L = bpy, phen, and biq, respectively. Only $[Ru(biq)_2(bicalutamide)_2]^{2+}$ displayed an increase in toxicity following 660 nm irradiation, with an IC_{50} of 15.8 ± 1.7 μM, lower than the IC_{50} of 47.4 ± 2.1 μM for free bicalutamide, and a PI value of 5.8. This latter complex provides an example of an effective PCT agent that can be activated within the PDT window.

4. Dual activity

A dual activity complex is one that is capable of undergoing both drug delivery through PCT together with the generation of singlet oxygen for PDT. A Jablonski diagram depicting the photophysical processes of a dual activity complex is shown in Fig. 13 and demonstrates a bifurcation of excited states as the ^{3}MLCT state either thermally populates a ^{3}LF state, leading to ligand dissociation, or undergoes internal conversion (IC) to a lower energy $^3\pi\pi^*$ state, where the latter is long-lived and can undergo energy transfer to generate $^1O_2$. The capability of a single molecule to undergo both PCT and PDT violates the "Laws of Photochemistry" derived for organic molecules, but inorganic complexes able to undergo both processes have been discovered. Such complexes are theorized to be more effective at treating cancer than a drug capable of performing either PCT or PDT alone.

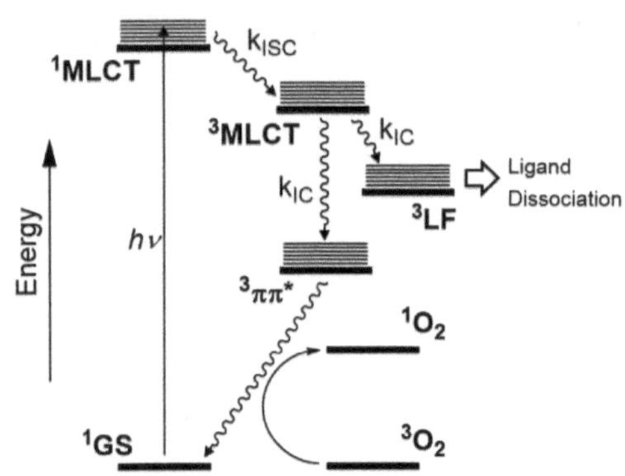

Fig. 13 Simplified Jablonski diagram for a generic Ru(II) polypyridyl complex that exhibits dual activity.

The first dual activity complex, published by the Turro group, is $[Ru(bpy)(dppn)(CH_3CN)_2]^{2+}$. This complex combines the ability to photodissociate CH_3CN ligands, releasing a ruthenium fragment able to coordinate to biomolecules, and the ability to sensitize $^1O_2$ from a long-lived $^3\pi\pi^*$ excited state localized on the dppn ligand.[97] The related complex $[Ru(bpy)_2(CH_3CN)_2]^{2+}$, which is only capable of undergoing a PCT process, exchanges one acetonitrile ligand with $\Phi_{400} = 0.21$ in H_2O, a value that shows that this photochemistry is more efficient than that for $[Ru(bpy)(dppn)(CH_3CN)_2]^{2+}$, which photodissociates CH_3CN with $\Phi_{400} = 0.002(3)$. As mentioned previously, the PDT analogue $[Ru(bpy)_2(dppn)]^{2+}$ does not undergo ligand dissociation upon irradiation, but possesses a lowest-energy $^3\pi\pi^*$ state with $\tau = 33$ μs and produces $^1O_2$ with $\Phi_\Delta = 0.88(2)$. These values in $[Ru(bpy)(dppn)(CH_3CN)_2]^{2+}$ were found to be $\tau = 20$ μs and $\Phi_\Delta = 0.72(2)$. It should be pointed out that the $^3\pi\pi^*$ state of $[Ru(bpy)(dppn)(CH_3CN)_2]^{2+}$ was shown using ultrafast spectroscopy to be populated from the ^{3}MLCT state with $\tau = 22$ ps.

Despite the lower quantum yields for each individual process, the combination of pathways for accessing PCT and PDT within the same complex proves to be beneficial for treating cancerous cells. In the dark, $[Ru(bpy)_2(dppn)]^{2+}$, $[Ru(bpy)_2(CH_3CN)_2]^{2+}$, and $[Ru(bpy)(dppn)(CH_3CN)_2]^{2+}$ displayed IC_{50} values against HeLa cells of 110 ± 28, 224 ± 23, and 334 ± 74 μM, respectively. Following irradiation, $[Ru(bpy)_2(CH_3CN)_2]^{2+}$ showed little difference in toxicity with $IC_{50} = 223 \pm 94$ μM, while the $^1O_2$ producing complexes displayed significantly higher toxicity, with $IC_{50} = 0.39 \pm 0.06$ μM for $[Ru(bpy)_2(dppn)]^{2+}$ and 0.47 ± 0.02 μM for $[Ru(bpy)(dppn)(CH_3CN)_2]^{2+}$. The PI value for $[Ru(bpy)_2(CH_3CN)_2]^{2+}$ against HeLa cells was 1.1 ± 0.4, indicating minimal enhancement in toxicity following irradiation. For $[Ru(bpy)_2(dppn)(CH_3CN)]^{2+}$ the PI value was 282 ± 69, a number which indicates greatly increased cytotoxicity following irradiation. The dual activity of $[Ru(bpy)(dppn)(CH_3CN)_2]^{2+}$ possessed $PI = 711 \pm 132$, greater than either the PCT or PDT complex and demonstrating the therapeutic advantage inherent in complexes capable of undergoing both processes.

While nitrile dissociation is important, the ability for the Ru(II) cages to undergo efficient pyridyl photodissociation was an important landmark for the delivery for therapeutics bearing aromatic nitrogen heterocycles. As such, the Turro group designed a new photocage that became the basis for many more dual activity PCT/PDT agents, $[Ru(tpy)(Me_2dppn)(py)]^{2+}$.[78] As demonstrated in previous publications, the Me_2dppn ligand, along with the tpy ligand, distorts the octahedral geometry around ruthenium center, such that it lowers the energy and permits the population of the dissociative ^{3}LF state, resulting in py dissociation upon irradiation. The extended π-system of the dppn ligand exhibits two different low-energy π^* molecular orbitals, one proximal and one distant to the coordinated Ru(II) center. The dppn-centered $^3\pi\pi^*$ state distal to the metal is lower in energy than the ^{3}MLCT state, such that $^1O_2$ is produced from the long-lived ligand-centered state. The addition of methyl groups pointing toward the Ru(II) center to the dppn ligand further distorts the pseudo-octahedral geometry around the metal, which was hypothesized to afford ligand exchange.

The resulting dual activity complex $[Ru(tpy)(Me_2dppn)(py)]^{2+}$, shown in Fig. 14, possesses a strong ^{1}MLCT transition at 486 nm, electronic absorption peaks throughout the visible region that tail out to 700 nm, and a long-lived $^3\pi\pi^*$ excited state with $\tau = 47$ μs, which was capable of producing $^1O_2$ with $\Phi_\Delta = 0.69(9)$. The analogous complex $[Ru(tpy)(dppn)(py)]^{2+}$ (Fig. 14) was also synthesized, lacking the steric distortion imposed by methyl groups, and was determined to have a similar electronic absorption spectrum, $\tau = 50$ μs, and $\Phi_\Delta = 0.98(6)$. When the ligand exchange quantum yields of the two complexes were measured, $[Ru(tpy)(Me_2dppn)(py)]^{2+}$ was found to exchange pyridine with $\Phi_{500} = 0.053(1)$ while the lack of methyl groups around Ru(II) in $[Ru(tpy)(dppn)(py)]^{2+}$ resulted in $\Phi_{500} < 0.0001$. These results indicate that the distortion of the geometry around the Ru(II) metal center, induced by the methyl groups on Me_2dppn in this case, is key to achieving dual activity.

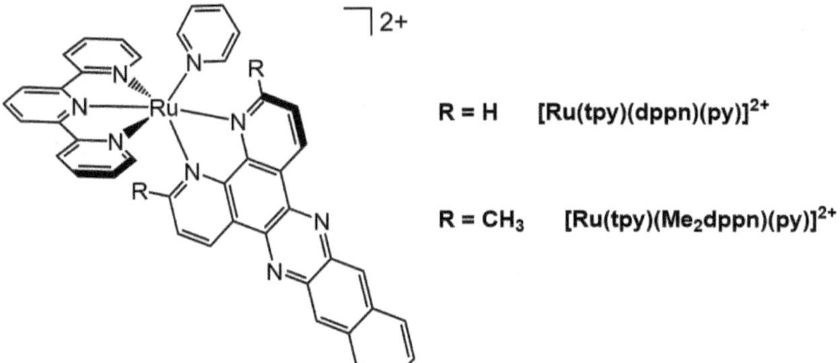

Fig. 14 Schematic representations of the structures of $[Ru(tpy)(dppn)(py)]^{2+}$ and $[Ru(tpy)(Me_2dppn)(py)]^{2+}$.

This conclusion is further supported by the photoinduced exchange of CH_3CN from the analogous $[Ru(tpy)(dppn)(CH_3CN)]^{2+}$. The bond to ruthenium from pyridine is typically stronger than from CH_3CN, which is reflected in the photoinduced ligand exchange quantum yields. In the case of $[Ru(tpy)(bpy)(py)]^{2+}$, $\Phi_{500} < 0.0001$ is observed, but under similar experimental conditions ligand dissociation in $[Ru(tpy)(bpy)(CH_3CN)]^{2+}$ occurs with several orders of magnitude larger quantum yield, $\Phi_{436} = 0.006$.[77,98] Yet for $[Ru(tpy)(dppn)(CH_3CN)]^{2+}$, the monodentate ligand exchanges with $\Phi_{450} = 0.00023(1)$, only slightly more efficient than the photodissociation of pyridine from $[Ru(tpy)(dppn)(py)]^{2+}$.[99]

The dual activity of $[Ru(tpy)(Me_2dppn)(py)]^{2+}$ provides it selective toxicity equivalent to or better than other complexes investigated by the Turro and Kodanko groups. When tested against the human triple-negative breast cancer cell line MDA-MB-231, $[Ru(tpy)(Me_2dppn)(py)]^{2+}$ displayed $EC_{50} = 4.6 \pm 0.5\ \mu M$ following irradiation and PI = 7.4.[100] A series of five complexes that photodissociate CH_3CN, $[Ru(bpy)_2(P(p\text{-}R\text{-}Ph)_3)(CH_3CN)]^{2+}$, where $P(p\text{-}R\text{-}Ph)_3$ represents *para*-substituted triphenylphosphine ligands, display no toxicity toward MDA-MB-231 cells when R = –OCH_3, –CH_3, and –F, but $EC_{50} = 7.0 \pm 1.4\ \mu M$ and PI > 4.2 for R = –H and $EC_{50} = 1.6 \pm 0.3\ \mu M$ and PI > 19 for R = –CF_3.[101] Regarding clinical suitability, $[Ru(tpy)(Me_2dppn)(py)]^{2+}$ has the advantage of absorbing well into the PDT window and ligand exchange quantum yields measured using 500 nm irradiation, while the $[Ru(bpy)_2(P(p\text{-}R\text{-}Ph)_3)(CH_3CN)]^{2+}$ series does not absorb beyond ~475 nm. The dual activity complex also kills cancer cells *via* a mechanism that is different than those typically observed in natural cell death, PCT, or PDT complexes, inducing necrosis in MDA-MB-231 cells as early as 4 h post-treatment, representing one of the first published examples of such a mechanism of achieving cancer cell death with ruthenium complexes.[100]

4.1 Derivatives of the prototypical $[Ru(tpy)(Me_2dppn)(py)]^{2+}$

Following the promising results of the dual activity complex $[Ru(tpy)(Me_2dppn)(py)]^{2+}$, modifications were made in an attempt to improve cell uptake and to individualize treatments by substituting a specific inhibitor for the model pyridine ligand. Synthetic modifications to the Me_2dppn ligand were made to vary the lipophilicity of the complex, producing three new Me_2dppn-containing ligands each with two aryl substituents (phenyl, 2,4-dimethyl phenyl, and 2,4-dimethoxy phenyl) at the 10- and 15-positions.[100] There is typically a strong correlation between the lipophilicity of a complex and the extent of its uptake by cells, such that the aryl modifications were expected to increase the lipophilicity as compared to Me_2dppn. Lipophilicity can be quantified by measuring the partition coefficients, P, of a given complex between water and octanol, such that negative LogP values indicate hydrophilicity while positive values point toward lipophilicity. As predicted, the prototypical dual activity complex $[Ru(tpy)(Me_2dppn)(py)]^{2+}$ was slightly hydrophilic, with LogP = − 0.27 ± 0.006, whereas the three arylated derivatives displayed varying levels of lipophilicity, with LogP values ranging from +0.11 to +0.32. In an unexpected result, the three lipophilic complexes were less effective at killing MDA-MB-231 cancer cells than the hydrophilic $[Ru(tpy)(Me_2dppn)(py)]^{2+}$, indicating there is no clear correlation between lipophilicity and cytotoxicity in these complexes.

The pyridine model ligand was later replaced with several different inhibitors to determine if dual activity was maintained with the new ligands and if the resulting complexes were effective at providing spatiotemporal control for treatment. One of the first inhibitors investigated by our group with the $[Ru(tpy)\text{-}(Me_2dppn)]^{2+}$ scaffold was the aforementioned imatinib, an important tyrosine kinase inhibitor.[85] The resulting complex $[Ru(tpy)(Me_2dppn)(imatinib)]^{2+}$ possesses a similar absorption spectrum to its parent complex, with an intense ^{1}MLCT transition at 487 nm and absorption trailing out to ~650 nm, produces $^1O_2$ with $\Phi_\Delta = 0.57(7)$, and exchanges imatinib for a solvent molecule with $\Phi_{500} = 0.073(1)$. The photoproduct formed by irradiation in H_2O, $[Ru(tpy)(Me_2dppn)(H_2O)]^{2+}$, was also determined to sensitize $^1O_2$ with $\Phi_\Delta = 0.22(2)$. These results demonstrate that the substitution of the model py ligand for a clinically relevant drug does not diminish the ability of the complex to exhibit dual activity and that the $[Ru(tpy)(Me_2dppn)(H_2O)]^{2+}$ photoproduct can remain useful for PDT treatment after a drug has been uncaged.

Cathepsin B (CTSB) is a cysteine protease localized on the cell surface whose expression is highly upregulated in a variety of cancers and is associated with growth, invasion, and metastasis of tumors.[102,103] Inhibition of CTSB has been shown to be an effective method of targeting cancerous cells and delivering anticancer drugs to their intended site of action. A derivative of the CTSB inhibitor CA-074, shown in Fig. 15, was synthesized and coordinated to ruthenium through a pyridyl moiety creating the

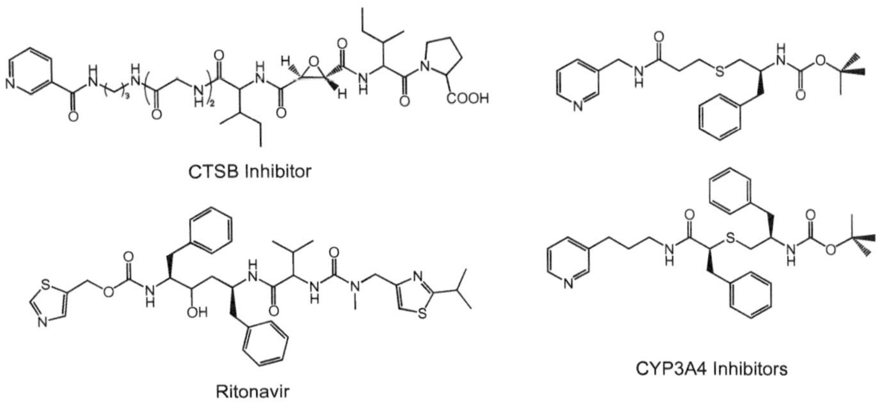

Fig. 15 Molecular structures of a cathepsin B (CTSB) inhibitor (top left), the CYP3A4 inhibitor Ritonavir (bottom left), and two derivatives of Ritonavir that function as CYP3A4 inhibitors (right).

complex [Ru(tpy)(Me_2dppn)(CTSB inhibitor)]$^{2+}$, which was determined to dissociate the inhibitor with $\Phi_{500}=0.11(1)$ in CH_3CN and produce $^1O_2$ with $\Phi_\Delta=0.58(3)$.[104] The free inhibitor and dual activity complex were both highly effective inhibitors, both irreversibly inhibiting CTSB activity at only 8.5 nM. In an unexpected result, [Ru(tpy)(Me_2dppn)(CTSB inhibitor)]$^{2+}$ was minimally toxic toward MDA-MB-231 cells in a 2D culture, a cell line that expresses CTSB at high levels on the cellular surface.[103] However, when the complex was evaluated against a 3D tumor model of MDA-MB-231 cells that better represents *in vivo* tumors, it was found to be highly toxic following irradiation, reducing cell viability to <10%, while remaining non-toxic in dark. The greatly increased toxicity in the 3D tumor model as compared to the 2D cell culture can be attributed to the greater surface area and subsequently greater surface CTSB concentration. The dual activity complex was also evaluated against healthy breast epithelial cells (MCF-10A) to investigate selectivity for cancerous over healthy cells. Importantly the complex did not reduce the viability of healthy MCF-10A cells in the dark or light, indicating great selectivity for cancerous cells.

A recent publication from the Turro and Kodanko groups focused on the coordination to ruthenium an inhibitor for CYP3A4, a P450 isoform that is most abundant in the liver and intestines and is vital to the metabolism of drugs.[105,106] Controlled inhibition of CYP3A4 is important for the treatment of a number of diseases as rapid metabolism of drugs by the enzyme can lead to reduced efficacy. The CYP3A4 inhibitor ritonavir, currently administered as part of a multidrug therapy for HIV and hepatitis C, was modified to contain a pyridyl group and then coordinated to the $[Ru(tpy)(Me_2dppn)]^{2+}$ fragment to create a dual activity complex.[107] As with other Ru(II) polypyridyl complexes, $[Ru(tpy)(Me_2dppn)(CYP3A4\ inhibitor)]^{2+}$ absorbs strongly in the visible region, possessing a ^{1}MLCT maximum at 485 nm and absorption features tailing to ~650 nm, produces $^1O_2$ with $\Phi_\Delta = 0.59(6)$, and photodissociates the CYP3A4 inhibitor with $\Phi_{500} = 0.024(4)$. Both the free inhibitor and the dual activity complex were able to inhibit CYP3A4 activity in the dark, with IC_{50} values of 1.54 μM and 250 nM, respectively. Surprisingly, the dual activity complex was more effective at inhibiting CYP3A4 activity in the dark than following irradiation, with a light IC_{50} value of 840 nM. A crystal structure of the related $[Ru(tpy)(Me_2bpy)(CYP3A4\ inhibitor)]^{2+}$ co-crystallized with CYP3A4 shows the intact complex bound within the active site, where the inhibitor is curled above the P450 heme, representing a likely method by which the complex inhibits CYP3A4 activity. The absence of an inhibitor-Fe bond in the crystal structure is an unexpected result, since free ritonavir and its derivatives inhibit CYP3A4 activity through coordination to the heme iron.[108]

The free ritonavir derivative and $[Ru(tpy)(Me_2dppn)\text{-}(CYP3A4\ inhibitor)]^{2+}$ were next evaluated against the DU-145 prostate cancer cell line in conjunction with vinblastine, a microtubule-destabilizing drug that has previously shown synergy with CYP3A4 inhibition for the treatment of prostate cancer.[109] The free CYP3A4 inhibitor displayed synergy with vinblastine in killing prostate cancer cells, reducing viability by ~40% in the presence of vinblastine as compared to only a ~10% decrease in the absence of vinblastine. When DU-145 cells were treated with vinblastine and $[Ru(tpy)(Me_2dppn)(CYP3A4\ inhibitor)]^{2+}$ in the dark, cell viability was reduced by only ~10%. Following irradiation, however, the viability was reduced by ~90%. $[Ru(tpy)(Me_2dppn)(py)]^{2+}$ was assayed against DU-145 cells alongside vinblastine as a control, as it produces $^1O_2$ with similar efficiency but releases pyridine instead of the CYP3A4 inhibitor. $[Ru(tpy)(Me_2dppn)(py)]^{2+}$ displayed some toxicity toward prostate cancer cells, but the effect was significantly more pronounced with $[Ru(tpy)(Me_2dppn)(CYP3A4\ inhibitor)]^{2+}$, indicating that CYP3A4 inhibition and PDT processes interact synergistically with vinblastine for the effective treatment of prostate cancer cells.

4.2 Other dual activity complexes from the Turro group

There have been several other dual activity complexes published by our group that do not utilize the $[Ru(tpy)(Me_2dppn)]^{2+}$ fragment as a scaffold. The tridentate analogue of dppn, pydppn, was coordinated to ruthenium to create $[Ru(pydppn)(biq)(py)]^{2+}$ in an effort to red-shift the absorption of the complex while maintaining dual activity and its structure is shown in Fig. 16.[110] The ^{1}MLCT transition shifted to 529 nm in acetone and singlet oxygen was produced with $\Phi_\Delta = 0.75(7)$, as compared to 486 nm and $\Phi_\Delta = 0.69(9)$ for $[Ru(tpy)(Me_2dppn)(py)]^{2+}$. The biq and pydppn ligands do not distort the geometry around ruthenium to the same extent as the Me_2dppn ligand, however, resulting in a lower efficiency of ligand exchange, $\Phi_{500} = 0.0070(4)$ for $[Ru(pydppn)(biq)(py)]^{2+}$ and $\Phi_{500} = 0.053(1)$ for $[Ru(tpy)(Me_2dppn)(py)]^{2+}$.

A dual activity complex without a dppn-type ligand was recently published by the Kodanko and Turro groups, which instead contained an iodinated py-BODIPY ligand that utilizes the heavy-atom effect to increase the population of a $^3\pi\pi^*$ state and "turn on" singlet oxygen production.[39] The protonated ligand, py-BODIPY-H, exhibits an excited state lifetime of <8 ns and does not produce $^1O_2$, while a long-lived ^{3}ES was observed for py-BODIPY-I with $\tau = 6.0$ μs and $\Phi_\Delta = 0.75(8)$. The coordination of this ligand to ruthenium to create $[Ru(tpy)(Me_2bpy)(py\text{-}BODIPY\text{-}I)]^{2+}$ (Fig. 16) reduced the lifetime to $\tau = 0.97$ μs and Φ_Δ to 0.23(4), but the complex was also capable of photodissociating the BODIPY-based ligand with

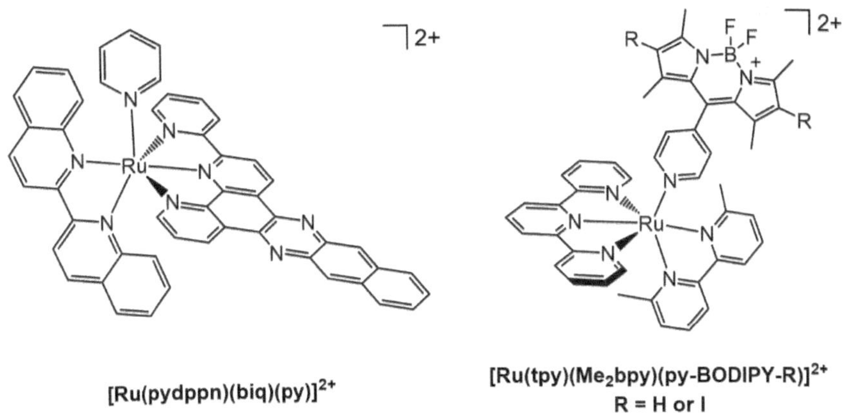

Fig. 16 Molecular structures of $[Ru(pydppn)(biq)(py)]^{2+}$ and $[Ru(tpy)(Me_2bpy)(py\text{-}BODIPY\text{-}R)]^{2+}$ (R = H, I).

$\Phi_{500} = 0.11(3)$. In contrast, the protonated analogue $[Ru(tpy)(Me_2bpy)(py\text{-}BODIPY\text{-}H)]^{2+}$ was unable to produce singlet oxygen but dissociated the ligand with $\Phi_{500} = 0.033(9)$. The py-BODIPY-I ligand alone is not toxic to MDA-MB-231 cells in the dark but following 520–530 nm irradiation achieved $EC_{50} = 70 \pm 30$ nM and a PI value of >730. $[Ru(tpy)(Me_2bpy)(py\text{-}BODIPY\text{-}I)]^{2+}$ was slightly less toxic following irradiation ($EC_{50} = 350 \pm 40$ nM), but was also nontoxic in the dark resulting in PI >140. Screening against healthy MCF-10A cells demonstrated that py-BODIPY-I had only moderate selectivity for cancerous cells while $[Ru(tpy)(Me_2bpy)(py\text{-}BODIPY\text{-}I)]^{2+}$ was more effective toward cancerous cells with a ratio of 6:1 over healthy cells.

The cathepsin K inhibitor Cbz-Leu-$NHCH_2CN$ was also bound to ruthenium to create the dual activity complex $[Ru(tpy)(dppn)\text{-}(Cbz\text{-}Leu\text{-}NHCH_2CN)]^{2+}$, which exchanges the inhibitor upon irradiation with $\Phi_{450} = 0.00014(3)$ and produced $^1O_2$ with $\Phi_\Delta = 0.64(3)$.[99] As with $[Ru(tpy)(dppn)(CH_3CN)]^{2+}$, the absence of methyl groups on the dppn ligand and the steric strain that they afford results in a small ligand exchange quantum yield, reducing efficacy of this complex as a dual activity agent. The Turro group has also published a rhodium-based dual activity complex, $[Rh_2(OCCH_3NH_2)_2(dppn)(CH_3CN)_2]^{2+}$, which is capable of dissociating one CH_3CN ligand with $\Phi_{450} = 0.0033(1)$ and producing $^1O_2$ with $\Phi_\Delta = 0.22(7)$.[111] The complex is capable of interacting with DNA, but its cytotoxicity toward cancerous cells remains to be determined.

4.3 Dual activity complexes from other research groups

The McFarland group published a series of Ru(II) polypyridyl complexes for dual activity employing Me_2bpy to induce steric distortion for ligand dissociation and an imidazo[4,5-*f*][1,10]phenanthroline ligand with $n = 1$, 2, or 3 thiophene moieties (IP-*n*T) to provide a ligand with an extended π-system on which a long-lived ^3ES could be localized.[112] $[Ru(Me_2bpy)(IP\text{-}n\text{T})]^{2+}$ (Fig. 17) produces $^1O_2$ with $\Phi_\Delta = 0.03$, 0.34, and 0.42 when $n = 1$, 2, or 3, respectively, and dissociates one of the Me_2bpy ligands following irradiation. All three complexes were assayed against the HL-60 human leukemia cell line and the complexes with two and three thiophene units were effective dual activity agents, possessing low dark toxicity ($EC_{50} = 115$–183 μM) and high toxicity following irradiation

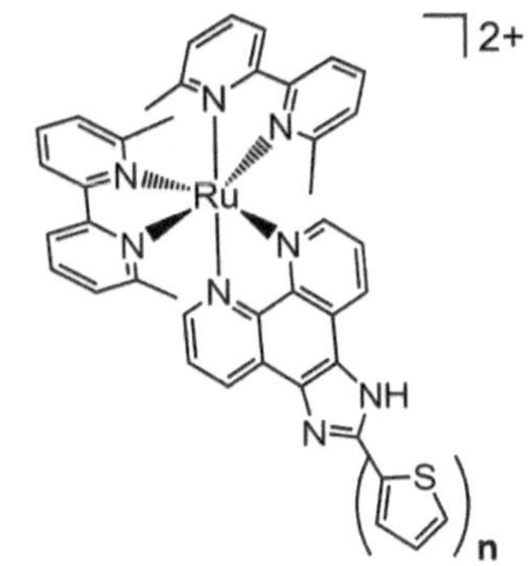

Fig. 17 Schematic representation of the McFarland dual activity complexes $[Ru(Me_2bpy)_2(IP\text{-}nT)]^{2+}$ where $n = 1, 2,$ or 3.

($EC_{50} = 0.2$–1.2 μM), with a significant PI = 750 for $[Ru(Me_2bpy)(IP\text{-}3T)]^{2+}$. The complex $n = 1$ was toxic to HL-60 in the dark, displaying $EC_{50} = 16 \pm 3$ after 16 h incubation and a PI = 15, making it ineffective for light-based therapy.

Another series of dual activity Ru(II) complexes containing the bidentate ligand 6,6′-dihydroxybipyridine (dhbp) employ hydroxyl groups to distort the geometry around the metal center and introduce a pH dependence on dual activity.[113] Complexes with the formula $[Ru(dhbp)(L)_2]^{2+}$ where L = bpy, phen, or dop (dop = 2,3-dihydro-[1,4]dioxino-[2,3-*f*][1,10]phenanthroline) were investigated for their ability to produce singlet oxygen and photodissociate the dhbp ligand in solutions at pH values of 5.0 and 7.5. At pH 5.0, with the hydroxyl group protonated, $\Phi_{450} = 0.0058(5)$, 0.0020(2), and 0.001(1) were reported for the complexes containing bpy, phen, and dop, respectively, while the singlet oxygen production for all complexes was in the 4–5% range. At pH 7.5, where the dhbp ligand contains deprotonated O^- groups, the ligand exchange quantum yields decreased by 4- to 50-fold, while the singlet oxygen production increased to $\Phi_\Delta = 0.18(2)$, 0.87(9), and 0.48(5) for the bpy, phen, and dop derivatives, respectively. DFT calculations show that the increased quantum yields at lower pH are likely due to the presence of a lowest energy ^{3}ES that is LF in nature, with significant anti-bonding character between ruthenium and the protonated dhbp ligand. At higher pH values, the ^{3}LF state is higher in energy and the ^{3}MLCT state significantly decreases in energy, making thermal population of the dissociative state very unlikely and increasing Φ_Δ while simultaneously decreasing Φ_{450}.

Ruthenium complexes containing arene groups have long been a focus of anticancer drug research due to their stability and tunability through synthetic modification.[114] A Ru(II) complex containing the η^6-arene *p*-cymene and the bidentate ligand 2,3-bis(2-pyridyl)benzo quinoxaline (dpb) with an expanded π-system was synthesized and characterized for its ability to produce $^1O_2$ and photodissociate a ligand.[115] The complex $[Ru(\eta^6\text{-}p\text{-cymene})(dpb)(py)]^{2+}$ sensitized singlet oxygen with $\Phi_\Delta = 0.25$ and unexpectedly exchanged both the dpb and py ligands following irradiation with visible light, $\lambda_{irr} > 400$ nm. The intact complex damages plasmid DNA following irradiation and was toxic to lung carcinoma cells (A549 cell line) with $IC_{50} = 27.6$ μM in the dark and $IC_{50} = 4.0$ μM following $\lambda_{irr} > 400$ nm irradiation, resulting in a PI value of 6.9.

5. Concluding remarks

Light-activated chemotherapy is a promising alternative cancer treatment that has the potential to prevent the systemic toxicity associated with many current chemotherapies by selectively activating anticancer drugs in an irradiated area. In this chapter, Ruthenium(II) PDT and PCT agents for the treatment of a variety of cancers have been reviewed and their selective activation with light, photochemistry, selectivity for cancerous cells, and toxicity toward cancer has been presented. In particular, the PDT agent TLD1433 has proven to be effective at treating bladder cancer and is currently in Phase II clinical trials. Dual activity complexes are capable of both photodissociating a ligand and producing $^1O_2$ upon irradiation, combining PDT and PCT processes into one molecule, which appears to improve on the effectiveness of complexes capable of only undergoing either PDT or PCT. Following the publication of the dual activity complex, $[Ru(tpy)(Me_2dppn)(py)]^{2+}$, the Turro and Kodanko groups have synthesized a variety of dual activity complexes with drugs that target a variety of cancer types, the most promising of which contains an inhibitor for the cytochrome P450 isoform CYP3A4, induces necrosis in cancerous cells, and displays drug cooperativity with vinblastine for the effective treatment of prostate cancer.

Acknowledgment

C.T. and J.J.K. are grateful for a grant from the National Science Foundation (CHE-2102508) for partial support of this work.

References

1. Kochanek, K. D.; Xu, J. Q.; Arias, E. In *National Center for Health Statistics Data Brief no. 395*; Hyattsvill, MD, 2020.
2. Horn, L.; Johnson, D. H. *J. Clin. Oncol.* **2010**, *28* (7), 1262–1268.
3. DeVita, V. T.; Rosenberg, S. A. *N. Engl. J. Med.* **2012**, *366* (23), 2207–2214.
4. Nielsen, J.; Strandberg, O. *Acta Radiol.* **1942**, *23* (3), 255–266.
5. Van Dyk, J.; Battista, J. J.; Almond, P. R. *Med. Phys. Int.* **2020**, *4*, 327–350.
6. Strebhardt, K.; Ullrich, A. *Nat. Rev. Cancer* **2008**, *8*, 473–480.
7. Erlich, P. *Dtsch. Med. Wochenschr.* **1910**, *17*, 1893–1896.
8. Gilman, A.; Philips, F. S. *Science (80-)* **1946**, *103* (2675), 409–415.
9. Gilman, A. *Am. J. Surg.* **1963**, *105*, 574–578.
10. Hirsch, J. *JAMA* **2006**, *296* (12), 1518–1520.
11. Goodman, L. S.; Wintrobe, M. M.; Dameshek, W.; Goodman, M. J.; Gilman, A.; McLennan, M. T. *JAMA* **1946**, *132*, 126–132.
12. Farber, S.; Diamond, L. K.; Mercer, R. D.; Sylvester, R. F., Jr.; Wolff, J. A. *N. Engl. J. Med.* **1948**, *238* (23), 787–793.
13. Dietzel, H. A.; Schwarz, R. H. *Arch. Surg.* **1966**, *92*, 301–303.
14. Huaizhi, Z.; Yuantao, N. *Gold Bull.* **2001**, *34* (1), 24–29.
15. Haller, J. S., Jr. *Pharm. Hist.* **1971**, *13* (2), 67–76.
16. Peyrone, M. *Ann. Chem. Pharm.* **1844**, *51*, 1–29.
17. Rosenberg, B.; VanCamp, L.; Trosko, J. E.; Mansour, V. H. *Nature* **1969**, *222*, 385–386.
18. Wiltshaw, E. *Platinum Met. Rev.* **1979**, *23* (3), 90–98.
19. Silverberg, E.; Holleb, A. I. *CA Cancer J. Clin.* **1975**, *25* (1), 2–7.
20. Yang, R.; Zhou, Y.; Wang, Y.; Du, C.; Wu, Y. *Ann. Transl. Med.* **2020**, *8* (24), 1671.
21. Ameta, R. K.; Malik, P. In *Advances in Metallodrugs*; Ul-Islam, S., Hashmi, A. A., Khan, S. A., Eds.; Scrivener Publishing LLC, 2020; pp. 203–245.
22. Galanski, M.; Jakupec, M. A.; Keppler, B. K. *Curr. Med. Chem.* **2005**, *12* (18), 2075–2094.
23. Kelland, L. *Nat. Rev. Cancer* **2007**, *7*, 573–584.
24. Paken, J.; Govender, C. D.; Pillay, M.; Sewram, V. *Semin. Hear.* **2019**, *40* (2), 108–121.
25. Santucci, N. M.; Garber, B.; Ivory, R.; Kuhn, M. A.; Stephen, M.; Aizenberg, D. *J. Otolaryngol. Head Neck Surg.* **2021**, *50* (19), 1–7.
26. Miller, R. P.; Tadagavadi, R. K.; Ramesh, G.; Reeves, W. B. *Toxins (Basel)* **2010**, *2*, 2490–2518.
27. Zhou, J.; Kang, Y.; Chen, L.; Wang, H.; Liu, J.; Zeng, S.; Yu, L. *Front. Pharmacol.* **2020**, *11* (343), 1–17.
28. Yao, Q.; Fan, J.; Long, S.; Zhao, X.; Li, H.; Du, J.; Shao, K.; Peng, X. *Chem* **2021**, *8*, 1–13.
29. Kanofsky, J. R.; Hoogland, H.; Wever, R.; Weiss, S. J. *J. Biol. Chem.* **1988**, *263* (20), 9692–9696.
30. Skovsen, E.; Snyder, J. W.; Lambert, J. D. C.; Ogilby, P. R. *J. Phys. Chem. B* **2005**, *109* (18), 8570–8573.
31. Fauci, A. S.; Harley, J. B.; Roberts, W. C.; Ferrans, V. J.; Gralnick, H. R.; Bjornson, B. H. *Ann. Intern. Med.* **1982**, *97* (1), 78–92.
32. Long, H.; Liao, W.; Wang, L.; Lu, Q. *Transfus. Med. Hemotherapy* **2016**, *43*, 96–108.

33. Foote, C. S.; Clennan, E. L. In *Active Oxygen in Chemistry*; Foote, C. S., Valentine, J. S., Greenberg, A., Liebman, J. F., Eds.; Blackie Academic & Professional: Glasgow, 1995; pp. 105–140.
34. Sies, H.; Menck, C. F. M. *Mutat. Res.* **1992**, *275*, 367–375.
35. Liang, P.; Kolodieznyi, D.; Creeger, Y.; Ballou, B.; Bruchez, M. P. *Front. Chem.* **2020**, *8*, 1–11.
36. Mroz, P.; Sharma, S. K.; Zhiyentayev, T.; Huang, Y. Y.; Hamblin, M. R. *Drug Deliv. Oncol. Basic Res. Cancer Ther.* **2011**, *3*, 1569–1603.
37. Anderson, E. M.; Kistiakowsky, G. B. *J. Chem. Phys.* **1969**, *51* (1), 182–188.
38. Wang, T.; Hou, Y.; Chen, Y.; Li, K.; Cheng, X.; Zhou, Q.; Wang, X. *Dalton Trans.* **2015**, *44*, 12726–12734.
39. Toupin, N. P.; Steinke, S. J.; Herroon, M. K.; Podgorski, I.; Turro, C.; Kodanko, J. J. *Photochem. Photobiol.* **2022**, *98*, 378–388.
40. B. Berman, N. Bhatia, D. Piacquadio, A. Houlihan, D. Davidson, and D. Siegel, *Photodiagnosis Photodyn. Ther.* **2020**, *32*, 102013.
41. Nishimura, T.; Hara, K.; Honda, N.; Okazaki, S.; Hazama, H.; Awazu, K. *Lasers Med. Sci.* **2020**, *35*, 1289–1297.
42. Knoll, J. D.; Turro, C. *Coord. Chem. Rev.* **2015**, *282–283*, 110–126.
43. Jones, L. R.; Grossweiner, L. I. *J. Photochem. Photobiol. B Biol.* **1994**, *26*, 249–256.
44. Gunaydin, G.; Gedik, M. E.; Ayan, S. *Front. Chem.* **2021**, *9* (686303), 1–26.
45. Garcia-Fresnadillo, D.; Georgiadou, Y.; Orellana, G.; Braun, A. M.; Oliverosb, E. *Helv. Chim. Acta* **1996**, *79*, 1222–1238.
46. Kaizu, Y.; Ohta, H.; Kobayashi, K.; Kobayashi, H.; Takuma, K.; Matsuo, T. *J. Photochem.* **1985**, *30* (1), 93–103.
47. Winkler, J. R.; Netzel, T. L.; Creutz, C.; Sutin, N. *J. Am. Chem. Soc.* **1987**, *109*, 2381–2392.
48. Abrahamsson, M.; Jäger, M.; Kumar, R. J.; Österman, T.; Persson, P.; Becker, H. C.; Johansson, O.; Hammarström, L. *J. Am. Chem. Soc.* **2008**, *130* (46), 15533–15542.
49. Sun, Y.; Joyce, L. E.; Dickson, N. M.; Turro, C. *Chem. Commun.* **2010**, *46*, 2426–2428.
50. Liu, Y.; Hammitt, R.; Lutterman, D. A.; Joyce, L. E.; Thummel, R. P.; Turro, C. *Inorg. Chem.* **2009**, *48* (1), 375–385.
51. Foxon, S. P.; Alamiry, M. A. H.; Walker, M. G.; Meijer, A. J. H. M.; Sazanovich, I. V.; Weinstein, J. A.; Thomas, J. A. *J. Phys. Chem. A* **2009**, *113*, 12754–12762.
52. Lifshits, L. M.; Roque, J. A., III; Ramasamy, E.; Thummel, R. P.; Cameron, C. G.; McFarland, S. A. J. *Photochem. Photobiol.* **2021**, *8*, 100067.
53. Sauvage, J. P.; Collin, J. P.; Chambron, J. C.; Guillerez, S.; Coudret, C.; Balzani, V.; Barigelletti, F.; De Cola, L.; Flamigni, L. *Chem. Rev.* **1994**, *94* (4), 993–1019.
54. Loftus, L. M.; Al-Afyouni, K. F.; Turro, C. *Chem. - A Eur. J.* **2018**, *24* (45), 11550–11553.
55. Al-Afyouni, M. H.; Rohrabaugh, T. N.; Al-Afyouni, K. F.; Turro, C. *Chem. Sci.* **2018**, *9* (32), 6711–6720.
56. Yin, H.; Stephenson, M.; Gibson, J.; Sampson, E.; Shi, G.; Sainuddin, T.; Monro, S.; McFarland, S. A. *Inorg. Chem.* **2014**, *53*, 4548–4559.
57. dos Santos, F. A.; de Almeida, D. R. Q.; Terra, L. F.; Baptista, M. S.; Labriola, L. *J. Cancer Metastasis Treat.* **2019**, *5* (25), 1–20.
58. Zhao, X.; Liu, J.; Fan, J.; Chao, H.; Peng, X. *Chem. Soc. Rev.* **2021**, *50* (6), 4185–4219.
59. Baskaran, R.; Lee, J.; Yang, S.-G. *Biomater. Res.* **2018**, *22* (25), 1–8.
60. Kamkaew, A.; Lim, S. H.; Lee, H. B.; Kiew, L. V.; Chung, L. Y.; Burgess, K. *Chem. Soc. Rev.* **2013**, *42*, 77–88.
61. Rademaker-Lakhai, J. M.; van den Bongard, D.; Pluim, D.; Beijnen, J. H.; Schellens, J. H. M. *Clin. Cancer Res.* **2004**, *10*, 3717–3727.

62. Hartinger, C. G.; Zorbas-Seifried, S.; Jakupec, M. A.; Kynast, B.; Zorbas, H.; Keppler, B. K. *J. Inorg. Biochem.* **2006**, *100*, 891–904.
63. Burris, H. A.; Bakewell, S.; Bendell, J. C.; Infante, J.; Jones, S. F.; Spigel, D. R.; Weiss, G. J.; Ramanathan, R. K.; Ogden, A.; Von Hoff, D. *ESMO Open* **2016**, *1*, 1–8.
64. Shi, G.; Monro, S.; Hennigar, R.; Colpitts, J.; Fong, J.; Kasimova, K.; Yin, H.; Decoste, R.; Spencer, C.; Chamberlain, L.; Mandel, A.; Lilge, L.; Mcfarland, S. A. *Coord. Chem. Rev.* **2015**, *282–283*, 127–138.
65. Monro, S.; Colon, K. L.; Yin, H.; Roque, J.; Konda, P.; Gujar, S.; Thummel, R. P.; Lilge, L.; Cameron, C. G.; McFarland, S. A. *Chem. Rev.* **2019**, *119*, 797–828.
66. Wolff, K.; Honigsmann, H. *Pharmacol. Ther.* **1981**, *12* (2), 381–418.
67. Pathak, M. A.; Fitzpatrick, T. B. J. *Photochem. Photobiol. B Biol* **1992**, *14*, 3–22.
68. Albert, A. *Nature* **1958**, *182* (4633), 421–423.
69. Rautio, J.; Meanwell, N. A.; Di, L.; Hageman, M. J. *Nat. Rev. Drug Discov.* **2018**, *17*, 559–587.
70. Rautio, J.; Kärkkäinen, J.; Sloan, K. B. *Eur. J. Pharm. Sci.* **2017**, *109*, 146–161.
71. Harris, A. L. *Nat. Rev. Cancer* **2002**, *2*, 38–47.
72. Fong, J.; Kasimova, K.; Arenas, Y.; Kaspler, P.; Lazic, S.; Mandel, A.; Lilge, L. *Photochem. Photobiol. Sci.* **2015**, *14*, 2014–2023.
73. Arenas, Y.; Monro, S.; Shi, G.; Mandel, A.; McFarland, S.; Lilge, L. *Photodiagnosis Photodyn. Ther.* **2013**, *10*, 615–625.
74. Vitaku, E.; Smith, D. T.; Njardarson, J. T. *J. Med. Chem.* **2014**, *57*, 10257–10274.
75. Cascioferro, S.; Parrino, B.; Spanò, V.; Carbone, A.; Montalbano, A.; Barraja, P.; Diana, P.; Cirrincione, G. *Eur. J. Med. Chem.* **2017**, *142*, 523–549.
76. Bonnet, S.; Collin, J. P.; Sauvage, J. P.; Schofield, E. *Inorg. Chem.* **2004**, *43* (26), 8346–8354.
77. Knoll, J. D.; Albani, B. A.; Durr, C. B.; Turro, C. *J. Phys. Chem. A* **2014**, *118* (45), 10603–10610.
78. Knoll, J. D.; Albani, B. A.; Turro, C. *Chem. Commun.* **2015**, *51*, 8777–8780.
79. Loftus, L. M.; Li, A.; Fillman, K. L.; Martin, P. D.; Kodanko, J. J.; Turro, C. *J. Am. Chem. Soc.* **2017**, *139* (50), 18295–18306.
80. Loftus, L. M.; Al-Afyouni, K. F.; Rohrabaugh, T. N.; Gallucci, J. C.; Moore, C. E.; Rack, J. J.; Turro, C. *J. Phys. Chem. C* **2019**, *123* (16), 10291–10299.
81. Druker, B. J.; Tamura, S.; Buchdunger, E.; Ohno, S.; Segal, G. M.; Fanning, S.; Zimmermann, J.; Lydon, N. B. *Nat. Med.* **1996**, *2* (5), 561–566.
82. Hochhaus, A.; Larson, R. A.; Guilhot, F.; Radich, J. P.; Branford, S.; Hughes, T. P.; Baccarani, M.; Deininger, M. W.; Cervantes, F.; Fujihara, S.; Ortmann, C.-E.; Menssen, H. D.; Kantarjian, H.; O'Brien, S. G.; Druker, B. J. *N. Engl. J. Med.* **2017**, *376* (10), 917–927.
83. Kantarjian, H.; O'Brien, S.; Jabbour, E.; Garcia-Manero, G.; Quintas-Cardama, A.; Shan, J.; Rios, M. B.; Ravandi, F.; Faderl, S.; Kadia, T.; Borthakur, G.; Huang, X.; Champlin, R.; Talpaz, M.; Cortes, J. *Blood* **2012**, *119* (9), 1981–1987.
84. De Giorgi, U.; Verweij, J. *Mol. Cancer Ther.* **2005**, *4* (3), 495–501.
85. Rohrabaugh, T. N.; Rohrabaugh, A. M.; Kodanko, J. J.; White, J. K.; Turro, C. *Chem. Commun.* **2018**, *54*, 5193–5196.
86. Huisman, M.; White, J. K.; Lewalski, V. G.; Podgorski, I.; Turro, C.; Kodanko, J. J. *Chem. Commun.* **2016**, *52* (85), 12590–12593.
87. Katunuma, N.; Murata, E.; Kakegawa, H.; Matsui, A.; Tsuzuki, H.; Tsuge, H.; Turk, D.; Turk, V.; Fukushima, M.; Tada, Y.; Asao, T. *FEBS Lett.* **1999**, *458*, 6–10.
88. Sudhan, D. R.; Siemann, D. W. *Pharmacol. Ther.* **2015**, *155*, 105–116.
89. McDonnell, A. M.; Dang, C. H. *J. Adv. Pract. Oncol.* **2013**, *4* (4), 263–268.
90. Auchus, M. L.; Auchus, R. J. *J. Invest. Med.* **2012**, *60* (2), 495–503.

91. Li, A.; Yadav, R.; White, J. K.; Herroon, M. K.; Callahan, B. P.; Podgorski, I.; Turro, C.; Scott, E. E.; Kodanko, J. J. *Chem. Commun.* **2017**, *53* (26), 3673–3676.
92. Lameijer, L. N.; Ernst, D.; Hopkins, S. L.; Meijer, M. S.; Askes, S. H. C.; Le Dévédec, S. E.; Bonnet, S. *Angew. Chem. Int. Ed.* **2017**, *56*, 11549–11553.
93. Gujar, A. D.; Le, S.; Mao, D. D.; Dadey, D. Y. A.; Turski, A.; Sasaki, Y.; Aum, D.; Luo, J.; Dahiya, S.; Yuan, L.; Rich, K. M.; Milbrandt, J.; Hallahan, D. E.; Yano, H.; Tran, D. D.; Kim, A. H. *Proc. Natl. Acad. Sci. U. S. A.* **2016**, *113* (51), E8247–E8256.
94. Sharma, P.; Xu, J.; Williams, K.; Easley, M.; Elder, J. B.; Lonser, R.; Lang, F. F.; Lapalombella, R.; Sampath, D.; Puduvalli, V. K. *Neuro Oncol.* **2022**, *24* (2), 229–244.
95. Goa, K. L.; Spencer, C. M. *Drugs Aging* **1998**, *12* (5), 401–422.
96. Zhao, J.; Liu, N.; Sun, S.; Gou, S.; Wang, X.; Wang, Z.; Li, X.; Zhang, W. *J. Inorg. Biochem.* **2019**, *196*, 110684.
97. Albani, B. A.; Peña, B.; Leed, N. A.; De Paula, N. A. B. G.; Pavani, C.; Baptista, M. S.; Dunbar, K. R.; Turro, C. *J. Am. Chem. Soc.* **2014**, *136*, 17095–17101.
98. Matsubara, Y.; Koga, K.; Kobayashi, A.; Konno, H.; Sakamoto, K.; Morimoto, T.; Ishitani, O. *J. Am. Chem. Soc.* **2010**, *132*, 10547–10552.
99. Rohrabaugh, T. N.; Collins, K. A.; Xue, C.; White, J. K.; Kodanko, J. J.; Turro, C. *Dalton Trans.* **2018**, *47* (34), 11851–11858.
100. Toupin, N. P.; Nadella, S.; Steinke, S. J.; Turro, C.; Kodanko, J. J. *Inorg. Chem.* **2020**, *59*, 3919–3933.
101. Lanquist, A. P.; Gupta, S.; Al-Afyouni, K. F.; Al-Afyouni, M.; Kodanko, J. J.; Turro, C. *Chem. Sci.* **2021**, *12*, 12056–12067.
102. Mohamed, M. M.; Sloane, B. F. *Nat. Rev. Cancer* **2006**, *6*, 764–775.
103. Roshy, S.; Sloane, B. F.; Moin, K. *Cancer Metastasis Rev.* **2003**, *22*, 271–286.
104. Arora, K.; Herroon, M.; Al-Afyouni, M. H.; Toupin, N. P.; Rohrabaugh, T. N.; Loftus, L. M.; Podgorski, I.; Turro, C.; Kodanko, J. J. *J. Am. Chem. Soc.* **2018**, *140* (43), 14367–14380.
105. Guengerich, F. P. *Annu. Rev. Pharmacol. Toxicol.* **1999**, *39*, 1–17.
106. Chen, Q.; Wei, D. *Adv. Exp. Med. Biol.* **2015**, *827*, 341–351.
107. Toupin, N.; Steinke, S. J.; Nadella, S.; Li, A.; Rohrabaugh, T. N.; Samuels, E. R.; Turro, C.; Sevrioukova, I. F.; Kodanko, J. J. *J. Am. Chem. Soc.* **2021**, *143* (24), 9191–9205.
108. Sevrioukova, I. F.; Poulos, T. L. *Proc. Natl. Acad. Sci. U. S. A.* **2010**, *107* (43), 18422–18427.
109. Blagosklonny, M. V.; Dixon, S. C.; Figg, W. D. *J. Urol.* **2000**, *163* (3), 1022–1026.
110. Loftus, L. M.; White, J. K.; Albani, B. A.; Kohler, L.; Kodanko, J. J.; Thummel, R. P.; Dunbar, K. R.; Turro, C. *Chem. - A Eur. J.* **2016**, *22* (11), 3704–3708.
111. Akhimie, R. N.; White, J. K.; Turro, C. *Inorg. Chim. Acta* **2017**, *454*, 149–154.
112. Sainuddin, T.; Pinto, M.; Yin, H.; Hetu, M.; Colpitts, J.; McFarland, S. A. *J. Inorg. Biochem.* **2016**, *158*, 45–54.
113. Qu, F.; Lamb, R. W.; Cameron, C. G.; Park, S.; Oladipupo, O.; Gray, J. L.; Xu, Y.; Cole, H. D.; Bonizzoni, M.; Kim, Y.; McFarland, S. A.; Webster, C. E.; Papish, E. T. *Inorg. Chem.* **2021**, *60*, 2138–2148.
114. Süss-Fink, G. *Dalton Trans.* **2010**, *39*, 1673–1688.
115. Chen, Y.; Lei, W.; Jiang, G.; Hou, Y.; Li, C.; Zhang, B.; Zhou, Q.; Wang, X. *Dalton Trans.* **2014**, *43*, 15375–15384.

CHAPTER NINE

Photochemical biosignaling with ruthenium complexes

Oscar Filevich[b] **and Roberto Etchenique**[a,*]

[a]Departamento de Química Inorgánica, Analítica y Química Física, INQUIMAE, Facultad de Ciencias Exactas y Naturales, Universidad de Buenos Aires, Intendente Güiraldes 2160, Ciudad Universitaria, Pabellón 2, Buenos Aires, Argentina

[b]Laboratorio de Neuroingeniería, Instituto de Tecnologías Emergentes y Ciencias Aplicadas (ITECA), UNSAM-CONICET, Escuela de Ciencia y Tecnología (ECyT), Universidad Nacional de San Martín, San Martín, Buenos Aires, Argentina

*Corresponding author: e-mail address: rober@qi.fcen.uba.ar

Contents

1. Introduction 322
2. Advantages of ruthenium caged compounds 324
3. Inorganic caged compounds and clinical applications 332
4. Value in diversity 333
5. The fifth ligand itch 337
6. Three-dimensional localization 341
7. *Cis, trans* and other structures 347
8. A brief gaze into the future 349
Acknowledgments 350
References 350

Abstract

Ruthenium coordination complexes bearing polypyridines have been studied for many years. While pursuing energy harvesting devices, considerable understanding about their photophysics and photochemistry was gained. This knowledge has recently been applied to a somewhat different application, namely the creation of a versatile scaffold which can release a molecule upon light absorption. The structural diversity, photochemical tunability, and clean photochemistry of this scaffold are of great advantage when designing new applications and as tools in biological research. In this chapter we describe the first incursions in the field of Ru-based phototriggers and the rationale backing their design. We highlight some of the recent applications in biology research, and note how the coupling with other emerging technologies aids in the sudden diversification of functions explored with Ru coordination complexes.

Advances in Inorganic Chemistry, Volume 80
ISSN 0898-8838
https://doi.org/10.1016/bs.adioch.2022.04.001

Abbreviations

1P	one-photon
2P	two-photon
4AP	4-aminopyridine
4pic	4-picoline
5HT	serotonin (5 hydroxy tryptamine)
5HT2A	serotonin receptor type 2A
AN	acetonitrile
BODIPY	4,4-difluoro-4-bora-3a,4a-diaza-s-indacene
bpy	2,2'bipyridine (or its analogues)
CA1	a specific area of the hippocampus in the brain
CB1	cannabinoid receptor 1
CDNI	4-carboxymethoxy-5,7-dinitroindolinyl
COM	commissural neuron
CPn	corticopontine pyramidal neuron
FRET	Forster Resonance Energy Transfer
GABA	gamma-aminobutyric acid
GM	Goeppert-Mayer unit
iGluSnFR	a genetically encoded glutamate sensor
LFP	local-field potential
LTD	long-term depression
LTP	long-term potentiation
Me_3[9]aneN_3	1,4,7-trimethyl 1,4,7-triazacyclononane
mGluR2/3	metabotropic glutamate receptor types 2 and 3 (group 2)
MLCT	metal-to-ligand charge transfer
MTG	methylthiogalactose
NMDA	N methyl D aspartate
NN	a generic bidentate ligand
NNN	a generic tridentate ligand
NO	nitric oxide
NOA60	Norland Optical Adhesive 60
PENB	3-(2-propyl)-4′-tris-ethoxy(methoxy)-4-nitrobiphenyl
PMe_3	trimethylphosphine
PPh_3	triphenylphosphine
Py	pyridine
RGH	arginine-glycine-histidine peptide sequence
Ru-bpy	complex containing Ru center and at least one 2,2′-bipyridine or analogue
Ru-tpy	complex containing a Ru center and at least one tpy or its analogues
tpy	2,2′:6′,2″-terpyridine (or its analogues)
UCNP	upconverting nanoparticles
φ_{PR}	quantum yield of photorelease

1. Introduction

During the most recent 20 years, neuroscience has become a very complex research field. The question of how the brain works has evolved beyond thought experiments into very specific perturbation and measuring techniques. Even so, we are reaching the limits of what our technology can do to help us. Simulations of the brain in different detail levels and temporal scales have been proposed, requiring yet unbuilt supercomputers running yet unwritten algorithms. Simple observation requires nanometer scale protein structure analysis. Experimental perturbation requires targeting anatomical features with nanometer precision and microsecond timing. The fact that the same molecule (say serotonin) has one behavioral effect when administered to one part of the brain, and a seemingly unrelated one when administered in some other brain structure precludes the classical all-organ drug application approach. Neurobiology has become a field that probes the functioning of the whole brain (which some consider the most complex structure in the universe) while making incredibly subtle and simple perturbations. The chemistry described in this chapter makes such experiments possible and opens a wide field of diverse applications.

Phototriggers, also known as caged-compounds, are photo-removable protecting groups which can be attached to a molecule of biological interest, hindering its potential interactions with cellular receptors or reaction partners. A particular chemical bond is broken when the phototrigger absorbs light of a given wavelength, yielding a non-interacting "cage" and a freely interacting biomolecule.

An ideal phototrigger does not have any biological effect at physiologically relevant concentrations until it is irradiated, follows a single photochemical pathway upon light absorption, and does so with high probability and a very fast time course.

Our group devised a new strategy in order to walk some steps toward the ideal caged compound: heterolytic metal-ligand photolysis. Based on the rich photochemistry of ruthenium bearing polypyridines, these inorganic caged compounds overcame several of the problems that traditional phototriggers had, extending the usability range well into the visible region and offering superb dark stability and clean and fast photoreactions. Since then, a variety of coordination caged compounds were developed and applied as tools in different biological systems.

Fine temporal control of the stimulus applied in a biological preparation has enabled testing hypotheses in unprecedented ways. Two examples are drug-based induction of gene expression, which allowed the understanding the effects of lethal-dominant gene mutations, and time-locked neural stimulation, which enabled testing hypotheses about neural dynamics, brain circuitry and perception. Derived clinical applications like deep brain stimulation and insulin pumps are changing the way some diseases are treated. In an analogous manner, fine spatial control of stimuli allows testing hypotheses in a wide range of spatial scales: from protein folding[1,2] and molecular interactions in ion channels[3] to whole-animal developmental effects of the product of a single gene.[4]

Many chemical platforms have been chosen or designed for use as photo-removable protecting groups. Nitrobenzyl and nitrophenyl compounds were the first well developed and used for uncaging amino acids, neurotransmitters, neuromodulators, molecules anchored to surfaces, peptides, bioactive drugs. Their major drawbacks were the high photon energy (~350 nm) needed to trigger the uncaging reaction and the associated biological damage, the rather complicated photolysis reaction pathways and the somewhat low stability of such phototriggers in biological media. Other platforms, based on different organic chromophores, were also devised. Among them, coumarins are perhaps the most investigated, due to the possibility to extend their sensitivity to visible light,[5,6] although p-hydroxyphenacyl, 4,4-difluoro-4-bora-3a,4a-diaza-s-indacene (BODIPY), quinolines, cyanines, benzoin and other photo-protecting groups have also been objects of research.[7–11]

On the other hand, inorganic caged compounds, mainly based in ruthenium-polypyridyl scaffolds, present a different and interesting chemistry. Octahedral coordination complexes allow the use of modular chemistry, since any of the multiple coordination sites can be manipulated individually in order to change chemical and photochemical properties. This tunability confers exceptionally broad caging capabilities, excellent photochemical properties and very good biocompatibility. Here we review recent advances in ruthenium based caged compounds in phototrigger design and some of their uses to answer interesting biosignaling questions.

2. Advantages of ruthenium caged compounds

In the mid 80's, Pinnick and Durham, based on previous works of Bosnich and Dwyer,[12] published a paper describing the way in which ruthenium-bipyridyl (Ru-bpy) complexes expel monodentate ligands upon irradiation with visible light. They found a clear dependence between the absorption maximum wavelength, the redox potential of the couple Ru^{2+}/Ru^{3+} in the complex, and its quantum yield of photolysis.[13] These findings were a continuation of work for synthetic purposes,[14] and established a rationale, profusely used in later works in which the Ru-bpy moiety was applied for harvesting solar energy. Obtaining complexes that

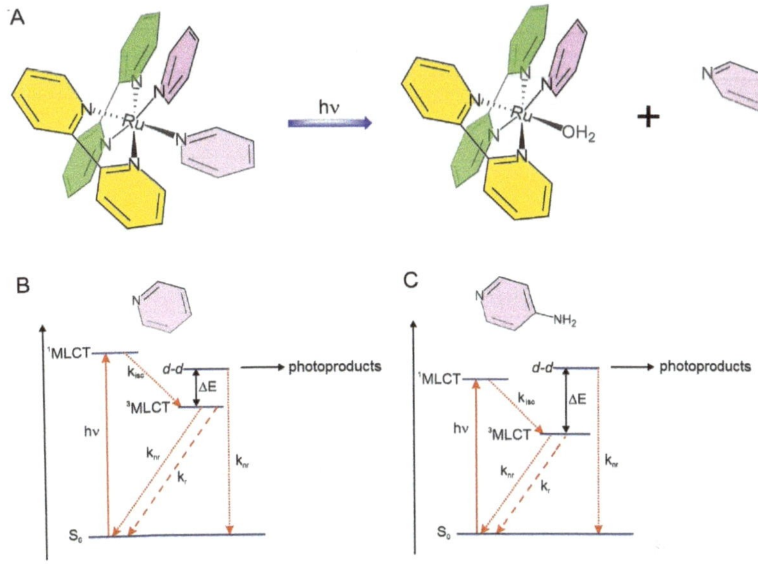

Fig. 1 (A) Scheme of the photolysis of the model complex $[Ru(bpy)_2Py_2]^{2+}$ in aqueous solutions. (B) States diagram for the complex depicted in (A). (C) States diagram for a similar complex bearing a stronger donor ligand (4-aminopyridine). In this case, the higher electron density on the Ru makes lowers the energy of the transition to the ^{1}MLCT, shifting the bands to longer wavelengths. The resulting higher gap between ^{3}MLCT and *d-d* states implies a lower quantum yield of photorelease.

present no degradation through photolysis, which is reduced but never zero for the bidentate ligands, was one of the goals for the use of Ru-bpy complexes as solar light collectors.[15,16]

A typical Ru^{II} octahedral complex bearing two bipyridines (*cis*-$[Ru(bpy)_2Py_2]^{2+}$) is depicted in Fig. 1A. Upon irradiation in aqueous solution, one of the pyridines is released, yielding the complex ion $[Ru(bpy)_2Py(H_2O)]^{2+}$ as the other photoproduct.

In the simplified model most often used, when a photon is absorbed charge transfer from the central atom to a bipyridine populates a singlet metal to ligand charge transfer state (^{1}MLCT), often formally described as Ru^{III}-bpy^{-}. Efficiently crossing to the triplet system gives a ^{3}MLCT state, which can emit phosphorescence, usually with low quantum yields, or populate a *d-d* state by thermal activation, as can be seen in Fig. 1B. It is from the latter state that the monodentate ligands are preferentially expelled. The energy gap between the ground state and the MLCT is modulated by the electron donor/acceptor properties of the ligands. In the case that the

^{1}MLCT state lays at a lower energy (Fig. 1C), a higher activation energy is needed to reach the dissociative *d-d* state, hence the quantum yield of photolysis diminishes.

It is important to note that this mechanism does not alter the oxidation state of the ligand nor the metal, it implies a non-redox dissociation. Conversely, Ru complexes with a similar structure but containing nitrosyl as a ligand are known to release nitric oxide (NO) upon irradiation with visible or UV light. In this case, despite its superficial similarity, the underlying mechanism is quite different, and even today is an object of some controversy. Ford, Santana da Silva and Malfant have focused their efforts to build NO photocages based on this mechanism and to elucidate it.[17,18] Other Ru-bpy complexes that photo-deliver NO without redox change were also devised (vide infra).

Aware of the growing interest in uncaging biomolecules using phototriggers, the inevitable Ru-bpy photo-degradation was seen as an advantage in our group. Instead of seeking the stability of the complexes, optimal control of the photochemical decomposition reaction was pursued. Fig. 2 shows some of the complexes and typical structures discussed in this paper.

The first useful Ru-based phototrigger developed in our lab was $[Ru(bpy)_2(4AP)_2]^{2+}$ (**i**) which releases the potassium-channel blocker 4-aminopyridine (4AP) upon light activation.[19] This compound displays maximum light absorption in the visible range at 489 nm with an estimated quantum yield of photosubstitution of $\varphi_{PR} = 0.02$. As 4AP is a K^+-channel blocker, the irradiation of a leech (*Hirudo medicinalis*) neuronal ganglion in the presence of (**i**) with green/blue light elicits a sudden increase in its neural activity, due to the depolarization induced by the 4AP freed to the medium. This complex, named RuBi4AP, was the first ruthenium-based caged compound, which started an ever-growing family of Ru-bpy and analogue phototriggers with several groups synthesizing and exploring new variants.

The good biological compatibility of RuBi4AP, and its ease of use in our biological preparations prompted the synthesis of caged versions of the major inhibitory and excitatory neurotransmitters in the central nervous system: two versions of RuBiGABA[20,21] (GABA = γ-aminobutyric acid) and RuBiGlutamate.[22] The structures of these three compounds (**viii**, **xi**, **xii**, respectively, in Fig. 2) include an auxiliary ligand, triphenylphosphine (PPh_3) or trimethylphosphine (PMe_3), which confers important stability against aquation, as will be discussed below. The availability of these core

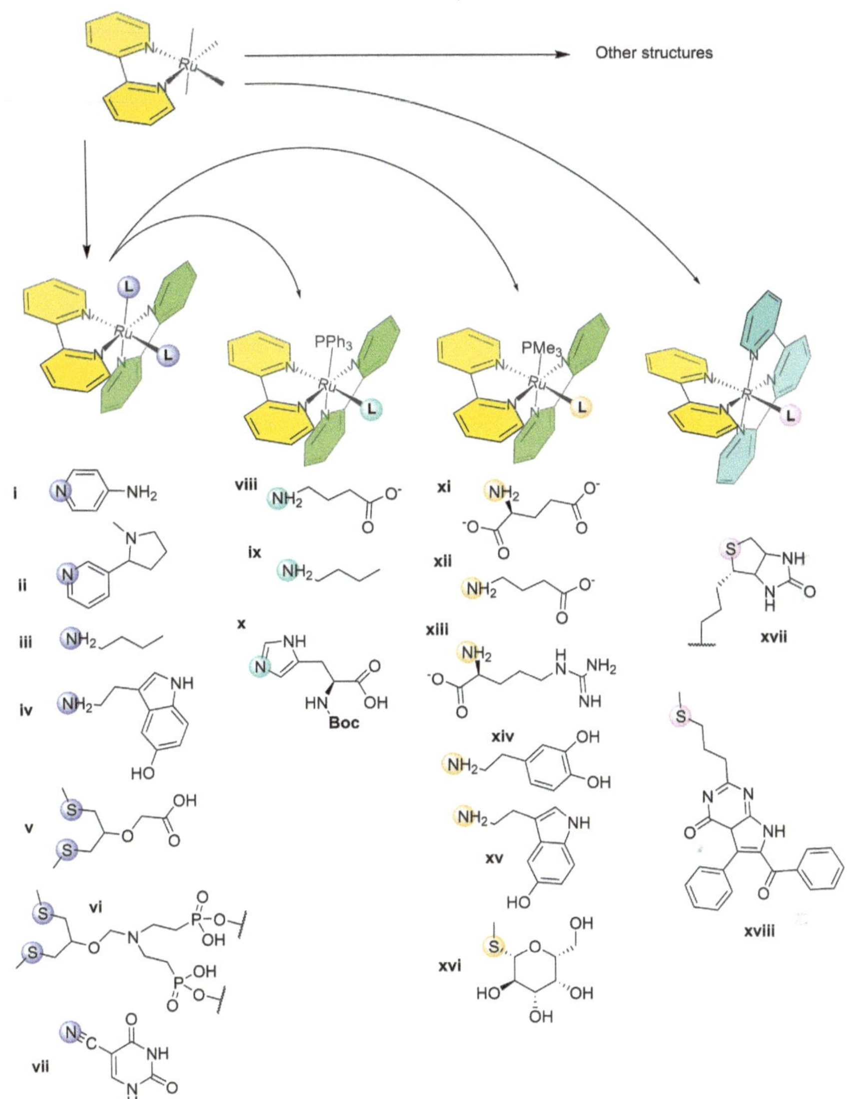

Fig. 2 A range of structures and complexes discussed in this chapter. Note that a bipyridine structure stands for any modified bipyridine or analogue (i.e., phenanthroline, dimethylbipyridines, etc.). The same cue is valid for terpyridine.

caged compounds raised interest in the neuroscientific community, especially from researchers studying circuitry and connectivity who understood the benefits of a water soluble, visible light-activatable caged neurotransmitter.

In an experimental electrophysiology lab studying circuitry, a recurring question is: "which neuron is presynaptic to the one I am monitoring?" i.e., "which circuit element is sending chemical/electrical signals to this other section". Yuste is one of the first neuroscientists to use RuBi-cages to answer this kind of biological question. He had experience with other (UV-activatable) caged neurotransmitters as his group pioneered the shift from the classical multielectrode-based electrophysiological techniques to a high throughput, automatable, whole-optical experimental configuration. The Yuste lab developed a technique to automatically detect every cell soma in a microscopy field, target each of them with an uncaging laser to trigger glutamate photo-release while recording the electrophysiological effect at the patched neuron, thus revealing which cells fire onto the patched cell.[23] This kind of work led to the concept of "cortical ensembles"[24] and enabled real-time behavioral control[25] by stimulating a few key neurons of an ensemble in a relatively unrestrained animal.[26]

In our lab, much more humble experiments involving optical influence of animal behavior were also performed. We used the complex $[Ru(bpy)_2(PMe_3)Arg]^{2+}$ (RuBiArginine, **iii** in Fig. 2) to trigger a feeding response in *Hydra vulgaris,* a simple cnidarian with a deceptively simple nervous system, capable of many behavioral "tricks."[27] Due to the high mechanical sensitivity of the tentacles, L-arginine delivery through more traditional methods, like "puffing" are prone to generating artifacts. The absence of mechanical stimulation upon photo-release ensures that the observed response is due to pure chemical stimulation. It also opens the door to truly understanding how exactly the feeding signaling chain works in this simple cnidarian, an issue which is not clear yet.

It is also remarkable that the combination of two emerging technologies, i.e. two-photon (2P) imaging of fluorescent probes and $[Ru(bpy)_2(PMe_3)Glu]^{2+}$ (RuBiGlutamate, **xi** in Fig. 2) uncaging, allows for very precise studies of the dynamics of nervous systems. Podgorski and others developed a two-photon imaging technique[28] which results in previously unimaginable field scanning speeds. They scan lines of excitation across a focal plane at multiple angles. Later, an algorithm computationally recovers high-resolution images, attaining frame rates over 1 kHz. This impressive imaging speed with 2P resolution is unprecedented, and wouldn't be as useful in the absence of a caged neurotransmitter with such fast photochemical kinetics.

Mappings of receptor distribution in a single cell or a whole brain are widespread applications of caged compounds. Although information related to the expression of receptor proteins can be obtained by immunohistochemistry techniques, information related to their function, the evoked currents and their overall role in the circuit, can only be obtained by activating the specific biosignaling pathways in a live preparation. Caged neurotransmitters are a fantastic tool for this, enabling drug release with sub-cellular resolution as members of our lab and others have shown.[29] Many groups have used RuBi-caged neurotransmitters to address key issues in current neurobiology. A brief and arbitrarily incomplete overview follows.

The Kind lab[30] has combined super-resolution microscopy studies with two-photon glutamate uncaging to study the fragile X syndrome, and found that spine morphology is conserved in model mice. This contrasts with the known hyperexcitability observed in nervous systems with this condition. So, this laboratory engaged in a functional analysis using RuBiGlutamate to reveal increased single-spine glutamate evoked currents, which explains the observed phenotypes and narrows the number of potential molecular actors involved.

A variety of mutants of the genetically encoded glutamate sensor iGluSnFR with improved affinity, stability and chromatic properties were reported by Marvin et al. and were characterized using RuBiGlutamate photorelease. These sensors allow "seeing" neural activity by imaging the glutamate concentration in a tissue like the brain.[31]

The Takahashi lab constructed a physical reservoir computing closed-loop experimental setup in which neural stimulation is executed by uncaging RuBiGlutamate.[32] Such experimental devices were a theoretical curiosity at first, but are now real models of neuromorphic computational systems.

Guo et al. used RuBiGlutamate uncaging to show that unipolar brush cells display a continuum of response profiles that reflect the graded expression of excitatory mGluR1 and inhibitory mGluR2/3 pathways (which are inversely correlated).[33] This is in contrast to what was previously believed: every neuron of a type would show a more or less fixed distribution of ion channel receptors.

In the Borgland lab, RuBiGlutamate was used directly on astrocytes, not neurons.[34] Those researchers showed that extended access to highly palatable, energy-dense food impairs astrocyte glutamate clearance. This in turn leads to a series of deregulations where GABA and extra-synaptic glutamate are involved, hindering the long-term plasticity mediated by endocannabinoids. This elegant study used RuBiGlutamate to directly affect astrocytes, as their role in regulation of neuronal activity became clearer in recent years. Astrocyte morphological features make a neuronal dendritic tree seem simple, even to the untrained eye. Neurons have a relatively fixed, stereotyped morphology. Compared to neurons, astrocyte physiological dynamics seem to be more complex. As a consequence, image analysis becomes more difficult. Also, the amount of information obtained by uncaging or mapping experiments (in any cell type) can be so large that it is not possible to analyze "by hand."

The Poskanzer lab designed a machine learning-based analysis software toolbox[35] to extract information of the wide-ranging fluorescent signals observed in most astrocyte imaging datasets. The result is an engine that does not require a predefined region of interest, instead automatically extracts information from the observed changes in the acquired images. In another study, Poskanzer and Yuste also showed that the stimulation of a single astrocyte, either with current injection or with RuBiGlutamate uncaging, activates other astrocytes in the local circuit and can trigger up state synchronizations of neighboring neurons.[36] This reinforces the evidence that astrocytes can regulate neural activity and addresses the issue of the real role of astrocytes in regulating behavior and neural computation.

One of the oldest substances known to have biosignaling functions is 5-hydroxy tryptamine (5HT), which is involved in many aspects of behavior. The 5HT phototrigger $[Ru(bpy)_2(PMe_3)(5HT)]^{2+}$ was used by Poskanzer's lab and members of our own lab to alter the firing modes of commissural and corticopontine pyramidal neurons (COM and CPn respectively), which change their behavior under 5HT, in the presence or absence of a 5HT2A receptor antagonist, thus showing that the effect of 5HT uncaging is indeed mediated by 5HT receptors.[37]

Araya and coworkers[38] established that single-spine stimulation protocols resulting in long-term potentiation (LTP) produce shrinkage of the activated spine neck and increase in synaptic strength, while protocols resulting in long-term depression (LTD) decrease synaptic strength without affecting spine shape. Notice the scales involved: a dendritic spine is smaller than 1 μm. These experiments were possible in a 2P microscope with 2P RuBiGlutamate release capabilities and superb spatial precision, which was described[39] for anyone to replicate, and addresses the issue about the computational capacity of the smallest established computational structure in the brain: the dendritic spine.

The two different caged compounds capable of delivering GABA that were developed, $[Ru(bpy)_2PPh_3GABA]^+$ (**viii**) and $[Ru(bpy)_2PMe_3GABA]^+$, (**xii**) differ in several aspects. The main difference is the respective lipophilicity, which make RuBiGABA-PPh_3 (**viii**) somewhat prone to enter into the membrane, having higher toxicity than its cousin RuBiGABA-PMe_3 (**xii**). On the other hand, the PPh_3 version has a higher quantum yield of photorelease, and a shorter maximum wavelength. Both RuBiGABAs were used to answer a broad spectrum of biological questions.

Zhe Pei et al.[40] demonstrated that more than one mechanism is involved in the direction selectivity of the retinal ganglion cell response. They used RuBiGABA to disrupt the stimulus-evoked GABA release from the presynaptic interneuron starburst amacrine cell and found that even without directional tuning of inhibitory inputs, direction selectivity is conserved in direction-selective ganglion cells.

Hong-Yuan Chu et al. used RuBiGABA-PPh_3 to study the interplay between GABA, NMDA and NO in the subthalamic nucleus of the basal ganglia.[41]

Cross-talk between different neurotransmitter systems is hard to dissect by just applying the drugs to the whole preparation. Diaspro et al. performed a systematic study of the experimental capabilities of RuBiGABA-PPh_3 in both linear and 2-photon regimes, in order to determine the optimal parameters for GABA photorelease in cell cultures.[42]

Two protecting groups are said to be orthogonal, if the wavelengths causing photorelease on one of them do not trigger photorelease of the other protecting group. This allows on to probe the biosignaling of two different molecules quickly, in the same preparation. The Kasai lab combined the use of two orthogonally-caged glutamate and GABA to gain understanding of the interplay of glutamate and GABA in the remodeling of dendritic spines.[43] For this, they used a mixed one- and two-photon approach to uncage RuBiGABA in linear regime and CDNI-glutamate (CDNI = 4-carboxymethoxy-5,7-dinitroindolinyl) in 2-photon regime independently and were able to manipulate the dendritic tree in real time.

In another subcellular compartmentalization study, the Losonczy group[44] uncaged both RuBiGABA and PENB-glutamate (PENB = 3-(2-propyl)-4′-tris-ethoxy(methoxy)-4-nitrobiphenyl). These researchers took advantage of the orthogonality of their excitation wavelengths, to dissect the origins of dendritic inhibition in CA1 pyramidal cells and its role in neural computation.

The functional organization of the brain is still being debated. One important question regarding the "unit of computation" remains unanswered: Which is the smallest anatomical structure capable of neural computation. Answering this question requires, among other pieces of information, knowing the spatial distribution of each neurotransmitter receptor type. These are known as "mapping" studies and caged neurotransmitters are superb tools for this.

Kwon et al. undertook a combined functional, molecular, and ultrastructural mapping of GABAergic inputs onto dendrites of pyramidal neurons using RuBiGABA in the 2P-regime.[45] They studied the upper layers of juvenile mouse somatosensory cortex, and found that most synapses in those neurons lack GABA receptors, despite the presence of receptors in spines, and concluded that those receptors must be extrasynaptic, and likely control clusters of excitatory inputs, defining functional zones on dendrites.

3. Inorganic caged compounds and clinical applications

The application of a drug to the whole brain is not desired, albeit it is sometimes the only available approach. Because of the brain's complex structure, and due to the fact that the same drug has different receptors in different brain areas with different behavioral effects, it would be an advantage to be able to delicately control both *when* and *where* in the brain a drug is released.

The question of whether RuBiGABA has potential in clinical applications was explored by Xiaofeng Yang at the Yuste and Rothman labs who used RuBiGABA in vivo to efficiently terminate epileptic seizures induced by free 4AP in rats as monitored by local field potentials (LFP).[46,47] Moreover, they found no cellular or histological damage in the sites of RuBiGABA uncaging as evaluated by immunohistochemistry.

Xinxia Cai and coworkers took this strategy further, enclosing RuBiGABA in antibody-coated nanoliposomes directed toward CB_1,[48] a cannabinoid receptor protein found highly expressed in the hippocampi of patients with epilepsy and in animal models with induced epilepsy. Neural activity was monitored by a microelectrode array which provides significantly better spatial resolution than LFP and thus a more accurate detection of epileptic events. This uncaging technique again resulted in effective seizure control. The ultimate clinical application in view is probably an autonomous closed-loop system, with an implantable light source and seizure detection algorithm which would trigger RuBiGABA when an epileptic seizure is detected.

The extremely fine spatial resolution attained by laser focusing coupled to caged-compounds allows probing subcellular organization of receptor distribution. Rowan et al. dissected the topological extent to which an axon is functionally compartmentalized during spike repolarization.[49] By using a voltage-sensitive dye to image axons transmembrane voltage and uncaging RuBi4AP to modify K^+-currents in a selected 6 μm section of an axonal branch, they were also able to assign a specific function to the axon initial segment thanks to the very fine spatial resolution attained for drug application.

Bub and coworkers combined RuBiNicotine (**ii** in Fig. 2) photo-uncaging and optogenetic neural stimulation with simple dye-free imaging of cardiac monolayers to measure electrical activity in an automated fashion and study the neuron-myocyte interaction,[50] illustrating the power and high throughput capability of such interrogations. This kind of experiments, in which one wishes to target a cell that grows on top of a moving (beating) monolayer, were technically defying with more traditional techniques. The availability of caged-nicotine makes this possible.[51]

A quite different biosignaling molecule is the stable radical NO (nitric oxide), a once elusive neurotransmitter/neuromodulator that acts on endothelial cells as a vasodilator. Santana da Silva, Slep and coworkers[52,53] have caged it in *cis*-$[Ru(Me_3[9]aneN_3)(bpy)(NO)]^-$ and *cis*-$[Ru(bpy)_2(4pic)(NO)]^{2+}$ from which NO is expelled upon irradiation with visible light in a clean manner, leaving the corresponding aquo complex as a sole other product. The exact nature of the photoactive state has not yet been elucidated. The non-innocent nature of the NO ligand introduces an additional complication in the mechanistic interpretation, although there is significant resemblance to the usual Ru based caged compounds.

Different kinds of caged-compounds have applications outside the field of neuronal probes. They have been used in bacterial ecological studies to create amino acid diffusion gradients, useful in chemotaxis experiments to probe responses to chemoattractants, another form of biosignaling. Carrara et al. uncaged amino acids by pointing the uncaging laser with a computer and shining spots of light over a bacterial culture.[54] They uncaged RuBiGlutamate and induced chemotaxis in the bacterium *Vibrio ordalii* in a very elegant setup which allows releasing the caged molecule to a predefined, desired concentration.

4. Value in diversity

Many different functional groups can be coordinated to Ru-bpy cages. Besides amines, nitriles, thiols and thioethers have been protected.

Thioethers and nitriles usually result in high quantum efficiencies of photo-release due to the weak electron-donor characteristics of the ligand. The particular molecule reported by Bonnet is a $Ru(bpy)_2$(bis-thioether) model molecule (**v**) which only releases its bidentate bis-thioether ligand after absorbing two photons,[55] with a high quantum yield of photorelease for the first photon, and a lower one for the second. The Bonnet group coordinated a biotin at the end of an extender molecule through the biotin sulfur atom obtaining a Ru(tpy)(bpy)-tagged biotin (**xvii**) which has a linker to another biotin that can still bind to avidin.[56] This shows that the avidin-biotin complex can still form when a biotin is tagged in this manner, which allows the use of the photosubstitution capacity of the Ru center for photochemotherapy.

Other biologically relevant thioether groups are those in amino acids side chains and some biosignaling molecules. In particular, the sugar methylthio galactose (MTG) is an S-substituted galactose. It triggers a required conformational change in the regulator protein of the *lac* operon allowing the expression of downstream encoded genes. Lac operon is a gene regulatory system first discovered in bacteria and now available for drug-induced gene regulation in mammals.[57] It is thus possible to induce gene expression by direct application of MTG into a desired zone. In our lab a caged version of MTG, the complex $[Ru(bpy)_2(PMe_3)MTG]^{2+}$ (**xvi**, RuBiMTG) was synthesized. A solution of RuBiMTG displays an absorption maximum at 422 nm, corresponding to a thioether coordination, and a high quantum efficiency $\varphi_{PR} = 0.30$. Fig. 3 shows the mechanism of gene activation through irradiation of RuBiMTG: A filter paper soaked in a suspension of *E.coli* capable of expressing the *lac* operon is placed on a Petri dish, and a picoprojector irradiates it with the image of a heart in a dark square for some seconds. Light photolyzes the caged-compound (1), which delivers MTG (2) to the medium. After entering bacteria, the free MTG is capable of binding the repressor (3) of *lac* operon (4), allowing mRNA transcription (5), which reaches the ribosomes (6) and causes β-galactosidase synthesis (7) the protein which cleaves lactose into its constituents glucose and galactose. The presence of this enzyme is developed with X-gal, a reagent which can be converted by β-galactosidase into a bright blue pigment (8). The result is shown in panel 9, where the blue color indicates the zones where the genes of *lac* operon were transcribed. In this way, it becomes feasible to express a gene or group of genes in a spatially controlled way, at the desired time.

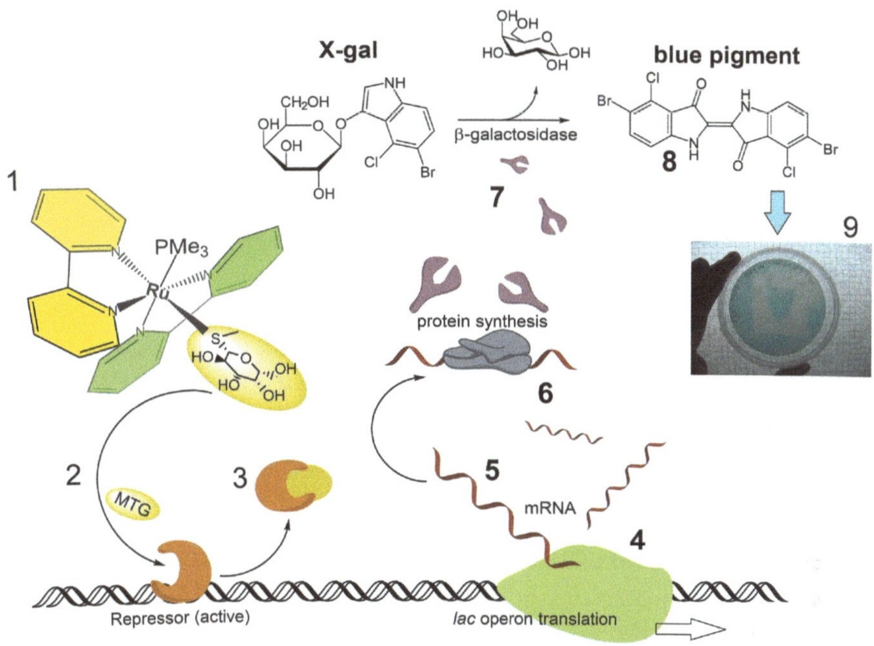

Fig. 3 Scheme of the caged-MTG gene inducing mechanism. After photolysis, the caged-compound (1) delivers MTG (2) to the medium. MTG enters into *E. coli.*, where it binds to the active repressor (3). Once inactivated, *lac* operon (4) transcribes to mRNA (5), which through ribosomes (6) synthesize β-galactosidase (7), an enzyme which cleaves lactose into glucose and galactose. X-gal, a reagent which can be converted by β-galactosidase into a bright blue pigment (8) is used to develop the image (9).

As the caged molecule becomes more and more complex, with a growing number of coordination sites exposed, the number of possible coordination products and structures quickly becomes huge. As an example, the amino acid histidine, which at first sight would be expected to behave as glutamate and coordinate the Ru atom through its imidazole nitrogen , as depicted in Fig. 4A, can also coordinate through its α-amino group as is shown in Fig. 4B.

If one wishes to cage a protein which contains some histidine, methionine and cysteine side chains, in addition to its α-amino end, the typical synthetic route is not efficient due to the multiple partially-coordinated products possible. Also, a protein caged with this strategy would need multiple photons for uncaging, and thus the apparent uncaging efficiency would be extremely low. The shortest route is not always the best. One can cage a

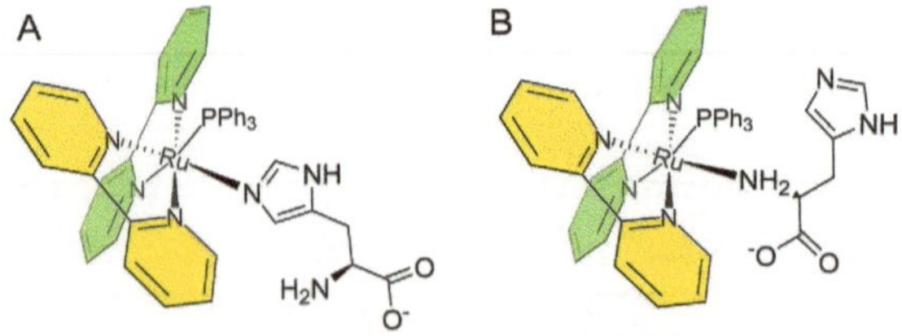

Fig. 4 Histidine can coordinate in two different ways to the Ru center: (A) through the imidazole nitrogen, (B) through the α-amino group.

single amino acid, and then (if the peptide is not too long) use solid-phase peptide synthesis to obtain a polypeptide that carries a single Ru caging group at a desired position in the sequence. To ensure the coordination does not involve the α-amino group of this amino acid so it can be linked to a polypeptide, a Boc-protected amino acid can be used. This is the approach taken by Mosquera et al. who report a light-sensitive histidine building block (**x**) for Fmoc/tBu solid-phase peptide synthesis in which the imidazole side chain is coordinated to a ruthenium complex.[57] These workers caged a peptide containing the RGH domain (Arg-Gly-His) which is known to coordinate a Ni(II) ion and act as a nuclease, cleaving DNA. In its caged form, the peptide shows very little nuclease activity, which is enhanced only after irradiation.

The concept of photoactivated chemotherapy (PACT) exploits the ability of antitumoral drugs to induce cell death in highly metabolic (rapidly dividing) cells with some degree of selectivity against normally dividing cells. By caging the antitumoral drug and photo-releasing ("photo-activating") it only in the neighborhood of the detected tumor, the overall free antitumoral concentration on the patient is lowered. This reduces unwanted side effects on other replicating cell types (like various epithelia and hair roots), while keeping a high effective concentration in light-irradiated sites, helps to avoid cell death induction in healthy tissues and reduces unwanted toxicity and side effects. One such application reports the caging of the sulfur atom of a rigidin analogue,[58] obtaining $[Ru(tpy)(bpy)(rig)]^{2+}$ (**xviii**) which releases rigidin, a cytotoxic molecule isolated from a marine tunicate that binds to and stabilizes microtubules, thus making cell division impossible.

Knowing that the complexes absorption maximum wavelength and the probability of photorelease (known as quantum efficiency) depend on the electron donating properties and sterical hindrance of the coordination sphere, we "tuned" these properties by experimenting with the behavior of dozens of ligands in very diverse combinations.

5. The fifth ligand itch

Ru^{II} complexes are usually six-coordinate. The photo-detachable ligands are nearly always monodentate, so, besides the position for the caged ligand, five positions on the octahedral moiety have to be occupied. This number can be fulfilled through several strategies. The simplest is to synthesize a traditional Ru-bpy bearing 2 bidentate ligands (phen or bpy) and add two copies of the bioactive ligand. The first column in Fig. 2 shows the basic structure of this subfamily. This approach has some advantages: the symmetric structure is easy to build and to characterize. In addition, the probability of releasing the desired bioactive ligand doubles. The first caged compound of the Ru family had this structure,[19] and also some other earlier phototriggers, structures **ii**, **iii** and **iv**.[51,59] Turro and coworkers have used this structure with the complex $[Ru(bpy)_2(5CNU)_2]^{2+}$ (**vii**) which is capable of delivering the bioactive compound 5-cyanouracyl upon irradiation, while the residue $[Ru(bpy)_2(H_2O)_2]^{2+}$ can bind to DNA, hence acting as a dual-action caged drug.[60]

On the other hand, this approach has some disadvantages: possibilities of tuning are lower, and unexpected behaviors can appear with this approach. As an example, the complex $[Ru(bpy)_2(GABA)_2]$ photo-delivers GABA upon irradiation.[59] However, coordinating two electron-rich amines to the Ru center red shifts the ^{1}MLCT band, which implies a low quantum yield of photolysis (see Fig. 1B/C). On the other hand, this labilizes the Ru center, and the thermal ligand exchange becomes much faster than for their PMe_3 or PPh_3 cousins. The direct effect is a noticeable release of GABA even in the dark, precluding this compound to be useful for most physiological experiments.

Attempts to synthesize the same $Ru(NN)_2L_2$ structure (where NN stands for the bipyridyl or analogue and L for the biorelevant ligand), for amino acids as glutamate yielded even worse results: although carboxylates do not establish stable bonds with Ru-bpy complexes in water environments, the α-carboxy group allows a bidentate coordination which impedes a second amino acid to enter the coordination sphere. Moreover, the now bidentate ligand, with a deprotonated oxygen coordinated to Ru, moves the ^{1}MLCT band even farther to the red, and this combination makes its photolysis extremely improbable. In conclusion, this simple structure just can be used for ligands bearing no additional coordinating groups in adjacent positions, and in cases where the fine tunability of the chemical properties is not needed.

If we number the bioactive molecule as "the sixth ligand," and we have at least two positions (1st and 2nd) occupied by a bidentate diimine ligand, there are different possibilities to fill the three remaining positions. One obvious approach is to use a tridentate ligand. Several tridentate ligands were in fact used in different studies to make photoactive complexes of the form [Ru(NN)(NNN)L]. Linear NNN = terpyridine or modified terpyridines were among the favorites, although some other ligands were also used.

Terpyridines (tpy) offer ease of synthesis and yield complexes that somewhat resemble the symmetric ones, except for the presence of an additional pyridine ring. Moreover, since tpy is a polypyridine, having a Ru-tpy ^{1}MLCT allows the use of innocent bidentate ligands instead of bpy, which can confer wanted or unwanted properties. However, nature is often tricky: the dimensions of tpy do not nicely match with those of the octahedral sites for Ru^{II} complexes. This fact translates into bond strain, and complexes having tpy have significantly less photodelivery efficiencies than their bis-bipyridine analogues. Several pathways can be taken to circumvent this problem. Electron-poor ligands, that are usually prone to photolysis (i.e., nitriles, thioethers), are quite photoreleasable in [Ru(tpy)(bpy)L] complexes. Based on the phenomenon of reverse FRET (vide infra) between Ru-bpy and Rhodamine,[61–63] Bonnet and coworkers have used complexes with tpy bearing a rhodamine antenna to increase the absorption cross section for thioether photodelivery.[64] Glazer has made an impressive and systematic analysis on these kind of structures, changing the electron donating properties of the ligands and their steric hindrance to shed light into a somewhat obscure landscape involving quantum yield, stability and biological compatibility/toxicity.[65] Extending the conjugation system of tpy, Turro et al. achieved red and even NIR active complexes that release the model molecule acetonitrile upon irradiation at 735 nm.[66] To our knowledge, this is the longest wavelength reported to cause ligand photo-exchange in a mononuclear Ru-polypyridine complex in the single photon excitation regime.

Back to the structures, in our lab we chose a third strategy: to use a traditional core $Ru(bpy)_2$ and occupy the fifth position with an auxiliary ligand, which can easily tune the properties of the whole complex. Among the ligands studied we can count NH_3, NH_2CH_3, Py, $P(Ph_2Me)$, $P(PhMe_2)$, $P(CH_2OH)_3$ and several other modified phosphines, although PPh_3 and PMe_3 were, by far, the most investigated and used.

In order to be a useful auxiliary fifth ligand, the key property that the entity must show is stability against irradiation. In principle, very weak donors such as nitriles, pirazines, etc. cannot be used, since their tendency is to be expelled from the complex upon irradiation, but also in the dark, being substituted by the solvent (usually water). The exception could be PPh_3, which despite its weak basicity (pKa ~ 11)[67] forms a notably stable bond with the metal center, due to its strong π-acceptor capability. Despite its usual stability, Steinke, Turro et al. recently reported the photochemical exchange of the PPh_3 ligand in a quite unusual Ru complex.[68]

Trimethylphosphine (PMe_3) combines two desirable properties, being an excellent π-acceptor, it is also a strong σ base, with $pK_b \sim 5$, and these confer superb stability as a fifth ligand. Although the complexes bearing PMe_3, as depicted in Fig. 2, structures **xi** to **xvi,** exhibit somewhat lower quantum yields of photo-delivery, the high thermal stability and low lipophilicity made this auxiliary ligand the one chosen for most biological applications.

As a rule of thumb, for two monodentate ligands in a $[Ru(bpy)_2L_1L_2]^{n+}$ complex, the one that will be expelled after irradiation is the one that has lower basicity (unless it is a phosphine). The complex $Ru(bpy)_2(NH_2CH_3)AN$ (AN = acetonitrile) will deliver AN upon irradiation, whereas the similar $Ru(bpy)_2(NH_2CH_3)PMe_3$ will expel the trimethylamine. In case both ligands have similar basicity, a proportion of each one will be delivered.

Regarding the absorption maxima, and therefore the quantum yields, basicity and π-acceptor properties of the fifth ligand can be used to tune them. More basic ligands will shift the band toward NIR, while poor bases, and strong π acceptors will move it to the UV. Both effects σ and π take place at the same time, and somewhat cancel out, as happens in the complexes bearing the strong σ base and π acceptor PMe_3. Fig. 5 shows the strong correlation between absorption maxima and quantum yield for complexes of the structure $[Ru(bpy)_2L_1L_2]^{n+}$. Although a general trend is easily visible, some interesting oddities appear. The complexes $[Ru(bpy)_2PMe_3L]^{n+}$ where L = Glutamate (**xi**), GABA (**xii**), Arginine (**xiii**), dopamine (**xiv**) and 5-hydroxytryptamine (**xv**), all of them coordinated through their primary amine group, show very different quantum yields of photo-delivery, although their absorption maxima are very close, all around 445 nm. (Comparison is somewhat difficult since the experiments which led to the values were done under somewhat different conditions.

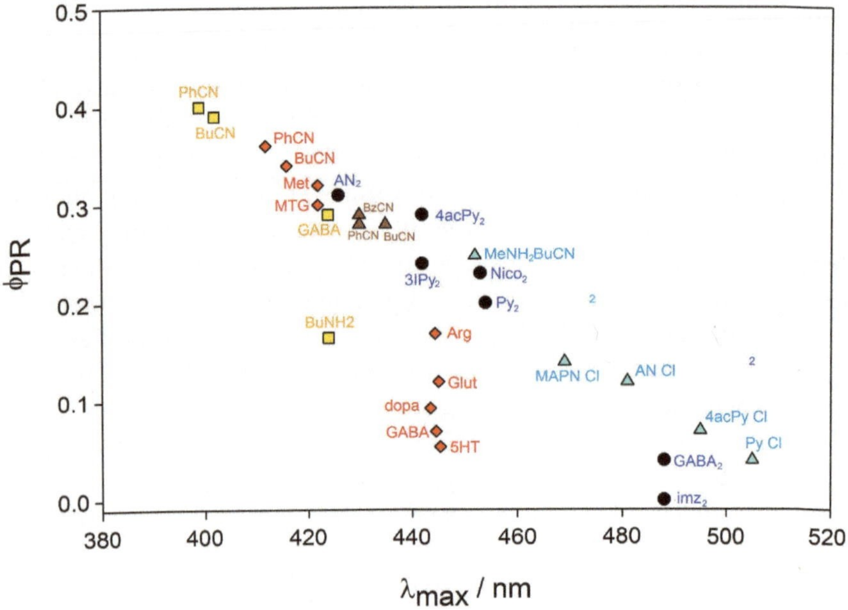

Fig. 5 A strong correlation is observed between the quantum yield of photolysis ϕ_{PR} and the wavelength of the maximum on the [1]MLCT band, although dependence with the exact nature of the ligand can significantly change these parameters.

In order to build this figure, we re-run some photolysis in distilled water, at 25 °C, using a unique irradiation wavelength of 405 nm). Arginine is delivered with $\phi_{PR} = 0.17$, serotonin is much more photostable, and presents a $\phi_{PR} = 0.05$, with Glutamate, dopamine and GABA in intermediate places. The overall structure of the ligand seems to play a role in the photo-delivery efficiency, even for complexes with very similar absorption spectra.

It is important to note that changes in the bipyridine-type ligands and changes in the auxiliary fifth ligand may modify the chemical properties in the same or in opposite ways. For instance, replacing the Py in the complex $[Ru(bpy)_2Py(AN)]^{2+}$ by isonicotinic acid introduces a carboxylate in position 4, which has several consequences. (1) The overall charge of the complex at physiological pH diminishes one unit, due ionization of the carboxylate group. (2) The hydrophilicity of the molecule increases. (3) The absorption maximum shifts toward UV, due to the acceptor nature of carboxylate which causes a lower donor ability at the pyridinic N. As shown in Fig. 5 the quantum yield of photorelease will probably be higher. (4) The redox potential increases, due to the same fact.

On the other hand, if a similar modification is made to one of the bipyridine rings, while properties 1, 2 and 4 change in the same way, and probably in almost the same magnitude, the position of the band (3) will be shifted in the opposite direction. Although a similar effect is also present on the Ru, now the –COOH, acceptor, is directly bonded to the bpy and shifts the MLCT to this bidentate ligand to lower energy. Hence, this is the main effect, diminishing the energy needed to reach the excited state (formally) Ru^{III}-bpy^{-}.

It is difficult to predict the outcome regarding the quantum yield given that the bipyridine ligands have changed. In practice, it is much more common to synthesize complexes bearing all the 4 rings with the same change (i.e., dicarboxybypyridines, dimethylbipyridines, phenanthrolines, biquinolines, etc.), so the differences are bigger, due to the simultaneous change of 4 coordination positions. Heteroleptic complexes of the type $[Ru(bpy_1)(bpy_2)L_1L_2]^{n+}$ have been synthesized several times, but their applications are much less visited than their homoleptic cousins. Finally, modifications in just one of the four rings are really scarce.

6. Three-dimensional localization

One of the most important characteristics that make caged compounds a useful tool, is the capability to deliver a drug precisely in time and space. Although the precise delivery in x and y dimensions (2D) is quite straightforward by using optical instruments, it is much more difficult to define the z position (depth) with high precision. The advantage of using multiphoton optics is shown in Fig. 6. The left panel (a) depicts a nonlinear (2 photon) focusing, while the right panel (b) shows standard linear focusing, both using the same microscope objective. In the latter case the total amount of light in any z-plane from the objective to the focal point and beyond is the same (regardless of absorption and/or dispersion). This fact means that the total amount of biomolecules uncaged in any plane will be also similar. Far from the focus this amount is somewhat diluted due to the higher area, but the total absolute uncaging above and below the focal plane is much more than in the desired place, and implies an important loss of precision in the z-axis.

One way to avoid this problem involves the use of the two-photon (2P) absorption phenomenon. When 2 photons of about one half of the energy required to reach an excited state reach a molecule in a quasi-simultaneous manner, it is possible to populate this state. However, in order to have a

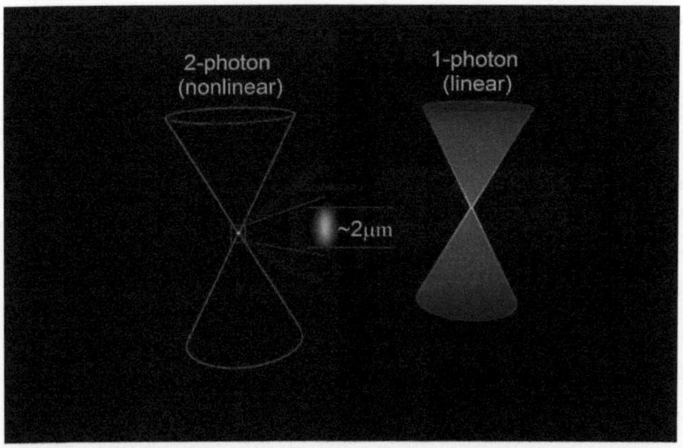

Fig. 6 Comparison of linear focusing (right) in a standard microscope with 2-photon excitation (left), obtained with a femtosecond pulsed NIR laser of very high instantaneous power. The quadratic response of the emission (or uncaging) implies a significant effect only in the very vicinity of the focal plane. Inset at the middle: an enlargement of the focal volume in 2P excitation.

measurable response, very high instantaneous power must be used. As the 2 photons must act simultaneously, the probability of absorption is not a linear function of the intensity as in the usual 1P absorption regime, but of the square of that intensity. This means that the focal point will have enough power density to elicit the excited state population (and uncaging) while far from it the total amount released will be near zero, as shown in Fig. 6 (left). For a good quality, high aperture objective, the precision in the z plane is typically near 2 μm, while in X and Y is usually a bit over the diffraction limit.

On this basis, one of the parameters that became important in the last generations of caged compounds is the capability to absorb NIR light in 2P regime. This capability, analogous to the molar absorptivity of an absorber, is usually given in terms of 2-photon absorption cross section, and measured in GM (1 GM $=10^{-50}$ cm^4 s/photon molecule).

RuBi4AP, the first member of the Ru based caged compounds, was demonstrated to be active in two photon regime[69] with 0.01 to 0.1 GM for action cross section (absorption cross section multiplied by the quantum yield of photo-delivery). This value is moderate, high enough to perform all the typical experiments of uncaging in biological specimens with high depth precision. The following GABA, Glutamate, Dopamine, Serotonin phototriggers were tested also active in 2P, indicating that the core structure of the complex presents 2P absorption capabilities somewhat regardless of the detachable ligand.

The 2P action cross-section is not easy to estimate. There are two main methods to measure the absorption and/or action cross section: z-scan[70] and 2P fluorescence comparison.[71] The latter is only useful when the substance undergoes radiative decay, emitting fluorescence or phosphorescence. The procedure needs a known fluorophore as standard to be compared with the emission of the sample. It is important to note that the focal zone or point-spread function (PSF) must be very well characterized. Multiphoton absorption is highly dependent on the light density, so a broader focus will promote much less emission than a sharp one. This fact introduces a difficult issue to solve in order to estimate the cross section. Even if these conditions are fulfilled, the method is quite useful to characterize fluorescent dyes, but no data of action cross section is obtained and the estimation must be done on the basis that both the quantum yield of photolysis and the relationship between emission and photolysis remain the same regardless of the excitation mechanism. *Z*-scan, on the other hand, is a technique that can be used with any absorber. The sample is moved through the optical axis of a focused laser beam, and the absorption is recorded. Although in theory this technique is straightforward, many confounding effects can appear, such as refraction issues due to the high instantaneous power of the irradiation, or excited state absorption by the sample, both of which lead to artifacts of very high absorption cross section. In this case, although no emission parameters are needed, the quantum yield of photolysis must be supposed to be equal to the parameter in the linear regime.

In order to characterize the capability of Ru-bpy complex to deliver the biomolecule upon 2P irradiation, we devised a new method that resembles the simplest way to measure the photoactivity in the linear regime: photolysis into a cuvette and measurement of spectral changes during a certain time. As the 2P irradiation only has effect in the close proximity of the focal point, the amount of delivered ligand is very small. A very small cuvette volume is needed to detect appreciable changes in its concentration. On a tiny volume, comparable with the typical field of a microscope, the direct measurement of the action cross section becomes straightforward. In our lab we built arrays of microcuvettes, in the tens of micrometers size range, which can be used to photolyze the complex using the same microscope setup that is used for the photo-physiological experiments. Fig. 7A shows

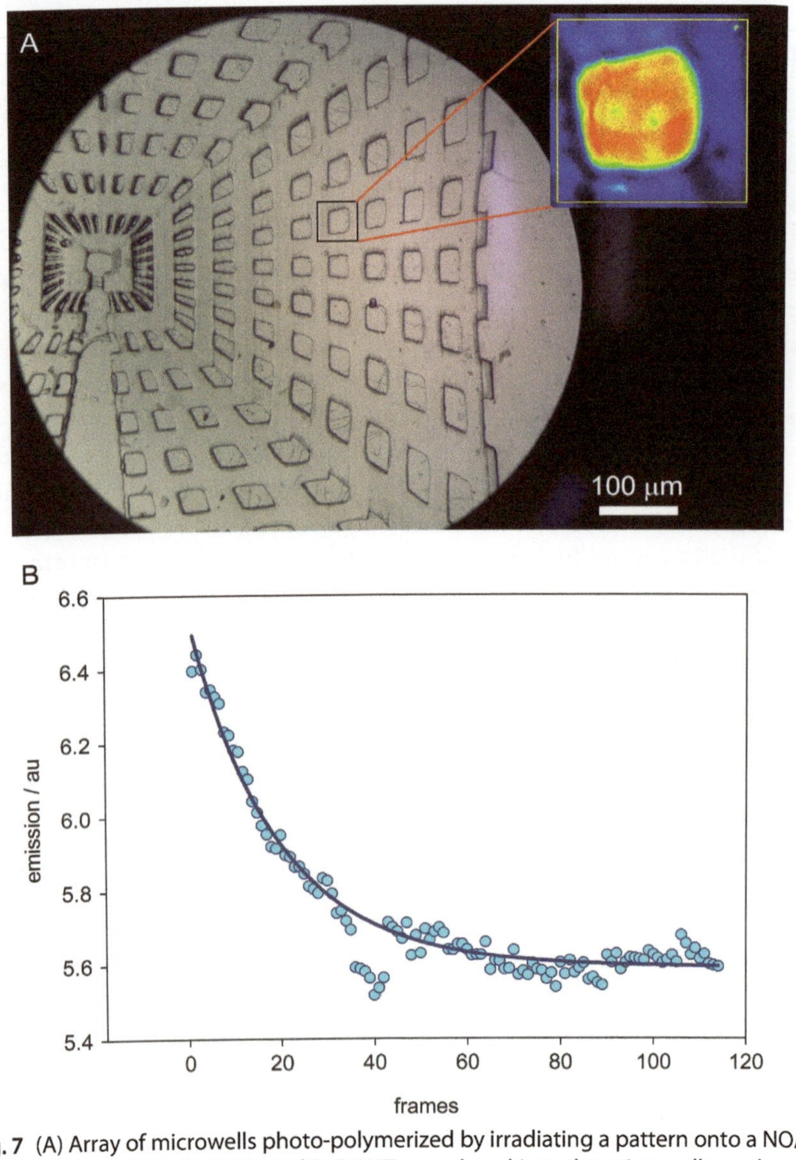

Fig. 7 (A) Array of microwells photo-polymerized by irradiating a pattern onto a NOA60 film on a glass slide. A solution of RuBi5HT was placed into the microwells, and one of them was irradiated with a 2P microscope. The inset shows a false color image of the complex emission during the 2P excitation. (B) Emission signal during irradiation, showing the aquation of the complex. (The aqua complex has lower emission.)

one of these cuvette assemblies, which are made by UV polymerization of an optical grade adhesive NOA60.[72]

The set of wells are polymerized by means of a 405 nm laser beam focused through a fluorescence microscope and controlled with XY galvanometers. The beam draws onto a sandwich of slide and coverglass of the desired thickness. The lateral dimensions, and thus area, of any of the wells can be easily measured by optical microscopy, while the depth of the resulting wells can be precisely determined measuring the absorbance of a dye solution of known molar absorptivity and concentration. Therefore, the volume of each well can be known and used as a microcuvette. A drop of a solution of the caged compound is placed onto the structure and a coverglass used as cap. The photolysis of the complex is performed in the same setup used for physiological experiments: a fluorescent microscope with femtosecond laser excitation, while recording the change in fluorescence (or absorption, if the sample is not fluorescent) with standard linear optics. Fig. 7A (inset) shows a false color picture of a well filled with a RuBi5HT solution and being photolyzed (and excited) at 777 nm. From the fitting of the curve of emission vs. time (Fig. 7B), and having a standard of known action cross section, the absorption/action parameters of the caged compound can be obtained. Some additional calculations may be needed to prevent artifacts arising from leakage between wells, which can be important when dealing with small dimensions. Measuring RuBi-caged compounds by using this technique yields values near 1 GM for the 2P absorption cross section, corresponding to 0.05–0.5 GM for 2P action cross section, depending on the quantum yield of photolysis.

Traditional multiphoton absorption with high power lasers is not the only mechanism by which two or more low energy NIR photons can be used to populate a high energy excited state, usually reached with visible or UV light. Upconversion is an alternative strategy, in which a ladder of long lived and somewhat equispaced excited states allows the sequential absorption of several photons in the microseconds range. This allows the use of very cheap laser diodes instead of expensive and large femtosecond lasers. Lanthanide ions usually exhibit these properties, being Er^{3+}, Tm^{3+} and Ho^{3+} the ones more widely used. After absorbing a number of photons, the system decays in a single step, emitting a visible or UV photon, with a high anti-Stokes shift. To achieve a high efficiency, it is useful to synthesize upconverting nanoparticles (UCNPs) containing the lanthanide ions in a suitable lattice[73]

As with traditional 2P absorption, upconversion processes give two advantages: use of low energy photons, which means less toxicity and more penetration, and z-axis sectioning, preventing the areas above and below the desired section from being contaminated with the released chemical. Although the first point was indisputable since the beginning, the latter was object of controversy. We were able to demonstrate that upconversion presents similar z-sectioning capability than traditional 2P, which was expected since the equations governing the intensity vs. power excitation are similar to those describing 2P behavior. The results[74,75] show that upconversion can be used to achieve true 3D localization.

Reaching the therapeutic window (600–1000 nm) is key in order to devise treatments based on the photo-delivery of drugs. Alternative to the use of UCNPs, upconversion can also be achieved by the Triplet-Triplet-Annihilation (TTA) mechanism. Bonnet and coworkers pioneered the use of Ru phototriggers under TTA-Upconversion[76] in liposomes, and showed that a thioether complex of the type $[Ru(tpy)(bpy)SR_2]^{n+}$ can be photolyzed using 630 nm light, although its ^{1}MLCT band lies around 450 nm. Salassa and coworkers followed the alternative pathway, and showed that the combination of Ru-bpy complexes and UCNPs can extend to the NIR the photosensitivity of the model complex $[Ru(bpy)_2Py_2]^{2+}$.[77] He et al. combined lanthanide-based upconverting nanoparticles (UCNP) with ruthenium-bipyridyl phototriggers to achieve an active drug photo-delivery upon NIR irradiation,[78] doxorubicin is released in an indirect way from a composite of a core-shell UCNP covered with silica and containing both the drug and a Ru-bpy complex bearing an aminosilane tail to be anchored to the silica. In a different approach, Meijer et al. tailored UCNPs with Ru-bpy complexes by means of a thioether-bridge-phosphonate construct.[79] Upon irradiation with 796 nm NIR light, the Ru complex itself (**vi**) is released as a bis-aqua complex.

Another possible way to elicit changes in chemical and biological systems would be to change the local pH in a controlled way. In our lab, we demonstrated that a Ru-bpy complex bearing a coordinated amine can be used to raise the pH by up to 5 units in a quasi-instantaneous way.[80] Wu extended the applicability of the method to the NIR (980 nm) by combining the complex and UCNPs within an hydrogel.[81] Upconversion-triggered caged compounds are yet a promising newborn child and their capabilities will surely be developed in the next few years.

A different technique that allows enhanced sensitivity to photons outside the maximum absorbance of the coordination complex is to include a highly absorptive fluorophore in the vicinity of the Ru coordination sphere. If the emission spectrum of the fluorophore overlaps with the absorption spectrum of the coordination complex, then the energy of photons captured by the fluorophore at its own absorption maximum can be transferred to the complex through FRET to trigger photo-release. A common mistake regarding the FRET mechanism consists in believing that for a Forster energy transfer to be efficient, the energy donor must have its emission spectrum at shorter wavelengths than those of the acceptor. Really, FRET mechanism can only occur when donor and acceptor energies are the same, and the requirement for a blue-shifted donor is just a convenient way to obtain a wide overlap between both spectra and thus high FRET at long distances. However, even if the donor is red-shifted, considering that both donor emission and acceptor absorption spectra are broad, the resulting overlap can be high enough. This kind of energy transfer, in which the donor emission band lays mainly at lower energies than the acceptor absorption band is commonly referred as "reverse FRET." In our case, the energy transfer can achieve even near 100% efficiency at the very short distances involved in a coordination bond.

We were able to show that this strategy actually works with a fluorescent compound which is quenched while coordinated to a Ru-bpy moiety, but after photorelease recovers its fluorescence.[61] Bonnet and coworkers extended the concept by showing that another ligand, which did not absorb the exciting photon, can instead be expelled. In this case, photolysis of the complex causes the photo-release of the most labile bond.[64] In our lab, we revisited the mechanism by which the energy capture and transfer and subsequent photorelease operates,[63] showing that the reverse-FRET efficiency can be as high as 0.98, even with a rather low spectrum overlap. The strategy of coordinating auxiliary fluorophores to the Ru-bpy center can be used to achieve two goals: Firstly, increase the molar absorptivity of the caged compound at a wavelength corresponding to an electronic transition of very low probability in the complex, but high probability in the fluorophore, extending the spectral sensitivity, and secondly, use the fluorophore absorption spectrum instead of that of the Ru complexes to establish best excitation wavelength, allowing a broader tunability and enhanced orthogonality of two or three Ru-based phototriggers.

7. *Cis, trans* and other structures

A characteristic of the vast majority of the Ru^{II} complexes bearing two bipyridines or its analogues is that these bidentate ligands are arranged in a *cis* configuration. There are several ways to synthesize the most common starting material, the neutral $[Ru(bpy)_2Cl_2]$, and all of them yield the thermodynamically favored *cis*-$[Ru(bpy)_2Cl_2]$.[82–84] The preparation of complexes with the *trans* configuration have hardly been studied. Early studies by Durham and Meyer on the photochemical properties of the aqua complex *cis*-$[Ru(bpy)_2(H_2O)_2]^{2+}$ showed that upon irradiation, the bis-aqua complex undergoes a transition to the *trans* form, which can be isolated due to its lower solubility as perchlorate salt.[85] To that day, it was thought that only the *cis* form could be obtained due to the steric hindrance between the hydrogens of *trans*-coordinated bipyridines. Although *trans* diaqua complexes can be prepared and even obtained as crystals, they revert to the *cis* form in solution in minutes, depending on the temperature. After this, some other *trans* complexes were prepared, bearing different ligands,[86–89] although no systematic attempt to compare them with their *cis* analogues were done.

In the interest of synthesizing new caged compounds, we noticed that some preparations presented a characteristic impurity in their NMR spectra. This impurity was consistently a *trans* form of the complex. Although some thermal synthesis of *trans* complexes were reported,[90] the usual way to obtain *trans* isomers is photochemically, so we decided to prepare the *trans* isomer of our starting complex $[Ru(bpy)_2PMe_3(H_2O)]^{2+}$, which would allow us to prepare a new family of phototriggers. *trans*-$[Ru(bpy)_2PMe_3(H_2O)]^{2+}$ was unexpectedly stable, its isomerization half-life in solution at room temperature is about 50 times longer than that of *trans*-$[Ru(bpy)_2(H_2O)_2]^{2+}$, while its lability toward monodentate ligand exchange was much higher than that of the *cis* isomer.[91]

Trans complexes exhibit a considerable bathochromic shift of their ^{1}MLCT band compared to their *cis* analogues. However, their quantum yields of photosubstitution are much higher than those for the corresponding *cis* forms. On the other hand, the thermal lability of *trans* complexes can be a drawback for applications in which a high stability in the dark is needed, i.e., drug delivery. Conversely, usage of *trans* Ru-bpy caged compounds in acute preparations, (e.g., typical neurophysiological experiments) can be one of the ways to achieve long wavelength sensitivity coupled with high quantum yields.

From the previous considerations, it is evident that just one bipyridine in the structure is enough to elicit the basic photochemical properties of Ru-bpy complexes. The complexes of the form $[Ru(bpy)L^1L^2L^3L^4]^{n+}$, in which L^1-L^4 are the same or different ligands often present good photolysis yield and therefore can be used for uncaging. Salassa, Sadler and coworkers showed that the complex $[Ru(bpy)(4AP)_4]^{2+}$ (4AP = 4-aminopyridine) can be photolyzed, yielding the *mer* and *trans* mono and di-aqua complexes in a two-step photoreaction.[92] A similar pathway is present in the complex

$[Ru(dppz)(4AP)_4]^{2+}$ (dppz = dipyrido[3,2-*a*:2′,3′-*c*]phenazine).[93] Our group had studied the similar complex $[Ru(bpy)AN_4]^{2+}$ (AN = acetonitrile), showing that it presents a strongly selective axial photocleavage toward *fac* aqua complexes in the first step.[94] The difference between the two behaviors lies in the very different basicity of AN and 4AP ligands compared with the bpy, which determines the orbital ordering and therefore the nature of the lowest lying ^{3}dd state responsible for the photochemistry. Sadler and coworkers have shown that Ru^{II}-arene complexes bearing a bipyrimidine or bipyridine can also be used as phototriggers. The complex $[Ru(bpm)(p\text{-cym})(py)]^{2+}$ (*p*-cymene, bpm = 2,2′-bipyrimidine) was the first example of a Ru-arene complex able to selectively dissociate a monodentate ligand upon irradiation.[95] Once pyridine is released, the aqua residue can coordinate a DNA base. Other compounds sharing the same structure were also presented. Although their absorptivity at visible wavelengths and quantum yields of photolysis are quite lower than the same figures for typical Ru-bpy complexes, some of these molecules can inhibit RNA synthesis upon irradiation,[96] a key feature to base an anti-cancer therapy.

8. A brief gaze into the future

No scientific breakthrough stands isolated. This family of caged biomolecules developed in-hand with optogenetics, and with the concept of "all-optical" experimental electrophysiology setups. Such setups combine multicolor or tunable lasers, confocal or even super-resolution microscopes and advances in fluorescent genetically encoded probes and light-activated ion channels.[97] They make it possible to measure biologically relevant variables, such as transmembrane voltage, pH and intracellular calcium concentration by imaging changes in the fluorescence of proteins expressed by the model animals themselves, usually induced by the injection of a viral vector. As may be noticed, traditional (mechanical) electrophysiology equipment was replaced or at least supplemented by advances in genetic engineering, protein design, tunable-laser technology and imaging sensors.[98] Timing of it all was key to the easy adoption of caged-compound technology.

In particular, Ru-based caged compounds came to fill some gaps in the usability of traditional organic phototriggers. The first one was long wavelength uncaging. Although some organic photoremovable groups were found useful to extend the range from the original UV to the blue and blue-green, Ru complexes can be used well beyond, with yellow, orange and even red light. On the other hand, phototriggering through usual organic caged compounds is frequently quite slow as compared with the nanosecond-scale of Ru-complexes. Thirdly, the modular nature of the complex synthesis allows a wider and richer set of modifications by which chemical and photophysical/photochemical properties can be easily determined. Redox potential, lipophilicity, efficiency, maximum absorption wavelength, quantum yield, are among the parameters that can be tuned by a proper and systematic selection of ligands.

These characteristics, together with the reverse-FRET excitation will allow researchers to devise orthogonal pairs of caged compounds, and to fine-tune orthogonal sets of optical actuator-probes, which are needed for high throughput neural stimulation and simultaneous activity recording by imaging.

Optogenetics is often stated as a technique that will overcome uncaging. However, the two procedures are rather complementary. Optogenetics needs previous molecular biology work, which is in cases expensive, often difficult or even impossible. Uncaging can be used not just for signaling but also as a way to deliver a drug, pursuing a medical objective. Anti-cancer drugs, usually exhibiting deleterious side effects in organisms, could be applied only in the affected tissues, even with cellular precision, at the required time. Medicine has not yet witnessed the possibilities of optochemistry as a tool, except in a couple of treatments. A long road will surely be built and paved in the next decades.

Acknowledgments

The authors thanks UBA, UNSAM, CONICET and ANPCyT for research funding. Fruitful talks with Dr. Leo Slep are acknowledged. R.E. and O.F. are members of CONICET.

References

1. Hansen, K. C.; Rock, R. S.; Larsen, R. W.; Chan, S. I. *J. Am. Chem. Soc.* **2000**, *122* (46), 11567–11568.
2. Chen, R. P.-Y.; Huang, J. J.-T.; Chen, H.-L.; Jan, H.; Velusamy, M.; Lee, C.-T.; Fann, W.; Larsen, R. W.; Chan, S. I. *Proc. Nat. Acad. Sci. U. S. A.* **2004**, *101*, 7305–7310.
3. Lester, H. A.; Mauri, E. K.; Krouse, M. E.; Nass, M. M.; Wasswermann, N. H.; Erlanger, B. F. *Nature* **1979**, *280*, 509–510.
4. (a) Bardhan, A.; Deiters, A.; Ettensohn, C. A. *Dev. Biol.* **2021**, *475*, 21–29; (b) Griepenburg, J. C.; Rapp, T. L.; Carroll, P. J.; Eberwineb, J.; Dmochowski, I. J. *Chem. Sci.* **2015**, *6*, 2342.
5. Shembekar, V. R.; Chen, Y.; Carpenter, B. K.; Hess, G. P. *Biochemistry* **2007**, *46*, 5479–5484.

6. Bao, C.; Fan, G.; Lin, Q.; Li, B.; Cheng, S.; Huang, Q.; Zhu, L. *Org. Lett.* **2012**, *14*, 572–575.
7. Josa Culleré, L.; Llebaria, A. *ChemPhotoChem* **2021**, *5*, 296–314.
8. López-Corrales, M.; Rovira, A.; Gandioso, A.; Bosch, M.; Nonell, S.; Marchán, V. *Chem. Eur. J.* **2020**, *26*, 16222–16227.
9. Lin, Q.; Yang, L.; Wang, Z.; Hua, Y.; Zhang, D.; Bao, B.; Bao, C.; Gong, X.; Zhu, L. *Angew. Chem., Int. Ed.* **2018**, *57*, 3722–3726.
10. Steinmetz, M. G.; Givens, R. S. *Photochem. Photobiol.* **2021**, *97*, 1168–1181.
11. Hennig, A.-L. K.; Deodato, D.; Asad, N.; Herbivo, C.; Dore, T. M. *J. Org. Chem.* **2020**, *85*, 726–744.
12. Bosnich, B.; Dwyer, F. P. *Aus. J. Chem.* **1966**, *19*, 2229–2233.
13. Pinnick, D. V.; Durham, B. *Inorg. Chem.* **1984**, *23*, 1440–1445.
14. Durham, B.; Walsh, J. L.; Carter, C. L.; Meyer, T. J. *Inorg. Chem.* **1980**, *19*, 860–865.
15. Kohle, O.; Grätzel, M.; Meyer, A. F.; Meyer, T. B. *Adv. Mater.* **1997**, *9*, 904–906.
16. Agrell, H. G.; Lindgren, J.; Hagfeldt, A. *Solar Energy* **2003**, *75*, 169–180.
17. Roose, M.; Tassé, M.; Lacroix, P. G.; Malfant, I. *New J. Chem.* **2019**, *43*, 755–767.
18. Pereira, F. L.; Cicillini, S. A.; Biazzotto, J. A.; Schiavon, M. A.; Mikhailovsky, A.; Burks, P.; Garcia, J.; Ford, P. C.; Santana da Silva, R. *J. Phys. Chem. A* **2014**, *118*, 12184–12191.
19. Zayat, L.; Calero, C.; Alborés, P.; Baraldo, L. M.; Etchenique, R. *J. Am. Chem. Soc.* **2003**, *125*, 882–883.
20. Zayat, L.; Noval, M. G.; Campi, J.; Calero, C. I.; Calvo, D. J.; Etchenique, R. *ChemBioChem* **2007**, *8*, 2035–2038.
21. Filevich, O.; Etchenique, R. *Photochem. Photobiol. Sci.* **2013**, *12*, 1565–1570.
22. Salierno, M.; Marceca, E.; Peterka, D. S.; Yuste, R.; Etchenique, R. *J. Inorg. Biochem.* **2010**, *104*, 418–422.
23. Peterka, D. S.; Nikolenko, V.; Fino, E.; Araya, R.; Etchenique, R.; Yuste, R. *Proc. SPIE* **2010**, 7548.
24. Carrillo-Reid, L.; Yang, W.; Bando, Y.; Peterka, D. S.; Yuste, R. *Science* **2016**, *353*, 691–694.
25. Carrillo-Reid, L.; Han, S.; Yang, W.; Akrouh, A.; Yuste, R. *Cell* **2019**, *178*, 1–11.
26. Carrillo-Reid, L.; Yuste, R. *Curr. Opin. Neurobiol.* **2020**, *64*, 89–95.
27. Rojas Perez, Y.; Etchenique, R. *Photochem. Photobiol. Sci.* **2019**, *18*, 208–212.
28. Kazemipour, A.; Novak, O.; Flickinger, D.; Marvin, J. S.; Abdelfattah, A. S.; King, J.; Borden, P. M.; Kim, J. J.; Al-Abdullatif, S. H.; Deal, P. E.; Miller, E. W.; Schreiter, E. R.; Druckmann, S.; Svoboda, K.; Looger, L. L.; Podgorski, K. *Nat. Meth.* **2019**, *16*, 778–786.
29. Araya, R.; Andino-Pavlovsky, V.; Yuste, R.; Etchenique, R. *ACS Chem. Neurosci.* **2013**, *4*, 1163.
30. Booker, S. A.; Domanski, A. P. F.; Dando, O. R.; Jackson, A. D.; Isaac, J. T. R.; Hardingham, G. E.; Wyllie, D. J. A.; Kind, P. C. *Nat. Comm.* **2019**, *10*, 4813–4826.
31. Marvin, J. S.; Scholl, B.; Wilson, D. E.; Podgorski, K.; Kazemipour, A.; Müller, J. A.; Schoch, S.; Quiroz, F. J. U.; Rebola, N.; Bao, H.; Little, J. P.; Tkachuk, A. N.; Cai, E.; Hantman, A. W.; Wang, S. S.; DePiero, V. J.; Borghuis, B. G.; Chapman, E. R.; Dietrich, D.; DiGregorio, D. A.; Fitzpatrick, D.; Looger, L. L. *Nat. Meth.* **2018**, *15*, 936–939.
32. Yada, Y.; Yasuda, S.; Takahashi, H. *Appl. Phys. Lett.* **2021**, *119*, 173701–173705.
33. Guo, C.; Huson, V.; Macosko, E. Z.; Regehr, W. G. *Nat. Comm.* **2021**, *12*, 5491–5503.
34. Lau, B. K.; Murphy-Royal, C.; Kaur, M.; Qiao, M.; Bains, J. S.; Gordon, G. R.; Borgland, S. L. *Cell Rep.* **2021**, *36*, 109563–109576.
35. Wang, Y.; DelRosso, N. V.; Vaidyanathan, T. V.; Cahill, M. K.; Reitman, M. E.; Pittolo, S.; Mi, X.; Yu, G.; Poskanzer, K. E. *Nat. Neurosci.* **2019**, *22*, 1936–1944.

36. Poskanzer, K.; Yuste, R. *Proc. Nat. Acad. Sci. U. S. A.* **2011**, *108*, 18453–18458.
37. Cabrera, R.; Filevich, O.; Garcia Acosta, B.; Athilingam, J.; Bender, K. J.; Poskanzer, K. E.; Etchenique, R. *ACS Chem. Neurosci.* **2017**, *8*, 1036–1042.
38. Tazerart, S.; Mitchell, D. E.; Miranda-Rottmann, S.; Araya, R. *Nat. Comm.* **2020**, *11*, 4276.
39. Mitchell, D. E.; Martineau, É.; Tazerart, S.; Araya, R. *Front. Syn. Neurosci.* **2019**, *11*, 19.
40. Pei, Z.; Chen, Q.; Koren, D.; Giammarinaro, B.; Acaron Ledesma, H.; Wei, W. *J. Neurosci.* **2015**, *35*, 13219–13232.
41. Chu, H.-Y.; Atherton, J. F.; Wokosin, D.; Surmeier, D. J.; Bevan, M. D. *Neuron* **2015**, *85*, 364–376.
42. Cozzolino, M.; Bazzurro, V.; Gatta, E.; Bianchini, P.; Angeli, E.; Robello, M.; Diaspro, A. *Sci. Rep.* **2020**, *10*, 13380–13389.
43. Hayama, T.; Noguchi, J.; Watanabe, S.; Takahashi, N.; Hayashi-Takagi, A.; Ellis-Davies, G. C. R.; Matsuzaki, M.; Kasai, H. *Nat. Neurosci.* **2013**, *16*, 1409–1418.
44. Lovett-Barron, M.; Turi, G. F.; Kaifosh, P.; Lee, P. H.; Bolze, F.; Sun, X.-H.; Nicoud, J.-F.; Zemelman, B. V.; Sternson, S. M.; Losonczy, A. *Nat. Neurosci.* **2012**, *15*, 423–430.
45. Kwon, T.; Merchán-Pérez, A.; Rial Verde, E. M.; Rodríguez, J. R.; DeFelipe, J.; Yuste, R. *Cereb. Cortex* **2019**, *29*, 2771–2781.
46. Yang, X.; Rode, D. L.; Peterka, D. S.; Yuste, R.; Rothman, S. M. *Ann. Neurol.* **2012**, *71*, 68–75.
47. Wang, D.; Yu, Z.; Yan, J.; Xue, F.; Ren, G.; Jiang, C.; Wang, W.; Piao, Y.; Yang, X. *Front. Neurol.* **2017**, *8*, 215.
48. Dai, Y.; Song, Y.; Xie, J.; Xiao, G.; Li, X.; Li, Z.; Gao, F.; Zhang, Y.; He, E.; Xu, S.; Wang, Y.; Zheng, W.; Jiang, X.; Qi, Z.; Meng, D.; Fan, Z.; Cai, X. *ACS Appl. Mater. Interf.* **2020**, *12*, 41148–41156.
49. Rowan, M. J. M.; Tranquil, E.; Christie, J. M. *J. Neurosci.* **2014**, *34*, 6611–6623.
50. Burton, R.-A. B.; Tomek, J.; Ambrosi, C. M.; Larsen, H. E.; Sharkey, A. R.; Capel, R. A.; Corbett, A. D.; Bilton, S.; Klimas, A.; Stephens, G.; Cremer, M.; Bose, S. J.; Li, D.; Gallone, G.; Herring, N.; Mann, E. O.; Kumar, A.; Kramer, H.; Entcheva, E.; Paterson, D. J.; Bub, G. *iScience* **2020**, *23*, 101334.
51. Filevich, O.; Salierno, M.; Etchenique, R. *J. Inorg. Biochem.* **2010**, *104*, 1248–1251.
52. Levin, N.; Codesido, N. O.; Bill, E.; Weyhermüller, T.; Segantin Gaspari, A. P.; da Silva, R. S.; Olabe, J. A.; Slep, L. D. *Inorg. Chem.* **2016**, *55*, 7808–7810.
53. Sauaia, M. G.; de Souza Oliveira, F.; de Lima, R. G.; Cacciar, I. A.; Tfouni, E.; da Silva, R. S. *Inorg. Chem. Commun.* **2005**, *8*, 347–349.
54. Carrara, F.; Brumley, D. R.; Hein, A. M.; Yawata, Y.; Salek, M. M.; Lee, K. S.; Sliwerska, E.; Levin, S. A.; Stocker, R. *J. Visualized Exp.* **2020**, 155.
55. Meijer, M. S.; Bonnet, S. *Inorg. Chem.* **2019**, *58*, 11689–11698.
56. Siewert, B.; Langerman, M.; Pannwitz, A.; Bonnet, S. *Eur. J. Inorg. Chem.* **2018**, 4117–4124.
57. Mosquera, J.; Sánchez, M. L.; Mascarenas, J. L.; Vázquez, M. E. *Chem. Comm.* **2015**, *51*, 5501–5504.
58. van Rixel, V. H. S.; Ramu, V.; Auyeung, A. B.; Beztsinna, N.; Leger, D. Y.; Lameijer, L. N.; Hilt, S. T.; Le Devédec, S. E.; Yildiz, T.; Betancourt, T.; Brenton Gildner, B.; Hudnall, T. W.; Sol, V.; Liagre, B.; Kornienko, A.; Bonnet, S. *J. Am. Chem. Soc.* **2019**, *141*, 18444–18454.
59. Zayat, L.; Salierno, M.; Etchenique, R. *Inorg. Chem.* **2006**, *45*, 1728–1731.
60. Garner, R. N.; Gallucci, J. C.; Dunbar, K. R.; Turro, C. *Inorg. Chem.* **2011**, *50* (19), 9213–9215.
61. del Mármol, J.; Filevich, O.; Etchenique, R. *Anal. Chem.* **2010**, *82*, 6259–6264.

62. Filevich, O.; García-Acosta, B.; Etchenique, R. *Photochem. Photobiol. Sci.* **2012**, *11*, 843–847.
63. Carrone, G.; Gantov, F.; Slep, L. D.; Etchenique, R. *J. Phys. Chem. A.* **2014**, *118*, 10416–10424.
64. Bahreman, A.; Cuello-Garibo, J.-A.; Bonnet, S. *Dalton Trans.* **2014**, *43*, 4494–4505.
65. Havrylyuk, D.; Stevens, K.; Parkin, S.; Glazer, E. C. *Inorg. Chem.* **2020**, *59*, 1006–1013.
66. Al-Afyouni, M. H.; Rohrabaugh, T. N., Jr.; Al-Afyounia, K. F.; Turro, C. *Chem. Sci.* **2018**, *9*, 6711–6720.
67. Henderson, W. A.; Streuli, C. A. *J. Am. Chem. Soc.* **1960**, *82*, 5791–5794.
68. Steinke, S. J.; Gupta, S.; Piechota, E. J.; Moore, E.; Kodanko, J. J.; Turro, C. *Chem. Sci.* **2022**, *13*, 1933–1945.
69. Nikolenko, V.; Yuste, R.; Zayat, L.; Baraldo, L. M.; Etchenique, R. *Chem. Comm.* **2005**, 1572–1574.
70. Belfield, K. D.; Bondar, M. V.; Hernandez, F. E.; Przhonska, O. V.; Yao, S. *J. Phys. Chem. B* **2007**, *111*, 12723–12729.
71. Wecksler, S. R.; Mikhailovsky, A.; Korystov, D.; Buller, F.; Kannan, R.; Tan, L. S.; Ford, P. C. *Inorg. Chem.* **2007**, *46*, 395–402.
72. Cabrera, R.; Gabriel, M.; Estrada, L.; Etchenique, R. *Anal. Chem.* **2019**, *91*, 5968–5972.
73. Garcia, J. V.; Yang, J.; Shen, D.; Yao, C.; Li, X.; Stucky, G. D.; Zhao, D.; Ford, P. C.; Zhang, F. *Small* **2012**, *8*, 3800–3805.
74. Hodak, J.; Chen, Z.; Wu, S.; Etchenique, R. *Anal. Chem.* **2016**, *88*, 1468–1475.
75. Sorbello, C.; Etchenique, R. *Chem. Commun.* **2018**, *54*, 1861–1864.
76. Askes, S. H. C.; Bahreman, A.; Bonnet, S. *Angew. Chem. Int. Ed.* **2013**, *53*, 1029–1033.
77. Ruggiero, E.; Habtemariam, A.; Yate, L.; Mareque-Rivas, J. C.; Salassa, L. *Chem. Comm.* **2014**, *50*, 1715–1718.
78. He, S.; Krippes, K.; Ritz, S.; Chen, Z.; Best, A.; Butt, H.-J.; Wu, S. *Chem. Comm.* **2015**, *51*, 431–434.
79. Meijer, M. S.; Natile, M. M.; Bonnet, S. *Inorg. Chem.* **2020**, *59*, 14807–14818.
80. Filevich, O.; Carrone, G.; Andino Pavlovsky, V.; Etchenique, R. *Anal. Chem.* **2012**, *84*, 5618–5624.
81. Chen, Z.; Xiong, Y.; Etchenique, R.; Wu, S. *Chem. Comm.* **2016**, *52*, 13959–13962.
82. Sullivan, B. P.; Salmon, D. J.; Meyer, T. J. *Inorg. Chem.* **1978**, *17*, 3334–3341.
83. Lay, P. A.; Sargeson, A. M.; Taube, H.; Chou, M. H.; Creutz, C. *Inorg. Synth.* **2007**, 291–299.
84. Viala, C.; Coudret, C. *Inorg. Chim. Acta* **2006**, *359*, 984–989.
85. Durham, B.; Wilson, S. R.; Hodgson, D. J.; Meyer, T. J. *J. Am. Chem. Soc.* **1980**, *102*, 600–607.
86. Cordes, A. W.; Durham, B.; Swepston, P. N.; Pennington, W. T.; Condren, S. M.; Jensen, R.; Walsh, J. L. *J. Coord. Chem.* **1982**, *11*, 251–260.
87. Bonneson, P.; Walsh, J. L.; Pennington, W. T.; Cordes, A. W.; Durham, B. *Inorg. Chem.* **1983**, *22*, 1761–1765.
88. Amouri, H.; Waern, J. B.; Caspar, R.; Barbieri, A.; Sabatini, C.; Zanelli, A.; Barigelletti, F. *Dalton Trans.* **2007**, *21*, 2179–2186.
89. Cordes, A. W.; Durham, B.; Pennington, W. T.; Kuntz, B.; Allen, L. *J. Cryst. Spec. Res.* **1992**, *22*, 699–704.
90. Rack, J. J.; Mockus, N. V. *Inorg. Chem.* **2003**, *42*, 5792–5794.
91. Rojas Pérez, Y.; Slep, L. D.; Etchenique, R. *Inorg. Chem.* **2019**, *58* (17), 11606–11613.
92. Salassa, L.; Garino, C.; Salassa, G.; Nervi, C.; Gobetto, R.; Lamberti, C.; Gianolio, D.; Bizzarri, R.; Sadler, P. *J. Inorg. Chem.* **2009**, *48*, 1469–1481.
93. Ruiu, T.; Garino, C.; Salassa, L.; Pizarro, A. M.; Nervi, C.; Gobetto, R.; Sadler, P. J. *Eur. J. Inorg. Chem.* **2010**, 1186–1195.

94. Petroni, A.; Slep, L. D.; Etchenique, R. *Inorg. Chem.* **2008**, *47*, 951–956.
95. Betanzos-Lara, S.; Salassa, L.; Habtemariam, A.; Sadler, P. J. *Chem. Commun.* **2009**, 6622–6624.
96. Betanzos-Lara, S.; Salassa, L.; Habtemariam, A.; Novakova, O.; Pizarro, A. M.; Clarkson, G. J.; Liskova, B.; Brabec, V.; Sadler, P. J. *Organometallics* **2012**, *31*, 3466–3479.
97. Rodriguez, E. A.; Campbell, R. E.; Lin, J. Y.; Lin, M. Z.; Miyawaki, A.; Palmer, A. E.; Tsien, R. Y. *Trends Biochem. Sci.* **2017**, *42*, 111–129.
98. Nikolaev, D. M.; Panov, M. S.; Shtyrov, A. A.; Boitsov, V. M.; Vyazmin, S. Y.; Chakchir, O. B.; Yakovlev, I. P.; Ryazantsev, M. N. *Progr. Photon Sci.* **2019**, 139–172.

CHAPTER TEN

Ruthenium phthalocyanines in nitric oxide modulation and singlet oxygen release: Selectivity and cytotoxic effect on cancer cell lines

Renata Galvão de Lima[a], Rafaella Rebecchi Rios[b], Antonio Eduardo da Hora Machado[c,d], and Roberto Santana da Silva[b,*]

[a]Instituto de Ciências Exatas e Naturais – ICENP, Universidade Federal de Uberlândia, Ituiutaba, MG, Brazil
[b]Faculdade de Ciências Farmacêuticas de Ribeirão Preto da Universidade de São Paulo, Ribeirão Preto – SP, Brazil
[c]Instituto de Física, Programa de Pós-Graduação em Ciências Exatas e Tecnológicas, Universidade Federal de Catalão, Catalão, GO, Brazil
[d]Instituto de Química, Laboratório de Fotoquímica e Ciência de Materiais, Universidade Federal de Uberlândia, Uberlândia, MG, Brazil
*Corresponding author: e-mail address: silva@usp.br

Contents

1. Introduction 356
2. Spectroscopic characteristics 359
3. Molecular orbital calculations 362
4. Light irradiation process 365
5. In vitro studies 373
6. Conclusions and future perspectives 377
Acknowledgment 377
References 377

Abstract

Light irradiation has been used for a long time in several countries as an approved therapy against cancer in a technique called photodynamic therapy (PDT). Essentially the process uses light, oxygen, and a photosensitizer (PS) to produce reactive oxygen species (ROS). However, PDT has been proven to be more effective for the early stages of cancer since many more developed tumors are hypoxic. It has driven scientists to search for new PSs to increase the selectivity against cancer. In this manuscript, we present a novel class of PSs based on ruthenium-phthalocyanine (Ru-Pc) complexes as agents for the production of ROS and reactive oxygen and nitrogen species (RONS). The cancer selectivity of the compounds could be achieved by targeting the Ru-Pc complexes

Advances in Inorganic Chemistry, Volume 80
ISSN 0898-8838
https://doi.org/10.1016/bs.adioch.2022.06.005

by inserting axial ligands "L" generating [RuLL'(Pc)] species. All synthesized Ru-Pc species are ROS producers and for L as nitrogen oxide derivative ligand, we also could generate nitric oxide (NO), which has been evaluated as an anticancer agent depending on its concentration. The synergism of ROS and NO are discussed, and the results suggest the treatment of this class of compounds as a possible metal-based drug to be clinically used in PDT application.

1. Introduction

Irradiation with light has been proposed as a suitable alternative to treat many types of diseases related to skin and eye conditions, as well as certain types of cancer. Photodynamic Therapy (PDT) is one of the most applied clinical processes involving irradiation with light. In oncology, PDT has been used to treat cancer, including deeper tumors.[1–3] In the clinical setting, PDT involves intravenous or topical administration of a photosensitizer (PS), which is absorbed by cells and activated by light in a therapeutic window, producing reactive oxygen species (ROS) and killing cancer cells.[1–3] On the basis of PDT established in mainstream clinical practice, several drugs have been employed in PDT protocols.[1–4] However, the wide application of PSs has been prevented by factors such as the need for subcellular PS localization and tissue oxygenation.[4] Thus, developing new PSs to improve the PDT outcome is crucial.

Phthalocyanines (Pcs) are macrocycles consisting of four isoindole units linked by nitrogen atoms in an azo position. Pcs have greater tendency to form radicals upon irradiation with light, so they have become an alternative to classic PSs used in PDT. Pcs generally present maximum absorption in the therapeutic window range, have high molar extinction coefficients, and produce singlet oxygen in high quantum yields[5–7] upon irradiation with light in the visible region. These macrocycles interact with metals, to form metal-coordinated complexes (M-Pc), some of which may produce hexacoordinated species with axial coordination. For biological purposes, M-Pc complexes can play a pivotal role. For example, they can exert effective cytotoxicity when their subcellular localization increases. Among M-Pc complexes, ruthenium phthalocyanines (Ru-Pcs) are worth highlighting: their chemistry is rich, and their therapeutic efficacy can be boosted by synergistic action of a bound spectator ligand such as nitric oxide (NO), favoring the use of Ru-Pcs in PDT.[8–13] Variations in the Pc structure have generated numerous Ru-(Pc-R) complexes. Their chemical and biological aspects, including energy transfer, charge transfer processes, and electrochemical and photochemical development have been investigated.[14–17]

Although several Ru-(Pc-R) complexes have been considered as anticancer agents, studies on the biochemical mechanisms related to the dose-dependent cytotoxicity of these complexes and consequent ROS-mediated cell death have not been fully exploited. The lack of Ru-Pc complexes might be partly due to synthetic difficulties.

We have recently reported the synthesis, characterization, and cytotoxic studies of a series of L_2Ru-Pc complexes (Fig. 1), obtained by different synthetic procedures. Some of those complexes display unique biological properties; for instance, cytotoxicity against cancer cells that cannot be achieved with typical free Pcs.

Inserting a ruthenium cation into the macrocyclic cavity is probably the major challenge. We have recently developed a simple and highly efficient method to prepare Ru-(Pc-R) complexes by the template method, as described in Scheme 1.[10] Perhaps this scheme will drive researchers through the synthesis of a series of M-(Pc-R) complexes like Ru-(Pc-R), consequently clarifying the structure-activity relationship.

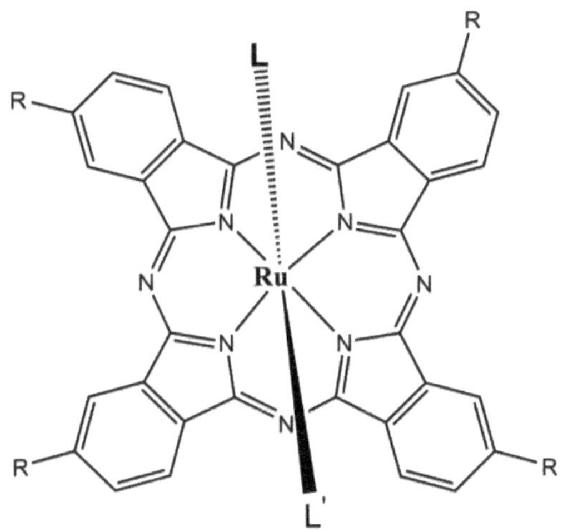

a) L = Cl-; L' = dimethylsulphoxide; $R_1 = R_2 = R_3 = R_4 = H$
b) L = NO; L' = NO_2^-; $R_1 = R_2 = R_3 = R_4 = H$
c) L = L' = 4-aaminopyridine; $R_1 = R_2 = R_3 = R_4 = H$
d) L = L' = pyrazine; $R_1 = R_2 = R_3 = R_4 = H$
e) L = L' = {RuNO(bpy)$_2$pz}; $R_1 = R_2 = R_3 = R_4 = H$
f) L = L' = Cl; $R_1 = COO^-$; $R_2 = R_3 = R_4 = H$;
g) L = Cl; L' = NO; $R_1 = R_2 = R_3 = R_4 = COO^-$

Fig. 1 Molecular structure of ruthenium-phthalocyanine complexes.

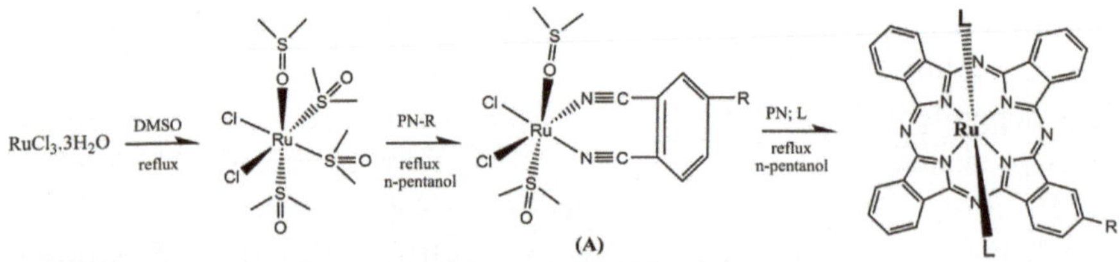

Scheme 1 Template synthesis for the [RuPc(L)$_2$] complex.

In the template method, the *cis*-[RuCl$_2$(DMSO)$_4$] complex is used as starting material. This complex can be easily obtained by following a previously published procedure.[18] Addition of different phthalonitrile (PN) compounds allows *cis*-[RuCl$_2$(DMSO)$_2$(phthalonitrile-R)] **(A)** to be synthesized, which is a key step in obtaining Ru(Pc-R) complexes. The most common ruthenium complexes present hexacoordinated structures. Therefore, when Ru is inserted into the Pc macrocycle, two spectator ligands can be attached at the axial positions. This is observed in the [RuPc(pz)$_2$] (pz = pyrazine) crystal structure,[19] where a pyrazine ligand is axially coordinated to the Ru(II) center, while Pc remains almost planar (Fig. 2).

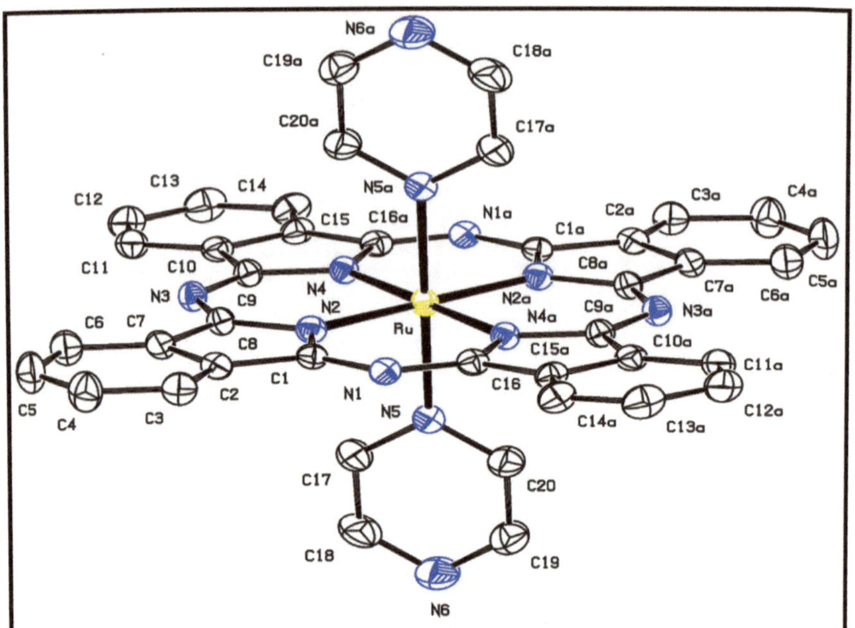

Fig. 2 X-ray crystal structure of [RuPc(pyrazine)$_2$].[19]

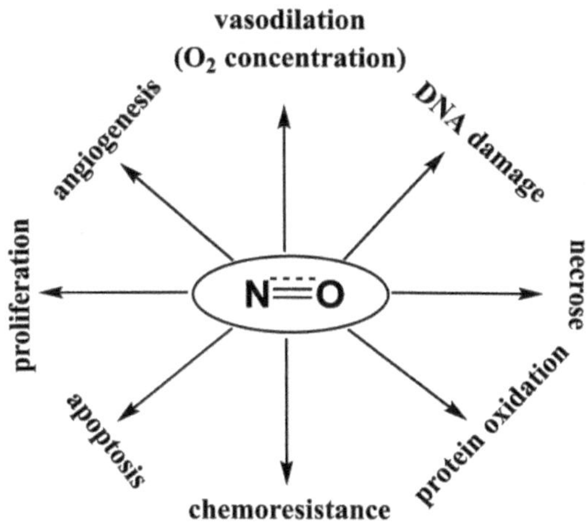

Fig. 3 Some direct and indirect actions of nitric oxide and derivatives in cancer.

An efficient strategy to enhance the cytotoxicity activity of [RuPc(L)(L')] complexes would consist of using a bioactive ligand such as nitric oxide (NO). The NO molecule is a short-lived diatomic signaling molecule that exerts multiple functions in biological systems,[20–27] as depicted in Fig. 3. In this work, we have used in some ruthenium complexes with NO axially bound to the Ru, which allowed free NO to be released by external stimulation.[8,9,13]

In cells, NO can either promote tumor growth or exert tumoricidal action depending on NO concentration, duration of exposure to NO, and cellular sensitivity to NO.[21,26,28,29] In general, NO at concentrations between 50 and 100 nM have a pro-tumor effect, whereas higher NO concentrations (>300 nM) exert a cytotoxic effect on neoplastic cells.[29] By further extending this knowledge, we have been able to acquire know-how about combining single-target drugs that can produce NO and ROS for synergistic therapy.[8,9,13]

2. Spectroscopic characteristics

The electronic spectral profile of Pcs is well known and has been widely described.[14,30–34] Typically, their UV-visible spectra present two electronic transitions between 350 and 500 nm, named B- or Soret-bands, as well as Q-bands between 500 and 700 nm; these bands have extinction coefficients around 10^5 M^{-1} cm^{-1}. However, interpreting the electronic spectra of Ru-(Pc-R) complexes is no easy task: bands due to electronic transitions involving interaction between the ligand and metal such as ligand field (LF), metal-ligand charge transfer (MLCT), and ligand-metal charge transfer (LMCT) bands may arise. Compared to the UV-visible spectrum of free PcH_2, the UV-visible spectra of Ru-(Pc-R) complexes (Table 1) are strikingly different. There is a single intense Q band and a shoulder much less intense in higher energy, whereas the UV-visible spectrum of free PcH_2 has two bands of roughly equal intensity in this region due to deviation from the D_{4h} symmetry caused by the two inner hydrogen atoms of free

Table 1 Comparison of the absorption bands, B and Q, and of the photophysical properties of free-PcH_2 and selected Ru-Pcs.

Compound	λ (nm)	λ_{em} (nm)	Φ_f	τ_f (ns)	Ref.
PcH_2	705, 643, 615, 340[a]	–	–	–	35
[RuPc]	632, 576, 318[b]	–	0.033[b]	–	19,36
[RuPc$(py)_2$]	622, 567, 375[c]	–	–	–	37
[RuPc(4-Mepy$)_2$]	622, 560, 375[c]	–	–	–	38
[RuPc$(pz)_2$]	640, 586, 442, 314[b]	–	–	–	19
[RuPc(4-Ampy$)_2$]	623, 385, 317[d]		0.011[c]	1.9; 3.1[c]	8
[RuPcCl(DMSO)]	640, 316[d]		0.007[c]	1.8; 3.2[c]	8
[Ru{$(tBu)_4$Pc}$(py)_2$]	627, 578, 376, 315[c]	870	0.0003	154	39
[Ru{$(tBu)_4$Pc}(4-mpy$)_2$]	627, 576, 378, 317[c]	–	–	–	39
[(Ru{$(tBu)_4$Pc}(4-NO_2py$)_2$]	645, 591, 382, 312[c]	–	–	–	39
[RuPc(NO)(NO_2)]	696, 608, 354[b]	696	0.29[b]	–	40
[RuPc(NO)(ONO)]	682, 616, 350[b]	690[b]	0.040[b]		36
[RuPc([Ru$(bpy)_2$(pz)NO$]^{3+})_2$]	634, 576, 318[d]	690	0.030[b]	–	12

[a]DMF.
[b]$CHCl_3$.
[c]CH_2Cl_2.
[d]DMSO.

PcH_2.[14,34] In the UV-visible spectra of Ru-(Pc-R) complexes, the Soret band (or B-band) appears in the region of 300–325 nm and is attributed to a $\pi \rightarrow \pi^*$ transition of the macrocycle. In Pcs, the maximum Q-band wavelength shifts to higher energy after coordination to the metal ion because the molecular symmetry increases, which also influences both the Soret band and Q-band extinction coefficients.

Axial ligands have been shown to affect Ru-Pc spectroscopic properties.[8–14] An axial ligand with higher π-acceptor character shifts the Q-band energy to lower values. For example, the UV-visible spectrum of the $[RuPc(pz)_2]$ complex displays Q-band at 640 nm, while the UV-visible spectra of the series $[RuPc(L)_2]$, where L=py, CH_3-py, or NH_2-py, have Q-bands around 622 nm. In addition, the latter spectra display absorption bands at 375–442 nm, ascribed to the $d_\pi(Ru^{II}) \rightarrow \pi^*$ (L) MLCT transitions.[8–10,12,13,19] When the macrocyclic chelates bear four peripheral tert-butyl groups, these groups have little influence on the position of the Q- and Soret bands.

Computational studies with {Ru-Pc-R} systems focusing on structural and spectroscopic properties have been performed by molecular modeling calculations. These studies have allowed the transition bands to be completely assigned and will possibly serve as fundamentals in the assignments of the UV-visible spectra of general Ru-(Pc-R) complexes (Fig. 4).

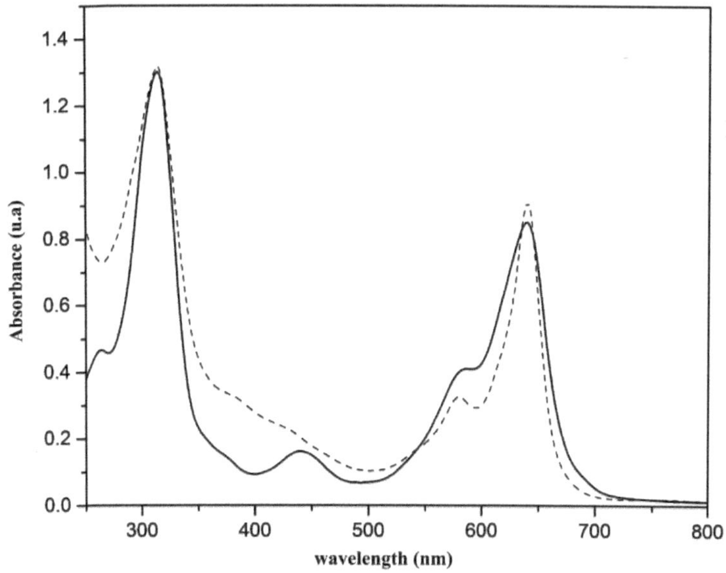

Fig. 4 UV-visible spectrum of 3.5 x 10^{-5} mol L^{-1} of $[Ru(Pc)(pz)_2]$ (solid line) and [RuPc] (dashed line) in dimethyl sulfoxide solution.[19]

3. Molecular orbital calculations

Molecular orbital calculations were performed for ruthenium complexes such as [RuPc(NO)(NO_2)], which were optimized and evaluated in terms of vibrational modes and UV-visible spectroscopy. The evaluations were carried out using density functional theory (DFT) with the CAM-B3LYP functional[41] combined with the TZP-DKH triple-zeta atomic basis set with polarization[42–44] and relativistic correction (DHK= Douglas-Kroll-Hess). Electronic spectra were obtained for the first 50 singlet states by calculation employing the time-dependent approach of the same theory (TD-DFT), with the same combination of functional and atomic basis set. To estimate the state diagram for Ru(II) phthalocyanine, the energies of the first six singlet and triplet states were calculated using the TZP-DKH basis set in combination with the hybrid functional B3LYP.[45] For the states of interest, the Natural Transition Orbitals (NTO)[46] were calculated in order to facilitate the analysis of the results. In all cases, the role of the solvent was considered by employing the polarizable dielectric continuum conductor-like continuum model (CPCM[47]) to describe the interaction of the dichloromethane solvent with the species studied. These calculations were performed using the program Gaussian 09 version E.01. We have examined the [RuPc(NO)(NO_2)] complex as a model to identify some general features involved in electronic transitions. The optimized geometry of [RuPc(NO)(NO_2)] (Fig. 5) was used in this work to calculate the UV-visible spectrum.

The optimized structure of [RuPc(NO)(NO_2)] complex shows that the group NO^+ on Ru(II) is axially distorted, when compared to the ligand

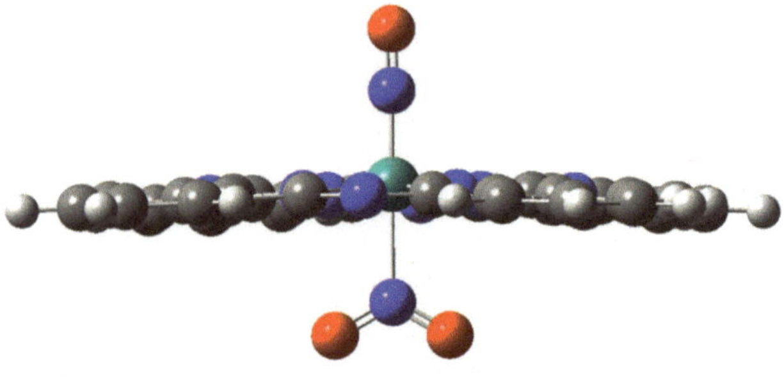

Fig. 5 Optimized structure of [RuPc(NO)(NO_2)] complex obtained using the CAM-B3LYP DFT (Density Functional Theory) functional combined with the triple-zeta TZP-DKH atomic basis set with polarization.

NO_2^-, in a diametrically opposite position. The UV-visible spectrum of [RuPc(NO)(NO_2)], from which the calculation is predicted, is in good agreement with the experimental one (Fig. 6).

The existence of a bipartite Q band in 600–660 nm region is characterized by two sets of singlet-singlet transitions each containing two transitions with expressive intensity. For the lowest energy experimental absorption band, that occurs with maximum centered at 677 nm, two transitions mostly π,π^* with charge transfer from the macrocycle to Ru(II) and axial ligands, are observed at 717 and 689 nm. The lowest energy transition (at 717 nm) was described as being composed by the first HOMO → LUMO (weighing 90.22%) and the second HOMO → LUMO + 2 (weighing 9.78%) (Fig. 7). Similar to that, the transition at 689.23 nm also presents two configurations associated to HOMO → LUMO + 1 (weighing 92.77%) and HOMO → LUMO + 3 (weighing 7.23%). The second pair of transitions, related to the experimental band centered at 607 nm, is also mostly $\pi \rightarrow \pi^*$, was theoretically described by two transitions. The first one, estimated at 552 nm, is 8.90% HOMO → LUMO and 91.10% HOMO → LUMO + 2. The second transition, at 540 nm, is described as being 6.40% HOMO → LUMO + 1

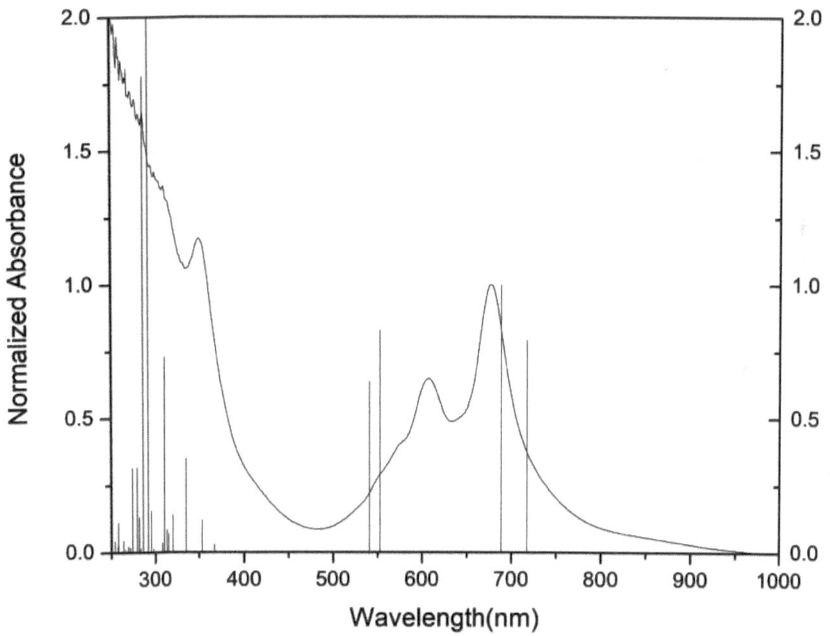

Fig. 6 Experimental UV-visible spectrum and calculated transition bars of [RuPc(NO)(NO_2)].

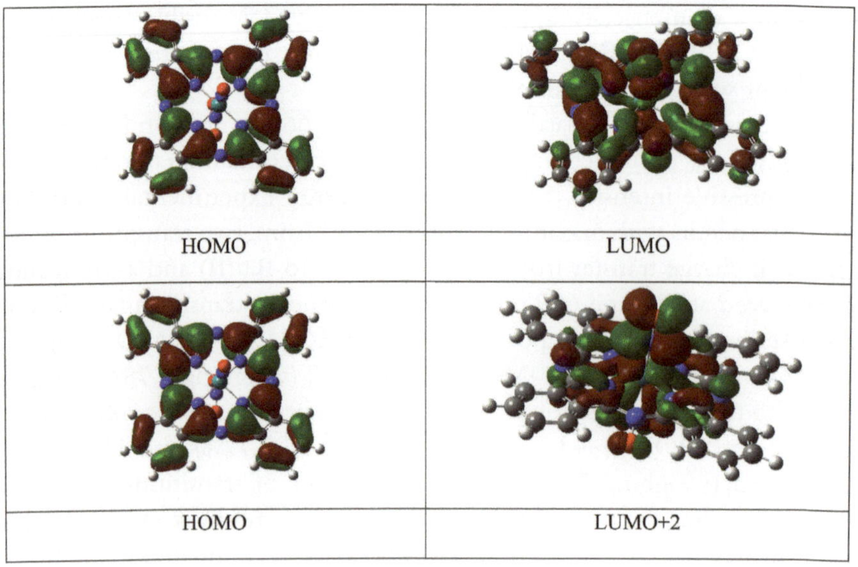

Fig. 7 Calculated HOMO/LUMO composition for the frontiers molecular orbital related to the transition at 717 nm of [RuPc(NO)(NO_2)] complex.

and 93.60% HOMO→LUMO+3. The deviations observed for the calculated transitions and experimental UV-visible spectrum of [RuPc(NO)(NO_2)] complex are inherent to the calculation methodology and do not significantly affect the prediction quality.

The existence of a bipartite Q band in 600–660 nm region is characterized by two sets each containing two singlet-singlet transitions with expressive intensity. For the lowest energy experimental absorption band, two transitions, also mostly π,π^*, with some charge transfer from the macrocycle to Ru(II), and axial ligands are observed at 717 and 689 nm, respectively. The lowest energy transition was described by the first HOMO→LUMO (90.22%) and the second HOMO→LUMO+2 (9.78%) (Fig. 7). Similar to that, the transition at 689 nm also presents two configurations associated to HOMO→LUMO+1 (92.77%) and HOMO→LUMO+3 (7.23%). The second pair of transitions, also mostly $\pi \rightarrow \pi^*$, is being described by two transitions. The first one estimated at 552 nm is due to HOMO→LUMO (8.9%) and HOMO→LUMO+2 (91.10%). A second transition at 540 nm is described by HOMO→LUMO+1 (6.40%) and HOMO→LUMO+3 (93.60%). The deviations observed for the calculated transitions and experimental UV-visible spectrum of the [RuPc(NO)(NO_2)] complex are inherent to the calculation methodology and do not significantly affect the prediction quality.

Experimentally the Ru-Pc-R complexes have been widely characterized by infrared spectroscopy, providing insight into metal–ligand interactions. The NH vibration observed in the infrared spectrum of free PcH_2 (ν_{N-H} 3200 and ν_{N-H} 1005 cm^{-1}) disappears upon Pc coordination to a metal ion. For the other Pc vibrations, Ru-Pc complexes display basically the same FTIR characteristics as free PcH_2 except for the decreased bond energies of the Pc systems in Ru-(Pc-R) complexes, a consequence of the presence of the metal ion. For {Ru-Pc} species, a relatively good similarity between the calculated and experimental infrared spectrum are indeed observed.

Considering the [RuPc(NO)(NO_2)] complex the infrared spectrum shows ν_{NO} stretching for the calculated (experimental) as 2002 (1890) cm^{-1}, while vibrations mode of bonded nitrite appears in 1521; 1513 (1640) and 1463; 1454 (1390) cm^{-1} among other bands attributed as ν_{CN} stretching. The band due to ν_{NO} stretching is a useful diagnostic band for the NO release process; this band disappears completely, as noted after irradiation with light or electrochemical stimulation.[11]

4. Light irradiation process

Under irradiation with visible light, some M-Pc complexes can produce singlet oxygen ($^1O_2$) efficiently, which makes these complexes promising PSs for PDT. The photophysical properties of M-Pc complexes depend strongly on the central atom and electronic absorption, fluorescence, phosphorescence, and triplet–triplet (T–T) absorption spectra.[48] Diamagnetic M-Pc complexes, including Ru(II)-Pc complexes, exhibit intense luminescence—the T_1 state is admixed with the S_1 state due to strong spin–orbit coupling between the central metal d_{xz} and d_{yz} orbitals.[8,49] Because of the high Φ_T and relatively long τ_T values of these complexes, they can be used as PSs for singlet oxygen generation (Table 2).

This information confirms that the axial ligand influences the $^3\pi\pi^*$ excited state and, consequently, the photophysical properties of M-Pc complexes.[8] For the [RuPcCl(DMSO)] and [RuPc(4-ampy)$_2$] (4-ampy = 4-aminopyridine) complexes, we have observed that their irradiation with light at 650 nm region generates the $^3\pi\pi^*$state transient spectra with λ_{max} of 500 and 540 nm, respectively (Fig. 8). The differences in the $^3\pi\pi^*$state spectra of these complexes clearly show that the axially coordinated ligands perturb the Pc electronic structure. Moreover, the $^3\pi\pi^*$state lifetime increases with the electron-withdrawing character of the ligand.

Table 2 Photophysical properties of some M-Pc in the excited states lifetime and singlet oxygen quantum yields.

Compound	τ_T	Φ_Δ	Ref.
[ZnPc]	0.30 μs[a]	0.67[a]	50
[RuPc]	0.13 μs[b]	0.39[b]	36
[RuPc(NO)(NO_2)]	0.50 μs[b]	–	–
[RuPc(NO)(ONO)]	–	0.22	36
[RuPc([Ru(bpy)$_2$(pz)NO]$^{3+}$)$_2$]	60 ns[b]	–	–
[RuPcCl(DMSO)]	3.68 μs	0.62	51
[RuPc(4-Mepy)$_2$]	3.7 μs	0.14	51
	90 ns[c]		38
[RuPc(py)$_2$]	135 ns	–	38

[a]Pyridine.
[b]$CHCl_3$.
[c]CH_2Cl_2.

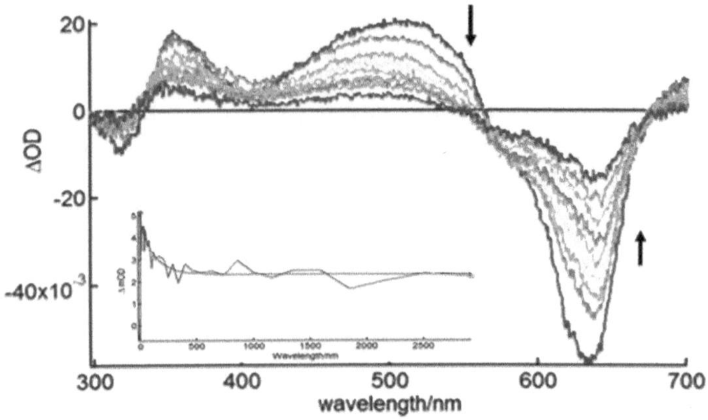

Fig. 8 Triplet state UV-vis transient absorption spectral change of [RuPcCl(DMSO)] in DMSO solution after excitation at λ_{ex}=650 nm (0–400 ns).

Table 2 showed that the nature of the axial ligands "L" in [RuPc(L)(L')] compounds affect the singlet oxygen quantum yield. The explanation maybe related to the intersystem crossing (ISC) of the ruthenium-phthalocyanine complexes. We have analyzed the [RuPc(NO)(NO_2)] complex as a model of [RuPc(L)(L')] systems by Spin-Orbit Coupling (SOC) approach. For SOC calculation, S_1 and T_1 state geometries were optimized using Orca software version 5.0.3.[52] The S_1 geometry was used to estimate the SOC for the first 10 excited singlet and triplet states. The atomic basis set was the same used previously (TZP-DKH), in combination, in this case, with the auxiliary basis set def2/J.[53] These calculations were performed using the B3LYP hybrid functional. The CPCM model was employed to account for the effect of the solvent (dichloromethane) on the electronic structure. The *Intersystem Coupling* between the S_1 state and the immediately adjacent lowest energy triplet state (T_2), can be qualitatively evidenced by inspection of the molecular orbitals involved in these transitions. In energetic terms, the phenomenon, which justifies the photodynamic activity presented by this ruthenium-phthalocyanine complex, is fully viable, since $E(S_1) > E(T_2)$ is an exothermic process, with an energy of the order of 42 kJ/mol (Fig. 9).

We have used natural transition orbitals (NTO) theory, which consider a hole-particle process to understand the effect of ISC as dependent on the axial ligands. Fig. 10 illustrates the hole-particle process involving in S_1 and T_2.

The NTOs designated as "Particle," for both S_1 and T_2, present a certain degree of orthogonality to each other with the MOs populated in both transitions, have different orbital symmetries. Furthermore, it is possible to verify the active participation of the metallic center in the process, an important aspect in the feasibility of the *Spin-Orbit Coupling* (SOC) and consequently of the *Intersystem Crossing* (ISC). Quantitatively, the calculated SOC matrix

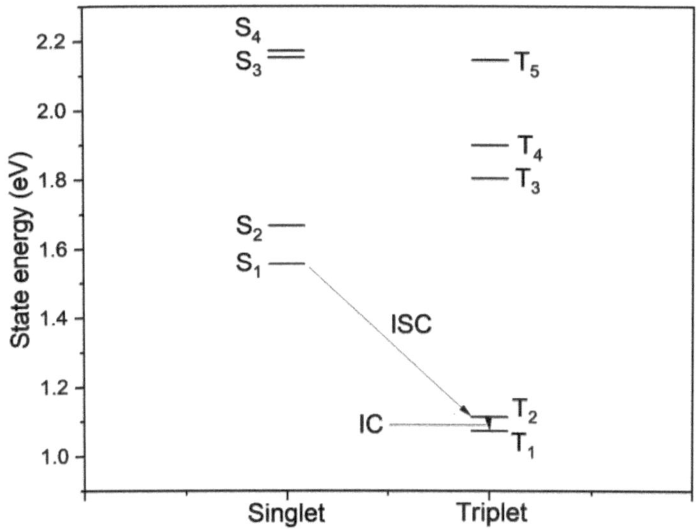

Fig. 9 State diagram considering non-relaxed excited states of Ru(II)-phthalocyanine.

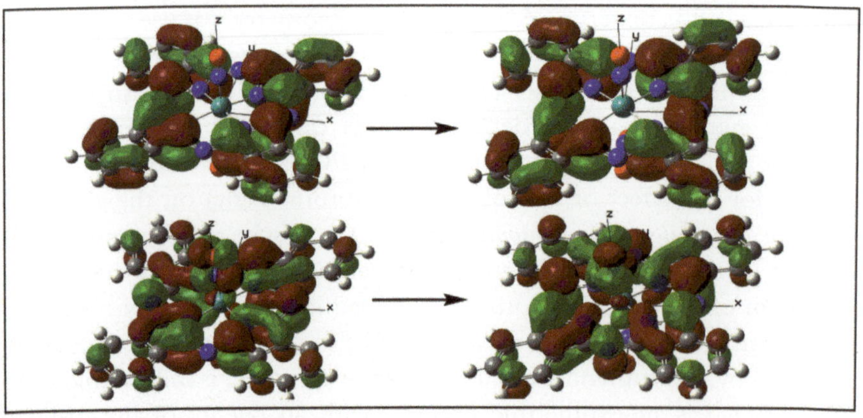

Fig. 10 Natural transition orbitals of [RuPc(NO)(NO_2)] complex illustrating the nature of optically active singlet and triplet orbitals.

element (SOCME) between S_1 and T_2, $< T_2 \,|HSO|\, S_1 >$, equal to 79 cm^{-1}, is consistent with a median value for SOC. This implies a moderate ISC, with consequence on the quantum yield production of singlet oxygen (Φ_Δ), resulting in a relatively low value, in agreement with the value experimentally estimated for the [RuPc(NO)(NO_2)] complex.[36] Expressive values of SOC are expected in systems whose involved states have high orthogonality, close energies and contribution of metallic centers that present high spin-orbit coupling constants.[54] In the ruthenium-phthalocyanine in question, unless there is orthogonality between states S_1 and T_2, the other conditions are observed. It is important to note that the SOC between S_1 and T_1 is unlikely, confirmed through the calculation, since the SOCME in this case is $< T_1 \,|HSO|\, S_1 > = 0.00$ cm^{-1}.

In general, irradiation of Pc or even some M-Pc complexes with visible light markedly decreases the Q-band, a consequence of the auto-oxidation process[8] elicited by the generated $^1O_2$, which efficiently attacks the Pc ring directly. We have found in contrast that, in solution, Ru-(Pc-R) complexes produce $^1O_2$ efficiently (Table 2) and are stable both in the dark and under visible light, as seen from the unaltered UV-visible spectrum upon irradiation with visible light. We have attributed this to strong Ru-Pc electronic coupling, which causes electron delocalization between Pc and Ru(II) orbitals, which was confirmed by molecular orbital calculations.

The [RuPc(NO)(NO_2)] complex is fascinating: it can produce NO by chemical or electrochemical reduction processes in non-aqueous or aqueous solvent, as observed in Fig. 11.

Similar results have been obtained upon irradiation of the [RuPc(NO)(NO_2)] complex with light at 377 nm. In this case, a photoactivatable

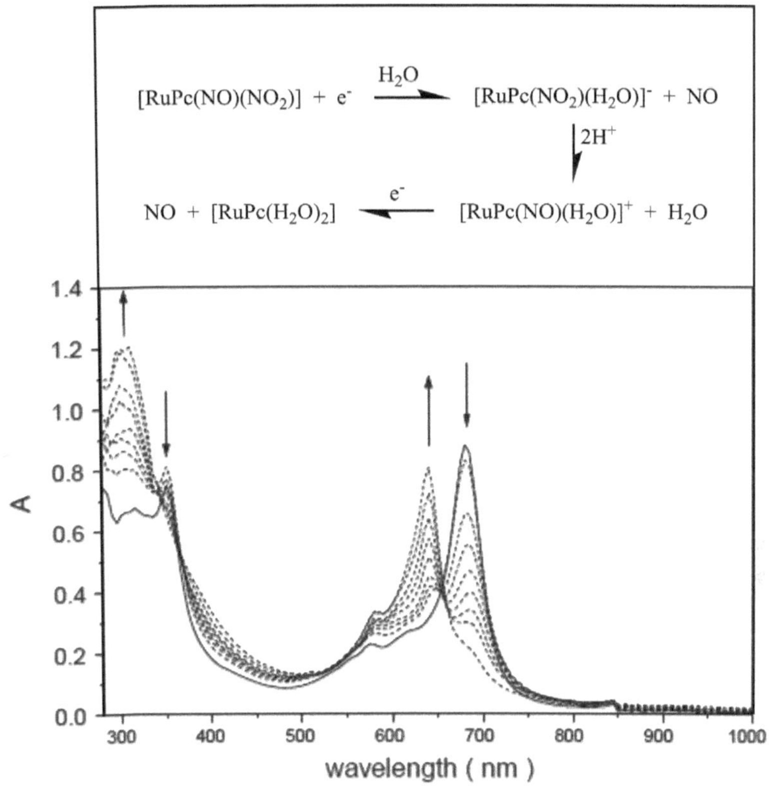

Fig. 11 UV-visible spectral changes for controlled reduction potential electrolysis of $[RuPc(NO)(NO_2)]$ at −0.20 V vs. SHE.[13] Inset: Electrochemical reduction process of $[RuPc(NO)(NO_2)]$ at −0.20 V vs. SHE.

NO-releasing ruthenium species is generated upon photo-oxidation of the metal ion, to give an intermediate Ru(III) species. Addition of ethanol or reducing agent immediately regenerates Q-band at higher energy consisting with the [RuPc] complex formation identified by electronic spectrum analysis (Fig. 12).

Comparison between how ruthenium(II) phthalocyanines and porphyrins behave with respect to NO photo-release as analyzed by the strength of the bond between the metal ion and macrocycle and axial ligands is interesting. The X-ray diffractogram of the $[Ru(porphyrin\text{-}R)NO(NO_2)]$ {porphyrin-R = tetraphenylporphyrinato dianion (TPP) and octaethylporphyrinato dianion (OEP)} species showed the planarity of the macrocycle ligand and the almost linear nitrosyl-ruthenium(II) bond, with nitrite in trans position.[55] For the ruthenium-porphyrin complex, ν_{NO} varied from 1835 (TPP) to 1854 cm^{-1} (OEP), whereas for $[RuPc(NO)(NO_2)$

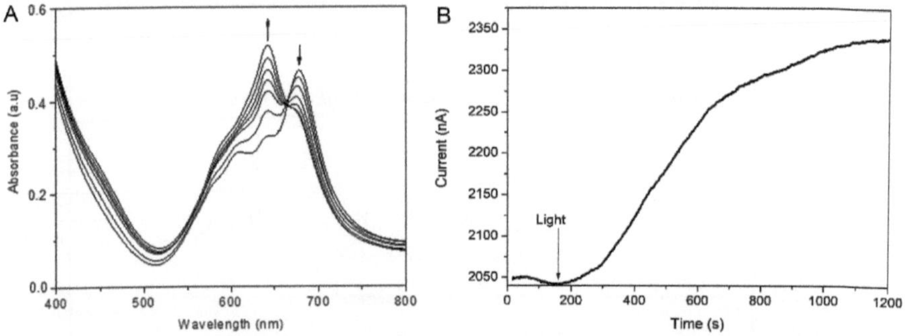

Fig. 12 Electronic spectral changes during light irradiation at 660 nm of 1.0×10^{-5} mol L^{-1} [RuPc(NO)(NO_2)] complex in ethanol solution (A). Chronoamperogram of nitric oxide released by photolysis at 377 nm of 1.0×10^{-5} mol L^{-1} [RuPc(NO)(NO_2)] complex in ethanol solution (B).

ν_{NO} appeared at 1890 cm^{-1}. The lower ν_{NO} energy indicated the presence of the Ru^{II}-π^*(NO) bond suggesting that ruthenium-porphyrin complexes are better electron donors than ruthenium-phthalocyanine complexes. Laser flash photolysis of [Ru(porphyrin-R)NO(NO_2)] at 355 nm in non-aqueous medium revealed NO and nitrite release with formation of transient pentacoordinate species [Ru(porphyrin-R)(ONO)] and [Ru(porphyrin-R)(NO)], as seen by the UV-visible analysis.[56] The starting material was regenerated upon NO addition. Photoreactivity showed that triplet ligand field states were close in energy to the MLCT state in these ruthenium-porphyrin complexes. This was quite different from [RuPc(NO)(NO_2)], for which light irradiation at 377 nm in aqueous solution produced exclusively NO release. The most probable explanation should involve the role played by the solvent, in which very fast conversion of {Ru^{II}-NO_2} to {Ru^{II}-NO^+} in the intermediate observed after NO release from the [RuPc(NO)(NO_2)] complex should occur as part of the process, as also observed by spectroelectrochemistry.[13] The absence of nitrite release indicated that the LUMO was essentially an MLCT transition in the [RuPc(NO)(NO_2)] complex, as described by our molecular orbital calculation.

The proof-of-concept to pursue the effect of nitrosyl ruthenium complexes on biological media and to establish a metal-based drug for NO delivery was effectively tested by evaluating endothelium dysfunction.[3,8,13,15,48,57–59] One effect is related to alteration in endogenous NO levels, which culminates in vasoconstriction, increasing blood pressure. Essentially, increased NO level and cGMP causes vasorelaxation by activation of a potassium channel by cGMP-dependent protein kinase.

Complexes such as [RuPc(NO)(NO_2)] can release NO and perform as vasodilator agents (Fig. 13).

As depicted in Fig. 13, vasorelaxation occurs upon reduction of the nitrosyl ruthenium complex and consequent NO release, affecting the biological pathways that lead to vasodilation. Therefore, we consider that nitrosyl ruthenium complexes can release NO in biological media. This property allows investigation into a synergistic effect between nitrosyl ruthenium complexes and a system that produces other radicals such as ROS. If such effect is verified, it could be used in cancer treatment.

All the Ru-Pc complexes described here produce ROS under photolysis at 650 nm, but the ratio between the resulting ROS is different depending on the Ru-Pc complex. This can be clearly observed when the [RuPcCl(DMSO)] and [RuPc(NO)(NO_2)]complexes are tested for the presence of hydrogen peroxide measured by Amplex™ Red[60] and hydroxyl radical measured by hydroxyphenyl fluorescein (HPF)[61] methods (Fig. 14), although they make up a major class of singlet oxygen producers (Table 2).

Therefore, we have hypothesized that such Ru-Pc complexes could be employed as models to evaluate synergistic effects during therapy with multiple releases of important agents such as NO and $^1O_2$. It has been observed that the nitrosyl ruthenium complex releases NO into cancer cells mostly by a mitochondrial NADH reduction process.[62] Subsequently, one of the possible mechanisms related to the NO anti-cancer activity is peroxynitrite formation, which is observed between NO and the superoxide produced

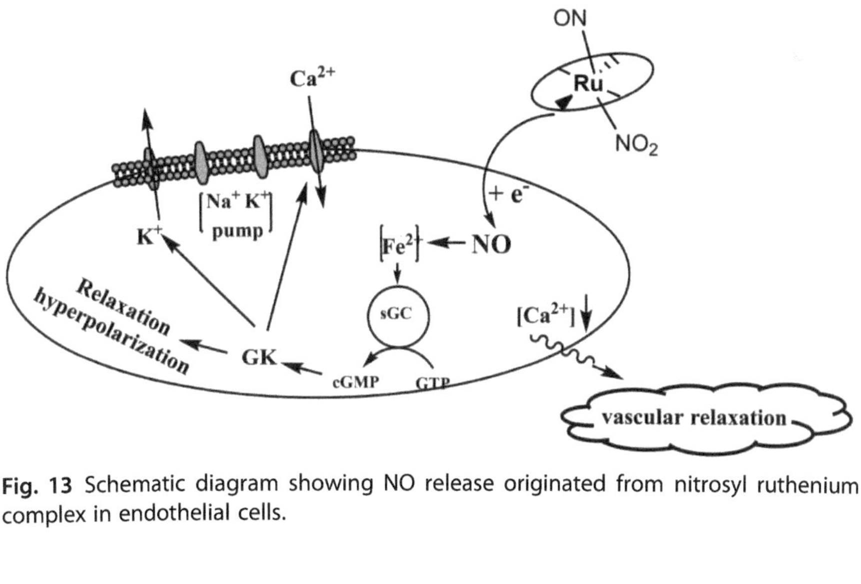

Fig. 13 Schematic diagram showing NO release originated from nitrosyl ruthenium complex in endothelial cells.

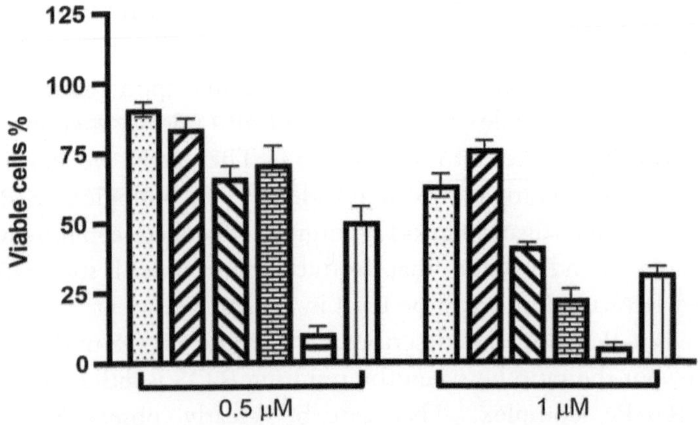

Fig. 14 Effect of ruthenium phthalocyanine complexes on cell viability of B16F10 cells. In the dark: [RuPc] ( ); [RuPcCl(DMSO)] ( ); [RuPc(4-ampy)$_2$] ( ); Under 650 nm light irradiation: [RuPc] ( ); [RuPcCl(DMSO)] ( ); [RuPc(4-ampy)$_2$] ( ).

by mitochondria that occurs near-diffusion limit, causing DNA damage via an oxidation process. Besides that, NO drives multiple pathways in cancer cell proliferation and has been pointed out as a possible therapy for cancer treatment. After the nitrosyl ruthenium complex such as [RuPc(NO)(NO$_2$)] is incubated, the resulting complex [RuPc] can produce ROS by light irradiation. Scheme 2 represents a simplified scheme of ROS and NO production by Ru-Pc complexes via a photochemical process.

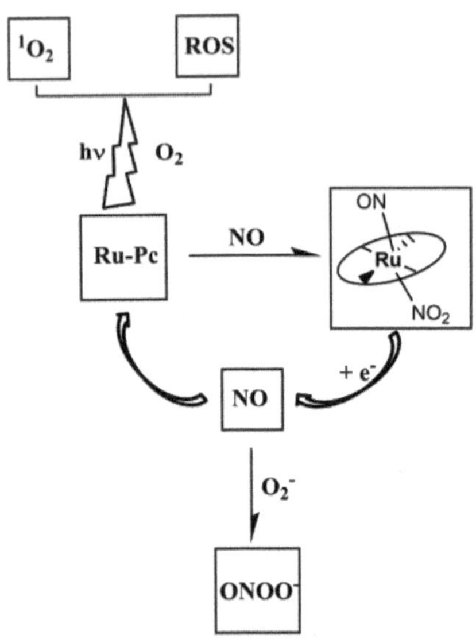

Scheme 2 Representative scheme for the ROS and NO generation based on ruthenium-phthalocyanine complexes.

5. In vitro studies

In vitro studies are the best way to understand the cytotoxicity mechanism of a bioactive compound. Ru-Pc complexes have potential clinical application—these complexes present high absorption molar coefficients in the therapeutic window and long-lived triplet excited states, which favor ROS production. The [RuPc], [RuPcCl(DMSO)], and [RuPc(4-ampy)$_2$] (4-ampy = 4-aminopyridine) complexes discussed here present dose-dependent cytotoxicity against the B16F10 cell line, as observed in Fig. 14.

The cytotoxicity of Ru-(Pc-R) complexes (as PS) against some cancer cell lines seems to depend on the $^1O_2$ generation quantum yield (Table 2) and on the chemical structure of these complexes. The [RuPcCl(DMSO)] complex is the most active against B16F10 cells, and this complex also shows low dark toxicity, which is consistent with high $^1O_2$ quantum yield. However, photolysis of this complex at 660 nm involves DMSO photo-release, which competes with ROS production. Knowing that DMSO has significant toxicity,[57] the observed photo-cytotoxicity may be due to a synergic effect (Scheme 3).

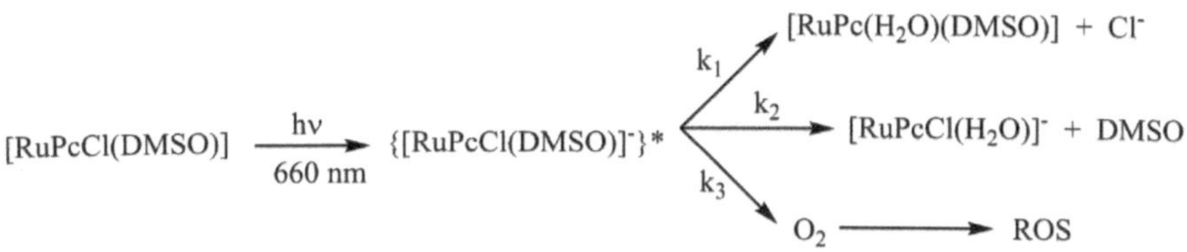

Scheme 3 Photochemical reactions of [RuPcCl(DMSO)] under light irradiation at 660 nm.

The dependence of Ru-(Pc-R) cytotoxicity on the Ru-(Pc-R) structure is closely related to the subcellular localization of the complex. In [RuPc(L)(L')] complexes, this localization is driven by ligands "L" containing functional group such as 4-aminopyridine in [Ru Pc(4-ampy)$_2$], which seems to confer cytotoxicity against the melanoma cell line B16F10 cell line. The 4-ampy ligand is known as a non-selective voltage-gated potassium channel (KV) blocker[8] that modulates the intracellular potassium concentration, affecting cell cycle progression, proliferation, and apoptosis.[48] Therefore, different structural modifications have been carried out in such complexes: axial ligands with specific functions such as PEG, carbohydrate units, or folic acid have been introduced to improve cellular uptake, thereby favoring cytotoxic activity.[15,58–60]

Martins et al.[10] made direct modifications to the Pc bound to Ru(II), to obtain the [Ru(Pc-DCBz)] (DCBz = 4-(3-carboxylicbenzyloxy)-phthalonitrile) complex (Fig. 15). Under stimulus with light ($\lambda = 660$ nm) and incubation with 10 μM [Ru(Pc-DCBz)] for 24 h, the latter complex proved to be as cytotoxic as [RuPc] against the B16F10 cell line. As for the MCF-7 breast cancer cell line, the [Ru(Pc-DCBz)] complex kills 90% of the cells, while Ru-Pc decreases this cell viability by only 15%.

Synergistic effects between the produced $^1O_2$ and NO have also been explored, providing enhanced cytotoxic activity.[12,13,34] The [RuPc(NO)(NO_2)] complex promotes cell death by apoptosis more effectively than its analogue Ru-Pc because the former can release NO together with $^1O_2$ after irradiation with light. Carneiro et al.[13] demonstrated that, upon irradiation with light at 660 nm (4 J cm^{-1}), both complexes reduce the cell viability of the murine melanoma tumor line, B16F10. Incubation with [RuPc(NO)(NO_2)] at 1 μM for 3 h provides the best result (24% cell death). In addition, both complexes are inactive in the dark given that neither NO nor $^1O_2$ is released without irradiation with light. Heinrich et al.[12] also explored the synergistic effect between these two species by using ruthenium complexes bearing nitrosyl as axial ligand. In this situation, NO is released from the $[Ru(Pc)(pz)_2\{Ru(bpy)_2NO\}_2]^{6+}$ complex by reduction in the biological medium, which favors cell death together with the oxygen released by the complex after its irradiation with light at 660 nm. Slightly higher concentrations of this complex (5 μM) are required to bring about the same cell death observed for the [RuPc(NO)(NO_2)] complex (Fig. 16). Notwithstanding, cell death is higher than that observed for the Ru-Pc complex under the same experimental conditions.

Negri et al.[9] showed the activity of these same complexes after photobiomodulation (PBM) at 850 nm, before PDT. Under these conditions, both complexes are more photocytotoxic probably because the ATP levels in the cell are higher, promoting greater cellular uptake of the complexes. These studies confirmed the great ability of Ru-Pc to kill cancer cells upon irradiation with light, especially in the presence of the synergistic effect elicited by NO release. These results can serve as basis for a structure-activity relationship and for the development of a promising antitumor compound.

The biochemical process involving the synergism between NO and $^1O_2$ is still under investigation. Some biological evaluations have shown that nitrosyl ruthenium complexes can release NO in membranes via a reduction process.[13] NO seems to affect the mitochondrial respiratory chain with a direct effect under ATP production—the amount of oxidant species increases, culminating in cell death (Scheme 4). Irradiation with visible light n following the reduction process activates $^1O_2$ production, which kills cells.

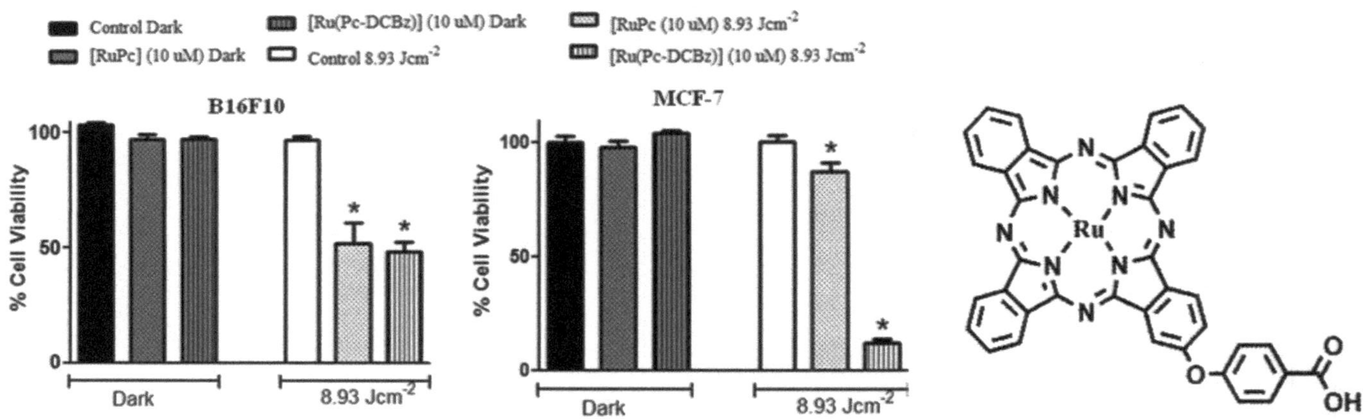

Fig. 15 Cell viability plot with or without light irradiation (660 nm, 8.93 J cm^{-2}, 10 min) of B16F10 and MCF-7 cancer cells treated with [RuPc] and [Ru(Pc-DCBz)] for 24 h.[34]

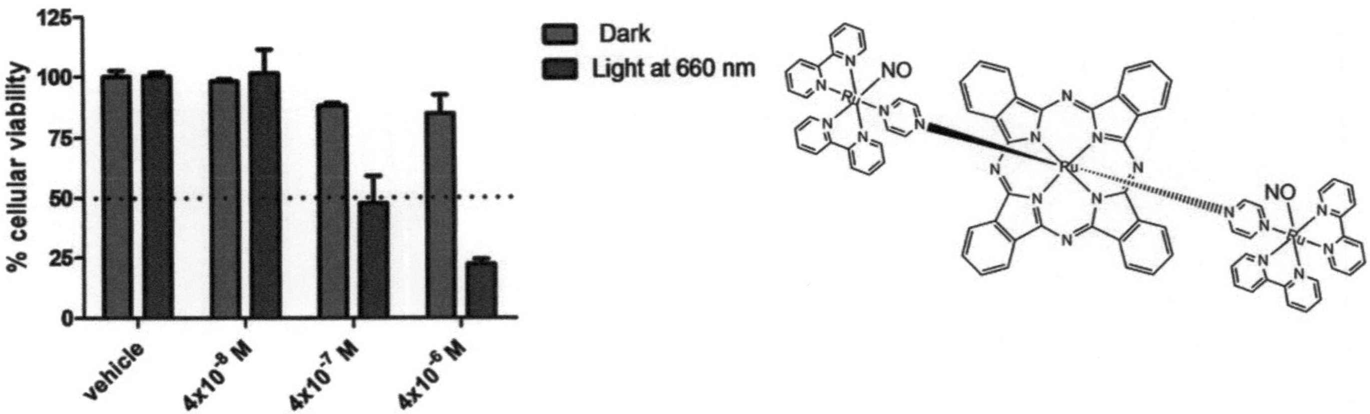

Fig. 16 Cell viability vs. concentration against the B16F10 cell line for the $[Ru(Pc)(pz)_2\{Ru(bpy)_2NO\}_2]^{6+}$ complexes.[12]

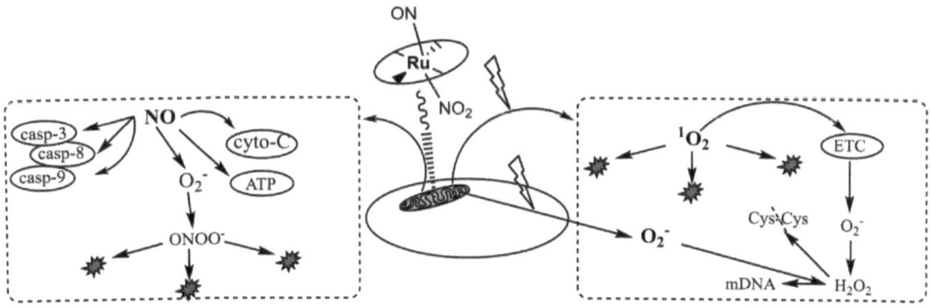

Scheme 4 Scheme summarizing the effect of NO and $^1O_2$ on cancer cell line.

6. Conclusions and future perspectives

The Ru-Pc complexes are an interesting class of anticancer species with potential application in PDT. Studies have shown that these compounds can have synergistic effects by producing cytotoxic molecules such as DMSO upon light irradiation or RONS depending on the axial coordination of ruthenium(II) followed by relative high singlet oxygen quantum yield production. The signaling pathways related to the cytotoxicity of Ru-Pcs indicate that cells die by apoptosis. A complementary approach for this system should involve the synthesis of new Ru-Pc complexes containing functional groups that specifically target cellular uptake. The synergistic effect of a binding ligand such as NO should also be considered because it could increase anti-tumor activity by affecting multiple signaling pathways that lead to cancer cell death. Besides the syntheses, evaluation of their in vivo behavior and studies related to their in vivo metabolism including their pharmacokinetics and pharmacodynamics should be considered for the intended clinical use. Using metal-phthalocyanine as PS in PDT is highly promising due to the cytotoxicity advantages that can be achieved over the conventional PSs.

Acknowledgment

The authors thank CAPES, CNPq and FAPESP (grant: 2019/19448-8).

References

1. Yanovsky, R. L.; Bartenstein, D. W.; Rogers, G. S.; Isakoff, S. J.; Chen, S. T. *Photodermatol. Photoimmunol. Photomed.* **2019**, *35* (5), 295–303.
2. Li, W.-P.; Yen, C.-J.; Wu, B.-S.; Wong, T.-W. *Biomedicines* **2021**, *9* (1), 69.
3. dos Santos, A. F.; de Almeida, D. R. Q.; Terra, L. F.; Baptista, M. S.; Labriola, L. *J. Cancer Metastasis Treatment* **2019**, *5*, 25.
4. Abrahamse, H.; Hamblin, M. R. *Biochem. J.* **2016**, *473* (4), 347–364.
5. Josefsen, L. B.; Boyle, R. W. *Theranostics* **2012**, *2* (9), 916–966.

6. Ogura, S.; Tabata, K.; Fukushima, K.; Kamachi, T.; Okura, I. *J. Porphyr. Phthalocyanin.* **2006**, *10* (09), 1116–1124.
7. Halaskova, M.; Rahali, A.; Almeida-Marrero, V.; Machacek, M.; Kucera, R.; Jamoussi, B.; Torres, T.; Novakova, V.; de la Escosura, A.; Zimcik, P. *ACS Med. Chem. Lett.* **2021**, *12* (3), 502–507.
8. Martins, T. J.; Negri, L. B.; Pernomian, L.; do Carmo Freitas Faial, K.; Xue, C.; Akhimie, R. N.; Hamblin, M. R.; Turro, C.; da Silva, R. S. *Front. Mol. Biosci.* **2021**, 7, 595830.
9. Negri, L. B.; Martins, J.; Santana Da Silva, R.; Hamblin, M. R. *J. Photochem. Photobiol. B Biol.* **2019**, *198*, 111564.
10. Negri, L. B.; Martins, T. J.; da Silva Gobo, N. R.; de Oliveira, K. T.; Hamblin, M. R.; da Silva, R. S. *Inorg. Chem. Commun.* **2019**, *99*, 60–63.
11. da Silva, R. S.; de Lima, R. G.; de Paula Machado, S. *Adv. Inorg. Chem.* **2015**, *67*, 265–294.
12. Heinrich, T. A.; Tedesco, A. C.; Fukuto, J. M.; da Silva, R. S. *Dalton Trans.* **2014**, *43* (10), 4021–4025.
13. Carneiro, Z. A.; de Moraes, J. C. B.; Rodrigues, F. P.; de Lima, R. G.; Curti, C.; da Rocha, Z. N.; Paulo, M.; Bendhack, L. M.; Tedesco, A. C.; Formiga, A. L. B.; da Silva, R. S. *J. Inorg. Biochem.* **2011**, *105* (8), 1035–1043.
14. Rawling, T.; Xiao, H.; Lee, S.-T.; Colbran, S. B.; McDonagh, A. M. *Inorg. Chem.* **2007**, *46* (7), 2805–2813.
15. Teles Ferreira, J.; Pina, J.; Alberto Fontes Ribeiro, C.; Fernandes, R.; Tomé, J. P. C.; Rodríguez-Morgade, M. S.; Torres, T. *J. Mater. Chem. B* **2017**, *5* (29), 5862–5869.
16. Charlesworth, P.; Truscott, T. G.; Brooks, R. C.; Wilson, B. C. *J. Photochem. Photobiol. B Biol.* **1994**, *26* (3), 277–282.
17. Enow, C. A.; Marais, C.; Bezuidenhoudt, B. C. B. *J. Porphyr. Phthalocyanin.* **2015**, *19* (04), 582–594.
18. Alessio, E.; Mestroni, G.; Nardin, G.; Attia, W. M.; Calligaris, M.; Sava, G.; Zorzet, S. *Inorg. Chem.* **1988**, *27* (23), 4099–4106.
19. Heinrich, T. A. *Aspectos químicos, fotoquímicos e fotobiológicos de complexo rutênio-nitrosilo como precursor de óxido nítrico. Princípios de aplicação como agente citotóxico em linhagens de células tumorais, Ph.D. Thesis*; Faculdade de Ciências Farmacêuticas de Ribeirão Preto, Universidade de São Paulo: Ribeirão Preto, Brazil, 2013.
20. Sant'Ana Dusse, L. M.; Vieira, L. M.; Carvalho, M.d. G. *J. Brasil. Patolog. Med. Lab.* **2003**, *39* (4), 343–350.
21. Ignarro Louis, J.; Bruce, F. *Nitric Oxide: Biology and Pathobiology*; Elsevier, 2017.
22. Wink, D. A.; Mitchell, J. B. *Free Rad. Biol. Med.* **1998**, *25* (4–5), 434–456.
23. Ghimire, K.; Altmann, H. M.; Straub, A. C.; Isenberg, J. S. *Am. J. Physiol. Cell Physiol.* **2017**, *312* (3), C254–C262.
24. Thomas, D. D.; Ridnour, L. A.; Isenberg, J. S.; Flores-Santana, W.; Switzer, C. H.; Donzelli, S.; Hussain, P.; Vecoli, C.; Paolocci, N.; Ambs, S.; Colton, C. A.; Harris, C. C.; Roberts, D. D.; Wink, D. A. *Free Rad. Biol. Med.* **2008**, *45* (1), 18–31.
25. Feldman, P. L.; Griffith, O. W.; Stuehr, D. J. *Chem. Eng. News* **1993**, *71* (51), 26–38.
26. Wink, D. *Carcinogenesis* **1998**, *19* (5), 711.
27. Butler, A. R.; Williams, D. L. H. *Chem. Soc. Rev.* **1993**, *22* (4), 233.
28. Korde Choudhari, S.; Chaudhary, M.; Bagde, S.; Gadbail, A. R.; Joshi, V. *World J. Surg. Oncol.* **2013**, *11* (11), 118.
29. Xu, W.; Liu, L. Z.; Loizidou, M.; Ahmed, M.; Charles, I. G. *Cell Res.* **2002**, *12* (5–6), 311.
30. Hush, N. S.; Woolsey, I. S. *Mol. Phys.* **1971**, *21* (3), 465.
31. Isago, H. *Optical Spectra of Phthalocyanines and Related Compounds A Guide for Beginners*; Springer Japan: Imprint: Springer, 2015. Tokyo.
32. Leznoff, C.; Lever, A. *Phthalocyanines Properties and Applications*; VCH Publishers: North York, Ontario, 1996.

33. Bonnett, R. *Chemical Aspects of Photodynamic Therapy*; CRC Press: London, 2000.
34. Negri, L. B. *Fotobiomodulação e óxido nítrico como estratégias para potencializar a ação anti-tumoral da terapia fotodinâmica mediada por rutênio-ftalocianinas: Aspectos químicos, bioquímicos e fotobiológicos. Ph.D. Thesis*; Faculdade de Ciências Farmacêuticas de Ribeirão Preto, Universidade de São Paulo: Ribeirão Preto, Brazil, 2019.
35. Ömeroğlu, İ.; Biyiklioğlu, Z. *Turk. J. Chem.* **2015**, *39* (2), 347.
36. Carneiro, Z. A. *Fotocitotoxicidade proveniente do sinergismo de oxigênio singlete e óxido nítrico gerado pelo complexo [Ru(NO)(ONO)(ftalocianina)]. Ph.D. Thesis*; Faculdade de Ciências Farmacêuticas de Ribeirão Preto, Universidade de São Paulo: Ribeirão Preto, Brazil, 2011.
37. Farrell, N. P.; Murray, A. J.; Thornback, J. R.; Dolphin, D. H.; James, B. R. *Inorg. Chim. Acta* **1978**, *144*.
38. Ferraudi, G. J.; Prasad, D. R. *J. Chem. Soc. Dalton Trans.* **1984**, *86* (10), 2137.
39. Guez, D.; Markovitsi, D.; Sommerauer, M.; Hanack, M. *Chem. Phys. Lett.* **1996**, *249* (5–6), 309.
40. da Rocha, Z. N.; de Lima, R. G.; Doro, F. G.; Tfouni, E.; da Silva, R. S. *Inorg. Chem. Commun.* **2008**, *11* (7), 737.
41. Yanai, T.; Tew, D. P.; Handy, N. C. *Chem. Phys. Lett.* **2004**, *393* (1–3), 51.
42. Campos, C. T.; Jorge, F. E. *Mol. Phys.* **2013**, *111* (2), 167.
43. Jorge, F. E.; Canal Neto, A.; Camiletti, G. G.; Machado, S. F. *J. Chem. Phys.* **2009**, *130* (6), 064108.
44. Pritchard, B. P.; Altarawy, D.; Didier, B.; Gibson, T. D.; Windus, T. L. *J. Chem. Inform. Model.* **2019**, *59* (11), 4814.
45. Stephens, P. J.; Devlin, F. J.; Chabalowski, C. F.; Frisch, M. J. *J. Phys. Chem.* **1994**, *98* (45), 11623.
46. Martin, R. L. *J. Chem. Phys.* **2003**, *118* (11), 4775.
47. Takano, Y.; Houk, K. N. *J. Chem. Theor. Comput.* **2005**, *1* (1), 70.
48. Bachmann, M.; Li, W.; Edwards, M. J.; Ahmad, S. A.; Patel, S.; Szabo, I.; Gulbins, E. *Front. Cell Dev. Biol.* **2020**, *8*, 611853.
49. Guo, J.-J.; Wang, S.-R.; Li, X.-G.; Yuan, M.-Y. *Dyes Pigments* **2012**, *93* (1–3), 1463.
50. Nunes, S. M. T.; Sguilla, F. S.; Tedesco, A. C. *Brazil. J. Med. Biol. Res.* **2004**, *37* (2), 273.
51. Martins, T. J. *Complexos rutênio-ftalocianinas como candidatos a fotossensibilizadores: estudos fotoquímicos, fotofísicos, fotobiológicos e avaliação do efeito "teranóstico"da interação compontos quânticos. Ph.D. Thesis*; Faculdade de Filosofia Ciências e Letras de Ribeirão Preto, Departamento de Química, Universidade de São Paulo: Ribeirão Preto, Brazil, 2017.
52. Neese, F.; Wennmohs, F.; Becker, U.; Riplinger, C. *J. Chem. Phys.* **2020**, *152* (22), 224108.
53. Weigend, F. *Phys. Chem. Chem. Phys.* **2006**, *8* (9), 1057.
54. Ramos, C.; Prado, F.; Carmo, M. E.; Farias, G.; Souza, B.; Machado, A. E.; Patrocinio, A. O. *J. Brazil. Chem. Soc.* **2022**, *33* (5), 425.
55. Miranda, K. M.; Bu, X.; Lorkovic, I.; Ford, P. C. *Inorg. Chem.* **1977**, *36*, 4838.
56. Lorkovic, I. M.; Miranda, K. M.; Lee, B.; Bernhard, S.; Schoonover, J. R.; Ford, P. C. *J. Am. Chem. Soc.* **1998**, *120*, 11674.
57. Nguyen, S. T.; Nguyen, H. T.-L.; Truong, K. D. *Biomed. Res. Ther.* **2020**, 7 (7), 3855.
58. Ferreira, J. T.; Pina, J.; Ribeiro, C. A. F.; Fernandes, R.; Tomé, J. P. C.; Rodríguez-Morgade, M. S.; Torres, T. *ChemPhotoChem* **2018**, *2* (7), 640.
59. Ferreira, J. T.; Pina, J.; Ribeiro, C. A. F.; Fernandes, R.; Tomé, J. P. C.; Torres, T.; Rodríguez-Morgade, M. S. *J. Porphyr. Phthalocyanin.* **2021**, *25* (10–12), 1193.
60. Karakuzu, O.; Cruz, M. R.; Liu, Y.; Garsin, D. A. *Bio. Protocol* **2019**, *9* (21), e3409.
61. Price, M.; Reiners, J. J.; Santiago, A. M.; Kessel, D. *Photochem. Photobiol.* **2009**, *85* (5), 1177.
62. Rodrigues, F. P.; Carneiro, Z. A.; Mascharak, P.; Curti, C.; da Silva, R. S. *Coord. Chem. Rev.* **2015**, *306*, 701.

CHAPTER ELEVEN

Photoactive organometallic compounds as antimicrobial agents

Ashwene Rajagopal[a,b], Jack Biddulph[a], Leila Tabrizi[a], Deirdre Fitzgerald-Hughes[b], and Mary T. Pryce[a,*]

[a]School of Chemical Sciences, Dublin City University, Dublin, Ireland
[b]Department of Clinical Microbiology, Royal College of Surgeons in Ireland, Dublin, Ireland
*Corresponding author: e-mail address: mary.pryce@dcu.ie

Contents

1. Introduction 382
2. Group 4 and 5 organometallic complexes 386
3. Group 6 organometallic complexes 387
4. Group 7 organometallic complexes 391
5. Group 8 organometallic complexes 398
6. Group 9 organometallic compounds 399
7. Group 10 and 11 organometallic complexes 401
8. Summary and future perspectives 404
Acknowledgment 405
References 405

Abstract

It is estimated that by 2050, 10 million lives will be at risk annually due to multidrug-resistant infections, if effective responses targeting antimicrobial resistance (AMR) are not established. The development of alternative therapeutics to manage AMR infections, and technologies to prevent the growth and spread of AMR is critical. Only two antimicrobial agents have been introduced to clinics in the last 40 years that have new targets within bacteria, daptomycin a lipopeptide that targets the cell membrane and linezolid, an oxazolidinone that targets early protein synthesis. Both antibiotics were developed for Gram-positive infections and resistance to both agents has already developed. There is a severely diminished pipeline for the development of new antimicrobial agents and an over-reliance on the development of agents with known targets within bacteria (the cell wall, nucleic acid synthesis, protein synthesis) which have already been compromised through the overuse of antibiotics. As we approach the post-antibiotic era, alternative strategies are required, and in this chapter the development of photoactive organometallic complexes for light assisted antimicrobial applications are discussed. These include compounds which can release CO or generate reactive oxygen species, such as singlet oxygen, for antimicrobial therapies. This article will focus on organometallic complexes that have been assessed for antimicrobial applications.

Advances in Inorganic Chemistry, Volume 80
ISSN 0898-8838
https://doi.org/10.1016/bs.adioch.2022.06.006

1. Introduction

The resistance of bacterial pathogens to antibiotics in the modern day is a global health crisis.[1] The World Health Organization (WHO) has reported that at least 700,000 people die every year due to drug resistant diseases. Without pro-active global responses to the antimicrobial resistance (AMR) crisis, it is estimated by 2050 that 10 million lives will be at risk annually due to untreatable infection.[2] Within a decade, AMR may have a significant impact on the economy, resulting in up to 24 million people facing financial difficulty and ultimately poverty. The WHO has designated the bacteria; *Enterococcus faecium, Staphylococcus aureus, Klebsiella pneumoniae, Acinetobacter baumannii, Pseudomonas aeruginosa,* and Enterobacter species as "priority pathogens" for which new antimicrobials are urgently needed. These pathogens, often referred to as 'ESKAPE' are associated with AMR and cause difficult-to-treat infections, especially in healthcare settings.[1] Gram-negative bacteria such as *P. aeruginosa, K. pneumoniae* present an additional challenge in terms of antimicrobial development, as in addition to acquired resistance (driven by genetic mutations and exchange) they also exhibit intrinsic resistance due the presence of an outer membrane which may impede the entry of antibiotics.[3] The ESKAPE pathogens, particularly *P. aeruginosa* and *S. aureus* form biofilms *in vivo* which add further complexity to infection management. Biofilms are communities of bacteria that form on abiotic surfaces, such as medical devices or damaged host tissues, such as wounds or the lungs of patients with cystic fibrosis.[4] In this mode cells produce an encasing extracellular matrix and become less metabolically active, factors which help them to avoid the immune responses and the effects of antibiotics, resulting in chronic infection and further antibiotic use.

Even the appropriate use of antimicrobials creates a selective pressure that favors microorganisms with inherent/acquired AMR. However, inappropriate and overuse of these medicines in human healthcare, food and agricultural sectors, continues to drive the world to an impending health, economic and societal crisis. Antibiotics are medicines which are used to prevent and treat infections in humans, animals and plants. However, since the era of antibiotic discovery in the 1930s–1960s, only two new antibiotic classes have reached the market with most discovery limited to derivatives of known classes.[5] The mechanisms of action of antibiotics in current use are therefore limited, involving disruption to cell wall synthesis (e.g. penicillin), nucleic acid synthesis (e.g. ciprofloxacin) or protein synthesis (e.g. gentamicin). Daptomycin, introduced to clinical use in 2003 targets the cell membrane. However, bacteria have evolved to develop sophisticated and sustainable AMR mechanisms that override attacks on these targets.[6] There is an urgent need for entirely novel therapeutics with different and multimodal mechanisms to combat AMR. Promising areas of research in this regard, pursued by our groups and the foci of this chapter are developments in two classes of photoactive organometallic materials for light assisted antimicrobial applications, compounds which can release CO or generate singlet oxygen by photodynamic therapy (PDT).

The use of PDT as a potential treatment strategy was proposed in the early 1900s. PDT requires light (ideally in the range 400–700 nm), oxygen and a photosensitizer (PS), Fig. 1.[7,8] This combination is required for the generation of reactive oxygen species (ROS), including singlet oxygen or hydrogen peroxide. Photosensitizers based on porphyrins, chlorins, bacteriochlorins and phthalocyanines are clinically approved for PDT in cancer diagnostics and chemotherapy, as a minimally invasive way of targeting malignant cells or tissues.[9,10] On the other hand, photodynamic inactivation (PDI) of microorganisms or antimicrobial photodynamic therapy (aPDT) has not been as extensively studied and holds much promise as a powerful tool in targeting microbial pathogens. ROS such as singlet oxygen or hydrogen peroxide, can induce cell mortality mediated by damaging interactions with proteins, DNA and cell membrane components of the bacteria (see Fig. 1).[11–13] Selective accumulation of the PS by bacteria and not the surrounding tissue

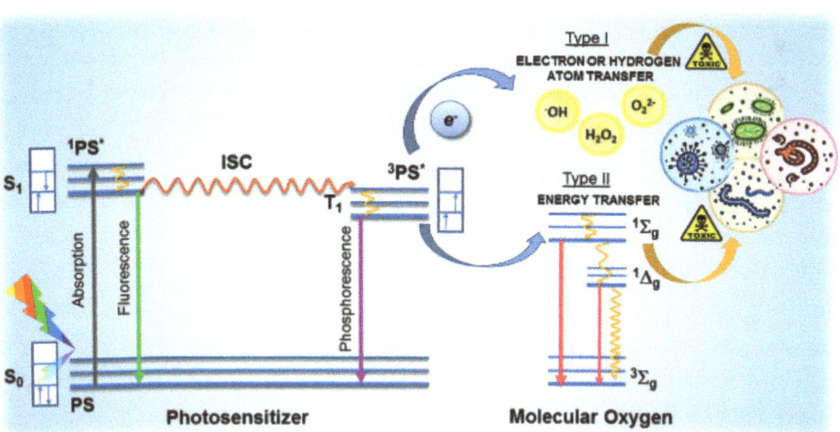

Fig. 1 Jablonski Diagram and the mechanisms (Type I and Type II) for the generation of ROS and subsequent inactivation of microorganisms. *Reproduced and adapted from Pucelik, B.; Dąbrowski, J. M.* Adv. Inorg. Chem. ***2022,*** *79, 65–103 with permission.*

limits cytotoxicity and the antimicrobial effects, and depending on the characteristics of the photosensitizer, it may be narrow or broad spectrum involving multiple intracellular targets (e.g. DNA damage, cell membrane disruption or inactivation of metabolic enzymes). The various mechanisms of resistance in microorganisms to antibiotics have recently been discussed in detail elsewhere.[13] aPDT was first introduced in the 1960s when toluidine blue was used against microorganisms, and following 30 min of irradiation at 632 nm, 99% of bacteria were killed, due to disruption of the cell membrane.[14,15] Other organic dyes including methylene blue and rose bengal, and cationic/anionic analogs were used as photosensitizers. It is only in more recent years that photosensitizers based on organometallic complexes have been identified for aPDT.

The second approach for antimicrobial inactivation, relies on the development of complexes for CO release. CO is formed in mammals from the oxidation of heme by heme oxygenase enzymes.[16] It participates in the mediation of essential biological functions and exhibits potential therapeutic properties e.g. anti-inflammatory and vasodilation. Under physiological conditions, it is essential that CO is delivered in a safe and controlled manner to specific sites due to the toxic attributes associated with over-exposure to CO. Hence, a safer method for CO delivery is via carbon monoxide-releasing molecules (CORMs), which have the potential to transport controlled, defined amounts of CO to target areas.[17,18] Organometallic complexes are excellent candidates for CO delivery systems due to the number of multiple CO releasing units bound to the metal center and in many cases, ease of synthesis. The photochemistry of metal carbonyl complexes has been studied for many years, however, it is only more recently that the therapeutic uses of these compounds has been studied. Various transition metals and ligands can be incorporated into the complexes to aid photochemical CO release, promote water solubility, and in some cases direct the CORM to target areas in the body. As metal carbonyl complexes can have multiple CO ligands attached, this may mean that lower doses of the CORM could be administered. The nature of the metal and co-ligands can influence the rate of CO release, which facilitates fast/slow CO release.[19] Metal carbonyls have the ability to generate precise amounts of CO and mimic the bioactivity of CO that is formed endogenously. Photo-assisted CO release (photoCORMs) provides the benefit of targeted applications and dosage control. Some metal-based CORMs contain trace elements present in the body such as Fe, Mn and Co, while other CORMs incorporate non-physiological metals such as W and Re.[20]

The ideal photoCORM should (i) not exhibit cytotoxic properties prior to and post CO release, (ii) should be stable for storage under ambient conditions (iii) be active and potent at the diseased tissues (iv) target specific sites (v) CO release should be controlled, and (vi) preferably water soluble or at least soluble in an aqueous based media.[17,21] However, recently, novel technologies have been developed to overcome the by-product toxicity and water solubility problems associated with CORMs by incorporating polymer conjugates, nanoporous non-woven and nanoparticles into the CORM sphere.[22]

Iron pentacarbonyl ($Fe(CO)_5$), and dimanganese decacarbonyl ($[Mn_2(CO)_{10}]$), the latter more commonly known as CORM-1 and the tricarbonyldichloro ruthenium(II)—dimer, $[Ru(CO)_3Cl_2]_2$, termed CORM-2, were the first complexes investigated for CO release, reported by Motterlini et al.[18] These complexes exhibited CO releasing properties in biological systems that mimicked the pharmacological effects of heme oxygenase—derived CO. The myoglobin (Mb) assay, where deoxy-Mb will be readily converted to Mb-CO in the presence "free CO" was used to quantify CO release in a protein-based aqueous media (at 37 °C and pH, 7.4).[23] The concentration of Mb-CO is measured by determining the change in absorbance spectra. Half-life ($t_{½}$) values are also used to compare CO release, which is the time taken for the concentration of a CORM to decrease by 50%, and can be determined using the myoglobin assay. Other approaches to quantifying CO loss include time resolved infrared spectroscopy and gas chromatography.[24,25]

Antimicrobial PDT and the use of CORMs can be effective against a wide range of organisms including bacteria, fungi and protozoa, but a significant advantage is the apparent limited tendency for resistance to develop after repeat sessions of therapy.[26] In the case of aPDT which is a rapid and efficient process, the bacteria lack sufficient time and agency to upregulate stress responses that normally mitigate the damaging effects of ROS.[27] Due to the alternative mechanisms of actions involving multiple targets, both approaches are promising methods to inactivate various microbes including AMR pathogens. In addition, investigations of these materials show potential in anti-biofilm activity. For example, CORM2, a ruthenium-containing CORM, showed attenuation of *P. aeruginosa* biofilms with additive effect in the presence of tobramycin.[28] A cationic Zn(II) phthalocyanine singlet oxygen generating dye disrupted and limited the regrowth of *S. aureus* and *P. aeruginosa* biofilms on titanium disks, a material used in prosthetics.[29]

2. Group 4 and 5 organometallic complexes

While organometallic complexes containing Ti, Zr and V, have been studied by a number of research groups for their antimicrobial activity, no studies have focused on photoactivation. A series of chlorocyclopentadienyl zirconium complexes were studied by El Khaldy et al., who investigated the biocidal effect of combining two bioactive components: the organozirconium moiety and the organophosphorus ligands attached (Fig. 2).[30] Five such complexes were synthesized and their antimicrobial activity assessed against both Gram-negative and Gram-positive bacterial strains (see Table 1). The antibacterial study conducted showed weak activity in all cases, however the Zr complexes did inhibit bacterial growth compared to the control (70% Ethanol, v/v). The weakest activity was observed against *S. aureus* with the lowest inhibition of 2.45%. The most susceptible bacterial strain tested was *Escherichia coli* with the greatest inhibition values observed between 5.0% and 11.1%. Despite the presence of the two bioactive moieties in the Zr- complexes only modest antibacterial activity was reported.

The only photoactive antimicrobial zirconium complexes in the literature, are based on a metallo-phthalocyanine, with 4-aminosalicylic or 4-aminophthalic acids at the axial position, which acted as bridging ligands for attaching to graphite oxide (GO) (Fig. 3).[31] The two complexes were screened for antibacterial activity using a series of Gram-negative and Gram-positive bacteria, and fungi. Following irradiation at >420 nm, weak antimicrobial activity was observed using concentrations of 5–10 mg/mL. Aggregation in aqueous solution, was suggested to contribute to the weak

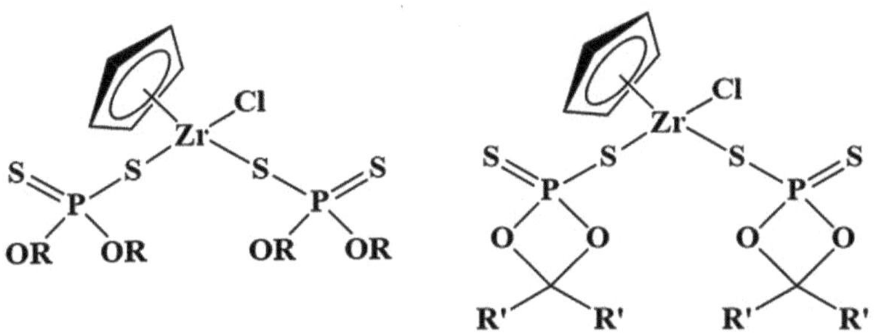

R = Et (1), Pr-n (2), or Pr-i (3) **R' = Me (4) or Et (5)**

Fig. 2 Complexes **1–3,** where R=Et, Pr-n, or Pr-i (left) and Complexes **4–5,** where R′=Me or Et (right).

Table 1 % Growth inhibition of Zr compounds (400 μg/mL) against bacterial strains.

	Gram-negative		Gram-positive	
Compound	***E. coli***	***P. vulgaris***	***S. aureus***	***B. subtilis***
$CpZrCl[S_2P(OEt)_2]_2$ (**1**)	11.1	5.1	2.45	10.3
$CpZrCl[S_2P(OPr\text{-}n)_2]_2$ (**2**)	5.0	5.0	6.4	5.0
$CpZrCl[S_2P(OPr\text{-}i)_2]_2$(**3**)	5.8	4.3	5.4	4.8
$CpZrCl[S_2POCH_2CMe_2CH_2O]_2$ (**4**)	8.8	5.9	3.33	6.7
$CpZrCl[S_2POCH_2CEt_2CH_2O]_2$ (**5**)	6.9	4.8	4.23	6.4

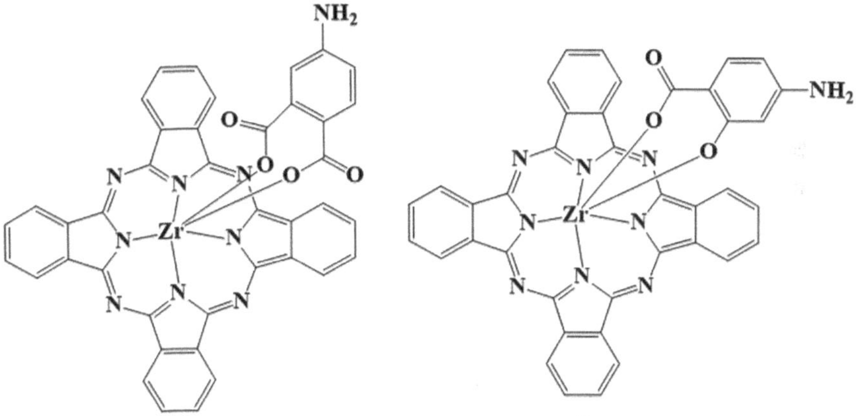

6 7

Fig. 3 Structure of PcZr(4-aminophthalate) [PcZr(APhA)] (**6**) and PcZr(4-aminosalicylate) [PcZr(APhA)] (**7**).

antimicrobial activity for the GO-PcZr composites. However, it should be noted for the GO-PcZr(ASA) or GO-PcZr(APhA) composites, the photoactive content, for the ROS generating component in the material is estimated at ~1%.

3. Group 6 organometallic complexes

The benefit of using metal carbonyls, as a CO delivery vector is that the timing, location and dosage can be controlled through light activation. Similar to many other research groups in recent years, we have attempted to design and identify photoactive complexes for antimicrobial applications. One of the main challenges in developing organometallic complexes is extending the absorption spectrum towards the therapeutic range.

If photo-activated CO release is limited to UV or near-UV light sources, the potential uses of these as therapeutics are restricted to topical applications. Clinical applications that require deeper tissue penetration necessitate longer wavelength excitation, ideally in the near-IR region to maximize penetration and to minimize harmful exposure to radiation. As a result, many of the photoCORMs discussed in this section, will require further study focused on extending the absorption desirable for physiological applications. In a recent review, Ford has described potential strategies to enable the use of longer wavelengths for CO-release, including two photon excitation, quantum dots, polymers, and metallic platforms, in addition to upconverting using nanoparticles antennae.[21] Numerous studies focus on CO-release and its quantification using an array of techniques, including the myoglobin assay.[32] However, in the following section we have focused solely on organometallic compounds that were assessed for antimicrobial activity.

The tetracarbonyl organometallic complexes shown in Fig. 4, which contain the conjugated 1,10-phenanthrolin-5-amine-*N*-(2-fluorenylmethylene) ligand were assessed for antibacterial and antifungal activity following irradiation at 365 nm.[33] Subsequently, Xian Lee et al. designed a similar

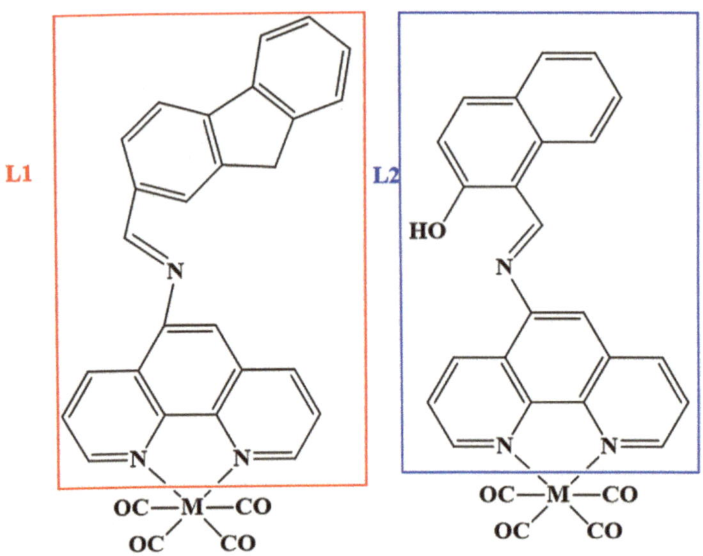

M = Cr (8), Mo (9), or W (10) **M = Cr (11), Mo (12), or W (13)**

Fig. 4 General structure of phenanthroline based group 6 photoCORMs Left: complexes **8–10**; $[M(C_{26}H_{17}N_3)(CO)_4]$, Right: complexes **11–13**; $[M(C_{23}H_{13}N_3OH)(CO)_4]$, where M = Cr, Mo, or W.

family of photoCORMs, again with a phenanthroline based ligand, but with a terminal naphthalene unit (Fig. 4).[34] All complexes displayed a strong absorption between 300 and 400 nm, together with a weak MLCT band in the visible region between 420 and 570 nm. Several Gram-negative and Gram-positive bacteria and fungi were tested, and the minimum inhibitory concentration (MIC) values obtained for each photoCORM are summarized in Table 2.

Surprisingly, for the fluorene containing complexes (**8–10**) no enhancement in antibacterial activity was observed following irradiation at 365 nm, (compared to studies performed in the absence of light) with the Cr photoCORM yielding some antimicrobial activity against the fungal strains. The activity observed for the Cr photoCORM (**8**) towards *Candida Spp.* was attributed to (i) the short half-life (14 mins for Cr vs. 52 mins for the tungsten analogue), and (ii) the lower molecular weight of the Cr complex, which aids diffusion across cell membranes. In the case of antibacterial inactivation, there was no apparent difference in activity with/without irradiation, indicating that photoactivated CO generation did not contribute to the antibacterial activity observed for these complexes.

When the fluorene unit in these complexes was replaced by a naphthalene moiety, the photoCORMs **11–13** displayed antibacterial activity. Again the Cr-photoCORM (**11**) had the highest activity, which was attributed to the short half-life (10 s (**11**) vs 34 mins for the W-complex (**13**), which results in faster CO release). However, the ligand alone had superior antimicrobial activity compared to corresponding complexes. For example, the MIC for the ligand (**L2**) was 3.13 mg/mL for *Bacillus cereus*, compared to 12.5 mg/mL for complex **12**. In contrast to complex **8**, the naphthalene-containing Cr photoCORM (**11**) showed a different response towards fungi, with no anti-fungal activity observed in the presence or absence of light.

We studied a tungsten Fischer carbene complex for photoactivated CO release and subsequent antimicrobial activity.[35] $[(CO)_5WC(NC_4H_8)Me]$ (Fig. 5) was regarded as an ideal CO-releasing antibacterial candidate due to the high photo quantum yield for CO release of 70%, which was determined from time resolved infrared experiments. The antibacterial activity for this complex was assessed against *S. aureus* (ATCC25923). In the absence of light, low levels of antibacterial activity were observed (10–20%). For all concentrations used (50–200 μM), increased activity was observed following photoactivation (375 nm) compared to that in the absence of a light source. The bactericidal activity was mainly mediated by CO release, which

Table 2 Minimum inhibitory concentration (MIC) values for Compounds **L1, L2**, and **8–13** (following irradiation at 365 nm).

	MIC (mg/mL)							
Microorganisms tested	**L1**	**8**	**9**	**10**	**L2**	**11**	**12**	**13**
Bacillus cereus ATCC14579	>100	100	100	>100	3.13	12.50	25	25
Bacillus subtilis ATCC8188	>100	100	>100	>100	6.25	12.50	25	25
Enterococcus faecalis ATCC29212	>100	>100	>100	>100	50	50	>100	>100
Vancomycin-resistant *E. faecalis* ATCC700802	>100	>100	>100	>100	50	>100	>100	>100
Methicillin-sensitive *S. aureus* ATCC29213	>100	>100	>100	>100	6.25	25	50	50
Methicillin-sensitive *S. aureus* ATCC33591	>100	>100	>100	>100	25	50	50	100
Methicillin-resistant *S. aureus* ATCC43300	>100	100	>100	>100	25	25	50	50
Methicillin-resistant *S. aureus* ATCC33519	>100	100	>100	>100	6.25	25	50	>100
Escherichia coli ATCC 25922	>100	100	>100	>100	6.25	50	50	100
Pseudomonas aeruginosa ATCC10145	>100	>100	>100	>100	25	>100	>100	>100
Klebsiella pneumoniae ATCC10031	>100	>100	>100	>100	>100	25	50	50
Shigella flexneri ATCC12022	>100	>100	>100	>100	12.50	50	100	100
Candida albicans IMR	>100	>100	>100	>100	>100	>100	>100	>100
Candida albicans clinical isolate	>100	25	>100	>100	>100	>100	>100	>100
Candida tropicalis clinical isolate	>100	25	>100	>100	>100	>100	>100	>100

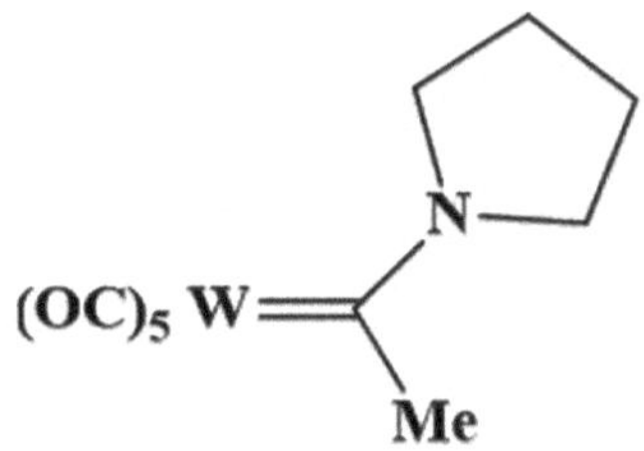

14

Fig. 5 Structure of the tungsten Fischer carbene complex **14**.

was confirmed by addition of the CO scavenger, bovine hemoglobin to the assays. Addition of bovine hemoglobin to the assays led to a decrease in bactericidal activity by 65%.

The photoCORMs described above are only active in the UV-region, and therefore any potential therapeutic applications are restricted to surfaces applications. Given the number of group 6 metal carbonyls that have been reported in literature, and many with high quantum yields for CO release (up to 0.72), it is somewhat surprising that so few complexes have assessed for antimicrobial applications.[36,37]

4. Group 7 organometallic complexes

Manganese is an essential bio-metal for human health. It is a required component of a number of mitochondrial enzymes, including Mn-dependent superoxide dismutase II, pyruvate carboxylase, and glutamine synthetase.[38–40] For more than a decade, manganese organometallic complexes have been recognized and studied for their antimicrobial properties. Coordination of well-established medicines like quinolones,[41–45] azoles,[46,47] polyether ionophores[48] to a manganese metal center shows enhanced antimicrobial activity towards Gram-positive and Gram-negative bacterial strains.

Manganese carbonyl complexes are widely studied for antimicrobial properties due to their rapid CO-releasing capabilities in solution. Different CO triggered methods such as redox-state,[49] thermal,[50] pH,[51] and enzymatic[52] in addition to photo-activation have been used to promote CO release in these complexes. Motterlini et al. in 2002 reported the photodissociation of dimanganese decacarbonyl $[Mn_2(CO)_{10}]$.[18] Due to its poor bioavailability and relatively limited options for further tuning the properties $[Mn_2(CO)_{10}]$ by chemical modifications, the direct use of $[Mn_2(CO)_{10}]$ for antimicrobial properties has not been well studied. PhotoCORMs based on $[Mn(CO)_3(tpa\text{-}\kappa^3N)]Br$ from $[MnBr(CO)_5]$ and tris(2-pyridylmethyl) amine (tpa) for the delivery of CO to biological systems, were developed and gave rise to inactivated CORMs (*iCORMs*) after the release of CO. Schatzschneider et al. showed that $[Mn(CO)_3(tpa\text{-}\kappa^3N)]Br$ (**15**) (Fig. 6) has long-term dark stability and antibacterial activity against *E. coli* was only observed following photoactivation at 365 nm.[53] Furthermore, the complex was active *in vitro* against the multidrug resistant uropathogenic strain of *E. coli*, EC958[54] and *Streptococcus pneumoniae*.[55] ICP-MS studies indicated that the complex was internalized by *E. coli* but remained nontoxic under dark conditions. Antibiotic synergy of the antibacterial activity of $[Mn(CO)_3(tpa\text{-}\kappa^3N)]Br$ with colistin and polymyxin B involving increased cytoplasmic membrane damage and with doxycycline involving reduce drug efflux pump expression have been shown. In addition, these authors showed that the breakdown products of $[Mn(CO)_3(tpa\text{-}\kappa^3N)]Br$, similarly to tetracyclines, activated iron chelation triggering an iron starvation, which could be abrogated by the addition of excess iron.[56]

More recently, Schatzschneider and colleagues, further modified the $[Mn(CO)_3(tpa\text{-}\kappa^3N)]Br$ complex [with bpqa = bis(2-pyridinylmethyl)(2-quinolinylmethyl)amine, bqpa = bis(2-quinolinylmethyl)(2-pyridinylmethyl)

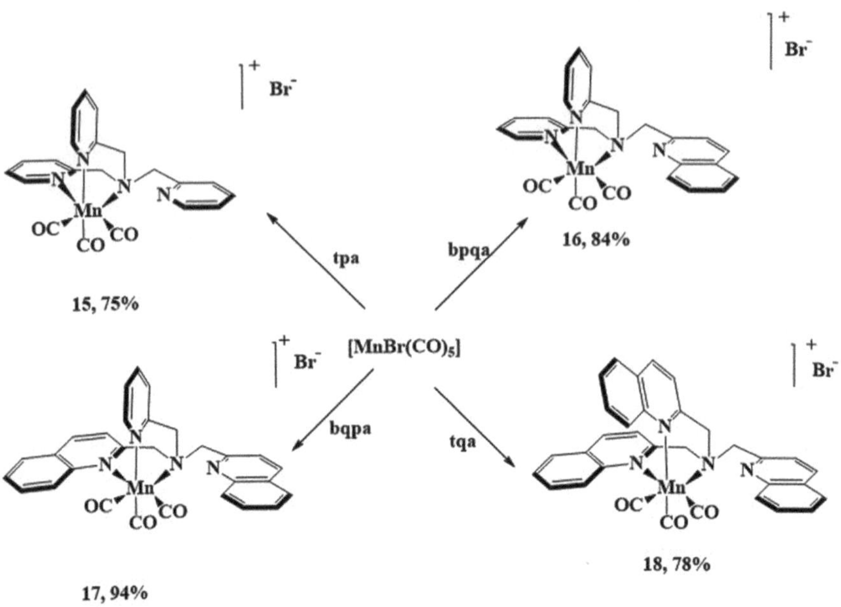

Fig. 6 $[Mn(CO)_3(X\text{-}\kappa^3N)]Br$ complexes (**15–18**) [X = bpqa = bis(2-pyridinylmethyl)(2-quinolinylmethyl)amine, bqpa = bis(2-quinolinylmethyl)(2-pyridinylmethyl)amine, tqa = tris(2-quinolinylmethyl)amine, and tpa = tris(2-pyridinylmethyl)amine].

amine, tqa = tris(2-quinolinylmethyl)amine, and tpa = tris(2-pyridinylmethyl) amine], (**15–18**) to further enhance the antibacterial potency of these Mn complexes (Fig. 6).[57] The bacterial killing activity of these Mn(I) complexes was assessed using 14 different multidrug resistant clinical isolates of *A. baumannii* and *P. aeruginosa*. The lipophilicity of the complexes is an important parameter in mediating passage of the complexes across the cell membrane and was determined to be in the order tpa < bpqa < bqpa ≈ tqa. The antimicrobial potency of the complexes correlated to the number of quinolinylmethyl groups, but was moderate, as evident from the MIC values of 100 μM for *A. baumannii* and 50 μM for *P. aeruginosa*. Bacterial cytoplasmic membrane damage was proposed as the antimicrobial mechanism for these Mn complexes, and these results demonstrate the importance of metal-CO ligand fragment, when assessing the biological activity of CORMs.

Ward and co-workers, prepared the tryptophan-derived Mn complexes (Fig. 7) which released CO using visible light.[58] Antibacterial effects of Trypto Mn-CORM were assessed against *E. coli*. The in situ degradation of the CORM under 400 nm irradiation was attributed to the bactericidal effect of the Trypto Mn-CORM. The effect against *E. coli* is associated with CO-release, and not a Mn species, and was confirmed using an i-CORM test. Additional toxicity studies were performed using Trypto Mn-CORM against *Neisseria gonorrhoeae* and *S. aureus* previously by the same group.[59] Trypto Mn-CORM is toxic to *N. gonorrhoeae*, even in the absence of a light source. As antimicrobial activity was abrogated in the presence of the CO-scavenger hemoglobin, it was concluded that bacterial toxicity was due to thermal release of CO. Trypto Mn-CORM exhibited a cytostatic

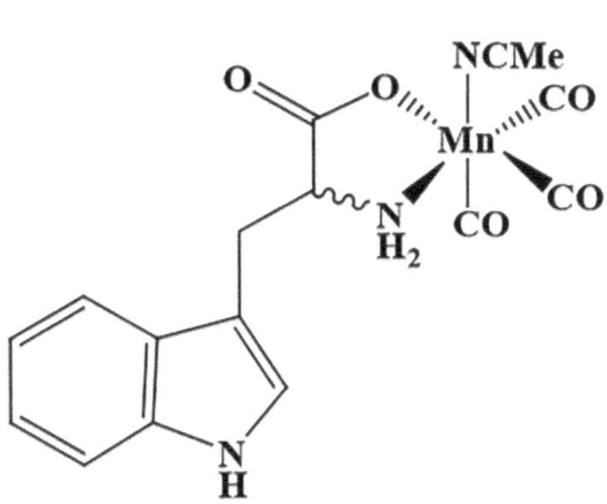

19

Fig. 7 Tryptophan-Mn CORM (complex **19**).

effect in the dark and cytotoxic effect when exposed to light against *S. aureus*. In addition, correlation between Trypto Mn-CORM activity with the bacterial culture age was highlighted. The mechanism was not investigated further, but greater sensitivity of older cultures, representing cells by the stationary phase of growth is interesting, given the slower metabolism of bacteria in biofilms, which contributes to the recalcitrance and longevity of these infections.

Rhenium: The *in vitro* anti-proliferative effects of rhenium complexes against tumor cells have been extensively studied over the last 20 years.[60] Metals such as iron, manganese, ruthenium and platinum had overshadowed rhenium in medicinal applications. However, the stability of rhenium compounds at room temperature, their capability of emitting light in the visible region, and suitable photophysical and photochemical properties, have made rhenium complexes more attractive in photoactivated chemotherapy and photoactivated antimicrobial therapy.[61]

In 2014, Meggers et al. reported the potency of rhenium(I) pyridocarbazole complexes toward light-induced antiproliferative activity in cancer cells.[62] A change in the ligand coordination within the rhenium tricarbonyl backbone had a marked physicochemical and biological impact. The generation of singlet oxygen was highest for complexes **22** and **23** (Fig. 8), using red light of $\lambda \geq 620$ nm. This publication highlighted the potential application of rhenium(I) tricarbonyl polypyridine complexes for antimicrobial PDT.

These studies provided the basis for the synthesis of light activated antibacterial rhenium bisquinoline complexes (Fig. 9) by Blaskovich et al.[63] ATCC strains of Gram-positive *S. aureus* and the Gram-negative *E. coli* were incubated with the rhenium complexes **24**–**26** and irradiated for 1 h at

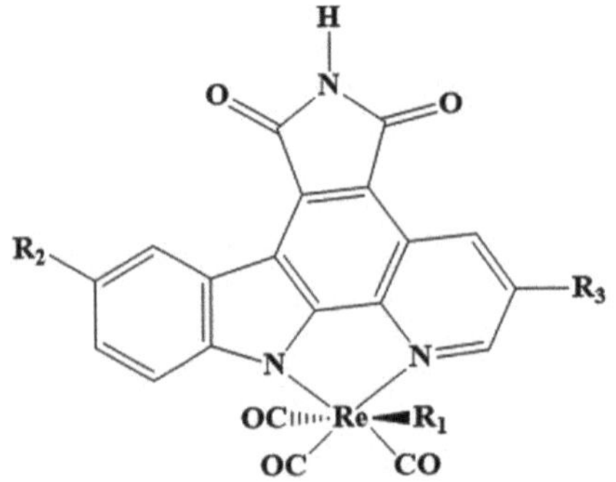

20	R_1: imidazole	R_2: H	R_3: H
21	R_1: imidazole	R_2: H	R_3: F
22	R_1: imidazole	R_2: H	R_3: CF_3
23	R_1: imidazole	R_2: OMe	R_3: F

Fig. 8 The structure of complexes **20–23**.

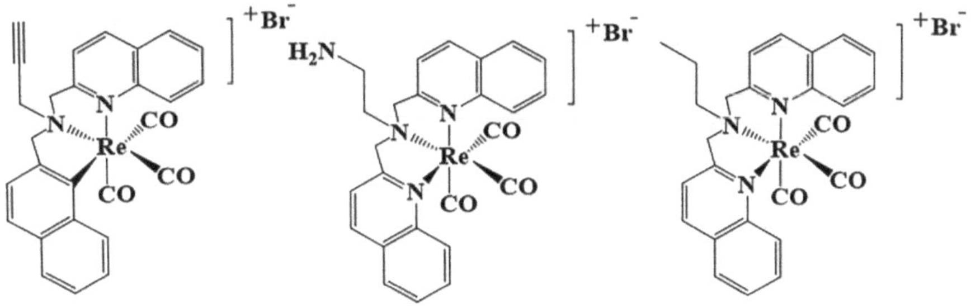

24 25 26

Fig. 9 Rhenium bisquinoline complexes **24–26.**

365 nm. The MIC values following irradiation were 4 to 16 times lower than in the absence of light, with MIC values between 0.25 and 64 μg/mL indicating the enhanced activity of the rhenium complexes upon irradiation (in Table 3). Compound **24** is the first rhenium compound reported that is active against Gram-positive and Gram-negative bacteria in addition to drug-resistant strains following irradiation. For these compounds, the mechanism of inactivation observed towards the bacterial strains used is attributed to the generation of ROS species and also an increase in aminoglycoside uptake.

In a recent study by Pizarro et al. facial rhenium(I) tricarbonyl complexes bearing P,N-bidentate pyridyl/phosphine ligands were synthesized with different phosphorus chelating rings and anions (RePNTfO and RePNNBr) (Fig. 10), were assessed for microbial inactivation.[64]

The *in vitro* antibacterial activity of RePNBr, RePNTfO and RePNNBr (**27–29**) was studied against *E. coli* and *S. aureus*. Following irradiation of the complexes with 365 nm, in the presence of *E. coli* and *S. aureus,* RePNBr (**27**) was most potent, inhibiting 99.9% of both bacterial strains. RePNTfO was more active against *S. aureus* at 50 μg/mL than toward *E. coli.* The singlet oxygen generation capabilities of the complexes are RePNBr > RePNTfO > RePNNBr, with singlet oxygen quantum yields of 0.0640, 0.0200 and 0.0015, respectively. This trend correlates with the 100-fold decrease in the $\log_{10}$ CFU of the bacterial strains. Infrared spectroscopy and DFT studies indicate photoisomerization of the *fac* to the *mer* isomer following photolysis.

Recent unpublished work from our laboratory based on rhenium N-heterocyclic metal carbonyl compounds, where we replaced Br or Cl halides, with a triflate ligand yielded a 100-fold increase in the antibacterial activity against multi-resistant *S. aureus* following irradiation at 375 nm. While this is a relatively small change, our studies demonstrate that subtle

Table 3 MIC values of **24–26** against *E. coli* and *S. aureus* (a), MIC values of **24** against methicillin-resistant *S. aureus* and colistin-resistant *E. coli* (b), MIC values of complex **24** against various mutant strains of *E. coli* (c), and against different mutants of *P. aeruginosa* (d).

	MICs against Gram(+) and Gram(−)			
	Dark		**365 nm**	
	(μg mL^{-1})	**(μM)**	**(μg mL^{-1})**	**(μM)**
S. aureus ATCC25923				
Vancomycin	1	0.7	1	0.7
24	8	11.6	0.5-1	0.72–1.45
25	32	46.2	4-8	5.8–11.6
26	2	2.9	0.25-0.5	0.36–0.72
E. coli ATCC 25922				
Polymyxin B	1	0.8	1	0.8
24	>64	>93.1	4-8	5.8–11.6
25	>64	>92.4	32	46.2
26	>64	>92.5	16	23.1
MICs against resistant Gram(+) and Gram(¯) *S. aureus* ATCC43300(MRSA)				
Vancomycin	2	1.4	2	1.4
24	4–8	5.8–11.6	2	2.9
E. coli (*mcr-1*)				
Polymyxin B	4	3.2	4	3.2
24	>64	>93.1	8	11.6
MICs against mutant *E. coli* strains				
E. coli (MB4827; control for mutants)				
Polymyxin B	0.13–0.25	0.1–0.2	0.13–0.25	0.1–0.2
24	>64	>93.1	8–16	11.6–23.3
E. coli (MB4902; lpxC)				
Polymyxin B	0.06	0.05	0.06	0.05
24	>64	>93.1	1–4	1.5–5.8

Table 3 MIC values of **24–26** against *E. coli* and *S. aureus* (a), MIC values of **24** against methicillin-resistant *S. aureus* and colistin-resistant *E. coli* (b), MIC values of complex **24** against various mutant strains of *E. coli* (c), and against different mutants of *P. aeruginosa* (d).—cont'd

	MICs against Gram(+) and Gram(−)			
	Dark		**365 nm**	
	(μg mL^{-1})	**(μM)**	**(μg mL^{-1})**	**(μM)**
E. coli (MB5747; tolC)				
Polymyxin B	0.13	0.1	0.13	0.1
24	8	11.6	2	2.9
E. coli (MB5746; lpxC, tolC)				
Polymyxin B	0.06	0.05	0.06	0.05
24	8	11.6	≤0.125	≤0.2
MICs against mutant *P. aeruginosa* strains				
P. aeruginosa (PAO1)				
Polymyxin B	1	0.8	1	0.8
24	>128	>186.2	64	93.1
P. aeruginosa (PAO397)				
Polymyxin B	1	0.8	1	0.8
24	128	186.2	4–8	5.8–11.6

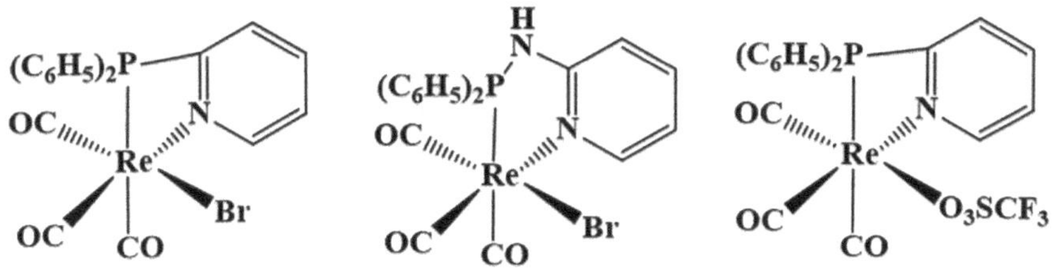

RePNBr, 27 **RePNNBr, 28** **RePNTfO, 29**

Fig. 10 Structure of facial rhenium(I) complexes with P,N-bidentate ligands.

changes led to a significant enhancement in antibacterial activity.[65] Other groups have also observed that Re(I) N-heterocyclic carbene complexes possess potent antibacterial activity against Gram-positive strains (MIC for *B. subtilis* = 0.7–1.3 μM), while no activity was observed towards

Gram-positive strains.[66] However, these studies were performed in the absence of a light source, and the activity was attributed to the positive charge which facilitates initial interactions of the NHC-Re compounds and the negatively charged phospholipids in the bacterial membrane.

A number of water soluble rhenium dicarbonyl compounds, $[Re(CO)_3(NN)(PR_3)]^+$, {where NN = a diimine ligand, and PR_3 = 1,3,5-triaza-7-phosphaadamantane (PTA), tris(hydroxymethyl)phosphine (THP), or 1,4-diacetyl-1,3,7-triaza-5-phosphabicylco[3.3.1]nonane (DAPTA)} were reported by Marker et al. These exhibited cytotoxic responses towards HeLa cells, and a number of ovarian cancer cell lines upon irradiation at 365 nm.[67] Minimal toxicity occurred in the absence of light, and the activity observed was attributed to both CO release and singlet oxygen. The rhenium compounds generated singlet oxygen with quantum yields up to 70%, and also undergo photosubstitution of CO ligand with quantum yields ranging from 1.1% to 5.5%. No antimicrobial studies however have been reported to date on these complexes.

5. Group 8 organometallic complexes

The polypyridine chelate complexes of ruthenium have been extensively studied for their PDT and antimicrobial PDT activity, but are not considered in this review, as they have been recently covered in a number of excellent reviews.[63] Few organometallic complexes of ruthenium, and none based on osmium have been studied for photoactivated antimicrobial activity. The polymeric dicarbonyldichlororuthenium(II) compound, $[Ru(CO)_2Cl_2]_n$ (Fig. 11, **30**) has been studied against *E. coli* and *S. aureus*.[68] This Ru-carbonyl precursor shows MIC at ng/mL, and is one of few photoactive organometallic

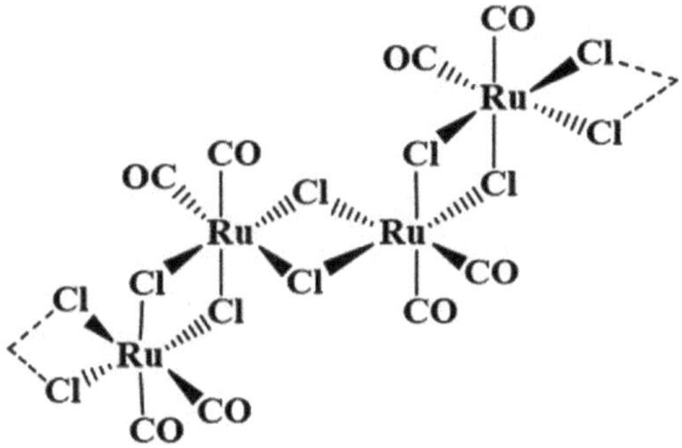

30

Fig. 11 Polymeric ruthenium precursor as a photoactivated antimicrobial agent.

antimicrobial polymers. The antimicrobial activity is markedly enhanced under photoirradiation, which can be ascribed to the generation of reactive oxygen species under UV irradiation. The activity of polymeric ruthenium as an antibacterial material is significant even at 6.6 ng/mL while remaining biocompatible to mammalian cells at much higher concentrations. $[Ru(CO)_2Cl_2]_n$ inhibits bacterial growth via the disruption of bacterial membranes and triggering upregulation of stress responses.

6. Group 9 organometallic compounds

Cobalt exists in very low abundance in the human body, but it constitutes an essential metal. It is found in Vitamin B_{12}, a cobaloxime, a cobalt metal center complexed with a glyoxime ligand.[69] Many studies focus on coordination of cobalt to ligands such as pyridine-amide,[70] imidazole, urea, thiourea, carbazide, acetamide or pyrazole, which in turn yield antibacterial, antifungal and antiviral activity against wide range of pathogens.[71] To the best of our knowledge, no light activated cobalt organometallic complexes for antimicrobial application have been published. Organometallic complexes of iridium have been extensively studied as anticancer agents, and a number as antimicrobial agents.[72,73] However, applications of these metal complexes as photo-activated antimicrobial agents is limited. Recently, a number of heteroleptic dipyrrinato Ir(III) complexes (Fig. 12) were assessed for antimicrobial photodynamic inactivation against *S. aureus* and *P. aeruginosa*.[74]

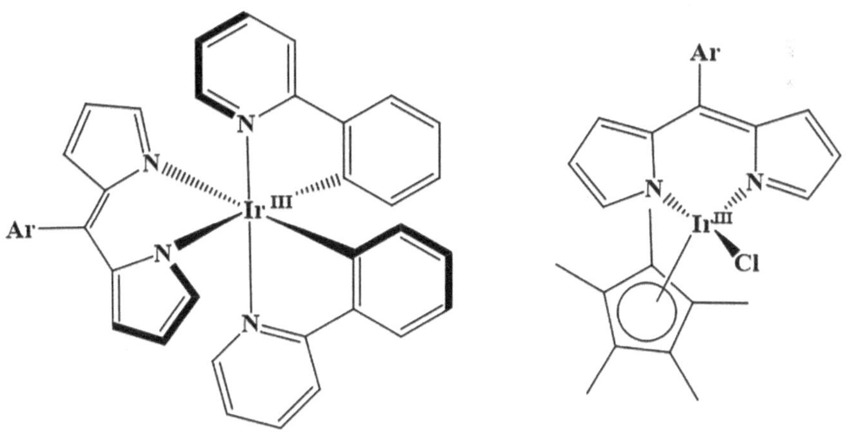

31 **32**

Fig. 12 Heteroleptic dipyrrinato complexes (for the R groups see Figs. 13 and 14) **31** and **32**.

The complexes were added to assays at a range of concentrations, 1, 10 and 100 μM, and irradiated (white light source). Most of the complexes led to >6 log reduction in *S. aureus* (CFU/mL) in the presence and absence of a light source, demonstrating the dark toxicity of these complexes (Figs. 13 and 14). Two of the complexes exhibited no dark or phototoxicity against *P. aeruginosa*. Antibacterial activity was retained against *S. aureus* in the presence of 10% serum while some compounds with alkyl substituted or non-substituted at the *meso* position showed strong phototoxic effects. In the case of the Gram-negative bacteria, the effectiveness of the iridium compounds as antibacterial agents was significantly reduced in the presence of 10% serum. Serum is added to the antimicrobial assays

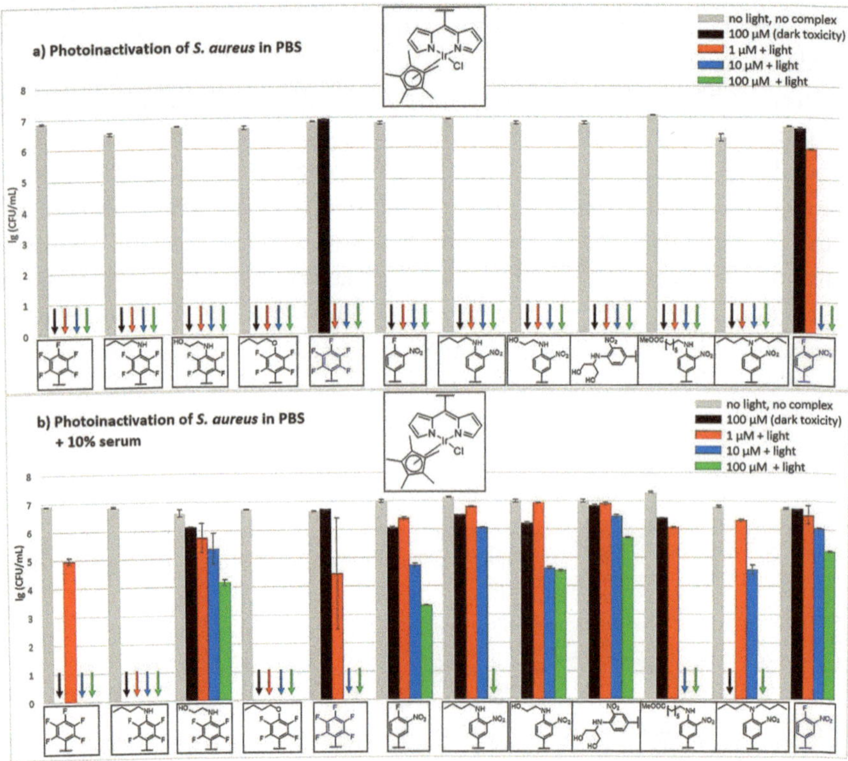

Fig. 13 Photoinactivation of *S. aureus* by chlorido(dipyrrinato)(pentamethylcyclopentadienyl)iridium(III) complexes (30 min incubation and irradiation with white light) in (A) phosphate-buffered saline (PBS) and (B) in PBS + 10% serum. The antibacterial toxicity is expressed as logarithm of the number of colony-forming units, lg (CFU mL^{-1}). Arrows indicate a suppression of bacterial growth below the detection limit. *Reproduced and adapted from Hohlfed, B.F.; Gitter, B.; Kingsbury, C.J.; Flanagan, K.J.; Steen, D.; Wieland, G.D.; Kulak, N.; Senge, M.O.; Wiehe, A.* Chem. Eur. J. ***2021,*** 27, *6440–6459 with permission.*

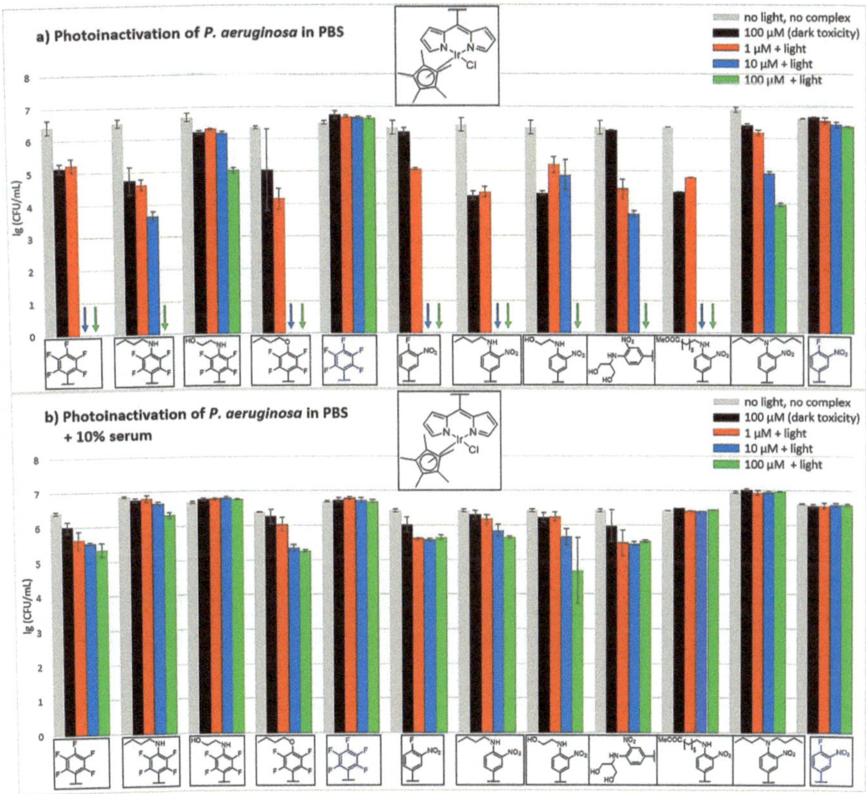

Fig. 14 Photoinactivation of *P. aeruginosa* by chlorido(dipyrrinato)(pentamethylcyclopentadienyl)iridium(III) complexes (30 min incubation and irradiation with white light) in (A) phosphate-buffered saline (PBS) and (B) in PBS + 10% serum. The antibacterial toxicity is expressed as logarithm of the number of colony-forming units/ml log (CFU mL^{-1}). Arrows indicate a suppression of bacterial growth below the detection limit. *Reproduced and adapted from Hohlfed, B.F.; Gitter, B.; Kingsbury, C.J.; Flanagan, K.J.; Steen, D.; Wieland, G.D.; Kulak, N.; Senge, M.O.; Wiehe, A.* Chem. Eur. J. ***2021,*** 27, *6440–6459 with permission.*

for the purpose of providing *in vivo* like conditions and this study demonstrates the importance of this aspect before progressing candidates to *in vivo* studies.

7. Group 10 and 11 organometallic complexes

Organometallic complexes of Pt, Ag, and Au are used widely in biomedical applications, particularly Pt cisplatin complexes which are well known for their anticancer activity. Antimicrobial studies have been carried out with Group 10 metals, with many based on NHC scaffolds.[75] Few studies however, have focused on light activated materials. Three new neutral luminescent cycloplatinated(II) photosensitizers of the type [Pt(dmba)(L)] (dmba = N,N-dimethylbenzylamine-*κN*, *κC*; L = 2-(benzo[d]oxazol-2-yl)-phenolato-κN,κO 1,2-(benzo[d]thiazol-2-yl)-phenolato-*κN*, *κO*2, and 2-(1-methyl-1H-benzo[d] imidazole-2-yl)phenolato-κN,κO_3) have been characterized and developed to potentially eliminate both resistant bacteria and lung cancer cells.[76] The phototherapeutic effects of complexes **33–35** were evaluated using low doses of blue light irradiation (40 min at 420 nm) against both Gram-positive (vancomycin-resistant *E. faecium* and methicillin-resistant *S. aureus* (MRSA)) and two Gram-negative strains *A. baumanii* and *P. aeruginosa* (see Table 4, Fig. 15). All three compounds were relatively inactive against the Gram-negative bacteria with MIC values >100. Only complex **34** containing sulfur, showed promising

Table 4 MIC values of the complexes **33–35** in the dark and upon 40 min of blue light irradiation (420 nm, 6 J cm^{-2}).[a]

	E. faecium		*S. aureus*	
	Dark	**Irradiated**	**Dark**	**Irradiated**
33	>100	>100	>100	>100
34	2	2	30	2
35	75	75	75	75

[a]Except for *E. Faecium*, t_{irr} = 2 h.

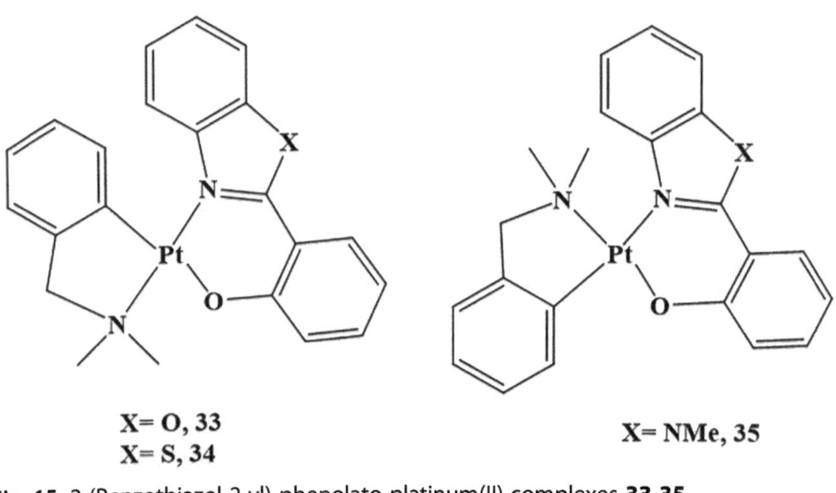

Fig. 15 2-(Benzothiazol-2-yl)-phenolato platinum(II) complexes **33-35**.

photoactivity against pathogenic Gram-positive bacteria strains of clinical interest, displaying a phototoxic index (PI) of 15 for MRSA, one of the major causes of hospital associated infections. Complex **34** was active in the dark, and when irradiation, indicating that the benzothiazole ligand plays a role in the antibacterial activity observed.

A tetraphenylethylene-based organoplatinum(II) metallacycle (**36**) photosensitizer which undergoes aggregation-induced emission has also been reported (Fig. 16).[77] The organoplatinum(II) metallacycle self-assembles with a transacting activator of transduction peptide-decorated virus coat protein (**37**) via electrostatic interactions. This assembly (**38**) was used as a photosensitizer for ROS generation. Electron spin resonance spectroscopy (ESR) was used to detect singlet oxygen, and the signal increased in intensity following irradiation of the self-assembled metallacycle (**38**) at 405 nm. UV-vis spectroscopy using the singlet oxygen probe, 9,10-anthracenediyl-bis(methylene) dimalonic acid was also used to confirm the photostability of the compounds and singlet oxygen generation. The assembled macrocycle (**38**) did not

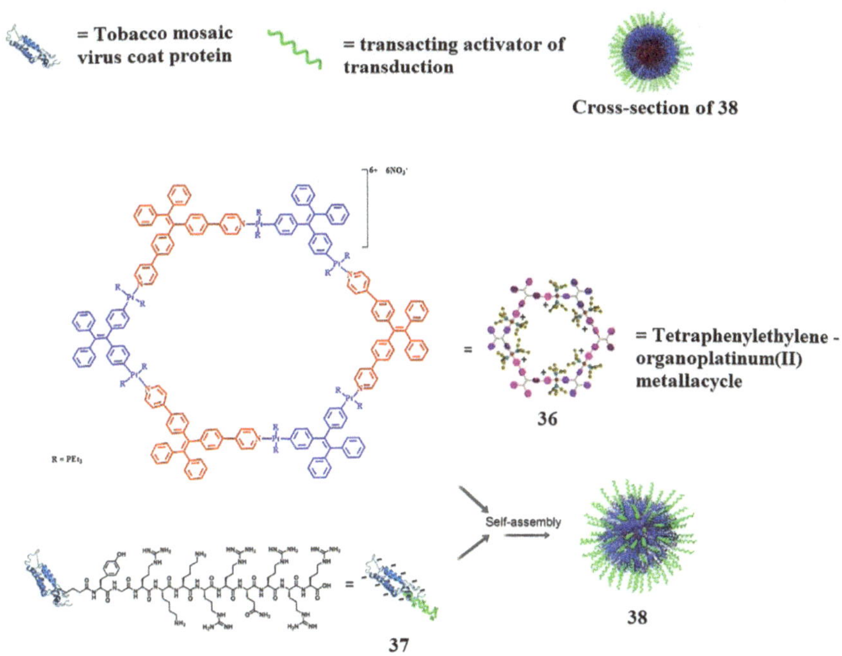

Fig. 16 Schematic illustration of the self-assembly of metallacycle **36** and protein **37** transacting activator of transduction. *Reproduced and adapted from Gao, S.; Xuzhou Yan, X.; Xie, G.; Zhu, M.; Ju, X.; Stang, P.J.; Ye Tian, Y.; Niu, Z.* Proc. Natl. Acad. Sci U.S.A. ***2019,*** 116, *23437–23443 with permission.*

generate as much singlet oxygen as the free metallacycle (**36**), and this was attributed to surrounding the photosensitizer with the protein shell (**37**), or reacting of the latter with the ROS species. Additionally, the assembled macrocycle has strong membrane-intercalating ability due to the peptide decorated peptide coat on the surface, which resulted in a significant enhancement in PDI against the bacterial strains used. The metallacycle assembly (**38**) led to a decrease in the survival rate of *E. coli* to almost zero using metallacycle concentrations of 20 and 40 μM (following 420 nm irradiation for 15 min). This decrease was attributed to intercalating of **38** with the bacterial cell membrane or entering the bacteria. In the case of *S. aureus* survival rates decreased to ~30% upon light irradiation (420 nm) using a 40 μM solution of **38**. This study has wide implications from the generation of multifunctional nanomaterials using a dual approach (ROS generation and membrane-intercalating) to the control bacterial infections, especially towards Gram-negative bacteria.

8. Summary and future perspectives

One recommendation that may advance the potential of metal complexes for clinical development as photo-active antimicrobials is evaluation of the conditions that are physiological and pathologically relevant. For example, determination of anti-biofilm activity is important in this regard and indeed CORMs and aPDT have demonstrated biofilm prevention and eradication properties.[78] We have found in general that PDT metal complexes, with high log killing (e.g. >4 log) of planktonic cells demonstrate anti-biofilm properties. As highlighted, it is also recommended to progress candidate complexes to antimicrobial evaluation in defined media relevant to the *in vivo* environment of infection, for example by the addition of human serum or simulated wound fluid. In the setting of wound infections, where photo-active complexes have the greatest potential, activity in iron-depleted media would also be appropriate and clinically relevant.

In terms of clinical and healthcare applications of aPDT, the management of wound infections and healthcare environment disinfection are at the forefront based on the shallower penetration afforded by shorter wavelengths of light. However, further extending PS activation to longer wavelengths or innovation in relation to PS modification and light source development has the potential to extend its antimicrobial applications, including exploration of theranostic applications.

In 2020 the global SARS-CoV2 pandemic necessitated healthcare systems re-organization and practice changes such as reduced AMR screening/surveillance, increased use antibiotics and particularly in the intensive care setting, unprecedented sustained staff absences, pressure on all healthcare professionals and reduction of in-person visits to the General Practioner and other services. There is no doubt that challenges in health care provision during the COVID-19 pandemic and future pandemics, have hastened the impending AMR crisis and reignited the urgent call to action.[79] Re-establishing and developing a fit-for-purpose pipeline is critical but remains an area of significant underinvestment. In 2019, there were 32 new antibiotics in clinical development, the majority of which were derivatives of current antibiotics for which bacteria are predicted to become resistant.[80] Challenges to substantial investment by the pharmaceutical industry in entirely innovative antimicrobials, such as those discussed here include: a regulatory environment which favors materials with known safety profiles and the sustainability of the market in a healthcare system which intends to spare these medicines. Nonetheless, as we show here, innovative antimicrobial discovery in the area of photoactive materials is actively pursued in academic settings and potential solutions exist that have several attractive features for development such as; activity against WHO priority pathogens, novel and multiple intracellular targets, on-target controllable activity and chemical modification capability including those that may enhance delivery for clinical applications. Encouragingly, it should be noted that PDT is an accepted and approved medicine in other areas such as cancer chemotherapy, dermatology and periodontology and cross disciplinary learning from the remarkable advances made in PS and light source development in these fields should be prioritized to advance aPDT to accepted medicines.

Acknowledgment

The authors thank the Health Research Board-Diabetes Ireland Research Alliance (HRBMRCG-2018-01 award (AR)) and SFI (19/FFP/6882 award (LT)) for financial support.

References

1. (a) Rossolini, G. M.; Arena, F.; Pecile, P.; Pollini, S. *Curr. Opin. Pharmacol.* **2014**, *18*, 56–60; (b) World Health Organization. *Global priority list of antibiotic resistant bacteria to guide research, discovery, and development of new antibiotics*; 2017. http://www.who.int/medicines/publications; (c) Boucher, H. W.; Talbot, G. H.; Bradley, J. S.; Edwards, J. E.; Gilbert, D.; Rice, L. B.; Scheld, M.; Spellberg, B.; Bartlett, J. *Clin. Infect. Dis.* **2009**, *48* (1), 1–12.
2. (a) WHO. *News Release: New Report Calls for Urgent Action to Avert Antimicrobial Resistance Crisis*; 2019. https://www.who.int; (b) Bassetti, M.; Poulakou, G.; Ruppe, E.; Bouza, E.; Van Hal, S. J.; Brink, A. *Intens. Care Med.* **2017**, *43* (10), 1464–1475;

(c) O'Neill, P. *Report of the International Review on Antimicrobial Resistance (AMR)*; 2014. https://amr-review.org.
3. (a) Walker, S. S.; Black, T. A. *Drug Discov. Today* **2021**, *26* (9), 2152–2158; (b) Gu, K.; Lin, G.; Zhu, Y.; Ji, X.; Li, J.; Dong, X.; Zhao, W. *Eur. J. Med. Chem.* **2020**, *199*, 112361.
4. Koo, H.; Allan, R. N.; Howlin, R. P.; Stoodley, P.; Hall-Stoodley, L. *Nat. Rev. Microbiol.* **2017**, *15* (12), 740–755.
5. Coates, A. R.; Halls, G.; Hu, Y. *Br. J. Pharmacol.* **2011**, *163* (1), 184–194.
6. Santajit, S.; Indrawattana, N. *Biomed. Res. Int.* **2016**, *2016*, 2475067.
7. Turksoy, A.; Yildiz, D.; Akkaya, E. U. *Coord. Chem. Rev.* **2019**, *379*, 47–64.
8. Castano, A. P.; Demidova, T. N.; Hamblin, M. R. *Photodiagn. Photodyn. Ther.* **2004**, *1* (4), 279–293.
9. Gunaydin, G.; Gedik, M. E.; Ayan, S. *Front Chem.* **2021**, *9*, 686303.
10. (a) Zhang, J.; Jiang, C.; Figueiró Longo, J. P.; Azevedo, R. B.; Zhang, H.; Muehlmann, L. A. *Acta Pharm. Sin. B* **2018**, *8* (2), 137–146; (b) Liu, M.; Li, C. *ChemPlusChem* **2020**, *85* (5), 948–957.
11. Agazzi, M. L.; Ballatore, M. B.; Durantini, A. M.; Durantini, E. N.; Tomé, A. C. *J. Photochem. Photobiol. C Photochem. Rev.* **2019**, *40*, 21–48.
12. Dryden, M. *Int. J. Antimicrob. Agents* **2018**, *51* (3), 299–303.
13. Pucelik, B.; Dąbrowski, J. M. *Adv. Inorg. Chem.* **2022**, *79*, 65–103.
14. Macmillan, J. D.; Maxwell, W. A. *Photochem. Photobiol.* **1966**, *555*. 5-65.
15. Bellin, J.; Lutwick, L.; Jonas, B. *Arch. Biochem. Biophys.* **1969**, *132*, 157–164.
16. Sjostrand, T. *Nature* **1949**, *164*, 580–581.
17. Mann, B. E.; Motterlini, R. *Chem. Commun.* **2007**, *41*, 4197–4208.
18. Motterlini, R.; Clark, J.; Foresti, R.; Sarathchandra, P.; Mann, B.; Green, C. *Circ. Res.* **2002**, *90*, E17–E24.
19. Schatzschneider, U. *J. Pharmacol.* **2015**, *172*, 1638–1650.
20. Heinemann, S. H.; Hoshi, T.; Westerhausen, M.; Schiller. *Chem. Commun.* **2014**, *50*, 3644–3660.
21. Ford, P. C. *Coord. Chem. Rev.* **2018**, (376), 548–564.
22. (a) Bruckmann, N. E.; Wahl, M.; Reiss, G. J.; Kohns, M.; Waetjen, W.; Kunz, P. C. *Eur. J. Inorg. Chem.* **2011**, *29*, 4571–4577; (b) Hasegawa, U.; van der Vlies, A. J.; Simeoni, E.; Wandrey, C.; Hubbell, J. *J. Am. Chem. Soc.* **2010**, *132*, 18273–18280; (c) Bohlender, C.; Glaeser, S.; Klein, M.; Weisser, J.; Thein, S.; Neugebauer, U.; Popp, J.; Wyrwa, R.; Schiller. *J. Mat. Chem. B* **2014**, *2*, 1454–1463; (d) Kunz, P. C.; Meyer, H.; Barthel, J.; Sollazzo, S.; Schmidt, A. M.; Janiak, C. *Chem. Commun.* **2013**, *49*, 4896–4898.
23. Atkin, A. J.; Lynam, J. M.; Moulton, B. E.; Sawle, P.; Motterlini, R.; Boyle, N. M.; Pryce, M. T.; Fairlamb, I. J. S. *Dalton Trans.* **2011**, *40*, 5755–5761.
24. Huber, W.; Linder, R.; Niesel, J.; Schatzschneider, U.; Spingler, B.; Kunz, P. C. *Eur. J. Inorg. Chem.* **2012**, *19*, 3140–3146.
25. (a) McMahon, S.; Rochford, J.; Halpin, Y.; Manton, J. C.; Harvey, E. C.; Greetham, G. M.; Clark, I. P.; Rooney, D.; Long, C.; Pryce, M. T. *Phys. Chem. Chem. Phys.* **2014**, *16*, 21230–21233; (b) Manton, J. C.; Carpentier, F. J. R.; Harvey, E. C.; Long, C.; Pryce, M. T. *Dalton Trans.* **2019**, *48*, 14642–14652.
26. Ghorbani, J.; Rahban, D.; Aghamiri, S.; Teymouri, A.; Bahador, A. *Laser Ther.* **2018**, *27* (4), 293–302.
27. Tim, M. *J. Photochem. Photobiol. B* **2015**, *150*, 2–10.
28. Murray, T. S.; Okegbe, C.; Gao, Y.; Kazmierczak, B. I.; Motterlini, R.; Dietrich, L. E.; Bruscia, E. M. *PLoS One* **2012**, 7 (4), e35499.
29. Vassena, C.; Fenu, S.; Giuliani, F.; Fantetti, L.; Roncucci, G.; Simonutti, G.; Romanò, C. L.; De Francesco, R.; Drago, L. *Int. J. Antimicrob. Agents* **2014**, *44* (1), 47–55.
30. El Khaldy, A. A. S.; Waddell, K.; Abu-Shanab, A. M. *J. Chem. Chem. Sci.* **2016**, *6* (9), 806–816.

31. Gerasymchuk, Y.; Tahershamsi, L.; Tomala, R.; Wedzynska, A.; Chernii, V.; Tretyakova, I.; Korona-Glowniak, I.; Rajtar, B.; Malm, A.; Piatek, D.; Lukowiak, A. *Chem. Pap.* **2021**, *75*, 5421–5433.
32. Marhenke, J.; Trevino, K.; Works, C. *Coord. Chem. Rev.* **2015**, *306*, 533–543.
33. Lee, S. X.; Tan, C. H.; Mah, W. L.; Wong, R. C. S.; Cheow, Y. L.; Sim, K. S.; Tan, K. W. *Inorgan. Chim. Acta* **2021**, *525*, 120491.
34. Xian Lee, S.; Hoe Tan, C.; Li Mah, W.; Chee Seng Wong, R.; Suhana Abdul Manan, N.; Lin Cheow, Y.; Shin Sim, K.; Wai Tan, K. *Inorg. Chim. Acta* **2022**, 120931.
35. McMahon, S.; Rajagopal, A.; Amirjalayer, S.; Halpin, Y.; Fitzgerald-Hughes, D.; Buma, W. J.; Woutersen, S.; Long, C.; Pryce, M. T. *J. Inorg. Biochem.* **2020**, *208*, 111071.
36. Kelly, J.; Long, C.; Bonneau, R. *J. Phys. Chem.* **1983**, *87*, 3344–3349.
37. (a) Rimmer, R. D.; Pierri, A. E.; Ford, P. C. *Coord. Chem. Rev.* **2012**, *256*, 1509–1519; (b) Zhang, W.; Atkin, A. J.; Fairlamb, I. J. S.; Whitwood, A. C.; Lynam, J. M. *Organometallics* **2011**, *30*, 4643–4654.
38. Chen, P.; Bornhorst, J.; Aschner, M. *Front Biosci (Landmark Ed).* **2018**, *23*, 1655–1679.
39. Smith, S. J.; Hadler, K. S.; Schenk, G.; Hanson, G. R.; Mitić, N. Manganese metalloproteins. In *Metals in Biology. Biological Magnetic Resonance*; Hanson, G., Berliner, L., Eds.; Vol. 29; Springer: New York, NY, 2010; pp. 273–341.
40. Li, L.; Yang, X. *Oxid. Med. Cell. Longev.* **2018**, *2018*, 7580707.
41. Firuzabadi, F. D.; Asadi, Z. *J. Mol. Struct.* **2021**, *1224*, 129248.
42. Zampakou, M.; Balala, S.; Perdih, F.; Kalogiannis, S.; Turel, I.; Psomas, G. *RSC. Adv.* **2015**, *5*, 11861.
43. Tan, Z.; Tan, F.; Zhao, L.; Li, J. *J. Cryst. Process. Technol.* **2012**, *2*, 55–63.
44. Barmpa, A.; Frousiou, O.; Kalogiannia, S.; Perdih, F.; Turel, I.; Psomas, G. *Polyhedron* **2018**, *145*, 166–175.
45. Zampakou, M.; Akrivou, M.; Andreadou, E. G.; Raptopoulou, C. P.; Psycharis, V.; Pantazaki, A. A.; Psomas, G. *J. Inorg. Biochem.* **2013**, *121*, 88–99.
46. Simpson, P. V.; Nagel, C.; Bruhn, H.; Schatzschneider, U. *Organometallics* **2015**, *34* (15), 3809–3815.
47. Iman, M.; Davood, A.; Gebbink, B. K.; Azerang, P.; Alibolandi, M.; Sardari, S. *Pharm. Chem. J.* **2014**, *48*, 513–519.
48. Dorkov, P.; Pantcheva, I. N.; Sheldrick, W. S.; Mayer-Figge, H.; Petrova, R.; Mitewa, M. *J. Inorg. Biochem.* **2008**, *102*, 26–32.
49. Aucott, B. J.; Ward, J. S.; Andrew, S. G.; Milani, J.; Whitwood, A. C.; Lynam, J. M.; Parkin, A.; Fairlamb, I. J. S. *Inorg. Chem.* **2017**, *56* (9), 5431–5440.
50. Yempally, V.; Kyran, S. J.; Raju, R. K.; Fan, W. Y.; Fschat, N.; Brothers, E. N.; Darenbourg, D. J.; Bengali, A. A. *Inorg. Chem.* **2014**, *53* (8), 4081–4088.
51. Zobi, F.; Degonda, A.; Schaub, M. C.; Bogdanova, A. Y. *Inorg. Chem.* **2010**, *49* (16), 7313–7322.
52. Stamellou, E.; Stroz, D.; Botov, S.; Ntasis, E.; Wedel, J.; Sollazzo, S.; Kraemer, B. K.; Van-Son, W.; Seelen, M.; Schmalz, H. G. *Redox Biol.* **2014**, *2*, 739–748.
53. Nagel, C.; McLean, S.; Poole, R. K.; Branuschweig, H.; Kramer, T.; Schatzschneider, U. *Dalton Trans.* **2014**, *24*, 765–780.
54. (a) Tinajero-Trejo, M.; Rana, N.; Nagel, C.; Jesse, H. E.; Smith, T. W.; Wareham, L. K.; Hippler, M.; Schatzschneider, U.; Poole, R. K. *Antioxid. Redox Sign.* **2016**, *24*, 765–780.
55. Liakopoulos, A.; La Ragione, R. M.; Nagel, C.; Schatzschneider, U.; Rozen, D. E.; Betts, J. W. *J. Glob. Antimicrob. Resist.* **2020**, *22*, 594–597.
56. Rana, N.; Jesse, H. E.; Tinajero-Trejo, M.; Butler, J. A.; Tarlit, J. D.; von Und Zur Muhlen, M. L.; Nagel, C.; Schatzschneider, U.; Poole, R. K. *Microbiology* **2017**, *163* (10), 1477–1489.
57. Guentzel, P.; Nagel, C.; Weigelt, J.; Betts, J. W.; Pattrick, C. A.; Southam, H. M.; La Ragione, R. M.; Poole, R. K.; Schatzschneider, U. *Metallomics* **2019**, *11*, 2033.

58. Ward, J. S.; Lynam, J. M.; Moir, J.; Fairlamb, I. J. S. *Chem. Eur. J.* **2014**, *20*, 1–9.
59. Ward, J. S.; Morgan, R.; Lynam, J. M.; Fairlamb, I. J. S.; Moir, J. W. B. *Med. Chem. Commun.* **2017**, *8*, 346–352.
60. Schindler, K.; Zobi, F. *Molecules* **2022**, *27*, 539–557.
61. (a) Frei, A. *Antibiotics* **2020**, *9*, 90; (b) Sovari, S. N.; Zobi, F. *Chemistry* **2020**, *2*, 418–452.
62. Wähler, K.; Ludewig, A.; Szabo, P.; Harms, K.; Meggers, E. *Eur. J. Inorg. Chem.* **2014**, *2014* (5), 807–811.
63. Frei, A.; Amado, M.; Cooper, M. A.; Blaskovich, M. A. T. *Chem. Eur. J.* **2020**, *26*, 2852–2858.
64. Acosta, A.; Antipán, J.; Fernández, M.; Prado, G.; Sandoval-Altamirano, C.; Günther, G.; Gutiérrez-Urrutia, I.; Poblete-Castro, I.; Vega, A.; Pizarro, N. *RSC Adv.* **2021**, *11*, 31959–31966.
65. Rajagopala, R.; Fitzgerald-Hughes, D.; Pryce, M. T. (unpublished results).
66. Siegmund, D.; Lorenz, N.; Gothe, Y.; Spies, C.; Geissler, B.; Prochnow, P.; Nuernberger, P.; Bandow, J. E.; Metzler-Nolte, N. *Dalton Trans.* **2017**, *46*, 15269–15279.
67. Marker, S. C.; MacMillan, S. N.; Zipfel, W. R.; Li, Z.; Ford, P. C.; Wilson, J. *J. Inorg. Chem.* **2018**, *57*, 1311–1331.
68. Ghosh, S.; Amariei, G.; Mosquera, M. E. G.; Rosal, R. *J. Hazard. Mater.* **2021**, *402*, 123788.
69. Chang, E. L.; Simmers, C.; Knight, D. A. *Pharmaceuticals* **2010**, *3* (6), 1711–1728.
70. Nagababu, P.; Latha, J. N. L.; Pallavi, P.; Harish, S.; Satyanarayana, S. *Can. J. Microbiol.* **2006**, *52*, 1247–1254.
71. Mishra, A.; Kaushik, N. K.; Verma, A. K. Gupta. R. *Eur. J. Med. Chem.* **2008**, *43*, 2189–2196.
72. Hearn, J. M.; Romero-Canelon, I.; Qamar, B.; Liu, Z.; Hands-Portmann, I.; Sadler, P. J. *ACS Chem. Biol.* **2013**, *8*, 1335–1343.
73. (a) Karpin, G. W.; Merola, J. S.; Falkinham, J. O., III. *Antimicrob. Agents Chemother.* **2013**, *57* (7), 3434–3436; (b) Simpson, P. V.; Schmidt, C.; Ott, I.; Bruhn, H.; Schatzschneider, U. *Eur. J. Inorg. Chem.* **2013**, *2013*, 5547–5554; (c) Pettinari, R.; Marchetti, F.; Pettinari, C.; Condello, F.; Petrini, A.; Scopelliti, R.; Riedel, T.; Dyson, P. J. *Dalton Trans.* **2015**, *44*, 20523–20531; (d) Karpin, G. W.; Morris, D. M.; Ngo, M. T.; Merola, J. S.; Falkinham, J. O., III. *Med. Chem. Commun.* **2015**, *6*, 1471–1478; (e) Lu, L.; Liu, L.-J.; Chao, W.-C.; Zhong, H.-J.; Wang, M.; Chen, X.-P.; Lu, J.-L.; Li, R.-N.; Ma, D.-L.; Leung, C.-H. *Sci. Rep.* **2015**, *5*, 14544–14553; (f) Novohradsky, V.; Zamora, A.; Gandioso, A.; Brabec, V.; Ruiz, J.; Marchan, V. *Chem. Commun.* **2017**, *53*, 5523–5526; (g) Chen, F.; Moat, J.; McFeely, D.; Clarkson, G.; Hands-Portman, I. J.; Furner-Pardoe, J. P.; Harrison, F.; Dowson, C. G.; Sadler, P. J. *J. Med. Chem.* **2018**, *61*, 7330–7344; (h) Chellan, P.; Avery, V. M.; Duffy, S.; Triccas, J. A.; Nagalingam, G.; Tam, C.; Cheng, L. W.; Liu, J.; Land, K. M.; Clarkson, J.; Romero-Canelon, I.; Sadler, P. J. *Chem. Eur. J.* **2018**, *24* (40), 10078–10090.
74. Hohlfed, B. F.; Gitter, B.; Kingsbury, C. J.; Flanagan, K. J.; Steen, D.; Wieland, G. D.; Kulak, N.; Senge, M. O.; Wiehe, A. *Chem. Eur. J.* **2021**, *27*, 6440–6459.
75. (a) Kostova, I. Platinum complexes as anticancer agents. *Recent Pat. Anticancer Drug Discov.* **2006**, *1* (1), 1–22; (b) Medici, S.; Peana, M.; Nurchi, V. M.; Lachowicz, J. I.; Crisponi, G.; Zoroddu, M. A. *Coord. Chem. Rev.* **2015**, *284*, 329–350; (c) Onar, G.; Gürses, C.; Karataş, M. O.; Balcıoğlu, S.; Akbay, N.; Özdemir, N.; Ateş, B.; Alıcı, B. *J. Organomet. Chem.* **2019**, *886*, 48–56; (d) Briguglio, I.; Piras, S.; Corona, P.; Gavini, E.; Nieddu, M.; Boatto, G.; Carta, A. *Eur. J. Med. Chem.* **2015**, *97*, 612–648; (e) Choo, K. B.; Lee, S. M.; Lee, W. L.; Cheow, Y. L. *J. Organomet. Chem.* **2019**, *898*, 120868; (f) Sim, W.; Barnard, R. T.; Blaskovich, M. A. T.;

Ziora, Z. M. *Antibiotics* **2018**, 7 (4), 93; (g) Liang, J.; Sun, D.; Yang, Y.; Li, M.; Li, H.; Chen, L. *Eur. J. Med. Chem.* **2021**, *224*, 113696; (h) Nomiya, K.; Morozumi, S.; Yanagawa, Y.; Hasegawa, M.; Kurose, K.; Taguchi, K.; Sakamoto, R.; Mihara, K.; Kasuga, N. C. *Inorg. Chem.* **2018**, *57* (18), 11322–11332; (i) Johnson, N. A.; Southerland, M. R.; Youngs, W. J. *Molecules* **2017**, *22* (8), 1263; (j) Lasmari, S.; Ikhlef, S.; Boulcina, R.; Mokrani, E. H.; Bensouici, C.; Gürbüz, N.; Dündar, M.; Karcı, H.; Özdemir, İ.; Koç, A.; Özdemir, I.; Debache, A. *J. Mol. Struct.* **2021**, *1238*, 130399.

76. Ortega, E.; Pérez-Arnaiz, C.; Rodríguez, V.; Janiak, C.; Busto, N.; García, B.; Ruiz, J. *Eur. J. Med. Chem.* **2021**, *222*, 113600.
77. Gao, S.; Xuzhou Yan, X.; Xie, G.; Zhu, M.; Ju, X.; Stang, P. J.; Ye Tian, Y.; Niu, Z. *Proc. Natl. Acad. Sci U.S.A.* **2019**, *116*, 23437–23443.
78. (a) Orlandi, V. T.; Rybtke, M.; Caruso, E.; Banfi, S.; Tolker-Nielsen, T.; Barbieri, P. *Biofouling* **2014**, *30* (8), 883–891.
79. Ukuhor, H. O. *J. Infect. Public Health* **2021**, *14* (1), 53–60.
80. Miethke, M.; Pieroni, M.; Weber, T.; Brönstrup, M.; Hammann, P.; Halby, L.; Arimondo, P. B.; Glaser, P.; Aigle, B.; Bode, H. B.; Moreira, R.; Li, Y.; Luzhetskyy, A.; Medema, M. H.; Pernodet, J. L.; Stadler, M.; Tormo, J. R.; Genilloud, O.; Truman, A. W.; Weissman, K. J.; Takano, E.; Sabatini, S.; Stegmann, E.; Brötz-Oesterhelt, H.; Wohlleben, W.; Seemann, M.; Empting, M.; Hirsch, A. K. H.; Loretz, B.; Lehr, C. M.; Titz, A.; Herrmann, J.; Jaeger, T.; Alt, S.; Hesterkamp, T.; Winterhalter, M.; Schiefer, A.; Pfarr, K.; Hoerauf, A.; Graz, H.; Graz, M.; Lindvall, M.; Ramurthy, S.; Karlén, A.; van Dongen, M.; Petkovic, H.; Keller, A.; Peyrane, F.; Donadio, S.; Fraisse, L.; Piddock, L. J. V.; Gilbert, I. H.; Moser, H. E.; Müller, R. *Nat. Rev. Chem.* **2021**, *5* (10), 726–749.

CHAPTER TWELVE

Mitochondria-targeting transition metal complexes

Ryan J. Morris and Massimiliano Massi*
School of Molecular and Life Sciences, Curtin University, Perth, Australia
*Corresponding author: e-mail address: m.massi@curtin.edu.au

Contents

1. Introduction 413
   1.1 Mitochondria structure and function 414
   1.2 Mitochondria and cell death 415
   1.3 Why target mitochondria? 416
   1.4 How are mitochondria targeted? 417
   1.5 Photodynamic therapy 418
2. Iridium 419
3. Platinum 464
4. Rhenium 470
5. Gold 476
6. Copper 479
7. Ruthenium 482
8. Osmium 486
9. Vanadium 489
10. Palladium 492
11. Rhodium 493
12. Zinc 495
13. Silver 496
14. Concluding remarks and future perspectives 497
Acknowledgments 499
References 499

Abstract

This chapter illustrates recent advances in mitochondria-targeting transition metal complexes and their potential use as diagnostic and/or therapeutic agents. The information presented in this chapter is arranged to systematically analyze data reported in recent literature that could be used to assess the structure-activity relationship for complexes of a variety of metallic elements. The diverse range of applications for transition metal complexes have led investigators toward optimizing such complexes to target mitochondria, elucidating structure and function of healthy mitochondria, as well as

Advances in Inorganic Chemistry, Volume 80
ISSN 0898-8838
https://doi.org/10.1016/bs.adioch.2022.06.003

identifying how these complexes can be used to evaluate disease pathophysiology. Furthermore, investigations into potential applications as metallo-drugs and metallo-prodrugs, biosensors, photodynamic therapy sensitizers, optical imaging agents, probes for correlative imaging, have all been explored in recent years with encouraging results.

Abbreviations

$^1O_2$ singlet oxygen
$^3O_2$ triplet oxygen
ADP adenosine 5′-diphosphate
AI(P)E aggregation induced (phosphorescent) emission
ATP adenosine 5′-triphosphate
CSC cancer stem cell
CVD β-cyclodextrin vesicles
Cys cysteine
DCA dichloroacetate
DLC delocalized lipophilic cation
DNA deoxyribonucleic acid
DPPZ dipyrido[3,2-a:2′,3′-c]phenazine
ETC electron transport chain
ER endoplasmic reticulum
Fc ferrocene
Hcys homocysteine
HDAC histone deacetylase
ISC intersystem crossing
MCS three-dimensional multi-cellular spheroids
MLCT metal-to-ligand charge transfer
MMP mitochondrial membrane potential
MMR mismatch repair
MPP mitochondria penetrating peptide
MPT mitochondrial permeability transition
mtDNA mitochondrial DNA
nDNA nuclear DNA
NHC *N*-heterocyclic carbene
NIR near-infrared
PAIDH 2-pyridyl-1H-anthra[1,2-d]imidazole-6,11-dione
PBN α-phenyl-N-tert-butylnitrone
PBS phosphate buffered saline
PDT photodynamic therapy
PEG poly(ethylene glycol)
PPG photolabile protecting group
RNA ribonucleic acid
ROS reactive oxygen species
SIM structured illumination microscopy
SRM super resolution microscopy
TD-DFT time-dependent density functional theory

TEM	transmission electron microscopy
TEMPO	2,2,6,6-tetramethylpiperidine-1-oxyl
TP-PLIM	two-photon phosphorescence lifetime imaging
TPP	triphenylphosphonium
TrxR	thioredoxin reductase

1. Introduction

The diverse structural chemistry and reactivity of transition metal elements have assisted advances in many areas, such as catalysis,[1] synthetic pigments,[2] organic light emitting diodes,[3] and solar cells,[4] for example. However, one field of research that has received extensive focus in past decades is life science. Transition metal complexes have been at the center of extensive investigation, with an emphasis on their use as diagnostic, therapeutic and theranostic (or theragnostic) agents. New applications have recently emerged where coordination and organometallic complexes of transition elements have demonstrated their potential as luminescent imaging agents, biological sensors, as well as prospective metallo-drugs and prodrugs. One particular focus area is the interaction between transition metal complexes and mitochondria, which is one of the key cellular organelles. The objective of this chapter is to systematically illustrate advances in mitochondria-targeting transition metal complexes that have been reported approximately within the last decade. For convenience, the discussion has been arranged by considering each individual transition metal, with occasional overlaps due to studies that compared different metal centers or featured molecular architectures comprising more than one metallic element. This chapter focuses on discrete mono or multinuclear complexes, excluding larger species such as nanoparticles and metal-organic frameworks. However, a few examples of complexes appended to polymeric chains have been included. The systematic information contained in this chapter is intended to assist in establishing a structure-activity relationship for the future design of mitochondria-targeting complexes, on the basis of rational planning rather than serendipitous discovery.

For ease of understanding, it is pertinent to list a variety of commonly encountered ligands incorporated in complexes that have been investigated for biological studies. Commonly incorporated cyclometalated ligands include: 2-phenylpyridine (**ppy**), 2-(2,4-difluorophenyl)pyridine (**dfppy**), 2-((1,10-biphenyl)-4-yl)pyridine (**pppy**), 2-phenylquinoline (**pq**), 1-phenylisoquinoline(**iso-pq**), 2-(2-pyridyl)benzothiophene (**btp**), 2-(benzo[b]thiophen-2-yl)quinoline (**btq**), 2-(1-naphthyl)benzothiazole (**btn**), 7,8-benzoquinoline (**bzq**), 2-phenylbenzothiazole (**pbt**) 2-((1, 1′biphenyl)-4-yl)benzothiazole (**ppbt**), dibenzo[f,h]quinoxaline (**dbq**), 6-(thieno[3,2-b]thiophen-2-yl)phenanthridine (**ttph**), 6-(benzo[b] thiophen-2-yl)phenanthridine (**btph**), 2-(thiophen-2-yl)pyridine (**tp**), and dibenzo[a,c]phenazine (**dbpz**). Commonly incorporated diimine ligands include 2,2′-bipyridine (**bipy**), 1,10-phenanthroline (**phen**), 4,4′-dimethyl-2,2′-bipyridine (**dmbpy**), 4,7-diphenyl-1,10-phenanthroline (**dip**), phenanthro[9,10-c][1,2,5]selenadiazole (**phen-Se**).

1.1 Mitochondria structure and function

Mitochondria are the primary producers of cellular energy, which is stored in the form of adenosine triphosphate (ATP).[5,6] Further to this essential role, mitochondria provide a number of other functions such as regulating Ca^{2+} homeostasis,[7,8] producing reactive oxygen species (ROS),[9,10] and playing a pivotal role in apoptosis, which is a form of programmed cell death.[11,12] Mitochondria are enclosed within two phospholipid membranes, which are termed outer and inner membranes. The outer membrane separates the mitochondrion from the cytosol. The folding inner membrane, which encloses the matrix, forms finger-like structures known as cristae. The aqueous matrix compartment contains intermediary metabolic enzymes, mitochondrial DNA (mtDNA), and the corresponding RNA necessary for translation.[13]

Mitochondria utilize a chemiosmotic mechanism to transduce energy via the oxidative phosphorylation pathway. Metabolic energy is used to pump protons out of the matrix, establishing an electrochemical gradient, which is commonly referred to as mitochondrial membrane potential (MMP). ATP synthase couples the translocation of protons, returning to the matrix, with the phosphorylation of adenosine diphosphate (ADP) to produce ATP.[13]

Whilst the anatomy of mitochondria is often presented as an idealized discrete bean-like form, their real structure is somewhat more nuanced. Individual mitochondria can fuse together to form long snake-like tubules, branched tree-like formations, and vast networks. The overall morphology of the network is determined by the balance of fusion and fission processes.[14,15]

A high fusion-to-fission ratio results in a long tubular highly interconnected network. On the other hand, a low fusion-to-fission ratio results in fragmented mitochondria that appear as small spheres or rod-like structures. Mitochondrial morphology is dynamic and can dramatically change in response to diverse events and stimuli, such as the current stage in the cell cycle, intracellular energy levels, and various cellular stresses.[14,15]

1.2 Mitochondria and cell death

Extensive reviews on our current understanding of the role that mitochondria play in cell death have been reported by Green et al.,[11] Wang et al.,[16] and Desagher et al.,[17] among others. Following a general classification, three mechanisms have been identified.

Mechanism 1 involves the disruption of the electron transport chain (ETC), oxidative phosphorylation, and ATP production. The disruption of electron transport chain has long been considered an early hallmark of cell death. The loss of electron transport results in a reduction of ATP production. Whilst it is true that reduced production of ATP is observed during apoptosis, it often occurs relatively late in the process. In fact, ATP is required for subsequent apoptosis-related events. Therefore, whilst mechanism 1 can result in cell death, it is unlikely that this mechanism results in the induction of apoptosis.[11]

Mechanism 2 involves the release of caspase-activating proteins. Caspases are a family of cysteine proteases that are the primary effectors during apoptosis that proteolytically dismantle many, if not most, cellular structures. During apoptosis, cytochrome *c* is released from mitochondria into the cytosol, resulting in the activation of caspase-9, which in turn activates other caspases. This cascade essentially orchestrates the death of the cell.[11,18,19] It is important to note that this is a very general and simplified overview of the process. The exact consequence of cytochrome *c* release is dependent on the cell type. For example, in cells where excess cytochrome *c* is present, enough cytochrome *c* may remain in the mitochondria to continue the production of ATP, while caspases concurrently dismantle cytosolic and nuclear structures, resulting in apoptosis. On the other hand, in cells that contain sufficient concentrations of endogenous caspase inhibitors, the release of cytochrome *c* may fail to induce caspase-dependent apoptosis, instead resulting in a necrotic process of cell death.[11]

Mechanism 3 involves alterations in cellular redox potential and production of ROS. These are short lived species that are highly reactive.

They are essential for the regulation of normal physiological functions such as signaling and cell cycle progression. The delicate balance of ROS is maintained by a variety of mechanisms involving antioxidant species. Excess levels of ROS can damage cellular structures such as proteins, lipids, and nucleic acids, and can lead to the activation of apoptosis.[20,21] Therefore, unregulated formation of ROS can induce apoptosis. In addition, mitochondria are the predominant source of the superoxide anion (O_2^-) via the ETC. Thus, a decrease of the coupling efficiency of the ETC can increase the production of highly reactive O_2^- species.[11,22]

1.3 Why target mitochondria?

Alterations in mitochondrial structure and function appear to be intrinsic to the pathophysiology of numerous afflictions, ranging from neurodegeneration[23–25] to metabolic disorders such as diabetes[26–28] and obesity.[28] Given these links, it is not only important to understand the anatomy and physiology of healthy mitochondria, but also how they can be affected by the onset of disease. The effective visualization of mitochondria via optical imaging often requires the use of luminescent molecular probes that can selectively interact with mitochondria and localize within these organelles.

Conventional luminescent probes for mitochondria are organic luminophores such as Rhodamine 123,[29] 10-N Nonyl-Acridine Orange,[30] and the most commonly employed mitochondrial probes known as MitoTrackers.[31] Whilst of widespread use, organic probes can suffer from certain drawbacks depending on the imaging conditions. These limitations include relatively fast photobleaching, concentration quenching, and potential difficulties in discriminating between the fluorescent signal of the probe and endogenous autofluorescence. Overcoming these drawbacks was one of the key motivations to explore luminescent transition metal complexes as complementary molecular probes. In general, transition metal luminophores display large Stokes shifts, relatively long excited state lifetimes, and resistance to both photobleaching and concentration quenching. These features assist in alleviating, and in some cases eliminating, the previously listed drawbacks of organic probes. Furthermore, the capacity to fine-tune photophysical and biological properties of transition metal luminophores via ligand design allows access to versatile probes for bioimaging. Some excellent summaries of phosphorescent heavy metal complex probes developed for bioimaging are available in the literature.[32–34]

Mitochondria-targeting is also an essential feature in the design of many therapeutic agents.[35–37] Therapeutic targets in mitochondria include the ETC, mitochondrial permeability transition (MPT), mitochondria-specific proteins, and mitochondrial DNA.[37,38] Potential applications include: (i) antioxidant delivery to prevent oxidative damage from neurodegenerative diseases, diabetes, as well as ischemia and reperfusion tissue injury[39]; (ii) drugs to inhibit the mitochondrial permeability transition in ischemia and reperfusion related tissue injury[40]; (iii) drugs to regulate ETC-protein coupling for the treatment of obesity and diabetes[41]; (iv) the delivery of cytotoxic drugs or proteins to trigger cell death in cancer therapies.[42] The potential of cancer therapies via targeting mitochondria is relevant to the context of this chapter, as many studies have been conducted exploring anticancer potential for mitochondria-targeting metal complexes.

The rapid proliferation and migration of cancer cells is an intensive energy demanding task. To fulfill this demand, cancer cells have been reported to reprogram their cellular activities to counteract metabolic stress during cancer progression.[43] In fact, mitochondria in cancerous cells can be altered to provide for energy requirements and macromolecular synthesis. This allows cells to interchangeably switch between glycolysis and oxidative phosphorylation as the means for producing energy for cancer cells.[44] The intricate relationship that mitochondria have with programmed cell death has resulted in this organelle becoming one of the primary targets for potential anticancer therapies. Consequently, a myriad of studies have focused on developing mitochondria-targeting anticancer compounds, including many examples of metal-based complexes.

1.4 How are mitochondria targeted?

Developing a structure-activity relationship that would guide the future design of coordination complexes, and molecular species in a more generalized context, is vital to advance the fields of diagnostics and therapy. Whilst the obvious answer resides in bioconjugation with mitochondria-targeting biological vectors, it is now evident that this approach is not the only available strategy. In fact, there are numerous examples of compounds that display efficient mitochondrial specificity without bioconjugation. It is therefore essential to try and identify trends that can assist in setting guidelines for the rational design of mitochondria-targeting species. At a very simplistic level, the key parameters are often identified as having a cationic charge to exploit potential gradients, and a lipophilic character to enhance cellular internalization and traverse membranes. The electrochemical gradient established by the MMP results in a net negative charge in the mitochondrial matrix. The electrical potential is generally reported to be within the range of 120–170 mV.[45] The value can be subject to fluctuations depending on health of the cell and the presence of specific pathologies. Importantly, this membrane potential is substantially larger than that of any other potential gradient established within the cell. As a result, ions featuring a positive charge are favored to penetrate mitochondria. Sufficiently lipophilic cations can pass through mitochondrial membranes and accumulate in the matrix.[45] A common method to achieve sufficient lipophilicity is to ensure the positive charge is delocalized via resonance. Combining these two concepts has resulted in investigating the so-called delocalized lipophilic cations (DLCs) as mitochondria-targeting species. In fact, one of the most common methods to promote mitochondrial internalization of a molecular species is to covalently bind the triphenylphosphonium cation (TPP).[39]

Regarding metal-complexes, an overview of the literature identifies three generic strategies to target mitochondria: (i) create a library of structurally related compounds to tune, rationally or serendipitously, the intrinsic biological properties of the complex and favor mitochondrial localization,[46] (ii) the incorporation of one or more mitochondria-targeting moieties such as TPP,[47] and (iii) the incorporation of one or more mitochondrial penetrating peptides (MPPs).[48] These strategies work to exploit the MMP ensuring the complex is cationic to promote mitochondrial affinity and retains a high enough lipophilicity to allow for penetration through the mitochondrial membranes. However, as reported in this chapter, there are also examples of cationic complexes that are efficient at targeting mitochondria even in the event of depolarization of the MMP. These examples clearly indicate that there is more that needs to be learned about the complex interaction of these species with biological environments.

1.5 Photodynamic therapy

Photodynamic therapy (PDT, also known as photochemotherapy) is one of the currently developed strategies for the treatment of cancers and other pathologies. Less than 30 years ago, Photofrin, an oligomer based on porphyrin units, was the first approved photosensitizer for use in clinical phototherapy.[49] PDT requires the presence of three essential components: (i) the photosensitizer, (ii) oxygen, and (iii) light. The absorption of light at an appropriate wavelength excites the photosensitizer from its ground state to an excited singlet state. After internal conversion, the excited singlet state rapidly undergoes intersystem crossing (ISC) to occupy a relatively long-lived excited triplet state. The excited triplet state then promotes the increase of ROS via sensitizing the formation of singlet oxygen ($^1O_2$) or alternative radicals. The produced ROS can induce severe oxidative damage to cellular structures, leading to cell death. It is important to note that, upon quenching, the photosensitizer returns to the ground state, meaning it is once again available for subsequent excitation. Hence, theoretically, only a catalytic amount of the photosensitizer is necessary for therapeutic effects.[50–52] Metal complexes have been extensively investigated as photosensitizers. Their intrinsic strong spin-orbit coupling, especially for those elements of the second and third transition periods, facilitates efficient intersystem crossing and population of excited states of triplet multiplicity (when the ground state is of singlet multiplicity).[53,54] These long-lived triplet excited states then result in interaction with molecular oxygen producing the desired PDT effects.[55–57]

2. Iridium

It would be an understatement to say that the majority of research toward developing mitochondria-targeting metal complexes in recent years have been focused on iridium-based complexes. Luminescent Ir(III) complexes are generally considered to be the 'gold standard' for subcellular imaging agents. Ir(III) complexes are typically characterized by long excited states, large Stokes shifts, appreciable biocompatibility (depending on their chemical structure) and resistance to photobleaching.[58] In recent years, there have also been strong indications that many Ir(III) complexes may display antiproliferative properties toward a variety of cancer cell types (*vide infra*), either intrinsically or following photoexcitation.[46,59] Mechanisms of anticancer activity for Ir(III) complexes have been found to differ from that of currently employed anticancer agents such as cisplatin,[60,61] as expected from their structural differences. Alternative mechanisms of action for anticancer therapies are indeed highly sought after to overcome issues related to platinum-based drug resistance of cancer cells.

A survey of the recent literature reveals that there are three predominant forms of mitochondria-targeting Ir(III) complex structures (see Fig. 1) synthesized in the past decade. Form 1 corresponds to the standard octahedral cyclometalated complexes with generic formulation $[Ir(\mathbf{C\hat{}N})_2(\mathbf{N\hat{}N})]^+$, where **C^N** represents bidentate cyclometalated ligands, whilst **N^N**

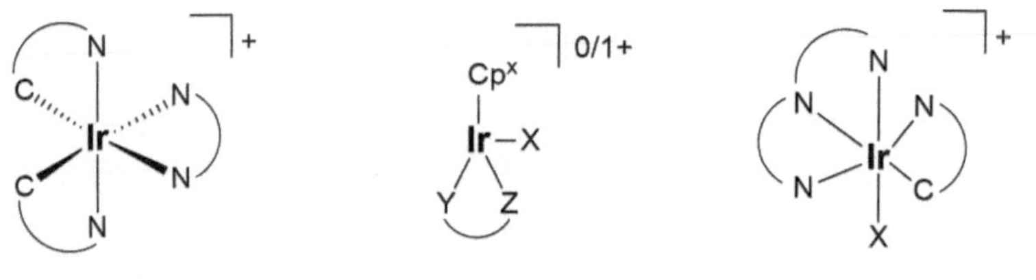

Fig. 1 Ir(III) complex general structures.

represents a bidentate diimine ligand. The structure of those Ir(III) complexes belonging to Form 2 is often referred to as half-sandwich, due to the similarity to metallocenes. These complexes have generic formulation $[Ir(\mathbf{Cp^x})(\mathbf{YZ})\mathbf{X}]^{+/0}$ where $\mathbf{Cp^x}$ is a π-bonded aromatic system such as the cyclopentadienyl anion, **YZ** is generally either a bidentate anionic cyclometalated **C^N** ligand or a neutral diimine **N^N** ligand, and **X** is an anionic halogen. Form 3 is the least common and comprises those complexes with formulation $[Ir(\mathbf{N^\wedge N^\wedge N})(\mathbf{C^\wedge N})\mathbf{X}]^+$, where **N^N^N** represents a tridentate ligand—such as a derivative of 2,2′:6′,2″-terpyridine, **C^N** represents a bidentate anionic cyclometalated ligand and **X** represents an anionic halogen. All iridium structures discussed in this section are shown in Fig. 2.

Li et al.[62] developed a class of luminescent Ir(III)-polyamine complexes featuring branched-poly(ethyleneimine) (**bPEI**) and ethylenediamine (**en**) ligands. It was shown that the photophysical and electrochemical properties of the complexes could be readily tuned via alterations in the cyclometalated ligands. *In vitro* investigations with HeLa human cervical carcinoma and HEK293T human embryonic kidney cell lines revealed that both the polyamine and the cyclometalated ligands modulate cellular uptake, lipophilicity and cytotoxicity. Upon internalization, **bPEI** complexes displayed affinity toward lysosomal compartments, whilst **en** complexes were found to localize within mitochondria. To improve biocompatibility, complex **Ir1** featuring **btn** ligands was grafted with poly(ethylene glycol) via ligand modification. The obtained copolymers retained the photophysical properties of the parent complex, **Ir1**, and exhibited improved water solubility, biocompatibility, and gene-delivery capacity *in vitro*. The study concluded that the copolymers developed serve as new luminescent transfection reagents and will likely provide new insights into understanding cellular uptake, intracellular trafficking, and DNA-delivery mechanisms of **bPEI**-based complexes.

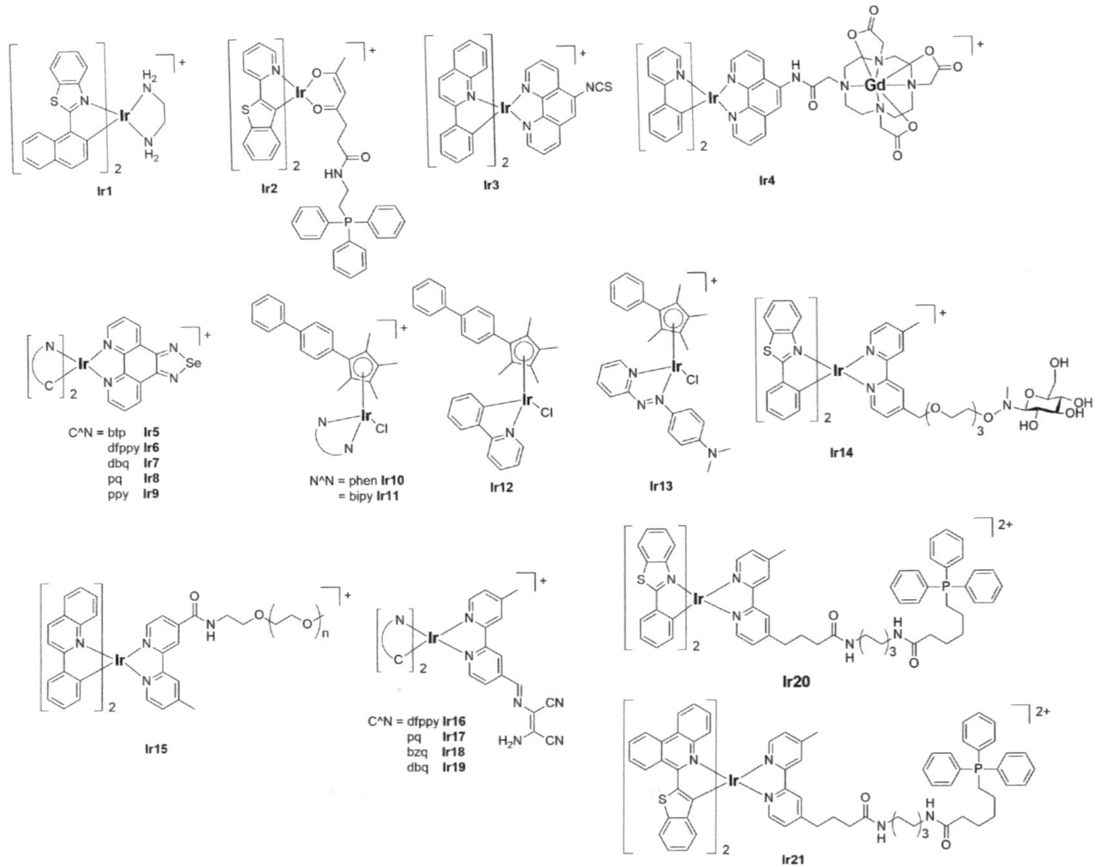

Fig. 2 (A) Mitochondria-targeting Ir(III) complexes **Ir1–21**. (B) Mitochondria-targeting Ir(III) complexes **Ir22–37**. (C) Mitochondria-targeting Ir(III) complexes **Ir38–65**. (D) Mitochondria-targeting Ir(III) complexes **Ir66–95**. (E) Mitochondria-targeting Ir(III) complexes **Ir121–142**. (F) Mitochondria-targeting Ir(III) complexes **Ir143–158**. (G) Mitochondria-targeting Ir(III) complexes **Ir143–158**. (H) Mitochondria-targeting Ir(III) complexes **Ir159–184.**

(Continued)

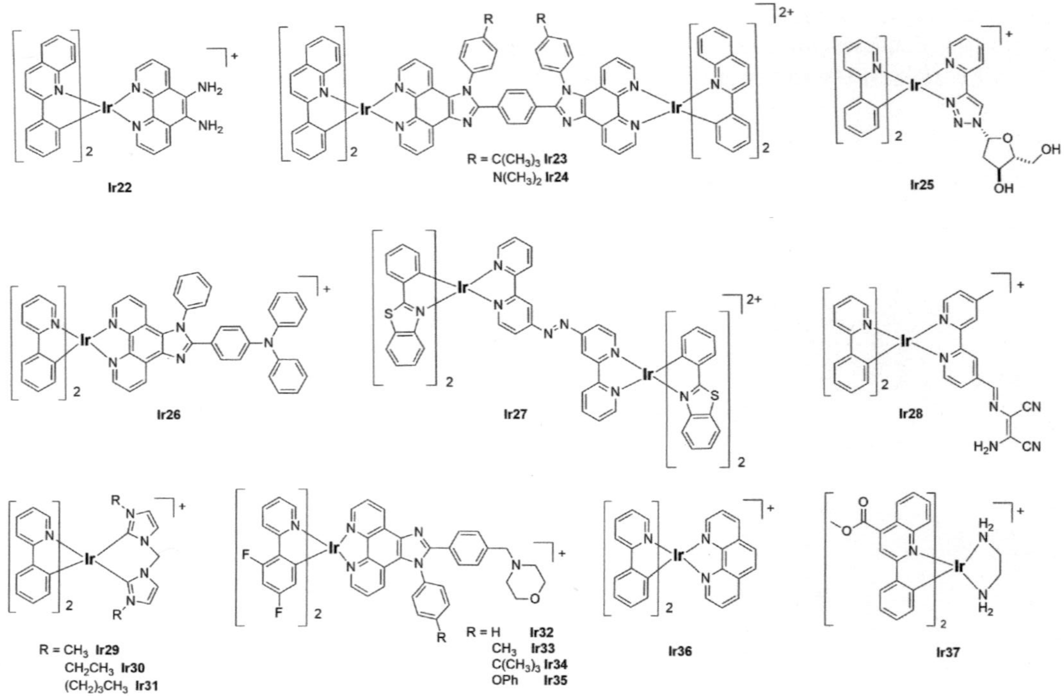
Ir22
R = C(CH3)3 Ir23
N(CH3)2 Ir24
Ir25
Ir26
Ir27
Ir28
R = CH3 Ir29
CH2CH3 Ir30
(CH2)3CH3 Ir31
R = H Ir32
CH3 Ir33
C(CH3)3 Ir34
OPh Ir35
Ir36
Ir37

Fig. 2—Cont'd

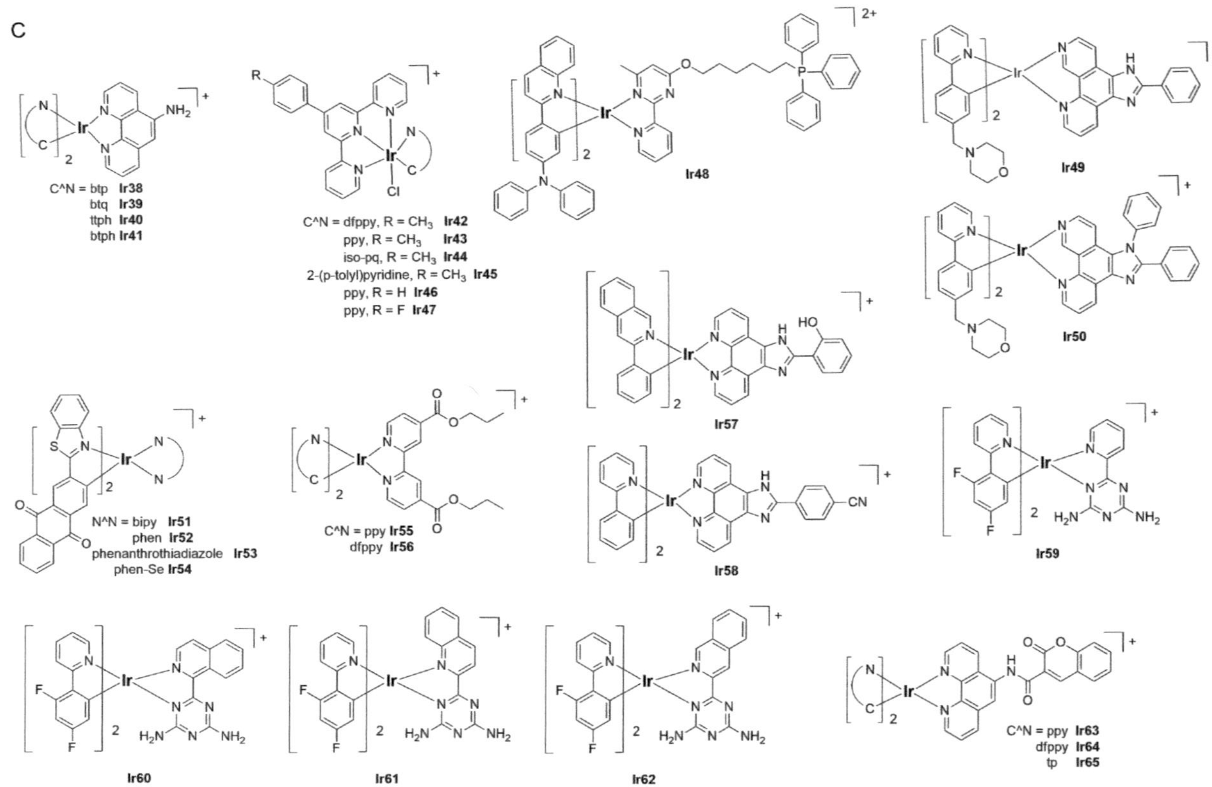
C
C^N = btp Ir38
btq Ir39
ttph Ir40
btph Ir41
C^N = dfppy, R = CH3 Ir42
ppy, R = CH3 Ir43
iso-pq, R = CH3 Ir44
2-(p-tolyl)pyridine, R = CH3 Ir45
ppy, R = H Ir46
ppy, R = F Ir47
Ir48
Ir49
Ir50
N^N = bipy Ir51
phen Ir52
phenanthrothiadiazole Ir53
phen-Se Ir54
C^N = ppy Ir55
dfppy Ir56
Ir57
Ir58
Ir59
Ir60
Ir61
Ir62
C^N = ppy Ir63
dfppy Ir64
tp Ir65

Fig. 2—Cont'd

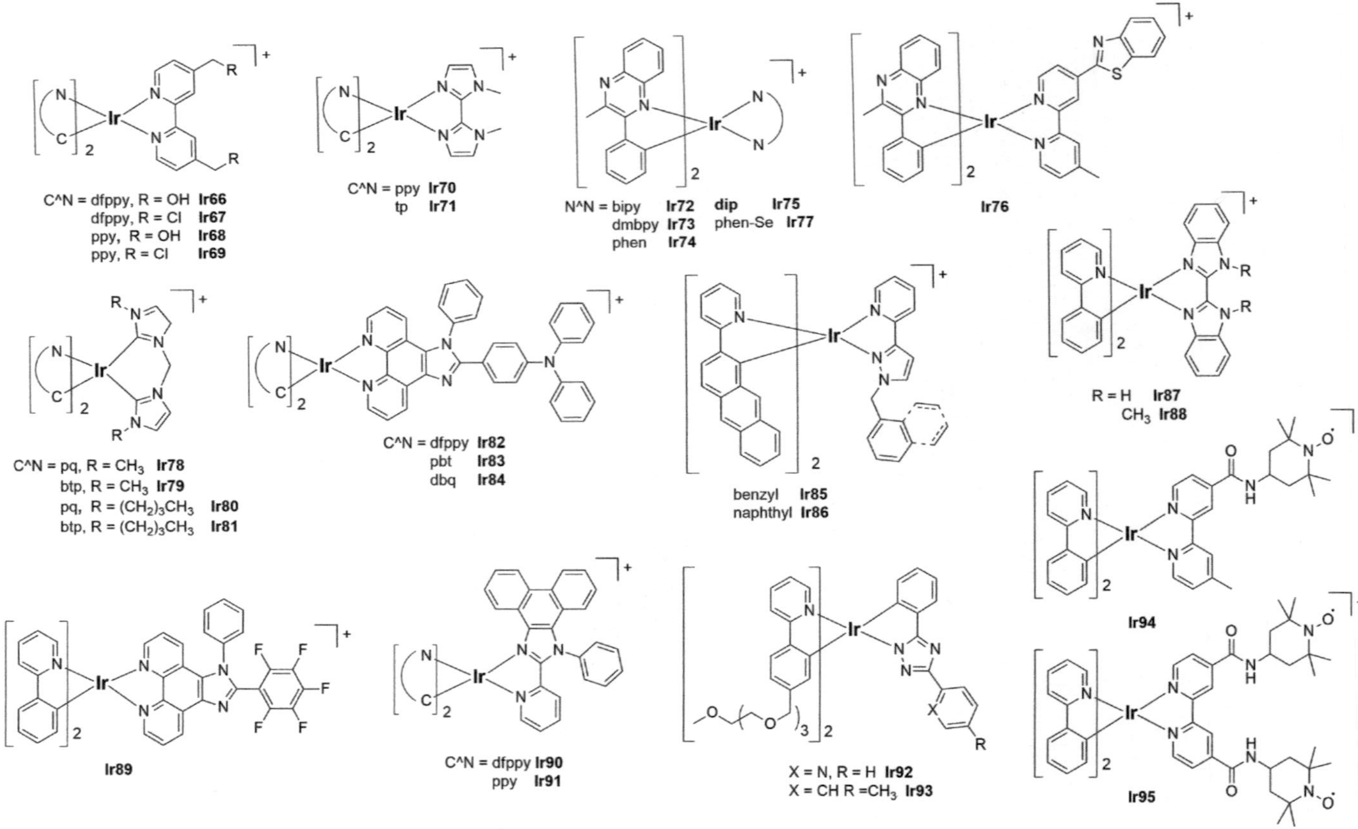
C^N = dfppy, R = OH Ir66
dfppy, R = Cl Ir67
ppy, R = OH Ir68
ppy, R = Cl Ir69
C^N = ppy Ir70
tp Ir71
N^N = bipy Ir72
dmbpy Ir73
phen Ir74
dip Ir75
phen-Se Ir77
Ir76
R = H Ir87
CH$_3$ Ir88
C^N = pq, R = CH$_3$ Ir78
btp, R = CH$_3$ Ir79
pq, R = (CH$_2$)$_3$CH$_3$ Ir80
btp, R = (CH$_2$)$_3$CH$_3$ Ir81
C^N = dfppy Ir82
pbt Ir83
dbq Ir84
benzyl Ir85
naphthyl Ir86
Ir94
Ir89
C^N = dfppy Ir90
ppy Ir91
X = N, R = H Ir92
X = CH R =CH$_3$ Ir93
Ir95

Fig. 2—Cont'd

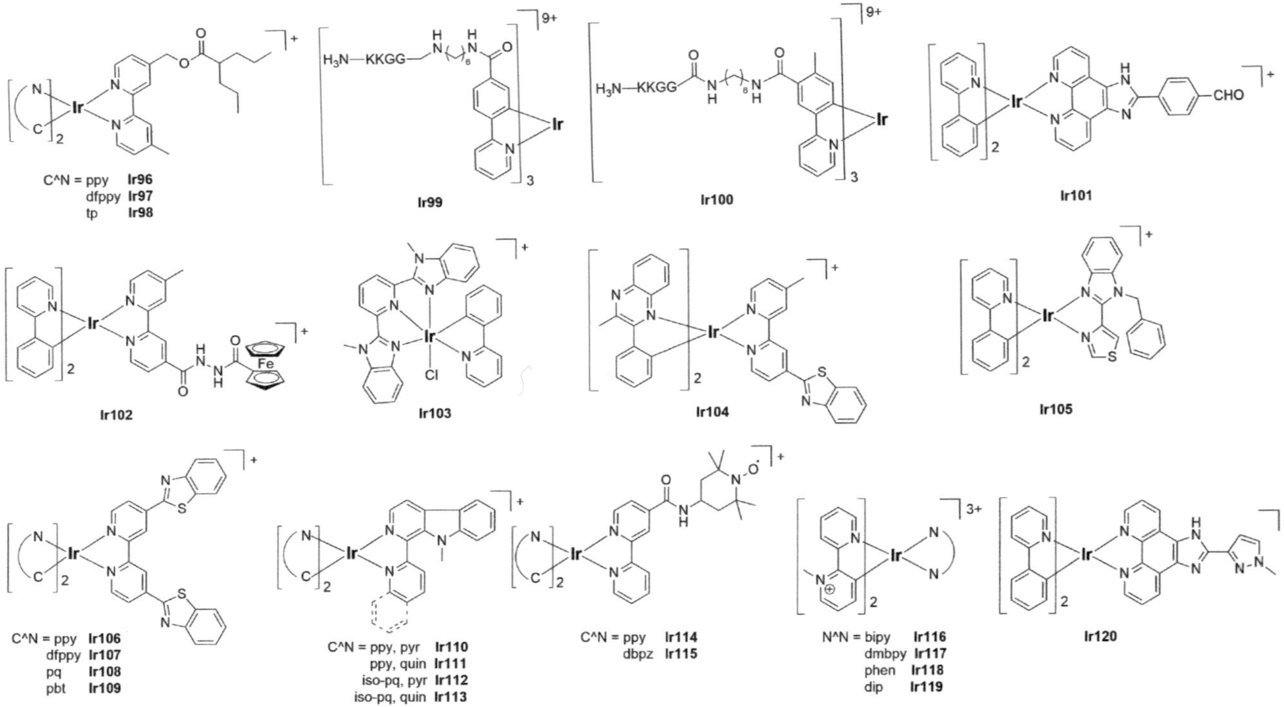
C^N = ppy Ir96
dfppy Ir97
tp Ir98
Ir99
Ir100
Ir101
Ir102
Ir103
Ir104
Ir105
C^N = ppy Ir106
dfppy Ir107
pq Ir108
pbt Ir109
C^N = ppy, pyr Ir110
ppy, quin Ir111
iso-pq, pyr Ir112
iso-pq, quin Ir113
C^N = ppy Ir114
dbpz Ir115
N^N = bipy Ir116
dmbpy Ir117
phen Ir118
dip Ir119
Ir120

Fig. 2—Cont'd

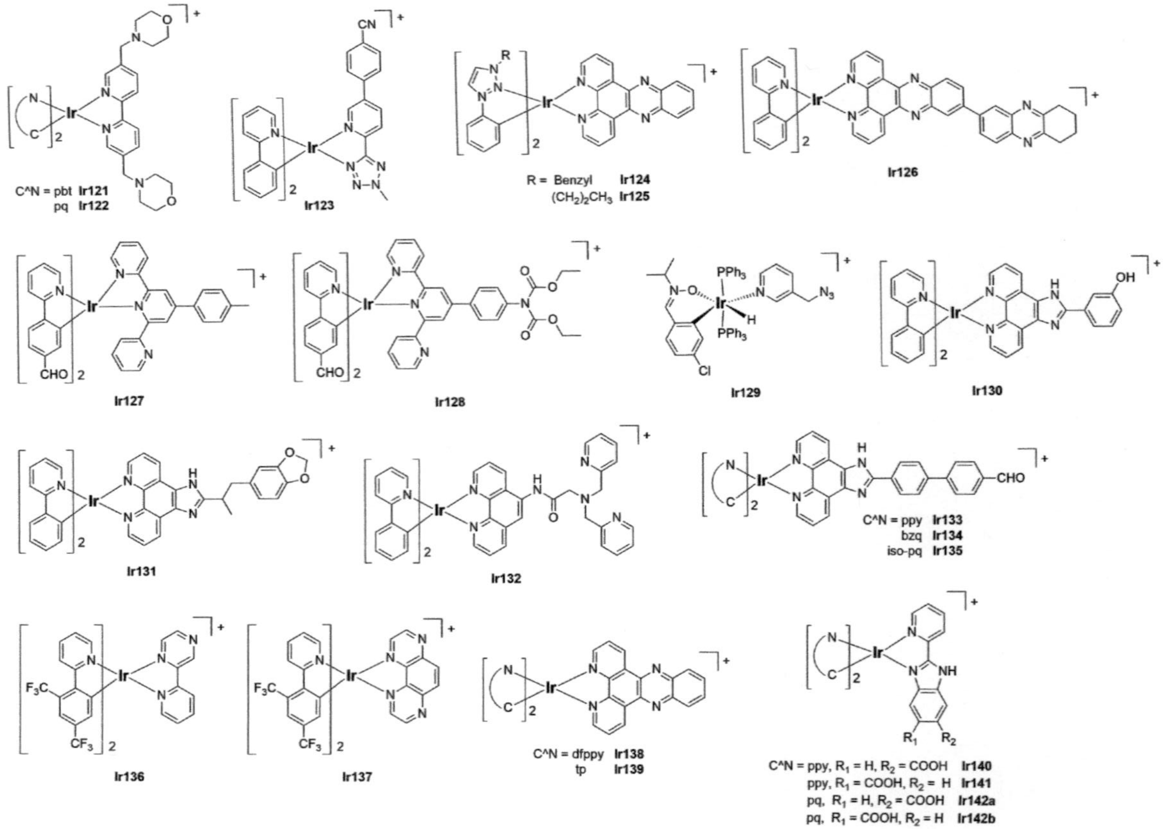
C^N = pbt Ir121
pq Ir122
Ir123
R = Benzyl Ir124
(CH2)2CH3 Ir125
Ir126
Ir127
Ir128
Ir129
Ir130
Ir131
Ir132
C^N = ppy Ir133
bzq Ir134
iso-pq Ir135
Ir136
Ir137
C^N = dfppy Ir138
tp Ir139
C^N = ppy, R1 = H, R2 = COOH Ir140
ppy, R1 = COOH, R2 = H Ir141
pq, R1 = H, R2 = COOH Ir142a
pq, R1 = COOH, R2 = H Ir142b

Fig. 2—Cont'd

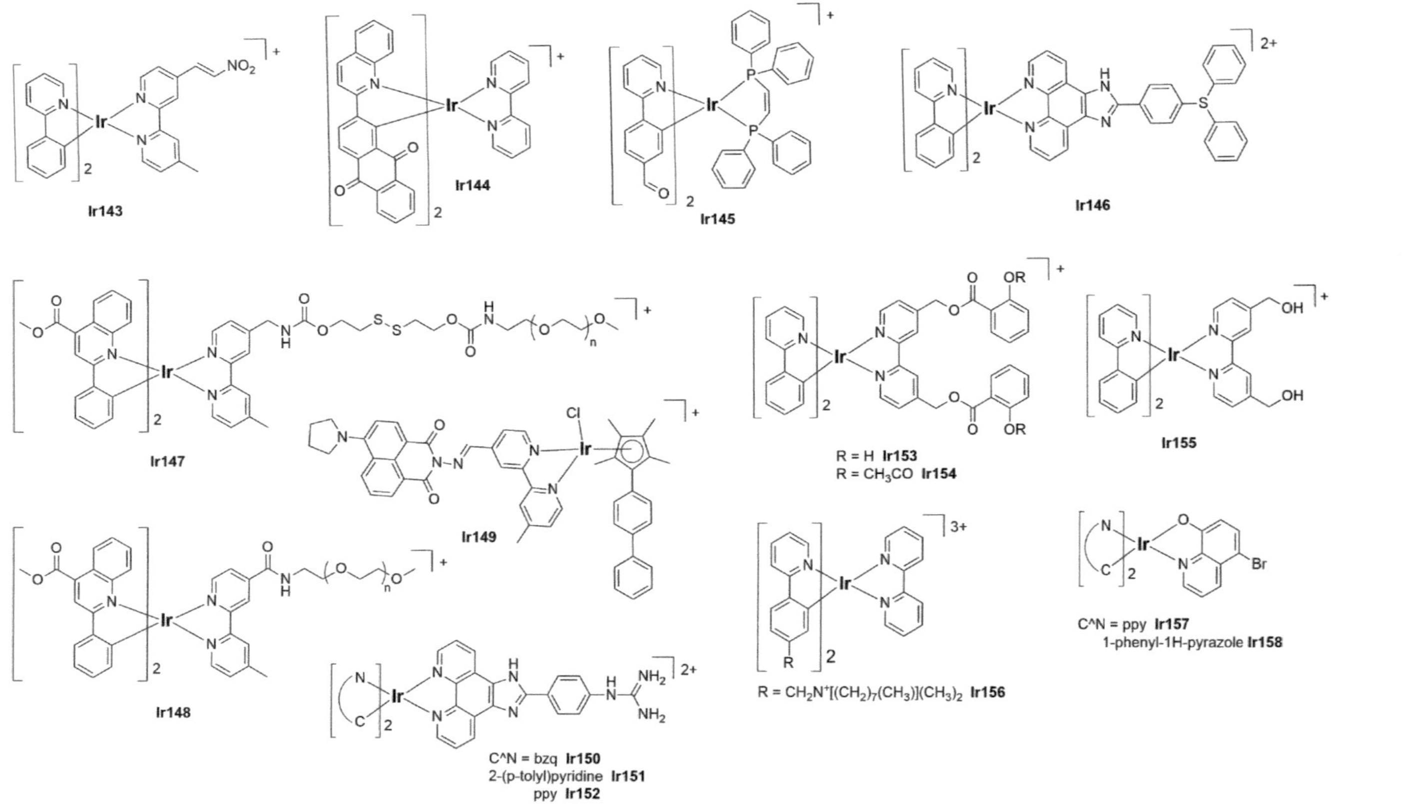
Ir143
Ir144
Ir145
Ir146
Ir147
Ir148
Ir149
C^N = bzq Ir150
2-(p-tolyl)pyridine Ir151
ppy Ir152
R = H Ir153
R = CH3CO Ir154
Ir155
R = CH2N+[(CH2)7(CH3)](CH3)2 Ir156
C^N = ppy Ir157
1-phenyl-1H-pyrazole Ir158

Fig. 2—Cont'd

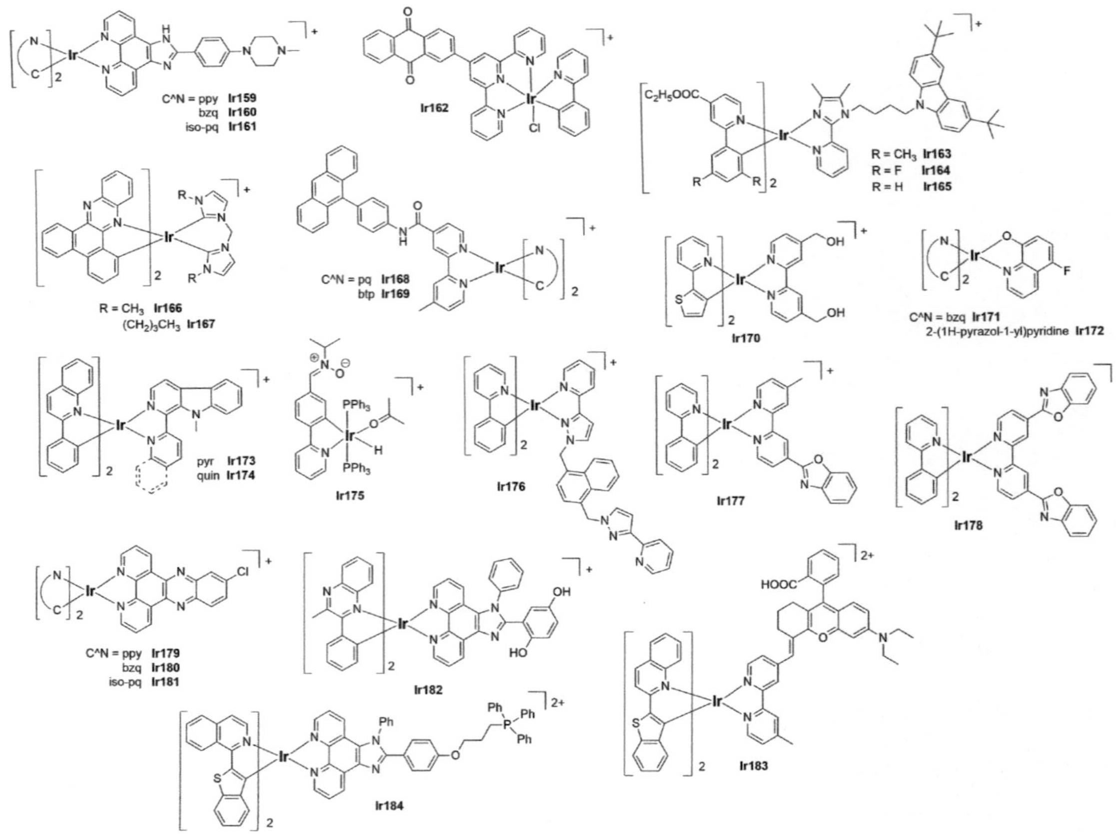
C^N = ppy Ir159
bzq Ir160
iso-pq Ir161
Ir162
R = CH3 Ir163
R = F Ir164
R = H Ir165
R = CH3 Ir166
(CH2)3CH3 Ir167
C^N = pq Ir168
btp Ir169
Ir170
C^N = bzq Ir171
2-(1H-pyrazol-1-yl)pyridine Ir172
pyr Ir173
quin Ir174
Ir175
Ir176
Ir177
Ir178
C^N = ppy Ir179
bzq Ir180
iso-pq Ir181
Ir182
Ir183
Ir184

Fig. 2—Cont'd

Murase et al.[47] developed a luminescent Ir(III) complex, **Ir2**, in the form $[Ir(\mathbf{btp})_2(\mathbf{L})]^+$ where **L** is acetylacetonato conjugated to TPP via an amide bridge. The phosphorescence of **Ir2** was observed to be effectively quenched in the presence of O_2, and the complex displayed mitochondrial selectivity in HeLa human cervical carcinoma cells. The complex showed excellent potential for use in lifetime-based oxygen-sensing measurements. The study concluded that **Ir2** represents the first mitochondrial-specific Ir(III) complex oxygen sensor appropriate for use in lifetime-based measurements.

Wang et al.[63] reported a novel approach combining microscopy, biochemical fractionation, and proteomic techniques for characterizing intracellular interactions of luminescent compounds. *In vitro* investigations with an Ir(III) complex (**Ir3**), in the form $[Ir(\mathbf{pq})_2(\mathbf{N^\wedge N})]^+$ where **N^N** is 5-isothiocyanate-1,10-phenanthroline, revealed mitochondrial specificity in 3T3 mouse embryonic fibroblast cells. **Ir3** was reported to permeate the plasma membrane via an energy-dependent pathway and covalently bind to primary amines of target proteins, switching on its emission within the mitochondria of live cells. Experiments with longer incubation times revealed more non-specific localization of **Ir3** throughout the cells, suggesting that a variety of cytoskeletal proteins also react with the complex. The study reports a novel approach to identify biological targets for a wide variety of luminescent probes.

Yang et al.[64] synthesized and conducted biological investigations with a heteronuclear Ir(III)-Gd(III) complex (**Ir4**) conjugated via an amide bridge. *In vitro* investigations indicated great potential for **Ir4** as an imaging agent for dual-mode phosphorescence and magnetic resonance imaging (MRI). **Ir4** displayed intense emission with a sub-microsecond lifetime in aqueous solution, low cytotoxicity, and good cellular penetration with subsequent mitochondrial localization. *In vivo* investigations with **Ir4** were able to produce high quality MRI images of murine livers, suggesting potential for applications into the diagnosis of liver disease. The design strategy might offer alternative pathways for development of more d-f heteronuclear systems that display both luminescent and magnetic properties appropriate for dual-modal imaging. The study concluded that the investigation will likely be useful for future research into the development of novel metal-based probes for dual mode imaging.

Chen et al.[65] developed an Ir(III) complex, **Ir5**, featuring aggregation-induced phosphorescent emission (AIPE) properties, and investigated its biological properties in HeLa human cervical carcinoma cells. It was found that **Ir5** displays mitochondrial specificity, excellent photostability, low cytotoxicity and resistance to loss of MMP. It was demonstrated that **Ir5** can be used for mitochondrial imaging and tracking, representing the first example of an AIPE-active Ir(III) complex for this use. The study facilitates new strategies for developing novel sub-cellular targeting imaging agents.

Chen et al.[66] then expanded on their previous work on **Ir5** to investigate the effects of altering the cyclometalated ligands. The new ligands for this family of complexes included **dfppy** (**Ir6**), **dbq** (**Ir7**), **pq** (**Ir8**) and **ppy** (**Ir9**). The complexes displayed luminescence in the visible region ranging from 528 to 589 nm, with relatively long-lived excited states. *In vitro* investigations revealed that all four complexes display a high degree of specificity toward mitochondrial localization in HeLa cells, both live and fixed, without the need for prior membrane permeabilization. Low cytotoxicity was observed for all complexes at concentrations appropriate for cellular imaging. Complexes **Ir7–9** were also found to exhibit superior resistance to photobleaching, relative to commercially available mitochondrial trackers. The results indicated that the complexes **Ir6–9** are well suited for mitochondria-targeting imaging agents. The information garnered in the investigation may be of benefit for future explorations into developing novel organelle-selective imaging agents.

Hearn et al.[67] synthesized four Ir(III) half-sandwich complexes in the form $[Ir(\mathbf{Cp^x})(\mathbf{XY})Cl]^n$ where $\mathbf{Cp^x}$ is a biphenyltetramethylcyclopentadienyl (**Ir10–12**) or phenyltetramethylcyclopentadienyl ligand (**Ir13**) and **XY** is **phen** (**Ir10**, $n = 1+$), **bipy** (**Ir11**, $n = 1+$), **ppy** (**Ir12**, $n = 0$) or N, N-dimethylphenylazopyridine (**Ir13**, $n = 1+$). The complexes were biologically screened to investigate potential anticancer activities. The complexes displayed mechanisms of action distinctly different from that of cisplatin, with relatively higher potency. Transmission electron microscopy (TEM) and biochemical assays found that DNA-damaging ROS production was the likely source of *in vitro* cytotoxicity, with mitochondria-targeting being observed and apoptosis induced within a 24-h period. The study concluded that cell selectivity and reactivity are strongly linked to the chelating ligand structure.

Law et al.[68] developed a class of Ir(III) bipyridine complexes appended to β-D-glucose or D-galactose, featuring the cyclometalated ligands **ppy**, **pbt**, **pq**, and **bzq**. The effect of varying the cyclometalated ligands and appending sugars on both photophysical and biological properties were explored. All the complexes were found to display long-lived excited states with emission from triplet excited states of metal-to-ligand charge-transfer (^{3}MLCT) [$d\pi(Ir) \rightarrow \pi^*(\mathbf{N}^\wedge\mathbf{N})$] or triplet intraligand (^{3}IL) ($\pi \rightarrow \pi^*$)($\mathbf{C}^\wedge\mathbf{N}$ and $\mathbf{N}^\wedge\mathbf{N}$) character. The biological properties in HeLa cells were assessed and showed that complex **Ir14**, featuring a β-D-glucose appended ligand and **pbt** bis-cyclometalated ligands, localized within mitochondria via an energy-dependent process such as endocytosis as well as an additional pathway mediated by glucose transporters. It was suggested that the photobleaching-resistant **Ir14** has potential for use as a luminescent glucose uptake indicator that is sensitive to hormonal and metabolic stimulation as well as inhibition. The study indicated that this class of Ir(III) complexes may serve to provide insights into the development of luminescent glucose uptake indicators to study cellular metabolism.

Li et al.[69] reported a class of luminescent Ir(III) complexes in the form $[Ir(\mathbf{C}^\wedge\mathbf{N})_2(\mathbf{N}^\wedge\mathbf{N})]^+$ where **N^N** is a PEG-appended bipyridine and **C^N** is **ppy**, **pppy**, **pq**, **pbt**, or **btn.** In the investigation, the PEG-functionalized complexes were compared with their PEG-free counterparts bound to 4-(N-ethylaminocarbonyl)-4′-methyl-2,2′-bipyridine. The complexes were found to display long-lived oxygen-sensitive excited states, with intense green to orange-red emissions depending on the structure of the cyclometalated ligand. Cell-based assays found that the PEG complexes were non-cytotoxic in the dark, but upon irradiation significant cytotoxicity was induced. The PEG complex **Ir15,** featuring the **pq** cyclometalated ligand, was found to localize within mitochondria in HeLa cells and induce rapid necrosis upon irradiation. The study indicated this family of luminescent Ir(III)-PEG complexes have potential for combined mitochondrial specific imaging and PDT applications.

Lo et al.[70] synthesized three phosphorescent Ir(III) polypyridine complexes appended with a dibenzocyclooctyne (DIBO) moiety. DIBO was selected as previous investigations have shown it is non-toxic and displays high reactivity toward azides. The potential for use as luminescent cellular imaging agents for targeting azide-modified biomolecules was investigated and assessed. The study indicated that the complexes displayed mitochondrial selectivity in murine ovarian cells and that the cytotoxicity could be effectively minimized by incorporating more polar constituents onto the ligands, such as PEG. Advantageously, the incorporation of different groups mostly preserved the photophysical properties. The study concluded that the first examples of phosphorescent bio-orthogonal labeling reagents for azide-modified biomolecules had been successfully identified.

Li et al.[71] developed four Ir(III) complexes featuring a 4-methyl-2,2′-bipyridine-4′-diaminomaleonitrile ligand. The diaminomaleonitrile moiety is a known quencher of luminescence and selectively reacts with hypochlorite. The family of complexes differentiated for the cyclometalated ligand, which included **dfppy** (**Ir16**), **pq** (**Ir17**), **bzq** (**Ir18**) and **dbq** (**Ir19**).

The study indicated a high degree of specificity for hypochlorite over other species. Upon reacting with hypochlorite, all complexes displayed luminescent emission in the visible range, spanning from green to red depending on the cyclometalated ligand. *In vitro* investigations revealed that complexes **Ir16–19** selectively localized into mitochondria in HeLa cells, where they effectively act as on-off phosphorescent sensors of hypochlorite. With low cytotoxicity, highly-selectivity and a range of emission wavelengths, complexes **Ir16–19** were concluded to be excellent candidates for mitochondrial specific hypochlorite imaging agents.

Zhang et al.[72] synthesized two Ir(III) complexes featuring a 4-methyl-2, 2′-bipyridine ligand appended to TPP via and amide bridges. The cyclometalated ligands for **Ir20** and **Ir21** were **pbt** and 6-(benzo[b] thiophen-2-yl)phenanthridine, respectively. The TPP moiety allowed for mitochondrial specificity, and the complexes were found to display strong resistance to photobleaching. The complexes were determined to show promising potential for use as mitochondrial specific luminescent cellular imaging agents.

Chen et al.[73] developed an Ir(III) complex (**Ir22**), in the form $[\mathrm{Ir}(\mathbf{pq})_2(\mathbf{N^\wedge N})]^+$ where **N^N** is 1,10-phenanthroline-5,6-diamine, with the aim of producing a two-photon luminescent probe for the detection of intracellular nitric oxide (NO). The adjacent amine groups on the phenanthroline ligand readily and selectively react with NO to form a triazole moiety, which results in a luminescent complex. *In vitro* investigations revealed low cytotoxicity and a high degree of specificity toward mitochondrial NO in HeLa cells. The study concluded that **Ir22** was the first mitochondrial NO specific on-off luminescent probe appropriate for two-photon microscopy investigations.

Chen et al.[74] synthesized seven dinuclear Ir(III) complexes, six of which displayed a high degree of specificity toward mitochondrial localization within HeLa cells and normal L02 human liver cells. The complexes displayed low cytotoxicity and resistance to loss of MMP. **Ir23** and **Ir24**, featuring appended tert-butyl and dimethyl amino moieties, respectively, were further optimized for tracking changes of mitochondrial morphology during the early stages of apoptosis. The study was reported to present the first examples of dinuclear Ir(III) complexes appropriate for mitochondrial imaging and tracking.

Choi et al.[75] developed a luminescent nucleoside-appended Ir(III) complex (**Ir25**) that displayed mitochondrial and nuclear accumulation in human epidermal KB3-1 carcinoma cells after a 24-h incubation period.

Investigations into the cytotoxic mechanism of action revealed that apoptosis occurred, resulting from the inhibition of cell cycle progression at the G_2/M phases. Mitochondrial depletion was also observed after longer incubation periods, which could potentially lead to additional cytotoxic effects. A key innovation in the research was the ability to use the luminescent properties of **Ir25** to directly visualize cellular uptake of the nucleoside to correlate its anti-cancer activity with intracellular localization. The study concluded that **Ir25** has the potential to provide new insights into understanding the complex biology of nucleoside metabolism.

Jin et al.[76] developed seven Ir(III) complexes, in the form [Ir(**ppy**)$_2$(**N**^**N**)]$^+$ where **N**^**N** represents various substituted imidazophenanthroline ligands, with the aim of developing new mitochondrial localizing imaging agents. The family of complexes were found to display mitochondrial selectivity, good photostability, excellent pH resistance, rapid cellular accumulation, and low cytotoxicity at concentrations appropriate for imaging. Overall, the complexes were deemed to be promising candidates for use in biomedical research.

Jin et al.[77] then expanded on their previous investigation by producing five analogous complexes, with the same generalized form of [Ir(**ppy**)$_2$(**N**^**N**)]$^+$ where **N**^**N** represents substituted imidazophenanthroline ligands, this time featuring an additional benzene ring. The new complexes were appropriate for use as two-photon microscopy imaging agents. These complexes were found to cross cellular membranes via non-endocytosis active transport and displayed resistance to loss of MMP. The study concluded that this new family of Ir(III) complexes is suitable for tracking mitochondrial morphology in the early stages of apoptosis with two-photon microscopy.

Further expanding on their previous investigations, Jin et al.[78] coupled a triphenylamine side group to their previously developed Ir(III) complexes. *In vitro* investigations found that these complexes also displayed mitochondrial selectivity in HeLa cells and were appropriate for confocal microscopy with only a relatively short incubation period required. Excellent photostability, relative to conventionally used organic probes, was observed, and attributed to the aggregation-induced emission (AIE) properties of the complexes. One complex (**Ir26**) was utilized alongside MitoTracker Green to successfully monitor artificially induced mitophagy. The investigation expands our current library of known mitochondrial-specific probes, with **Ir26** showing particular promise for dynamic studies into the process of mitophagy.

Li et al.[79] developed four dinuclear Ir(III) complexes and explored their potential use as luminescent probes for *in vivo* detection of sulfite and bisulfite.

Complex **Ir27** was concluded to have great potential for use as a sulfite probe. Each iridium center was cyclometalated to 2-phenylbenzo[d]thiazole and bridged via an (E)-1,2-di([2,2′-bipyridin]-4-yl)diazene (**BPD**) ligand. The azo group in the **BPD** ligand was incorporated as the sulfite-active moiety. **Ir27** is only weakly emissive in aqueous solutions, but, upon reacting with sulfite, the emission intensity increases 18-fold. *In vitro* investigations revealed that **Ir27** displays preferential mitochondrial localization in HepG2 cells, where it rapidly and selectively reacts with sulfite and bisulfite, allowing for their direct imaging using one- and two-photon microscopy. **Ir27** was subsequently used in two-photon confocal imaging of the *in situ* generation of sulfite in *Caenorhabditis elegans* and *in vivo* monitoring of sulfite distribution in rat hippocampal tissue. These findings demonstrated that **Ir27** is an effective tool for one- and two-photon imaging of biological sulfite and bisulfite in subcellular organelles and live tissues.

Li et al.[80] also investigated the potential use of the Ir(III) complex **Ir28**, as a two-photon microscopy probe. **Ir28**, in the form $[Ir(\mathbf{ppy})_2(\mathbf{N^\wedge N})]^+$, where **N^N** is a reactive hypochlorite-specific diaminomaleonitrile-appended bipyridine ligand, displayed a high degree of mitochondrial specificity in human HeLa cells and murine RAW264.7 cells. The study demonstrated that **Ir28** is appropriate for use as a hypochlorite-sensitive mitochondrial specific two-photon probe *in vivo* using zebra fish as an animal model.

Li et al.[81] developed three luminescent Ir(III) complexes (**Ir29–31**), in the form $[Ir(\mathbf{ppy})_2(\mathbf{N^\wedge N})]$ where **N^N** is a bis-*N*-heterocyclic carbene (NHC) ligand appended by a hydrocarbon chain of varying length. The complexes were investigated for their anticancer activities and PDT potential. All three complexes localized within mitochondria and were found to display greater cytotoxicity than cisplatin to all the screened cancer cell lines, by inducing a cascade of events associated with mitochondrial dysfunction. The luminescent properties of the complexes allowed for monitoring morphological changes to mitochondria. Up to three orders of magnitude increase in cytotoxicity was observed upon irradiation. It was reported that this is one of the highest photocytotoxic responses for Ir(III) complexes. The results indicated that complexes **Ir29–31** have great potential for use as potent mitochondria-targeting theranostic PDT agents for cancer therapy.

Qiu et al.[82] synthesized four Ir(III) complexes (**Ir32–35**) in the form $[Ir(\mathbf{dfppy})_2(\mathbf{N^\wedge N})]^+$, where **N^N** is an imidazophenanthroline derivative appended with a morpholine group, and investigated their potential use as one- and two-photon microscopy probes. Curiously, the morpholine moiety was attributed to the high specificity for all the complexes toward mitochondrial localization. Generally, morpholine is predicted to result in lysosomal accumulation.[83] The complexes were found to display strong two-photon absorption properties, low cytotoxicity, high resistance to photobleaching, and retained the ability to emit intense phosphorescence regardless of changes to the MMP. **Ir32–35** were demonstrated to be applicable for tracking mitochondrial morphology during the early stages of apoptosis and were further developed for mitochondrial imaging in 3D multicellular spheroids. **Ir32–35** are expected to find future use as mitochondrial-specific two-photon imaging agents for use in biological research.

Takizawa et al.[84] identified that the ability of **Ir36** to photosensitize the production of singlet oxygen is highly dependent on the local environment. This sensitivity to the local environment could potentially influence results obtained from a multitude of singlet oxygen photosensitized experiments in cells. In particular, mechanistic studies of photosensitized singlet oxygen-mediated cell death require care when interpreting results. These findings suggest that the photophysical properties of **Ir36** in bulk solutions may not accurately represent intracellular activity and *in vivo* quantification of singlet oxygen dosage can be difficult to accurately measure.

Tang et al.[85] synthesized four complexes in the form $[\mathrm{Ir}(\mathbf{C^\wedge N})_2(\mathbf{en})]^+$. All the complexes were investigated for their potential use as bioimaging agents. Complex **Ir37** was observed to efficiently internalize in HeLa cells, wherein it accumulates within mitochondria. The complex displayed good biocompatibility in the dark and, upon irradiation, generated singlet oxygen, thus resulting in a high degree of cytotoxicity. Complex **Ir37** also displayed an increase in emission intensity upon binding to bovine serum albumin, which was tentatively attributed to hydrophobic interactions. The study indicated that the identity and polarity of the cyclometalated ligands play a pivotal role in lipophilicity, which consequently affects subcellular localization.

With the aim of developing oxygen sensitive luminescent probes, Yoshihara et al.[86] synthesized a luminescent iridium complex (**Ir38**) in the form $[\mathrm{Ir}(\mathbf{btp})_2(\mathbf{phen\text{-}NH_2})]^+$ (**phen-NH$_2$** = 5-amino-1,10-phenanthroline) and analogous complexes featuring extended π-conjugated cyclometalated ligands: $[\mathrm{Ir}(\mathbf{btq})_2(\mathbf{phen\text{-}NH_2})]^+$ (**Ir39**), $[\mathrm{Ir}(\mathbf{ttph})_2(\mathbf{phen\text{-}NH_2})]^+$ (**Ir40**) and $[\mathrm{Ir}(\mathbf{btph})_2(\mathbf{phen\text{-}NH_2})]^+$ (**Ir41**). It was observed that increasing the degree of π-conjugation results in a remarkable red-shifting of absorption and emission spectra, with the emission bands of **Ir40** and **Ir41** lying in the NIR region. *In vitro* investigations found that this family of complexes displays selective accumulation within mitochondria in HeLa cells. Dimethylation and diethylation of the amine group found progressive enhancement of cellular uptake efficiency. All the synthesized complexes were also observed to display an emission intensity that was sensitive to the presence of oxygen in HeLa cells. Overall, the research revealed great potential for **Ir38** and its derivatives for use as mitochondrial-selective oxygen sensors for use in cell imaging.

Zhang et al.[87] developed a series of Ir(III) complexes in the form $[Ir(\mathbf{C\hat{}N})_2(\mathbf{N\hat{}N})]^+$ where **C^N** is either **pppy**, N-(4-(pyridin-2-yl)benzyl)butan-1-amine (**ppy-C4**) or N-((4′-(pyridin-2-yl)-[1,1′-biphenyl]-4-yl)methyl)butan-1-amine (**pppy-C4**) and **N^N** represents varying diimine ligands. Complexes featuring **ppy-c4 and pppy-C4** were found to display dual emissions. TD-DFT calculations indicated that these properties were the result of an amine-to-ligand triplet state disrupting internal conversion from the ^{3}IL state to the ^{3}CT state. The structure of both cyclometalated and diimine ligands were systematically altered to investigate the effect that substituents had on the dual emissive properties. The data indicated that the emission profile of each complex is strongly dependent on both the structure of coordinating ligands and the nature of its local environment. *In vitro* investigations revealed that the complexes localized within mitochondria, cell-membranes, or both in HeLa cells. Intracellular localization appeared to correlate with the lipophilicity of the ligands, with the most lipophilic species showing mitochondrial selectivity. In addition, it was demonstrated that the inclusion of a biotin moiety allows for the complexes to be used as ratiometric sensors for avidin. Although, it is important to note that the addition of the polar biotin group significantly reduced selectivity for mitochondria, while maintaining interaction with the cell membrane. The study concluded that this investigation will likely contribute to future efforts in developing novel organelle-selective dual-emissive complexes for use as ratiometric probes and biomedical imaging agents.

Huang et al.[88] developed five Ir(III) complexes in the form $[Ir(\mathbf{N\hat{}N\hat{}N})(\mathbf{C\hat{}N})Cl]^+$ where **N^N^N** is 4′-(*p*-tolyl)-2,2′:6′,2″-terpyridine and **C^N** is **dfppy** (**Ir42**), **ppy** (**Ir43**), **iso-pq** (**Ir44**), 2-(*p*-tolyl)pyridine, or **pbt** (**Ir45**). The complexes synthesized were subsequently investigated for their potential use as two-photon imaging agents. Complexes **Ir42–45** were all found to display high quantum yields, large two-photon absorption cross sections, and selective accumulation within mitochondria. The complexes were demonstrated to produce clear images of mitochondria in the inner structure of 3D multicellular spheroids with two-photon microscopy. Complex **Ir43** was used in real-time one- and two-photon imaging of live *Caenorhabditis elegans*, where it was successfully utilized to track mitochondrial fusion and fission processes. It was concluded that all complexes synthesized, particularly **Ir43**, can potentially find use for mitochondrial imaging both in vitro and in vivo.

Huang et al.[89] then extended their prior investigation to explore the effects of altering the tridentate ligand on **Ir43**. The methyl group on the 4′-(*p*-tolyl)-2,2′:6′,2″-terpyridine ligand was substituted for a hydrogen or fluorine atom in complexes **Ir46** and **Ir47**, respectively. The two complexes were found to retain the desirable photophysical and biological properties of the parent complex. **Ir47** was concluded to display superior properties than **Ir43** and **Ir46** for mitochondrial two-photon imaging. **Ir47** was subsequently used to monitor mitochondrial dynamics and mitophagy in live cells in real-time. The study concluded that **Ir47** can likely find use in future two-photon microscopy investigations to further our knowledge of key events associated with mitochondrial fission, fusion and mitophagy.

Lv et al.[90] developed the Ir(III) complex **Ir48** appended with a TPP moiety, along with the TPP-free analogous complex. **Ir48** was observed to localize within mitochondria, whilst the TPP-free analogue displayed accumulation within lysosomes. With this observation, the investigation aimed to evaluate PDT effects of lysosomes- and mitochondria-targeting photosensitizers under both normoxic and hypoxic conditions. The results indicated that **Ir48** exhibits greater therapeutic effects, particularly under normoxic conditions. The proposed explanation was that **Ir48** potentially inhibited mitochondrial respiration, resulting in higher intramitochondrial oxygen content. This research may prove to be beneficial for future explorations into developing specific organelle-targeting PDT agents.

Qiu et al.[91] developed a family of complexes which included **Ir49** and **Ir50**, functionalized with a morpholine group. Both complexes were observed to target mitochondria, whereas the other complexes in the family, displaying less lipophilic ligands, were observed to target lysosomes. Intriguingly, **Ir49** and **Ir50** were found to enter the cells using entirely different mechanisms. **Ir49** entered cells via an energy-dependent pathway, whilst **Ir50** used an energy-independent pathway as observed for the lysosomes-targeting complexes. All complexes synthesized were found to display low toxicity and excellent optical properties within a pH range of approximately 4–8. The results obtained from this investigation present a method for targeting mitochondria or lysosomes by regulating the lipophilicity of the coordinated ligands. This may be beneficial for future explorations into developing novel organelle-targeting diagnostic or therapeutic agents.

Sun et al.[92] developed four mitochondrial specific luminescent Ir(III) complexes for imaging in hypoxic tumor cells. The complexes were in the form $[Ir(\mathbf{C^\wedge N})_2(\mathbf{N^\wedge N})]^+$ where **C^N** represents 2-(benzo[d]thiazol-2-yl)anthracene-9,10-dione and **N^N** is **bipy** (**Ir51**), **phen** (**Ir52**), phenanthro[9,10-c][1,2,5]thiadiazole (**Ir53**), or **phen-Se** (**Ir54**). The **C^N** ligand incorporates the hypoxic-sensitive anthraquinone moieties. Anthraquinone groups effectively quench the phosphorescence of the iridium(III) complexes. However, under conditions of high reductive stress, such as in hypoxic tumor tissues, the anthraquinone groups can convert to hydroquinone. In this form, intense phosphorescence is restored, resulting in an on-off mitochondrial specific phosphorescence with minimal background interference. The reduced complexes showed significant two-photon absorption cross-sections and were demonstrated to distinguish hypoxic mitochondria in adherent cells, 3D multicellular spheroids, and live zebrafish. It was concluded that complexes **Ir51–54** have potential application for tracking mitochondrial morphology, in real-time and under hypoxic conditions. Proposed applications include recognizing mitochondria-related physiological and pathological changes, improving targeted cancer therapies, overcoming therapeutic resistance induced by hypoxia and possibly in the development of novel diagnostic strategies.

Tso et al.[93] developed two mitochondria-targeting PEG polymers featuring Ir(III)-based pendants and investigated their potential for use as PDT agents. The Ir(III) moiety was attached to the PEG framework via a nitrobenzyl photolabile protecting group (PPG) linker. The linking group was attached to suppress the cytotoxicity of the Ir(III) moiety. Upon irradiation, the Ir(III) moiety is photocleaved from the protecting group and its cytotoxicity is restored. The PPG-PEG pendant was demonstrated to suppress dark cytotoxicity and allow for photoactivatable oxygen-dependent cytotoxicity. The study concluded that photoactivatable PPG-PEG transition metal complexes may have a future in the development of mitochondria-targeting anticancer therapeutics.

Wang et al.[94] developed a series of ten luminescent complexes containing 2,2′-bipyridine-4,4′-dicarboxylic acid and various diester derivatives. The complexes were investigated for their potential anticancer activity, and it was found that their cytotoxicity correlates well with the length of the ester hydrocarbon chain. Complexes **Ir55** and **Ir56,** featuring propyl esters, were found to be highly active against various cancer cell lines, including cisplatin resistant lines. *In vitro* mechanistic investigations revealed that esterases hydrolyze the ester bonds. In turn, this leads to mitochondrial accumulation and induction of a series of cell-death related events, including simultaneous caspase-dependent apoptosis and pro-death autophagy. The study concluded that ester modification is a simple and effective method for increasing anticancer activities of Ir(III) complexes, and opens the potential for synergistic effects via ester linkage to other drugs.

Wang et al.[95] developed an Ir(III) complex (**Ir57**) in the form $[Ir(\mathbf{pq})_2(\mathbf{N}^\wedge\mathbf{N})]^+$ where **N^N** is 2-(2-hydroxyphenyl)imidazo[4,5-f]-1, 10-phenanthroline. Investigations in live cells found that **Ir57** displayed excellent mitochondrial specificity, resistance to loss of MMP, and negligible cytotoxicity at imaging concentrations. **Ir57** displayed excellent photostability, better than commercially available mitochondrial probes, which could potentially allow for tracking mitochondrial morphology over long time periods. The study concluded that **Ir57** is well suited for use as a mitochondrial selective imaging agent.

Following from this work, Wang et al.[96] developed an Ir(III) complex (**Ir58**) in the form $[Ir(\mathbf{ppy})_2(\mathbf{N}^\wedge\mathbf{N})]^+$ where **N^N** is 2-(4-cyanophenyl) imidazo[4,5-f]-1,10-phenanthroline. *In vitro* investigations found that **Ir58** displays dose-dependent inhibition of the proliferation, migration, and invasion of human breast cancer MDA-MB-231 cells by down-regulating AKT/ERK signal pathways. Confocal microscopy revealed that **Ir58** localizes within mitochondria and triggers intrinsic-pathway apoptosis. The study concluded that **Ir58** displays both antineoplastic and antimetastatic properties, suggesting it may find use in therapies for the treatment of breast tumor metastasis.

Xiong et al.[97] synthesized four Ir(III) complexes in the form $[Ir(\mathbf{dfppy})_2(\mathbf{N}^\wedge\mathbf{N})]^+$ where **N^N** is 6-(pyridin-2-yl)-1,3,5-triazine-2, 4-diamine (**Ir59**), 6-(isoquinolin-1-yl)-1,3,5-triazine-2,4-diamine (**Ir60**), 6-(quinolin-2-yl)-1,3,5-triazine-2,4-diamine (**Ir61**), or 6-(isoquinolin-3-yl)-1,3,5-triazine-2,4-diamine (**Ir62**). *In vitro* investigations found that the complexes display a cytotoxic mechanism of action distinctly different from that of cisplatin. **Ir59–62** displayed mitochondrial selectivity and induced intrinsic-pathway apoptosis in a variety of cancer cell lines, including cisplatin-resistant A549R cells. The complexes featured a high degree of selective toxicity toward tumor cells over normal cells and possessed greater cytotoxicity than cisplatin. Interestingly, the mechanism of cellular uptake varied depending on the cell line. Uptake occurred via an energy-independent mechanism for normal L02 human liver cells and an energy-dependent mechanism for cisplatin-resistant A549R human lung adenocarcinoma cells. The study suggested that the greater cytotoxicity observed in cancer cells was a result of the different uptake mechanisms. Though curiously, despite utilizing different uptake mechanisms, the intracellular distribution was comparable in both cell lines. The study offered new insights into the mechanisms behind Ir(III) complex cellular uptake and may assist future exploration into the development of improved chemo-therapeutics for human cancers.

Ye et al.[98] developed three luminescent Ir(III) complexes in the form $[Ir(\mathbf{C^\wedge N})_2(\mathbf{N^\wedge N})]^+$ where **N^N** is **phen** appended with coumarin via an amide bridge and **C^N** is **ppy** (**Ir63**), **dfppy** (**Ir64**), or **tp** (**Ir65**). All three complexes were found to display mitochondrial specificity and greater antiproliferative activity than that of cisplatin in various cancer cell lines including HeLa, human pulmonary carcinoma (A549), cisplatin-resistant A549 (A549R) and human hepatocellular liver carcinoma (HepG2) cell lines. The luminescent properties and *in vitro* cytotoxicity may potentially allow for theranostic applications, where changes in mitochondrial morphology can be simultaneously induced and monitored. Mechanistic investigations revealed the *in vitro* cytotoxicity of complexes **Ir63–65** results from the induction of a cascade of events related to mitochondrial dysfunction, including ROS production, caspase activation and apoptosis. The study concluded that the investigated Ir(III) complexes display great potential for the development of novel anticancer theranostics.

Cao et al.[99] developed the four complexes **Ir66–69. Ir67** and **Ir69** have reactive chloromethyl groups that can bind to free thiols in mitochondria allowing for mitochondrial fixation. Whilst all complexes displayed mitochondrial localization, **Ir67** and **Ir69** displayed cytotoxicity greater than that of their non-fixable analogues in the cancer cells including human cervical carcinoma (HeLa), human lung carcinoma (A549), cisplatin-resistant A549 (A549R), human breast cancer (MDB-MA-231), human prostate carcinoma (PC3) and human normal liver (L02) cells. Notably, **Ir69** showed 11-fold higher selectivity for cancerous A549 cells over noncancerous L02 cells. Mechanistic investigations found that the *in vitro* cytotoxicity of **Ir67** and **Ir69** is a result of the induction of caspase-dependent apoptosis via cellular ATP depletion, mitochondrial respiration inhibition, and ROS elevation. The luminescent properties of **Ir67** and **Ir69** allowed for *in vitro* monitoring of perinuclear clustering of mitochondria in real time. The work provided a strategy for developing potential mitochondria-targeting anticancer theranostics through the rational modification of Ir(III) complexes.

Caporale et al.[46] developed a series of five neutral Ir(III) tetrazolato complexes and their methylated cationic analogues. Substituents and degree of conjugation of π-systems were varied so that the structure-activity relationship for cellular internalization, intracellular localization, cytotoxicity and photophysical properties could be explored. The work showed that methylation of the complexes resulted in a systematic red-shift of emission profiles. The neutral complexes predominantly accumulated in the endoplasmic reticulum and lipid droplets, with a degree of partition somewhat dependent on the chemical identity of the ligands. On the other hand, mitochondrial accumulation was observed for the cationic analogues. The cationic species displayed more potent cytotoxicity that their neutral counterparts in long term incubation experiments. No direct and clear links between lipophilicity and cellular uptake were observed, suggesting diverse uptake mechanisms may be at play. Notably, the most lipophilic cationic complex displayed mitochondrial accumulation in fixed tissue even in the event of depolarization of the MMP.[100] The study concluded that the neutral iridium(III) complexes are promising building blocks for designing novel endoplasmic reticulum- or lipid droplet-targeting luminescent probes for cellular imaging at long timescales.

Chen et al.[101] synthesized two luminescent Ir(III) complexes in the form $[Ir(\mathbf{C^\wedge N})_2(\mathbf{N^\wedge N})]^+$ where **N^N** is a methylated bisimidazole and **C^N** is **ppy** (**Ir70**) or **tp** (**Ir71**). *In vitro* investigations incubating A549 cells with **Ir70** and **Ir71** found that both complexes selectively target mitochondria, inducing mitophagy via the reduction of MMP, lowering cellular ATP levels, perturbing mitochondrial metabolism, and inducing oxidative stress. **Ir70** was demonstrated to be a potential theranostic agent that can simultaneously induce mitophagy while allowing for visualization of the mitochondrial morphology throughout the process. The study concluded that both **Ir70** and **Ir71** have potential for use as theranostic agents that may find application in mitophagy investigations.

Jin et al.[102] developed a series of luminescent Ir(III) complexes in the form $[Ir(\mathbf{C^\wedge N})_2(\mathbf{N^\wedge N})]^+$ where **C^N** is 2-methyl-3-phenylquinoxaline and **N^N** is **bipy** (**Ir72**), **dmbpy** (**Ir73**), **phen** (**Ir74**), **dip** (**Ir75**), 2-(4′-methyl-[2,2′-bipyridin]-4-yl)benzo[d]thiazole (**Ir76**), or **phen-Se** (**Ir77**). Complexes **Ir72–77** were all found to emit in the NIR region with appreciable quantum yields under two-photon excitation. With lifetimes ranging within 210–240 ns, the complexes were determined to be appropriate for two-photon phosphorescence lifetime and time-gated imaging protocols. *In vitro* investigations found that all the complexes localize within the mitochondria of HeLa cells. Complex **Ir76**, which displays the most promising photophysical properties, was successfully utilized in multimodal luminescence imaging of cell cultures, multicellular spheroids, and murine hippocampus tissue. The study will likely inspire the design of future NIR-emitting Ir(III) complexes for multimodal imaging.

Li et al.[103] synthesized four bis-*N*-heterocyclic carbene (NHC) Ir(III) complexes in the form $[Ir(\mathbf{C\hat{}N})_2(\mathbf{C\hat{}C})]^+$, where **C^C** is a methylated NHC and **C^N** is **pq** (**Ir78**) and **btp** (**Ir79**). **Ir80** and **Ir81** are the respective butylated analogues. *In vitro* investigations found that the complexes efficiently enter A549 cells via an energy-dependent pathway and accumulate within mitochondria. In the dark, complexes **Ir78–81** displayed cytotoxic potency greater than that of cisplatin. Upon irradiation with visible light, the cytotoxicity increased by an average of 2 orders of magnitude. The antiproliferative activity and photocytotoxicity were found to be dependent on the identity of the cyclometalated **C^N** ligand. Mechanistic investigations revealed that the *in vitro* cytotoxicity of complexes **Ir78–81** originates from a series of events related to mitochondrial damage including ROS production, MMP depolarization, caspase activation and apoptosis. The study suggested that these complexes could potentially find use as mitochondria-targeting anticancer and PDT agents.

Liu et al.[104] developed three mitochondria-targeting Ir(III) complexes characterized by AIE, **Ir82–84**. *In vitro* cytotoxicity experiments with HeLa human cervical cancer cells and normal L02 human liver cells found low dark cytotoxicity for all complexes. However, upon irradiation, the cytotoxicity dramatically increased. The increase was determined to be primarily a result of singlet oxygen generation. The complexes displayed a greater degree of photocytotoxicity toward HeLa cells, over L02 cells. The study concluded that **Ir82** is the most efficient singlet oxygen generator and displays the greatest lethality, suggesting it may be a promising candidate for future investigations into PDT agents. Liu et al.[105] synthesized a series of mitochondria-targeting Ir(III) complexes incorporating a dichloroacetate (DCA) moiety, a known mitochondrial pyruvate dehydrogenase kinase 1 inhibitor. The complexes were in the form $[Ir(\mathbf{ppy})_2(\mathbf{N\hat{}N})]^+$ where **N^N** is a DCA-appended derivative of imidazophenanthroline. The complexes were observed to accumulate in mitochondria and selectively alter the metabolism of HeLa cells, sensitizing them for two-photon PDT. This selective-sensitization allowed for cancer-specific enhanced two-photon PDT in hypoxic multi-cellular tumor spheroids. The study concluded that the complexes offer promising routes for developing cancer-specific two-photon PDT agents.

Liu et al.[106] synthesized two Ir(III) complexes in the form [Ir(**C^N**)$_2$(**N^N**)]$^+$ where **C^N** is 2-(anthracen-2-yl)pyridine and **N^N** is 2-(1-benzylpyrazol-3-yl)-pyridine (**Ir85**) or its naphthalen-1-yl analogue (**Ir86**). The biological properties of **Ir85** and **Ir86** were investigated in HeLa cells, revealing the complexes display mitochondrial selectivity, excellent photostability, low cytotoxicity, and resistance to changes in MMP and pH. The NIR-emitting complexes were successfully utilized to monitor changes in mitochondrial morphology over extended time periods. The study concluded that **Ir85** and **Ir86** have great potential for use as mitochondrial imaging agents and encourage further investigations into Ir(III) complexes featuring 2-(anthracen-2-yl)pyridine ligands for use as NIR-emitting luminescent probes.

McKenzie et al.[107] synthesized the Ir(III) complexes **Ir87** and **Ir88**. The complexes were found to display high photostability and two-photon absorption cross-sections. *In vitro* investigations found **Ir87** and **Ir88** localize within mitochondria and lysosomes in U2OS osteosarcoma cells, with the former displaying relatively low dark cytotoxicity. Both complexes were observed to be efficient photosensitizers using one-photon irradiation, inducing apoptotic cell death with relatively low dosage of light irradiation. Complex **Ir87** was also found to be an efficient photosensitizer under two-photon excitation, displaying photocytotoxicity with NIR light. The study concluded that **Ir87** shows great potential for use as a two-photon PDT agent.

Ouyang et al.[108] developed a family of luminescent Ir(III) complexes in the form [Ir(**ppy**)$_2$(**N^N**)]$^+$ where **N^N** is a progressively fluorinated derivative of a functionalized imidazophenanthroline. The biological properties of the complexes were investigated in five separate cancer cell lines, where they were found to localize rapidly and preferentially within mitochondria. The complexes displayed antiproliferative activities greater than that of cisplatin. A direct correlation between the degree of fluorination and cytotoxicity was observed, with the most fluorinated complex, **Ir89**, displaying the greatest antiproliferative ability. *In vitro* incubation of the complex in HeLa, HepG2, BEL-7402, A549, and A549R cancer cells resulted in greater cytotoxicity than normal L02 liver cells. Mechanistic investigations suggested that the cytotoxicity of **Ir89** was a result of MMP depolarization, caspase activation, and the induction of intrinsic-pathway apoptosis. The study concluded that **Ir89** is a very promising candidate for further *in vivo* evaluation as a mitochondria-targeting anticancer theranostic agent.

Qiu et al.[109] developed two Ir(III) complexes that display aggregation-induced two-photon emission properties, **Ir90** and **Ir91**. *In vitro* investigations with HeLa cells and multicellular spheroids found the lipophilic complexes aggregate in mitochondria. The aggregates were observed to display photostability and high singlet oxygen production under two-photon excitation. Complexes **Ir90** and **Ir91** were successfully employed as mitochondria-targeting imaging and PDT agents. The study concluded that aggregation-induced emission can effectively overcome issues associated with reduction in singlet oxygen generation and luminescence quenching associated with the accumulation of photosensitizers in targeted organelles.

Shewring et al.[110] developed two mitochondria-targeting Ir(III) complexes, **Ir92** and **Ir93**, that provide excellent image quality and contrast for both optical microscopy and TEM. Advantageously, incubation of the complexes at concentrations around 50 μM was found to be suitable for imaging with both techniques. The complexes are among the first that have found simultaneous compatibility with both super-resolution and TEM imaging. The excited state lifetime of the complexes was appreciably reduced by the presence of oxygen, allowing for mapping of mitochondrial O_2 concentration. The compatibility of the complexes with both electron and luminescence microscopy suggests the potential for these imaging agents to be applied to correlative light-electron microscopy. The study concluded that the organelle-selectivity and high photostability of complexes **Ir92** and **Ir93** makes them excellent starting points for developing future mitochondria-targeting correlative imaging probes.

Venkatesh et al.[111] synthesized two mitochondria-targeting Ir(III) complexes in the form $[Ir(\mathbf{ppy})_2(\mathbf{N^\wedge N})]^+$, where **N^N** is a bipyridine ligand incorporating one (**Ir94**) or two (**Ir95**) 2,2,6,6-tetramethylpiperidine-1-oxyl (TEMPO) groups, appended by amide bridges. Electron paramagnetic resonance showed spin-spin interactions between TEMPO moieties in **Ir95**. **Ir94** and **Ir95** displayed intense emission with relatively long-lived excited states. A monoexponential decay was observed for **Ir94**, whilst a biexponential decay was observed for **Ir95**, indicating there may likely be more than one energetically accessible conformation. *In vitro* investigations revealed that **Ir95** displays greater antiproliferative activity than **Ir94** toward several cancer cell lines. **Ir95** displayed particularly high potency toward PC3 human prostate cancer cells, approximately 8 times greater than that of cisplatin. It also displayed greater cytotoxicity toward various cancer cells relative to normal MRC5 lung fibroblast cells. The study concluded that this class of Ir(III) complexes shows favorable potential for use as mitochondria-targeting anticancer agents.

Ye et al.[112] developed three luminescent Ir(III) complexes in the form $[Ir(\mathbf{C^{\wedge}N})_2(\mathbf{N^{\wedge}N})]^+$ where **N^N** is a bipyridine ligand appended to valproic acid and **C^N** is **ppy** (**Ir96**), **dfppy** (**Ir97**), or **tp** (**Ir98**). Complexes **Ir96–98** were found to display excellent photophysical properties, favorable for two-photon imaging. Free valproic acid is a known histone deacetylase (HDAC) inhibitor. *In vitro* investigations with normal L02 human liver cells and various cancer cell lines found that, upon cellular uptake, esterases quickly hydrolyze the complexes **Ir96–98**, resulting in inhibition of HDAC. **Ir96–98** were observed to display potent *in vitro* toxicity toward cisplatin-resistant A549R adenocarcinoma cells, in a range 50–80 times greater than that of cisplatin. Mechanistic investigations found efficient cellular uptake for complexes **Ir96–98**, whereupon they localized within mitochondria and induced a series of mitochondrial related cell death events. This included ROS production, cell cycle arrest, MMP depolarization, caspase activation and apoptosis. The study concluded that multifunctional anticancer drugs could potentially be achieved by conjugating metal complexes to existing organic drugs via ester linkages.

Yokoi et al.[113] developed a series of cationic cyclometalated Ir(III) complexes appended with oligopeptides. The nonacationic complexes **Ir99** and **Ir100** feature KKGG oligopeptides (K = lysine, G = glycine) at the 4′ and 5′ positions of the **ppy** moiety, respectively, linked by secondary amide groups and a hexamethylene chain. *In vitro* investigations with Jurkat human T lymphocyte cells revealed that complexes **Ir99** and **Ir100** effectively induced cell death through a Ca^{2+}-dependent pathway accompanied by membrane disruption. Co-staining experiments suggested that **Ir99** and **Ir100** bind to Ca^{2+}-calmodulin and are localize within mitochondria during necrosis-like cell death. The study may provide valuable information on mitochondrial related cell death.

Zhang et al.[114] synthesized the luminescent Ir(III) complex **Ir101**. The complex displayed no dark cytotoxicity toward the cell lines screened. However, upon irradiation, **Ir101** displayed potent antiproliferative activity toward SGC-7901 human gastric cancer cells, approximately an order of magnitude greater than that of normal L02 human liver cells. Mechanistic investigations suggested that the photocytotoxicity originates as a result of inducing a cascade of mitochondrial related cell death pathways. These included MMP depolarization, caspase activation, ROS generation and intrinsic-pathway apoptosis. The study concluded that **Ir101** shows potential for the design of new PDT agents based on iridium complexes.

Zhang et al.[115] developed an Ir(III) complex, **Ir102**, in the form $[Ir(\mathbf{ppy})_2(\mathbf{N^\wedge N})]^+$, where **N ^ N** is a bipyridine ligand appended by a ferrocene (Fc) unit via a hydrazine linker. The linker is sensitive to the presence of hypochlorous acid. Fc is an effective emission quencher, however, HOCl reacts with the hydrazine group cleaving the Fc unit from the complex and restoring its luminescence. The complex **Ir102** was therefore investigated as a photo- and electrochemical luminescent sensor for HOCl. Investigations on the cytotoxicity of **Ir102** in HepG2 cells suggested good biocompatibility. One- and two-photon microscopy and lifetime imaging were employed to visualize the distribution of HOCl in biological samples. Furthermore, **Ir102** was employed to visualize the generation of HOCl in the liver of Zebrafish and mouse models, demonstrating *in vivo* applicability. This study opens new avenues for the application of chemosensors in diagnostics and monitoring.

Zheng et al.[116] developed a series of mixed ligand Ir(III) complexes and investigated potential applications as PDT agents. Complex **Ir103** was found to specifically target mitochondria in A549 adenocarcinoma cells. Cytotoxicity and PDT activity were explored in various cancer cell lines and normal L02 human liver cells. The results indicated that **Ir103** displayed moderate dark cytotoxicity and relatively high photocytotoxicity, particularly in A549 cells. The relatively high photocytotoxicity was also evident in cisplatin-resistant A549R cells. Mechanistic investigations suggested that the cytotoxicity results from inducing ROS-dependent and caspase-mediate apoptosis. The study concluded that whilst **Ir103** shows potential for PDT applications, the short excitation wavelength required may limit its use to superficial tumors.

Chen et al.[117] investigated potential applications in super-resolution microscopy of the Ir(III) complex **Ir104**. The complex was demonstrated to track mitochondrial dynamics in live cells using structured illumination microscopy (SIM) to a resolution of approximately 80 nm. The complex displayed excellent mitochondrial specificity, photostability and appreciable cell permeability. **Ir104** was successfully employed to visualize mitochondrial cristae and record dynamic approach and separation of mitochondria. The study concluded that **Ir104** can be used to quantitatively investigate interactions between mitochondria and lysosomes.

Perez-Arnaiz et al.[118] synthesized two Ir(III) complexes with thiabendazole ligands and their Rh(III) analogues to investigate anticancer and PDT agent potentials. The effect on both biological and photophysical properties, resulting from altering ancillary ligands and metal centers, was explored.

The complex **Ir105**, featuring **ppy** and a N-benzyl-thiabendazole ligands, displayed selective localization in mitochondria. All four complexes were found to be an order of magnitude more cytotoxic than cisplatin toward human A549 lung and SW480 colon carcinoma cell lines, without the need for photoactivation. Mechanistic investigations showed that all complexes interfere with mitochondrial functions, resulting in the activation of apoptotic cell death pathways. Both Ir(III) complexes also displayed an increase in cytotoxicity when irradiated with blue light, due to photo-oxidation of sulfur-containing amino acids. This suggests that oxidative stress could be selectively promoted via photocatalysis. Conversely, the Rh(III) complexes did not display enhanced cytotoxicity upon irradiation, as they are unable to induce photo-oxidation. The study concluded that the knowledge of varying PDT behaviors, depending on the identity of the metal center, will likely aid in the design of future metal-based photosensitizers.

Guan et al.[119] developed a series of Ir(III) mitochondria-targeting complexes, **Ir106–109**. These complexes displayed potent cytotoxicity toward various cancer cell lines, including cisplatin resistant lines, whilst displaying a lower cytotoxic potency toward normal HL-7702 human liver cells and HEK 293 human kidney cells. Mechanistic investigations found the complexes induced oncosis in cisplatin resistant A549R adenocarcinoma cells. The study represents the first metallo-components to induce oncosis in drug-resistant cancer cells.

He et al.[120] synthesized four Ir(III) complexes and conducted investigations into their potential anticancer activity. *In vitro* investigations found the complexes displayed efficient uptake by A549 cells wherein they localized specifically within mitochondria. Complexes **Ir110–113** displayed cytotoxicity greater than cisplatin against all cancer cell lines screened, including cisplatin-resistant lines. An approximate fourfold reduction of *in vitro* cytotoxicity was observed when normal LO2 human liver cells were incubated with the same complexes. Mechanistic investigations revealed that **Ir111** induced paraptosis in tandem with excessive mitochondria-derived cytoplasmic vacuoles, affects the ubiquitin–proteasome system and mitogen-activated protein kinase signaling pathways, and induces a series of mitochondria-related cell death events. The predominant cause of cell death was identified as caspase-independent non-apoptotic paraptosis. The study offers valuable insights into the design and mechanisms of action of novel metal-based anticancer agents.

Jing et al.[121] developed two luminescent Ir(III) complexes in the form $[Ir(\mathbf{C}\hat{}\mathbf{N})_2(\mathbf{N}\hat{}\mathbf{N})]^+$ where **N^N** is a **phen** ligand incorporating a

TEMPO group appended by an amide bridge, and **C^N** is **ppy** (**Ir114**) or **dbpz** (**Ir115**). The mitochondria-targeting complexes were investigated as functional theranostic agents, capable of inducing mitochondrial oxidative stress under two-photon excitation and simultaneously monitor morphological changes. Interestingly, the PDT efficacy could be evaluated by monitoring changes in phosphorescence intensity and excited state lifetime throughout the process. The study concluded that Ir(III)-nitroxide conjugates have very high potential for use as PDT agents.

Laws et al.[122] developed a series of four tricationic Ir(III) complexes that were reported to display potent cytotoxicity toward cancer stem cells (CSCs). The complexes were in the form [Ir(**C^N**)$_2$(**N^N**)]$^{3+}$ where **C^N** is 1-methyl-[2,2′-bipyridin]-1-ium, and **N^N** was **bipy** (**Ir116**), **dmbpy** (**Ir117**), **phen** (**Ir118**), or **dip** (**Ir119**). Complex **Ir118** was particularly noteworthy as it displayed equivalent *in vitro* cytotoxic potency toward bulk breast cancer cells and breast CSCs, suggesting it may have the potential to eliminate heterogenous breast cancer populations with a single dose. Mechanistic investigations revealed that the *in vitro* cytotoxicity of **Ir118** results from the inhibition of cytochrome *c* oxidase, promoting MMP depolarization, elevating ROS levels, and triggering caspase-dependent apoptosis. The study was the first to explore anti-CSC properties of iridium complexes. It is suggested that previously identified mitochondria-targeting Ir(III) complexes should be used in anti-CSC investigations as that may potentially provide leads for the development of novel anti-CSC compounds.

Liang et al.[123] developed the Ir(III) complex **Ir120** and investigated its potential anticancer activity. *In vitro* studies indicated that **Ir120** is biologically compatible with all cells screened, with the exception of HepG2 carcinoma cells, where it induced moderate cytotoxicity. The complex displayed mitochondrial accumulation in HepG2 cells and induced apoptosis via the inhibition of the P13K/AKT/mTOR pathway. In addition, **Ir120** was observed to increase ROS levels, decrease MMP and cause autophagy. The study will likely aid in future design of novel Ir(III) anticancer agents.

Qiu et al.[124] developed two luminescent Ir(III) complexes and investigated their potential as mitochondria-targeting pH sensors. The complexes were in the form [Ir(**C^N**)$_2$(**N^N**)]$^+$, where **N^N** is 5,5′-bis (morpholinomethyl)-2,2′-bipyridine and **C^N** is **pbt** (**Ir121**) or **pq** (**Ir122**). Emission intensities for both **Ir121** and **Ir122** were found to be linearly correlated with pH values, encouraging investigations into the potential use of the complexes as pH probes. *In vitro* studies found **Ir121** and **Ir122** display mitochondrial selectivity in HeLa cells, low cytotoxicity, and effectively act as two-photon pH sensors. The mitochondrial selectivity is notable given that the morpholine moiety is commonly exploited to provide selectivity for the lysosomes. The two-photon absorption properties of the complexes allowed for the illumination of mitochondria deep inside multicellular spheroids, and consequently enabled monitoring of pH fluctuations during induced apoptosis. The results showed that both **Ir121** and **Ir122** hold great promise for studying mitochondrial pH.

Schibilla et al.[125] synthesized a range of luminescent Ir(III) complexes in the form [Ir(**C^N**)$_2$(**N^N**)]$^+$ where **C^N** is **ppy** or **dfppy,** and **N^N** is a **bipy** ligand appended by one or two adamantyl groups. This family of complexes was designed to bind to β-cyclodextrin vesicles (CDV) with high affinity. The Ir(III)-CDV adduct results in enhanced luminescence intensity and a blue-shift of the emission profile, attributed to the less polar and more viscous environment of the CDV. *In vitro* investigations found the Ir(III)-CDV adducts are readily internalized into HUVEC cells, where they are transported to mitochondria via the endocytic pathway. In contrast, Ir(III) complexes free from CDV showed inefficient cellular uptake, demonstrating the delivery vehicle capability of CDV. The complexes proved to be effective imaging agents for use in live cells. The study concluded that CDVs decorated with luminescent Ir(III) complexes are promising candidates for bioimaging agents, and that CDVs offer a potential platform for developing future diagnostic and theranostic agents.

Sorvina et al.[126] investigated the imaging potential of an Ir(III) complex, **Ir123**. The complex, in the form [Ir(**ppy**)$_2$(**N^N**)]$^+$ where **N^N** is 5-(5-(4-cyanophen-1-yl)pyrid2-yl)tetrazolate, was found to be compatible with selectively imaging mitochondria in both live and fixed tissue samples. Compatibility with fixed samples, where the MMP is no longer sustained, suggested that its affinity for mitochondria results from more than just the cationic charge. It was suggested that the proposed staining protocol could effectively overcome limitations in existing techniques for fixed tissue staining that rely on immunohistochemistry. The study represents the first example of a mitochondrial targeting luminescent probe that is compatible with both live and fixed tissue samples and offers a potential starting point for the development of alternative imaging agents.

Sreedharan et al.[127] developed two Ir(III) complexes and investigated their potential as cellular imaging agents. The complexes, in the form [Ir(**C^N**)$_2$(**N^N**)]$^+$, where **N^N** is dipyrido[3,2-a:2′,3′-c]phenazine (DPPZ) and **C^N** is a benzyl (**Ir124**) or propyl (**Ir125**) functionalized phenyl 1,2,3-triazole ligand, were found to localize within mitochondria in live cancer cell lines. Investigations into the DNA binding properties of the complexes indicated that the DPPZ ligand exhibits appreciable binding affinity. Preliminary investigations indicated that the cytotoxicity appears to be variable, depending on the cell type. **Ir124** and **Ir125** displayed switch-on luminescence upon binding with DNA, potentially allowing for selective DNA imaging. The study concluded that, given the wide range of triazole-based ligands available, **Ir124** and **Ir125** may potentially open new avenues for developing novel Ir(III)-based DNA-specific luminescent probes and phototherapeutics.

Tang et al.[128] developed a mitochondria-targeting Ir(III) complex, **Ir126**, and investigated its potential for use as an anticancer agent. The complex was found to display potent cytotoxicity and cell growth inhibition toward a range of cancer cells, and low *in vitro* cytotoxicity toward normal NIH/3T3 murine fibroblast cells. Mechanistic investigations indicated **Ir126** raises the concentration of ROS, decreases MMP, enhances intracellular Ca^{2+} concentration, releases cytochrome *c*, and activates apoptosis and autophagy. The study concluded that **Ir126** has potential applications as an anticancer therapeutic for B16 cells.

Wang et al.[129] synthesized two Ir(III) complexes, **Ir127** and **Ir128**, and investigated their potential as luminescent probes for imaging cysteine (Cys) and homocysteine (HCys). Complex **Ir128** was determined to have greater potential than **Ir127** for selectively imaging Cys and Hcys, as it displayed a turn-on luminescent response to both species. *In vitro* investigations demonstrated that **Ir128** can selectively accumulate in the mitochondria of HepG2 cells and light up upon interaction with Cys and Hcys. The NIR emission of **Ir128** allowed for two-photon imaging of mitochondria deep in multicellular spheroids. **Ir128** was then successfully applied to imaging Cys/Hcys in living tissue and animal models. The study opens new opportunities for developing novel two-photon Ir(III) luminescent probes for detecting mitochondrial Cys/Hcys *in vivo*.

Wang et al.[130] developed a series of Ir(III) complexes and investigated their potential anticancer properties. The mitochondria-targeting complex **Ir129**, in the form $[Ir(PPh_3)_2(\mathbf{C^{\wedge}O})(\mathbf{L})H]^+$, where **C^O** is α-phenyl-N-tert-butylnitrone (PBN) and **L** is 3-(azidomethyl)pyridine, displayed the most potent antiproliferative activity toward A2780 human ovarian cancer cells. It was suggested that the *in vitro* cytotoxicity of **Ir129** may be due to the complex causing an imbalance of intracellular redox and catalytic reaction processes. The study concluded that complexes with PBN ligands and azide functional groups may offer a general strategy to probe the mechanisms of action of metal-based anticancer drugs.

Xiao et al.[131] developed a luminescent Ir(III) complex, **Ir130**, and investigated its potential as a mitochondria-targeting anticancer agent. **Ir130** was observed to display more potent cytotoxicity than cisplatin against a range of cancer cell lines, including cisplatin-resistant cells, during *in vitro* investigations. The mitochondrial selectivity of the complex in HeLa cells allowed for imaging and tracking morphological changes to mitochondria in real-time, demonstrating its theranostic potential. Mechanistic investigations indicated that **Ir130** elevates ROS concentrations, activates caspases, induces changes in MMP, and triggers intrinsic-pathway apoptosis. The study concluded that **Ir130** holds potential for use as a mitochondria-targeting theranostic anticancer agent.

Yi et al.[132] investigated the antiproliferative capacity of the Ir(III) complex **Ir131**. The complex localized within mitochondria and was found to display superior cytotoxic potency than cisplatin against A549 cells, as well as potent cytotoxicity toward several other cancer cells. *In vivo* investigations showed that **Ir131** inhibits tumor growth by over 63% relative to the control. Mechanistic investigations revealed that **Ir131** increase ROS concentrations, decreases MMP, inhibits cell invasion and promotes apoptosis and autophagy. The results demonstrated that **Ir131** induces apoptosis in A549 cells via mitochondrial dysfunction and PI3K/AKT/mTOR signaling pathways. The study concluded that **Ir131** displays potential for use as a potent anticancer therapeutic.

Zhang et al.[133] developed the mitochondria-targeting Ir(III) complex **Ir132**. The phenathroline ligand is functionalized with a tetradentate active site that shows specific Zn^{2+}-induced luminescence enhancement and elongation of excited state lifetime decay. The lifetime was found to be unaffected by environmental factors such as lipophilicity, viscosity, polarity and ambient oxygen. **Ir132** was demonstrated as a reversible tracker of exogenous labile Zn^{2+} using phosphorescence and phosphorescence lifetime imaging. The study concluded that **Ir132** shows great promise for use in luminescent imaging of mitochondrial labile Zn^{2+}.

Zhang et al.[134] synthesized a series of four Ir(III) complexes in the form $[Ir(\mathbf{ppy})(\mathbf{N^\wedge N^\wedge N})Cl]^+$ where **N^N^N** represents progressively fluorinated derivatives of 4′-phenyl-2,2′:6′,2″-terpyridine. All four complexes were found to exhibit a high degree of specificity for mitochondrial localization in both live and fixed cells, with low cytotoxicity and resistance to photobleaching. The degree of fluorination could be used to tune the photophysical properties of the complexes, with more fluorinated complexes displaying emission at longer wavelengths and shorter excited state lifetimes. The complexes were successfully used to monitor mitochondrial morphology during apoptosis. The study concluded that this family of mitochondria-targeting complexes are suitable for luminescence imaging techniques.

Zhang et al.[135] developed three mitochondria-targeting Ir(III) complexes, **Ir133–135**, and investigated their potential anticancer activities. *In vitro* studies revealed that the complexes display no dark cytotoxicity toward any of the various cancer cell lines screened. However, upon irradiation, all three complexes displayed potent cytotoxicity toward BEL-7402 hepatoma cells. Mechanistic investigations showed that **Ir133–135** increases intracellular ROS and Ca^{2+} concentrations, reduces MMP, activates caspases, induces autophagy and apoptosis, releases cytochrome *c*, inhibits polymerization of α-tubulin, and regulates the expression of Beclin-1 and LC3 proteins. Inhibition of BEL-7402 cell growth was determined to result from ROS-mediated mitochondrial dysfunction and targeting tubules pathways. The study illustrated new insights into the mechanisms of action and design of Ir(III) anticancer complexes.

Bevernaegie et al.[136] synthesized two Ir(III) complexes capable of photo-oxidizing biomolecules under oxygen-free conditions. The complexes were in the form $[Ir(\mathbf{C^\wedge N})_2(\mathbf{N^\wedge N})]^+$ where **C^N** is 2-(3,5-bis(trifluoromethyl)phenyl)pyridine and **N^N** is 2-(pyridin-2-yl)pyrazine (**Ir136**) or pyrazino[2,3-f]quinoxaline (**Ir137**). *In vitro* investigations revealed the complexes rapidly enter cells and localize in mitochondria, with the capacity to penetrate deep into multicellular spheroids. Upon irradiation, the complexes induced rapid apoptotic cell death. It was determined that complex **Ir137** displays the greater potential as an anticancer photosensitizer, since under hypoxic conditions it is able to yield a modest production of singlet oxygen and initiate type 1 oxygen-independent processes. **Ir137** was observed to achieve complete destruction of multicellular spheroids at concentrations as low as 10 μM, with two irradiation doses within a 24-h interval. The study concluded that both **Ir136** and **Ir137** warrant further *in vivo* investigation.

Cao et al.[137] synthesized a family of Ir(III) complexes and explored their potential as anticancer agents. *In vitro* investigations revealed the complexes localize within mitochondria of FaDu cells and display cytotoxicity more potent than cisplatin in a range of cancer cells screened. The complexes **Ir138** and **Ir139** were determined to intercalate within mitochondrial

DNA *in situ* and induce mitochondrial DNA damage. Mechanistic investigations revealed that **Ir138** and **Ir139** decrease MMP, negate ATP synthesis, and disrupt mitochondrial energetic and metabolic processes, resulting in mitophagy, G_0 and G_1 phase cell cycle arrest, and apoptosis. *In vivo* investigations revealed that **Ir139**, featuring **tp** cyclometalated ligands, can effectively inhibit tumor xenograft growth in murine models. The study concluded that targeting the mitochondrial genome may prove to be an effective strategy for overcoming platinum-drug resistance.

Chen et al.[138] synthesized a small family of Ir(III) complexes with the aim of developing luminescent probes for monitoring mitochondrial pH fluctuations. The pK_a values for complexes **Ir140** and **Ir141** were found to be ideal (7.49 and 7.41, respectively). They displayed an enhanced luminescence intensity and a shift in emission in response to pH fluctuations in a range of 6.0 to 8.0, allowing for monitoring of pH in live cells. *In vitro* investigations found **Ir140–142** localize within mitochondria of HepG2 cells. Upon irradiation, the complexes generated singlet oxygen and displayed potent photocytotoxicity. The study concluded that the pH-sensitive phosphorescence of the complexes could be employed to monitor mitochondrial pH and induce mitochondrial dysfunction via PDT.

Du et al.[139] developed a mitochondria-targeting Ir(III) complex, **Ir143**, in the form $[Ir(\mathbf{ppy})_2(\mathbf{N^\wedge N})]^+$, where **N^N** is (E)-4-methyl-4′-(2-nitrovinyl)-2,2′-bipyridine. The nitroolefin moiety on the **N^N** ligand acts as effective quencher for the emission of the complex. However, upon selective reaction with Cys, the luminescence is restored in a switch-on manner. *In vitro* investigations revealed that **Ir143** displays high selectivity toward intracellular cysteine and good biocompatibility with MCF-7 cells. The long-lived excited state of **Ir143** allowed for time-gated luminescence analysis of Cys in cells and human sera. Furthermore, *in vivo* experiments demonstrated that **Ir143** can effectively image Cys distribution in *Daphnia magna*, zebrafish, and mouse models. The study concluded that the development of **Ir143** provides a new approach for detecting mitochondrial Cys *in vivo*. The new probe is expected to contribute to future research related to inflammation-based diseases.

Guo et al.[140] developed four Ir(III) complexes, in the form $[Ir(\mathbf{C^\wedge N})_2(\mathbf{bipy})]^+$, where **C^N** is a cyclometalated ligand incorporating an anthraquinone moiety, and investigated their potential use as PDT agents. *In vitro* investigations revealed that complex **Ir144**, featuring 2-(quinolin-2-yl)anthracene-9,10-dione cyclometalated ligands, shows the greatest PDT potential. It was found to localize within mitochondria in various human cancer cell lines. **Ir144** displayed a high singlet oxygen yield, low dark cytotoxicity, and displays excellent therapeutic effects under both normoxic and hypoxic conditions. The study concluded that **Ir144** shows great potential for PDT applications and results garnered in this investigation might aid future explorations into developing novel PDT agents.

Hao et al.[141] synthesized six Ir(III) complexes, in the form $[Ir(\mathbf{C^\wedge N})_2(\mathbf{N^\wedge N})]^+$, where **C^N** is **ppy** or 4-(pyridin-2-yl)benzaldehyde, and **N^N** is bis(diphenylphosphino)methane, 1,2-bis(diphenylphosphino) ethane, or (Z)-1,2-bis(diphenylphosphino)ethene. The mitochondria-targeting complexes contain rotatable groups that result in environment-sensitive emission intensity and excited state lifetime, allowing for use as quantitative probes for viscosity. The viscosity-dependent photophysical properties are attributed to conformational changes in the phosphine ligands, as evidenced by ^{1}H-NMR spectra. *In vitro* investigations revealed complex **Ir145** displays mitochondrial accumulation in A549 cells, potent cytotoxic activity and induces caspase-dependent apoptotic cell death. It was demonstrated that **Ir145** can induce mitochondrial dysfunction and at the same time quantitatively monitor mitochondrial viscosity in A549 cells via two-photon phosphorescence lifetime imaging (TP-PLIM). Upon incubation with **Ir145**, mitochondria displayed a time-dependent increase in viscosity, heterogeneity and loss of MMP. The study demonstrated the potential of multifunctional luminescent anticancer organometallic complexes to induce and precisely monitor microenvironmental changes at subcellular levels using TP-PLIM.

He et al.[142] developed a mitochondria-targeting photoacid generator, **Ir146**, featuring a sulfonium cation, and investigated its anticancer potential. Photoactivation of **Ir146** results in photoacid generation, disturbing the pH balance in mitochondria. Photolytic by-products were also demonstrated to be effective singlet oxygen generators. Hence, treatment with **Ir146** results in dual mode oxygen-dependent and oxygen-independent photodynamic mitochondrial damage, effectively killing cancer cells in both normoxic and hypoxic conditions. The study concluded that **Ir146** offers a promising strategy for developing novel dual-mode anticancer PDT agents for the treatment of hypoxic tumors.

Li et al.[143] synthesized a family of five Ir(III) complexes in the form $[Ir(\mathbf{ppy})_2(\mathbf{N^\wedge N})]^+$, where **N^N** is an anthraquinone-appended imidazophenathroline ligand, and investigated their potential as anticancer agents. The anthraquinone moiety was incorporated to allow for PDT treatment in hypoxic cancer cells. *In vitro* investigations revealed the complexes efficiently respond to hypoxia with a turn-on yellow emission, successfully allowing for the detection of hypoxia in multicellular tumor spheroids. The complexes were found to display potent photocytotoxicity toward a range of cancer cells, including cisplatin-resistant cell lines. Mechanistic investigations revealed the complexes predominantly localize within mitochondria, where they induce mitochondrial dysfunction and caspase-dependent apoptotic cell death. The study concluded that this family of Ir(III) complexes show great promise for use as theranostic anticancer agents for studying hypoxic cancers.

Li et al.[144] developed **Ir147** and **Ir148**, featuring appended PEG chains to the **bipy** ligand. The complexes are differentiated by the presence of a disulfide functional group in **Ir147**. Both complexes were found to localize within mitochondria in HeLa cells. The complex **Ir147** was found to display greater cytotoxicity than its disulfide-free counterpart. This was determined to be a result of intracellular reactions with selected sulfhydryl-containing compounds including H_2S, Cys and reduced glutathione. The intracellular reaction detaches the PEG pendant and subsequently releases the Ir(III) complex. The study concluded that disulfide-containing transition metal complexes may provide more sophisticated systems for sensing and therapeutic applications.

Ma et al.[145] developed a series of five Ir(III) and Ru(II) half-sandwich complexes and investigated their anticancer potential. Complex **Ir149** was found to localize within mitochondria in A549 cells. **Ir149** displayed greater cytotoxicity than cisplatin, inducing a variety of cell death mechanisms including apoptosis and autophagy, and displayed good antimetastatic properties. Complex **Ir149** was found to cause cell arrest in the S and G_2/M phases, induce both early and late cell decay, decrease MMP, and strongly increase intracellular ROS concentrations. In addition, it was observed that complex **Ir149** does not have a significant effect on the NADH binding reaction and displays moderate binding affinity to bovine serum albumin. The study concluded that the naphthalimide modified half-sandwich Ir(III) and Ru(II) complexes have strong potential for use as antitumor agents, and may find use in future research into anticancer treatments.

Song et al.[146] developed three Ir(III) complexes, **Ir150–152**, and investigated their potential for use as cellular imaging and PDT agents. Complexes **Ir150–152** were found to localize within mitochondria and display potent photocytotoxic effects in a range of cancer cells screened. **Ir150** showed negligible dark cytotoxicity in HepG2 cells, then, upon irradiation for 10 min, the cytotoxicity levels increased over 380-fold.

Further mechanistic investigations indicated that **Ir150** could induce apoptosis via ROS-mediated mitochondrial signaling pathways with or without irradiation. *In vivo* investigations demonstrated that **Ir150** could effectively inhibit HepG2 tumor growth in xenograft bearing mouse models under light irradiation. The study concluded that **Ir150** shows great potential for use as an anticancer PDT agent and that the overall anticancer drug design methodology shows promise for developing future novel Ir(III)-based PDT agents.

Wu et al.[147] developed two Ir(III) complexes in the form [Ir(**ppy**)$_2$(**N^N**)]$^+$ where **N^N** is bipyridine appended to two salicylic acid (**Ir153**) or aspirin (**Ir154**) moieties. The complexes can be readily hydrolyzed to form **Ir155**. *In vitro* investigations revealed that the complexes localize within mitochondria in PC3 cells and exhibit greater antitumor and anti-inflammatory properties than an equivalent mixture of **Ir155** and aspirin. The addition of the aspirin and salicylic acid moieties was also observed to improve cellular uptake, relative to that of **Ir155**. The luminescent properties of **Ir153** and **Ir154** allowed for tracking morphological changes in mitochondria, and phosphorescence lifetime imaging revealed that both complexes can aggregate during mitochondrial dysfunction. Predictably, immunomodulatory activity was observed by regulating the activity of immune factors and inhibiting COX-2. Mechanistic investigations showed that **Ir153** and **Ir154** induce caspase-dependent apoptosis, caspase-independent paraptosis and inhibit metastasis. *In vivo* investigations found that both complexes effectively inhibit xenograft tumor growth in mouse models. The study concluded that the combination of mitochondria-targeting and immunomodulatory activities is an effective strategy for developing multifunctional metal-based anticancer agents.

Yi et al.[148] developed a series of amphiphilic gemini Ir(III) complexes in the form [Ir(**C^N**)$_2$(**N^N**)]$^{3+}$ and investigated their potential as theranostic imaging and PDT agents. Gemini refers to compounds containing two hydrophilic heads. In the context of this study, it refers to the presence of two quaternary ammonium groups appended to each **C^N** cyclometalated ligands or both appended to the **N^N** ancillary ligand. Spectroscopic analysis indicated that the introduction of ammonium groups to **C^N** ligands mitigates aggregation-induced emission quenching. One such complex **Ir156**, featuring a C_8-hydrocarbon chain, displayed efficient cellular penetration, mitochondrial accumulation, outstanding photostability and impressive photocytotoxicity. **Ir156** was determined to show satisfactory performance in mitochondrial imaging and PDT of tumor cells. *In vivo* investigations with **Ir156** demonstrated excellent antitumor activity and inhibition of HepG2 cell growth with PDT treatment. The study concluded that this strategy for developing multifunctional Ir(III) complexes for theranostic imaging and PDT applications holds great promise.

Bi et al.[149] developed four Ir(III) complexes, in the form $[Ir(\mathbf{ppy})_2(\mathbf{N\hat{}N})]^+$ where **N^N** are derivatives of (E)-2-(4-styrylphenyl)-imidazo [4,5-f][1,10]phenanthroline, and investigated their potential as two-photon imaging agents and NIR PDT agents. Encouraged by the high singlet oxygen yield and impressive two photon absorption, *in vitro* experiments were conducted, revealing mitochondrial accumulation in A549 cells with low dark cytotoxicity. Upon irradiation, the complexes were observed to induce apoptosis with high phototherapeutic indexes. The study concluded that the complexes display great potential for use as both mitochondrial specific luminescent probes and two-photon PDT agents.

Chen et al.[150] synthesized two Ir(III) complexes, **Ir157** and **Ir158**, and investigated their potential anticancer activities. *In vitro* investigations revealed **Ir157** and **Ir158** display potent antineoplastic activity and greater *in vitro* cytotoxic potency toward HeLa carcinoma cells than toward normal HL-7702 human liver cells. Mechanistic investigations showed that **Ir157** and **Ir158** decrease MMP, release cytochrome *c*, alter mitochondrial related proteins, and trigger caspase-dependent apoptosis. The study concluded that **Ir157** and **Ir158** show promising anticancer activity and the results garnered from the investigation may aid the future design for Ir(III)-based anticancer compounds.

Gu et al.[151] developed three Ir(III) complexes, **Ir159–161**, and investigated their potential anticancer activity in a range of cancer cells. The complexes were found to display potent cytotoxicity toward A549 cells. Mechanistic investigations showed that complexes **Ir159–161** lower MMP, increase cellular ROS concentrations and induce caspase-dependent apoptosis. *In vivo* investigations of **Ir161** revealed moderate tumor growth inhibition of A549 cells in mouse models with minimal adverse effects. The study concluded that **Ir161** shows potential for use as an anticancer therapeutic and may provide new insights into anticancer mechanisms and Ir(III)-based anticancer complex design.

Kuang et al.[152] developed a mitochondria-targeting Ir(III) complex (**Ir162**) in the form $[Ir(\mathbf{ppy})(\mathbf{N\hat{}N\hat{}N})Cl]^+$, where **N^N^N** is 2-([2,2′:6′,2″-terpyridin]-4′-yl)anthracene-9,10-dione, for two-photon PDT under hypoxic conditions. **Ir162** was observed to display hypoxia-specific reduction by cellular reductases, after which its emissive properties are effectively switched on. Upon irradiation, **Ir162** generates carbon radicals, increases ROS concentration, and decreases MMP. The complex was demonstrated for use as a two-photon PDT agent in multicellular spheroids and in mouse models bearing A549 carcinoma xenografts. The study concluded that **Ir162** shows great potential for theranostic imaging and PDT applications. **Ir162** represents one of the few examples of metal-based theranostic agents to generate carbon radicals for oxygen-independent two-photon PDT.

Li et al.[153] synthesized a series of five Ir(III) complexes, in the form $[Ir(\mathbf{C^\wedge N})_2(\mathbf{N^\wedge N})]^+$, where **N^N** is 2-(1-(4-bromobutyl)-4,5-dimethyl-imidazol-2-yl)pyridine appended with a ditert-butylcarbazole group and **C^N** represents various esterified derivatives of **ppy**. Of the five complexes, **Ir163**, **Ir164**, and **Ir165** were found to localize within mitochondria. These complexes displayed greater photostability than common commercially available mitochondrial trackers. All three complexes displayed significant AIE, attributed to restricted intramolecular motion upon aggregation. Studies of their cytotoxicity revealed negligible effects at concentrations appropriate for imaging. The study concluded that **Ir163–165** hold excellent potential for use as mitochondria-targeting imaging agents and highlighted new insights into the structure-activity relationship of Ir(III)-based complexes.

Li et al.[154] developed two Ir(III)-NHC complexes in the form $[Ir(\mathbf{dbpz})_2(\mathbf{C^\wedge C})]^+$ where **C^C** represents methylated (**Ir166**) and butylated (**Ir167**) NHCs, and explored their potential anticancer activity. *In vitro* investigations indicated that complexes **Ir166** and **Ir167** localize within mitochondria in A549 cells. The complexes displayed greater cytotoxicity than cisplatin, which could be further enhanced via irradiation. Mechanistic investigations indicated that the complexes increase cellular ROS concentrations, reduce MMP, and induce a degree of early apoptosis. Further analysis revealed that **Ir166** can induce mitophagy in the dark and necrocytosis upon irradiation. *In vivo* investigations revealed that **Ir166** can effectively inhibit tumor growth in A549 xenografts in mouse models. The study concluded that **Ir166** has great potential for use as an anticancer PDT agent. The study provides insights into the design of Ir(III) complexes for use as mitochondrial targeted anticancer PDT agents.

Liu et al.[155] developed two Ir(III) complexes, **Ir168** and **Ir169**, and explored their potential for sensing and imaging singlet oxygen. The complexes display only low intensity emission. However, upon selectively reacting with singlet oxygen, they form intensely emitting endoperoxide derivatives. *In vitro* investigations showed that both **Ir168** and **Ir169** display low cytotoxicity toward HeLa, A549 and 3T3 cells. At concentrations lower than 20 μM, **Ir168** was reported to be slightly less cytotoxic than **Ir169**. Interestingly, the complexes displayed similar levels of cytotoxicity upon irradiation, despite analogous complexes displaying appreciable photo-cytotoxic properties. This could be potentially due to the fact that the complexes themselves are able to trap singlet oxygen. Co-localization studies revealed the complexes were localized within mitochondria. Their reactivity toward singlet oxygen limited photocytotoxic properties, as cellular oxidative damage was reduced. The study concluded that both **Ir168** and **Ir169** have potential application as mitochondria-targeting luminescent imaging agents.

Peng et al.[156] synthesized two Ir(III) complexes, in the form $[Ir(\mathbf{C}^\wedge\mathbf{N})_2(\mathbf{N}^\wedge\mathbf{N})]^+$, where **N^N** is [2,2′-bipyridine]-4,4′-diyldimethanol and **C^N** is **ppy** (**Ir155**) or **tp** (**Ir170**), and investigated their potential antitumor properties against glioma stem cells (GSCs). *In vitro* investigations revealed that **Ir155** and **Ir170** efficiently penetrate GSCs, localize within mitochondria, and selectively kill GSCs with low IC_{50} values. Mechanistic investigations showed that the complexes elevate ROS levels, reduce MMP, activate caspases, induce mitochondrial mediated apoptosis and effectively suppress proliferation. In addition, **Ir155** and **Ir170** were observed to inhibit GSC colony formation. The study concluded that **Ir155** and **Ir170** have great potential for GSC treatment applications.

Qin et al.[157] developed four Ir(III) complexes, in the form $[Ir(\mathbf{C}^\wedge\mathbf{N})_2(\mathbf{L})]^+$, where **L** is 5-fluoroquinolin-8-ol. Complexes **Ir171** and **Ir172**, featuring **bzq** and 1-phenylpyrazole cyclometalated ligands, respectively, were observed localize within mitochondria. **Ir171** and **Ir172** displayed the highest degree of cytotoxicity in HeLa cells, with IC_{50} values in the nanomolar range. Mechanistic investigations revealed the complexes promote MMP depolarization and trigger caspase-dependent apoptosis. The complexes both displayed antineoplastic activity in HeLa and SK-OV-3/DPP cells. The study represents the first investigation to explore the anticancer activity of Ir(III)-5-fluoro-8-quinolinol complexes and highlights the expanding potential of mitochondria-targeting anticancer agents.

Qin et al.[158] developed two mitochondria-targeting Ir(III) complexes, **Ir173** and **Ir174**, and explored their potential anticancer properties. Both complexes displayed potent cytotoxicity toward the range of cancer cells screened. Irradiation was shown to greatly enhance cytotoxicity, particularly in A549 cells, with photocytotoxicity index of 120 and 93, respectively. Mechanistic investigations of complex **Ir173** revealed that the complex increases ROS concentrations, reduces ATP production, damages mitochondrial DNA, increases lipid peroxidation levels, and inhibits proteasomal activity. **Ir173** also displayed theranostic potential, as it was successfully utilized to monitor morphological changes in mitochondria with two-photon microscopy during its own therapeutic effect. The study concluded that **Ir173** and **Ir174** display potential for use in anticancer theranostic applications. Results garnered from the investigation offer insights into the design of novel Ir(III) chemotherapeutic and PDT agents.

Wang et al.[159] developed the mitochondria-targeting Ir(III) complex **Ir175**, where the **ppy** ligand is functionalized by a nitrone group, and investigated its potential anticancer activity. *In vitro* studies found that complex **Ir175** display potent cytotoxicity toward a range of cancer cell lines, in particular A549 cells. Mechanistic investigations suggested that the predominant mechanism of cell death was HMOX1-mediated ferroptosis. Increased expressions of proteins HMOX1, LC3, ferritin FTL, and NCOA4 were also observed. The study represents the first observation of an Ir(III)-anticancer complex inducing ferroptosis. The results indicate that targeting ferroptosis pathways may be a promising strategy for developing novel anticancer metallodrugs.

Deng et al.[160] synthesized the complex **Ir176** and investigated its potential as a luminescent detector of intracellular Cu^{2+} and Cys. **Ir176** displayed reversible on-off sequential photoluminescence in response to Cu^{2+} and Cys. The free bidentate pyrazolyl-pyridine ligand on **Ir176** binds selectively to Cu^{2+} forming a non-luminescent **Ir176**-Cu(II) assembly. The Cu(II) moiety can then be efficiently released via a selective reaction with Cys, restoring the luminescence of **Ir176**. The on-off process could be accomplished at very fast time frames (within 30 s) and repeated a minimum of five times without loss of luminescence intensity. *In vitro* investigations revealed **Ir176** targets mitochondria in HeLa cells, display good permeability and low cytotoxicity. *In vivo* explorations found that complex **Ir176** can be used to effectively image endogenous Cys in zebrafish. The study presents a novel approach to imaging and detecting Cu^{2+} and Cys *in vitro* and *in vivo*.

Guan et al.[161] developed the complexes **Ir177** and **Ir178**, and investigated their potential anticancer properties. *In vitro* investigations revealed the complexes localize within mitochondria and display potent cytotoxicity in a range of cancer cells, including cisplatin resistant A549R cells. Mechanistic investigations revealed that the cytotoxicity of **Ir177** and **Ir178** was a result of oxidative stress, loss of MMP, and induction of necroptosis. Necroptosis was confirmed by the activation of receptor-interacting serine-threonine kinase 3 (RIPK3) and mixed lineage kinase domain-like pseudokinase (MLKL). **Ir177** and **Ir178** were also observed to interfere with the progression of the cell cycle, causing cell cycle arrest in the G_0 and G_1 phase. The study concluded that **Ir177** and **Ir178** are promising candidates for cisplatin-resistant anticancer therapeutics.

Hao et al.[162] developed three mitochondria-targeting Ir(III) complexes bound to a chlorodipyridophenazine ligand and **ppy** (**Ir179**), **bzq** (**Ir180**), or **iso-pq** (**Ir181**), and investigated their potential anticancer activities. *In vitro* investigations found that the complexes displayed cytotoxicity toward a range of cancer cells screened, with complex **Ir180** displaying the highest potency, particularly for SGC-7901 cells. Mechanistic investigations revealed that **Ir179–181** elevate ROS concentrations, lower MMP, release cytochrome *c*, and induce intrinsic apoptosis. In addition, the complexes were observed to arrest the cell cycle at the G_0/G_1 phases, suppress cell colony formation and migration, and influence the expression of proteins associated with apoptosis. The study concluded that complexes **Ir179–181** are promising candidates for further investigations into novel Ir(III) anticancer therapeutics.

Wu et al.[163] developed the Ir(III) complex **Ir182** and investigated its potential use as a luminescent probe for detecting and imaging intracellular peroxynitrite ($ONOO^-$)/glutathione (GSH) redox cycles. The complex displayed NIR emission, relatively high quantum yield, and excellent selectivity toward $ONOO^-$/GSH in aqueous solutions. *In vitro* investigations revealed that complex **Ir182** displays excellent permeability, low cytotoxicity, and mitochondrial localization in HepG2 cells. **Ir182** was demonstrated to effectively visualize and monitor induced changes in intracellular $ONOO^-$/GSH levels both *in vitro* and *in vivo* in live mouse models. The study concluded that **Ir182** has excellent potential as a luminescent probe for detecting endogenous $ONOO^-$/GSH redox cycles, as well as for evaluating hepatotoxicity induced by drugs.

Wu et al.[164] developed the Ir(III) complex **Ir183**, featuring a **bipy** ligand appended with a xanthene derivative, and investigated its potential for use as a PDT agent. The complex was observed to display NIR-emission and a high degree of photostability. *In vitro* investigations compared **Ir183** with its xanthene-free analogue, and it was found that **Ir183** displays improved singlet oxygen generation and an overall greater phototherapeutic effect via the induction of apoptosis. **Ir183** featured negligible dark cytotoxicity and localized within mitochondria in MCF-7 cells. *In vivo* investigations showed that **Ir183** could effectively act as a PDT agent in MCF-7 tumor-bearing mouse models. The study offers a novel method for improving PDT efficiency of photosensitizers for anticancer therapies.

Yuan et al.[165] synthesized **Ir184**, featuring a functionalized imidazophenathroline ligand appended to a TPP group. The complex was found to localize within mitochondria. *In vitro* investigations found that the complex exhibit type 1 PDT processes, resulting in the induction of ferroptosis in hypoxic tumor cells. Type 1 PDT processes involve the production of oxygen and hydroxide radicals, which are likely responsible for the induction of ferroptosis. Ferroptosis was characterized by lipid peroxide accumulation, glutathione peroxidase 4 downregulation, ferrostatin-1 inhibited cell death, and mitochondrial shrinkage. **Ir184** was also observed to induce loss of MMP, suppress ATP production, and induce apoptosis. The dual action of inducing ferroptosis and apoptosis provides a synergistic effect, resulting in antiproliferative activity being observed in a range of cancer cell lines. The dual cell death mode of apoptosis and ferroptosis offers a new opportunity for overcoming hypoxia-induced PDT-resistance in anticancer therapies.

Research over the past decade has generated many examples of mitochondria-targeting luminescent Ir(III)-based complexes displaying potential application for use as biological imaging probes, with emission maxima spanning visible to NIR light. This includes early examples of dinuclear Ir(III) complexes that are appropriate for mitochondrial imaging and tracking.[74,79] Many Ir(III) complexes have displayed properties that allow for the use of two photon excitation, potential use in correlative imaging,[110] and compatibility with super-resolution microscopy techniques.[110,117] A variety of investigations demonstrated the excellent photostability of Ir(III) complexes, allowing for long-term luminescence studies with minimal photodegradation.[66,68,72,82,107,134] Several examples of Ir(III) complexes that display AIPE characteristics were developed in recent years.[65,78,104,109,153] These complexes tend to display resistance to concentration-dependent luminescence and singlet oxygen yield quenching, and may provide an avenue for overcoming commonly encountered issues with luminescent probes and PDT agents.

Explorations into biosensing applications for Ir(III) complexes have produced a library of potential sensors. The results have indicated potential sensors for the targeted detection and monitoring of mitochondrial oxygen,[47,86] azide-modified biomolecules,[70] hypochlorite[71] and hypochlorous acid,[115] nitric oxide,[73] sulfite and bisulfite,[79] cysteine,[129,139,160] viscosity,[141] peroxynitrite and glutathione redox cycles,[163] Zn^{2+} and Cu^{2+},[133,160] and pH.[124,138] Most probes were focused on developing on-off emission for target detection. However, sensors based on emission intensity, wavelength, and lifetime were also investigated. This diverse range demonstrates the versatility of Ir(III) complexes for sensing applications, while retaining the ability to localize in mitochondria. With so many avenues of exploration available, Ir(III)-based probes will indubitably find use in many future biological investigations and stand to illuminate currently unknown aspects of mitochondrial structure and function.

Applications for mitochondria-targeting Ir(III) complexes as photodynamic therapeutics have been thoroughly explored since 2011, with exciting prospects. Many Ir(III) complexes displayed intrinsic cytotoxicity toward cancerous cells, often comparable or even greater than that of cisplatin. These activities were, in general, improved upon irradiation, occasionally to a staggering degree.[81,103] The most promising candidates were those that displayed low to negligible dark cytotoxicity, then upon illumination promoted strong anticancer activity, thus allowing for targeted phototherapy.[85,107,114,146,149] Very encouraging advances have been found for PDT activity in hypoxic tumor cells. As expressed previously, an important factor in the PDT process is the presence of oxygen to initiate type II reactions. Hence, photodynamic treatment of hypoxic tumors is a challenging task. Several Ir(III) complexes were found to display PDT effects under hypoxic conditions.[140,142,143,152] Kuang et al.,[152] Guo et al.,[140] and Li et al.[143] developed potential theranostic compounds containing anthroquinone moieties that are reduced to form luminescent species under hypoxic conditions and display appreciable PDT effects. The phototherapeutic effect of **Ir144,** developed by He et al.,[142] follows an alternative pathway. Upon irradiation, dissociation of the Ir(III) moiety produces photoacid. Photolytic products can then further sensitize singlet oxygen, allowing the compound to kill cancer cells under both normoxic and hypoxic conditions.

Another excellent advancement is the observation that Ir(III) complexes can display selective toxicity toward cancerous cells.[97,105,108,111,114,119,128] Complexes **Ir59–62**, **Ir89**, **Ir101**, **Ir106–109** and **Ir126** were all found to display greater cytotoxic potency toward various cancerous cells than normal cells. Alternatively, an investigation by Liu et al.[105] featured Ir(III)-DCA complexes that selectively altered cancer cell metabolism, enhancing the efficacy of PDT. This method of cancer selectivity is also observed to be effective in hypoxic environments.

Encouraging results have demonstrated a variety of cell death mechanisms in cancer cells. The predominant mechanism of cell death was apoptosis. However, paraptosis,[120,147] necrosis,[69,113] mitophagy,[78,89,101,137,154] and first examples of Ir(III) complexes to induce oncosis in cisplatin-resistant cancer cells were also observed.[119] One interesting example was **Ir166**, which displayed alternative cytotoxic mechanisms dependent on the presence or absence of light.[154] Devoid of light, the complex displayed potent cytotoxicity via the induction of mitophagy. However, upon photoexcitation the mechanism of cell death converted to necrocytosis. Selected examples also were observed to induce ferroptosis in cancer cells,[159,165] including **Ir184** which was found to simultaneously induced apoptosis and ferroptosis. These investigations carry great significance, as alternative mechanisms of cell death may provide a pathway to overcoming platinum-drug resistance.

3. Platinum

Since cisplatin was first observed to inhibit cell division in the 1960s, platinum-based metallodrugs have been extensively studied for their use in anticancer therapies.[166] Mechanistic investigations have revealed that cisplatin targets DNA, forming intra- and inter-DNA crosslinks that disrupt DNA transcription and replication, consequently inducing apoptosis in cancer cells.[167] Whilst cisplatin has proved to be a valuable tool in cancer therapeutics, there are currently a number of cell lines that display intrinsic or developed resistance to cisplatin. Resistant cell lines and harsh side effects, including nephrotoxicity and neurotoxicity,[167] have guided researchers toward developing novel metallodrugs that utilize alternative mechanisms of action to combat cancer proliferation. One approach is to focus on alternative sub-cellular targets, such as mitochondria.[168] This research effort, combined with the intrinsic photophysical properties of Pt(II) complexes,[169] has generated a library of compounds that can be considered useful building blocks for therapy and imaging. Many of these complexes display mitochondrial specificity. All platinum structures discussed in this section are shown in Fig. 3.

Guo et al.[170] investigated the potential as cellular imaging agents for a series of oligo(ethylene glycol)-substituted luminescent Pt(II) complexes. Complexes **Pt1** and **Pt2**, substituted with morpholine and piperidine, respectively, were found to preferentially localize within mitochondria with a high degree of specificity in HeLa cells. Conversely, the five other Pt(II) complexes synthesized were found to localize within the cytoplasm and display weaker emission, despite the similar level of cellular uptake. The study showed that the emission of this family of complexes is quenched upon cellular uptake, then either restored or enhanced upon accumulation within hydrophobic regions. All complexes were found to have potent cytotoxic effects in HeLa cells, with IC_{50} values being approximately an order of magnitude lower than that of cisplatin. This study may provide valuable considerations for the design of future Pt(II) imaging agents that show selective mitochondrial accumulation.

Li et al.[171] investigated a series of Pt(II)-based oligo(ethylene glycol)-substituted complexes in a diverse range of cancer cell lines. The complex **Pt2** (previously mentioned in investigations by Guo et al.[170]) was found to trigger substantial alterations in mitochondrial dynamics involving the depolarization of MMP, release of cytochrome *c* and consequent cell necrosis. The work details the role of **Pt2** in both cell adhesion and effective inhibitory activities of cell invasion/migration. The dual functions of **Pt2** and

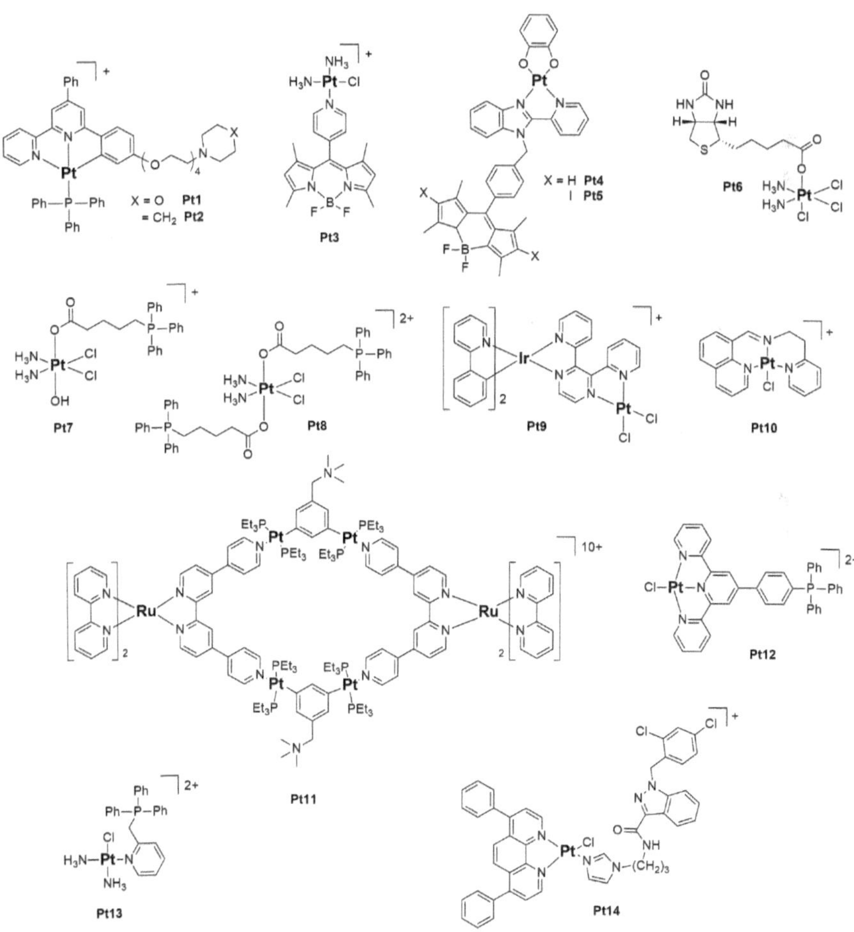

Fig. 3 Mitochondria-targeting platinum complexes **Pt1–14.**

insights into their mechanism of action represent promising starting points for the development of mitochondria-targeting Pt(II)-based complexes for anticancer therapies.

Sun et al.[172] developed a luminescent BODIPY-Pt(II) conjugate, **Pt3**, which was shown to accumulate within mitochondria. *In vitro* investigations of **Pt3** found strong antiproliferative activity in MCF-7 and HeLa cancer cell lines, with localization being sensitive to the MMP. The cytotoxicity was mildly lower than that of cisplatin. However, it was also possible to image platinum distribution in living cells due to the presence of the BOPIDY group. **Pt3** therefore shows great potential for a multifunctional agent able to fulfill both cytostatic and imaging roles.

Mitra et al.[173] developed two Pt(II) catecholate complexes, **Pt4** and **Pt5,** featuring BODIPY-appended 2-(2-pyridyl)benzimidazole ligands. These complexes displayed affinity for the mitochondria. The bidentate catecholate ligand was selected to allow for the slow release of the active Pt(II) species, capable of mtDNA crosslinking. Both **Pt4** and **Pt5** displayed strong emissive properties. The incorporation of two iodine atoms in the BODIPY moiety of **Pt5** promoted more efficient triplet excited-state population, resulting in greater singlet oxygen generation, and hence a lower IC_{50} value than **Pt4** upon light exposure. Importantly, no cytotoxic effects were observed in the dark, suggesting the possibility that these complexes could be suitable as PDT agents. Mechanistic studies revealed light-dependent generation of ROS resulting in MMP depolarization and induction of intrinsic apoptosis in HaCaT cells. The results highlight the ability to tune the photophysical properties of BODIPY-appended platinum complexes to induce mtDNA damage. The targeting of mtDNA may offer alternative measures for countering resistance to traditional Pt-based anticancer agents, with the potential for minimizing side effects.

Zhao et al.[174] developed and explored the anticancer activities of three Pt(IV)-based complexes that feature a biotin ligand and various ancillary ligands. Of these, complex **Pt6** was found to display cytotoxicity 2–10 times more potent than cisplatin in a range of human cancer cells. Whilst the exact subcellular localization was not identified, it was determined that **Pt6** affected mitochondria via the expression of Bax, an apoptosis regulating protein. The expression of Bax induced the release of cytochrome *c* and activated caspase-3, thus resulting in apoptosis. It was concluded that biotinylated Pt(IV) complexes show potential for the targeted delivery of platinum-based metallodrugs.

Jin et al.[175] synthesized two Pt(IV) prodrugs, **Pt7** and **Pt8**, featuring mono- and bis-coordinated pentanoic acid ligands appended with TPP, respectively. The biological properties of the prodrugs were investigated to determine the effect that mitochondria-targeting groups have on cytotoxicity and bioreactivity. Whilst both complexes were found to enter mitochondria, **Pt7** displayed greater mitochondrial accumulation than **Pt8**. This may seem counterintuitive, as **Pt8** features two mitochondria-targeting TPP moieties. An explanation can be offered based on similar observations previously reported by Muhammad et al.,[176] which involve the different rate of reduction of **Pt7** and **Pt8** *in cellulo* to form cisplatin. The study concluded that the cytotoxicity of subcellular targeting Pt(IV) complexes are dependent on both reduction potential and the nature of the targeting group.

Ouyang et al.[177] synthesized a hetero-dinuclear Pt(II)-Ir(III) complex, **Pt9**, and investigated its anticancer properties. The results indicated that **Pt9** displays potent *in vitro* cytotoxicity toward cisplatin-resistant A549R tumor cells. The complex induced cell death via mitochondrial dysfunction resulting from superoxide accumulation and alteration of metabolism. The cell death pathway was determined to be necrosis, which was found to be more effective than apoptosis. The findings are expected to facilitate new strategies for the development of novel antitumor agents.

Wang et al.[178] synthesized two Pt(II) complexes with 8-substituted quinoline derivatives as ligands. The complex **Pt10** was found to predominantly localize within mitochondria, display effective anti-proliferative activity in several cancer cell lines and lower systemic toxicity against non-cancerous HL-7702 cells than cisplatin. **Pt10** induced a series of events associated with mitochondrial dysfunction, including MMP depolarization and ATP depletion. ROS generation and thioredoxin reductase (TrxR) inhibition resulted in an apparent endoplasmic reticulum stress response. Mechanistic studies indicated **Pt10** can simultaneously induce intrinsic pathway-dependent apoptosis and apoptosis-dependent pro-death autophagy in A549 cells. In a mouse model, **Pt10** displayed better anticancer activity and lower systemic toxicity than cisplatin. Overall, the results indicate that **Pt10**, or derivatives thereof, may prove to be useful in the development of novel anticancer drugs.

Yilmaz et al.[179] developed a series of Pt(II) saccharinate complexes bound to phosphine ligands. The ligands range from triphenylphosphine to tricyclohexylphosphine, by sequentially substituting phenyl with cyclohexyl substituents. The anticancer activity of the complexes was investigated using various human cancer cell lines and normal BEAS-2B human bronchial epithelial cells. Of the synthesized complexes, only those of *trans* configuration displayed promising activity. All the complexes showed localization predominantly within the cytoplasm. However, the results indicated that the antiproliferative mechanism involves apoptotic pathways through activation of caspases, externalization of phosphatidylserine, excessive production of ROS, mitochondrial dysfunction, and genomic DNA damage. The study suggests that the *trans* complexes of this series can be considered potential anticancer agents that act through mitochondrial and DNA damage. The results garnered valuable insights into the future design of Pt(II)-based anticancer agents.

Zhou et al.[180] synthesized a decacationic heteronuclear Ru(II)-Pt(II) polypyridine metallocycle, **Pt11**, and investigated its potential as a two-photon PDT agent. **Pt11** was found to display NIR emission, two-photon absorption, and high singlet oxygen generation efficiency. Upon incubation, **Pt11** was found both within mitochondria and nuclei of A549 cells. **Pt11** exhibits only weak cytotoxicity in the dark; however, upon photoexcitation the generation of singlet oxygen simultaneously damages mitochondria and nuclei, resulting in a high degree of photocytotoxicity. *In vivo* on A549 tumor-bearing mouse model studies demonstrated that **Pt11** ablates cancer tissues with low dosages of NIR light. The synthetic pathway of **Pt11** provides multiple avenues for facile alterations in the structure. Future studies could explore tuning properties to allow for sub-cellular targeting or even allow multiple imaging modalities to be used in tandem. Findings suggest that **Pt11** shows very promising potential for use as a two-photon PDT agent.

Wang et al.[181] synthesized a TTP-appended terpyridine Pt(II) complex, **Pt12**, that was designed to inhibit TrxR in cancer cells. *In vitro* investigations revealed **Pt12** displayed caspase-3-independent cytotoxicity toward cisplatin-resistant human ovarian cancer cells and showed preferential inhibition of mitochondrial TrxR. Levels of both mitochondrial and cellular ROS were decreased and mitochondrial function and morphology were severely damaged. Significant impairment of glycolytic and mitochondrial bioenergetics resulted in the cancer cells entering a hypometabolic state, eventually starving to death. The results suggest that targeting redox homeostasis is an effective way to modulate metabolic pathways in cancer cells.

Zhu et al.[182] developed three Pt(II) complexes, featuring two ammonia ligands, one chlorido ligand and a pyridine ring appended with a TPP moiety to promote mitochondrial accumulation. The complexes differed on the basis of the position of the TPP moiety. It was found that the position of the TPP moiety significantly influences cellular uptake, cellular distribution and cytotoxicity. Notably, the *ortho*-substituted complex, **Pt13**, was found to exhibit excellent *in vitro* cytotoxicity toward various cancer cell lines, whilst displaying lower toxicity to normal HL-7702 human liver cells than cisplatin. **Pt13** displayed several mechanisms of action, including the release of cytochrome *c*, glycolysis inhibition and mtDNA/nDNA interruption. *In vivo* investigations with mouse models showed promising tumor growth inhibition properties. The study provided great insight into our current understanding of anticancer mechanisms of action for platinum complexes.

Muhammad et al.[183] synthesized three Pt(II) complexes, featuring a lonidamine-appended imidazole ligand and either two ammonia ligands, one 1,10-phenathronline ligand or one 4,7-diphenyl-1,10-phenanthroline ligand. The biological activity of each complex was evaluated and compared to determine the effect of the ligand variation. The molecule featuring the 4,7-diphenyl-1,10-phenanthroline ligand, **Pt14**, was found to preferentially accumulate within mitochondria, causing significant mtDNA damage and inducing mitochondrial de-energization via disruption of mitochondrial bioenergetics. Mechanistic investigations showed that **Pt14** inhibit the cell cycle in the G0 and G1 phases, and induce caspase-dependent apoptosis. Interestingly, **Pt14** displayed greater *in vitro* cytotoxicity toward a variety of cancerous cells than normal MCF-10A human breast epithelial cells. The study provides a novel strategy to produce new anticancer Pt(II)-based compounds via the conjugation with mitochondrial-specific regulators, that may work on resistant cancer cell lines.

Platinum-based metallodrugs designed for targeting mitochondria have taken remarkable strides in recent years. Some of the investigated complexes are bound to functional groups such as BODIPY,[172,173] TPP,[175,182] and oligo(ethylene glycol).[170] Pt(II) complexes have demonstrated their potential to serve both as imaging probes and therapeutic agents, especially in the treatment of cancer. The assembly **Pt11**, reported by Zhou et al.,[180] offers particularly exciting potential as a multifunctional imaging and therapeutic agent, given its two-photon absorption and NIR emitting properties, coupled with its light-induced cytotoxicity.

Most of the investigations involving Pt(II) complexes suggested cell death was a result of inducing apoptosis through a number of mechanisms, including oxidative stress by raising ROS levels, depolarization of MMP, and mtDNA and/or nDNA damage. However, studies conducted by Li et al.,[171] Wang et al.,[178] and Ouyang et al.,[177] indicated that mitochondria-targeting Pt(II) complexes are also capable of inducing necrotic or autophagic cell death.

Overall, the work done in recent years has shown that despite the challenges faced by traditional platinum-drug resistant cell lines, Pt-based therapeutics that target mitochondria may have a significant role to play in future anticancer therapies.

4. Rhenium

Facial rhenium(I) tricarbonyl complexes are well known for their desirable photophysical properties, as they generally feature relatively long-lived excited states, large Stokes shifts, environment-sensitive emission and good photobleaching resistance.[184] These properties have resulted in a large number of investigations into the potential of Re(I) complexes as cellular imaging agents. A review by Lo et al.[185] provides an excellent summary of investigations focused on the use of Re(I) complexes in imaging prior to 2011. Re(I) complexes, particularly cationic complexes with lipophilic ligands, have previously been observed to display mitochondrial selective localization.[186] In recent times, a number of studies have indicated that Re(I) organometallic complexes may have potential for use as anticancer agents.[187] All rhenium structures discussed in this section are shown in Fig. 4.

Kitanovic et al.[188] conducted a series of biological experiments on the previously investigated complex **Re1**,[189] which displayed a high degree of cytotoxicity but could not be observed in cancer cells by optical microscopy. **Re1** features ligands in the form of $[Re(CO)_3(\mathbf{N^\wedge N^\wedge N})]^+$ where **N^N^N** is 2-azido-N,N-bis[(quinolin-2-yl)methyl]ethanamine. The study indicated that **Re1** displays concentration-dependent emission, which is likely responsible for quenching in live cells. MCF-7 breast cancer cells showed reduced cellular respiration and an enhancement in glycolysis upon incubation with **Re1**. This indicated a shift in cellular metabolism and inhibition of mitochondrial activity. Analysis of isolated mitochondria showed that mitochondrial respiration was a direct target of **Re1**.

Kowalski et al.[190] developed a neutral Re(I) tricarbonyl complex, **Re2**, featuring a 1,10-phenanthroline ligand and a chromone-appended alkynyl ligand. The complex was found to exhibit orange-red luminescence assigned to a triplet ^{3}MLCT excited state resulting of d(Re)-π^* character. *In vitro* investigations showed mitochondrial, cytoplasmic and endoplasmic reticulum localization of **Re2** in murine C2C12 cancer cells. However, uptake was limited due to poor aqueous solubility. **Re2** displayed impressive photostability in cellular environments, with no loss of emission observed upon prolonged irradiation. It was proposed that complexes featuring

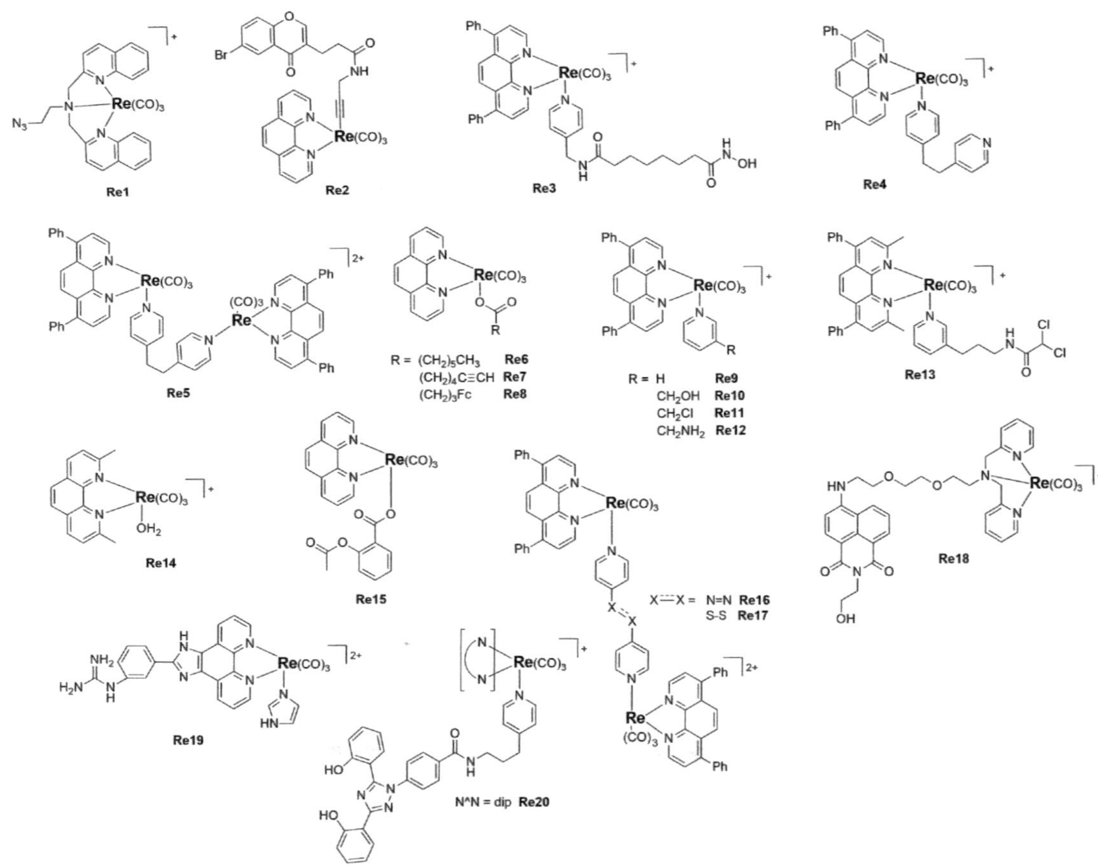

Fig. 4 Mitochondria-targeting rhenium complexes **Re1–20.**

analogous ligands could be expected to preserve the characteristic emission, whilst modulating the biological properties. Future explorations into developing Re(I) complexes with more selective affinity to cytoplasmic organelles may benefit from the knowledge garnered from this investigation.

Ye et al.[191] developed a histone deacetylase (HDAC) targeting luminescent tricarbonyl Re(I) complex, **Re3**. Human HDAC7 has previously been reported to localize within the mitochondrial inner-membrane space and plays a vital role in apoptosis regulation and cancer cell proliferation. **Re3** features a (N^1-hydroxy-N^8-pyridin-4-ylmethyl)octanediamide ligand, a 4-(aminomethyl)pyridine derivative of a previously reported HDAC inhibitor. Colocalization experiments revealed that **Re3** accumulates within mitochondria in HeLa cells, whereby it inhibits HDACs and induces caspase-independent paraptosis. **Re3** allows for cellular imaging as a result of its luminescent properties. In combination with its cytotoxicity, **Re3** can simultaneously induce paraptosis and monitor consequent changes in mitochondrial morphology. Its theranostic capabilities may provide new insights into links between mitochondrial morphology and paraptosis.

Ye et al.[192] developed a series of luminescent tricarbonyl Re(I) complexes including **Re4**, featuring **dip** and 1,2-di(pyridin-4-yl)ethane ligands, as well as its dinuclear counterpart, **Re5**. The complex **Re4** and **Re5** were found to localize within mitochondria, with the latter depolarizing MMP and inducing caspase-independent paraptosis. The study reveals useful insights for the rational design of mitochondria-targeting anticancer mono- and multinuclear Re(I) complexes.

Skiba et al.[193] developed three neutral Re(I) tricarbonyl complexes in the form *fac*-$[Re(CO)_3(\mathbf{phen})(\mathbf{L})]^+$: **Re6**, **Re7** and **Re8**. Complexes **Re6** and **Re7** were initially investigated for use as luminescent probes in HeLa cells. Findings revealed that both complexes selectively accumulate within mitochondria. To investigate the mechanism of mitochondrial staining for these complexes, **Re8**, containing a luminescence quenching ferrocene moiety was synthesized. The emission of **Re8** is only restored when the ferrocene appended ligand dissociates to produce a species containing the luminescent $[Re(CO)_3(\mathbf{phen})]^+$ core. Confocal microscopy revealed mitochondrial localization for all three complexes, including **Re8**. Based on these results, it was proposed that, upon entering the cell, carboxylate ligands dissociate to produce Re(I)-based luminescent complexes, which then accumulate within mitochondria. These complexes were found to induce photocytotoxicity via the production of ROS. Overall, the results indicated that this family of complexes feature a dissociation-driven mechanism in cells with excellent potential for biological applications.

Yang et al.[194] investigated a family of tricarbonyl Re(I) complexes of varying lipophilicities and investigated their potential for use as anticancer agents. The complexes were in the form $[Re(CO)_3(\mathbf{N^\wedge N})(\mathbf{L})]^+$, where **N^N** is **phen** or **dip** and **L** is pyridine, pyridine-3-ylmethanol, 3-(chloromethyl)pyridine, or pyridin-3-ylmethanamine. The complexes containing **phen** ligands displayed poor cellular uptake, whereas the **dip** complexes, **Re9–12,** were found to penetrate into A549 cells quickly and efficiently. They displayed mitochondrial localization and greater cytotoxicity than that of cisplatin in various cancer cell lines, including resistant lines. **Re9–12** induced several events upon mitochondrial accumulation including ATP depletion, ROS species elevation, mitochondrial respiration inhibition and caspase-dependent apoptosis. Of particular interest was complex **Re11**, designed for mitochondrial immobilization. This complex displayed greater cytotoxicity and more effective repression of mitochondrial metabolism in cancer cells. **Re11** displayed efficient phosphorescence and an O_2 sensitive lifetime, allowing for the real-time tracking of mitochondrial morphology, suggesting it constitutes a valuable tool to understand anticancer mechanisms. Work done in the investigation showed that targeting mitochondrial metabolism is an effective method for anticancer therapy. The study concluded that Re(I) complexes designed for mitochondrial immobilization can improve cytotoxicity and selectivity toward cancer cells.

Yang et al.[195] also developed a luminescent tricarbonyl Re(I) complex, **Re13**, for theranostic applications by conjugating a known metabolic modulator, dichloroacetate (DCA), to a known mitochondria-targeting Re(I) complex featuring a 2,9-dimethyl-4,7-diphenyl-1,10-phenanthroline and pyridine ligands. **Re13** retains its selective accumulation within mitochondria and reverses cancer cell metabolism from glycolysis to glucose oxidation at pharmacologically relevant DCA concentrations. The efficient phosphorescence and O_2-sensitive excited state of **Re13** allowed for phosphorescent lifetime imaging of oxygen consumption during the process and real-time tracking of mitochondrial morphology. *In vitro* investigations revealed that **Re13** also displays selective killing of NCI-H1229 human lung cancer cells in a co-culture model with normal L02 human liver cells. The study reveals that the design of Re(I)-based metallodrug theranostic agents may provide new insights into cancer therapies.

Konkankit et al.[196] conducted biological studies on **Re14**, a previously investigated tricarbonyl Re(I) complex in the form $[Re(CO)_3(H_2O)(\mathbf{L})]^+$ where **L** is 2,9-dimethyl-1,10-phenanthroline. A high degree of mitochondrial accumulation was observed in A2780 and cisplatin resistant A2780CP70 cells. It was also observed that the luminescence of **Re14** is sensitive to local pH and coordination environment. *In vivo* studies with mouse models bearing ovarian cancer tumor xenografts showed that **Re14** exhibits favorable antitumor properties with minimal discernible side effects. The promising results support future exploration into Re(I)-based anticancer drug candidates and potentially pH-sensitive luminophores.

Skiba et al.[197] synthesized four Re(I) tricarbonyl complexes in the form of $[Re(CO)_3$(**phen**)(NSAID)] where NSAID is a non-steroidal anti-inflammatory drug, either aspirin, indomethacin, (S)-(+)-naproxen, or (S)-(+)-ibuprofen. The aim of the investigation was to gain insights into mechanisms of anticancer activity for Re(I)-NSAID conjugates and screen for potential antibacterial properties. Results indicated that only $[Re(CO)_3$ (**phen**)(aspirin)$]^+$, **Re15**, display *in vitro* cytotoxicity toward HeLa cells. **Re15** was also shown to be inactive toward non-cancerous mouse L929 cells. Upon uptake into Hela cells, **Re15** dissociates the carboxylate ligand and predominantly localizes within mitochondria, raises ROS levels, and triggers apoptotic cell death. Meanwhile, the dissociated aspirin ligand retains its characteristics as a COX-2 inhibitor. The work concurs with previous studies by Skiba et al.,[193] highlighting the possibility to exploit dissociation of carboxylate ligands from Re(I) complexes in cells for combined imaging and therapeutic functions.

Wang et al.[198] synthesized two dinuclear Re(I) tricarbonyl complexes, **Re16** and **Re17**, in the form $[Re_2(CO)_6$(**dip**)$_2$(L)$]^{2+}$, where L=4,4′-azopyridine (**Re16**) or 4,4′-dithiodipyridine (**Re17**). *In vitro* investigations of the complexes revealed that both **Re16** and **Re17** target mitochondria, induce oxidative stress, slow down bioenergetic rate, and interfere with glutathione metabolism in HeLa cells. **Re16** and **Re17** were found to simultaneously induce both caspase-dependent apoptosis and necroptosis. Finally, **Re16** and **Re17** were demonstrated to inhibit tumor growth in nude mice bearing carcinoma xenografts. The study shows that dinuclear Re(I) complexes can exert *in vitro* cytotoxicity via irreversible oxidative stress and disruption of the cellular redox balance.

Day et al.[199] developed a series of six tricarbonyl Re(I) complexes featuring a tridentate dipicolylamine chelator conjugated to 1,8 naphthalimide with a range of different linker units. By varying the linker units, the overall lipophilicity of the complex could be tuned. One water-soluble complex, **Re18**, containing two ether groups in the linker unit, was shown to localize in mitochondria of MCF-7 cells. A ^{99m}Tc(I) analogue was synthesized and investigated for its potential as a mitochondria-targeting radiotracer. The ^{99m}Tc(I) analogue was found to display excellent mitochondrial uptake characteristics, and good quality images were able to be produced in MCF-7 cells. The study demonstrated that variations in linker units can be utilized to tune solubility, lipophilicity and sub-cellular accumulation in this family of complexes.

He et al.[200] synthesized Re(I) complexes featuring guanidinium-functionalized ligands and investigated their anticancer potential. Complex **Re19** was found to selectively accumulate in mitochondria and induce caspase-dependent apoptosis via the elevation of ROS, ATP reduction, and cell cycle arrest at the S phase. **Re19** was also found to effectively depress the cell migration, cell invasion and colony formation. The work provides a greater insight for the design of novel Re(I) anticancer agents.

Pan et al.[201] developed three tricarbonyl Re(I) complexes in the form $[Re(CO)_3(\mathbf{N^\wedge N})(\mathbf{L})]^+$ where **L** is a pyridine ligand appended with deferasirox, and **N^N dip** (**Re20**), 4,7-dimethyl-1,10-phenanthroline, or **phen**. Deferasirox is a clinically used iron chelating agent and was incorporated to simultaneously influence iron homeostasis and mitochondrial metabolism in cancer cells. **Re20** was found to relocate iron to mitochondria and alter key metabolic species related to epigenic modification by down-regulating the expression of Fe(II)/2-oxoglutarate-dependent demethylases. It was observed that **Re20** concurrently elevates methylation levels of DNA, RNA, and histone, consequently affecting events related to RNA polymerases, apoptosis, and T-cell receptor signaling. **Re20** was shown to induce immunogenic apoptosis and displayed potent anticancer activity and low systemic toxicity *in vivo*. The study proposed a novel strategy for recoding the cancer epigenome and offers new insights into the development of treatments for tumors resistant to traditional therapies.

Whilst *facial* Re(I) tricarbonyl complexes are indeed known for their excellent luminescent properties, mitochondria-targeting Re(I) complexes are certainly promising building blocks for the development of anticancer metallodrugs. In particular, Re(I) scaffolds have demonstrated valuable multifunctional features. These complexes have found excellent applicability toward anticancer agents capable of monitoring their own effects on mitochondrial morphology while inducing cell death mechanisms such as apoptosis and paraptosis. Examples of Re(I) complexes inducing necroptosis and immunogenic apoptosis were also reported. With complexes displaying appreciable cytotoxicity in platinum-drug resistant cancer cell lines and mechanisms of cell death distinctly different from that of cisplatin, mitochondria-targeting Re(I) complexes may potentially find use as potent anticancer therapeutics and theranostic agents.

5. Gold

Since Auranofin found clinical use as an antirheumatic drug in the 1980s, exploration into the medicinal use of gold complexes has become quite popular. Recent decades have predominantly focused on the use of gold-nanoparticles as therapeutic agents, drug delivery systems, imaging probes and more.[202] Whilst gold-nanoparticles have shown impressive versatility, a number of interesting investigations into small-molecule Au-based complexes have also been reported. Regarding mitochondria-targeting Au complexes, most studies appear to have been focused on developing novel inhibitors of thioredoxin reductase (TrxR), an enzyme responsible for cellular homeostasis that is a known target for the treatment of diseases such as rheumatoid arthritis and cancer.[203] Au(I) and Au(III) complexes have previously shown potent and selective inhibition of TrxRs catalytic functions,[204] suggested to be a result of covalent binding of the Au ion to the C-terminus of TrxR.[205] Upon the inhibition of TrxR, regular redox balance is disrupted, resulting in oxidative stress and cell death.[206] With mechanisms of action distinctly different to that of cisplatin,[166] gold-complexes offer an alternative approach for therapies focused on platinum-resistant cancers. All gold structures discussed in this section are shown in Fig. 5.

Rubbiani et al.[207] synthesized and investigated the cationic complexes **Au1** and **Au2**, bound to the NHC ligand 1,3-diethylbenzimidazol-2-ylidene. **Au1** is bound to the NHC ligand and triphenylphosphine, whereas **Au2** is bound to the same two NHC ligands. As a comparison, the neutral complex **AuCl**, bound to the NHC and chloro ligands, was also prepared. The study aimed to compare the efficacy of each complex for their use as TrxR inhibitors. It was concluded that the pharmacodynamic effects of complexes **AuCl, Au1** and **Au2** was a result of multiple mechanisms related to TrxR inhibition and potentially unrelated effects on mitochondrial biochemistry. Complexes **Au1** and **Au2** were found to have greater mitochondrial accumulation and cytotoxicity than **AuCl**, likely due to their overall positive charge and bond dissociation energies. The complexes were found to have remarkable antiproliferative activity and to strongly induce apoptosis via the enhancement of ROS species formation, offering promising potential for use as anticancer chemotherapeutics.

Marmol et al.[208] evaluated the antitumor potential of a previously synthesized alkynyl Au(I) complex, **Au3**. The investigation was performed using a colorectal adenocarcinoma Caco-2 cell line. The results suggested

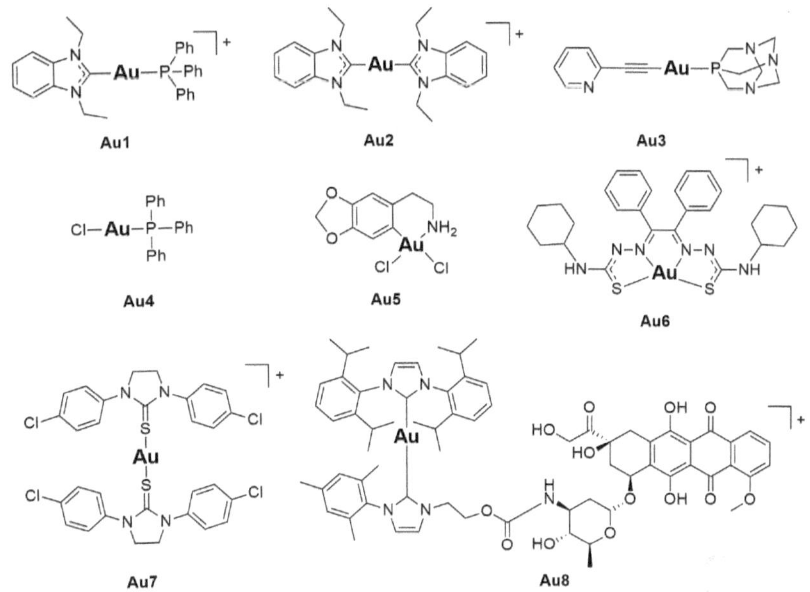

Fig. 5 Mitochondria-targeting gold complexes **Au1–8.**

that **Au3** passes through the mitochondrial membrane and alters the mitochondrial function to increase ROS levels, triggers tumor necrosis factor induced necroptosis, dependent on NF-κB signaling, and activates RIP-1. Since the mechanism of cell death is via necroptosis, **Au3** may find use in the treatment of apoptosis resistant tumors. This study represents the first report of an alkynyl Au(I) complex inducing necroptosis in cancer cells and may provide new insights into the use of gold complexes as substitutes for platinum-based chemotherapeutics.

Chrysouli et al.[209] investigated the mechanism of action for the known metallodrug chloro(triphenylphosphine)Au(I) (**Au4**). It was found that **Au4** exhibits more potent *in vitro* cytotoxicity than cisplatin against MCF-7 cells. An apoptotic mechanism of cell death was confirmed by the observation of *in vitro* permeabilization of the mitochondrial membrane and the *ex vivo* catalytic oxidation of linoleic acid. Strong LOX inhibitory activity and poor DNA binding affinity further suggested that apoptosis is the mechanism of cell death. It was concluded that **Au4** is a promising candidate for the development of new metallodrugs for breast cancer treatment.

Huang et al.[210] developed two Au(III) complexes, with structures described as being similar to functionalized tetrahydroisoquinolines, and investigated their anticancer potential. **Au5** was determined to display enhanced antiproliferative activity against various cancer cell lines, including cisplatin resistant cells, relative to normal HL-7702 human liver cells. It was determined that **Au5** induces a number of events that are indicative of mitochondrial dysfunction, including MMP depolarization, an increase in ROS levels, and ATP depletion. Intrinsic apoptosis and pro-death autophagy were also implicated in the cytotoxicity, with the level of autophagy showing an inverse correlation with the level of apoptosis. In a mouse model, **Au5** displayed greater anticancer activity than cisplatin, with lower levels of systemic toxicity. The study suggests that **Au5**, or derivatives thereof, may prove to be potential anticancer drug candidates.

Rodriguez-Fanjul et al.[211] developed the Au(I) complex **Au6** as part of a small family of complexes. A tetradentate chelating ligand was chosen to retain structural stability when compared to analogous complexes. *In vitro* investigations into the anticancer activities of each complex were conducted using various cancer cell lines including MCF-7 human breast cell lines. Despite the structural similarities, the complexes were found to display different mechanisms of cytotoxicity. Treatment of MCF-7 cells with **Au6** displayed in a high degree of accumulation within mitochondria and cytosol, resulting in ROS-dependent cell death via the inhibition of TrxR. **Au6** was found to have similar *in vitro* cytotoxicity to that of cisplatin and may represent the first complex in a new class of TrxR inhibitors for use in anticancer research.

Yu et al.[212] developed the complex, **Au7**. The study indicated that **Au7** triggers mitochondrial MPT and decreases membrane fluidity. **Au7** was also found to decrease MMP, increase ROS generation and increase the permeability of hydrogen and potassium cations as a result of MPT. In addition, **Au7** was found to severely affect mitochondrial respiration and metabolism. The cytotoxicity of **Au7** was determined to be a result of both impairment of the mitochondrial respiratory chain and MPT. The study provides important supplementary information for understanding Au(I) complex cytotoxicity mechanisms.

Sen et al.[213] synthesized a series of heteroleptic Au(I) complexes coordinated to NHC ligands. One particular complex, **Au8**, containing an anthracycline doxorubicin moiety, was found to target mitochondria. This is particularly interesting as free doxorubicin, a chemotherapy agent used in the treatment of various cancers, is known to concentrate in the nucleus. It was proposed that this redirection could be the cause of the slight reduction of cytotoxic potency, relative to the free doxorubicin. Conversely, it was also proposed that mitochondrial accumulation could potentially reduce or even eliminate cardiotoxicity associated with nuclear targeting. This research presents encouraging results for reducing the toxicity of doxorubicin-based chemotherapeutics.

Research in mitochondria-targeting gold complexes over the last decade has predominantly focused on understanding the mechanisms of cytotoxicity and intracellular localization. Advancements of current understanding of mitochondria-targeting gold-based complexes were mostly found in the form of structure-activity relationships and cytotoxic mechanisms. Complexes characterized by potent cytotoxicity, comparable to or greater than that of cisplatin, and relatively low levels of systemic toxicity have been reported. With assorted mechanisms of cytotoxicity in various cancer cells reported, including necrosis, apoptosis, MPT and impairment of the mitochondrial respiratory chain, gold-based mitochondrial anticancer complexes will likely continue to find use in future investigations for treatments of apoptotic and cisplatin resistant cancers.

6. Copper

Copper is a vital trace element in all eukaryotes, with an essential role in aerobic respiration, due in part to its redox properties that allow for the interconversion between Cu(I) and Cu(II).[214] Copper is naturally present in mitochondria, being an intrinsic part of cytochrome *c* oxidase, the terminal enzyme of the mitochondrial respiratory chain.[215] By virtue of their redox properties, many copper complexes also display potential anticancer activities via the generation of ROS.[216–218] This has resulted in a number of studies investigating the potential anticancer properties of novel copper-based complexes. All copper structures discussed in this section are shown in Fig. 6.

Zhou et al.[219] developed a cationic Cu(II) complex, **Cu1**, featuring a tridentate TPP-appended terpyridine ligand. *In vitro* investigations revealed that **Cu1** displayed a high degree of cytotoxicity toward a variety of cancer cells, including in cisplatin-resistant tumor cells. The cytotoxicity was originating from multiple mechanisms, including the generation of ROS, alteration of DNA conformation, scission of DNA strands, and dissipation of MMP. The study concluded that mitochondria-targeting copper complexes could find use as effective anticancer drugs in cisplatin resistant cell lines.

Bhattacharyya et al.[220] developed two BODIPY appended Cu(II) complexes, featuring curcumin ligands, **Cu2** and **Cu3**. These complexes displayed relatively low dark toxicity and remarkable PDT effect upon irradiation, with **Cu3** being comparable to Photofrin. Photocytotoxicity was

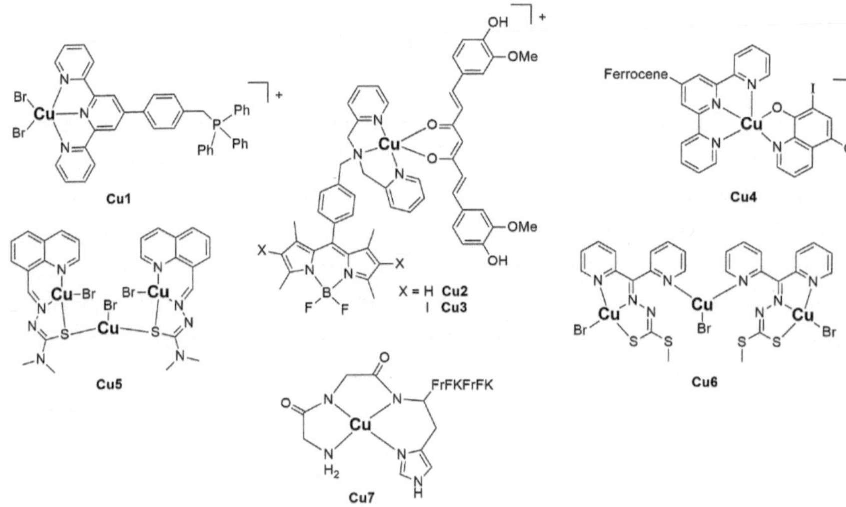

Fig. 6 Mitochondria-targeting copper complexes **Cu1–7.**

determined to be a result of ROS generation. The exemplary PDT effect of **Cu3** was proposed to be a result of the diiodinated BODIPY ligand. The heavy atom effect facilitates efficient ISC, resulting in tandem photosensitization from both the curcumin and BODIPY-based ligands. **Cu2**, being more emissive than its iodinated counterpart, was used for localization studies revealing accumulation within mitochondria. Mechanistic studies indicated light induced generation of singlet oxygen within mitochondria results in the activation of apoptosis. The work shows excellent potential for the use of BODIPY-based Cu(II) complexes for use as PDT agents.

Deka et al.[221] synthesized six Cu(II) complexes and conducted *in vitro* investigations to determine any potential anticancer applications. One complex featuring a terpyridine ligand appended with ferrocene and a dihalogenated 8-hydroxyquinoline ligand, **Cu4**, was determined to primarily localize within mitochondria of HeLa cells. **Cu4** was found to display remarkable cytotoxicity toward cancer cells lines such as MCF-7 and HeLa, whilst displaying significantly reduced cytotoxicity toward normal MCF-10A cells. Cytotoxicity was determined to be originating from the generation of ROS, resulting in the induction of apoptosis. The study represents the first example of ferrocene-appended Cu(II) complexes featuring hydroxyquinoline ligands and displaying selective cytotoxicity toward cancerous cells.

Gu et al.[222] synthesized the multinuclear mixed-valence complex **Cu5** and investigated its anticancer activity. The complex was found to accumulate within mitochondria and displayed significantly greater cytotoxicity than cisplatin against a panel of human cancer cells. It was determined that the *in vitro* cytotoxicity of **Cu5** may involve the mitochondrial mediated apoptotic pathway, likely due to the generation of ROS and reduction of MMP. **Cu5** was also found to display cytostatic effects in 3D HeLa spheroid models. The luminescent properties of **Cu5** could also be exploited for bioimaging in cells. The study concluded that the encouraging results present opportunities for further investigations on **Cu5** and analogous complexes as potential mitochondria-targeting anticancer agents coupled with imaging potential.

Guo et al.[223] synthesized three luminescent copper-based complexes, including the trinuclear mixed-valent Cu(I)/Cu(II) complex, **Cu6**. *In vitro* investigations with **Cu6** revealed cytotoxic potency greater than that of cisplatin in three types of human pancreatic cancer cells. Mechanistic investigations revealed cell death likely occurs through multiple pathways, including the mitogen-activated protein kinase pathway, cytokine-cytokine receptor interaction pathway, and ferroptosis. **Cu6** was demonstrated to be effective for use in both *in vitro* and *in vivo* imaging applications. Given pancreatic cancer cell lines generally display resistance to apoptosis-inducing anticancer agents, **Cu6** and analogous complexes present promising candidates for preclinical studies aimed at the treatment of chemoresistant tumors.

Li et al.[224] developed a Cu(II) complex incorporating a MPP, FrFKFrFK-CONH2 (Phe-r-Phe-Lys-Phe-r-Phe-Lys-CONH2, where r = D-arginine), **Cu7**, that allowed for mitochondria-targeting in cancer cells. In mitochondria, **Cu7** induced Fenton-type reactions with endogenous hydrogen peroxide to produce the highly reactive hydroxyl radicals. The production of ROS depolarizes the MMP, blocks the cell cycle and induces apoptosis in cancer cells. Results suggest a novel system for copper-mediated hydroxyl radical generation via intra-mitochondrial Fenton reactions in response to the microenvironment of cancer cells. The study offers promising new avenues for treatments in cancer therapies.

The small number of mitochondria-targeting copper complexes investigated in recent times have displayed excellent potential for combating cancers. With complexes that display multifunctional imaging and anticancer activity, powerful PDT potential, potential cancer-selective cytotoxicity, strong cytotoxicity toward cisplatin-resistant cancers, and unique anticancer mechanisms, copper-based species represent a fertile ground for further research into anticancer metallodrugs.

7. Ruthenium

Ru(II) polypyridine complexes are well known for their unique photophysical and photochemical properties. Interest in Ru(II) complexes likely stems from their reversible redox processes, excited state reactivities, luminescence properties and relatively long-lived excited state lifetimes.[225] Mitochondria are known targets for many Ru(II) complexes. Recent interest in the development of novel Ru(II) complexes stems from the established anticancer properties of ruthenium-containing compounds.[226,227] In general, the structure of Ru(II) mononuclear complexes can be divided into two main classes (Fig. 7). Class 1 complexes are dicationic and in the form $[Ru(^{1}\mathbf{N}^\wedge\mathbf{N})_2(^{2}\mathbf{N}^\wedge\mathbf{N})]^{2+}$ where $^{1}\mathbf{N}^\wedge\mathbf{N}$ are two identical bidentate ligands, forming the main structure of the compound, and $^{2}\mathbf{N}^\wedge\mathbf{N}$ is a unique bidentate ancillary ligand. Both ligands can be altered to tune the photophysical and biological properties. Class 2 complexes are monocationic and feature a π-bonded arene group, typically benzene, *p*-cymene or derivatives thereof, a halogen and a bidentate ligand. This class of complexes is often referred to as half-sandwich due to structural similarities with metallocenes. All ruthenium structures discussed in this section are shown in Fig. 8.

Qian et al.[228] synthesized four Ru(II) complexes all featuring 2-pyridyl-1H-anthra[1,2-d]imidazole-6,11-dione (PAIDH) as the ancillary ligand. PAIDH is an imidazole-fused anthraquinone derivative that is known to display potent inhibitory activity on topo I in cancer cell lines. However, the low solubility in water of PAIDH limits biological applications, hence the desire to form metallic complexes that allow tuning of solubility. To explore the effect that altering lipophilicity has on biological properties in cancer cell lines, the complexes were systematically altered to feature **bipy**, **phen**, **dmp** and **dip** (**Ru1**) ligands. Findings suggested that all complexes display potent anti-proliferative activity toward HeLa, A549, BEL-7402, and HepG2 carcinoma cell lines and feature relatively lower *in vitro*

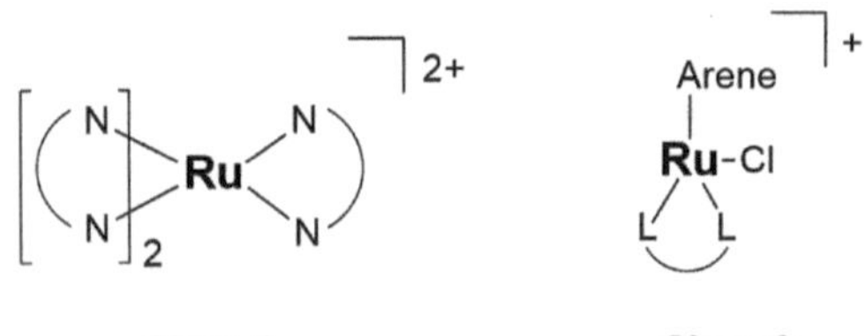

Fig. 7 Mitochondria-targeting Ru(II) complexes general structures.

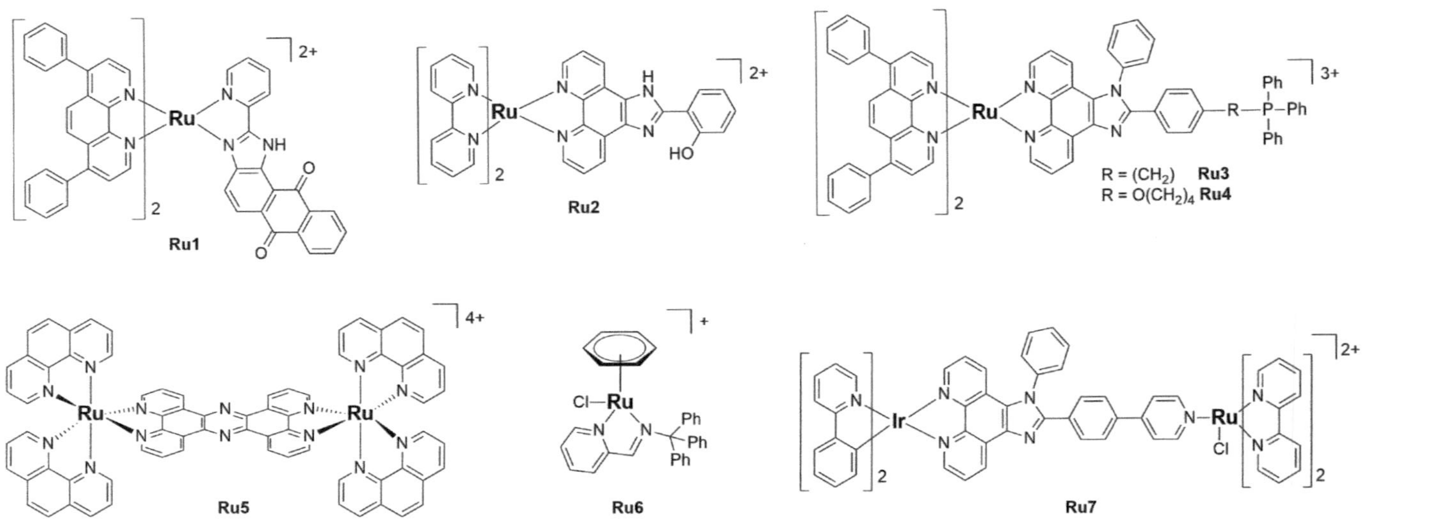

Fig. 8 Mitochondria-targeting ruthenium complexes **Ru1–7**.

cytotoxicity toward normal L02 human liver cells. **Ru1** was determined to be the most active complex as it exhibits a higher degree of cytotoxicity than cisplatin in HeLa cells. **Ru1** preferentially accumulated in mitochondria of HeLa cells and induced apoptosis via the mitochondrial pathway, suggesting that **Ru1** is a strong candidate for future investigations into developing novel chemotherapeutic agents for human cancers.

Martin et al.[229] developed a heptacationic peptide-bridged dinuclear Ru(II) complex for dynamically monitoring local O_2 concentration. The MPP incorporated was FrFKFrFK-$CONH_2$ (r = D-arginine). The use of a MPP allowed for precise accumulation within mitochondria. A series of *in vitro* calibration studies indicated the probe responds dynamically to changes in O_2 concentration in a temperature-sensitive manner. The complex was moderately cytotoxic in the dark and displayed significant photocytotoxicity under continuous irradiation. The parent complex was also conjugated to a known cell penetrating peptide, octaarginine, which displayed a wider distribution within the cytoplasmic region of the cell and showed no evidence of specific mitochondrial localization. Interestingly, the complex was found to be more cytotoxic than the MPP conjugated complex under the same conditions. The results highlight the value of combining the luminescent properties of metal complexes to the selectivity offered by MPPs. Despite the toxicity over longer time periods, the probe offers a potentially powerful tool for use in short-term dynamic studies of mitochondrial function in disease progression.

Wang et al.[230] synthesized six Ru(II) complexes featuring **bipy** and various derivatives of imidazophenanthroline as ancillary ligands. All complexes were evaluated to determine their antiproliferative activity against various cancer cell lines. ICP-MS was used to determine preferential mitochondrial localization of all six complexes in HeLa cells. **Ru5** was of particular interest, given it was found to have IC_{50} values close to that of cisplatin, whilst displaying a greater degree of cytotoxicity toward HeLa, HepG2, BEL-7402, MG-63, MCF-7, A549, and A549-CP/R cancer cell lines than normal L02 human liver cells. Additionally, the cisplatin resistant cell line A549-CP/R was found to be more sensitive to **Ru2** than cisplatin. Further investigations showed that the cytotoxicity of **Ru2** is exerted through the mitochondrial apoptotic pathway, accompanied by the regulation of Bcl-2 family members and caspase activation. The results indicate that this family of complexes may have potential in the development of selective chemotherapeutics for cisplatin-resistant tumors.

Liu et al.[231] produced a small library of Ru(II) complexes, including **Ru3** and **Ru4** appended to TPP moieties. The complexes were found to exhibit very high singlet oxygen generation in methanol solutions. *In vitro* investigations were conducted using both monolayer cells and three-dimensional multicellular spheroids. All complexes were found to display mitochondrial accumulation and a deep penetration depth. The complexes displayed very low dark cytotoxicity and were found to generate sufficient singlet oxygen using both one- and two-photon irradiation to induce cell death. **Ru4** was found to have a very low IC_{50} value (<2 μM) in two-photon PDT of 3D MCSs and displayed cytotoxicity an order of magnitude greater than that of cisplatin. All complexes were found to show potential for use as efficient two-photon PDT agents, with **Ru4** showing the most promise.

Sreedharan et al.[232] investigated a dinuclear tetracationic Ru(II) complex, **Ru5**, for its potential use in super-resolution microscopy (SRM). Previous investigations had indicated that **Ru5** is extremely photostable, and emits in the NIR with an appreciable Stokes shift. The complex displayed mitochondrial specificity at low concentrations and nuclear specificity at concentrations greater than 20 μM. This feature makes this complex extremely attractive for use in SRM. It was found to be very well suited for SRM techniques such as two-color structured illumination microscopy, stimulated emission depletion (STED) microscopy and 3D-STED microscopy in both live and fixed cells. Its vastly superior photostability, relative to conventional SRM probes, allowed imaging of nuclear DNA at an unprecedented resolution.

Li et al.[233] developed a series of half-sandwich Ir(III) and Ru(II) complexes and investigated their anticancer properties. The study reported that the complexes exhibited potent *in vitro* cytotoxicity toward a range of cancer cells. **Ru6** was found to localize within mitochondria of A549 human lung cancer cells. The complex entered cells through active transport pathways and induced excess production of ROS, resulting in a notable decrease in MMP. **Ru6** also deactivated CDK 4/cyclin D1 and arrested the cell cycle at the G0 and G1 phases. The work was reported as instrumental for the future understanding of the mechanism of action for this class of complexes, which hold great potential as theranostic agents.

Du et al.[234] developed a one-pot synthetic plan for producing five-coordinated half-sandwich aminoimine Ru(II) and Ir(III) complexes. All complexes showed a high degree of cytotoxicity toward HepG2, A549, and HeLa human cancer cells. Investigations on their mechanism of action revealed that Ir(III) cytotoxicity occurs as a result of cell necrosis, whilst Ru(II) cytotoxicity is associated with the induction of apoptosis. The observation of an increase in ROS levels also contributes to the *in vitro* cytotoxicity. The Ru(II) and Ir(III) complexes both displayed photophysical properties compatible with cellular imaging. Confocal microscopy showed that these two complexes are quickly taken into A549 cells via an energy-independent pathway and localized predominantly within mitochondria and lysosomes. The study represents the first evaluation of the biological properties of stable five-coordinated half sandwich Ru(II) and Ir(III) complexes.

Zhang et al.[235] synthesized a hetero-dinuclear Ru(II)-Ir(III) complex, **Ru7**, and conducted *in vitro* analyses into its potential anticancer properties. Results indicated that **Ru7** exhibits potent antitumor activity toward cisplatin resistant cell lines. **Ru7** enters cells through active transport mechanisms and localizes within mitochondria. Irradiation resulted in the activation of apoptosis via mitochondrial dysfunction and damage to mtDNA. The Ir(III) moiety provides the molecule with its mitochondria-targeting ability. Upon irradiation, the Ir(III) moiety photodissociates and forms a photosensitizer for PDT. On the other hand, the Ru(II) moiety acts as a photoactivated chemotherapeutic (PACT), inducing mtDNA damage. This study represents the first example of mitochondria-targeting Ru(II)-Ir(III) PACT agent. The study concluded that the combination of PDT and PACT dual therapy results in a more effective therapeutic outcome than either modality alone.

Ruthenium-based mitochondria-targeting complexes seem to have received a modest amount of attention in recent years, when compared to other transition metal elements. However, the reported examples clearly illustrate the potential that these mitochondria-targeting complexes have as multimodal imaging and therapeutic agents. With complexes displaying cytotoxicity in cisplatin resistant cell lines, potential selective cytotoxicity toward cancer cell lines, and a class of five-coordinated half-sandwich complexes, it is clear that the research into Ru(II) metallodrugs and theranostic agents still holds great promise.

8. Osmium

In general, the investigation of osmium coordination compounds as potential diagnostic and/or therapeutic agents is significantly more limited with respect to the other d^6 metal complexes. Perhaps this lack of investigation might be linked to the well-known acute toxicity of OsO_4. Nonetheless, exploration into biological applications for Os(II) polypyridine complexes has garnered interest in the past few decades due to their attractive photophysical properties, kinetic inertness and anticancer activity.[60,236–238] Investigations into mitochondria-targeting Os(II) complexes seem to be quite rare, however, the potential for luminescent nonplatinum-based anticancer theranostic agents has led a few researchers to interesting conclusions in this field. All osmium structures discussed in this section are shown in Fig. 9.

Omar et al.[239] synthesized a luminescent Os(II) complex, **Os1**, coordinated to triazole-based tridentate ligands. The complex was reported to have a relatively high quantum yield and exhibited a significant degree of oxygen-dependent quenching, with a 43-fold reduction of luminescence intensity in air-equilibrated acetonitrile, relative to the degassed solution. The relatively high quantum yield and large degree of oxygen quenching suggest **Os1**, or derivatives thereof, could be suitable candidates for singlet oxygen sensitizers for PDT applications. Preliminary investigations showed favorable uptake of **Os1** in HeLa and U2OS cell lines, with low toxicity ($LD_{50} > 100$ μM in HeLa cells) and localization within mitochondria. The study showcases the potential for novel theranostic agents based on **Os1**.

Sanchez-Cano et al.[240] investigated the subcellular distribution of **Os2** in human ovarian cancer cells using synchrotron X-ray fluorescence, TEM and ICP-MS. Results suggested mitochondrial localization of **Os2** accompanied by ER-mitochondrial Ca^{2+} transfer, a known signaling event for apoptotic cell death. **Os2** is able to induce rapid bursts of ROS, in particular superoxide formed in the first step of mitochondrial O_2 reduction. The metabolic targeting of **Os2** differs from that of platinum-based drugs, suggesting **Os2** and analogous compounds could find potential use as anticancer therapeutics for cisplatin resistant cancer cell lines as well as correlative microscopy techniques.

Gkika et al.[241] synthesized an Os(II) complex featuring two tridentate terpyridyl ligands appended with benzoic acid. The complex was then bisconjugated with MPPs via amide coupling. The resulting conjugate displayed a relatively intense NIR-emission and excellent photostability, making it an attractive potential probe for cellular imaging. Preliminary *in vitro* investigations in HeLa and MCF-7 cells showed the bis-MPP Os(II) complex localizes in mitochondria at concentrations below 50 μM. However, at higher concentrations the complex was found to display less specific subcellular distribution, including penetration of the nuclear envelope and localization within nucleoli structures. Extensive cell death was confirmed with DRAQ 7 staining. Mitochondrial depolarization was

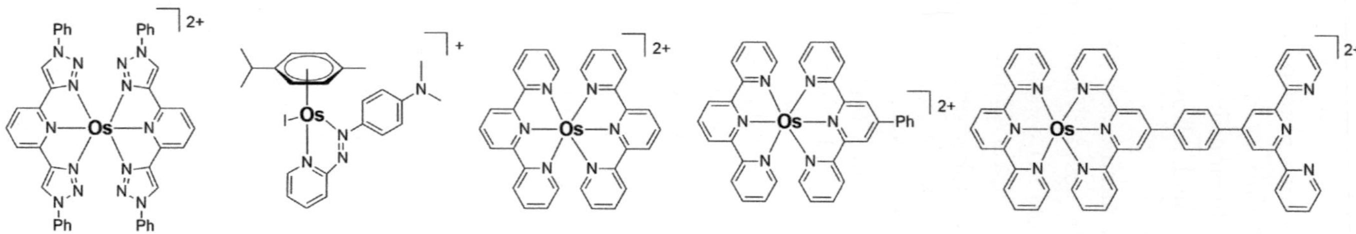

Fig. 9 Mitochondria-targeting osmium complexes **Os1–5.**

inferred at a 30 μM concentration with moderate caspase activity. At a 100 μM concentration, caspase activity increased and mitochondrial depolarization decreased, indicating a change in the mechanism of cell death with more non-specific distribution of the bis-MPP Os(II) complex.

Ge et al.[242] developed the three Os(II) complexes **Os3**, **Os4** and **Os5**. The complexes all displayed emission maxima in the NIR region and specific staining of mitochondria in living Hep-G2 cells, indicating potential use as mitochondrial cellular imaging probes. The complexes were also shown to produce singlet oxygen, oxidize NADH to NAD^+ and display a large degree of photocytotoxicity against cancer cells. Results indicate that terpyridyl-based Os(II) complexes have potential for the development of new NIR-PDT agents.

Whilst the biological properties of mitochondria-targeting Os(II) polypyridine complexes have not been investigated in detail to date, these somewhat preliminary results indicate the potential of this element and its complexes for both diagnostic and therapeutic applications.

9. Vanadium

With the ubiquitous distribution of vanadium in air, soil and water, it is not surprising that this element has found use in biological processes and is considered an essential element in most organisms.[243] Vanadium complexes have been widely investigated for medicinal uses in the treatment of cancer, diabetes, and diseases caused by parasites.[243] In regard to mitochondria-targeting vanadium complexes, the main body of work originates from the Chakravarty's, Kondaiah's and Nagaraju's research groups. In this area, the investigated complexes predominantly feature curcumin ligands. Curcumin is a major component of turmeric and has previously been found to display anticancer activity due to its ability to selectively induce apoptosis in cancer cells without significant cytotoxicity toward healthy cells.[244] All vanadium structures discussed in this section are shown in Fig. 10.

Prasad et al.[245] developed a oxidovanadium(IV) complex, **V1**, featuring an anthracenyl fluorophore and a dipeptide moiety appended to an imidazophenathroline ligand. The complex was found to specifically target mitochondria in HeLa and HaCaT cells. It displayed remarkable phototoxicity using NIR light, comparable to that of Photofrin. The phototoxicity originated as a result of apoptosis caused by ROS generation on exposure to light. The results are expected to open new avenues for developing mitochondria-targeting metal-based photochemotherapeutic agents based on vanadium.

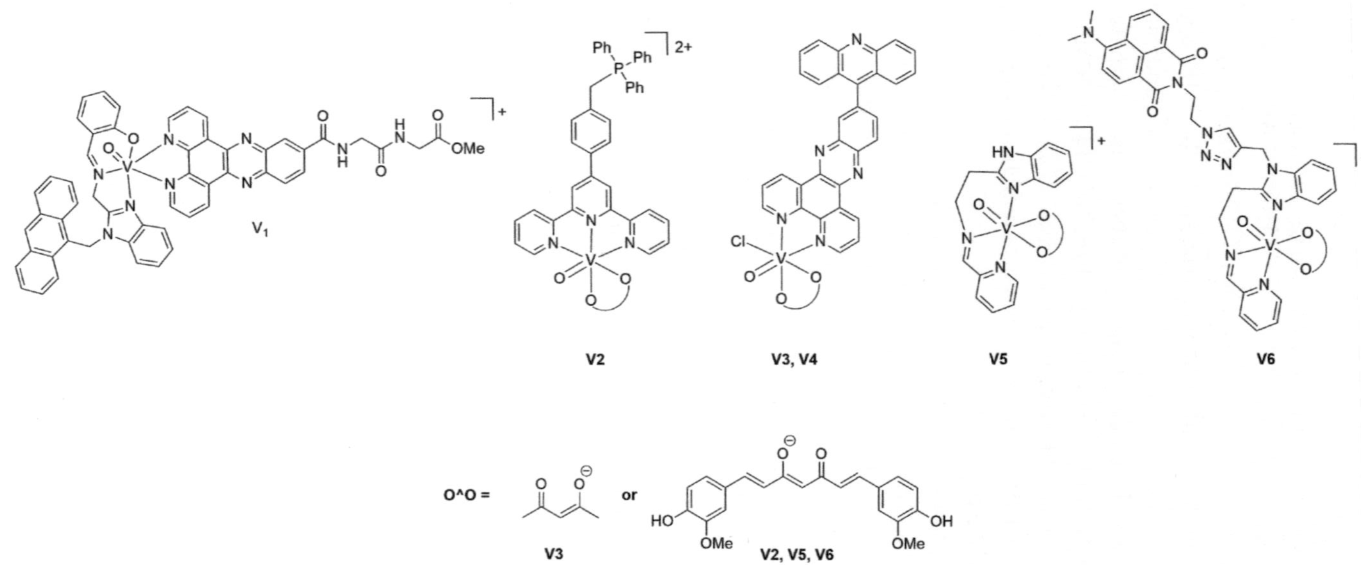

Fig. 10 Mitochondria-targeting vanadium complexes **V1–6.**

Banik et al.[246] investigated further oxovanadium(IV) complexes for use as mitochondria-targeting PDT agents. The complex **V2** was designed to overcome the limitations of curcumin, namely its low bioavailability and instability in biological environment. The degradation of curcumin in biological environments has previously been shown to be arrested when bound to a metal ion. The inclusion of a TPP moiety offers mitochondrial selectivity, whilst the terpyridine group guarantees an appropriate framework for vanadium complexation. Control analogues of each of the complexes featured acetylacetonato ligands in place of curcumin. The control complexes showed low DNA photocleavage activity in visible light, whereas the curcumin analogues displayed significant PDT activity via the formation of hydroxyl radicals. The IC_{50} value of **V2** is comparable to that of Photofrin. The work presents a novel method for the delivery of photocytotoxic agents into mitochondria.

Banerjee et al.[247] reported curcumin-based oxidovanadium(IV) complexes that induce photocytotoxicity with the use of visible and NIR light, whilst displaying low dark toxicity. The phototoxicity was reported to be greater than that of Photofrin. Mechanistic studies indicated DNA photocleavage occurs via singlet oxygen pathway that targets mitochondria, whilst cellular damage occurs via apoptotic pathway. Green emission maxima (*ca.* 520 nm) for complexes **V3** and **V4** in aqueous media allowed for confocal imaging to assess cellular uptake and localization. Complexes **V3** and **V4** were diffusely localized in the cytosol of HaCaT and HeLa human cancer cells, however they also displayed significant mitochondrial uptake. The results suggest this work offer a new generation of vanadium-based mitochondria-targeted PDT and tumor imaging agents.

The desire to increase the bioavailability and hydrolytic stability of curcumin led Prasad et al.[248] to the development of two oxidovanadium complexes, **V5** and **V6**, that exhibit selective mitochondrial localization and PDT properties. To increase the efficacy of vanadium-based curcumin complexes, **V6** features an appended naphthalimide moiety. Naphthalimide is a known DNA intercalator, fluorescence tag and photosensitizer. The naphthalimide-bearing complex displayed DNA binding and photocleavage greater than that of **V5**. Complexes **V5** and **V6** were deemed to be potential PDT agents given their photocytotoxicity in visible light and lack of toxicity in the dark. Fluorescence microscopy with complex **V6** displayed a large degree of specificity toward mitochondrial localization. Combined with the functional similarity to Photofrin, this complex is unique in the chemistry of metal-based PDT agents.

The preliminary work on vanadium complexes has opened new avenues toward the development of novel PDT agents. The work so far has shown encouraging results for alternative vanadium-based metallotherapeutics that synergistically work with curcumin.

10. Palladium

Palladium microspheres and nanoparticles have previously been shown to induce depropargylation, deallylation or even Suzuki-Miyaura cross-coupling reactions in biological settings,[249–251] an exciting prospect when it comes to the development of pro-drugs. If discrete small-molecule Pd catalysts could be used for the same purpose, these would offer the possibility of customizing ligands to alter stability, solubility, intracellular targeting, and reactivity. However, previous studies indicated that commercially available discrete Pd catalysts generally perform poorly in complex biological environments with issues related to toxicity, solubility, bioorthogonal reactivity, and decomposition to inactive Pd species. Whilst the prospects are indeed exciting, to date there are only few examples of mitochondria-targeting palladium complexes. All palladium structures discussed in this section are shown in Fig. 11.

A study by Serrano et al.[252] investigated the anticancer activity of the cyclopalladated dinuclear complex **Pd1** in melanoma cells. This complex was determined to interact with thiol groups on mitochondrial membrane proteins, resulting in the dissipation of the MMP, inducing Bax translocation from the cytosol to mitochondria, significantly decreasing ATP levels and increasing the concentration of cytosolic Ca^{2+} sourced from intracellular compartments. Chromatin condensation, DNA degradation and the activation of effector caspases suggested that **Pd1** activated the intrinsic apoptotic pathway in murine melanoma cells. Preclinical *in vitro* studies showed that **Pd1** induces cell death in multiple human cancer cells, including cisplatin-resistant cells. The study concluded that **Pd1** is an effective

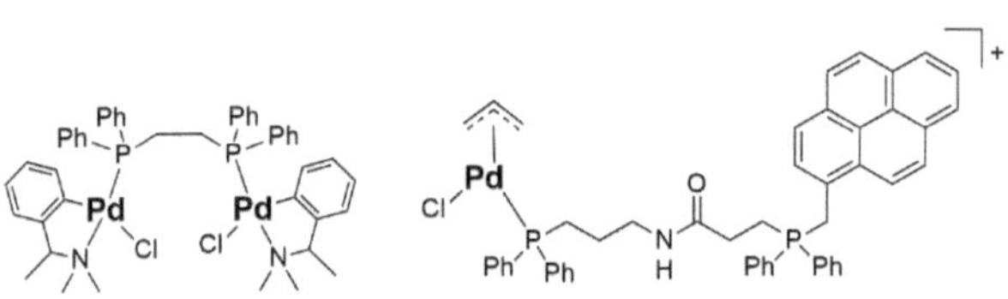

Pd1 **Pd2**

Fig. 11 Mitochondria-targeting palladium complexes **Pd1–2.**

chemotherapeutic anticancer compound against primary and metastatic tumors in both human and murine samples, including cisplatin resistant cells, via the induction of intrinsic pathway apoptosis.

Inspired by the potential use of palladium small-molecule catalysts in biological media, Martinez-Calvo et al.[252] investigated a series of palladium catalysts with customized phosphine ligands. The complexes were designed to mediate intracellular deprotection reactions, and their efficiency was compared to the more traditional palladium catalysts, such as $Pd(OAc)_2$. Their findings confirmed that, in general, traditional Pd(0) and Pd(II) catalysts are less effective in cellular media than in water and PBS. However, the complexes featuring appropriately designed phosphine ligands were able to catalyze depropargylation and deallylation reactions in cell lysates. The same complexes were able to be incubated into living mammalian cells and still function as catalysts for these reactions. The work also demonstrated preferential mitochondrial accumulation for the complex **Pd2,** where a phosphine ligand is tethered to a phosphonium and pyrene groups. The demonstrated catalytic feature *in cellulo* could potentially open new opportunities for selective prodrug activation.

11. Rhodium

Rhodium complexes are well known for their catalytic activities[253] and, to a lesser extent, their biological applications.[254,255] The observation of Rh(III) complexes binding to biomolecules has led to their use as small molecule sensors as well as cellular imaging agents.[256,257] In recent years, several studies have shown mitochondrial localization for unique Rh(III) complexes, with a number of potential applications. All rhodium structures discussed in this section are shown in Fig. 12.

Komor et al.[258] investigated the effect of lipophilicity on the cellular selectivity for a family of rhodium metalloinsertors, which include the

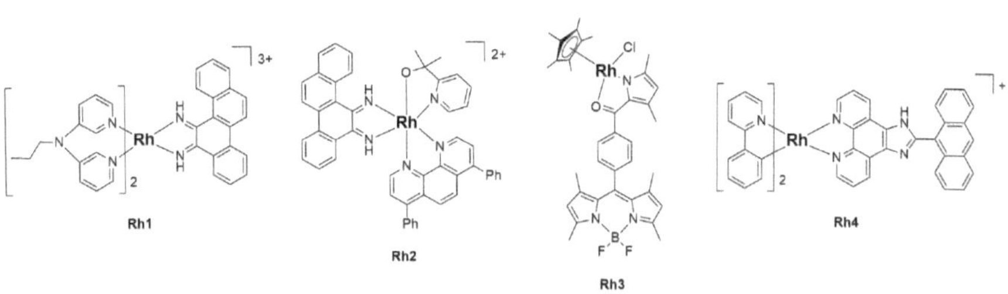

Fig. 12 Mitochondria-targeting rhodium complexes **Rh1–4.**

complex **Rh1**. Rhodium complexes are known to bind selectively DNA mismatches *in vitro*. All the complexes synthesized were found within the nucleus with sufficient concentration to be able to detect the presence of DNA mismatches. Interestingly, an increase in lipophilicity resulted in enhanced accumulation within mitochondria, which also corresponded to a decreased affinity for DNA mismatch repair (MMR) deficient cell lines. This work elegantly illustrates the fine balance between lipophilicity, which is usually increased to facilitate membrane penetration, and potentially altered biological properties and cellular selectivity.

Following from the previous work, Boyle et al.[259] prepared a family of second generation metalloinsertors that includes the complex **Rh2**. This family was shown to have a greater potency and selectivity toward MMR-deficient cells than traditional metalloinsertors, regardless of structural changes to ancillary ligands and degree of lipophilicity. This work overcomes the previously noted limitation of rhodium metalloinsertors whereby an increase in lipophilic results in enhanced mitochondrial specificity and decreased selectivity for MMR-deficient cells, making this new class of rhodium complexes promising for the development of anticancer therapies.

Gupta et al.[260] developed a bidentate ligand appended to a BODIPY moiety and coordinated it to half-sandwich Rh(III) (**Rh3**), Ru(II), and Ir(III) complexes. All complexes displayed efficient cellular uptake, localizing within mitochondria with a high degree of specificity. The complexes also display high photostability and their specificity did not decrease upon loss of MMP. Furthermore, they displayed low cytotoxicity at concentrations appropriate for imaging. These findings expand the versatility of molecular architectures bearing BODIPY and a metal center for mitochondrial selectivity.

Peng et al.[261] developed a small family of Rh(III) complexes, including **Rh4**. All complexes displayed cytotoxic potency greater than that of cisplatin against various cancer cell lines. Mechanistic studies revealed mitochondrial accumulation, subsequent elevation of ROS levels and the release of cytochrome *c*, eventually resulting in cellular apoptosis. **Rh4** was observed to sufficiently destroy the mitochondrial membrane and confirmed that the Rh(III) complexes presented in the study induce apoptosis via mitochondrial dysfunction. The ligand fluorescence could also be exploited for real-time tracking of mitochondrial morphology via confocal microscopy. The work offers a series of novel mitochondria-targeting Rh(III) complexes with greater cytotoxicity that cisplatin and strong potential for non-platinum chemotherapeutics.

There are only a handful of reports in recent years that have observed mitochondrial localization of Rh(III) complexes. These complexes display efficient mitochondrial specificity without the need of functional groups such as TPP cations, and in some cases the requirement of active MMP. These structures are definitely promising building blocks for the development of alternative non-platinum chemotherapeutics.

12. Zinc

Zinc is one of the most important trace elements in the human body and is essential for many biological functions, including metabolism, enzymatic roles, wound healing, protein and DNA synthesis, as well as cell division.[262,263] Recently, a growing body of evidence has suggested that intracellular accumulation of zinc has a strong impact on the function of mitochondria and can result in increasing ROS levels and membrane potential depolarization.[264–266] To understand the role of zinc in mitochondrial function, a number of organic switch on fluorophores have been developed.[267–269] However, only a few examples of zinc-based metal complexes designed to target mitochondria have been reported in recent years. All zinc structures discussed in this section are shown in Fig. 13.

Liu et al.[270] synthesized three terpyridine-based Zn(II) probes for two-photon monitoring of apoptosis in cells. Of the three probes, **Zn1** displayed mitochondrial localization, good biocompatibility, and an appreciable two-photon absorption cross section in the NIR region. With the use of two-photon microscopy and TEM, it was observed that **Zn1** migrated from the mitochondria to the nucleolus during apoptosis, and potentially bound to RNA. The study concluded not only that **Zn1** offers the opportunity for correlative imaging, but potentially also acts as a real-time indicator for apoptosis at a subcellular level.

Shen et al.[271] developed **Zn2** as a luminescent mitochondria-targeting thiophene-based terpyridine Zn(II) complex. The complex displays high

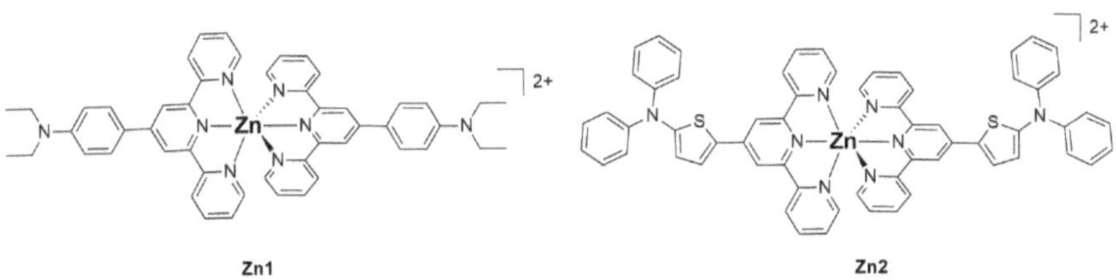

Fig. 13 Mitochondria-targeting zinc complexes **Zn1–2.**

biocompatibility, resistance to photobleaching, an appreciable two-photon absorption cross section, and was found to interact with mitochondrial DNA. **Zn2** was successfully used as an imaging agent in stimulated emission depletion (STED) experiments in live cells. This allowed for mapping of mitochondrial DNA distribution within cristae and the inner matrix at an unprecedented resolution. The results will likely provide a pathway for the design of future mitochondria-targeting bioimaging probes with super-resolution imaging capabilities.

Whilst examples of mitochondria-targeting zinc complexes are not abundant, the applications of these complexes are indeed promising. Both examples demonstrate the potential use of Zn(II) complexes as selective and efficient molecular probes. Future investigations into alternative functionalities of mitochondria-targeting Zn(II) complexes are definitely warranted.

13. Silver

Silver salts, organometallic complexes and nanoparticles have all found use as efficient antibacterial agents, typically in the form of topical metallotherapeutics for infections.[272,273] Several silver complexes have also been previously shown to display anticancer activity, including bipyridine, carboxylate and phosphane complexes.[274] The majority of research into mitochondria-targeting silver compounds appears to be focused on silver nanoparticles, which is beyond the scope of this chapter. However, there are sparse examples of research into small-molecule silver compounds that may target mitochondria. All silver structures discussed in this section are shown in Fig. 14.

Eloy et al.[275] investigated fourteen Ag(I) complexes bearing *N*-heterocyclic carbene (NHC) ligands and evaluated their anticancer activities. It is worth noting that these complexes should only be tentatively classified as mitochondria-targeting. Mitochondrial localization for this family of complexes was inferred by synthesizing the analogous complex **Ag1**, bound to

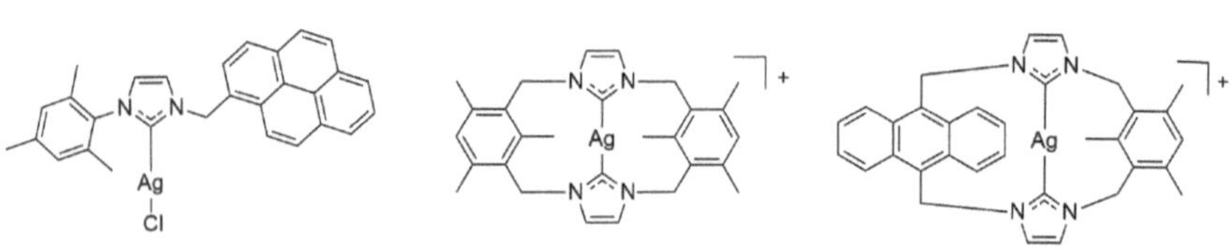

Ag1 Ag2 Ag3

Fig. 14 Potential mitochondria-targeting silver complexes **Ag1–3.**

a pyrene moiety as a fluorescence tag, and performing co-localization experiments with MitoTracker Red. Whilst the experiments did indeed reveal a strong correlation with MitoTracker Red, further investigations would be necessary to confirm that the entire family of complexes exhibit similar localization patterns. The study was the first of its kind to explore the mechanism of cell death induced by Ag(I)-NHC complexes, and it was found that apoptosis is activated without the involvement of primary necrosis. Human cancer cells treated with the Ag(I)-NHC complexes resulted in a lowering of the MMP, which likely releases mitochondrial proteins that elicit early apoptosis. There was no evidence for the generation of ROS nor the activation of caspase-3. Rather, the translocation of caspase-12 and mitochondrial apoptosis-inducing factor (AIF) proteins to the nucleus promote DNA fragmentation, resulting in cell death. This pathway for inducing cell death showed no modification in the cell-cycle distribution, hence Ag(I)-NHC complexes were confirmed to be non-genotoxic. The complexes displayed IC_{50} values in the nanomolar range and strong indication that Ag(I)-NHC can exhibit potent antiproliferative activity toward human cancer cell lines, including cisplatin resistant lines. The findings suggest that Ag(I)-NHC complexes could likely lead to alternative methods for the therapeutic treatment of cancer.

In a study conducted by Li et al.,[276] two Ag(I)-NHC complexes derived from cylcophanes (**Ag2** and **Ag3**) and two Au(I)-NHC analogues were synthesized and analyzed for their antitumor properties. Despite the complexes featuring identical ligand structures, they were found to exert cytotoxicity through different mechanisms. The Ag(I) complexes were found to induce early apoptosis at relatively high concentrations through caspase- and ROS-independent pathways, whilst the Au(I) analogues were more inclined to induce both early and late apoptosis at lower concentrations through caspase- and ROS-dependent pathways. Intracellular localization experiments found that all complexes localize in mitochondria, however, the Au(I) analogues displayed a greater degree of specificity toward mitochondria. The findings may open new avenues to better understand the effect of the metal center on antitumor mechanisms.

14. Concluding remarks and future perspectives

The large body of investigations carried out in the past decade into mitochondria-targeting metal complexes demonstrates how important and largely active this area of research continues to be. Ir(III)-based complexes are clearly at the forefront and by far have garnered the most attention. General mitochondria-targeting luminescent probes have been developed spanning the visible light to NIR range of the electromagnetic spectrum. More specialized probes for selective interaction with a variety of mitochondrial targets have also been synthesized and heavily investigated. Luminescent Ir(III) complexes have been modulated and honed for use in super-resolution microscopy, transition electron microscopy, correlative imaging, and lifetime-based imaging techniques. Whilst their optical properties are indeed impressive, the potential of mitochondria-targeting Ir(III) complexes is not limited to imaging techniques. Many of the complexes investigated display cytotoxicity toward cancer cell lines at a comparable level, if not better, than that of cisplatin. This combined research is leading to the design of novel anticancer therapeutics with a wide variety of cell death mechanisms including apoptosis, paraptosis, mitophagy, necrosis, ferroptosis, and oncosis. In addition, efficient intersystem crossing allows for quenching of their excited states by endogenous oxygen and biomolecules, promoting the generation of ROS. Many Ir(III) complexes have been observed to display impressive PDT applications in both normoxic and hypoxic environments. The biological properties of Ir(III) complexes have been tuned to target mitochondria so effectively that they can even be used as an appended targeting group in multinuclear complexes. A range of cancer cell specific Ir(III) complexes have been reported and exploited for therapeutic effects. A vast array of Ir(III) structural motifs were presented in this chapter, including neutral and cationic species, half-sandwich structures, homogenous and heterogenous dinuclear complexes, polymers, and peptides. Needless to say, the future for iridium-based mitochondria-targeting complexes looks very bright.

Platinum-based complexes are mostly known for their anticancer properties by targeting DNA. However, in recent years many complexes have been tuned to interact with mitochondria as both an alternative target for anticancer therapeutics and for dual functional therapeutics and imaging agents. Whilst examples of intrinsically mitochondria-targeting platinum complexes were reported, in most cases the assistance of mitochondria-targeting moieties, such as TPP, was necessary. Investigations included the exploration of the effects of altering the structure of platinum prodrugs, overcoming platinum drug resistance, the effects of regioselective addition of targeting groups, and the targeted delivery of metallodrugs. Overall, the findings revealed a variety of mechanisms for platinum drug toxicity, including mitophagy, necrosis, as well as mitochondrial and nuclear DNA damage.

Platinum PDT agents were developed, BODIPY appendages were able to image the intracellular distribution of platinum, and appreciable cytotoxicity toward traditional platinum-drug resistant cancer cell lines was observed. Platinum complexes still have high standing as therapeutics against cancer, and now mitochondria are classified as a new target.

Rhenium is another up-and-coming element for mitochondria-targeting anticancer research. The favorable optical properties of Re(I) complexes have led to the development of mitochondrial imaging agents and multifunctional agents that can monitor their own effects while delivering potent cytotoxic actions. Re(I) complexes were observed to selectively kill cancer cells with multiple cell death mechanisms. It is easy to see why interest in developing novel rhenium complexes is on the rise.

The interest in mitochondria-targeting gold complexes seems to be focused on developing alternatives to platinum-based drugs, to overcome drug resistance and reduce systemic toxicity. These investigations are warranted as various mechanisms of cell death have been reported in a variety of cancer cell lines. Mechanisms reported in the past decade include necrosis, apoptosis, mitochondrial permeability transition, and impairment of the mitochondrial respiratory chain.

Whilst other transition metal elements have not received the same amount of attention as those previously discussed, these more isolated findings are clearly starting to pave the way toward their application in life science. This chapter has made it clear that every discussed transition element offers unique opportunities to design and prepare molecular species for targeted applications. Whilst this area has been under investigation for a longer period than the advances covered in this chapter from the last decade, it is clear the field is still wide open for further investigation.

Acknowledgments

The provision of an Australian Government Research Training Program Scholarship to R.J.M. is kindly acknowledged.

References

1. Brandsma, L.; Vasilevsky, S. F.; Verkruijsse, H. D., Application of Transition Metal Catalyts in Organic Synthesis, Springer Science & Business Media: 2012.
2. Vishnu, V. S.; Jose, S.; Reddy, M. L. *J. Am. Ceram. Soc.* **2011**, *94* (4), 997–1001.
3. Yersin, H. *Triplet Emitters for OLED Applications. Mechanisms of Exciton Trapping and Control of Emission Properties*; Springer Berlin Heidelberg, **2004**; pp. 1–26.
4. Gerling, L. G.; Mahato, S.; Morales-Vilches, A.; Masmitja, G.; Ortega, P.; Voz, C.; Alcubilla, R.; Puigdollers, J. *Sol. Energy Mater. Sol. Cells* **2016**, *145*, 109–115.
5. Scheffler, I. E., Mitochondria, John Wiley & Sons: 2011.

6. Friedman, J. R.; Nunnari, J. *Nature* **2014**, *505* (7483), 335–343.
7. Pizzo, P.; Drago, I.; Filadi, R.; Pozzan, T. *Pflugers Arch* **2012**, *464* (1), 3–17.
8. Luongo, T. S.; Lambert, J. P.; Gross, P.; Nwokedi, M.; Lombardi, A. A.; Shanmughapriya, S.; Carpenter, A. C.; Kolmetzky, D.; Gao, E.; van Berlo, J. H.; Tsai, E. J.; Molkentin, J. D.; Chen, X.; Madesh, M.; Houser, S. R.; Elrod, J. W. *Nature* **2017**, *545* (7652), 93–97.
9. Starkov, A. A. *Measurement of Mitochondrial ROS Production*; Humana Press, 2010; pp. 245–255.
10. Andreyev, A. Y.; Kushnareva, Y. E.; Starkov, A. A. *Biochemistry* **2005**, *70* (2), 200–214.
11. Green, D. R.; Reed, J. C. *Science* **1998**, *281* (5381), 1309–1312.
12. Shi, Y. *Nat. Struct. Mol. Biol.* **2001**, *8* (5), 394–401.
13. Frey, T. G.; Mannella, C. A. *Trends. Biochem. Sci.* **2000**, *25* (7), 319–324.
14. Yaffe, M. P. *Nat. Cell. Biol.* **1999**, *1* (6), E149–E150.
15. Wang, C.; Du, W.; Su, Q. P.; Zhu, M.; Feng, P.; Li, Y.; Zhou, Y.; Mi, N.; Zhu, Y.; Jiang, D.; Zhang, S.; Zhang, Z.; Sun, Y.; Yu, L. *Cell. Res.* **2015**, *25* (10), 1108–1120.
16. Wang, C.; Youle, R. J. *Annu. Rev. Genet.* **2009**, *43* (1), 95–118.
17. Desagher, S.; Martinou, J. C. *Trends Cell Biol.* **2000**, *10* (9), 369–377.
18. Cai, J.; Yang, J.; Jones, D. P. *Biochim. Biophys. Acta* **1998**, *1366* (1–2), 139–149.
19. Garrido, C.; Galluzzi, L.; Brunet, M.; Puig, P. E.; Didelot, C.; Kroemer, G. *Cell Death Differ.* **2006**, *13* (9), 1423–1433.
20. Redza-Dutordoir, M.; Averill-Bates, D. A. *Biochim. Biophys. Acta* **2016**, *1863* (12), 2977–2992.
21. Simon, H. U.; Haj-Yehia, A.; Levi-Schaffer, F. *Apoptosis* **2000**, *5* (5), 415–418.
22. Brand, M. D.; Affourtit, C.; Esteves, T. C.; Green, K.; Lambert, A. J.; Miwa, S.; Pakay, J. L.; Parker, N. *Free. Radic. Biol. Med.* **2004**, *37* (6), 755–767.
23. Beal, M. F. *Ann. Neurol.* **2005**, *58* (4), 495–505.
24. Federico, A.; Cardaioli, E.; Da Pozzo, P.; Formichi, P.; Gallus, G. N.; Radi, E. *J. Neurol. Sci.* **2012**, *322* (1–2), 254–262.
25. Moreira, P. I.; Zhu, X.; Wang, X.; Lee, H. G.; Nunomura, A.; Petersen, R. B.; Perry, G.; Smith, M. A. *Biochim. Biophys. Acta* **2010**, *1802* (1), 212–220.
26. Szendroedi, J.; Phielix, E.; Roden, M. *Nat. Rev. Endocrinol.* **2012**, *8* (2), 92–103.
27. Kelley, D. E.; He, J.; Menshikova, E. V.; Ritov, V. B. *Diabetes* **2002**, *51* (10), 2944–2950.
28. Ritov, V. B.; Menshikova, E. V.; He, J.; Ferrell, R. E.; Goodpaster, B. H.; Kelley, D. E. *Diabetes* **2005**, *54* (1), 8–14.
29. Johnson, L. V.; Walsh, M. L.; Chen, L. B. *Proc. Natl. Acad. Sci. U. S. A.* **1980**, 77 (2), 990–994.
30. Maftah, A.; Petit, J. M.; Ratinaud, M. H.; Julien, R. *Biochem. Biophys. Res. Commun.* **1989**, *164* (1), 185–190.
31. Buckman, J. F.; Hernandez, H.; Kress, G. J.; Votyakova, T. V.; Pal, S.; Reynolds, I. J. *J. Neurosci. Methods* **2001**, *104* (2), 165–176.
32. Zhao, Q.; Huang, C.; Li, F. *Chem. Soc. Rev.* **2011**, *40* (5), 2508–2524.
33. Lo, K. K.; Choi, A. W.; Law, W. H. *Dalton Trans.* **2012**, *41* (20), 6021–6047.
34. Lo, K. K. W.; Li, S. P. Y. *RSC Adv.* **2014**, *4* (21), 10560–10585.
35. Heller, A.; Brockhoff, G.; Goepferich, A. *Eur. J. Pharm. Biopharm.* **2012**, *82* (1), 1–18.
36. Wongrakpanich, A.; Geary, S. M.; Joiner, M. L.; Anderson, M. E.; Salem, A. K. *Nanomedicine* **2014**, *9* (16), 2531–2543.
37. Armstrong, J. S. *Br. J. Pharmacol.* **2007**, *151* (8), 1154–1165.
38. Dong, L.; Neuzil, J. *Cancer. Commun.* **2019**, *39* (1), 63.
39. Murphy, M. P. *Biochim. Biophys. Acta* **2008**, *1777* (7–8), 1028–1031.
40. Walters, A. M.; Porter, G. A., Jr.; Brookes, P. S. *Circ. Res.* **2012**, *111* (9), 1222–1236.

41. Zhang, C.-Y.; Baffy, G.; Perret, P.; Krauss, S.; Peroni, O.; Grujic, D.; Hagen, T.; Vidal-Puig, A. J.; Boss, O.; Kim, Y.-B.; Zheng, X. X.; Wheeler, M. B.; Shulman, G. I.; Chan, C. B.; Lowell, B. B. *Cell* **2001**, *105* (6), 745–755.
42. Chari, R. V. J. *Acc. Chem. Res.* **2008**, *41* (1), 98–107.
43. Wang, Y.; Xia, Y.; Lu, Z. *Cancer Commun.* **2018**, *38* (1), 65.
44. Jia, D.; Park, J.; Jung, K.; Levine, H.; Kaipparettu, B. *Cells* **2018**, 7 (3), 21.
45. Murphy, M. P. *Trends Biotechnol.* **1997**, *15* (8), 326–330.
46. Caporale, C.; Bader, C. A.; Sorvina, A.; MaGee, K. D. M.; Skelton, B. W.; Gillam, T. A.; Wright, P. J.; Raiteri, P.; Stagni, S.; Morrison, J. L.; Plush, S. E.; Brooks, D. A.; Massi, M. *Chem. Eur. J.* **2017**, *23* (62), 15666–15679.
47. Murase, T.; Yoshihara, T.; Tobita, S. *Chem. Lett.* **2012**, *41* (3), 262–263.
48. Horton, K. L.; Stewart, K. M.; Fonseca, S. B.; Guo, Q.; Kelley, S. O. *Chem. Biol.* **2008**, *15* (4), 375–382.
49. Sternberg, E. D.; Dolphin, D.; Bruckner, C. *Tetrahedron* **1998**, *54* (17), 4151–4202.
50. Imberti, C.; Zhang, P.; Huang, H.; Sadler, P. J. *Angew. Chem. Int. Ed.* **2020**, *59* (1), 61–73.
51. Dougherty, T. J.; Gomer, C. J.; Henderson, B. W.; Jori, G.; Kessel, D.; Korbelik, M.; Moan, J.; Peng, Q. *J. Natl. Cancer Inst.* **1998**, *90* (12), 889–905.
52. Dolmans, D. E.; Fukumura, D.; Jain, R. K. *Nat. Rev. Cancer.* **2003**, *3* (5), 380–387.
53. Li, J.; Chen, T. *Coord. Chem. Rev.* **2020**, *418*, 213355.
54. Koziar, J. C.; Cowan, D. O. *Acc. Chem. Res.* **1978**, *11* (9), 334–341.
55. Heinemann, F.; Karges, J.; Gasser, G. *Acc. Chem. Res.* **2017**, *50* (11), 2727–2736.
56. McFarland, S. A.; Mandel, A.; Dumoulin-White, R.; Gasser, G. *Curr. Opin. Chem. Biol.* **2020**, *56*, 23–27.
57. Felder, P. S.; Keller, S.; Gasser, G. *Adv. Ther.* **2019**, *3* (1), 1900139.
58. Flamigni, L.; Barbieri, A.; Sabatini, C.; Ventura, B.; Barigelletti, F. *Topics in Current Chemistry: Photochemistry and Photophysics of Coordination Compounds: Iridium*; Springer Berlin Heidelberg, **2007**; pp. 143–203.
59. Huang, H.; Banerjee, S.; Sadler, P. J. *Chembiochem* **2018**, *19* (15), 1574–1589.
60. Konkankit, C. C.; Marker, S. C.; Knopf, K. M.; Wilson, J. J. *Dalton Trans.* **2018**, *47* (30), 9934–9974.
61. Sharma, S. A.; Pete, S.; Roy, N.; Paira, P. *Inorg. Chim. Acta* **2020**, *513*, 119925.
62. Li, S. P.; Tang, T. S.; Yiu, K. S.; Lo, K. K. *Chem. Eur. J.* **2012**, *18* (42), 13342–13354.
63. Wang, B.; Liang, Y.; Dong, H.; Tan, T.; Zhan, B.; Cheng, J.; Lo, K. K.; Lam, Y. W.; Cheng, S. H. *Chembiochem* **2012**, *13* (18), 2729–2737.
64. Yang, H.; Ding, L.; An, L.; Xiang, Z.; Chen, M.; Zhou, J.; Li, F.; Wu, D.; Yang, S. *Biomaterials* **2012**, *33* (33), 8591–8599.
65. Chen, Y.; Qiao, L.; Yu, B.; Li, G.; Liu, C.; Ji, L.; Chao, H. *Chem. Commun.* **2013**, *49* (94), 11095–11097.
66. Chen, Y.; Qiao, L.; Ji, L.; Chao, H. *Biomaterials* **2014**, *35* (1), 2–13.
67. Hearn, J. M.; Romero-Canelon, I.; Qamar, B.; Liu, Z.; Hands-Portman, I.; Sadler, P. J. *ACS Chem. Biol.* **2013**, *8* (6), 1335–1343.
68. Law, W. H.; Lee, L. C.; Louie, M. W.; Liu, H. W.; Ang, T. W.; Lo, K. K. *Inorg. Chem.* **2013**, *52* (22), 13029–13041.
69. Li, S. P.; Lau, C. T.; Louie, M. W.; Lam, Y. W.; Cheng, S. H.; Lo, K. K. *Biomaterials* **2013**, *34* (30), 7519–7532.
70. Lo, K. K.; Chan, B. T.; Liu, H. W.; Zhang, K. Y.; Li, S. P.; Tang, T. S. *Chem. Commun.* **2013**, *49* (39), 4271–4273.
71. Li, G.; Lin, Q.; Ji, L.; Chao, H. *J. Mater. Chem. B* **2014**, *2* (45), 7918–7926.
72. Zhang, Q.; Cao, R.; Fei, H.; Zhou, M. *Dalton Trans.* **2014**, *43* (44), 16872–16879.

73. Chen, X.; Sun, L.; Chen, Y.; Cheng, X.; Wu, W.; Ji, L.; Chao, H. *Biomaterials* **2015**, *58*, 72–81.
74. Chen, Y.; Xu, W. C.; Zuo, J. R.; Ji, L. N.; Chao, H. *J. Mater. Chem. B* **2015**, *3* (16), 3306–3314.
75. Choi, J. S.; Maity, A.; Gray, T.; Berdis, A. J. *J. Biol. Chem.* **2015**, *290* (15), 9714–9726.
76. Jin, C.; Liu, J.; Chen, Y.; Li, G.; Guan, R.; Zhang, P.; Ji, L.; Chao, H. *Dalton Trans.* **2015**, *44* (16), 7538–7547.
77. Jin, C.; Liu, J.; Chen, Y.; Zeng, L.; Guan, R.; Ouyang, C.; Ji, L.; Chao, H. *Chem. Eur. J.* **2015**, *21* (34), 12000–12010.
78. Jin, C. Z.; Liu, J. P.; Chen, Y.; Guan, R. L.; Ouyang, C.; Zhu, Y. J.; Ji, L. N.; Chao, H. *Sci. Rep.* **2016**, *6* (1), 1–13.
79. Li, G. Y.; Chen, Y.; Wang, J. Q.; Wu, J. H.; Gasser, G.; Ji, L. N.; Chao, H. *Biomaterials* **2015**, *63*, 128–136.
80. Li, G.; Lin, Q.; Sun, L.; Feng, C.; Zhang, P.; Yu, B.; Chen, Y.; Wen, Y.; Wang, H.; Ji, L.; Chao, H. *Biomaterials* **2015**, *53*, 285–295.
81. Li, Y.; Tan, C. P.; Zhang, W.; He, L.; Ji, L. N.; Mao, Z. W. *Biomaterials* **2015**, *39*, 95–104.
82. Qiu, K.; Huang, H.; Liu, B.; Liu, Y.; Zhang, P.; Chen, Y.; Ji, L.; Chao, H. *J. Mater. Chem. B* **2015**, *3* (32), 6690–6697.
83. Wan, Q.; Chen, S.; Shi, W.; Li, L.; Ma, H. *Angew. Chem. Int. Ed.* **2014**, *53* (41), 10916–10920.
84. Takizawa, S. Y.; Breitenbach, T.; Westberg, M.; Holmegaard, L.; Gollmer, A.; Jensen, R. L.; Murata, S.; Ogilby, P. R. *Photochem. Photobiol. Sci.* **2015**, *14* (10), 1831–1843.
85. Tang, T. S. M.; Leung, K. K.; Louie, M. W.; Liu, H. W.; Cheng, S. H.; Lo, K. K. W. *Dalton Trans.* **2015**, *44* (11), 4945–4956.
86. Yoshihara, T.; Murayarria, S.; Masuda, T.; Kikuchi, T.; Yoshida, K.; Hosaka, M.; Tobita, S. *J. Photochem. Photobiol. A* **2015**, *299*, 172–182.
87. Zhang, K. Y.; Liu, H. W.; Tang, M. C.; Choi, A. W. T.; Zhu, N. Y.; Wei, X. G.; Lau, K. C.; Lo, K. K. W. *Inorg. Chem.* **2015**, *54* (13), 6582–6593.
88. Huang, H.; Yang, L.; Zhang, P.; Qiu, K.; Huang, J.; Chen, Y.; Diao, J.; Liu, J.; Ji, L.; Long, J.; Chao, H. *Biomaterials* **2016**, *83*, 321–331.
89. Huang, H.; Zhang, P.; Qiu, K.; Huang, J.; Chen, Y.; Ji, L.; Chao, H. *Sci. Rep.* **2016**, *6* (1), 20887.
90. Lv, W.; Zhang, Z.; Zhang, K. Y.; Yang, H.; Liu, S.; Xu, A.; Guo, S.; Zhao, Q.; Huang, W. *Angew. Chem. Int. Ed.* **2016**, *55* (34), 9947–9951.
91. Qiu, K.; Liu, Y.; Huang, H.; Liu, C.; Zhu, H.; Chen, Y.; Ji, L.; Chao, H. *Dalton Trans.* **2016**, *45* (41), 16144–16147.
92. Sun, L.; Chen, Y.; Kuang, S.; Li, G.; Guan, R.; Liu, J.; Ji, L.; Chao, H. *Chem. Eur. J.* **2016**, *22* (26), 8955–8965.
93. Tso, K. K.; Leung, K. K.; Liu, H. W.; Lo, K. K. *Chem. Commun.* **2016**, *52* (24), 4557–4560.
94. Wang, F. X.; Chen, M. H.; Hu, X. Y.; Ye, R. R.; Tan, C. P.; Ji, L. N.; Mao, Z. W. *Sci. Rep.* **2016**, *6* (1), 1–15.
95. Wang, J. Q.; Hou, X. J.; Jin, C. Z.; Chao, H. *Chin. J. Chem.* **2016**, *34* (6), 583–588.
96. Wang, J. Q.; Hou, X. J.; Zhao, Z. Z.; Bo, H. B.; Chen, Q. Z. *Inorg. Chem. Commun.* **2016**, *67*, 40–43.
97. Xiong, K.; Chen, Y.; Ouyang, C.; Guan, R. L.; Ji, L. N.; Chao, H. *Biochimie* **2016**, *125*, 186–194.
98. Ye, R. R.; Tan, C. P.; Ji, L. N.; Mao, Z. W. *Dalton Trans.* **2016**, *45* (33), 13042–13051.
99. Cao, J. J.; Tan, C. P.; Chen, M. H.; Wu, N.; Yao, D. Y.; Liu, X. G.; Ji, L. N.; Mao, Z. W. *Chem. Sci.* **2017**, *8* (1), 631–640.

100. Sorvina, A.; Bader, C. A.; Caporale, C.; Carter, E. A.; Johnson, I. R. D.; Parkinson-Lawrence, E. J.; Simpson, P. V.; Wright, P. J.; Stagni, S.; Lay, P. A.; Massi, M.; Brooks, D. A.; Plush, S. E. *Oncotarget* **2018**, *9* (85), 35541–35552.
101. Chen, M. H.; Wang, F. X.; Cao, J. J.; Tan, C. P.; Ji, L. N.; Mao, Z. W. *ACS Appl. Mater. Interfaces* **2017**, *9* (15), 13304–13314.
102. Jin, C.; Guan, R.; Wu, J.; Yuan, B.; Wang, L.; Huang, J.; Wang, H.; Ji, L.; Chao, H. *Chem. Commun.* **2017**, *53* (75), 10374–10377.
103. Li, Y.; Liu, B.; Lu, X. R.; Li, M. F.; Ji, L. N.; Mao, Z. W. *Dalton Trans.* **2017**, *46* (34), 11363–11371.
104. Liu, J.; Jin, C.; Yuan, B.; Liu, X.; Chen, Y.; Ji, L.; Chao, H. *Chem. Commun.* **2017**, *53* (12), 2052–2055.
105. Liu, J. P.; Jin, C. Z.; Yuan, B.; Chen, Y.; Liu, X. G.; Ji, L. N.; Chao, H. *Chem. Commun.* **2017**, *53* (71), 9878–9881.
106. Liu, Y.; Zhang, P.; Fang, X.; Wu, G.; Chen, S.; Zhang, Z.; Chao, H.; Tan, W.; Xu, L. *Dalton Trans.* **2017**, *46* (14), 4777–4785.
107. McKenzie, L. K.; Sazanovich, I. V.; Baggaley, E.; Bonneau, M.; Guerchais, V.; Williams, J. A.; Weinstein, J. A.; Bryant, H. E. *Chem. Eur. J.* **2017**, *23* (2), 234–238.
108. Ouyang, M.; Zeng, L. L.; Huang, H. Y.; Jin, C. Z.; Liu, J. P.; Chen, Y.; Ji, L. N.; Chao, H. *Dalton Trans.* **2017**, *46* (20), 6734–6744.
109. Qiu, K.; Ouyang, M.; Liu, Y.; Huang, H.; Liu, C.; Chen, Y.; Ji, L.; Chao, H. *J. Mater. Chem. B* **2017**, *5* (27), 5488–5498.
110. Shewring, J. R.; Cankut, A. J.; McKenzie, L. K.; Crowston, B. J.; Botchway, S. W.; Weinstein, J. A.; Edwards, E.; Ward, M. D. *Inorg. Chem.* **2017**, *56* (24), 15259–15270.
111. Venkatesh, V.; Berrocal-Martin, R.; Wedge, C. J.; Romero-Canelon, I.; Sanchez-Cano, C.; Song, J. I.; Coverdale, J. P. C.; Zhang, P.; Clarkson, G. J.; Habtemariam, A.; Magennis, S. W.; Deeth, R. J.; Sadler, P. J. *Chem. Sci.* **2017**, *8* (12), 8271–8278.
112. Ye, R. R.; Cao, J. J.; Tan, C. P.; Ji, L. N.; Mao, Z. W. *Chem. Eur. J.* **2017**, *23* (60), 15166–15176.
113. Yokoi, K.; Hisamatsu, Y.; Naito, K.; Aoki, S. *Eur. J. Inorg. Chem.* **2017**, *2017* (44), 5295–5309.
114. Zhang, C.; Lai, S. H.; Yang, H. H.; Xing, D. G.; Zeng, C. C.; Tang, B.; Wan, D.; Liu, Y. J. *RSC Adv.* **2017**, *7* (29), 17752–17762.
115. Zhang, F.; Liang, X.; Zhang, W.; Wang, Y. L.; Wang, H.; Mohammed, Y. H.; Song, B.; Zhang, R.; Yuan, J. *Biosens. Bioelectron.* **2017**, *87*, 1005–1011.
116. Zheng, Y.; He, L.; Zhang, D. Y.; Tan, C. P.; Ji, L. N.; Mao, Z. W. *Dalton Trans.* **2017**, *46* (34), 11395–11407.
117. Chen, Q.; Jin, C.; Shao, X.; Guan, R.; Tian, Z.; Wang, C.; Liu, F.; Ling, P.; Guan, J. L.; Ji, L.; Wang, F.; Chao, H.; Diao, J. *Small* **2018**, *14* (41), e1802166.
118. Perez-Arnaiz, C.; Acuna, M. I.; Busto, N.; Echevarria, I.; Martinez-Alonso, M.; Espino, G.; Garcia, B.; Dominguez, F. *Eur. J. Med. Chem.* **2018**, *157*, 279–293.
119. Guan, R.; Chen, Y.; Zeng, L.; Rees, T. W.; Jin, C.; Huang, J.; Chen, Z. S.; Ji, L.; Chao, H. *Chem. Sci.* **2018**, *9* (23), 5183–5190.
120. He, L.; Wang, K. N.; Zheng, Y.; Cao, J. J.; Zhang, M. F.; Tan, C. P.; Ji, L. N.; Mao, Z. W. *Dalton Trans.* **2018**, *47* (20), 6942–6953.
121. Jing, Y.; Cao, Q.; Hao, L.; Yang, G. G.; Hu, W. L.; Ji, L. N.; Mao, Z. W. *Chem. Commun.* **2018**, *54* (3), 271–274.
122. Laws, K.; Eskandari, A.; Lu, C. X.; Suntharalingam, K. *Chem. Eur. J.* **2018**, *24* (57), 15205–15210.
123. Liang, Z. H.; Wan, D.; Yi, Q. Y.; Zhang, W. Y.; Liu, Y. J. *Transit. Met. Chem.* **2018**, *43* (3), 243–257.

124. Qiu, K.; Ke, L.; Zhang, X.; Liu, Y.; Rees, T. W.; Ji, L.; Diao, J.; Chao, H. *Chem. Commun.* **2018**, *54* (19), 2421–2424.
125. Schibilla, F.; Holthenrich, A.; Song, B.; Linard Matos, A. L.; Grill, D.; Rota Martir, D.; Gerke, V.; Zysman-Colman, E.; Ravoo, B. J. *Chem Sci* **2018**, *9* (40), 7822–7828.
126. Sorvina, A.; Bader, C. A.; Darby, J. R. T.; Lock, M. C.; Soo, J. Y.; Johnson, I. R. D.; Caporale, C.; Voelcker, N. H.; Stagni, S.; Massi, M.; Morrison, J. L.; Plush, S. E.; Brooks, D. A. *Sci. Rep.* **2018**, *8* (1), 8191.
127. Sreedharan, S.; Sinopoli, A.; Jarman, P. J.; Robinson, D.; Clemmet, C.; Scattergood, P. A.; Rice, C. R.; Smythe, C. G. W.; Thomas, J. A.; Elliott, P. I. P. *Dalton Trans.* **2018**, *47* (14), 4931–4940.
128. Tang, B.; Wan, D.; Wang, Y. J.; Yi, Q. Y.; Guo, B. H.; Liu, Y. J. *Eur. J. Med. Chem.* **2018**, *145*, 302–314.
129. Wang, H.; Hu, L.; Du, W.; Tian, X. H.; Hu, Z. J.; Zhang, Q.; Zhou, H. P.; Wu, J. Y.; Uvdal, K.; Tian, Y. P. *Sens. Actuators B Chem.* **2018**, *255*, 408–415.
130. Wang, X.; Zhu, M.; Gao, F.; Wei, W.; Qian, Y.; Liu, H. K.; Zhao, J. *J. Inorg. Biochem.* **2018**, *180*, 179–185.
131. Xiao, Q. M.; Zhao, Z. Z.; Lin, K.; Wang, J. Q. *Inorg. Chem. Commun.* **2018**, *94*, 75–79.
132. Yi, Q. Y.; Wan, D.; Tang, B.; Wang, Y. J.; Zhang, W. Y.; Du, F.; He, M.; Liu, Y. J. *Eur. J. Med. Chem.* **2018**, *145*, 338–349.
133. Zhang, C.; Liu, M.; Liu, S.; Yang, H.; Zhao, Q.; Liu, Z.; He, W. *Inorg. Chem.* **2018**, *57* (17), 10625–10632.
134. Zhang, C.; Qiu, K.; Liu, C.; Huang, H.; Rees, T. W.; Ji, L.; Zhang, Q.; Chao, H. *Dalton Trans.* **2018**, *47* (37), 12907–12913.
135. Zhang, W. Y.; Yi, Q. Y.; Wang, Y. J.; Du, F.; He, M.; Tang, B.; Wan, D.; Liu, Y. J.; Huang, H. L. *Eur. J. Med. Chem.* **2018**, *151*, 568–584.
136. Bevernaegie, R.; Doix, B.; Bastien, E.; Diman, A.; Decottignies, A.; Feron, O.; Elias, B. *J. Am. Chem. Soc.* **2019**, *141* (46), 18486–18491.
137. Cao, J. J.; Zheng, Y.; Wu, X. W.; Tan, C. P.; Chen, M. H.; Wu, N.; Ji, L. N.; Mao, Z. W. *J. Med. Chem.* **2019**, *62* (7), 3311–3322.
138. Chen, H.; Ge, C.; Cao, H.; Zhang, X.; Zhang, L.; Jiang, L.; Zhang, P.; Zhang, Q. *Dalton Trans.* **2019**, *48* (46), 17200–17209.
139. Du, Z.; Zhang, R.; Song, B.; Zhang, W.; Wang, Y. L.; Liu, J.; Liu, C.; Xu, Z. P.; Yuan, J. *Chem. Eur. J.* **2019**, *25* (6), 1498–1506.
140. Guo, S.; Han, M. P.; Chen, R. Z.; Zhuang, Y. L.; Zou, L.; Liu, S. J.; Huang, W.; Zhao, Q. *Sci. China Chem.* **2019**, *62* (12), 1639–1648.
141. Hao, L.; Li, Z. W.; Zhang, D. Y.; He, L.; Liu, W.; Yang, J.; Tan, C. P.; Ji, L. N.; Mao, Z. W. *Chem. Sci.* **2019**, *10* (5), 1285–1293.
142. He, L.; Zhang, M. F.; Pan, Z. Y.; Wang, K. N.; Zhao, Z. J.; Li, Y.; Mao, Z. W. *Chem. Commun.* **2019**, *55* (70), 10472–10475.
143. Li, J.; Chen, H. M.; Zeng, L. L.; Rees, T. W.; Xiong, K.; Chen, Y.; Ji, L. N.; Chao, H. *Inorg. Chem. Front.* **2019**, *6* (4), 1003–1010.
144. Li, S. P.-Y.; Shum, J.; Lo, K. K.-W. *Dalton Trans.* **2019**, *48* (26), 9692–9702.
145. Ma, W.; Zhang, S.; Tian, Z.; Xu, Z.; Zhang, Y.; Xia, X.; Chen, X.; Liu, Z. *Eur. J. Med. Chem.* **2019**, *181*, 111599.
146. Song, X. D.; Chen, B. B.; He, S. F.; Pan, N. L.; Liao, J. X.; Chen, J. X.; Wang, G. H.; Sun, J. *Eur. J. Med. Chem.* **2019**, *179*, 26–37.
147. Wu, X. W.; Zheng, Y.; Wang, F. X.; Cao, J. J.; Zhang, H.; Zhang, D. Y.; Tan, C. P.; Ji, L. N.; Mao, Z. W. *Chem. Eur. J.* **2019**, *25* (28), 7012–7022.
148. Yi, S.; Lu, Z.; Zhang, J.; Wang, J.; Xie, Z.; Hou, L. *ACS Appl. Mater. Interfaces* **2019**, *11* (17), 15276–15289.

149. Bi, X. D.; Yang, R.; Zhou, Y. C.; Chen, D. M.; Li, G. K.; Guo, Y. X.; Wang, M. F.; Liu, D. D.; Gao, F. *Inorg. Chem.* **2020**, *59* (20), 14920–14931.
150. Chen, Z. L.; Zou, B. Q.; Qin, Q. P.; Wang, Z. F.; Tan, M. X.; Huang, X. L.; Liang, C. J.; Liang, H. *Inorg. Chem. Commun.* **2020**, *115*, 107854.
151. Gu, Y.; Wen, H.; Bai, L.; Zhou, Y.; Zhang, H.; Tian, L.; Zhang, Y.; Hao, J.; Liu, Y. *J. Inorg. Biochem.* **2020**, *212*, 111215.
152. Kuang, S.; Sun, L.; Zhang, X.; Liao, X.; Rees, T. W.; Zeng, L.; Chen, Y.; Zhang, X.; Ji, L.; Chao, H. *Angew. Chem. Int. Ed.* **2020**, *59* (46), 20697–20703.
153. Li, Y.; Ma, B.; Long, Y. G.; Song, Z.; Su, J. Y.; Wang, Y.; Liu, R.; Song, G. L.; Zhu, H. J. *J. Mater. Chem. C* **2020**, *8* (7), 2467–2474.
154. Li, Y.; Wang, K. N.; He, L.; Ji, L. N.; Mao, Z. W. *J. Inorg. Biochem.* **2020**, *205*, 110976.
155. Liu, X.; Dai, P.; Gu, T.; Wu, Q.; Wei, H.; Liu, S.; Zhang, K. Y.; Zhao, Q. *J. Inorg. Biochem.* **2020**, *209*, 111106.
156. Peng, W.; Hegazy, A. M.; Jiang, N.; Chen, X.; Qi, H. X.; Zhao, X. B.; Pu, J.; Ye, R. R.; Li, R. T. *J. Inorg. Biochem.* **2020**, *203*, 110909.
157. Qin, L. Q.; Zou, B. Q.; Qin, Q. P.; Wang, Z. F.; Yang, L.; Tan, M. X.; Liang, C. J.; Liang, H. *New J. Chem.* **2020**, *44* (19), 7832–7837.
158. Qin, W. W.; Pan, Z. Y.; Cai, D. H.; Li, Y.; He, L. *Dalton Trans.* **2020**, *49* (11), 3562–3569.
159. Wang, X. X.; Chen, F.; Zhang, J. Y.; Sun, J. X.; Zhao, X. Y.; Zhu, Y. L.; Wei, W.; Zhao, J.; Guo, Z. J. *Sci. China Chem.* **2020**, *63* (1), 65–72.
160. Deng, P.; Pei, Y.; Liu, M.; Song, W.; Wang, M.; Wang, F.; Wu, C.; Xu, L. *RSC Adv.* **2021**, *11* (13), 7610–7620.
161. Guan, R. L.; Xie, L. N.; Wang, L. L.; Zhou, Y.; Chen, Y.; Ji, L. N.; Chao, H. *Inorg. Chem. Front.* **2021**, *8* (7), 1788–1794.
162. Hao, J.; Zhang, H.; Tian, L.; Yang, L.; Zhou, Y.; Zhang, Y.; Liu, Y.; Xing, D. *J. Inorg. Biochem.* **2021**, *221*, 111465.
163. Wu, W.; Liao, X.; Chen, Y.; Ji, L.; Chao, H. *Anal. Chem.* **2021**, *93* (22), 8062–8070.
164. Wu, Y.; Wu, J.; Wong, W. Y. *Biomater. Sci.* **2021**, *9* (14), 4843–4853.
165. Yuan, H.; Han, Z.; Chen, Y.; Qi, F.; Fang, H.; Guo, Z.; Zhang, S.; He, W. *Angew. Chem. Int. Ed.* **2021**, *60* (15), 8174–8181.
166. Dasari, S.; Tchounwou, P. B. *Eur. J. Pharmacol.* **2014**, *740*, 364–378.
167. Galluzzi, L.; Senovilla, L.; Vitale, I.; Michels, J.; Martins, I.; Kepp, O.; Castedo, M.; Kroemer, G. *Oncogene* **2012**, *31* (15), 1869–1883.
168. Armstrong, J. S. *Br. J. Pharmacol.* **2006**, *147* (3), 239–248.
169. Williams, J. A. G. *Topics in Current Chemistry: Photochemistry and Photophysics of Coordination Compounds: Platinum*; Springer Berlin Heidelberg, **2007**; pp. 205–268.
170. Guo, Z.; Tong, W. L.; Chan, M. C. *Chem. Commun.* **2014**, *50* (14), 1711–1714.
171. Li, J.; He, X.; Zou, Y.; Chen, D.; Yang, L.; Rao, J.; Chen, H.; Chan, M. C. W.; Li, L.; Guo, Z.; Zhang, L. W.; Chen, C. *Metallomics* **2017**, *9* (6), 726–733.
172. Sun, T.; Guan, X.; Zheng, M.; Jing, X.; Xie, Z. *ACS Med. Chem. Lett.* **2015**, *6* (4), 430–433.
173. Mitra, K.; Gautam, S.; Kondaiah, P.; Chakravarty, A. R. *ChemMedChem* **2016**, *11* (17), 1956–1967.
174. Zhao, J.; Hua, W.; Xu, G.; Gou, S. *J. Inorg. Biochem.* **2017**, *176*, 175–180.
175. Jin, S.; Hao, Y.; Zhu, Z.; Muhammad, N.; Zhang, Z.; Wang, K.; Guo, Y.; Guo, Z.; Wang, X. *Inorg. Chem.* **2018**, *57* (17), 11135–11145.
176. Muhammad, N.; Sadia, N.; Zhu, C.; Luo, C.; Guo, Z.; Wang, X. *Chem. Commun.* **2017**, *53* (72), 9971–9974.
177. Ouyang, C.; Chen, L.; Rees, T. W.; Chen, Y.; Liu, J.; Ji, L.; Long, J.; Chao, H. *Chem. Commun.* **2018**, *54* (49), 6268–6271.

178. Wang, F. Y.; Tang, X. M.; Wang, X.; Huang, K. B.; Feng, H. W.; Chen, Z. F.; Liu, Y. N.; Liang, H. *Eur. J. Med. Chem.* **2018**, *155*, 639–650.
179. Yilmaz, V. T.; Icsel, C.; Turgut, O. R.; Aygun, M.; Erkisa, M.; Turkdemir, M. H.; Ulukaya, E. *Eur. J. Med. Chem.* **2018**, *155*, 609–622.
180. Zhou, Z.; Liu, J.; Rees, T. W.; Wang, H.; Li, X.; Chao, H.; Stang, P. J. *Proc. Natl. Acad. Sci. U. S. A.* **2018**, *115* (22), 5664–5669.
181. Wang, K.; Zhu, C.; He, Y.; Zhang, Z.; Zhou, W.; Muhammad, N.; Guo, Y.; Wang, X.; Guo, Z. *Angew. Chem. Int. Ed.* **2019**, *58* (14), 4638–4643.
182. Zhu, Z. Z.; Wang, Z. H.; Zhang, C. L.; Wang, Y. J.; Zhang, H. M.; Gan, Z. J.; Guo, Z. J.; Wang, X. Y. *Chem. Sci.* **2019**, *10* (10), 3089–3095.
183. Muhammad, N.; Tan, C. P.; Muhammad, K.; Wang, J.; Sadia, N.; Pan, Z. Y.; Ji, L. N.; Mao, Z. W. *Inorg. Chem. Front.* **2020**, 7 (20), 4010–4019.
184. Kirgan, R. A.; Sullivan, B. P.; Rillema, D. P. *Topics in Current Chemistry: Photochemistry and Photophysics of Coordination Compounds: Rhenium*; Springer Berlin Heidelberg, **2007**; pp. 45–100.
185. Lo, K. K. W.; Zhang, K. Y.; Li, S. P. Y. *Eur. J. Inorg. Chem.* **2011**, *2011* (24), 3551–3568.
186. Fernandez-Moreira, V.; Thorp-Greenwood, F. L.; Amoroso, A. J.; Cable, J.; Court, J. B.; Gray, V.; Hayes, A. J.; Jenkins, R. L.; Kariuki, B. M.; Lloyd, D.; Millet, C. O.; Williams, C. F.; Coogan, M. P. *Org. Biomol. Chem.* **2010**, *8* (17), 3888–3901.
187. Leonidova, A.; Gasser, G. *ACS Chem. Biol.* **2014**, *9* (10), 2180–2193.
188. Kitanovic, I.; Can, S.; Alborzinia, H.; Kitanovic, A.; Pierroz, V.; Leonidova, A.; Pinto, A.; Spingler, B.; Ferrari, S.; Molteni, R.; Steffen, A.; Metzler-Nolte, N.; Wolfl, S.; Gasser, G. *Chem. Eur. J.* **2014**, *20* (9), 2496–2507.
189. Gasser, G.; Pinto, A.; Neumann, S.; Sosniak, A. M.; Seitz, M.; Merz, K.; Heumann, R.; Metzler-Nolte, N. *Dalton Trans.* **2012**, *41* (8), 2304–2313.
190. Kowalski, K.; Szczupak, L.; Bernas, T.; Czerwieniec, R. *J. Organomet. Chem.* **2015**, *782*, 124–130.
191. Ye, R. R.; Tan, C. P.; Lin, Y. N.; Ji, L. N.; Mao, Z. W. *Chem. Commun.* **2015**, *51* (39), 8353–8356.
192. Ye, R. R.; Tan, C. P.; Chen, M. H.; Hao, L.; Ji, L. N.; Mao, Z. W. *Chem. Eur. J.* **2016**, *22* (23), 7800–7809.
193. Skiba, J.; Bernas, T.; Trzybinski, D.; Wozniak, K.; Ferraro, G.; Marasco, D.; Merlino, A.; Shafikov, M. Z.; Czerwieniec, R.; Kowalski, K. Molecules. **2017**, *22* (5), 809.
194. Yang, J.; Zhao, J. X.; Cao, Q.; Hao, L.; Zhou, D.; Gan, Z.; Ji, L. N.; Mao, Z. W. *ACS Appl. Mater. Interfaces* **2017**, *9* (16), 13900–13912.
195. Yang, J.; Cao, Q.; Zhang, H.; Hao, L.; Zhou, D.; Gan, Z.; Li, Z.; Tong, Y. X.; Ji, L. N.; Mao, Z. W. *Biomaterials* **2018**, *176*, 94–105.
196. Konkankit, C. C.; King, A. P.; Knopf, K. M.; Southard, T. L.; Wilson, J. J. *ACS Med. Chem. Lett.* **2019**, *10* (5), 822–827.
197. Skiba, J.; Kowalczyk, A.; Staczek, P.; Bernas, T.; Trzybinski, D.; Wozniak, K.; Schatzschneider, U.; Czerwieniec, R.; Kowalski, K. *New J. Chem.* **2019**, *43* (2), 573–583.
198. Wang, F. X.; Liang, J. H.; Zhang, H.; Wang, Z. H.; Wan, Q.; Tan, C. P.; Ji, L. N.; Mao, Z. W. *ACS Appl. Mater. Interfaces* **2019**, *11* (14), 13123–13133.
199. Day, A. H.; Domarkas, J.; Nigam, S.; Renard, I.; Cawthorne, C.; Burke, B. P.; Bahra, G. S.; Oyston, P. C. F.; Fallis, I. A.; Archibald, S. J.; Pope, S. J. A. *Dalton Trans.* **2020**, *49* (2), 511–523.
200. He, S. F.; Pan, N. L.; Chen, B. B.; Liao, J. X.; Huang, M. Y.; Qiu, H. J.; Jiang, D. C.; Wang, J. J.; Chen, J. X.; Sun, J. *J. Biol. Inorg. Chem.* **2020**, *25* (8), 1107–1116.

201. Pan, Z. Y.; Tan, C. P.; Rao, L. S.; Zhang, H.; Zheng, Y.; Hao, L.; Ji, L. N.; Mao, Z. W. *Angew. Chem. Int. Ed.* **2020**, *59* (42), 18755–18762.
202. Balfourier, A.; Kolosnjaj-Tabi, J.; Luciani, N.; Carn, F.; Gazeau, F. *Proc. Natl. Acad. Sci. U. S. A.* **2020**, *117* (37), 22639–22648.
203. Holmgren, A.; Lu, J. *Biochem. Biophys. Res. Commun.* **2010**, *396* (1), 120–124.
204. Bindoli, A.; Rigobello, M. P.; Scutari, G.; Gabbiani, C.; Casini, A.; Messori, L. *Coord. Chem. Rev.* **2009**, *253* (11–12), 1692–1707.
205. Pratesi, A.; Gabbiani, C.; Ginanneschi, M.; Messori, L. *Chem. Commun.* **2010**, *46* (37), 7001–7003.
206. Mahmood, D. F.; Abderrazak, A.; El Hadri, K.; Simmet, T.; Rouis, M. *Antioxid. Redox Signal* **2013**, *19* (11), 1266–1303.
207. Rubbiani, R.; Can, S.; Kitanovic, I.; Alborzinia, H.; Stefanopoulou, M.; Kokoschka, M.; Monchgesang, S.; Sheldrick, W. S.; Wolfl, S.; Ott, I. *J. Med. Chem.* **2011**, *54* (24), 8646–8657.
208. Marmol, I.; Virumbrales-Munoz, M.; Quero, J.; Sanchez-de-Diego, C.; Fernandez, L.; Ochoa, I.; Cerrada, E.; Yoldi, M. J. R. *J. Inorg. Biochem.* **2017**, *176*, 123–133.
209. Chrysouli, M. P.; Banti, C. N.; Kourkoumelis, N.; Panayiotou, N.; Markopoulos, G. S.; Tasiopoulos, A. J.; Hadjikakou, S. K. *J. Inorg. Biochem.* **2018**, *179*, 107–120.
210. Huang, K. B.; Wang, F. Y.; Tang, X. M.; Feng, H. W.; Chen, Z. F.; Liu, Y. C.; Liu, Y. N.; Liang, H. *J. Med. Chem.* **2018**, *61* (8), 3478–3490.
211. Rodriguez-Fanjul, V.; Lopez-Torres, E.; Mendiola, M. A.; Pizarro, A. M. *Eur. J. Med. Chem.* **2018**, *148*, 372–383.
212. Yu, B.; Ma, L.; Jin, J.; Jiang, F.; Zhou, G.; Yan, K.; Liu, Y. *Toxicol. Res.* **2018**, 7 (6), 1081–1090.
213. Sen, S.; Li, Y.; Lynch, V.; Arumugam, K.; Sessler, J. L.; Arambula, J. F. *Chem. Commun.* **2019**, *55* (71), 10627–10630.
214. Vest, K. E.; Hashemi, H. F.; Cobine, P. A. *The Copper Metallome in Eukaryotic Cells*; Springer Netherlands, 2013; pp. 451–478.
215. Horn, D.; Barrientos, A. *IUBMB Life* **2008**, *60* (7), 421–429.
216. Marzano, C.; Pellei, M.; Tisato, F.; Santini, C. *Anticancer Agents Med. Chem.* **2009**, *9* (2), 185–211.
217. Santini, C.; Pellei, M.; Gandin, V.; Porchia, M.; Tisato, F.; Marzano, C. *Chem. Rev.* **2014**, *114* (1), 815–862.
218. Tardito, S.; Marchio, L. *Curr Med Chem* **2009**, *16* (11), 1325–1348.
219. Zhou, W.; Wang, X. Y.; Hu, M.; Zhu, C. C.; Guo, Z. J. *Chem. Sci.* **2014**, *5* (7), 2761–2770.
220. Bhattacharyya, A.; Dixit, A.; Mitra, K.; Banerjee, S.; Karande, A. A.; Chakravarty, A. R. *MedChemComm* **2015**, *6* (5), 846–851.
221. Deka, B.; Sarkar, T.; Banerjee, S.; Kumar, A.; Mukherjee, S.; Deka, S.; Saikia, K. K.; Hussain, A. *Dalton Trans.* **2017**, *46* (2), 396–409.
222. Gu, S.; Yu, P.; Hu, J.; Liu, Y.; Li, Z.; Qian, Y.; Wang, Y.; Gou, Y.; Yang, F. *Eur. J. Med. Chem.* **2019**, *164*, 654–664.
223. Gou, Y.; Chen, M.; Li, S.; Deng, J.; Li, J.; Fang, G.; Yang, F.; Huang, G. *J. Med. Chem.* **2021**, *64* (9), 5485–5499.
224. Li, X.; Hao, S. J.; Han, A. L.; Yang, Y. Y.; Fang, G. Z.; Liu, J. F.; Wang, S. *J. Mater. Chem. B* **2019**, 7 (25), 4008–4016.
225. Campagna, S.; Puntoriero, F.; Nastasi, F.; Bergamini, G.; Balzani, V. *Topics in Current Chemistry: Photochemistry and Photophysics of Coordination Compounds: Ruthenium*; Springer Berlin Heidelberg, **2007**; pp. 117–214.
226. Bratsos, I.; Jedner, S.; Gianferrara, T.; Alessio, E. *Chimia* **2007**, *61* (11), 692–697.
227. Levina, A.; Mitra, A.; Lay, P. A. *Metallomics* **2009**, *1* (6), 458–470.

228. Qian, C.; Wang, J. Q.; Song, C. L.; Wang, L. L.; Ji, L. N.; Chao, H. *Metallomics* **2013**, *5* (7), 844–854.
229. Martin, A.; Byrne, A.; Burke, C. S.; Forster, R. J.; Keyes, T. E. *J. Am. Chem. Soc.* **2014**, *136* (43), 15300–15309.
230. Wang, J. Q.; Zhang, P. Y.; Qian, C.; Hou, X. J.; Ji, L. N.; Chao, H. *J. Biol. Inorg. Chem.* **2014**, *19* (3), 335–348.
231. Liu, J. P.; Chen, Y.; Li, G. Y.; Zhang, P. Y.; Jin, C. Z.; Zeng, L. L.; Ji, L. N.; Chao, H. *Biomaterials* **2015**, *56*, 140–153.
232. Sreedharan, S.; Gill, M. R.; Garcia, E.; Saeed, H. K.; Robinson, D.; Byrne, A.; Cadby, A.; Keyes, T. E.; Smythe, C.; Pellett, P.; Bernardino de la Serna, J.; Thomas, J. A. *J. Am. Chem. Soc.* **2017**, *139* (44), 15907–15913.
233. Li, J.; Guo, L.; Tian, Z.; Zhang, S.; Xu, Z.; Han, Y.; Li, R.; Li, Y.; Liu, Z. *Inorg. Chem.* **2018**, *57* (21), 13552–13563.
234. Du, Q.; Guo, L. H.; Ge, X. X.; Zhao, L. P.; Tian, Z. Z.; Liu, X. C.; Zhang, F. J.; Liu, Z. *Inorg. Chem.* **2019**, *58* (9), 5956–5965.
235. Zhang, C.; Guan, R. L.; Liao, X. X.; Ouyang, C.; Rees, T. W.; Liu, J. P.; Chen, Y.; Ji, L. N.; Chao, H. *Chem. Commun.* **2019**, *55* (83), 12547–12550.
236. Kumaresan, D.; Shankar, K.; Vaidya, S.; Schmehl, R. H. *Topics in Current Chemistry: Photochemistry and Photophysics of Coordination Compounds: Osmium*; Springer Berlin Heidelberg, **2007**; pp. 101–142.
237. Hanif, M.; Babak, M. V.; Hartinger, C. G. *Drug Discov. Today* **2014**, *19* (10), 1640–1648.
238. Needham, R. J.; Sanchez-Cano, C.; Zhang, X.; Romero-Canelon, I.; Habtemariam, A.; Cooper, M. S.; Meszaros, L.; Clarkson, G. J.; Blower, P. J.; Sadler, P. J. *Angew. Chem. Int. Ed.* **2017**, *56* (4), 1017–1020.
239. Omar, S. A.; Scattergood, P. A.; McKenzie, L. K.; Bryant, H. E.; Weinstein, J. A.; Elliott, P. I. *Molecules* **2016**, *21* (10), 1382.
240. Sanchez-Cano, C.; Romero-Canelon, I.; Yang, Y.; Hands-Portman, I. J.; Bohic, S.; Cloetens, P.; Sadler, P. J. *Chem. Eur. J.* **2017**, *23* (11), 2512–2516.
241. Gkika, K. S.; Byrne, A.; Keyes, T. E. *Dalton Trans.* **2019**, *48* (47), 17461–17471.
242. Ge, C.; Zhu, J. Y.; Ouyang, A.; Lu, N.; Wang, Y.; Zhang, Q. L.; Zhang, P. Y. *Inorg. Chem. Front.* **2020**, 7 (20), 4020–4027.
243. Pessoa, J. C.; Etcheverry, S.; Gambino, D. *Coord. Chem. Rev.* **2015**, *301*, 24–48.
244. Hsu, C.-H.; Cheng, A.-L. In *Clinical Studies With Curcumin*; Aggarwal, B. B., Surh, Y.-J., Shishodia, S., Eds.; Springer US: Boston, MA, **2007**; pp. 471–480.
245. Prasad, P.; Khan, I.; Kondaiah, P.; Chakravarty, A. R. *Chem. Eur. J.* **2013**, *19* (51), 17445–17455.
246. Banik, B.; Somyajit, K.; Nagaraju, G.; Chakravarty, A. R. *Dalton Trans.* **2014**, *43* (35), 13358–13369.
247. Banerjee, S.; Prasad, P.; Khan, I.; Hussain, A.; Kondaiah, P.; Chakravarty, A. R. *Z. Anorg. Allg. Chem.* **2014**, *640* (6), 1195–1204.
248. Prasad, P.; Pant, I.; Khan, I.; Kondaiah, P.; Chakravarty, A. R. *Eur. J. Inorg. Chem.* **2014**, *2014* (14), 2420–2431.
249. Yusop, R. M.; Unciti-Broceta, A.; Johansson, E. M.; Sanchez-Martin, R. M.; Bradley, M. *Nat. Chem.* **2011**, *3* (3), 239–243.
250. Weiss, J. T.; Dawson, J. C.; Macleod, K. G.; Rybski, W.; Fraser, C.; Torres-Sanchez, C.; Patton, E. E.; Bradley, M.; Carragher, N. O.; Unciti-Broceta, A. *Nat. Commun.* **2014**, *5* (1), 3277.
251. Wang, J.; Cheng, B.; Li, J.; Zhang, Z. Y.; Hong, W. Y.; Chen, X.; Chen, P. R. *Angew. Chem. Int. Ed.* **2015**, *54* (18), 5364–5368.

252. Serrano, F. A.; Matsuo, A. L.; Monteforte, P. T.; Bechara, A.; Smaili, S. S.; Santana, D. P.; Rodrigues, T.; Pereira, F. V.; Silva, L. S.; Machado, J., Jr.; Santos, E. L.; Pesquero, J. B.; Martins, R. M.; Travassos, L. R.; Caires, A. C.; Rodrigues, E. G. *BMC Cancer* **2011**, *11* (1), 296.
253. van Leeuwen, P. W. N. M.; Roobeek, C. F. *J. Organomet. Chem.* **1983**, *258* (3), 343–350.
254. Harlos, M.; Ott, I.; Gust, R.; Alborzinia, H.; Wolfl, S.; Kromm, A.; Sheldrick, W. S. *J. Med. Chem.* **2008**, *51* (13), 3924–3933.
255. Wirth, S.; Rohbogner, C. J.; Cieslak, M.; Kazmierczak-Baranska, J.; Donevski, S.; Nawrot, B.; Lorenz, I. P. *J. Biol. Inorg. Chem.* **2010**, *15* (3), 429–440.
256. Wang, J.; Angnes, L. *Anal. Chem.* **2002**, *64* (4), 456–459.
257. Wang, J.; Nie, J. J.; Guo, P.; Yan, Z.; Yu, B.; Bu, W. *J. Am. Chem. Soc.* **2020**, *142* (6), 2709–2714.
258. Komor, A. C.; Schneider, C. J.; Weidmann, A. G.; Barton, J. K. *J. Am. Chem. Soc.* **2012**, *134* (46), 19223–19233.
259. Boyle, K. M.; Barton, J. K. *J. Am. Chem. Soc.* **2018**, *140* (16), 5612–5624.
260. Gupta, G.; Kumari, P.; Ryu, J. Y.; Lee, J.; Mobin, S. M.; Lee, C. Y. *Inorg. Chem.* **2019**, *58* (13), 8587–8595.
261. Peng, Y. B.; He, W.; Niu, Q.; Tao, C.; Zhong, X. L.; Tan, C. P.; Zhao, P. *Dalton Trans.* **2021**, *50* (26), 9068–9075.
262. Chasapis, C. T.; Loutsidou, A. C.; Spiliopoulou, C. A.; Stefanidou, M. E. *Arch. Toxicol.* **2012**, *86* (4), 521–534.
263. Osredkar, J.; Sustar, N. *J. Clinic. Toxicol. S* **2011**, *3* (2161), 0495.
264. Sensi, S. L.; Ton-That, D.; Sullivan, P. G.; Jonas, E. A.; Gee, K. R.; Kaczmarek, L. K.; Weiss, J. H. *Proc. Natl. Acad. Sci. U. S. A.* **2003**, *100* (10), 6157–6162.
265. Dineley, K. E.; Votyakova, T. V.; Reynolds, I. J. *J. Neurochem.* **2003**, *85* (3), 563–570.
266. Dineley, K. E.; Richards, L. L.; Votyakova, T. V.; Reynolds, I. J. *Mitochondrion* **2005**, *5* (1), 55–65.
267. Masanta, G.; Lim, C. S.; Kim, H. J.; Han, J. H.; Kim, H. M.; Cho, B. R. *J. Am. Chem. Soc.* **2011**, *133* (15), 5698–5700.
268. Chyan, W.; Zhang, D. Y.; Lippard, S. J.; Radford, R. J. *Proc. Natl. Acad. Sci. U. S. A.* **2014**, *111* (1), 143–148.
269. Xue, L.; Li, G.; Yu, C.; Jiang, H. *Chem. Eur. J.* **2012**, *18* (4), 1050–1054.
270. Liu, D. D.; Zhang, M. Z.; Du, W.; Hu, L.; Li, F.; Tian, X. H.; Wang, A. D.; Zhang, Q.; Zhang, Z. P.; Wu, J. Y.; Tian, Y. P. *Inorg. Chem.* **2018**, *57* (13), 7676–7683.
271. Shen, Y.; Shao, T.; Fang, B.; Du, W.; Zhang, M. Z.; Liu, J. J.; Liu, T. Y.; Tian, X. H.; Zhang, Q.; Wang, A. D.; Yang, J. X.; Wu, J. Y.; Tian, Y. P. *Chem. Commun.* **2018**, *54* (80), 11288–11291.
272. Chernousova, S.; Epple, M. *Angew. Chem. Int. Ed.* **2013**, *52* (6), 1636–1653.
273. Atiyeh, B. S.; Costagliola, M.; Hayek, S. N.; Dibo, S. A. *Burns* **2007**, *33* (2), 139–148.
274. Zhu, H.-L.; Chen, Q.; Peng, W.-L.; Qi, S.-J.; Xu, A.-L.; Chen, X.-M. *Chin. J. Chem.* **2010**, *19* (3), 263–267.
275. Eloy, L.; Jarrousse, A. S.; Teyssot, M. L.; Gautier, A.; Morel, L.; Jolivalt, C.; Cresteil, T.; Roland, S. *ChemMedChem* **2012**, 7 (5), 805–814.
276. Li, Y.; Liu, G. F.; Tan, C. P.; Ji, L. N.; Mao, Z. W. *Metallomics* **2014**, *6* (8), 1460–1468.

Index

Note: Page numbers followed by "*f*" indicate figures, "*t*" indicate tables, and "*s*" indicate schemes.

A

Abiraterone (ABI), 303–304, 303*f*
Antimicrobial resistance (AMR), 382–383
Azide, 52–54

B

β-CD branched polymer, 196–197, 198–199*f*
Bicalutamide, 305–306, 305*f*
Bimetallic Au-Ag core-shell NPs, 195–196, 196*f*
Bioorthogonal chemical reporters
  azide, 52–55
  definition, 52
  nitrone, 77–80, 78–79*f*
  PDT treatment, 54–55
  pretargeting approach for cancer, 76
  sydnones, 80–82, 81*f*
  tetrazine moiety, 82–88, 86–87*f*
Biotin receptors, 42–44
Biotinylated photoCORM, 224, 224*f*
Block copolymers (BCP), 117–118, 117*f*

C

Caged-compounds. *See* Phototriggers
Carbon dots (CDs), 119, 119*f*, 184–186, 185*f*, 187*f*
Carbonic anhydrase (CA), 48–49
Carbon monoxide
  heme degradation, 131–132
  photo-activated CO releasing moieties (photoCORM) (*see* Uncaging of bioactive compound)
Cathepsin B (CTSB), 310–311, 311*f*
Cathepsin K inhibitor Cbz-Leu-$NHCH_2CN$, 314
Cathepsin L (CTSL), 303, 303*f*
Cisplatin, 287–288, 287*f*
Copper complexes, 479–481, 480*f*
Cyclooxygenase (COX), 50
CYP3A4 inhibitor, 311–312

D

Diimine $Re(CO)_3$ complexes, 8–10, 8*f*, 10*f*
Dopamine receptors, 45–46

E

Epidermal growth factor receptor (EGFR), 48
ESKAPE pathogens, 382

F

Fluoroquinolones (FQs), 193–194, 194*f*
Folic acid (FA)/meso-tetrakis(4-carboxyphenyl)porphyrin (TCPP) supramolecular assembly, 180–182, 181*f*

G

Gasotransmitters, 131–132, 206
Gastrin-releasing peptide (GRP), 46
Glucose transporters, 39–42, 41*f*
Gold-nanoparticles, 476–479, 477*f*
G-protein-coupled receptors, 45–47
Graphene oxide (GO), 186–189, 188*f*

H

Hematoporphyrin derivatives, 96
Heterolytic metal-ligand photolysis, 323
Hollow gold nanoshells (HGN), 160–161, 161*f*
Hydrogels, 116–117
Hypereosinophilic syndrome, 290

I

Imatinib, 302–303, 303*f*, 310
Indomethacin, 50
Integrins, 44–45
Iridium
  anticancer therapeutics, 438–440, 445, 447, 450, 452, 455, 457, 461
  biological targets, 429
  bPEI complexes, 420
  cell death, 463–464

Iridium (*Continued*)
chemosensors, 446
chemotherapeutics, 459–460
correlative imaging, 429, 444
dinuclear Ir(III) complexes, 432
hypochlorite imaging agents, 431–432, 434
hypoxic tumors, 454–455, 463
inflammation-based diseases, 453
intracellular Cu^{2+} and Cys detection, 460
luminescent glucose uptake indicators, 430–431
luminescent Ir(III) complex, 420, 429, 431, 451
mitochondrial pH, 448–449, 453
mitochondria-targeting Ir(III) complex structures, 419–420, 420–428*f*
mitophagy, 433, 437
nucleoside metabolism, 432–433
organelle-selective imaging agents, 430, 436–437, 444
oxygen sensors, 435–436
phosphorescent bio-orthogonal labeling reagents, 431
subcellular imaging agents, 419, 429–430, 432, 435, 454
theranostic PDT agents, 434, 441–443, 445, 447–449, 455–457
two-photon imaging agents (luminescent probe), 432–435, 458–459
Isonitrile-containing $Re(CO)_3$ complexes
advantage, 14–15
co-localization, 15–16, 16*f*
elemental distribution, 15–16, 16*f*
in vivo antitumor efficacy, 21, 22*f*
mechanism of action, 16–17, 18*f*, 19–21, 20*f*
perfusion imaging agent, 14–15
P-glycoprotein (Pgp), 17–19
π-acceptor ligands, 14–15
resistance mechanisms, 17–19
structure, 15–16, 15*f*
structure-activity relationships (SARs), 19–21, 19–20*f*

L

Langmuir-Schaefer (LS) technique, 192–193
Laser biostimulation. *See* Photobiostimulation
Layer-by-layer (LbL) approach, 189–191, 191*f*
Levulan, 292, 297–298, 297*f*
Light-triggered unconventional therapy. *See also* Unconventional phototherapy
approaches, 173
modalities, 173
non-metal and metal-based nanoscaffolds, 173–174, 174*f*
Liposomes, 159, 160*f*

M

Macrophage Trojan Horses, 161–163, 162–163*f*
Manganese carbonyl complexes with *fac*-$[Mn(CO)_3]$ moiety
biological studies, 206, 206*f*
chemical and photochemical properties, 207–208, 207*f*
chemotherapeutic applications, 216–217, 217*f*
CO delivery at target site, 220–222, 221*f*
CO donor visualization, 217–220, 218–219*f*
CO-releasing materials and platforms
aluminosilicate mesoporous particles, 225, 225*f*
biocompatibility, 224–225
glycero-phosphocholine lipid vesicle, 227, 227*f*
non-toxic zeolites, 225
peptide-based porous materials, 226–227
photosensitive CO-releasing MOF framework, 226, 226*f*
future research, 227–228
ligand substitution reactions, 207–208, 207*f*
low dose inhalation, 206
photoCORMs with coligands
antibacterial activity, 222–223, 223*f*
antibody-tethered photoCORM, 224, 224*f*
biotinylated photoCORM, 224, 224*f*
HDAC activity, 223–224
one-two punch effect, 222
silent killer, 206
smart design

coligands, 209–210, 209*f*
coligands with α,α′-diimine functionality, 214–215, 214*f*
CO release on light exposure, 214–216, 215*f*
electronic absorption spectra, 209–210, 210*f*
HOMO/LUMO energy diagram, 210–212, 211*f*, 215–216
Jablonski diagram, 213–214, 214*s*
ligand frame structures on photoband shift, 212–213, 212*f*
moderate sensitivity to visible light, 208–209, 208*f*
π-acceptor properties, 213–214, 214*s*
quantum yield, 215
trackable photoCORM
cytotoxicity, 220
fluorescent photoCORM, 217–219, 218*f*
therapeutic coligands, 219, 219*f*
turn-ON photoCORM, 220–222, 221*f*
water-soluble photoCORM, 216–217, 217*f*
Mesoporous silica nanoparticles (MSNPs), 179–180, 180*f*
Metal-to-ligand charge transfer (MLCT), 4–6, 5*f*
Mitochondria-targeting transition metal complexes
copper complexes, 479–481, 480*f*
future research, 497–499
gold-nanoparticles, 476–479, 477*f*
iridium
anticancer therapeutics, 438–440, 445, 447, 450, 452, 455, 457, 461
biological targets, 429
bPEI complexes, 420
cell death, 463–464
chemosensors, 446
chemotherapeutics, 459–460
correlative imaging, 429, 444
dinuclear Ir(III) complexes, 432
hypochlorite imaging agents, 431–432, 434
hypoxic tumors, 454–455, 463
inflammation-based diseases, 453
intracellular Cu^{2+} and Cys detection, 460
luminescent glucose uptake indicators, 430–431
luminescent Ir(III) complex, 420, 429, 431, 451
mitochondrial pH, 448–449, 453
mitochondria-targeting Ir(III) complex structures, 419–420, 420–428*f*
mitophagy, 433, 437
nucleoside metabolism, 432–433
organelle-selective imaging agents, 430, 436–437, 444
oxygen sensors, 435–436
phosphorescent bio-orthogonal labeling reagents, 431
subcellular imaging agents, 419, 429–430, 432, 435, 454
theranostic PDT agents, 434, 441–443, 445, 447–449, 455–457
two-photon imaging agents (luminescent probe), 432–435, 458–459
mitochondria
and cell death, 415–416
structure and function, 414–415
targeting strategy, 417–418
therapeutic targets, 416–417
osmium complex, 486–489, 488*f*
palladium complex, 492–493, 492*f*
photodynamic therapy (PDT), 418–419
platinum-based metallodrugs, 464–469, 465*f*
rhenium complexes, 470–475, 471*f*
rhodium complexes, 493–495, 493*f*
Ru(II) polypyridine complexes, 482, 482–483*f*, 484–486
silver complexes, 496–497, 496*f*
vanadium complex, 489, 490*f*, 491–492
zinc complexes, 495–496, 495*f*
Molecular systems
molecular metal complexes, 252–258, 256*f*
molecular nitric oxide generating systems, 246–247, 248–249*f*, 249–250
molecular photosensitizers, 251–252, 252–253*f*
multimodal systems
carbon nitride-carbon dot composite nanoparticles, 273–274
micelles formation, 271–273, 273*f*

Molecular systems (*Continued*)
molecular hybrid based on β-cyclodextrin, 271, 272*f*
NO/PDT combined anticancer therapy, 269–271, 271*f*
ruthenium-based phototherapeutics, 269–271, 271*f*
simultaneous photo-induced NO and $^1O_2$ generation, 269, 270*f*
supramolecular systems, 271–273
tetrapyrrolic photosensitizers and NO photodonors, 271–273, 273*f*
phototriggered NO release, 250–251

N

Nano-carriers
photo-uncaging
N-doped graphene QD, 158–159
water soluble nanocarrier, 157–158, 158*f*
platinum anticancer complexes
block copolymers (BCP), 117–118, 117*f*
carbon dots, 119, 119*f*
hydrogels, 116–117
polymetallodrugs, 118–119, 118*f*
polysaccharides, 115–116, 116*f*
proteins, 114–115, 115*f*
up-converting nanoparticles (UCNPs), 120–122, 120–122*f*
Nanostructured systems
multimodal systems, 274–275
noble metals nanoparticles, 259–261
NO delivery systems, 258–259, 258*f*
ROS generation, 259
semiconductor nanocatalysts
cancer cell degradation, 262–263
intracellular redox environment, 266–267
photochemotherapeutic (PCT) agent, 265–266
photodynamic inactivation (PDI) of bacteria, 264–265
phototoxicity, 264
quantum yield, 265
ROS generation, 261–262, 263*f*
Nitric oxide photodynamic therapy (NO-PDT), 176–177, 177*f*
photo-activated NO releasing moieties (photoNORM) (*see* Uncaging of bioactive compound)
photochemistry, 131
signaling pathways, 131–132

O

Optogenetics, 238–240
Organometallics, photoactive
antimicrobial resistance (AMR), 382–383
CO delivery systems, 384
future research, 404–405
group 4 and 5 organometallic complexes, 386–387, 386–387*f*, 387*t*
group 6 organometallic complexes, 387–391, 388*f*, 390*t*, 391*f*
group 7 organometallic complexes, 391–398, 392–395*f*, 396–397*t*, 397*f*
group 8 organometallic complexes, 398–399, 398*f*
group 9 organometallic complexes, 399–401, 399–401*f*
group 10 and 11 organometallic complexes, 401–404, 402*t*, 402–403*f*
iridium complexes
antibacterial activity, 400–401, 400–401*f*
heteroleptic dipyrrinato complexes, 399, 399*f*
manganese complexes
antibacterial properties, 391–393, 392*f*
manganese carbonyl complexes, 391–393, 392*f*
tryptophan-derived Mn complexes, 393–394, 393*f*
metal carbonyls
benefits, 387–388
fluorene containing complexes, 389
minimum inhibitory concentration (MIC) values, 388–389, 390*t*
naphthalene moiety, 389
photoCORM, 388–389, 388*f*
tetracarbonyl organometallic complexes, 388–389, 388*f*
tungsten Fischer carbene complex, 389–391, 391*f*
photo-assisted CO release (photoCORMs), 384–385

photodynamic therapy (PDT), 383–384, 383*f*
platinum complexes
2-(benzothiazol-2-yl)-phenolato platinum(II) complexes, 401–403, 402*f*
MIC values, 401–403, 402*t*
tetraphenylethylene-based organoplatinum(II) metallacycle photosensitizer, 403–404, 403*f*
rhenium complexes
facial rhenium(I) tricarbonyl complexes, 395, 397*f*
in vitro anti-proliferative effects, 394
minimum inhibitory concentration (MIC) values, 396–397*t*
rhenium bisquinoline complexes, 394–395, 395*f*, 396–397*t*
rhenium(I) pyridocarbazole complexes, 394, 394*f*
rhenium N-heterocyclic metal carbonyl compounds, 395–398
water soluble rhenium dicarbonyl compounds, 398
ROS generation mechanism, 383–384, 383*f*
ruthenium complexes, 398–399, 398*f*
zirconium complexes
antimicrobial activity, 386, 387*t*
biocidal effect, 386, 386*f*
metallo-phthalocyanine, 386–387, 387*f*
Osmium complex, 486–489, 488*f*
Ototoxicity, 288

P

Palladium complex, 492–493, 492*f*
Photoactivated chemotherapy (PACT), 336
Photo-assisted CO release (photoCORMs), 384–385
Photobiostimulation, 238, 240*f*
Photocatalysis
definition, 244
heterogenous photocatalysis, 244–246, 245*f*
f-block metal-based/doped nanostructures, 246
metallic nanoparticles, 246
principles, 244, 245*f*
semiconductor nanoparticles, 244–246
homogenous photocatalysis, 244, 244*f*
Photochemical release. *See* Uncaging of bioactive compound
Photochemical release rate, 132
Photochemotherapy (PCT)
advantage, 299
agents
abiraterone (ABI), 303–304, 303*f*
bicalutamide, 305–306, 305*f*
cathepsin L (CTSL), 303, 303*f*
imatinib, 302–303, 303*f*
STF-31, 303*f*, 304–305
concept, 298–299
Jablonski diagram, 299–300, 300*f*
vs. PDT, 299
photocaging, 298–299
photodissociation of ligand, 299–300, 300*f*
Ru(II) polypyridyl complexes design strategy, 300–302, 301*f*
singlet oxygen sensitization, 299
Photocleavage, 240, 241*f*
Photodynamic therapy (PDT), 175, 176*f*
Photofrin®, 96, 292, 297–298, 297*f*
Photopharmacology, 228
Photosensitization
concept, 240–241
endogenous and exogenous photosensitizers, 241
photosensitized ROS generation
molecular oxygen, 241, 242*f*
Wilkinson mechanism, 242–243, 243*f*
Photothermal therapy (PTT), 177–178, 178*f*
Phototriggers, 323
Plasma membrane proteins, 38–39
Platinum anticancer complexes, photoactivated, 2–3, 3*f*, 464–469, 465*f*
axial derivatives of Pt(IV) complexes
multi-action derivatives, 110–111, 110*f*, 112*f*
targeting vectors, 108–109, 109*f*
azide-alkyne cycloaddition, 112–114, 113*f*
cisplatin, 97–98, 97*f*

Platinum anticancer complexes, photoactivated (*Continued*)
click reaction, 112–114, 113*f*
cyclometalated Pt(II) complexes, 98
diazido-Pt(IV) diamine complexes
^{14}N NMR spectra, 102, 103*f*
gram-scale syntheses, 102
LMCT bands, 100–101, 100*f*
methyl substituent, 102
photoactive prodrugs, 100, 100*f*
photodecomposition pathways, 101
pyridine ligand, 101–102
trans complex, 101
diiodido-Pt(IV) complexes, 99–100, 99*f*
kinetically-inert Pt(IV) complexes, 98
mechanism of action
DNA target, 105–106, 107*f*
immunogenic cell death, 108
peptide and protein targets, 107
radical and ROS generation, 104–105, 104–105*f*
metal-ligand bond, 96–97
nanocarriers
block copolymers (BCP), 117–118, 117*f*
carbon dots, 119, 119*f*
hydrogels, 116–117
polymetallodrugs, 118–119, 118*f*
polysaccharides, 115–116, 116*f*
proteins, 114–115, 115*f*
up-converting nanoparticles (UCNPs), 120–122, 120–122*f*
oxaliplatin, 98–99
platinum coordination complexes, 97, 97*f*
Pt(IV) azide–acetylene derivatives, 112–114, 113*f*
tissue penetration depth, 96, 97*f*
Polymetallodrugs, 118–119, 118*f*
Polysaccharides, 115–116, 116*f*
Proteins, 114–115, 115*f*
Psoralen, 96
Pyridylimine $Re(CO)_3$ complexes
with alkyl chain substituents, 12–14, 14*f*
canonical necrotic cell death, 11, 13*f*
combinatorial synthetic strategy, 11, 12*f*
dose-response curves, 12–14, 14*f*
in vitro anticancer activity, 11
lipophilicity, 12–14, 14*f*

R

Reactive small inorganic molecules (RSIMs)
biomedical applications, 235, 236*f*
future research, 275–277
molecular systems
molecular metal complexes, 252–258, 256*f*
molecular nitric oxide generating systems, 246–247, 248–249*f*, 249–250
molecular photosensitizers, 251–252, 252–253*f*
phototriggered NO release, 250–251
multimodal systems
molecular systems, 269–274, 270–273*f*
nanostructured systems, 274–275
nanostructured systems
noble metals nanoparticles, 259–261
NO delivery systems, 258–259, 258*f*
ROS generation, 259
semiconductor nanocatalysts, 261–269, 263*f*
photogeneration
light-controlled processes, 236–237, 237*f*
light management strategies, 237–238, 239*f*
optogenetics, 238–240
photobiostimulation, 238, 240*f*
photocatalysis, 244–246, 244–245*f*
photocleavage, 240, 241*f*
photosensitization, 240–243, 242–243*f*
physiological and pathological processes, 234–235, 235*f*
regulation, 236*f*
Rhenium complexes, 470–475, 471*f*
Rhenium(I) tricarbonyl complexes
diimine $Re(CO)_3$ complexes, 8–10, 8*f*, 10*f*
future research, 27–28
imaging and therapeutic applications, 3–4
isonitrile-containing $Re(CO)_3$ complexes
advantage, 14–15
co-localization, 15–16, 16*f*
elemental distribution, 15–16, 16*f*
in vivo antitumor efficacy, 21, 22*f*

mechanism of action, 16–17, 18*f*, 19–21, 20*f*
perfusion imaging agent, 14–15
P-glycoprotein (Pgp), 17–19
π-acceptor ligands, 14–15
resistance mechanisms, 17–19
structure, 15–16, 15*f*
structure-activity relationships (SARs), 19–21, 19–20*f*
Jablonski diagram, 5*f*
metal-to-ligand charge transfer (MLCT), 4–6, 5*f*
photoactivation
$^1O_2$ generation, 22–23
CisPtRe and OxaliPtRe, 25–27
cytotoxicity, 23–25, 24*f*
diimine/quinolylamine ligands, 22–23
Frontier Kohn-Sham molecular orbitals, 27*f*
ligand photosubstitution process, 23
multimetallic complexes, 25–27, 26–27*f*
$[Re(CO)_2(OH_2)(NN)(PR_3)]^+$, 23–25
$[Re(CO)_3(phen)(DAPTA)]^+$ (DAPTA-1), 25
TRIP-suc, 25–27
photoluminescent imaging agents, 6–8, 6–7*f*
vs. platinum-based drugs, 3–4
properties and applications, 4–6, 5*f*
pyridylimine $Re(CO)_3$ complexes
with alkyl chain substituents, 12–14, 14*f*
canonical necrotic cell death, 11, 13*f*
combinatorial synthetic strategy, 11, 12*f*
dose-response curves, 12–14, 14*f*
in vitro anticancer activity, 11
lipophilicity, 12–14, 14*f*
theranostic anticancer agents, 8–10, 8*f*, 10*f*
Rhodium complexes, 493–495, 493*f*
Ritonavir, 311–312
Ruthenium complexes
advantages
dendritic spine computation, 330–331
direction-selective ganglion cells, 331
glutamate sensor, 329–330
mapping studies, 331–332
neurotransmitters, 326–329
nitric oxide (NO) release, 326
optical influence, 328
photolysis, 325–326, 325*f*
Ru-based phototrigger, 326, 327*f*
two-photon imaging technique, 328–329
brain stimulation, 322–324
cis, trans and other structures, 347–349
clinical applications
bacterial ecological studies, 333
epileptic seizures, 332
optogenetics, 333
vasodilator, 333
coumarins, 324
drug-based induction of gene expression, 323–324
fifth ligand
absorption maxima, 339–340, 340*f*
bioactive ligand, 337
bipyridine (bpy), 337
chemical property, 340
terpyridines (tpy), 338
trimethylphosphine (PMe_3), 339
future research, 349–350
inorganic caged compounds, 324
nitrobenzyl and nitrophenyl compounds, 324
phototriggers, 323
three-dimensional localization
absorptive fluorophore, 346–347
cross section estimation, 343
cuvette assemblies, 343–344, 344*f*
linear focusing, 341, 342*f*
local pH, 346
multiphoton optics, 341, 342*f*
set of wells, 344–345
therapeutic window, 346
two-photon (2P) absorption phenomenon, 341–342
upconversion, 345–346
time-locked neural stimulation, 323–324
value in diversity
coordination sites, 335, 336*f*
methylthio galactose (MTG) gene inducing mechanism, 334, 335*f*
photoactivated chemotherapy (PACT), 336

Ruthenium complexes (*Continued*)
protein uncaging, 335–336
quantum efficiency, 333–334
thioethers and nitriles, 333–334
Ruthenium(II) polypyridyl complexes, 482, 482–483*f*, 484–486
cancer, 286
chemotherapy, 286–287
vs. cisplatin, 287–288, 287*f*
dual activity complex, 288–289
^{1}MLCT transition, 308
arene groups, 316
bifurcation of excited state, 306, 306*f*
cancer cell death, 309
cathepsin B (CTSB), 310–311, 311*f*
cathepsin K inhibitor Cbz-Leu-$NHCH_2CN$, 314
CH_3CN ligand photodissociation, 307
CYP3A4 inhibitor, 311–312
cytotoxicity, 307
hydroxyl groups, 315–316
imatinib, 310
iodinated py-BODIPY ligand, 313–314
ligand exchange quantum yields, 309
lipophilicity, 309–310
McFarland dual activity complexes, 314–315, 315*f*
photophysical processes, 306, 306*f*
prototypical $[Ru(tpy)(Me_2dppn)(py)]^{2+}$, 309–312, 311*f*
pyridyl photodissociation, 307–308, 308*f*
ritonavir, 311–312
$[Ru(pydppn)(biq)(py)]^{2+}$ and $[Ru(tpy)(Me_2dppn)(py)]^{2+}$, 313, 313*f*
selective toxicity, 309
vinblastine, 312
light-based treatment, 288–289
photochemotherapy (PCT)
agents, 302–306, 303*f*, 305*f*
design strategy, 300–302, 301*f*
mechanism, 298–300, 300*f*
photodynamic therapy (PDT)
agents, 297–298, 297*f*
criteria, 289
reaction mechanism, 289–290
singlet oxygen (*see* Singlet oxygen production)
visible light absorption, 288
Ruthenium phthalocyanines
direct and indirect actions of nitric oxide, 359, 359*f*
future research, 377
in vitro studies
cell death, 374, 377*s*
cell viability plot, 374, 375–376*f*
cytotoxicity mechanism, 373
photobiomodulation, 374
photo-cytotoxicity/photochemical reactions, 373, 373*s*
subcellular localization, 373
synergistic effects, 374, 376*f*
light irradiation
cell viability, 371–372, 372*f*
electronic spectrum analysis, 368–369, 370*f*
intersystem crossing (ISC), 366–368
natural transition orbitals (NTO), 367–368, 368*f*
non-relaxed excited states, 366–367, 367*f*
NO photo-release, 369–371, 371*f*
photophysical properties, 365, 366*t*
ROS and NO generation, 371–372, 372*s*
spin-orbit coupling (SOC), 366–368
triplet state UV-vis transient absorption spectra, 365, 366*f*
UV-visible spectra, 368, 369*f*
molecular orbital calculations
calculated and experimental infrared spectrum, 365
calculated HOMO/LUMO composition, 363–364, 364*f*
CAM-B3LYP density functional theory (DFT), 362, 362*f*
experimental UV-visible spectrum and calculated transition bars, 362–363, 363*f*
molecular structure, 356–357, 357*f*
spectroscopic characteristics
absorption bands, 359–361, 360*t*
photophysical properties, 359–361, 360*t*
UV-visible spectra, 361, 361*f*
template synthesis, 357, 358*s*
X-ray crystal structure, 358, 358*f*

S

Sialylated glycoproteins and glycolipids, 51–55, 52–53*f*
Silver complexes, 496–497, 496*f*
Silver nanoparticles, 192, 192*f*
Single photon excitation (SPE), 149–150, 149–150*f*
Singlet oxygen production
  cell death, 290–291
  efficiency, 291–292
  energy diagram, 290, 290*f*
  Jablonski diagram, 292–293*f*
  ligand-centered $^3\pi\pi^*$ states, 294–295, 295*f*
  localization, 291
  quantum yield, 291–292
  steady state electronic absorption, 295–297, 296*f*
  triplet metal-to-ligand charge transfer (^{3}MLCT) state, 293–294, 293–294*f*
  type II PDT agents, 291–292, 292*f*
Small molecule bioregulators (SMBs), 131–132
  release (*see* Uncaging of bioactive compound)
Somatostatin (SST), 46–47
STF-31, 303*f*, 304–305

T

Talaporfin, 297–298
"Three-bullets" MSNPs, 182–184, 182*f*
TLD1433, 297*f*, 298
Trackable photoCORM
  cytotoxicity, 220
  fluorescent photoCORM, 217–219, 218*f*
  therapeutic coligands, 219, 219*f*
Transition metal complexes, photofunctional, 2–4
  bioorthogonal chemical reporters
    nitrone, 77–80, 78–79*f*
    pretargeting approach for cancer, 76
    sydnones, 80–82, 81*f*
    tetrazine moiety, 82–88, 86–87*f*
  design strategies, 38
  DNA mismatch repair (MMR), 76
  enzyme activity
    azoreductases, 72–73
    nitroreductases, 73–74
    quinone reductases, 74–75
  future research, 88–90
  hypoxia, tumor
    2-(2-pyridyl)benzothiophene-based iridium(III) acetylacetonate complex, 56–58
    dual phosphorescence, 58–59, 59*f*
    iridium(III) complexes with non-conjugated amino groups, 58–59, 59*f*
    oxygen supply, 56
    platinum(II) porphyrins, 60–61
  optical imaging, 37–38
  pH
    cancer- and lysosome-targeted imaging and PDT, 63–65
    tumor acidosis and hypoxia, 62–63
  photosensitizers, 37–38
  polarity, cellular, 65
  real-time imaging, 37–38
  redox homeostasis
    glutathione (GSH), 67–71, 68*f*
    nicotinamide adenine dinucleotide (NADH), 71–72
    ROS generation and elimination, 67
  for targeted cancer diagnosis and therapy
    biotin receptors, 42–44
    carbonic anhydrase (CA), 48–49
    cyclooxygenase (COX), 50
    epidermal growth factor receptor (EGFR), 48
    glucose transporters, 39–42, 41*f*
    G-protein-coupled receptors, 45–47
    integrins, 44–45
    plasma membrane proteins, 38–39
    sialylated glycoproteins and glycolipids, 51–55, 52–53*f*
  viscosity, intracellular, 66–67
Trojan Horse mechanism, 161–163, 162–163*f*
Turn-ON photoCORM, 220–222, 221*f*
Two photon excitation (TPE)
  $Mn(CO)_3(\eta^2$-TPYOH)X, 153, 153*f*
  Roussin's red ester, 151–152, 152*f*
  simplified models, 150–151, 151*s*
Tyrphostin 1, 48

U

Uncaging of bioactive compound
  biological targeting
    hollow gold nanoshells (HGN), 160–161, 161*f*

Uncaging of bioactive compound (*Continued*)
liposomes, 159, 160*f*
macrophage Trojan Horses, 161–163, 162–163*f*
definition, 132
multiphoton excitation, 132–133
photo-activated CO releasing moieties (photoCORMs)
compound location, 141–142
design goals, 139
metal carbonyls, 139
photoreaction, 140–141, 141*s*
quantification, 140
$Re(PR_3)(CO)_3(NN)^+$ complexes, 142, 143*s*
Schlenk cell, 140, 141*f*
stability, 139–140
theranostics, 141–142
transition metal photoCORMs, 139–140
$W(CO)_5(TPPTS)^{3-}$, 140–141, 141*s*
photo-activated NO releasing moieties (photoNORM)
absorptivity, 138–139
chromium(III) nitrito cation, 137–139, 138*s*
flash photolysis studies, 134–135, 134*s*
formulas, 133–134, 134*f*
photo-labilization, 134–135, 134*s*
quantification, 135, 135*f*
Roussin's red salt esters, 135–136, 136*f*
ruthenium nitrosyls, 137
Sievers Nitric Oxide Analyzer (NOA), 135, 135*f*
photochemical release rate, 132
phototherapeutic window
definition, 148–149
illustration, 148–149, 148*f*
nano-carriers, 157–159, 158*f*
simultaneous two photon excitation, 150–153, 151*s*, 152–153*f*
single photon excitation (SPE) strategies, 149–150, 149–150*f*
upconverting nanoparticles (UCNPs), 153–157, 154–156*f*
small molecule bioregulator (SMB) release rates enhancement
anth-CrONO and pyr-CrONO, 145–146, 145*f*
CdSe/ZnS core/shell quantum dots, 146–147, 147*f*
charge transfer mechanism, 147–148
graphene QDs (GQDs), 148
illustration, 143, 144*s*
organic chromophore antennas, 143–146, 144–146*f*
PPIX-RSE complex, 143–145, 144*f*
ruthenium photoCORM with pendent BODipy dye antenna, 146, 146*f*
semiconductor core/shell quantum dots (QDs), 146–148, 147*f*
target delivery, 132
target site, 132
toxicity, 132
two-photon excitation (TPE), 132–133
Unconventional phototherapy
advantages, 175
future research, 199
minimally invasive therapeutic modalities, 175
nitric oxide (NO) photodynamic therapy (NO-PDT), 176–177, 177*f*
noble metal-based NPs
benzophenone (BP), 195, 196*f*
β-CD branched polymer, 196–197, 198–199*f*
bimetallic Au-Ag core-shell NPs, 195–196, 196*f*
dual-function gold nanoparticles, 189–191, 190*f*
fluoroquinolones (FQs), 193–194, 194*f*
Langmuir-Schaefer (LS) technique, 192–193
layer-by-layer (LbL) approach, 189–191, 191*f*
localized surface plasmon resonance (LSPR) in VIS/NIR region, 196–197, 198–199*f*
plasmonic NPs, 194–195
silver nanoparticles, 192, 192*f*
nonmetallic inorganic NPs
carbon dots (CDs), 184–186, 185*f*, 187*f*

folic acid (FA)/meso-tetrakis (4-carboxyphenyl)porphyrin (TCPP) supramolecular assembly, 180–182, 181f
graphene oxide (GO), 186–189, 188f
mesoporous silica nanoparticles (MSNPs), 179–180, 180f
"three-bullets" MSNPs, 182–184, 182f
photodynamic therapy (PDT), 175, 176f
photoprecursors, 175
photothermal therapy (PTT), 177–178, 178f
Upconverting nanoparticles (UCNPs)
photo-activated platinum anticancer complexes, 120–122, 120–122f
photo-uncaging
biocompatible polymer disks (PDs), 155–156
illustration, 153–154, 154f
PetACrONO, 156–157, 156f
visible PL spectrum, 154–155, 155f

V

Vanadium complex, 489, 490f, 491–492

W

Water-soluble photoCORM, 216–217, 217f

Z

Zinc complexes, 495–496, 495f

FSC
www.fsc.org
MIX
Papier | Fördert gute Waldnutzung
FSC® C083411

Kolibri 360 GmbH
Ferdinand-Jühlke-Straße 7
99095 Erfurt, Deutschland
produktsicherheit@kolibri360.de